Prentice Hall Canada is proud to present a truly **Canadian** Environmental Science textbook for introductory courses. To hail this event, Prentice Hall Canada is delighted to support the work of The Nature Conservancy of Canada through the sale of this textbook.

The Nature Conservancy of Canada is the only national organization dedicated to preserving Canadian biodiversity through the acquisition and protection of ecologically significant natural areas. Information from scientific sources, including a network of Conservation Data Centres, guides its programs to secure critical habitat. Since 1962, with the support of individuals, corporations, foundations, and governments, The Conservancy has helped protect over 464 000 hectares across Canada. To learn more about The Nature Conservancy of Canada, e-mail **Nature@NatureConservancy.ca**

Promoting environmentally responsible publishing in Canada, Prentice Hall Canada has printed this text on recycled paper with vegetable-based biodegradable inks.

ENVIRONMENTAL SCIENCE
A CANADIAN PERSPECTIVE

BILL FREEDMAN
Dalhousie University

Prentice Hall Canada
Scarborough, Ontario

Canadian Cataloguing in Publication Data
Freedman, Bill
Environmental science: a Canadian perspective
Includes index.
ISBN 0-13-524240-1

1. Environmental sciences - Canada. I. Title.

GE160.C3F74 1998 363.7 C97-931272-8

Prentice-Hall, Inc., Upper Saddle River, New Jersey
Prentice-Hall International (UK) Limited, London
Prentice-Hall of Australia, Pty. Limited, Sydney
Prentice-Hall Hispanoamericana, S.A., Mexico City
Prentice-Hall of India Private Limited, New Delhi
Prentice-Hall of Japan, Inc., Tokyo
Simon & Schuster Asia Private Limited, Singapore
Editora Prentice-Hall do Brasil, Ltda., Rio de Janeiro

ISBN 0-13-524240-1

Publisher: Pat Ferrier
Acquisitions Editor: Sarah Kimball
Senior Marketing Manager: Ann Byford
Developmental Editor: Maurice Esses
Production Editor: Susan James
Copy Editor: Rosemary Tanner
Production Coordinator: Deborah Starks
Permissions/Photo Research: Karen Taylor
Cover Design: Kyle Gell
Cover Image: Trudy Woodcock's Image Artwork Inc.
Page Layout: Hermia Chung

Visit the Prentice Hall Canada Web site! Send us your comments, browse our catalogues, and more. **www.phcanada.com** Or reach us through e-mail at **phcinfo_pubcanada@prenhall.com**

1 2 3 4 5 CC 02 01 00 99 98

Printed and bound in the United States of America.

Every reasonable effort has been made to obtain permissions for all articles and data used in this edition. If errors or omissions have occurred, they will be corrected in future editions provided written notification has been received by the publisher.

This book is printed on recycled paper.

There is absolutely no inevitability
as long as there is a willingness
to contemplate what is happening.

Marshall McLuhan,
in *The Medium is the Message* (1967)

ABOUT THE AUTHOR

BILL FREEDMAN is a professor of Biology at Dalhousie University in Halifax, Nova Scotia, where he has taught for the past 16 years. He earned his M.Sc. and his Ph.D. from the Department of Botany at the University of Toronto. Bill has researched many aspects of environmental science including the ecological effects of forestry, acidification, toxic metals, sulfur dioxide, pesticides, and nutrient inputs to fresh water. Bill is a Trustee and Science Advisor to the Nature Conservancy of Canada and a member on the Board of Directors of the Tree Canada Foundation. In addition to writing this book, the author has written a book entitled Environmental Ecology (Academic Press '95), co-edited a book on arctic ecology, and written over 200 scientific papers and research reports.

In his spare time, Bill enjoys natural history, especially birding and botanizing in wild places, collecting antiques and folk art, and playing squash. Bill and his spouse, George-Anne Merrill, have two teen-aged kids, Jonathan and Rachael.

BRIEF CONTENTS

CONTENTS

V Environmental Damages

List of Feature Boxes

List of Case Studies

PREFACE

ENVIRONMENTAL LITERACY

Today, an understanding of environmental science is crucial. Humans are presently engaged in a frenzy of destructive behaviour that is causing enormous damage to the ecosystems that sustain both our species and Earth's legacy of biodiversity. All around us, we see collapsing fisheries, deforestation, soil degradation, species extinctions, pollution, and other damages.

Nevertheless, we need not be totally pessimistic. If we take constructive actions now, it is not too late to prevent or repair these environmental problems, which threaten the welfare of humans and most other species. Within limits, humans are prescient creatures, and our societies are capable of designing and implementing sustainable economic systems that would support human livelihoods as well as healthy ecosystems.

It is clear, however, that sustainable economies will mean very different ways of doing business and will require fundamental changes in the lifestyles of many people, especially in wealthy industrialized countries such as Canada. Ultimately, such socioeconomic transformations will have to involve substantially less use of energy, materials, and other resources, in comparison with what many of us take for granted today. A more respectful attitude toward nature will also be required.

Achieving such transformations will depend on a sound understanding of environmental issues by national citizenries. Any imposition of substantial restrictions on access to resources will initially be uncomfortable for many people. I believe, however, that people will be more willing to soften their lifestyles if they understand the reasons for those changes in the context of economic sustainability and the provision of livelihoods for future generations. With such an understanding, most individuals will support economic and social changes that conserve the quality of their environment and that of other species.

Environmental literacy will be a key requirement if a country such as Canada is to achieve the difficult transition into an ecologically sustainable economy. I hope that this book will make it easier for more Canadians to become literate about important environmental issues.

A CANADIAN TEXT

This textbook is intended to provide the core elements of a curriculum for teaching environmental science at the introductory level in Canadian colleges and universities. Such classes are normally taught in first year, or sometimes in second year. This book is suitable for students starting in environmental science or environmental studies programs, for arts students who require a science elective, and for science students who require a non-major elective.

To date, most introductory textbooks in environmental science have failed to include much Canadian content. It is regrettable that many Canadian students in environmental studies and ecology have had to use books that are so lacking in information about their own country. Canada has unique national and regional contexts and perspectives, which should be understood by Canadian students.

One of the main goals of this book is to address the needs of Canadian students. Canadian information and examples are integrated throughout the text, along with more general international data. Special Canadian Focus Boxes illustrate Canadian applications of important concepts. And special Canadian Case Studies provide additional examples in a Canadian context.

To help make a difference, Prentice Hall is donating some of the proceeds from the sale of this book to The Nature Conservancy of Canada. The Nature Conservancy of Canada is a charitable organization dedicated to preserving Canadian biodiversity through the acquisition and protection of ecologically significant natural areas. By purchasing a copy of this new textbook, you too are taking a constructive step towards establishing Canadian sustainability!

APPROACH AND ORGANIZATION

Environmental science draws on knowledge and methods from many fields of the sciences and social sciences, including biology, chemistry, geography, geology, physics, statistics, medicine, economics, ethics, political science, and sociology. Many environmental scientists adopt an interdisciplinary approach to integrate these different kinds of knowledge in order to help understand and prevent environmental damage.

This book also adopts an interdisciplinary approach by drawing on different disciplines. At the same time, the choice of topics and the offered interpretations reflect my own world view as an ecologist.

The book is organized into six main Parts. Part I (Ecosystems and Humans) consists of one chapter that serves as an introduction. It defines environmental science and ecology, explains the principles of the ecosystem approach, briefly considers the environmental stressors caused by human activities, and describes various world views.

Part II (The Biosphere: Characteristics and Dynamics) consists of eight chapters that provide a scientific foundation for much of what follows. It discusses the scientific approach (Chapter 2), the geological, hydrological, and atmospheric dynamics of planet Earth (Chapter 3), energy (Chapter 4), the flows and cycles of nutrients (Chapter 5), evolution (Chapter 6), biodiversity and the classification of organisms (Chapter 7), biomes and ecozones (Chapter 8), and ecology (Chapter 9).

As its name implies, Part III (Human Populations) deals with the structure, growth, and implications of the human population. It consists of two chapters: one on global populations (Chapter 10) and one on the Canadian population (Chapter 11).

Part IV (Resources) consists of three chapters that deal with the natural resources that humans and all other species require to sustain their livelihoods. It discusses the relation between resources and sustainable development (Chapter 12), the limited supplies of non-renewable natural resources (Chapter 13), and potentially renewable natural resources (Chapter 14).

Part V (Environmental Damages) consists of ten chapters dealing with the damage caused by humans (individually and collectively) during the course of their activities. It discusses pollution and disturbance as environmental stressors (Chapter 15), gaseous air pollution (Chapter 16), atmospheric gases and climatic change (Chapter 17), toxic elements (Chapter 18), acidification (Chapter 19), eutrophication (Chapter 20), pesticides (Chapter 21), oil spills (Chapter 22) , the ecological effects of forestry (Chapter 23), and the biodiversity crisis (Chapter 24).

Part VI (Ecologically Sustainable Development) consists of one chapter that serves as a conclusion. It discusses the process of assessing environmental impacts and it considers the prospects for Canada and for spaceship Earth.

FEATURES

A special effort has been made with this book to incorporate features that will facilitate learning and enhance an understanding of environmental science.

- **Chapter Objectives** at the beginning of each chapter summarize the skills and knowledge to be learned in that chapter.

ECOSYSTEMS AND HUMANS

1

ECOSYSTEMS AND HUMANS

CHAPTER OUTLINE

Environmental science
The larger context: Earth, life, and ecosystems
The relationship between species and ecosystems
Environmental stressors and ecological responses
Human activities: important environmental stressors
Environmental ethics and world views
The environmental crisis

CHAPTER OBJECTIVES

After completing this chapter, you will be able to:

1. Define environmental science and ecology and distinguish between them.
2. Outline a hierarchical framework of the universe that includes consideration of life on Earth at various scales.
3. Identify important principles of the ecosystem approach.
4. Compare humans and other species in terms of their abilities to cope with environmental constraints.
5. Describe how environmental stressors and disturbances can affect species and ecosystems.
6. List three direct ways in which humans influence their environment.
7. Identify four broad classes of environmental values.
8. Describe five important world views.
9. Classify the issues of the environmental crisis into three categories and give examples of each.
10. Discuss environmental effects of humans as a function of both population and lifestyle.

- **Key terms** are boldfaced where they are defined in the text, and they are listed near the end of each chapter. For easy reference, all the key terms and their definitions are collated in a **Glossary** near the end of the book.

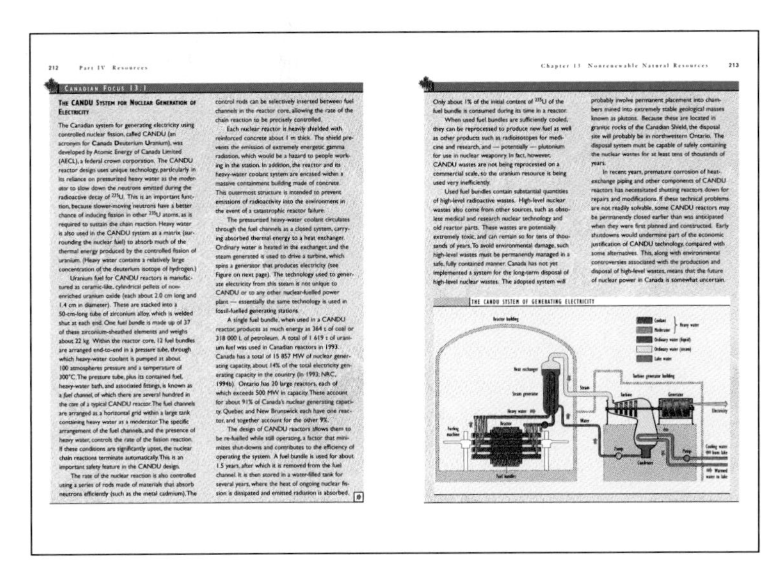

- **Canadian Focus Boxes** illustrate the application of important concepts to the Canadian environment.

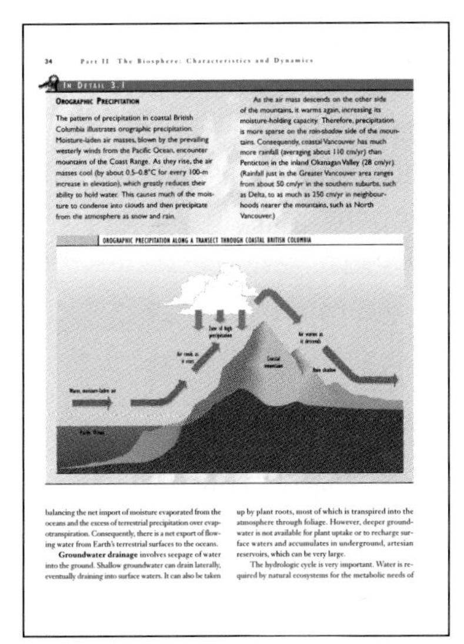

- **In Detail Boxes** provide additional technical information.

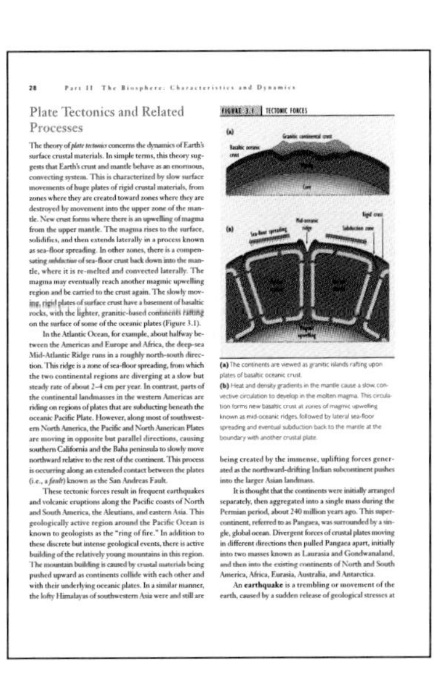

- **Figures, Tables, and colour Photos** throughout the book each carry explanatory captions.

- **Questions for Discussion** near the end of each chapter provide thought-provoking questions which help stimulate careful reflection and class discussion.
- **References** are listed at the end of each chapter to help guide users to further reading.

254 Part IV Resources

CBC CANADIAN CASE 4

THE SEAL HUNT

Marine mammals, such as seals and whales, have long been hunted as natural resources — as sources of meat, oil, and other useful products. Although marine mammals can potentially be utilized as renewable resources, until recently most species were over-hunted, being harvested at rates that exceeded their rate of regeneration. The populations of some species of seals and whales were endangered by this overexploitation, and several became globally extinct. This is similar to the "mining" of a nonrenewable natural resource.

Extirpation is not, however, the present fate of the harp seal of the northwestern Atlantic Ocean. Although the populations of this species have been somewhat depleted by harvesting, the species was never endangered by the commercial hunts for its fur and meat. Today, in fact, the harp seal is one of the world's most populous large wild animals, numbering more than five million.

The intense controversy about seal hunting in eastern Canada has focused mainly on ethical concerns, rather than on issues associated with resource depletion. The most important economic commodities gained from the hunt of harp seals in Atlantic Canada have been furs, especially those of baby seals known as "whitecoats," and more recently penises, which are valued in traditional medicine in eastern Asia. The meat of hunted seals is also used, but this is not the primary objective of the modern commercial hunt — the pelts and penises are.

Although the products of the harp-seal harvest are valuable, they are not particularly crucial to the workings of modern economies. In fact, many people believe that wild animals should not be subjected to commercial hunts for relatively frivolous commodities such as furs, ivory, horns, or penises. Moreover, many people are outraged by the brutality of seal hunting, and by the fact that baby seals are killed in large numbers.

In contrast, proponents of the seal hunt in eastern Canada point out that they are attempting to engage in a sustainable exploitation of a renewable natural resource. They also contend that seals eat a great deal of fish (some of which could be caught and eaten by people), and that the slaughter of domestic animals in agriculture is also a brutal enterprise. In view of these considerations, why, they ask, pick on the sealers?

Questions for Discussion

1. Do you think that seals represent a renewable natural resource that can be commercially exploited in a sustainable fashion? What about other marine mammals, such as whales? Or terrestrial animals, such as deer and furbearing mammals?
2. Many people believe that the cultivation and slaughter of domestic animals in modern, industrial agriculture is a brutal and, from the biocentric perspective, dubiously ethical enterprise. What do you think about this issue? Is it appropriate to justify the hunting of wild animals on the basis of the methods that we use in industrial agriculture?
3. Do you think that it is ethical to kill wild animals to obtain non-essential materials, such as furs, ivory, or horns? Make a list of some plausible reasons for and against this kind of resource harvesting.

Video Resource: "Seal Hunt." *News in Review* (April, 1996)

- **Canadian Case Studies**, at the end of each of Parts II, III, IV, V and VI, provide further examples of important applications and include their own Questions for Discussion. A special CBC video segment supplements each Canadian Case Study.

- Appendix I lists some standard Units of Measurement and Their Conversions
- Appendix II presents an annotated list of Environmental Organizations in Canada and abroad. Web sites are listed where available.
- Appendix III lists Canadian and International Journals, Magazines, and Periodicals on the Environment.
- Four Indices (Subject, Geographical, Chemical, and Species) make looking up items easy.
- For convenience, a Map showing the Ecozones of Canada (i.e., Figure 8.3) is restated on the inside of the front cover.

SUPPLEMENTS

The following supplements have been carefully prepared to accompany this new book:

- An *Instructor's Manual*, which includes suggested answers to all the Questions for Discussion at the end of each chapter and at the end of each Canadian Case Study.
- *Transparency Masters* of all the Figures and Tables in the text.
- A *Test Item File* of questions, organized by chapter, with the level of difficulty (i.e., easy, moderate, or difficult) indicated for each question.
- *PH Custom Test (For Windows)*, a special computerized version of the Test Item File, which enables instructors to edit existing questions, add new questions, and generate tests.
- *CBC/Prentice Hall Canada Video Library for Environmental Science*, is a special compilation of 8 video segments from news magazines of the Canadian Broadcasting Corporation. Each video segment has been carefully selected to accompany one of the Canadian Case Studies in the book.
- *College Newslink*, a service available for a fee, that selects articles by discipline from some of the leading North American newspapers and electronically transmits them daily.

ACKNOWLEDGEMENTS

A number of acknowledgements are in order. Numerous colleagues provided an extremely valuable service by informally reviewing one or more draft chapters of this book, and by making important ideas and information available to me. Inevitably, I was not able to incorporate all of the criticisms and suggestions, sometimes because they did not correspond with my own views or interpretations of the subject matter. However, the great majority of suggestions and criticisms offered by these people resulted in changes in preliminary manuscripts, and they improved the quality, and sometimes the accuracy, of the material. These helpful colleagues are: Gordon Beanlands, Christine Beauchamp, Stephen Beauchamp, Marian Binkley, Chris Corkett, Ray Cote, Roger Cox, Les Cwynar, Roger Doyle, William Ernst, Tracy Fleming, George Francis, David Gauthier, William Gizyn, Patricia Harding, Chris Harvey-Clarke, Owen Hertzman, Jeff Hutchings, Joseph Kerekes, Allan Kuja, Brian Le, Judy Loo, Annette Luttermann, Paul Mandell, Ian McLaren, Chris Miller, Pierre Mineau, Gunther Muecke, Neil Munro, David Nettleship, David Patriquin, Allan Pinder, Stephen Price, Robert Scheibling, Donald Stewart, Torgney Viegerstad, Richard Wassersug, Peter Wells, Mary-Anne White, Hal Whitehead, Martin Willison, Stephen Woodley, and Vince Zelazney.

In addition, I am grateful to the following instructors for providing formal reviews of part or all of the manuscript:

Danny Blair (University of Winnipeg)
Deborah Boersma (St. Clair College)
Thomas Clair (Environment Canada)
L. Gordon Goldsborough (University of Manitoba)
Peeter Kruus (Carleton University)
Lawrence E. Licht (York University)
M. Anne Naeth (University of Alberta)
Roger Saint-Fort (Mount Royal College)
Ted J. Spicer (Conestoga College)
James Michael Waddington (McMaster University)

A number of the professional staff at Prentice-Hall have been of great assistance in the preparation of this book, especially Sarah Kimball, Tracey Hawken, Maurice Esses, Susan James, Rosemary Tanner, and Deborah Starks.

Several personal acknowledgements are also in order. I thank my spouse, George-Anne Merrill, for her patient and uncomplaining tolerance of my work habits and lifestyle, and for being my best friend in spite of everything I do and don't do. Also, my children, Jonathan and Rachael, for mysterious motivations that succeeding generations engender in their parents, for tolerating my often busy days, and for not minding too much that I am frequently away from home.

If readers of this book have any comments, suggestions, or criticisms about the material or the approach, I will be very grateful if you would communicate your ideas to me for future improvements.

Bill Freedman,
Department of Biology
Dalhousie University
Halifax, Nova Scotia
B3H 4J1
(E-mail address: billfree@is.dal.ca)

ECOSYSTEMS AND HUMANS

Chapter Outline

Environmental science
The larger context: Earth, life, and ecosystems
The relationship between species and ecosystems
Environmental stressors and ecological responses
Human activities: important environmental stressors
Environmental ethics and world views
The environmental crisis

Chapter Objectives

After completing this chapter, you will be able to:

1. Define environmental science and ecology and distinguish between them.
2. Outline a hierarchical framework of the universe that includes consideration of life on Earth at various scales.
3. Identify important principles of the ecosystem approach.
4. Compare humans and other species in terms of their abilities to cope with environmental constraints.
5. Describe how environmental stressors and disturbances can affect species and ecosystems.
6. List three direct ways in which humans influence their environment.
7. Identify four broad classes of environmental values.
8. Describe five important world views.
9. Classify the issues of the environmental crisis into three categories and give examples of each.
10. Discuss environmental effects of humans as a function of both population and lifestyle.

Environmental Science

Environmental science investigates questions related to the rapidly increasing human population, the use and abuse of resources, damage caused by pollution and disturbance, and the endangerment and extinction of species and natural ecosystems. Typical questions that might be examined in environmental science include the following:

1. What might the human population be in Canada, or on Earth, in fifty years or in two centuries?
2. How can fossil fuels, a non-renewable resource, be integrated into a sustainable economy?
3. How can we harvest cod in Atlantic Canada, wild salmon in British Columbia, or wheat in the Prairie Provinces without degrading the resource?
4. What kinds of ecological damage are caused by acid rain, pesticides, or other kinds of pollution, and how can these be prevented or repaired?
5. How quickly are species and natural ecosystems becoming endangered or extinct in Canada and globally, and how can these ecological calamities be prevented?

Planet Earth is the third closest plant to the sun, and is the only place in the universe known to sustain life.

Photo: NASA/Karen Taylor

Environmental scientists examining these questions use several disciplines in their study. The most relevant of these scientific disciplines are atmospheric science, biology, ecology, geography, chemistry, geology, oceanography, physics, mathematics, statistics, computer science, and medical science. Ecology and geography themselves are also interdisciplinary fields.

Environmental science involves just the science-related aspects of *environmental studies*, which is also a highly *interdisciplinary* field. Environmental studies encompasses a wide diversity of kinds of knowledge, all relevant to the study of the effects of humans and their activities on environmental quality. Important non-science disciplines include anthropology, business, economics, ethics, law, literature, philosophy, political studies, psychology, religion, and sociology. The interdisciplinary nature of environmental studies is illustrated in Figure 1.1.

This book deals mainly with environmental science, although some relevant non-science topics are also discussed, such as environmental ethics, philosophy, and economics.

Of all the academic disciplines, ecology is the most relevant to environmental science, and in fact the two terms are commonly confused by many people and the popular media. **Ecology** can be simply defined as the study of the relationships among organisms and between organisms and their environment. Ecology is mostly a biological field of study, but knowledge of chemistry, computer science, mathematics, physics, geology, and other fields is also important. Figure 1.2 highlights the interdisciplinary nature of ecology.

Most academic ecologists do not study the relationships between human activities and environmental quality. Rather, they research and write about the structures and functions of natural ecosystems and the factors that influence these. However, increasing numbers of ecologists are focusing on the study of human (or *anthropogenic*) influences on ecosystems. This applied specialization is sometimes called *environmental ecology*: the ecological ef-

FIGURE 1.1 | THE INTERDISCIPLINARY NATURE OF ENVIRONMENTAL STUDIES

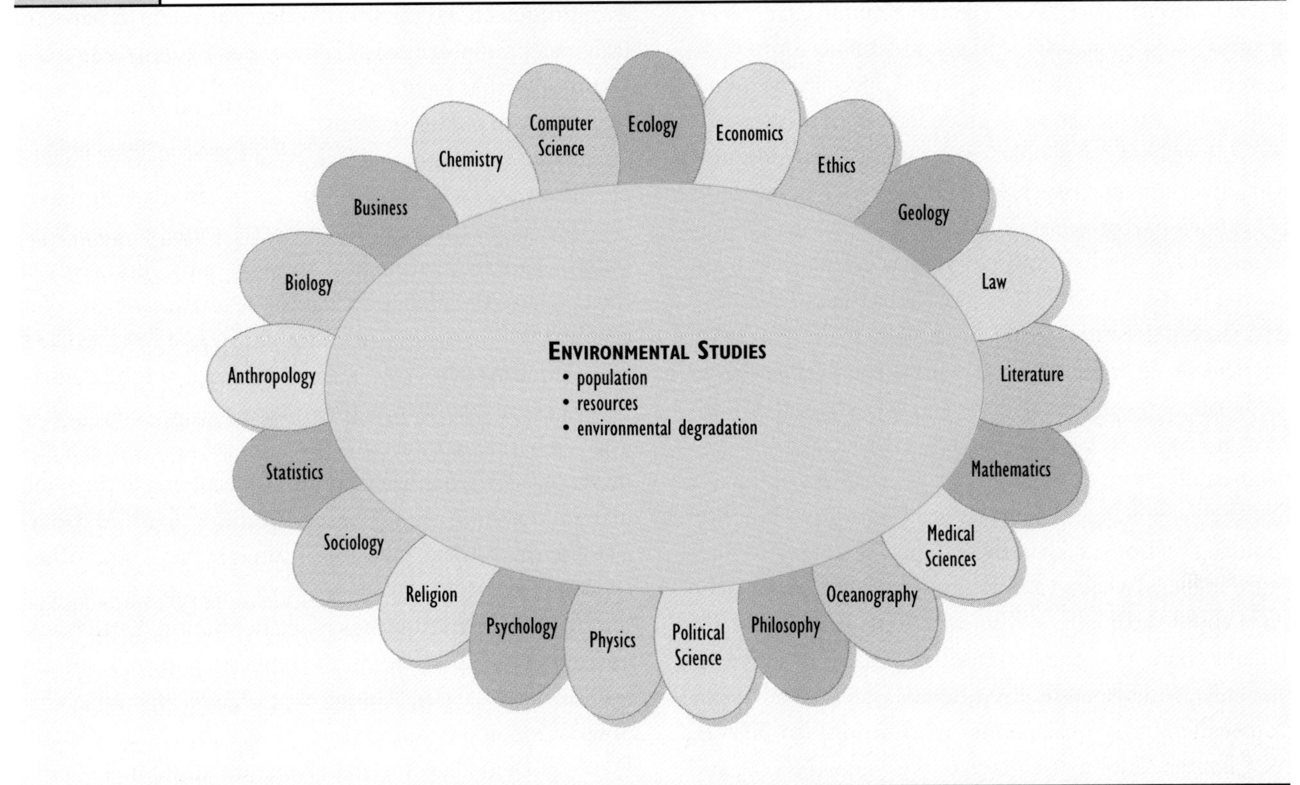

fects of pollution and disturbance. Major subject areas of environmental ecology are:

1. the management of biological resources, such as those important in agriculture, fisheries, and forestry;
2. the prevention or mitigation of ecological problems, such as those related to endangered biodiversity, restoration of degraded land or water, and management of greenhouse gases (such as carbon dioxide); and
3. the management of ecological processes such as productivity, nutrient cycling, hydrology, and erosion.

Ecologists are specialists who study relationships among organisms and their environment. *Environmental scientists*, on the other hand, are generalists, using any science-related knowledge relevant to environmental quality, such as air or water chemistry, climate modelling, or the ecological effects of pollution or disturbance. For example, three famous Canadian environmental scientists are David Schindler,

FIGURE 1.2 | ECOLOGY IS ALSO AN INTERDISCIPLINARY FIELD

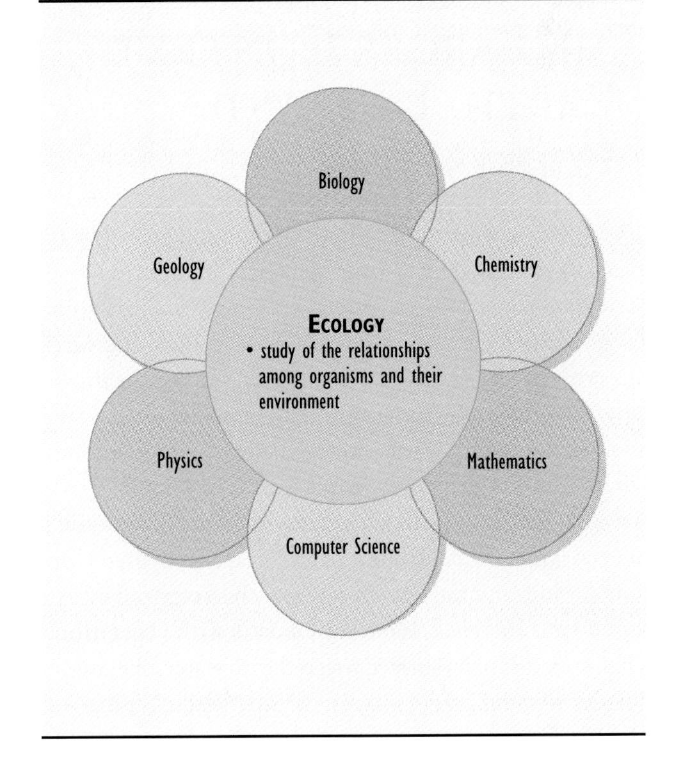

University of Alberta, who studies the effects of pollution and climate change on lakes; Hamish Kimmins, University of British Columbia, who works in sustainable forest management; and Tom Hutchinson, Trent University, who studies how human influences degrade terrestrial ecosystems.

A third group of people with a significant involvement with environmental issues, especially in an advocacy sense, are called **environmentalists**. Advocacy involves taking strong public stances for or against particular environmental issues. David Suzuki is perhaps the most famous environmentalist in Canada, because he so effectively influences the attitudes of Canadians through the popular media of television, newspapers, and books. Another well-known Canadian is Paul Watson, a direct-action environmentalist who operates under the auspices of the Sea Shepherd Society and is involved in non-governmental "policing" actions, such as the sabotage of enterprises engaged in illegal whaling and fishing. However, anyone who cares about the quality of the environment, and who advocates change for the better, can be called an environmentalist. Environmentalists often pursue their advocacy in so-called non-government organizations (or NGOs; see Chapter 25 for a discussion of the role of NGOs, and Appendix II for a list of Canadian organizations).

The Larger Context: Earth, Life, and Ecosystems

The universe consists of billions of stars and probably an even larger number of associated planets. Our Earth is one particular planet, located within a seemingly ordinary solar system, which consists of the sun, nine planets, and various orbiting comets and asteroids. Our Earth is the third-closest planet to the sun, orbiting that medium-sized star about every 365 days at an average distance of 149 million kilometres, and revolving on its axis every 24 hours. Earth is a spherical body with a diameter of about 12 700 kilometres. Approximately 70% of its surface is covered with liquid water, and the remaining area, characterized by exposed land and rock, is covered mostly with vegetation. With so much of its surface covered with water, one might wonder why our planet was not named "Water" instead of "Earth."

Like any other planet in the universe, Earth has its own unique characteristics. What makes Earth particularly exceptional are certain qualities (or factors) of its environment that have led to the genesis and subsequent evolution of living organisms and ecosystems. These favourable environmental factors include aspects of Earth's chemistry, surface temperature, and strength of gravity.

The beginning of life on Earth occurred about 3.5 billion years ago, within perhaps only one billion years of the origin of the planet during the formation of the solar system. We do not know exactly how life first evolved from inanimate matter, although almost all biologists today believe that life began spontaneously. In other words, genesis happened naturally, as a direct result of appropriate physical-chemical conditions occurring in the same place at the same time. These conditions involved chemistry, temperature, solar radiation, pressure, and other critical factors.

Aside from the musings of science fiction, Earth is celebrated as the only place in the universe known to sustain life and its associated ecological processes. Of course, this observation simply reflects our present state of knowledge. We do not actually know that living organisms do not exist elsewhere — only that life or its signals have not yet been discovered anywhere else in the universe. In fact, many (but not all) scientists believe that, because of the extraordinary diversity of environments that must exist among the innumerable planets that occur in the multitudinous solar systems of the universe, it is statistically likely that life forms have developed elsewhere. Nevertheless, the fact remains that Earth is the only planet definitely known to support organisms and ecosystems. This makes Earth an extraordinarily special place in the greater scheme of things.

We can consider the universe at various hierarchical levels (Figure 1.3). The scale ranges from the extremely small, such as sub-atomic particles and photons (an energy unit), to the fantastically large, such as galaxies and ultimately the universe itself.

Life on Earth occupies several intermediate levels of this hierarchy. The realm of ecology encompasses:

1. **individual organisms**, which are living entities that are genetically and physically discrete;
2. **populations**, individuals of the same species which occur together in time and space;
3. **communities**, populations of various species, also co-occurring at the same time and place;

FIGURE 1.3 HIERARCHICAL ORGANIZATION OF THE UNIVERSE

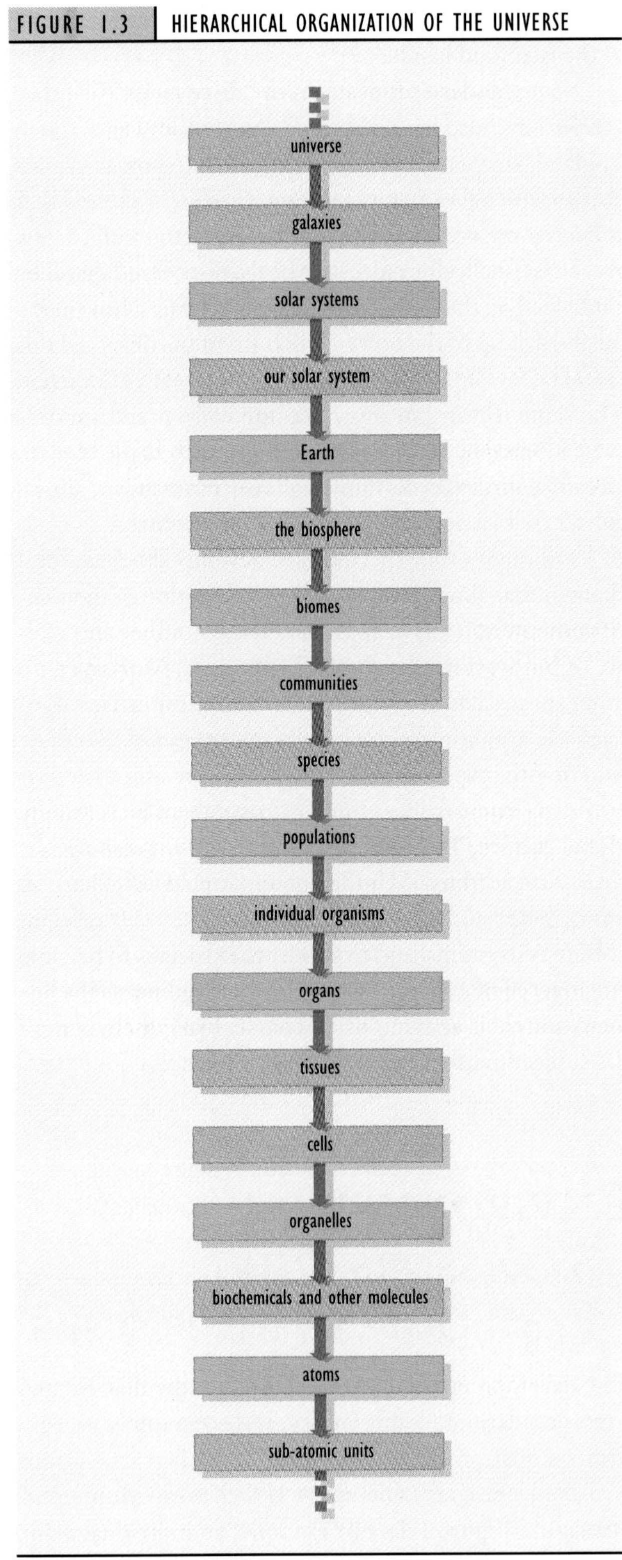

This diagram focuses on Earth, its ecosystems, and life, and suggests a linear structure to the organization of the universe. However, it is important to understand that all elements are intrinsically interconnected.

4. **landscapes** and **seascapes**, spatial integrations of various communities over large areas, known as landscapes in terrestrial environments and seascapes in marine environments; and
5. the entire **biosphere**, composed of all life and ecosystems on Earth.

The Relationship between Species and Ecosystems

A **species** is an aggregation of individuals and populations that can potentially interbreed and produce fertile offspring (see Chapter 7). The word "species" is both singular and plural. The word **ecosystem** is a generic term used to describe one or more communities of organisms that are interacting with their environment as a defined unit. As such, ecosystems can themselves be organized in a hierarchical fashion. Ecosystems can range from small units occurring in discrete microhabitats (for example, an aquatic ecosystem contained in a pitcher plant or a garden surrounded by pavement), to larger units such as landscapes and seascapes. Even the biosphere can ultimately be viewed as a single ecosystem.

Ecological interpretations of the natural world consider the diverse, web-like interconnections among the many components of ecosystems in a holistic manner. This **ecosystem approach** does not simplistically view each ecosystem as a random grouping of populations, species, communities, and environments — it confirms all of these as intrinsically connected and mutually dependent, although in varying degrees.

Another important ecological principle is that all species are sustained by environmental resources: the "goods and services" provided by ecosystems. All organisms require specific necessities of life, such as inorganic nutrients, food, and habitat with particular biological and physical qualities. Green plants, for example, need an adequate supply of moisture, inorganic nutrients such as nitrate and phosphate, sunlight, and space. Animals require suitable foods of plant or animal biomass (that is, organic matter), along with other habitat requirements that are different for each species.

It is important to understand that humans are no different from other species. We also depend on environmental resources, such as food, energy, shelter, and clean water, to sustain ourselves and our economies. Like all other species, humans are intimately connected with the ecosystems that sustain them. The reliance of humans on natural goods and services is an undeniable fact, even though this dependence may not always be apparent in our daily lives.

It follows that the development and growth of individual humans, their populations, and their societies and cultures are all limited to some degree by environmental factors. Examples of such constraints include excessively cold or dry climatic conditions, inhospitable terrain, and other factors that influence the rate of food production by agriculture or hunting. However, humans are often able to manipulate their environmental circumstances favourably. For example, crop productivity may be increased by irrigating or fertilizing agricultural lands or managing pests. In fact, humans can overcome their environmental constraints much more than any other species can. This is a distinguishing characteristic of the human species.

The human species is labelled by the scientific term *Homo sapiens*, a two-worded name (or *binomial*) which is Latin for "wise man." Indeed, humans are the most intelligent of all the species on Earth, with an enormous cognitive ability (that is, an aptitude for solving problems). When humans and their societies perceive an environmental constraint, such as scarcity of resources, they have commonly been able to understand which factors are causing the scarcity. Then they manipulate the environment accordingly. The clever solutions have generally involved management of the environment or other species to the benefit of humans, or the development of social systems and technologies that allow a more efficient exploitation of natural resources. Of course, not all problems have solutions. Where possible, however, humans have generally been able to discover short-term resolutions to their problems, and to utilize this knowledge to their advantage.

Humans are not the only species that can cope with ecological constraints in clever ways. Other species have learned to use rudimentary tools to exploit the resources of their environment more efficiently. For example, the woodpecker finch of the Galapagos Islands uses cactus spines to pry its food of insects out of fissures in bark and rotting wood. Chimpanzees modify twigs and use them to extract termites, a favoured food, from termite mounds. Egyptian vultures pick up stones in their beaks and drop them on ostrich eggs, breaking them and allowing access to the rich food inside.

Some modern innovations or "discoveries" by other species have also been observed. About fifty years ago in England, whole milk was hand-delivered to homes in glass bottles which had a bulbous compartment at the top that collected the cream as it separated from the milk. A few great tits (small, chickadee-like birds) discovered that they could feed on the nutritious cream by tearing a hole in the cardboard cap of the bottle. Other great tits observed this useful behavioural novelty and enthusiastically adopted it. The same feeding strategy became widespread and was even adopted by several other species, such as the blue tit. Cream-eating was certainly a clever innovation, allowing access to a novel and valuable food resource.

Although other species have developed behavioural changes that allow more efficient exploitation of their environment, none have approached the number and variety of innovations developed by humans. Moreover, no other species has developed a cumulative expertise for intensively exploiting such a broad range of useful resources. And no other species has managed to spread these capabilities as extensively as humans have, in an increasingly global culture. The human ability to exploit resources is clearly extraordinary. Unfortunately, humans also have an unparalleled ability to degrade resources, to disturb and pollute ecosystems, and to cause other species to become endangered or extinct. The intense damage caused by humans and their activities is, of course, a major element of the subject matter of environmental science.

Environmental Stressors and Ecological Responses

The development and productivity of individual organisms, populations, communities, and ecosystems are naturally constrained by environmental factors. These environmental constraints can be viewed as **environmental stressors** (Figure 1.4). For example, an individual plant may be stressed by inadequate nutrition, perhaps caused by infertile soil or by competition with nearby plants for scarce resources. Less-than-optimal access to nutrients, water, or sunlight results in physiological stress, which

FIGURE 1.4 CONCEPTUAL MODEL OF THE ECOLOGICAL EFFECTS OF ENVIRONMENTAL STRESS

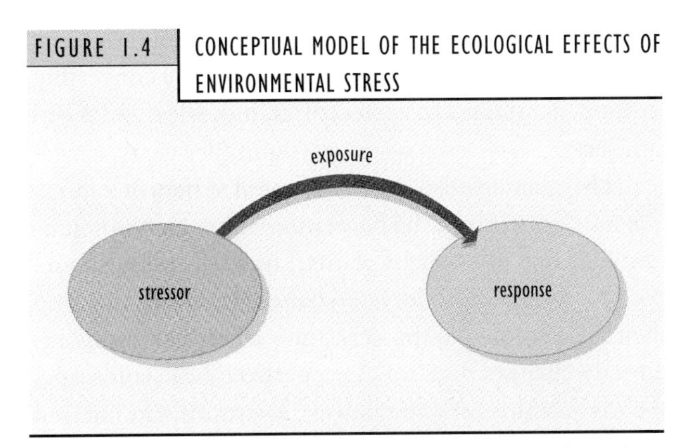

Environmental stressors are present at some variable intensity, which is referred to as "exposure." Effects on individual organisms, species, communities, or larger ecological units occur in response to exposures to one or more environmental stressors. See text for additional discussion.

causes the plant to be less productive than it is genetically capable of being. One result of this stress-response relationship is that the plant may develop relatively few seeds during its lifetime. Since evolutionary success is related to the number of progeny an organism leaves to carry on its genetic lineage, the evolutionary success of this individual plant is less than its potential.

Similarly, the development and productivity of an animal (including any human) are constrained by the environmental conditions under which it lives. For instance, an individual may have to deal with environmental stresses caused by food shortage or related to interactions with other animals through predation, parasitism, or competition for scarce resources.

The most benign (or least stressful) natural environments on Earth are characterized by conditions in which environmental factors such as moisture, nutrients, and temperature are not unduly constraining to biological processes; while extensive disturbances associated with wildfire, windstorm, disease, or other cataclysms are rare. These relatively benevolent conditions allow Earth's most complex and biodiverse ecosystems to develop, namely old-growth rainforests and coral reefs. Other environments, however, are characterized by conditions that are more stressful, and ecological development is therefore limited to less complex ecosystems, such as prairie, tundra, or desert.

All of Earth's ecosystems change profoundly, and quite naturally, over time. Many ecosystems are relatively dynamic, regularly experiencing large changes in their component species, quantities of biomass, and rates of functional properties such as productivity and nutrient cycling (these are described in following chapters). For example, ecosystems that occur in strongly seasonal climates usually have a discrete growing season, followed by a dormant period when little or no growth occurs. To varying degrees, all of the natural ecosystems of Canada are seasonally dynamic: a warm growing season is followed by a dormant period characterized by cold, wintry temperatures during which no plant productivity or growth occurs. Animals may survive the difficult time of winter by migrating, hibernating, or feeding on plant biomass remaining from the previous growing season.

Ecosystems that have been stressed by a recent **disturbance** (that is, an episode of destruction) are particularly dynamic, because they are undergoing a process of community-level, ecological recovery known as **succession**. Succession occurs in response to changes associated with natural disturbances such as wildfire, windstorms, and insect and disease epidemics. These cataclysmic factors can kill many of the dominant organisms in an ecosystem. This creates temporary opportunities for relatively short-lived species, which may dominate the earlier years of the post-disturbance, successional recovery.

Natural disturbance dynamics can be far-reaching, in some cases affecting extensive landscapes. For example, in most years millions of hectares of the conifer-dominated boreal forest of northern Canada are disturbed by wildfires. Similarly, millions of hectares of forest in eastern Canada may be affected by sudden increases in abundance of spruce budworm, a moth whose caterpillars can kill most of the mature trees in vulnerable, fir-spruce stands. An even more extensive example occurred about 15 000 years ago, when glaciation covered virtually all of Canada with enormous ice-sheets, several kilometres thick in places. All of today's ecosystems in Canada have developed since the melt-back of that latest glacial advance.

In other cases, disturbances can be quite local in scale. For example, the death of a large tree within an otherwise intact forest creates a local zone of damage, referred to as a *microdisturbance*. This small-scale disturbance induces a local *microsuccession* of vigorously growing plants that attempt to achieve individual success by occupying the newly available gap in the forest canopy.

Even Earth's most stable ecosystems change inexorably over time. However, the dynamics of change in tropical rainforests and in the communities of the deepest regions of the oceans are generally slow. Catastrophic disturbances may affect these stable ecosystems, but they are rare under natural conditions. Nevertheless, as with all of Earth's ecosystems, these relatively stable types are profoundly influenced by widespread changes in climate and by other long-term dynamics, such as evolution.

In fact, natural environmental and ecological changes have caused the extinctions of almost all of the species that have ever lived on Earth since life began about 3.5 billion years ago. Many of these extinctions occurred because species could not cope with the demands of changes in such environmental factors as climate or in such biological interactions as predation, competition, or disease. However, many of the extinctions appear to have occurred synchronously (i.e., at about the same time), and were presumably caused by unpredictable, catastrophic disturbances, such as a large meteorite colliding with the Earth. (See Chapters 7 and 24 for discussions of natural extinctions and those caused by human influences.)

Environmental stressors and disturbances have always been an important, natural context for life on Earth. So, too, have been the resulting ecological responses, including changes in species and the dynamics of their communities and ecosystems.

Human Activities: Important Environmental Stressors

These days, of course, the character and development of ecosystems are influenced not only by "natural" environmental stresses. In many situations, human influences are the major constraining factor on the productivity of species and on the structure and function of ecosystems more generally. Throughout the history of *Homo sapiens*, the scale, frequency, and intensity of environmental stressors and disturbances have been altered through the activities of humans. These anthropogenic influences have intensified enormously in modern times.

Anthropogenic influences on the environment can be direct or indirect. Humans directly affect ecosystems and species in three ways: a) by harvesting economically valuable biomass, such as trees and hunted animals; b) by causing toxicity through pollution; and c) by converting natural ecosystems to agricultural, industrial, or urban land-uses.

These actions also engender a great variety of indirect effects. For example, the harvesting of trees alters conditions for the diversity of plants, animals, and microorganisms that require forested habitat, thereby affecting their populations. At the same time, forest harvesting indirectly changes functional properties of the landscape, such as quantities of water flowing in streams, erosion, and productivity. Both the direct and indirect effects of humans on ecosystems are important.

Humans have always left "footprints" in nature; that is, to some degree they have always influenced the ecosystems of which they are a component. During most of the more than one million years of evolution of *Homo sapiens*, that footprint was relatively shallow and obscure. This was because the capability of humans for exploiting their environment was not much different from that of other similarly abundant, large species. However, during the cultural evolution of our species, the ecological changes associated with human activities intensified progressively. This process of **cultural evolution** has been characterized by increasingly more sophisticated methods, tools, and social organizations used to secure resources by exploiting the environment and other species.

Certain important advances during the cultural evolution of humans represented great increases of capability. Because of their enormous influence on human success, these advances are referred to as "revolutions." Examples of earlier technological revolutions include:

1. methods for fashioning weapons to hunt animal prey;
2. domestication of the dog, which greatly facilitated hunting;
3. methods for the domestication of fire, which provided warmth and cooked, more-digestible foods;
4. methods for cultivating and domesticating plants and livestock, resulting in huge increases in food availability; and
5. techniques for smelting metal-containing minerals and working the raw metals into tools, which were much better than those made of wood, stone, or bone.

The rate of new discoveries has increased enormously over time. More recent technological revolutions include:

1. methods of utilizing fuels and machines to accomplish work previously done by humans or draught animals,
2. further advances in the domestication and cultivation of plants and animals,
3. discoveries in medicine and sanitation, and
4. extraordinary strides in communications and information-processing technologies.

These revolutionary innovations led to substantial, and sometimes enormous, increases in the abilities of humans to exploit the resources of their environment and to achieve population growth (see Chapter 10). Unfortunately, enhanced exploitation has rarely been accompanied by the development of a compensating cultural ethic — that is, one that encourages conservation of the dwindling resources required for survival. Even the early hunting societies, more than eight thousand years ago, caused exterminations of species that they hunted too efficiently (see Chapter 24). Enthusiastic over-harvesting of resources continues today, and much more efficiently than in prehistoric times.

The diverse effects of human activities on environmental quality are vital issues in environmental science, and we will discuss them in detail later. For now, we emphasize the message that substantial environmental stresses are associated with diverse human activities, and these are forcing environmental and ecological changes on Earth. Many of these changes are reducing the ability of the environment and ecosystems to sustain humans and their societies. Human activities are also causing enormous degradation of natural ecosystems in general, including the habitats required to support other species.

In fact, the environmental and ecological damage caused by humans is so severe that an appropriate metaphor for the human enterprise is that of a malignancy or cancer. This is a sobering image. It is useful to dwell on it so that its meaning does not escape our understanding. **Humans and their activities are endangering species and natural ecosystems at such a tremendous scale and rate that the integrity of Earth's life-support systems is at risk.** From the ecological perspective, the pace and intensity of these changes is truly staggering. Moreover, the damage will become substantially worse, before corrective actions are (hopefully) undertaken to stop and reverse the degradation, allowing an ecologically sustainable human enterprise to become possible. From a pessimistic perspective, however, it may prove to be beyond the capability of human societies to act effectively to reverse the damage and to design and implement solutions for sustainability.

Modern consumerism results in huge demands for material and energy resources to manufacture and operate machines.
Photo: Tom Tracy/Tony Stone Images

These are, of course, only opinions, albeit the highly informed, expert opinions of many environmental scientists. Anticipating the future is always uncertain, and things may turn out to be less grim than is now commonly predicted. For example, we might be wrong about the amounts of resources needed to sustain future generations of humans. Still, the clear indications from recent patterns of change are that the environmental crisis is severe, and that it will worsen in the foreseeable future.

The belief that humans are causing grievous damage to the life-support systems of Earth, and the metaphor of human activities as a malignancy, should not be dismissed as the opinions of misanthropic individuals. These beliefs are remarkably common among ecologists and environmental scientists. These opinions result from objective interpretations of recent trends in environmental quality, and of reasonably anticipated changes in the future. If many ecologists and environmental scientists have predicted correctly, the future may, indeed, be quite grim. The causes of this future predicament are environmental damage and resource shortages associated with human activities.

All this damage is not, however, inevitable. There is hope that human societies will yet make appropriate adjustments and will choose to pursue options that are more sustainable than those now being followed.

Environmental Ethics and World Views

Human actions can influence environmental quality by affecting the availability of resources, by causing pollution, and by endangering species and natural ecosystems. Decisions influencing environmental quality are influenced by two types of considerations: *knowledge* and *ethics*.

In the sense meant here, *knowledge* refers to information and understanding about the natural world, and *ethics* refers to the perception of right and wrong and the appropriate behaviour of people toward each other, other species, and nature. Humans can, of course, choose to interact with the environment and ecosystems in various ways. On the one hand, knowledge provides guidance about the consequences of alternative choices, including damage that might be caused and actions that might be taken to avoid that damage. On the other hand, ethics provides guidance about which alternative actions should be favoured or even allowed to occur.

According to the biocentric and ecocentric world views, all species have equal intrinsic value. This does not, however, mean that one species cannot exploit another.

Photo: Wagne R. Bilenduke/Tony Stone Images

The influence of both knowledge and ethics on choices is critical, because modern humans have enormous powers to utilize and damage the environment. And we can choose from among various alternatives. For example, individual people can choose whether to have children, purchase an automobile, or eat meat, while society can decide whether to allow hunting of whales, clear-cutting of forests, or constructing nuclear-power plants. All of these choices have implications for resource consumption, pollution, biodiversity, and other elements of environmental quality.

Perceptions of *value* (that is, of merit or importance) profoundly influence interpretations of the consequences of human actions. **Environmental values** can be divided into four broad classes: utilitarian, ecological, aesthetic, and moral.

Utilitarian values are based on the known importance of something to the welfare of humans (see also the discussion of the anthropocentric world view, below). Accordingly, components of environmental quality are only considered important if they are resources necessary to sustain humans, that is, if they bestow economic benefits and provide livelihoods.

Ecological values are based on the needs of humans as well as on those of other species and natural ecosystems. Ecological values are broader than utilitarian values and take a longer-term view.

Aesthetic values are based on an appreciation of the beauty of nature. They are subjective and influenced by cultural perspectives. Aesthetics might value natural wilderness over human-dominated ecosystems, and free-living whales over whale meat. Maintenance of aesthetic values can provide substantial cultural, social, psychological, and economic benefits.

Moral values are based on the belief that components of the natural environment (such as species and natural ecosystems) have intrinsic value and a right to exist, regardless of any positive, negative, or neutral relationships with humans. Under this system, it would be wrong for humans to treat other organisms cruelly, or to take actions that cause other natural entities to become endangered or extinct, or to fail to prevent such an occurrence.

As noted previously, ethics concerns the perception of right and wrong, and the values and rules that should govern human conduct. Clearly, ethics of all kinds depend upon the values that people believe are important. **Environmental ethics** deals with the responsibilities of the present human generation to future generations of people and other species, to ensure that the world will continue to provide adequate resources and livelihoods (this is a key element of sustainable development; see Chapter 12). The environmental values discussed above underlie this system of ethics. Applying environmental ethics often means analyzing and balancing standards that may conflict, since utilitarian, ecological, aesthetic, and moral values rarely coincide with each other (see In Detail 1.1).

Values and ethics in turn support larger systems known as world views. *World views* are comprehensive philosophies of human life and the universe, and of the relation-

IN DETAIL 1.1

OLD-GROWTH FOREST: A CASE OF VALUES IN COMPETITION

Ethics and values are substantially influenced by cultural attitudes. Because the attitudes of people vary considerably, proposals to exploit natural resources as economic commodities often give rise to controversy. Consider, for example, the case of old-growth rainforests on Vancouver Island.

Old-growth forests in the coastal zone of British Columbia contain many ancient trees, some of which are hundreds of years old and of gigantic height and girth (see Chapter 23). The cathedral-like aesthetics of old-growth forests are inspiring to many people, providing them with a deeply natural, even religious experience. Major elements of the culture of the First Nations of coastal British Columbia are based on values associated with old-growth forests. Whatever their culture, however, few people fail to be inspired by a walk through a tract of old-growth forest on Vancouver Island.

Old-growth forests are also a special kind of natural ecosystem, different from other forests, and supporting species that cannot survive in other kinds of habitat. These ecological characteristics give coastal B.C.'s old-growth forests an intrinsic value that is not replicated elsewhere in Canada. They represent a distinct element of our natural heritage.

Old-growth forests are also an extremely useful natural resource, because they contain large trees of desirable species that can be harvested and manufactured into lumber or paper. If utilized in this manner, old-growth forests can provide livelihoods for people and revenues for local, provincial, and national economies. Old-growth forests also support other economic values, including deer and salmon that can be harvested by hunters and fishers, and birds and wildflowers that entice ecotourists to visit these special ecosystems. Intact old-growth forests also provide other valuable services, such as yielding a supply of clean stream water and helping to regulate atmospheric concentrations of such gases as oxygen and carbon dioxide.

At one time, old-growth forests were common on Vancouver Island, but they are now endangered there and almost everywhere else in Canada. This has happened largely because old-growth forests have been extensively harvested and replaced by younger, second-growth forests. The second-growth forests are themselves harvested as soon as they become economically mature, which happens long before they can develop into old-growth forests. The net result is rapidly diminishing areas of old-growth forests and endangerment of the ecosystem type and some of its dependent species.

Obviously, the different values concerning old-growth forests on Vancouver Island are in severe conflict. Proposals to harvest old-growth forests to supply raw materials to sawmills and pulp mills are incompatible with other proposals to protect these special ecosystems in parks and ecological reserves. The conflicting perceptions of values have resulted in emotional confrontations between loggers and preservationists, in some cases resulting in civil disobedience, arrests, and jail terms. Ultimately, these controversies can only be resolved by finding a balance between the utilitarian, ecological, aesthetic, and moral values of old-growth forests, and ensuring that all these values are sustained into the future.

ship between humans and the natural world. World views include traditional religions as well as other belief systems. Three important environmental world views are known as anthropocentric, biocentric, and ecocentric, while the frontier and sustainability world views are more directly related to the use of resources.

The **anthropocentric world view** considers humans as the centre of the universe. Humans are seen as more worthy than any other species and as being uniquely disconnected from the rest of nature. Thus, the anthropocentric world view judges the importance and worthiness of anything, including other species and ecosystems, in terms of the implications for human welfare.

The **biocentric world view** considers all species (and individuals) as having equal intrinsic value. Humans are seen as a unique and special species, but no more important or worthy than any other species. Thus, the biocentric world view rejects discrimination against other species, known as *speciesism* (a term similar to racism or sexism).

The **ecocentric world view** considers the direct and indirect connections among species within ecosystems to be invaluable. It incorporates the biocentric world view but goes beyond it by stressing the importance of interdependent ecological functions, such as productivity and nutrient cycling.

The **frontier world view** asserts that humans have a right to exploit nature by consuming natural resources in boundless quantities. Based on utilitarian values and derived from the anthropocentric world view, the frontier world view claims that humans are superior and have a right to "subdue" and exploit nature. The supply of resources to sustain humans is considered to be limitless, because new resources or substitutes can always be discovered. The consumption of resources is considered to be good because it enables economies to grow. Nations and individuals should be allowed to consume resources aggressively, as long as no humans are hurt in the process.

The **sustainability world view** acknowledges that humans must have access to vital resources, but the exploitation of those resources should be governed by appropriate ecological, aesthetic, and moral values. The sustainability world view can assume various forms. The **spaceship world view**, for example, is quite anthropocentric. It focuses only on sustaining those resources needed by people, and it assumes that humans can exert a great degree of control over natural processes and can pilot "spaceship Earth." In contrast, **ecological sustainability** is more ecocentric. It considers humans within an ecological context, and focuses on sustaining all components of Earth's life-support system by preventing human actions that would degrade them. In an ecologically sustainable economy, natural goods and services should be utilized only in ways that do not compromise their future availability and do not endanger the survival of species or natural ecosystems.

The attitudes of humans and their societies toward other species, natural ecosystems, and resources have enormous implications for environmental quality. Extraordinary damages have been legitimized by attitudes based on a belief in the inalienable right of humans to harvest whatever they desire from nature, without considering pollution, threats to species, or the availability of resources for future generations of people. One of the keys to resolving the environmental crisis is to achieve a widespread adoption of ecocentric and sustainability world views.

The Environmental Crisis

The current environmental crisis embraces many issues. In large part, we can classify the issues into three categories: population, resources, and environmental quality.

Population

In 1997, the human population numbered about 6 billion globally, including about 30 million in Canada. The human population has been increasing because of changes in the balance between certain *demographic* parameters, particularly the excess of the human *birth rate* over the *death rate.* The recent explosive population growth, and the poverty of so many people, is the root cause of much of the environmental crisis. Directly or indirectly, population growth leads to extensive deforestation, expanding deserts, erosion, water shortages, climate change, endangerment and extinction of species, and other problems. Considered together, these environmental degradations represent changes in the character of Earth's biosphere that are as cataclysmic as major geological events, such as glaciation.

We will discuss human population in more detail in Chapters 10 and 11.

Resources

We can distinguish between two types of natural resources. A **non-renewable resource** is present on Earth in a finite quantity. These resources can only be mined, so as they are used, they are available in smaller and smaller quantities for future generations. Non-renewable resources include metals and fossil fuels such as petroleum and coal.

A **renewable resource** can regenerate after harvesting, and if managed suitably, can provide a supply that is sustainable virtually forever. To be renewable, the ability of the resource to regenerate cannot be compromised by environmental damage or inappropriate management practices. Examples of renewable resources include fresh water; the biomass of trees, agricultural plants, and livestock; and hunted animals such as fish and deer. Ultimately, sustainable economies can be supported only by renewable resources. Too often, however, potentially renewable resources are not used wisely, thereby impairing their renewal and representing a type of mining. Resources will be discussed in Chapters 12, 13, and 14.

Environmental Quality

Environmental quality deals largely with anthropogenic pollution and disturbance, and the effects of these on humans, their economies, other species, and natural ecosystems. Pollution can be caused by pesticides, gases emitted by power plants and smelters, and hot water discharged by factories into lakes. Examples of disturbance include clear-cutting, fishing, and forest fires. We will study the consequences of pollution and disturbance in terms of their effects on biodiversity, climate change, resource availability, risks to human health, and other aspects of environmental quality in Chapters 15 to 24.

Impact of Humans on the Environment

In general terms, the cumulative impact of humans on the biosphere is a function of two factors: the size of the human population and the per capita environmental impact. The human population varies greatly among and within countries, as does the per capita impact, which depends on the nature and degree of technological development.

Places where people live, work, grow food, and harvest natural resources are affected by many kinds of anthropogenic stressors. These result in ecosystems that are not very natural in character, such as the pavement and grassy edges of a major highway in Toronto.

Photo: B. Freedman

How does Canada's impact on the environment compare with that of much more populous countries, such as China and India? Let us use energy use and gross domestic product (or GDP, the annual value of all goods and services produced by a country) as simple *indicators* of the degree of environmental impact of both individual people and national economies. Energy use is a useful indicator because energy is required in virtually all activities in an industrial society, including driving automobiles, heating homes, mining metal ores and fossil fuels, harvesting forests and fish, manufacturing industrial products, running computers, and watching television. Gross domestic product represents all of the economic activities in a country, each of which results in some degree of environmental damage.

Canada, with a population of 28.5 million in 1995, has a much smaller population than China (1238 million) or India (931 million), which are the two most populous countries in the world (Table 1.1). However, as indicated by per capita energy use and per capita GDP, individuals in

Canada affect the environment much more intensely than do individuals in China or India. Overall, in terms of national energy use and national GDP, the environmental impacts of Canada, China, and India are rather similar.

This is a remarkable observation. Although the population of Canada is relatively small, the intensive resource use by Canadians means that their aggregate effect on the environment is large, and comparable to that of more populous India or China. We can conclude that **the environmental crisis is due to both overpopulation and excessive resource consumption**.

TABLE 1.1 | THE RELATIVE ENVIRONMENTAL IMPACTS OF CANADA, CHINA, AND INDIA

The environmental impacts of Canada, China, and India, and of their individual citizens, can be roughly compared using energy use and gross domestic product as simple indicators. National energy use equals population multiplied by per capita energy use, and national GDP equals population times per capita GDP.

		ENERGY USE		GROSS DOMESTIC PRODUCT	
COUNTRY	POPULATION (MILLIONS)	PER CAPITA (10^9 JOULE)	NATIONAL (10^{18} J)	PER CAPITA (10^3 \$US)	NATIONAL (10^9 \$US)
Canada	28.5	324	9.23	20.7	591
China	1238	23	28.5	0.364	451
India	931	9	8.37	0.330	307

Note: Population data are for 1995; energy-use and GDP data are for 1991.

Source: Data from World Resources Institute (1995)

Key Terms

- environmental science
- ecology
- environmentalist
- individual organism
- population
- community
- landscape
- seascape
- biosphere
- ecosystem
- ecosystem approach
- species
- environmental stressors
- disturbance
- succession
- cultural evolution
- environmental values
- environmental ethics
- anthropocentric world view
- biocentric world view
- ecocentric world view
- spaceship world view
- sustainability world view
- frontier world view
- ecological sustainability
- non-renewable resources
- renewable resources

Questions for Discussion

1. Identify the important environmental stressors that you think may be affecting a familiar ecosystem in your area (such as a local park). Make sure that you consider both natural and anthropogenic stressors.
2. List your connections with ecosystems that occur through the resources you consume (as food, energy, and materials) as well as your recreational activities. Which of these linkages could you do without?
3. How are your own ethical standards related to utilitarian, ecological, aesthetic, and moral values? Think about your world view, and describe how it relates to the anthropocentric, biocentric, and ecocentric models of world views.
4. According to information presented in this chapter, Canada might be considered as overpopulated as India and China. Do you believe that this is a reasonable conclusion? Justify your answer.

References

Botkin, D. 1990. *Discordant Harmonies. A New Ecology for the 21st Century.* New York: Oxford University Press.

Callicott, J.B. 1988. *In Defence of the Land Ethic: Essays in Environmental Philosophy.* Albany, NY: State University of New York Press.

Devall, B. and G. Sessions. 1985. *Deep Ecology: Living as if Nature Mattered.* Salt Lake City, UT: Peregrine Smith Books.

Evernden, L.L.N. 1985. *The Natural Alien: Humankind and Environment.* Toronto, ON: University of Toronto Press.

Evernden, L.L.N. 1992. *The Social Creation of Nature.* Baltimore, OH: Johns Hopkins Press.

Freedman, B. 1995. *Environmental Ecology: The Ecological Effects of Pollution, Disturbance, and Other Stresses.* San Diego, CA: Academic Press.

Hargrove, E.C. 1989. *Foundations of Environmental Ethics.* Englewood Cliffs, NJ: Prentice Hall.

Leopold, A. 1949. *A Sand County Almanac.* New York: Oxford University Press.

Livingston, J.A. 1994. *Rogue Primate: An Exploration of Human Domestication.* Toronto, ON: Key Porter Books.

Miller, G.T. 1990. *Living in the Environment.* Belmont, CA: Wadsworth Pub. Co.

Nash, R.F. 1988. *The Rights of Nature: A History of Environmental Ethics.* Madison, WI: University of Wisconsin Press.

Regan, T. 1984. *Earthbound: New Introductory Essays in Environmental Ethics.* New York: Random House.

Rowe, J.S. 1990. *Home Place: Essays on Ecology.* Edmonton, AB: NeWest Pub.

Schumacher, E.F. 1973. *Small Is Beautiful.* New York: Harper & Row.

Shrader-Frechette, K. 1986. Environmental ethics and global imperatives. pp. 97–127 in: *Global Possible: Resources, Developments and the New Century.* (R. Repetto, ed.) New Haven, CT: Yale University Press.

Singer, P. 1990. *Animal Liberation.* New York: New York Review of Books.

Wackernagle, M. and E.E. Rees. 1996. *Our Ecological Footprint: Reducing Human Impact on the Earth.* Gabriola Island, BC: New Society Publishers.

White, L. 1967. The historical roots of our ecologic crisis. *Science* **155**: 1203–1207.

Wilson, E.O. 1984. *Biophilia.* Cambridge, MA: Harvard University Press.

World Resources Institute. 1995. *World Resources 1994–95.* New York, NY: Oxford University Press.

CBC CANADIAN CASE 1

BANFF NATIONAL PARK: WILDERNESS OR DEVELOPMENT?

Human welfare is intrinsically connected to the welfare of other species and ecosystems. People and their economies cannot survive without access to resources obtained from the environment — as sources of food, materials, energy, and recreation. However, the exploitation of natural resources always causes some degree of damage to other species, ecosystems, and environmental quality. This damage is said to be "unsustainable" when it exceeds the level that the system can tolerate without suffering irretrievable destruction.

A specific example of unsustainable use of a potentially renewable natural resource can be seen in Banff National Park. The ability of the ecosystems in Banff and its surrounding area to deliver a variety of "goods and services" to Canadian society is being seriously hampered. The two most important "goods and services" are the recreational opportunities for millions of people each year and corridors for crucial elements of the national transportation system (the Trans-Canada Highway and the Canadian Pacific Railway). These uses support the livelihoods of thousands of people who live in and near the park. They deliver enormous tax revenues to regional, provincial, and national economies. Unfortunately, Banff National Park cannot sustain all of these uses by humans, while continuing to maintain populations of such key species as grizzly bear and timber wolf and protecting desirable natural values such as wilderness.

Established in 1885, Banff National Park was the first of Canada's national parks. Because of its rugged beauty and outstanding natural values, Banff is considered to be the "jewel in the crown" of the Canadian system of national parks. Unfortunately, the special values of this internationally acclaimed natural area are being severely threatened by excessive development. Ecologists believe that the limits of sustainable use of Banff National Park and its surrounding area have already been exceeded. There is not a social consensus on this issue, however, because mainstream business and political interests believe that much more economic development can and should be undertaken in the region. This dichotomy about sustainable use is the source of controversy.

Questions for Discussion

1. Make a list of the kinds of environmental damages being caused by the different human uses of Banff National Park. Compare these damages with those being caused to a protected area near where you live.
2. Should national parks be expected to sustain viable populations of grizzly bear, timber wolf, and other large mammals that sometimes have conflicts with humans? Why or why not?
3. Is it appropriate for national parks to contain sizeable commercial towns (such as the town of Banff and the village of Lake Louise)? What about golf courses, hotels and lodges, extensive campgrounds, ski hills, and other ecologically damaging infrastructure that support outdoor recreation? List arguments both for and against these kinds of economic developments in national parks.

Video Resource: "Jewel in the Crown," *The National Magazine* (September 11, 1996)

2

SCIENCE AS A WAY OF UNDERSTANDING THE NATURAL WORLD

CHAPTER OUTLINE

The nature of science
Inductive and deductive logic
Goals of science
Facts, hypotheses, and experiments
Uncertainty

CHAPTER OBJECTIVES

After completing this chapter, you will be able to:

1. Describe the nature of science and its usefulness in explaining the structure and function of the natural world.
2. Distinguish between facts, hypotheses, and theories.
3. Outline the methodology of science, including the critical importance of tests designed to disprove hypotheses.
4. Discuss the importance of uncertainty in many scientific predictions and the relevance of this to environmental controversies.

The Nature of Science

Science can be defined as the systematic study of the character and behaviour of the natural, physical world. Science is also a rapidly enlarging body of knowledge about the natural world. A goal of science is to discover the simplest general principles that explain the enormous complexity of the natural world. These principles can then be used to gain insights about the structure and function of nature and to make predictions about future changes.

Science is a relatively recent way of learning about natural phenomena, having largely replaced other, less objective methods and world views. The major alternatives to science are belief systems that are influential in all cultures, including those based on religion, morality, and aesthetics. Modern science evolved from natural philosophy, a way of learning developed by classical Greeks that was concerned with rational investigations of existence, knowledge, and phenomena. Compared with modern science, however, analyses in natural philosophy use relatively unsophisticated technologies and methods and are not very quantitative, sometimes involving only the application of logic.

Modern science began with the systematic investigations of such famous 16th- and 17th-century scientists as:

1. Nicolaus Copernicus (1473–1543), a Polish astronomer who first conceived the modern theory of the solar system;
2. Galileo Galilei (1564–1642), an Italian who conducted research on the physics of objects in motion as well as astronomy;
3. William Harvey (1578–1657), an Englishman who described the circulation of the blood;
4. William Gilbert (1544–1603), an Englishman who worked on magnetism; and
5. Isaac Newton (1642–1727), an Englishman who made important contributions to understanding gravity, laws of motion, the nature of light, and the mathematics of calculus.

Inductive and Deductive Logic

The English philosopher Francis Bacon (1561–1626) was also highly influential in the development of modern science. Bacon was not an actual practitioner of science, but he was a strong proponent of its emerging methodologies. He promoted the usefulness of **inductive logic**, in which conclusions are objectively developed from the accumulating evidence of experience and the results of experiments. Inductive logic can lead to unifying explanations based on large bodies of data and observations of natural phenomena. Consider the following illustration of inductive logic, applied to an environmental topic:

Observation 1: marine mammals off the Atlantic coast of Canada have large residues of DDT and other chlorinated hydrocarbons in their fat and other body tissues.

Observation 2: so do marine mammals off British Columbia.

Observation 3: as do marine mammals in the Arctic Ocean, although in lower concentrations.

Inductive conclusion: There is widespread contamination of marine mammals with chlorinated hydrocarbons. Further research may demonstrate a global contamination. This suggests a potentially important environmental problem.

In contrast, the application of **deductive logic** involves making initial assumptions and then drawing logical conclusions from those assumptions. Consequently, the truth of deductive conclusions depends entirely on the truth of the original assumptions. If those assumptions are based on false information or on incorrect supernatural belief, then any deduced conclusions about natural phenomena are likely to be wrong. Consider the following illustration of deductive logic:

Assumption: TCDD, an extremely toxic chemical in the dioxin family, is poisonous when present in even the smallest concentrations in food and water — even a single molecule of this substance can cause toxicity.

Deductive conclusion 1: No exposure to TCDD is safe.

Deductive conclusion 2: No emissions of TCDD should be allowed.

The two conclusions are consistent with the original assumption. However, scientists disagree about that assumption. Many toxicologists believe that exposures to TCDD (and other potentially toxic chemicals) must exceed a "threshold of biological tolerance" before any poisoning will result (see Chapter 15). In contrast, other scientists believe that even the smallest exposure to TCDD carries some degree of toxic risk. In this and many other cases in the environmental field, the science issues are not yet resolved, and may never be.

In general, inductive logic plays a much stronger role in modern science than deductive logic does. In both cases, however, the usefulness of any conclusion depends greatly on the accuracy of the observations and data on which it is based. Poor data may lead to an inaccurate conclusion through the application of inductive logic, as will inappropriate assumptions in deductive logic.

Goals of Science

The goals of science are to understand natural phenomena and to explain any changes in them. To achieve those goals, scientists undertake scientific investigations which are based on information, inferences, and conclusions developed through a systematic application of inductive logic. Scientists observe natural phenomena and conduct experiments.

A higher goal of scientific research is to formulate laws that describe the workings of the universe in general terms. (See, for example, Chapter 4 for a discussion of the laws of thermodynamics, which describe the transformations of energy among its various states.) Universal laws, along with *theories* and *hypotheses*, are used to understand natural phenomena. Many natural phenomena, however, are extremely complex and may never be fully understood in terms of physical laws. This may be particularly true of the ways that ecosystems work and are organised.

Scientific investigations can be *pure* or *applied*. Pure science is the unfettered search for knowledge and understanding, without regard for its usefulness in human welfare. Applied science is more goal-oriented, and deals specifically with practical difficulties and problems of one sort or another. Applied science might examine how to improve the technological instruments available to society, to solve problems in the management of natural resources, or to reduce pollution or manage other kinds of environmental damage associated with human activities.

Facts, Hypotheses, and Experiments

A **fact** is an event or thing that is known to have happened, to exist, and to be true. Facts are based on experience and scientific evidence. In contrast, a **hypothesis** is a proposed explanation of the occurrence or cause of a phenomenon. Scientists formulate hypotheses as statements and test them through experiments and other forms of research. Hypotheses are developed using logic, inference, and mathematical arguments to explain the nature of observed phenomena. It must always be possible to refute scientific hypotheses, and they can never ultimately be confirmed (see discussion of *null hypotheses*, below). Thus, the hypothesis that "cats are so intelligent that they prevent humans from discovering it" cannot be logically refuted, and so is not a scientific hypothesis.

Theory is a broader term that refers to a set of laws, rules, and explanations. These are supported by a large body of experimental and observational evidence, all leading to robust, internally consistent conclusions. Some of the most famous theories in science are:

1. the theory of evolution by natural selection, published simultaneously in 1858 by the English naturalists Charles Darwin (1809–1882) and Alfred Russel Wallace (1823–1913);
2. the theory of gravitation, first proposed by Isaac Newton (1642–1727); and
3. the theory of relativity of the Swiss physicist Albert Einstein (1879–1955).

Celebrated theories such as these are strongly supported by large bodies of evidence, and they will likely persist for a long time. It cannot, however, be said that these (or any other) theories are absolute truth — some future experiment may yet falsify even these famous theories.

The **scientific method** begins with the identification of a question involving the structure or function of the natural world, usually developed using inductive logic (Figure 2.1). The question is interpreted in terms of existing theory, and specific hypotheses are formulated to explain the character and causes of a natural phenomenon. Observations in nature and experiments are then designed and conducted to disprove (rather than to prove) the various hypotheses. Most hypotheses are rejected because their predictions are not borne out during the course of research. Any viable hypotheses are further examined through additional research, again largely involving experiments designed to disprove each hypothesis. Once a large body of evidence accumulates in support of a hypothesis, it can be used to support the original theory.

The scientific method can be used to investigate only those questions that can be critically examined through observation and experiment. Consequently, science cannot

Charles Darwin (1809–1882), regarded by some as the greatest naturalist of all time, is best known for the theory of evolution by natural selection. This theory was announced in 1858 by Darwin and Alfred Russel Wallace (1823–1913) simultaneously.
Photo: Corbis-Bettman

FIGURE 2.1 DIAGRAMMATIC REPRESENTATION OF THE SCIENTIFIC METHOD

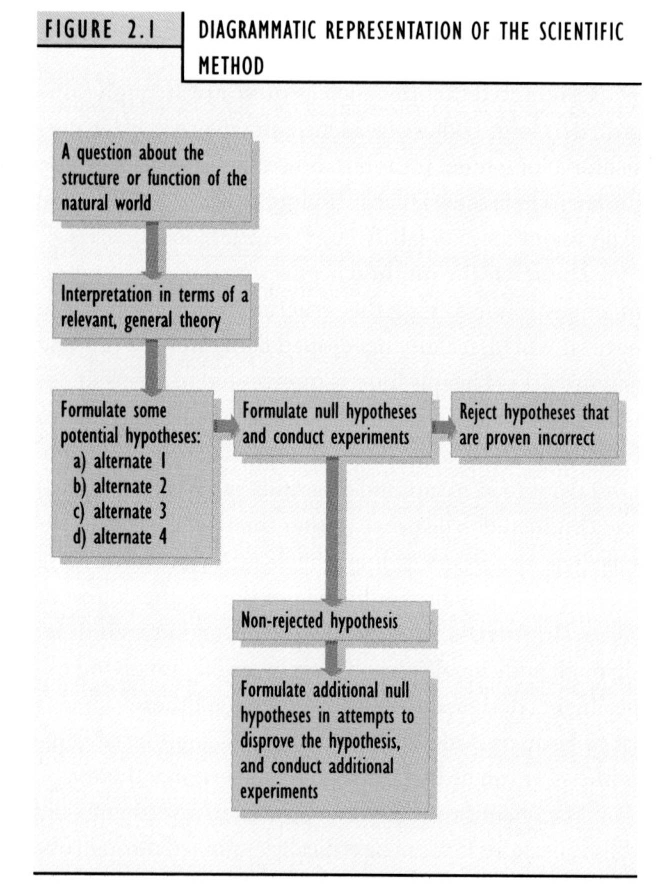

Source: Modified from Raven and Johnson (1992)

help resolve value-laden questions such as the meaning of life, good versus evil, or the existence and nature of God or any other supernatural being or force.

An **experiment** is a test or investigation designed to provide evidence for, or preferably against, a hypothesis. **Natural experiments** are conducted by observing variations of phenomena in nature, and then developing explanations for these through analysis of possible causal mechanisms. **Manipulative experiments** involve deliberate alterations of the factors that are hypothesized to influence phenomena. These manipulations are carefully planned and controlled in order to determine whether predicted responses will occur, thereby uncovering causal relationships.

By far the most useful working hypotheses in scientific research are **null hypotheses**, which seek to disprove rather than to support a hypothesis. This is a very important aspect of scientific investigation. A hypothesis might, for example, be supported by a large number of confirming experiments or observations. This does not, however, serve to "prove" the hypothesis, only to support its conditional acceptance. As soon as a clearly defined hypothesis is falsified by an appropriately designed and well-conducted experiment, it is disproved for all time. This is why experiments designed to disprove hypotheses are a key element of the scientific method. Great advances in understanding can occur when important hypotheses are rejected. For example, once it was discovered that the Earth is not flat, it became possible to sail beyond the visible horizon without fear of falling off the planet.

Factors believed to influence natural phenomena are called **variables**. For example, a scientist might hypothesize that the productivity of a wheat crop was potentially limited by such variables as the availability of water or of nutrients such as nitrogen and phosphorus. Some of the most powerful scientific experiments involve manipulating *key* (or controlling) variables and comparing the results with those of a **control** treatment which was not manipulated. In the example just described, the specific variable that controls wheat productivity could be identified by providing various amounts of water, nitrogen, and phosphorus, alone and in combination, and comparing the results with a non-manipulated control treatment.

In some respects, the explanation of the scientific method offered above is a bit uncritical. It perhaps suggests a too-orderly progression in terms of logical, objective experimentation and comparison of alternative hypotheses. These are, in fact, critical components of the scientific method. However, it is also important to understand that the insights and personal biases of scientists are significant as well in the progression of science. In most cases, scientists design experiments that they think will "work," that is, will yield useful results that will contribute to the orderly advancement of knowledge in their field. Karl Popper (1902–1994), a famous European philosopher, noted that scientists tend to use their "imaginative preconception" of the workings of the natural world, and to design experiments based on those informed insights. This means that effective scientists must be more than knowledgeable and technically skilled — they must also be capable of a certain degree of insightful creativity when forming their ideas, hypotheses, and research.

An experiment is a controlled investigation designed to provide evidence for, or preferably against, a hypothesis about the working of the natural world. This experiment exposed test populations of a grass to different concentrations of toxic chemicals in the laboratory.

Photo: B. Freedman

UNCERTAINTY

Much scientific investigation involves the collection of observations and data by measuring and monitoring phenomena in the natural world. Another important aspect of science involves making predictions about the future values of variables. Such projections require some degree of understanding of the relationships among variables and their influencing factors, and of recent patterns of change. However, many types of scientific information and predictions are subject to *inaccuracy*. Frequently, measured data are only approximations of the true values of phenomena, and predictions are rarely fulfilled exactly. The accuracy of observations and predictions is influenced by various factors, including the following:

Phenomena Can Be Predictable or Uncertain

Some phenomena are highly predictable, but most are uncertain. A few phenomena are considered to have a universal character and are consistent wherever and whenever they are accurately measured. One of the best examples, called a universal constant, is the speed of light, which always has a value of 2.998×10^8 m/s, regardless of where it is measured or of the speed of the body from which the light is emitted. Similarly, certain relationships describing transformations of energy and matter, known as the laws of thermodynamics (Chapter 4), always give reliable predictions.

Most natural phenomena, however, are not universally consistent — depending on circumstances, there are exceptions to general predictions about these phenomena. This circumstance is particularly true of ecology, a field of science in which virtually all general predictions have exceptions. We have not yet discovered any inviolate laws or unifying principles of ecology, in contrast to the several esteemed laws and universal constants of physics. For this reason, ecologists have great difficulties in making accurate predictions about ecological responses to environmental change. This is why ecologists are sometimes said to have "physics envy."

In large part, the inaccuracies of ecology occur because ecological functions are controlled by complexes of poorly understood, and sometimes unidentified, environmental influences. Consequently, predictions about future values of ecological variables or the causes of ecological changes are seldom accurate. For example, even though population ecologists in eastern Canada have been monitoring the population size of spruce budworm (an important pest of conifer forests) for some years, they cannot accurately predict its future abundance in particular stands of forest or in larger regions. This is because the abundance of this moth is influenced by a complex of environmental factors, including tree-species composition, age of the forest, abundance of its predators and parasites, quantities of its preferred foods, weather at critical times of year, and insecticide use to reduce its populations (see Chapter 21).

Variable Phenomena

Many natural phenomena vary in space and time. This is true of physical and chemical variables as well as of biological and ecological ones. Within a forest, for example, the amount of sunlight reaching the ground varies depending on time of day and season of the year. It also varies spatially, depending on the density of foliage over the place where sunlight is being measured. Similarly, the density of a particular species of fish within a river typically varies greatly, in response to variations of habitat conditions and other influences. Most fish populations also vary greatly over time, especially migratory species such as salmon. In environmental science, replicated (that is, independently repeated) measurements and statistical analyses are used to measure and account for these kinds of spatial and temporal variations.

Accuracy, Precision, and Significant Figures

Accuracy refers to the degree to which a measurement or observation reflects the actual, or true, value of the subject. For example, the insecticide DDT and the metal mercury are potentially toxic chemicals that occur in trace concentrations in all organisms, but their small residues are difficult to analyze chemically. Some of the analytical methods used to determine DDT and mercury concentrations are more accurate than others, and therefore provide more useful and reliable data compared with less accurate methods. In fact, analytical data are often approximations of the real values — exact accuracy is rarely attainable.

Precision is different from accuracy, and is related to the degree of *repeatability* of a measurement or observation. For example, suppose that the actual number of caribou in a particular migrating herd is 10 246 animals. A wildlife ecologist might estimate that there were about 10 000 animals in that herd, which for practical purposes is a reasonably accurate approximation of the actual number of caribou. If other ecologists also estimate the size of the herd at about 10 000 caribou, there is a good degree of precision among the estimations. If, however, some systematic bias existed in the methodology used to count the herd, giving consistent estimates of 15 000 animals (remember, the actual population is 10 246 caribou), these estimates would be considered precise, but not particularly accurate.

Precision is also related to the number of digits with which data are reported. If you were using a flexible tape to measure the lengths of ten large, live snakes, you would probably measure the animals only to the nearest centimetre. The strength and squirminess of the subjects make more precise measurements impossible. The reported average length of the ten snakes should reflect the original measurements, and might be given as 204 cm, and not a value such as 203.8759. The latter number might be displayed as a calculated average by a calculator or computer, but it is unrealistically precise.

Significant figures are also related to precision, and can be defined as the number of digits used to report data from analyses or calculations. Significant figures are best understood by examples. The number 179 has three significant figures, as does the number 0.0849 and also 0.000 794 (i.e., the zeros preceding the significant integers do not count). However, the number 195 000 000 has nine significant figures (i.e., the following zeros are meaningful), although the same number written as 195×10^6 is considered to have only three significant figures (see also Appendix I). It is rarely useful to report environmental or ecological data with more than 2–4 significant figures. This is because any more would generally exceed the precision of the methodology used in the estimation and would therefore be unrealistic. For example, the approximate population of Canada in 1997 was 30.2 million people (or 30.2×10^6; both of these notations have three significant figures). However, this should not be reported as 30 200 000 people, which implies an unrealistic precision of eight significant figures.

The Need for Scepticism in Environmental Science

Environmental science is filled with examples of uncertainty, in present values and future changes of environmental variables as well as in predictions of ecological responses to those changes. To some degree the difficulties associated with scientific uncertainty can be mitigated by developing and using improved technologies for analysis, and by modelling and studying changes occurring in ecosystems in different parts of the world. The latter enhances our understanding by providing convergent evidence about the occurrence and causes of natural phenomena.

Scientific information and understanding will always, however, be subject to some degree of uncertainty, and therefore predictions will always be to some extent inaccurate. Uncertainty must always be considered in understanding and dealing with the causes and consequences of environmental and ecological changes. Therefore, all information and predictions in environmental science must be critically interpreted with uncertainty in mind. This should be done whenever one is learning about an environmental issue, whether by listening to a speaker in a classroom or conference or on television or radio, or when reading an article in a newspaper, magazine, textbook, or respected scientific journal. Because of the uncertainty of many predictions in science, and particularly in environmental science and ecology, a certain amount of scepticism and critical analysis is always useful.

Environmental issues are acutely important to the welfare of humans and other species. Science and its methods allow critical and objective identification of important issues, the investigation of their causal mechanisms, and some degree of understanding of the ecological consequences of environmental change. Scientific information influences decision-making about environmental issues, including whether to pursue expensive strategies to avoid further, but often uncertain, damage.

Scientific information is, however, only one consideration for decision-makers, who are also concerned with the political, economic, and cultural contexts of environmental problems (see Chapter 25). In fact, when deciding how to deal with the causes and consequences of environmental changes, decision-makers may give greater weight to non-scientific considerations than to scientific ones, especially when there is uncertainty about the latter. Some of the most important decisions about environmental issues are made by politicians and senior bureaucrats in government, rather than by environmental scientists. Decision-makers typically worry about the short-term implications of their decisions on their chances for re-election and continued employment, and for the economic activity of society at large, as much as they do about the ecological consequences of environmental changes.

Key Terms

science
inductive logic
deductive logic
scientific method
fact
hypothesis
theory
working hypothesis
experiment
natural experiment
manipulative experiment
null hypothesis
variables
parameters
control
accuracy
precision
significant figures

Questions for Discussion

1. Outline the reasons why science is a rational way of understanding the natural world.
2. What factors result in legitimate scientific controversies about environmental issues? Contrast these with environmental controversies that exist because of differing values and world views.
3. Identify an environmental question of interest. Suggest some useful hypotheses for investigation, the null hypotheses that would be examined, and experiments that might be conducted.

References

AAAS. 1989. *Science for all Americans.* Washington, DC: American Association for the Advancement of Science (AAAS).

Barnes, B. 1985. *About Science.* London, U.K.: Blackwell Ltd.

Giere, R.N. 1979. *Understanding Scientific Reasoning.* New York: Holt, Reinhart & Winston.

McCain, G. and E.M. Siegal. 1982. *The Game of Science.* Boston, MA: Holbrook Press Inc.

Popper, K. 1979. *Objective Knowledge: An Evolutionary Approach.* Oxford, U.K.: Clarendon Press.

Raven, P.H. and G.B. Johnson. 1992. *Biology.* Toronto: Mosby Year Book.

3

THE PHYSICAL WORLD

CHAPTER OUTLINE

CHAPTER OBJECTIVES

After completing this chapter, you will be able to:

1. Explain the geological structure and dynamics of planet Earth.
2. Describe the importance of glaciation and other geological forces in modifying the landscapes of Canada.
3. Outline the four major elements of Earth's water cycle.
4. Describe Earth's atmosphere and its circulation.
5. Explain the elements of climate and weather.

Introduction

In this chapter we describe various aspects of the physical world, including the origin of Earth and the nature and dynamics of its physical and geological attributes. Understanding these subjects is important in environmental science, because they provide context for interpreting the environmental changes being caused through human activities.

As recently as about 15 000 years ago, virtually all of Canada was covered by glaciers. Remnants still occur, such as this icecap on Baffin Island.

Photo: Victor Last/Geographical Visual Aids

Planet Earth

The sun is an ordinary star, one of billions that exist in the universe. The universe is thought to have originated as many as 15–20 billion years ago during an immense cataclysmic event known as the "big bang." Initially, virtually all of the mass of the universe consisted of the two lightest elements, hydrogen and helium, which existed as a diffuse, extremely dispersed, gaseous mass. Eventually, under the influence of gravity, the hydrogen and helium aggregated and was compressed under enormously high pressures and temperatures into developing stars. This caused the formation of heavier elements through nuclear fusion reactions, accompanied by the release of tremendous quantities of energy. Because of these processes, there are now 87 naturally occurring elements. Hydrogen and helium are, however, still the most abundant elements, comprising more than 99.9% of the mass of the universe.

The sun, its nine orbiting planets, miscellaneous comets, meteors, asteroids, and other local materials (such as space dust) are known as **the solar system**. This particular region of the universe is organized and held together by a balance of the attractive force of gravitation and repulsive forces associated with rotation and orbiting (these same forces also organize the universe). The age of the solar system (and Earth) is at least 4.5 billion years.

Earth is a dense planet, as are the other so-called terrestrial planets located relatively close to the sun, namely, Mercury, Venus, and Mars. These planets are composed almost entirely of heavier elements such as iron, nickel, magnesium, aluminum, and silicon. These inner planets were formed by a selective condensing of heavier elements out of the primordial planetary nebula (i.e., the disk of gases and other matter that slowly rotated around the sun during the early stages of formation of the solar system). This happened because the inner planets were subjected to relatively intense heating by solar radiation, which allowed heavier elements to liquify and solidify, while much of the lighter gases such as hydrogen and helium ended up in the outer planets. Consequently, the more distant planets in the solar system, such as Jupiter and Saturn, are relatively large, gaseous, and diffuse in character. Most of their volume is composed of an extensive atmosphere of hydrogen and helium, although these planets may contain heavier elements in their cores.

Earth, the third-closest planet to the sun, is the only place in the universe definitely known to sustain life. It is quite possible, however, that other places in the cosmos also sustain life. Although there is no direct evidence on this, many scientists nevertheless consider it likely that life has evolved elsewhere. One estimate suggests that the universe contains 10^{20} (i.e., 10^{11} billion) stars, with perhaps 10% of these having planetary systems (i.e., 10^{19} systems). It is highly probable that at least one other of the billions of planetary systems in the universe supports suitable conditions for the genesis of life.

Earth is a spherical body with a diameter of about 12 800 km. It revolves around the sun in an elliptical orbit, at an average distance of about 149 million km, completing an orbit in 365.26 days, or one year. Earth also rotates on its axis every 24 hours, or one day. Earth's single moon has a diameter of about 3500 km, and a mass about 1.2% that of Earth. The Moon revolves around Earth in an elliptical orbit at an average distance of about 385 000 km, completed every 27.3 days (the lunar month).

Earth's sphere is thought to be composed of four layers — the core, mantle, lithosphere, and crust — arranged in concentric layers like an onion. Earth's massive **core** has a diameter of about 3500 km, and is made up mostly of hot, molten metals, particularly iron and nickel. The internal heat of Earth is generated by the slow, radioactive decay of unstable isotopes of such elements as uranium.

The **mantle** is a less-dense region enclosing the core, about 2800 km thick and composed of minerals in a plastic, semi-liquid state known as *magma*. The mantle contains large quantities of relatively light elements, notably silicon, oxygen, and magnesium, occurring as various compounds. Magma from the upper mantle sometimes erupts to the Earth's surface at mountainous vents known as volcanoes and is usually spewed from the surface as lava, which cools as basaltic rock. The next layer, the **lithosphere**, is only about 80 km thick and made up of rigid, relatively light rocks, especially basaltic, granitic, and sedimentary rocks. These rocks contain elements found in the mantle plus enriched quantities of aluminum, carbon, calcium, potassium, sodium, sulphur, and other lighter elements.

The outermost layer is known as **crust**. Oceanic crust is typically thin, averaging about 10–15 km, while continental crust is 20–60 km thick. Earth's crust has an extremely complex mineralogical composition, in contrast with the mantle and especially the core, which are thought to be relatively uniform in their structure and constitution. The most abundant elements in the crust are oxygen (45%), silicon (27%), aluminum (8.0%), iron (5.8%), calcium (5.1%), magnesium (2.8%), sodium (2.3%), potassium (1.7%), titanium (0.86%), vanadium (0.17%), hydrogen (0.14%), phosphorus (0.10%), and carbon (0.032%).

The rocks forming the crust can be grouped into three basic types: igneous, sedimentary, and metamorphic. **Igneous rocks** include basalt and granite, which are formed by the cooling of molten magma. The mineral forms depend on the rate of cooling and other factors. Basaltic rocks are the major constituent of oceanic crust, while granitic rocks dominate the continental crust.

Sedimentary rocks include limestone, dolomite, shale, sandstone, and conglomerates. These form from particles eroded from other rocks or from precipitated minerals such as calcite ($CaCO_3$), which become lithified (turned into stone) under great pressure in deep oceanic sediment. Sedimentary rocks typically overlie basaltic or granitic rocks.

Metamorphic rocks are formed from igneous or sedimentary rocks that were changed under the influences of geological heat and pressure. These conditions are encountered when the primary rocks are carried deeper into the lithosphere by crustal movements, such as those associated with mountain building (described below). Gneiss, for example, is a metamorphic rock derived from granite, while marble is derived from limestone and slate from shale.

About 30% of Earth's surface is covered by the solid substrates of continents and islands. The other 70% of Earth's surface is liquid water, virtually all of which is oceanic. In addition, Earth's dense sphere is immersed in a gaseous envelope known as the atmosphere, which extends to a distance of about 1000 km. However, about 99% of the mass of the atmosphere occurs within 30 km of the planet's surface.

Geological Dynamics

Throughout its history, Earth has been subject to enormous geological forces which have greatly affected its mineralogical composition and surface features. The predominant forces are **tectonic forces**, which are associated with crustal movements and other processes that cause structural deformation of rocks and minerals. Geological forces also cause the continents and their underlying plates to slowly move about Earth's surface, much like rafts of solid rock riding upon a sea of magma, building mountain ranges where crustal plates collide and push up surface rocks.

Earthquakes and *volcanoes* are also tectonic phenomena, influencing Earth's crust and surface with extremely powerful, sometimes disastrous events. Other massive geological forces include rare, cataclysmic strikes of our planet by *meteorites* and extensive *glaciation* associated with climatic cooling. Less cataclysmic but more pervasive geological forces are *erosion*, caused by water, wind, and gravity, and *weathering*, the fracturing of rocks and dissolution of minerals.

Over geological time, these physical processes have profoundly influenced the present character of Earth. Geological forces continue to have enormous influences on Earth and its ecosystems, over both short- and longer-term time scales. Environmental changes associated with Earth's geological dynamics provide a naturally occurring context for the substantial changes that humans are now causing through their various economic activities.

Meteorites

Earth is frequently struck by fast-moving, rocky or metallic objects from space known as **meteorites**. Although meteorites are relatively small objects (by planetary standards), they have immense momentum because of their speed, which typically ranges from 10 to 100 km/s. The smallest, most numerous meteorites reaching Earth typically burn up or explode in the atmosphere because of heat generated by friction, but larger ones can survive to strike Earth's surface. It has been estimated, for example, that each day a meteorite weighing at least 100 g impacts somewhere in Canada.

Very large meteorites are extremely rare, but can cause enormous damage. The impact site is typically obliterated, and a large crater is created when vast quantities of crustal materials are ejected into the atmosphere. Immense *sea waves* can also be caused by a meteorite impact. The largest of the several dozen large meteorite craters known in Canada are an ovoid depression with a diameter of 140 km near Sudbury, Ontario, and a doughnut-shaped lake with a diameter of 100 km at Manicouagan, Quebec. These were created by collisions with meteorites more than 570 million years ago. These extraordinary events must have caused tremendous damage to the species and ecosystems of the time.

Earth's evolutionary history has been punctuated by a number of catastrophic events of mass extinction, when most of the existing biota disappeared in a short period of time, to be later replaced by new species (see Chapter 6). Paleontologists recognize these cataclysms by rapid changes in Earth's fossil record, which point to the transitions between stages in the geological time scale (Table 3.1). According to surviving evidence, the most intense mass-extinction event occurred 245 million years ago at the end of the Permian period, when an astonishing 96% of marine species may have become extinct. Another set of mass extinctions occurred 65 million years ago, at the end of the Cretaceous period, when perhaps 76% of species became extinct, including the last of the dinosaurs.

TABLE 3.1 THE GEOLOGICAL TIME SCALE

Divisions between geological time stages are largely assigned on the basis of rapid changes in species composition of the fossil record. These are related to events of mass extinction, followed by the evolutionary radiation of new species and families. The record is most detailed for relatively recent times, because recent fossil records are more complete.

Time is given in millions of years, and indicates the beginning of each time stage (e.g., the Holocene epoch ranges from 10 000 years ago to the present; the Pleistocene ranges from 1.6 million years ago to 10 000 years ago).

ERA	PERIOD	EPOCH	TIME (× 10^6 YEARS)
Cenozoic	Quaternary	Holocene (recent)	0.01 (i.e., 10 000 y)
		Pleistocene	1.6
	Tertiary	Pliocene	5
		Miocene	26
		Oligocene	38
		Eocene	54
		Paleocene	65
Mesozoic	Cretaceous		140
	Jurassic		210
	Triassic		245
Paleozoic	Permian		290
	Carboniferous		365
	Devonian		413
	Silurian		441
	Ordovician		504
	Cambrian		570
Precambrian	Proterozoic		2400
	Archaean		>4500

According to one popular theory, the extinctions at the end of the Cretaceous were caused when a 10–15-km-wide meteorite impacted Earth. It is thought that huge quantities of fine dust were spewed into the upper atmosphere, resulting in a climatic cooling that large animals and many ecosystems could not tolerate. Some geologists believe that the impact site was near the coast of the Yucatan Peninsula in Mexico, where a buried, 170-km-wide ring structure exists, dated as about 65 million years old. Although controversial, the theory of rare, meteorite-caused catastrophes has also been used to explain other rapid changes of species in the geological record.

Plate Tectonics and Related Processes

The theory of *plate tectonics* concerns the dynamics of Earth's surface crustal materials. In simple terms, this theory suggests that Earth's crust and mantle behave as an enormous, convecting system. This is characterized by slow surface movements of huge plates of rigid crustal materials, from zones where they are created toward zones where they are destroyed by movement into the upper zone of the mantle. New crust forms where there is an upwelling of magma from the upper mantle. The magma rises to the surface, solidifies, and then extends laterally in a process known as sea-floor spreading. In other zones, there is a compensating *subduction* of sea-floor crust back down into the mantle, where it is re-melted and convected laterally. The magma may eventually reach another magmic upwelling region and be carried to the crust again. The slowly moving, rigid plates of surface crust have a basement of basaltic rocks, with the lighter, granitic-based continents rafting on the surface of some of the oceanic plates (Figure 3.1).

In the Atlantic Ocean, for example, about halfway between the Americas and Europe and Africa, the deep-sea Mid-Atlantic Ridge runs in a roughly north-south direction. This ridge is a zone of sea-floor spreading, from which the two continental regions are diverging at a slow but steady rate of about 2–4 cm per year. In contrast, parts of the continental landmasses in the western Americas are riding on regions of plates that are subducting beneath the oceanic Pacific Plate. However, along most of southwestern North America, the Pacific and North American Plates are moving in opposite but parallel directions, causing southern California and the Baha peninsula to slowly move northward relative to the rest of the continent. This process is occurring along an extended contact between the plates (i.e., a *fault*) known as the San Andreas Fault.

These tectonic forces result in frequent earthquakes and volcanic eruptions along the Pacific coasts of North and South America, the Aleutians, and eastern Asia. This geologically active region around the Pacific Ocean is known to geologists as the "ring of fire." In addition to these discrete but intense geological events, there is active building of the relatively young mountains in this region. The mountain building is caused by crustal materials being pushed upward as continents collide with each other and with their underlying oceanic plates. In a similar manner, the lofty Himalayas of southwestern Asia were and still are being created by the immense, uplifting forces generated as the northward-drifting Indian subcontinent pushes into the larger Asian landmass.

FIGURE 3.1 TECTONIC FORCES

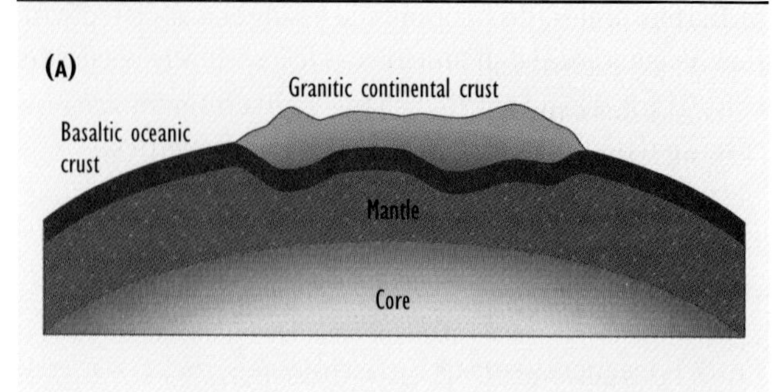

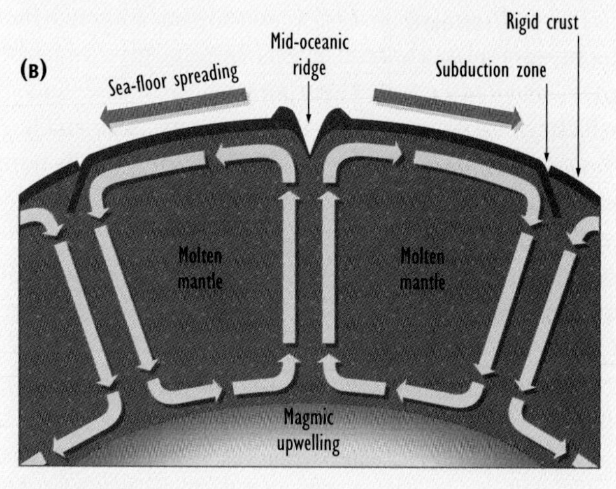

(a) The continents are viewed as granitic islands rafting upon plates of basaltic oceanic crust.
(b) Heat and density gradients in the mantle cause a slow, convective circulation to develop in the molten magma. This circulation forms new basaltic crust at zones of magmic upwelling known as mid-oceanic ridges, followed by lateral sea-floor spreading and eventual subduction back to the mantle at the boundary with another crustal plate.

It is thought that the continents were initially arranged separately, then aggregated into a single mass during the Permian period, about 240 million years ago. This supercontinent, referred to as Pangaea, was surrounded by a single, global ocean. Divergent forces of crustal plates moving in different directions then pulled Pangaea apart, initially into two masses known as Laurasia and Gondwanaland, and then into the existing continents of North and South America, Africa, Eurasia, Australia, and Antarctica.

An **earthquake** is a trembling or movement of the earth, caused by a sudden release of geological stresses at some point within the crust or upper mantle. Earthquakes are most commonly caused when crustal plates slip across or beneath each other at their intersections, known as faults. Earthquakes can also be caused by a volcanic explosion. Although their seismic energy can affect large areas, earthquakes have a spatial focus, known as an epicentre and defined as the surface position lying above the deep point of energy release. Intense earthquakes can cause great damage to buildings, and the collapsing structures, fires, and other destruction can take a toll of human lives.

In 1556 an earthquake struck Shanxi Province in China and caused about 830 000 deaths, making it the most deadly earthquake in recorded history. The most famous catastrophic earthquake in North America was the San Francisco event in 1906, which killed 503 people and resulted in tremendous physical damage. This earthquake was caused by slippage along the San Andreas Fault. Other 20th-century earthquakes have resulted in greater losses of human lives, including one in 1976 that killed 242 000 people in Tangshan, China; another in 1927 that killed 200 000 in Nan-Shan, China; and one in Tokyo-Yokohama, Japan, that killed 200 000 in 1926.

The events in San Francisco and Tokyo affected large cities. The powerful tremors caused great damage, partly because of weak architectural designs that were unable to withstand the strong forces. In both cases, however, about 90% of the actual destruction resulted from fires. Earthquakes can also cause soil to lose some of its mechanical stability, resulting in destructive landslides and subsidence (i.e., sinking) of land and buildings.

Undersea earthquakes can trigger fast-moving, sea-surface phenomena known as *tsunami*, seismic sea waves, or tidal waves. A tsunami is barely noticeable at sea, but it can become gigantic when the wave reaches shallow water and piles up to heights that can swamp coastal villages and towns. In 1912, for example, an earthquake off eastern Canada generated a seismic sea wave that killed people in Newfoundland and Nova Scotia. In 1946, a large earthquake, centred on Umiak Island in the Aleutian Islands, caused a tsunami to strike Hawaii, 4500 km away, with an 18-m crest.

Volcanoes are vents in the Earth's surface that spew molten lava onto the ground and eject molten, solid, and gaseous materials into the atmosphere. The largest volcanic eruptions can literally explode mountains, ejecting immense quantities of material into the environment and causing enormous damage and loss of life. For example, an eruption of Mount Vesuvius in the year 79 (CE) buried the Roman city of Pompeii, killing most of its inhabitants. A 1902 explosion of Mont Pelee on the Caribbean island of Martinique killed 30 000 people.

The greatest eruption of modern times involved Tambora, a volcano in Indonesia which exploded in 1815 and blew more than 300 km^3 of material into the atmosphere, including the top 1300 m of the mountain. Some of the finer particulates of this massive eruption were blown into the upper atmosphere (the stratosphere), causing an increase in Earth's reflectivity which resulted in global cooling. The year 1816 became known as the "year without a summer" in Europe and North America because of its unusually cool and wet weather, including frost and snowfall during the summer.

Another famous Indonesian eruption was that of Krakatau in the Sunda Strait in 1883, which ejected 18–21 km^3 of material as high as 50–80 km into the atmosphere. The 30-m-high tsunami associated with the eruption of Krakatau killed about 36 000 people in coastal towns and villages.

Large volcanic eruptions can also disturb great expanses of forest and other ecosystems. For instance, the 1980 explosion of Mount St. Helen's in Washington State blew down about 21 000 ha of coniferous forest, killed another 10 000 ha of forest by heat injury, and otherwise damaged another 30 000 ha. Mudslides also devastated large areas and a vast area was covered by particulate debris (known as tephra) that settled from the atmosphere to a depth of up to 50 cm.

Some volcanoes produce chronic lava flows and venting of gases. These volcanoes tend to form distinctive, cone-shaped mountains from their accumulated lava, which solidifies into finely crystalline, glassy rocks. An active example of this spectacular process is Mount Kilauea in Hawaii, which often erupts continuously for years. The slowly flowing lava from these volcanoes can destroy buildings and vegetation that are directly impacted, but it is not otherwise dangerous because people and animals can avoid the molten streams.

Glaciation

Glaciers, persistent sheets of ice, are common features in high-latitude environments of the Arctic and Antarctic. Glaciers also occur at high altitude on mountains, even in some tropical countries such as equatorial New Guinea. Glaciers are formed from deep, persistent snowpacks, which are compressed into ice as their weight accumulates. Most glaciers occur on land, but some extend onto the ocean or exist as extensive oceanic ice shelves. At the present time, about 10% of the land surface of Earth (about 14.9 million km^2) is covered with glaciers, the largest of which are the continental ice sheets of Antarctica. The largest glaciers in the Northern Hemisphere are in Greenland, but parts of Baffin and Ellesmere Islands in the Canadian Arctic are also extensively covered with glacial ice, as are some mountainous areas in western Canada.

Glaciation refers to an extensive geological change characterized by advancing ice sheets associated with extended periods of global climatic cooling, sometimes known as ice ages. There have been a number of glacial periods during Earth's history. However, significant details are known only about the most recent glaciation, during the Pleistocene epoch, because most traces of earlier events have been obliterated. The most recent glacial period, known as the Wisconsin, began about 100 000 years ago and ended about 10 000 years ago. (Note that this should not be referred to as the "last" glaciation, which has not yet happened!)

At the height of the Wisconsin glaciation, ice covered about 30% of Earth's land surface, including virtually all of what is now Canada, and extensive areas of the continental shelf which are now beneath the ocean. (Sea level was about 100 m lower during the Wisconsin glaciation, because so much water was tied up in ice on land.) The greatest ice mass in Canada was the Laurentide Ice Sheet, which reached a thickness of about 4 km. The Cordilleran Ice Sheet of the western mountains contained ice up to 2 km thick.

The Holocene (Recent) epoch, relatively warm and ice-free, is referred to as an *interglacial* stage. Climate has not, however, been uniformly warm during the present interglacial. For example, the period 1450 to 1850 is known as the Little Ice Age because of its relatively cool temperatures. During that period there was a moderate expansion of glaciers and snowfields in many parts of the world, including the Arctic and western mountains of Canada.

Glaciers are extremely erosive forces, crushing, scouring, and excavating the underlying terrain by their massive weight. Glaciers also move huge quantities of excavated debris around the landscape. These solid materials are eventually deposited when the glaciers melt, tumbling from the ablating (i.e., melting) ice mass or carried away by running meltwater. Extensive deposits of glacial debris are common over virtually all of Canada, often occurring as distinctive landforms, such as the following:

1. moraines, which are long, mounded hills, usually lying perpendicular to the glacier's flow, containing mixed rocky debris known as till;
2. drumlins, which are teardrop-shaped hills that are elongated in the direction of movement of the glacier and composed of a mixture of rocky materials;
3. eskers, which are long, serpentine mounds of mixed debris deposited by rivers running beneath glaciers;
4. rounded boulders known as erratics, which can be incongruously scattered over the landscape;
5. long, U-shaped valleys in mountainous terrain, carved from pre-existing river valleys by the erosive forces of glaciers;
6. fiords, which are long, narrow, steep-sided inlets of the ocean;
7. outwash plains, which contain a mixture of rocky materials that were deposited over a relatively wide area by streams and rivers fed by glacial meltwaters; and
8. the former basins of extensive lakes of glacial meltwater, which today are characterized by flat, fine-grained, often fertile plains. For example, southern Manitoba has extensive former lakebeds of postglacial Lake Agassiz, while in southern Ontario and Quebec, flat areas were once part of the more-extensive basins of the Great Lakes and St. Lawrence River.

The tremendous ice sheets that once obliterated virtually all of Canada were largely gone by 8000–10 000 years ago, although glacial remnants occur on islands in the High Arctic and on mountains of western Canada. The Canadian landscape has been profoundly shaped by the impressive geological signatures of the advance and retreat of the immense continental glaciers. Since then, the terrain and landforms have been substantially modified by other geological forces, such as erosion and weathering, and by the redevelopment of ecosystems after the retreat of the great ice sheets. However, these forces have had a relatively small influence on the enduring, essentially glacial character of the Canadian landscape.

Weathering and Erosion

Meteorite impacts, earthquakes, volcanic explosions, and glaciation are all tremendous environmental forces, capable of obliterating both natural and anthropogenic ecosystems. Other, less forceful geological dynamics are also important, although they exert their influences more pervasively, by operating relatively slowly over longer time scales, rather than as extremely destructive events.

Weathering refers to the physical and chemical processes by which rocks and minerals are broken down by environmental agents. Non-biological agents include rain, wind, and temperature changes (especially freeze-thaw cycles), and biological influences include the rock-cracking forces that can be exerted by plant roots. Weathering proceeds by the physical fracturing of rocks and by the chemical decomposition (i.e., solubilization) of minerals by acidic rainwater and corrosive solutions secreted by plant roots and microorganisms. Weathering is an *in situ* (i.e., "in place") phenomenon — the weathered rocks and minerals are not necessarily transported elsewhere.

Erosion refers to the physical removal of rocks and soil through the actions of flowing water, ice, wind, and gravity. Erosion is a pervasive geological process, occurring at various rates in all environments. Usually it is gradual, occurring as particles are slowly removed by flowing water or blowing wind, or as dissolved minerals are carried away by underground and surface flows of water. It also occurs as mass events, such as landslides and mudslides in steep terrain. Over extremely long periods of time, erosion tends to create a relatively flat and homogeneous landscape, known as a *peneplain*.

Even geological features as immense as mountains are slowly eroded away, with their enormous mass gradually deposited in lower regions. For instance, the Precambrian Shield that is so extensive in regions of Canada is composed of the granitic basement rocks of ancient mountains that were slowly eroded away by the actions of water, wind, and glaciers. The somewhat less ancient hills of the Appalachians of eastern North America, which extend into New Brunswick, Nova Scotia, and Newfoundland, are also the eroded relics of a once-great mountain range. The youngest mountain range in North America, the Rocky Mountains, extends from the western United States north into Alberta, British Columbia, the Yukon, and the western Northwest Territories. The Rockies still have many towering, sharp peaks because they have not yet been much reduced by the mass-wasting forces of erosion.

Rates of natural erosion are influenced by many factors, including the hardness of rocks, degree of consolidation of soil and sediment, amount of vegetation cover, rate of water flow, slope of the land, speed and direction of winds, and frequency of storm events and other types of disturbances. Some of these factors can be greatly influenced by human actions. When we disturb vegetation, for example, its moderating influence on erosion rates is reduced or eliminated. In fact, human activities associated with agriculture, forestry, and road-building have greatly increased rates of erosion in almost all regions of the world. In many cases, the increased losses of soil have had serious consequences for the productivity of agricultural lands and for natural biodiversity. We will discuss anthropogenic influences on erosion in Chapters 14 and 23.

Materials eroded from mountains and other uplands must, of course, go somewhere. These materials are carried to lower altitudes and much of the mass is eventually deposited in the oceans, settling to the bottom in a process known as **sedimentation**. Over extremely long periods of time (that is, tens or more millions of years), as the mass of sedimented materials builds up, sufficient pressures are exerted on underlying materials to cause them to aggregate, become more densely packed, and cement into sedimentary rocks in a process called **lithification**. Common examples of sedimentary rocks are sandstone, mudstone, shale, limestone, and mixtures of these known as conglomerates (the latter may also contain eroded, non-sedimentary rocks such as granites and basalts).

Eventually, under the influence of tectonic forces, enormously slow and powerful collisions of crustal plates can cause areas of deep-oceanic, sedimentary rocks to *uplift*, sometimes raising them to great altitudes and forming new mountain ranges underwater or on the continents. Geological uplift is the means by which oceanic rocks and marine fossils can find their way to the tops of Earth's highest mountains. Uplift and mountain-building are important stages in the geological recycling of some of the continental mass that was wasted down-slope during millions of years of erosion.

The Hydrosphere

Earth's **hydrosphere** is the portion of the planet that contains water (H_2O), including the oceans, the atmosphere, the land surface, and underground. The **hydrologic** or **water cycle** refers to the rates of movement (or **fluxes**) of water among these various reservoirs (or **compartments**). The hydrologic cycle is a global phenomenon, although it also operates on local scales. The major elements of the hydrologic cycle are illustrated in Figure 3.2.

Each compartment of the hydrologic cycle has input and output fluxes, and the sum of all of these comprises the cycle. If the rate of input equals the rate of output, then a compartment is in a flow-through equilibrium and its size does not change. Of course, if input exceeds output, the compartment increases in size over time (and decreases if input is less than output).

On the global scale, the major compartments of the hydrological cycle are in a long-term equilibrium condition. This is not, however, generally true of local scales, particularly over shorter intervals of time. For example, areas may temporarily flood or dry out. In addition, local hydrological conditions can change over the long term. Glaciation, for example, stores immense quantities of solid water on land, and excessive use of groundwater can deplete artesian reservoirs.

We can distinguish four major compartments of the hydrological cycle (see Table 3.2):

1. **The oceans** are the largest hydrological compartment, accounting for about 97.4% of all water on the planet.
2. **Surface waters** occur on Earth's landmasses, and account for 2.3% of global water. Virtually all surface water is tied up in glaciers, mostly in Antarctica. Lakes, ponds, rivers, streams, and other surface bodies containing liquid water amount to only 0.002% of global water.
3. **Groundwater** accounts for 0.32% of Earth's water. Groundwater can occur in relatively shallow soil horizons, where it is accessible for uptake by plants, or can drain laterally into surface waters such as lakes and streams. Deeper groundwater is inaccessible for these purposes, and forms *artesian reservoirs* in spaces within fractured bedrock.
4. **Atmospheric moisture** accounts for only about 0.001% of Earth's water. Atmospheric water can

FIGURE 3.2 | MAJOR ELEMENTS OF THE HYDROLOGIC CYCLE

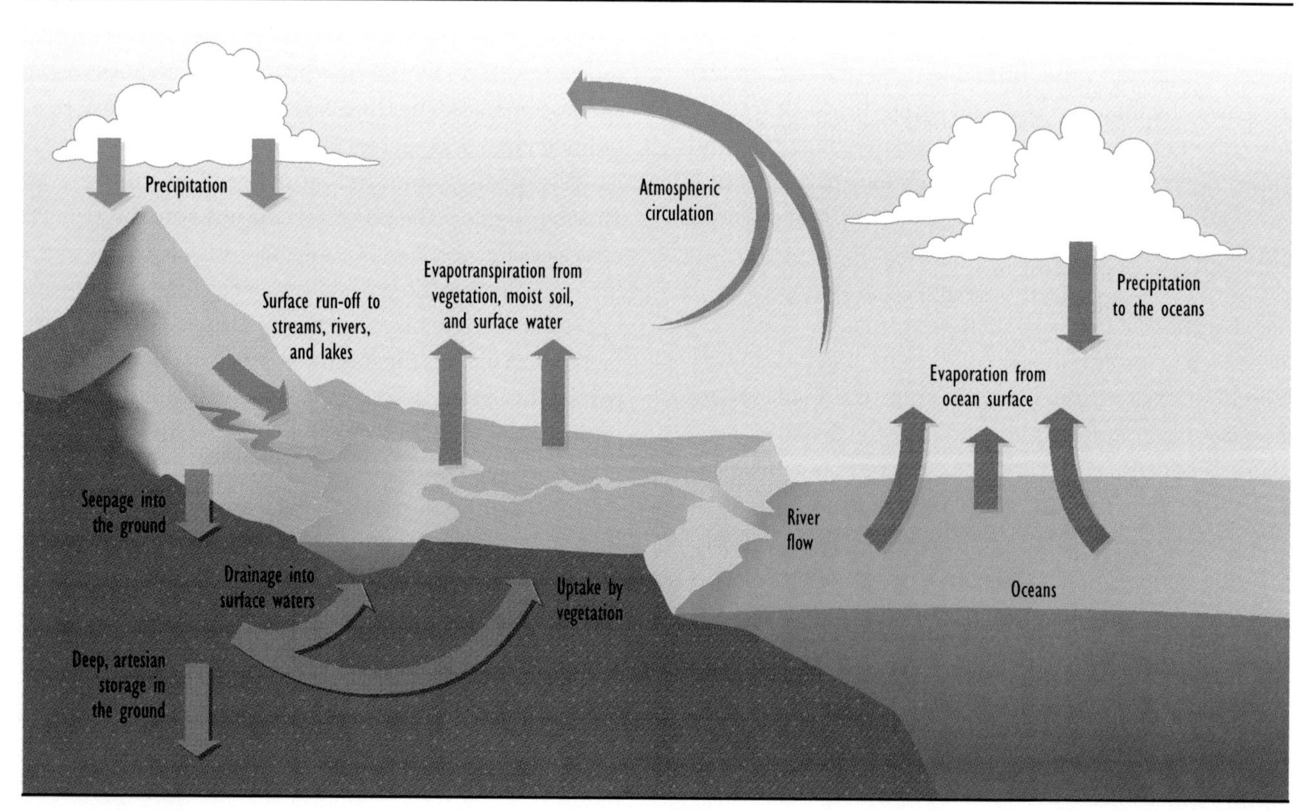

TABLE 3.2 | THE HYDROLOGIC CYCLE

Estimates of the sizes of major compartments and fluxes of the global hydrologic cycle.

COMPARTMENT	QUANTITY (10^{14} tonnes)	PERCENTAGE OF TOTAL
oceans	12 300	97.4
glaciers	286	2.3
groundwater (to depth of 0.8 km)	40	0.32
inland waters	0.25	0.002
atmosphere	0.13	0.001

FLUX	QUANTITY (10^{14} t/year)
evaporation	
from oceans	3.8
from land surfaces	0.6
total evaporation	4.4
precipitation	
to oceans	3.4
to land surfaces	1.0
total precipitation	4.4
atmospheric export from oceans to terrestrial	0.4
surface runoff from land	0.2

Sources: Odum (1983) and Botkin and Keller (1995)

occur as a gas, vapour (tiny, suspended droplets), or solid (ice crystals), all of which are highly variable over space and time. Clouds are dense aggregations of liquid or solid water in the atmosphere, while gaseous water is invisible. (Note that the maximum amount of water a volume of atmosphere can hold is highly dependent on temperature, with warmer air having a much greater water-storage capacity than cold air. The term **humidity** refers to the actual concentration of water in the atmosphere (measured in g/m^3), while **relative humidity** expresses actual humidity as a percent of the *saturation value* for a particular temperature.)

The six major fluxes among the four major compartments of the hydrological cycle are listed in Table 3.2. The more important fluxes are described below.

Evaporation is a change of state of water from a liquid to a gas or from a solid, such as snow or ice, directly to a gas (more properly referred to as sublimation). Globally,

The global hydrologic cycle involves water movement through the atmosphere, on the surface, and underground, as well as storage in oceans, lakes, glaciers, and groundwater. This river in Labrador represents a flow to the ocean of water deposited to the landscape as precipitation.

Photo: B. Freedman

about 86% of evaporation is from the oceans, and the rest is from terrestrial surfaces. On terrestrial landscapes water can evaporate from bodies of surface waters, from moist soil and rocks, and from vegetation. **Transpiration** refers specifically to the evaporation of water from plants, while **evapotranspiration** refers to all sources of evaporation from a landscape.

Precipitation is the deposition of water from the atmosphere as liquid rain or as solid snow or hail. In addition, vapour-phase water in the atmosphere can condense or freeze onto surfaces as dew or frost. As previously noted, most global evaporation is from the oceans, much of which precipitates back to that surface. Some, however, is transported by moving air masses over the continents, resulting in a net import of evaporated water from Earth's oceans to its land surfaces. Precipitation volumes can be especially large in hilly or mountainous areas. This phenomenon is known as **orographic precipitation** (In Detail 3.1).

Surface flows involve water that is transported in brooks, streams, and rivers. (In contrast, lakes and ponds are relatively static, storage reservoirs.) Surface flows move in response to gravitational gradients associated with altitude (in other words, water flows downhill). Most surface flows ultimately carry water to the oceans, thereby

In Detail 3.1

Orographic Precipitation

The pattern of precipitation in coastal British Columbia illustrates orographic precipitation. Moisture-laden air masses, blown by the prevailing westerly winds from the Pacific Ocean, encounter mountains of the Coast Range. As they rise, the air masses cool (by about 0.5–0.8°C for every 100-m increase in elevation), which greatly reduces their ability to hold water. This causes much of the moisture to condense into clouds and then precipitate from the atmosphere as snow and rain.

As the air mass descends on the other side of the mountains, it warms again, increasing its moisture-holding capacity. Therefore, precipitation is more sparse on the *rain-shadow* side of the mountains. Consequently, coastal Vancouver has much more rainfall (averaging about 110 cm/yr) than Penticton in the inland Okanagan Valley (28 cm/yr). (Rainfall just in the Greater Vancouver area ranges from about 50 cm/yr in the southern suburbs, such as Delta, to as much as 250 cm/yr in neighbourhoods nearer the mountains, such as North Vancouver.)

OROGRAPHIC PRECIPITATION ALONG A TRANSECT THROUGH COASTAL BRITISH COLUMBIA

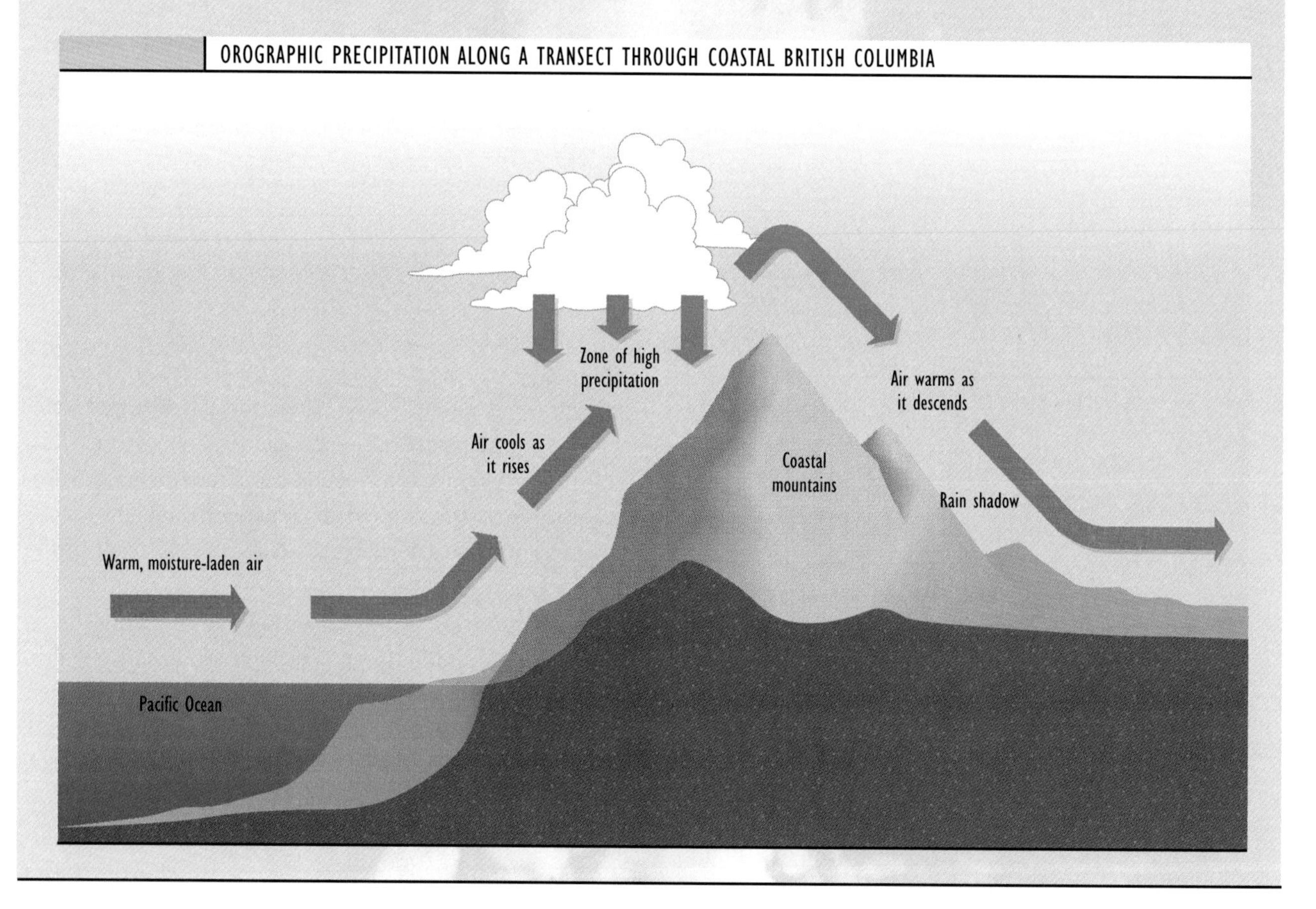

balancing the net import of moisture evaporated from the oceans and the excess of terrestrial precipitation over evapotranspiration. Consequently, there is a net export of flowing water from Earth's terrestrial surfaces to the oceans.

Groundwater drainage involves seepage of water into the ground. Shallow groundwater can drain laterally, eventually draining into surface waters. It can also be taken up by plant roots, most of which is transpired into the atmosphere through foliage. However, deeper groundwater is not available for plant uptake or to recharge surface waters and accumulates in underground, artesian reservoirs, which can be very large.

The hydrologic cycle is very important. Water is required by natural ecosystems for the metabolic needs of organisms, for cooling, and as a ubiquitous solvent that allows water-soluble nutrients to be absorbed by organisms. Water is also required by humans for use in agriculture, industry, and recreation. In many regions, unfortunately, water and its biological resources (such as fish) have been used excessively, and water quality has been degraded through pollution. These damages to water and its resources are common but regrettable themes in many chapters in this book.

The Atmosphere

Earth's **atmosphere** is the envelope of gases that surrounds the planet and is held in place by the attractive forces of gravity. The density of the atmospheric mass is greater at lower altitudes and decreases rapidly with increasing altitude.

The atmosphere consists of four layers, of which the boundaries are not precisely defined and vary over time and space.

1. The **troposphere** contains about 85–90% of the atmospheric mass and extends from the surface to 8–17 km (being thinner at high latitudes and thicker at equatorial latitudes, as well as varying seasonally). Air temperature typically decreases with increasing altitude within the troposphere, and convective air currents (or winds) are common. Consequently, the troposphere is sometimes called the "weather layer."
2. The **stratosphere** extends from above the troposphere to as much as about 50 km, depending on season and latitude. Air temperature varies little with altitude within the stratosphere, and there are few convective air currents.
3. The **mesosphere** extends beyond the stratosphere to about 75 km.
4. The **thermosphere** extends to 450 or more km.

Beyond the atmosphere is **outer space**, a region where the Earth's atmosphere exerts no detectable chemical or thermal influences.

About 78% of the mass of the atmosphere is nitrogen gas (N_2), while 21% is oxygen (O_2), 0.9% argon (Ar), and 0.035% carbon dioxide (CO_2). The rest is various trace gases, including potentially toxic chemicals such as ozone (O_3) and sulphur dioxide (SO_2) (see Chapter 16). The atmosphere also contains highly variable concentrations of water vapour, which can range from only 0.01% in frigid winter air in the Arctic to 5% in warm, humid, tropical air. On average, the total weight of Earth's atmospheric mass exerts a pressure at sea level of around 1.0×10^5 pascals, or *one atmosphere*, approximately equivalent to the weight of a 1.0-kg mass per square centimetre. Air pressure is, however, variable over space and time.

Earth's atmosphere is a highly dynamic medium. This is particularly true of the troposphere, within which temperature and energy gradients are most pronounced. To even out these energy gradients, there is a streaming of atmospheric mass from regions of relatively high pressure to those with lower pressure. These more-or-less lateral atmospheric movements are known as **winds**, the vigour and speed of which can range from barely perceptible to several hundred km/h in extremely turbulent conditions, such as a tornado or hurricane. In general, winds are caused when air heated by the sun becomes less dense and rises in altitude, to be replaced at the surface by an inflow of cooler, denser air. Simply interpreted, this movement of atmospheric mass represents an enormous, gaseous convective cell. These atmospheric movements occur on both local and global scales and are extremely variable over space

The atmosphere is composed of a mixture of gases, fine particulates, and water vapour, often occurring as clouds. This view of a cloudy atmosphere, mountains, and montane forest was taken in Jasper National Park, Alberta.

Photo: B. Freedman

FIGURE 3.3 | ATMOSPHERIC CIRCULATION IN THE WESTERN HEMISPHERE

Note that air masses tend to circulate from areas of relatively high pressure to areas with lower pressure. In such cases, surface winds blow from high pressure towards low pressure and are replaced by a higher-altitude flow of air in the opposite direction, as is suggested by the diagrams around Earth's sphere.

Source: Modified from Botkin and Keller (1995)

and time. At the global level, however, a broadly general pattern of circulation is discernible (Figure 3.3).

As noted above, wind directions are influenced by the relative locations of areas having high or low atmospheric pressures. Wind directions are also influenced by Earth's **Coriolis effect**, caused by the west-to-east rotation of the planet. The Coriolis effect makes winds in the Northern Hemisphere deflect to the right and those in the Southern Hemisphere to the left. Local patterns of wind flow are also influenced by surface topography — mountains are barriers that deflect winds upward or around, while valleys can channel wind flow.

Prevailing winds blow more-or-less continuously in a dominant direction. There are three major classes of prevailing winds: *trade winds* are tropical airflows that blow from the northeast (that is, to the southwest) in the Northern Hemisphere and from the southeast in the Southern Hemisphere; *westerlies* are mid-latitude winds that blow from the southwest in the Northern Hemisphere and from the northwest in the Southern Hemisphere; and *polar easterlies* blow from the northeast at high northern latitudes, and from the southeast near Antarctica.

Climate and Weather

Climate refers to the prevailing, long-term, atmospheric conditions of temperature, precipitation, humidity, wind speed and direction (together, these are wind velocity), insolation (i.e., incoming solar radiation), visibility, fog, and cloud cover in a place or region. Climatic data are usually calculated as statistics (such as averages or ranges of values) using data obtained from several decades or more of monitoring (the preferred period for the calculation of "normal" climatic parameters is at least 30 years).

In contrast, **weather** refers to relatively short-term, day-to-day or instantaneous meteorological conditions (the latter is sometimes referred to as "real-time" weather). Because weather is related to short-term conditions, it is much more variable over time and space than climate.

Many aspects of Earth's climate are functions of solar insolation and how this incoming electromagnetic energy is absorbed, reflected, and re-radiated by the atmosphere, oceans, and terrestrial surfaces. It is not practical here to discuss this complex subject in detail. Nevertheless, we can describe several ecologically important aspects (see also the discussion of physical energy budgets in Chapter 4).

Give Thanks to the Sun

If it were not for the warming influence of solar radiation, the temperature of Earth's surface and atmosphere would approach the coldest that is physically possible — absolute zero, or –273°C (equal to 0 on the Kelvin scale). Although Earth has a limited ability to generate its own heat energy (by the decay of radioactive elements in its core), this is insufficient to provide much warming at the surface. Solar energy is critical to maintaining Earth's surface temperature within a range that organisms can tolerate.

Reflection and Absorption in the Atmosphere

Conditions in Earth's atmosphere have a great influence on climatic variables. For instance, cloud cover and tiny atmospheric particulates are highly reflective of many wavelengths of incoming visible radiation and therefore have a cooling effect on the lower atmosphere and surface. In addition, Earth's atmosphere contains trace concentrations of gases that effectively absorb some of the infrared radiation that the planet radiates to cool itself of the heat obtained from solar radiation. Most important in this regard are water vapour, carbon dioxide, and methane. Because of the influence of these so-called "greenhouse gases," Earth's surface temperature is maintained at an average of about 15°C, or 33° warmer than the –18°C it would be without this moderating effect (see Chapters 4 and 17).

Night and Day

At any place on Earth's surface, the input of solar radiation is relatively high during the day and small at night. (At night, the only radiation inputs are from distant stars and from solar radiation reflected from atmospheric particulates and the moon — inputs known as "skylight.") The daily, 24-hour (or *diurnal*) variations in energy input result in large changes in weather during the 24-hour period. This effect varies greatly between tropical and polar latitudes. Tropical regions have approximately equal day- and night-lengths of about 12 hours each, which do not vary much during the year. In contrast, polar latitudes are much more seasonal in this respect, with virtually continuous light during part of the summer and unremitting night during part of the winter. Temperate latitudes are intermediate, with longer day-lengths during the summer and shorter ones during winter.

Effects of Latitude

Places at tropical latitudes tend to face incoming solar radiation on a relatively perpendicular angle (i.e., closer to 90° at noon-time). Polar latitudes have a more oblique angle of solar incidence, and temperate latitudes are intermediate in this regard. The more perpendicular the angle of incidence of solar radiation, the smaller the surface area over which the incoming energy is distributed, and the more intense is the resulting heating. The angle of solar incidence has a strong influence on differences in unit-area solar radiation received at various latitudes, and is a major reason why the tropics are warmer than polar regions.

Seasons

Earth's axis tilts at a 23.5° angle relative to the angle of incidence of solar radiation. Consequently, during Earth's annual revolution around the sun, there are seasonal differences in energy received between the Northern and Southern Hemispheres. In the Northern Hemisphere, the angle of incidence is closer to perpendicular from March 21 to September 22, giving relatively warmer conditions, while the angle is more oblique from September 22 to March 21, resulting in cooler conditions. These seasons are reversed in the Southern Hemisphere. Because Earth's orbit is elliptical, climatic seasons are also influenced by the varying distance from the sun. This effect is relatively small, however, compared with that of the inclination of Earth's axis.

Aspect and Slope

On a local scale, the direction that a slope faces (known as **aspect**) has a substantial influence on the amount of solar radiation received. In the Northern Hemisphere, south- and to a lesser degree west-facing slopes are relatively warm, while north- and east-facing slopes are cooler. (In the Southern Hemisphere, north-facing slopes are warmer).

The degree of **slope** also affects the amount of energy received. The closer the slope approximates a perpendicular angle to incoming solar radiation, the greater the energy input per unit of surface area. In the Northern Hemisphere, this effect is greatest on south-facing slopes.

Soil and Vegetation Cover

Darker surfaces absorb much more solar radiation (particularly visible radiation) than do lighter surfaces. This is the reason why a black asphalt surface gets much hotter during the day than a lighter-coloured, cement surface. Vegetation canopies also vary greatly in their absorption and reflection characteristics, depending on the colour of the dominant foliage and on the angle at which the foliage is oriented to incoming solar radiation. Major changes in the character of vegetation, as occur when forests are converted into agricultural or urban land-uses, can affect local, and sometimes regional, weather and climate.

Snow and Ice Cover

Because snow and ice are highly reflective to solar radiation, relatively little insolation is absorbed by snow- or ice-covered surfaces. The melting of snow cover in the springtime exposes a much more absorptive ground surface, and warming accelerates greatly.

Evaporation of Water

Moist surfaces can be cooled through the evaporation of water, a process that absorbs thermal energy. Therefore, the transpiration of water from plant foliage has a cooling effect, as does the evaporation of sweat from the body surface of a human.

The above influences on the input, reflection, absorption, and dissipation of solar radiation lead to great variations of air, water, and surface temperatures over Earth's surface. The resulting energy gradients result in global processes that attempt to distribute the energy more evenly, by movements of air masses in the atmosphere (winds) and currents of water in the oceans. In addition, prevailing wind directions can interact with oceanic currents to generate circular flows known as *gyres*. Subtropical gyres rotate clockwise in the Northern Hemisphere and counter-clockwise in the Southern Hemisphere, while subpolar gyres rotate in the opposite directions.

Climate has an important influence on the character of ecological development in any region or place. Climatic conditions can vary on a large scale, called **macroclimate**, and thereby affect the nature of ecosystems over an extensive area. Climatic conditions can also vary on much smaller scales, called **microclimates**, caused, for example, by topography, proximity to the ocean or a large lake, or understorey conditions beneath a dense canopy of tree foliage.

Four important climatic factors particularly affect the development of ecosystems. Of these, variations of precipitation and temperature generally have the greatest influence.

Precipitation volumes are greatly affected by the flow of prevailing winds, the humidity of air masses, and the influence of topography (through the orographic effect; see In Detail 3.1). Dry climates can support only desert vegetation, whereas wetter conditions may allow old-growth forests and wetlands to develop.

Temperature is relatively warm in tropical latitudes and at lower altitude in mountainous terrain, and cooler at high latitude and high altitude. In general, places with cold temperatures develop a tundra vegetation, whereas warmer temperatures may support forests. Temperate and polar latitudes have large, seasonal fluctuations in temperature. Tropical forests develop in moist regions where temperature remains uniformly warm, while temperate and boreal forests are dominated by tree species that can tolerate cold temperatures during the winter.

Wind can also have a substantial ecological influence, although this is typically less important than those of precipitation and temperature. Very windy locations may not be able to support forest, even though precipitation and temperature are otherwise favourable. This occurs in many coastal habitats in Canada, where windy conditions result in shrub-dominated ecosystems developing, rather than the forests that occur farther inland.

Extreme events of weather, such as drought, flooding, hurricanes, and tornadoes, can also be important. Severe disturbances influence ecological development, especially where they occur frequently. For example, frequent drought or severe windstorms may restrict the development of forest in some regions, even though the average climatic conditions may be favourable.

Major elements of the climates of Canada are described in Canadian Focus 3.1. We will discuss their relationship with ecological development in Chapter 8.

CANADIAN FOCUS 3.1

CLIMATES OF CANADA

Canada is a huge country, the second-largest in the world after Russia. The wide range of latitude means that Canadian climates range from cold polar in the High Arctic to warm temperate in southern Ontario and southern British Columbia. Canada also has extremely varied topography, with extensive regions at sea level and many areas at high elevation, particularly in the mountains of Labrador, the eastern Arctic Islands, and the Rocky Mountains. Thus, Canada's regions are characterized by huge differences in climate. The data below reflect "normal" or average values calculated for the period 1961 to 1990.

	ST. JOHN'S NFLD.	HALIFAX N.S.	MONTREAL QUE.	TORONTO ONT.	WINNIPEG MAN.	REGINA SASK.	EDMONTON ALTA.	PENTICTON B.C.	VANCOUVER B.C.	YELLOWKNIFE N.W.T.	RESOLUTE N.W.T.
TEMPERATURE (°C)											
Annual average	4.8	6.1	6.5	7.3	2.2	2.2	3.1	8.9	9.8	−5.4	−16.6
Coldest	−23.3	−26.1	−37.8	−31.1	−45.0	−50.0	−48.3	−27.2	−17.8	−51.1	−52.2
Warmest	30.6	34.5	37.8	38.3	40.6	43.3	34.4	40.6	33.3	32.2	18.3
GROWING DEGREE-DAYS (warmer than 5°C)	1196	1694	2113	2127	1785	1677	1560	2136	1994	1027	33
PRECIPITATION (cm/y)											
Annual total	152	149	95	77	53	40	49	29	111	28	13
Annual rainfall	116	122	72	64	41	29	35	22	105	15	5
Annual snowfall	36	27	23	13	12	11	14	7	6	13	8
Average windspeed (km/h)	24	18	16	15	19	21	14	13	12	16	22
Bright sunshine (h/y)	1497	1885	2054	2045	2321	2331	2264	2032	1920	2277	1505
Days with fog (d/y)	124	122	20	35	20	29	17	1	45	21	62
Days with rain (d/y)	156	125	114	99	72	59	70	78	156	46	20
Days with snow (d/y)	88	64	62	47	57	58	59	29	15	82	82
Days with wind >63 km/h	23	3	1	<1	1	9	0	<1	<1	<1	25
Frost-free period (days)	131	155	157	149	121	109	140	148	216	111	9

Source: Data are from Phillips (1990)

Key Terms

the solar system
Earth's core
mantle
lithosphere
crust
igneous rocks
sedimentary rocks
metamorphic rocks
tectonic forces
meteorites
earthquakes
volcanoes
glaciers (glaciation)
weathering
erosion
sedimentation
lithification
hydrosphere
hydrologic (or water) cycle
fluxes
compartments
ocean
surface water
groundwater
atmospheric water
humidity (relative humidity)
evaporation
transpiration
evapotranspiration
precipitation
orographic precipitation
surface flows
deep drainage
atmosphere
troposphere
stratosphere
mesosphere
thermosphere
outer space
wind
Coriolis effect
prevailing winds
climate
weather
aspect
slope
macroclimate
microclimate

Questions for Discussion

1. What major geological forces have influenced landscape features in the region where you live?
2. Where does your drinking water come from? Trace its origins and disposal in terms of the hydrologic cycle.
3. Outline the differences between climate and weather and discuss their influences on your daily and annual life.
4. Describe the various layers of Earth's solid sphere and atmosphere.

References

Alvarez, W., E.G. Kauffman, F. Surlyk, L.W. Alvarez, F. Asaro, and H.V. Michel. 1984. Impact theory of mass extinctions and the invertebrate fossil record. *Science*, **223**: 1135–1141.

Botkin, D.B. and E.A. Keller. 1995. *Environmental Science: Earth as a Living Planet.* New York: J. Wiley & Sons.

Cowen, R. 1995. *History of Life.* London, U.K.: Blackwell Sci. Pub.

Flint, R.F. and B.J. Skinner. 1987. *Physical Geology.* New York: J. Wiley & Sons.

Jablonski, D. 1991. Extinctions: A paleontological perspective. *Science*, **253**: 754–757.

Margulis, L. and L. Olendzenski (eds.). 1992. *Environmental Evolution.* Cambridge, MA: MIT Press.

Phillips, D. 1990. *The Climates of Canada.* Ottawa, ON: Environment Canada.

Pielou, E.C. 1991. *After the Ice Age: The Return of Life to Glaciated North America.* Chicago, IL: University of Chicago Press.

Raup, D.M. 1986. *The Nemesis Affair.* New York: Norton Pub. Co.

Schneider, S.H. 1989. The changing climate. *Scientific American*, **261 (3)**: 70–79.

Van Andel, T.H. 1985. *New Views on an Old Planet: Continental Drift and the History of the Earth.* Cambridge, U.K.: Cambridge University Press.

4

ENERGY AND ECOSYSTEMS

Chapter Outline

Chapter Objectives

After completing this chapter, you will be able to:

1. Describe the nature of energy, its various forms, and the laws that govern its transformations.
2. Discuss how Earth is a flow-through system for solar energy.
3. Identify the three major components of Earth's energy budget.
4. Describe energy relationships within ecosystems, including the fixation of solar energy by primary producers and the passage of that fixed energy through other components of the ecosystem.
5. Explain why the trophic structure of ecological productivity is pyramid-shaped and why ecosystems cannot support many top predators.
6. Compare the feeding strategies of humans living a hunting and gathering lifestyle and those of modern cultures living in cities and towns.

Introduction

Earth's ecosystems are not closed, self-sustaining systems. In fact, without continuous access to an external source of energy, ecosystems would quickly deplete their quantities of stored energy and cease to function. The external source of energy is solar energy, which is stored mainly as biomass and heat.

Solar energy is absorbed by green plants and algae, and utilized to *fix* carbon dioxide and water into simple sugars, through a process known as *photosynthesis*. Ultimately, this biological fixation of solar energy provides the energy basis for most organisms and ecosystems (the few exceptions are described later).

Energy is critical to the functioning of physical processes throughout the universe and of ecological processes on Earth. In this chapter we will examine the physical nature of energy, the laws that govern its behaviour and transformations, and the role of energy in ecosystems.

The energy from the sun is derived from nuclear fusion reactions involving hydrogen nuclei. The reactions generate enormous quantities of thermal and electromagnetic energy. Solar electromagnetic energy is the most crucial source of energy sustaining ecological and biological processes on Earth.

Source: Corbis-Bettmann

The Nature of Energy

Energy is a fundamental physical entity, and is defined simply as the capacity of a body or system to accomplish work. In physics, **work** is defined as the result of a force being applied over a distance. In all of the following examples of work, energy is transformed and some measurable outcome is achieved.

- a hockey stick strikes a puck, causing it to speed toward a target
- a book is picked up from the floor, lifted, and then laid on a table
- a vehicle is driven along a road
- heat from a stove is absorbed by water in a kettle, causing it to become hotter and eventually boil
- the photosynthetic pigment chlorophyll absorbs sunlight, converting the electromagnetic energy into a form that plants and algae can utilize to synthesize sugars.

Forms of Energy

Energy can exist in various states, each of which is fundamentally different from the others. Under suitable conditions, however, any energy state can be converted into another through various physical or chemical transformations. The states can be grouped into three categories of energy: electromagnetic, kinetic, and potential.

Electromagnetic energy

Electromagnetic energy (or electromagnetic radiation) is associated with *photons*, which are entities with properties of both particles and waves and travel through space at a constant speed of 3×10^8 meters per second (that is, at the speed of light). Electromagnetic energy exists in a continuous spectrum of wavelengths, which (ordered from the shortest to longest wavelengths) are known as gamma, X-ray, ultraviolet, visible light, infrared, microwave, and radio (Figure 4.1). The human eye can perceive electro-

FIGURE 4.1 | THE ELECTROMAGNETIC SPECTRUM

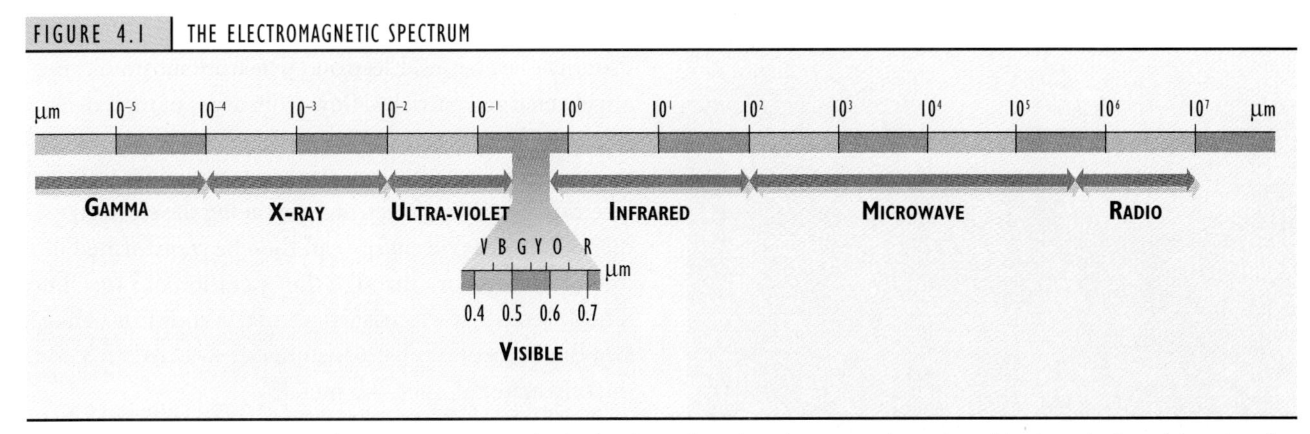

The spectrum is divided into major components on the basis of wavelength and presented on a logarithmic scale ($\log_{10}$) in units of micrometers (μm = 10^{-3} mm = 10^{-6} m). Note the expansion of the visible component, and the wavelength ranges for red, orange, yellow, green, blue, and violet colours.

magnetic energy with wavelengths between about 0.4–0.7 micrometers (1 micrometer, or 1 μm, is 10^{-6} m; see Appendix I), a range commonly referred to as visible radiation or light.

Electromagnetic energy is given off (radiated) by all objects with a surface temperature greater than absolute zero (i.e., greater than –273°C). The rate and spectral quality of the emitted radiation is determined by the surface temperature of the body. Compared with cooler bodies, relatively hotter ones have much greater emission rates and their radiation is dominated by shorter, higher-energy wavelengths. The sun has an extremely hot surface temperature, about 6000°C (the interior of the sun is much hotter still), and most of its radiation is ultraviolet (0.2–0.4 μm), visible (0.4–0.7 μm), and near infrared (0.7–2 μm). Because the surface temperature of Earth averages only about 15°C, it radiates much smaller quantities of longer-wavelength energy (peaking at a wavelength of about 10 μm).

Kinetic energy

Kinetic energy is associated with motion. We can distinguish two classes of kinetic energy: mechanical and thermal.

Mechanical kinetic energy is associated with objects in motion, for example, a baseball flying through the air, a deer running through a forest, water flowing in a stream, or a planet moving through space. The quantity of mechanical kinetic energy is determined by the mass of the object and its speed.

Thermal kinetic energy is associated with the rate of vibration of atoms or molecules. Such vibrations are frozen at –273°C (i.e., absolute zero), but they are progressively more vigorous at higher temperatures, reflecting a larger content of thermal kinetic energy. Thermal energy is sometimes referred to as *heat*.

Potential energy

Potential energy is the stored ability to perform work. To actually perform work, potential energy must be transformed into electromagnetic or kinetic energy. There are various types of potential energy, associated with gravity, chemicals, compressed gases, electrical potential, magnetism, and the sub-atomic organization of matter.

Gravitational potential energy results from the attractive forces between objects (that is, gravity). For example, water stored at some height above sea level contains gravitational potential energy. This can be converted into kinetic energy if a pathway allows the water to flow downhill in response to gravity. Gravitational potential energy can be converted into electrical energy through the technology of hydroelectric power plants.

Chemical potential energy is stored in the bonds between atoms within molecules. Chemical potential energy can be liberated by *exothermic reactions* (i.e., those which lead to a net release of thermal energy). For example:

- Chemical potential energy is stored in the molecular bonds of sulphide minerals (such as iron sulphide, FeS_2), and some of this energy is released

Organic matter and fossil fuels contain potential chemical energy, which is released during combustion to generate heat and electromagnetic radiation. This forest fire near Dawson City in the Yukon was ignited naturally by lightning and burned the organic matter of boreal-forest vegetation.

Source: T. Keith

when the sulphides are oxidized. Specialized bacteria can metabolically use the chemical potential energy of sulphide minerals to support their own productivity, through a process known as *chemosynthesis* (discussed later in this chapter).

- The ionic bonds of salts also store chemical potential energy. For example, when table salt (sodium chloride, or NaCl) is dissolved in water, ionic potential energy is released as heat, slightly increasing the temperature of the water.
- Hydrocarbons store energy in the bonds between their hydrogen and carbon atoms (hydrocarbons contain only these atoms). The chemical potential energy of gasoline, a mixture of liquid hydrocarbons, is liberated in an internal combustion engine to achieve the kinetic energy of vehicular motion.
- Organic compounds (biochemicals) produced metabolically by organisms also store large quantities of potential energy in their inter-atomic bonds. Carbohydrates typically contain about 16.8 kilojoules/gram, proteins about 21.0 kJ/g, and lipids or fats about 38.5 kJ/g. Many organisms store their energy reserves as fats because these biochemicals have such a high energy density.

Electrical potential energy results from differences in the quantity of electrons. Electrons, which are subatomic, negatively charged particles, flow from areas of high density to areas of low density. When an electrical switch is used to complete a circuit connecting two areas with different electrical potentials, electrons flow along the electron gradient. The electric energy can then be transformed into light, heat, or work through the operation of a machine. The difference in electrical potential is known as voltage, and the current of electrons must flow through a conducting material, such as a metal.

Compressed gases also store potential energy which can do work when the gases are allowed to expand. This type of potential energy is present in a cylinder containing compressed or liquified gas.

Nuclear potential energy results from the extremely strong binding forces that exist within atoms. This is by far the densest form of energy. Huge quantities of electromagnetic and kinetic energy are liberated when nuclear energy is released by processes that convert matter into energy. **Fission reactions** involve the splitting of isotopes of certain heavy atoms, such as uranium-235 and plutonium-239, to generate smaller atoms plus enormous amounts of energy. Fission reactions occur in nuclear explosions and, under controlled conditions, in nuclear reactors, such as those used to generate electricity.

Fusion reactions involve the combining of certain light elements, such as hydrogen, to form heavier atoms under conditions of extremely high temperature and pressure, while liberating huge quantities of energy. Fusion reactions involving hydrogen occur in stars, and are responsible for the unimaginably large amounts of energy that these celestial bodies generate and radiate into space. It is thought that all heavy atoms in the universe were produced by fusion reactions occurring in stars (see Chapter 3). Fusion reactions also occur in a type of nuclear explosion, known as a hydrogen bomb. A technology has not yet been developed to allow controlled fusion reactions; if and when available, controlled fusion could be used to generate virtually unlimited amounts of electricity (see Chapter 13).

Energy Units

As we explained above, energy can exist in various forms, all of which can be measured in the same or equivalent units.

The *SI* recommended unit (or Système Internationale d'Unités, the internationally accepted system for scientific units) is the **joule** (abbreviation J). A joule is defined as the energy required to accelerate 1 kg of mass at 1 m/s^2 (1 m per second per second) for a distance of 1 m.

A **calorie** (or gram-calorie, abbreviation cal) is another unit of energy. A calorie, defined as being equivalent to 4.184 J, is equal to the amount of energy required to raise the temperature of one gram of pure water by one degree Centigrade (specifically, from 15°C to 16°C). Note that the dietician's "Calorie" is equivalent to 1000 calories (i.e., 1 Calorie = 1 kcal). However, the energy content of many food products are now listed in kJ in countries using the SI system of units, such as Canada.

Energy Transformations and the Laws of Thermodynamics

As was previously noted, energy can be transformed among its various states. For example, when solar electromagnetic radiation is absorbed by a dark object, it is transformed into thermal kinetic energy and the absorbing body increases in temperature. The gravitational potential energy of water stored at a height is converted into the kinetic energy of flowing water at a waterfall, or it may be utilized in hydroelectric technology to spin a turbine and generate electrical energy. As well, visible wavelengths of solar radiation are absorbed by chlorophyll, a green pigment in the foliage of plants, and some of the absorbed energy is converted into chemical potential energy of sugars through the biochemistry of photosynthesis.

All transformations of energy must behave according to certain physical principles, known as the **laws of thermodynamics**. These are universal principles, meaning they are always true, regardless of the circumstances.

The First Law of Thermodynamics

The **first law of thermodynamics**, also known as the law of conservation of energy, can be stated: *energy can undergo transformations among its various states but it is never created or destroyed; thus, the energy content of the universe remains constant.* A consequence of this law is that there is always a zero balance among the energy inputs to a system, any net storage within the system, and the energy output from the system.

Consider the case of an automobile driving along a highway. The vehicle consumes gasoline, an energy input that can be measured. The potential energy of the fuel is converted into various other kinds of energy, including kinetic energy embodied in forward motion of the automobile, electrical energy powering the lights and windshield wipers, heat from friction between the vehicle and the atmosphere and road surface, and the hot exhaust gases (thermal energy) and unburned fuel (chemical potential energy) that are vented through the tailpipe. Overall, in accordance with the first law of thermodynamics, an accurate accounting of all of these transformations would find that while the energy of the gasoline was converted into various other forms, the *total* amount of energy was *conserved* (remained constant).

The Second Law of Thermodynamics

We can express the **second law of thermodynamics** as follows: *transformations of energy can occur spontaneously only under conditions in which there is an increase in the entropy of the universe.* **Entropy** is a physical attribute related to disorder, and is associated with the degree of randomness in the distributions of matter and energy. As the randomness (i.e., disorder) increases, so does entropy. A decrease in disorder is referred to as negative entropy.

Consider, for example, an inflated balloon. Because of the potential energy of its compressed gases, that balloon may slowly leak its contents to the surrounding atmosphere; alternatively, it may burst. Either of these events can readily occur spontaneously, because both processes would represent increases in the entropy of the universe. This is because compressed gases are more highly ordered than gases dispersed in the atmosphere. In contrast, the dispersed gases in the atmosphere would never spontaneously relocate to inflate a balloon. A balloon can be inflated only if energy is expended through a local application of work, for example, by a person blowing into the balloon. In other words, energy must be expended to locally decrease entropy in the system. Note that this energy cost itself gives rise to an increase in the entropy of the universe. For instance, the effort of the balloon blower involves additional respiration, which uses biochemical energy and results in heat being expelled into the environment.

Another example concerns planet Earth. Earth continuously receives solar radiation, almost all of which is comprised of visible and near-infrared wavelengths in the range of about 0.4–2.0 µm. Some of this electromagnetic energy is absorbed and converted to thermal energy, heating Earth's atmosphere and surface. The planet cools itself in various ways, but ultimately Earth dissipates all of the absorbed solar radiation by radiating its own electromagnetic energy to outer space as longer-wave infrared radiation (of a spectral quality that peaks at a wavelength of 10µm). In this case, relatively short-wavelength solar radiation is ultimately transformed into the longer-wavelength radiation emitted by Earth, a process that represents a spontaneous degradation in quality of the energy and an increase in the entropy of the universe.

An important corollary (or secondary proposition) of the second law of thermodynamics is that energy transformations can never be completely efficient — some of the initial content of energy must always be converted to heat, so that entropy increases. This helps to explain why, even when using the best available technology, only about 30% of the potential energy of gasoline can be converted into the kinetic energy of a moving automobile, and no more than about 40% of the energy of coal can be transformed into electricity in a generating station. There are also thermodynamic limits to the efficiency of photosynthesis, by which plants convert visible radiation into biochemicals, even under ideal ecological conditions with optimal amounts of nutrients, water, and light.

A superficial assessment might suggest that life in general appears to contradict the second law of thermodynamics. Plants, for example, absorb visible wavelengths of electromagnetic radiation and use this highly dispersed form of energy to fix simple inorganic molecules (carbon dioxide and water) into extremely complex and energy-dense biochemicals. The plant biomass may then be consumed by animals and microbes, who synthesize their own complex biochemicals. These various biological syntheses represent energy transformations that greatly decrease local entropy, because relatively dispersed electromagnetic energy and inorganic compounds are actively converted into the complex, highly ordered biochemicals of organisms. Do these biological transformations contravene the second law of thermodynamics?

This seeming paradox of life can be successfully resolved using the following logic: the localized bio-concentrating of negative entropy can occur because the system (i.e., life on Earth) receives a constant input of energy in the form of solar radiation. If this external source of energy were terminated, all the organisms and organic materials would quickly and spontaneously degrade, releasing simple inorganic molecules and heat and thereby increasing entropy in the universe. Life and ecosystems cannot survive without continual inputs of solar energy, which are required to organize and maintain their negative entropy. In this sense, the biosphere can be viewed as representing an island (system) of negative entropy, localized in space and time, and continuously fuelled by the sun as an external source of energy.

Earth: An Energy Flow-through System

Solar electromagnetic radiation is by far the major input of energy that drives Earth's ecosystems. Solar energy heats the planet, circulates its atmosphere and oceans, evaporates its water, and sustains almost all its ecological productivity. Eventually, all solar energy absorbed by Earth is reradiated back to space in the form of electromagnetic radiation with a longer wavelength than what was originally absorbed. Therefore, Earth maintains a virtually perfect balance between these inputs and outputs of electromagnetic energy. In other words, Earth is a **flow-through system**, with an input of solar energy, an output of reradiated energy, and no net storage of energy.

In addition, Earth's ecosystems depend absolutely on solar radiation as the source of energy that photosynthetic organisms (such as green plants) can utilize to synthesize simple organic compounds (such as sugars) from inorganic molecules (such as carbon dioxide and water). Plants then use the chemical potential energy in these sugars, plus inorganic nutrients, to synthesize a huge diversity of biochemicals through various metabolic reactions. Plants grow and reproduce by using these biochemicals and their potential energy. Moreover, plant biomass is used as food by the enormous numbers of organisms that are incapable of photosynthesis. These organisms include *herbivores* that eat plants directly, *carnivores* that eat other animals, and *detritivores* that feed on dead biomass. (The energy relationships within ecosystems are described later in more detail.)

Less than 0.02% of the solar energy received at Earth's surface is absorbed and fixed by photosynthetic plants and algae. Although this represents a quantitatively trivial component of Earth's energy budget, it is extremely important qualitatively. This biologically absorbed and fixed energy is the foundation of ecological productivity.

Ultimately, however, the solar energy fixed by plants and algae is released to the environment again as heat and is eventually radiated back to outer space. This reinforces the idea of Earth as a flow-through system for energy.

EARTH'S PHYSICAL ENERGY BUDGET

An **energy budget** of a given system describes the rates of input and output of energy as well as any internal transformations of energy among its various states, including changes in stored quantities. Figure 4.2 illustrates key aspects of the budget of Earth's physical energy.

The rate of input of solar radiation to Earth averages about 8.36 joules per cm^2 per minute (equivalent to 2.00 $cal/cm^2 \cdot min$), measured at the outer limit of the planet's atmosphere. About 1/2 of this energy input is visible radiation and 1/2 is near-infrared. The output of energy from Earth also occurs at a rate of about 8.36 $J/cm^2 \cdot min$. Because the rates of energy input and output are equal, there is no

FIGURE 4.2 | IMPORTANT COMPONENTS OF EARTH'S PHYSICAL ENERGY BUDGET

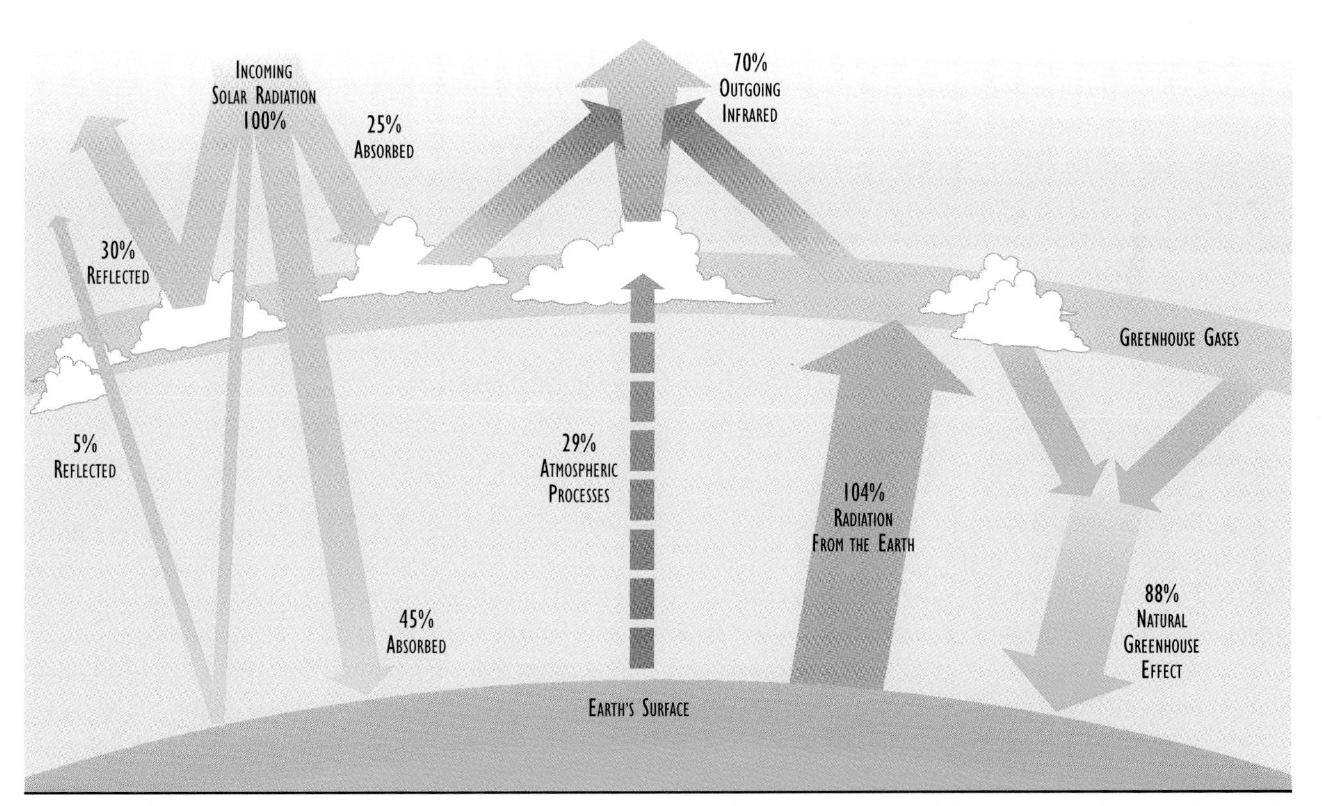

About 30% of the incoming solar radiation is reflected by atmospheric clouds and particulates or by Earth's surface. The remaining 70% is absorbed and then dissipated in various ways. Much of the absorbed energy heats the atmosphere and terrestrial surfaces, and most is reradiated as long-wave infrared radiation. Atmospheric moisture and greenhouse gases interfere with this process of reradiation, keeping Earth's surface warmer than it would be otherwise (see Chapter 17). The numbers refer to percentage of incoming solar radiation.

Source: Modified from Schneider (1989)

net storage of energy, and Earth's average surface temperature remains rather stable. Therefore, as was previously noted, the energy budget of Earth can be roughly characterized as a zero-sum, flow-through system.

However, the above is not exactly true. Over extremely long, geological time scales, a small storage of solar energy has occurred through the accumulation of undecomposed biomass that eventually transforms into fossil fuels. Minor long-term fluctuations in Earth's surface temperature also occur, representing an important element of climate change. These are, however, quantitatively trivial exceptions to the statement that Earth is a zero-sum, flow-through system for solar energy.

Even though the quantity of energy emitted by Earth eventually equals the amount of solar radiation that is absorbed, many ecologically important transformations occur between the initial absorption and eventual reradiation. These are the internal elements of the planet's physical energy budget (see again Figure 4.2), and the most important components are described below.

Reflection

On average, Earth's atmosphere and surface reflect about 30% of incoming solar energy back to outer space. Earth's reflectivity (or *albedo*) is strongly influenced by such factors as the angle of the incoming solar radiation (which varies during the day and over the year), amounts of cloud cover and atmospheric particulates (also highly variable), and the character of the surface, especially the types and amounts of water (including snow and ice) and vegetation.

Absorption by the Atmosphere

About 25% of incident solar radiation is absorbed by gases, vapours, and particulates in the atmosphere, including clouds. The rate of absorption is wavelength-specific, with portions of the infrared range being intensively absorbed by the so-called "greenhouse" gases (especially carbon dioxide; see Chapter 17). The absorbed energy is converted to heat and reradiated as infrared radiation of a longer wavelength.

Absorption by the Surface

On average, about 45% of incoming solar radiation passes through the atmosphere and is absorbed at Earth's surface by living and non-living materials, increasing the temperature of the absorbing surfaces. This figure of 45% is highly variable, however, depending on atmospheric conditions, especially cloud cover, and also on whether the incident light has passed through plant canopies.

Over the longer term (that is, years), and even the medium term (i.e., days), the global net storage of heat is virtually zero. In some places, however, there may be substantial changes in the net storage of thermal energy within the year. This occurs everywhere in Canada because of the strong seasonality of its climate (that is, Canadian environments are warmer during summer than during winter). Most of the absorbed energy is eventually dissipated by reradiation from the surface as long-wave infrared.

Evaporation of Water

Some of the thermal energy of surfaces causes water to evaporate from living and non-living surfaces in a process known as **evapotranspiration.** Evapotranspiration has two components: **evaporation** of water from lakes, rivers, streams, moist rocks, soil, and other non-living substrates, and **transpiration** of water from any living surface, particularly from plant foliage, but also from moist body surfaces and lungs of animals.

Melting of Snow and Ice

Absorbed thermal energy can also cause ice and snow to melt, representing an additional energy transformation associated with a change of state of water (that is, from a solid to a liquid form).

Wind and Water Currents

Thermal energy on Earth's surface has a highly uneven distribution, with some regions being quite cold (e.g., the Arctic) and others relatively warm. Because of this irregular allocation of heat, Earth's surface develops processes to diminish the energy gradients by transporting mass around the globe, such as by winds, water currents, and waves on the surface of waterbodies (see also Chapter 3).

Biological Fixation

A very small but ecologically critical portion of incoming solar radiation (globally averaging less than 0.02%) is absorbed by chlorophyll in plants and algae and used to drive photosynthesis. This biological fixation allows some of the solar electromagnetic energy to be temporarily stored as potential energy in biochemicals, thereby serving as the energetic basis for ecological productivity and life on Earth.

Energy and Energy Budgets in Ecosystems

Ecological energy budgets focus on the absorption of energy by photosynthetic organisms and the transfer of that fixed energy through the *trophic levels* of ecosystems ("trophic" refers to the means of organic nutrition). Ecologists classify organisms in terms of the sources of energy that they utilize.

Autotrophs are capable of synthesizing their complex biochemical constituents using simple inorganic compounds and an external source of energy to drive the process. Most autotrophs are **photoautotrophs**, which use sunlight as their external source of energy. Photoautotrophs capture the solar radiation using photosynthetic pigments, the most important of which is chlorophyll. Green plants are the most abundant examples of photoautotrophic organisms, but algae and some bacteria are also photoautotrophic.

A much smaller number of autotrophs are **chemoautotrophs**, which harness some of the energy content of certain inorganic chemicals to drive *chemosynthesis*. The bacterium *Thiobacillus thiooxidans*, for example, oxidizes sulphide minerals to sulphate and uses some of the energy liberated during this reaction to chemosynthesize organic molecules.

Because autotrophs are the biological foundation of ecological productivity, ecologists refer to them as **primary producers**. The total fixation of solar energy by all the primary producers within an ecosystem is known as **gross primary production** (GPP). Primary producers use some of this production for their own **respiration** (R), that is, for the physiological functions needed to maintain themselves in a healthy condition. Respiration, the metabolic oxidation of biochemicals, requires a supply of oxygen and releases carbon dioxide and water as waste products. **Net primary production** (NPP) refers to the fraction of gross primary production that remains after primary producers have used some of their GPP for their own respiration. In other words: NPP = GPP – R.

Plant productivity is sustained by solar energy which is fixed by chlorophyll in the plant and used to combine carbon dioxide, water, and other simple inorganic compounds into the complex molecular structures of organic matter. These ecologists are studying the productivity of a plant community on Sable Island, Nova Scotia.

Source: B. Freedman

The energy fixed by primary producers is the basis for the productivity of all other organisms, known as heterotrophs. **Heterotrophs** rely on other organisms, living or dead, to supply the energy they need; they cannot fix solar energy themselves. Animal heterotrophs that feed on plants are known as **herbivores** (or **primary consumers**), three familiar examples being deer, geese, and grasshoppers. Heterotrophs that consume other animals are known as **carnivores** (or **secondary consumers**), for example, timber wolves, peregrine falcons, sharks, and spiders. Some species feed on both plant and animal biomass and are known as **omnivores** — the grizzly bear is a good example of this feeding strategy, as is our own species. Many other heterotrophs feed primarily on dead organic matter and are called **decomposers** or **detritivores**, examples being vultures, earthworms, and most fungi and bacteria.

Productivity is production expressed as a rate function, that is, per unit of time and area. Productivity in terrestrial ecosystems is often expressed in units such as kilograms of dry biomass (or its energy equivalent) per hectare per year (e.g., kg/ha·y or kJ/ha·y), while aquatic productivity is often given as $kg/m^3 \cdot y$.

Numerous studies have been made of the productivity of the different trophic levels in various types of ecosystems. For example, studies of a natural oak-pine forest in New York found that the total fixation of solar energy by the vegetation (i.e., the annual gross primary productivity) was equivalent to 48 100 kilojoules per hectare per year (4.81×10^4 kJ/ha·y) (Odum, 1993). This fixation rate was equivalent to less than 0.01% of the annual input of solar radiation to the forest. Because the plants used 2.72×10^4 kJ/ha·y during respiration, the net primary productivity in the ecosystem was 2.09×10^4 kJ/ha·y, represented mainly by the accumulating biomass of the trees. The various heterotrophic organisms in the forest used 1.26×10^4 kJ/ha·y to support their respiration. Ultimately, the net accumulation of biomass by all the organisms in the ecosystem (referred to as the **net ecosystem productivity**) was equivalent to 0.84×10^4 kJ/ha·y.

The primary productivities of the world's major classes of ecosystems are summarized in Table 4.1. Note that the rate of production is greatest in tropical forests, wetlands, coral reefs, and estuaries. The production for each ecosystem type is calculated as its productivity multiplied by its

TABLE 4.1 PRIMARY PRODUCTION OF EARTH'S MAJOR ECOSYSTEMS

Productivity is the rate of production, standardized to area and time, while production is the total amount of biomass (in dry tonnes) produced by the total global area of each ecosystem. Ecosystems are arranged in order of net primary productivity. See Chapter 8 for descriptions of biomes.

ECOSYSTEM	AREA ($\times 10^6$ km^2)	NET PRIMARY PRODUCTIVITY (t/ha·y)	GLOBAL NET PRODUCTION ($\times 10^9$ t/y)
Wetlands	2.0	30.0	6.0
Tropical rain forest	17.0	22.0	37.4
Tropical seasonal forest	7.5	16.0	12.0
Temperate evergreen forest	5.0	13.0	6.5
Temperate deciduous forest	7.0	12.0	8.4
Savannah	15.0	9.0	13.5
Boreal forest	12.0	8.0	9.6
Open woodland	8.5	7.0	6.0
Cultivated land	14.0	6.5	9.1
Temperate grassland	9.0	6.0	5.4
Lake and stream	2.0	4.0	0.8
Tundra, arctic and alpine	8.0	1.4	1.1
Desert and semidesert scrub	18.0	0.9	1.6
Extreme desert	24.0	<0.1	0.1
TOTAL CONTINENTAL	149.0	7.8	117.5
Reefs and estuaries	2.0	18.0	3.7
Shelf and upwelling	27.0	3.6	9.8
Open ocean	332.0	1.3	41.5
TOTAL MARINE	361.0	1.5	55.0
WORLD TOTAL	510.0	3.4	172.5

Source: Whittaker and Likens (1975)

area. The largest amounts of production occur in tropical forests and the open ocean. Note that the open ocean has a relatively small productivity, but its global production is large because of its enormous area.

Ecological **food chains** are hierarchical models of feeding relationships among species within an ecosystem. An example of a simple food chain in the northern forests and tundra of Canada is: lichens and sedges, eaten by caribou, eaten by wolf.

Food webs are more complex models of feeding relationships, describing the connections among all food chains within an ecosystem. Wolves, for instance, are opportunistic predators that feed on snowshoe hare, voles, lemming, beaver, birds, and other prey in addition to their usual prey of deer, moose, and caribou. Therefore wolves participate in various food chains within their ecosystem. However, no natural predators feed on wolves, which are therefore referred to as *top carnivores* or *top predators*.

Figure 4.3 illustrates important elements of the food web of Lake Erie, one of the Great Lakes. In this lake, shallow-water environments support aquatic plants, while phytoplankton occur throughout the upper water column. The shallow-water plants are consumed by ducks, muskrat, and other herbivores, while phytoplankton are consumed by tiny crustaceans (or zooplankton) and bottom-living filter-feeders such as clams. Zooplankton are eaten by small fish such as smelt, which are fed upon by larger fish, which may eventually be eaten by cormorants, bald eagles, or humans. Dead biomass from any level of the food web may settle to the bottom, where it enters a detrital food web. There it is eaten by small animals and ultimately decomposed by bacteria and fungi.

In accordance with the second law of thermodynamics, the transfer of energy in food webs is always inefficient because some of the fixed energy must be degraded into heat. For example, when a herbivore consumes plant biomass, only some of the energy content can be assimilated and transformed into herbivore biomass. The rest is excreted in faeces or utilized in respiration (Figure 4.4). Consequently, in all ecosystems the amount of productivity by autotrophs is always much greater than that of herbivores, which in turn is always greater than that of their predators. As a broad generalization, there is about a 90% loss of energy at each transfer stage. In other words,

FIGURE 4.3 | MAJOR ELEMENTS OF THE FOOD WEB IN LAKE ERIE

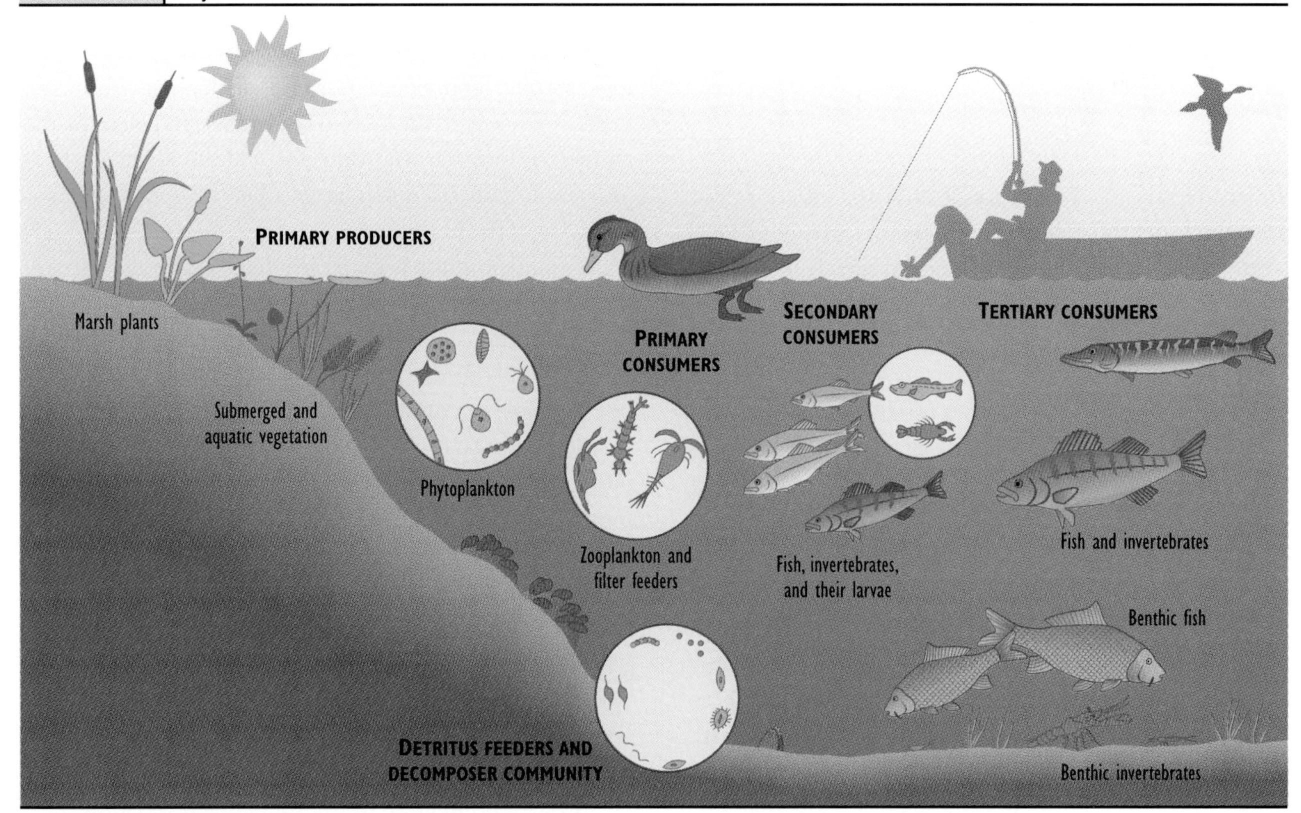

FIGURE 4.4 | MODEL OF ENERGY TRANSFER IN AN ECOSYSTEM

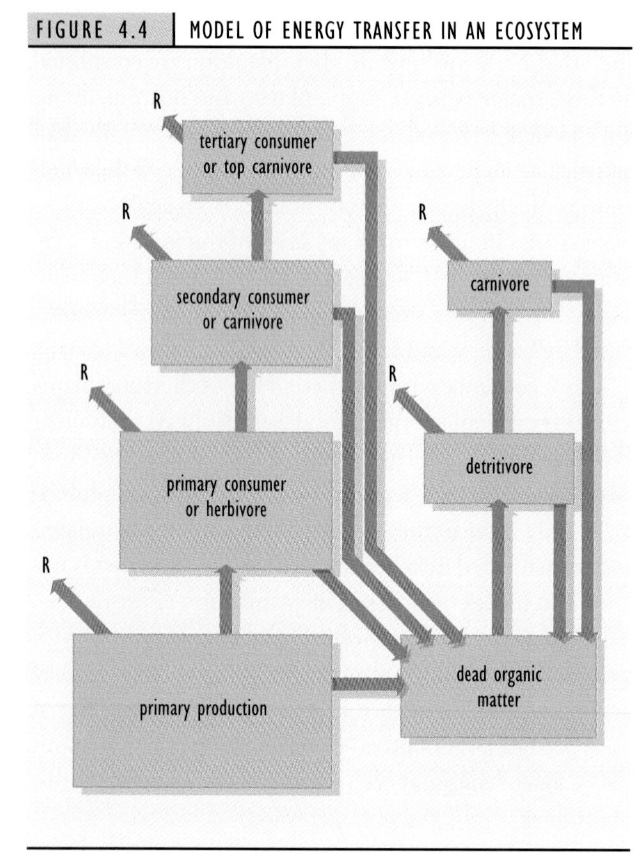

Note that lower levels of the food web always have a greater productivity than higher levels, so the trophic structure is roughly pyramidal. Although not drawn to an appropriate scale, this relationship is roughly indicated by the areas of the boxes. "R" indicates respiration.

the productivity of herbivores is only 10% that of their plant food, and the productivity of the first carnivore level is only 10% that of the herbivores that they feed upon.

These productivity relationships can be displayed graphically, using so-called **ecological pyramids** to represent the **trophic structure** of an ecosystem. Ecological pyramids are organized with plant productivity on the bottom, that of herbivores above the plants, and carnivores above the herbivores. If the ecosystem sustains top carnivores, they are represented at the apex of the pyramid. The sizes of the trophic boxes in Figure 4.4 suggest the pyramid-shaped structure of ecosystem productivity.

The second law of thermodynamics applies to ecological productivity, a function that is directly related to energy flow. The second law does not, however, directly apply to the accumulated biomass of the ecosystem. Consequently, it is only the productivity trophic structure that is pyramid shaped. In some ecosystems, other variables may have a pyramid-shaped trophic structure, such as quantities of biomass (or *standing crop*) present at specific times, or the sizes or densities of populations. These particular variables are not, however, pyramid shaped in all ecosystems.

For example, in the open ocean, phytoplankton are the primary producers, but they often maintain a biomass similar to that of the small zooplankton that feed upon them. The phytoplankton cells are relatively short-lived and their biomass turns over quickly because of their high productivity. In contrast, the individual zooplankton are longer lived and much less productive than the phytoplankton. Consequently, the productivity of the phytoplankton is much larger than that of the zooplankton, even though at any particular time both of these trophic levels may have a similar biomass.

Some ecosystems may even have an inverted pyramid of biomass, characterized by a smaller biomass of plants than of herbivores. This sometimes occurs in grasslands, in which the dominant plants are relatively small, herbaceous species that can be quite productive but do not maintain a large biomass. In comparison, some of the herbivores that feed on the plants are relatively large, long-lived animals, which may maintain a larger total biomass than the vegetation. Some temperate and tropical grasslands have such an inverted biomass pyramid, especially during the dry season when there may be large populations (and biomass) of long-lived herbivores such as deer, bison, antelope, gazelle, hippopotamus, rhino, or elephant. However, in accordance with the second law of thermodynamics, the annual (or long-term) productivity of the plants in these grasslands is always much larger than the long-term productivity of the herbivores.

In addition, the population densities of animals are not necessarily smaller than those of the plants that they eat. For instance, insects are the most important herbivores in many forests and they commonly maintain large populations. In contrast, the numbers of trees are much smaller, because each individual plant is large and occupies a great deal of space. Forests typically maintain many more herbivores than trees and other plants, so the pyramid of numbers is inverted in shape. As in all ecosystems, however, the pyramid of forest productivity is much wider at the bottom than at the top.

Because of the inefficiency of the energy transfer between trophic levels, there are energetic limits to the number of top carnivores (such as wolves, killer whales, eagles, and sharks) that can be sustained by an ecosystem. To sustain a viable population of top predators, there must be a suitably high productivity of prey that these animals can exploit. This prey must in turn be sustained by an appropriately high plant productivity. Because of these ecological constraints, only extremely productive or very extensive ecosystems can support top predators.

Of all Earth's terrestrial ecosystems, none supports more species of higher-order carnivores than the savannas and grasslands of Africa. The most prominent of these top predators are cheetah, lion, leopard, spotted and striped hyaena, and wild dog. This unusually high richness of top predators can be sustained because these African ecosystems are extensive and rather productive of vegetation, except during years of drought. In contrast, the tundra of northern Canada can support only one natural species of top predator, the wolf, because, although extensive, tundra is a relatively unproductive ecosystem.

Some pre-industrial human populations functioned as top predators, including aboriginal peoples of Canada such as the arctic Inuit and many sub-arctic Indian cultures. As an ecological consequence of their higher-order feeding strategy within the food web, these cultures were not able to maintain very large populations. In most modern economies, however, humans interact with ecosystems in an omnivorous manner — we harvest an extremely wide range of foods and other biomass products of microbes, fungi, algae, plants, and invertebrate and vertebrate animals.

Key Terms

energy
work
electromagnetic energy
kinetic energy
potential energy
fission reactions
fusion reactions
joule
calorie
first law of thermodynamics
second law of thermodynamics
entropy
flow-through system
energy budget
evapotranspiration (evaporation, transpiration)
autotrophs (photoautotrophs, chemoautotrophs)
primary producers
gross primary production
respiration
net primary production
heterotrophs
herbivores (primary consumers)
carnivores (secondary consumers)
omnivores
decomposers (detritivores)
net ecosystem production
food chains
food webs
ecological pyramids
trophic structure

Questions for Discussion

1. Identify the forms of energy and the ways in which each can be transformed into other forms.
2. Describe the first and second laws of thermodynamics and how they govern transformations of energy.
3. Outline the major elements of Earth's physical energy budget.
4. Explain why the trophic structure of ecological productivity is pyramid shaped.
5. Discuss why it would be more ecologically efficient for humans to be vegetarians.

References

Botkin, D.B. and E.A. Keller. 1995. *Environmental Science: Earth as a Living Planet.* New York: J. Wiley & Sons.

Freedman, B. 1995. *Environmental Ecology. 2nd Edition* San Diego, CA: Academic Press.

Gates, D.M. 1985. *Energy and Ecology.* New York: Sinauer.

Odum, E.P. 1983. *Basic Ecology.* New York: Saunders College Publishing.

Priest, J. 1991. *Energy.* New York: Addison-Wesley.

Schneider, S.H. 1989. The changing climate. *Sci. Amer.*, **261 (3)**: 70–79.

Whittaker, R.H. and G.E. Likens. 1975. The biosphere and man. pp. 305–328 in: *Primary Productivity of the Biosphere.* (H. Lieth and R.H. Whittaker, eds.). New York: Springer-Verlag.

5

FLOWS AND CYCLES OF NUTRIENTS

CHAPTER OUTLINE

Nutrients
Nutrient flows and cycles
The soil ecosystem
The carbon cycle
The nitrogen cycle
The phosphorus cycle
The sulphur cycle

CHAPTER OBJECTIVES

After completing this chapter, you will be able to:

1. Explain what nutrients are, giving examples.
2. Discuss the concept of nutrient cycling and describe important compartments and fluxes.
3. Describe the factors that affect the development of major soil types.
4. Describe key aspects of the cycles of carbon, nitrogen, phosphorus, and sulphur.

NUTRIENTS

Nutrients are any chemicals required for the proper functioning of organisms. We can distinguish two basic types of nutrients: inorganic chemicals required by autotrophic organisms for use in photosynthetic reactions and metabolism, and organic compounds ingested by heterotrophic organisms. This chapter deals with the inorganic nutrients.

Plants absorb a wide range of inorganic nutrients from their environment, typically as inorganic compounds. For example, most plants obtain their carbon as gaseous carbon dioxide (CO_2) from the atmosphere, their nitrogen as the ions nitrate (NO_3^-) or ammonium (NH_4^+), their phosphorus as phosphate (PO_4^{-3}), and their calcium and magnesium as simple ions (Ca^{2+} and Mg^{2+} respectively). The ions are obtained from soil water. Plants utilize these nutrients in photosynthetic reactions and other metabolic processes to manufacture all of the biochemicals required for growth and reproduction.

Some inorganic nutrients, referred to as *macronutrients*, are required by plants in relatively large quantities. These are carbon, oxygen, hydrogen, nitrogen, phosphorus, potassium, calcium, magnesium, and sulphur. Carbon and oxygen are needed in the largest quantities, because carbon typically comprises about 45–50% of the dry weight of plant biomass and oxygen slightly less. Hydrogen accounts for about 6% of dry plant biomass, while nitrogen and potassium occur in concentrations of 1–2%. The concentrations of calcium, phosphorus, magnesium, and sulphur are about 0.1%–0.5%. *Micronutrients*, required by plants in much smaller quantities, include boron, chlorine, copper, iron, manganese, molybdenum, and zinc. Each micronutrient accounts for less than 0.01% of plant biomass and as little as a few ppm (i.e., parts per million or 10^{-6}; 1 ppm is equivalent to 0.0001%; see Appendix I).

Heterotrophs obtain the nutrients they require by eating plants and/or other heterotrophs. This ingested biomass contains nutrients in various organically bound forms, such as biochemicals. The animals digest the organic forms of nutrients and assimilate them as simple organic or inorganic compounds, which they use to synthesize their own necessary biochemicals through various metabolic processes.

The productivity of a natural ecosystem is often limited by the supply of nutrients, a phenomenon that can be investigated by artificially fertilizing the system. In this case, nitrogen fertilizer was added to a meadow on the arctic tundra, resulting in increased productivity. The experimental plot is a darker colour.
Source: B. Freedman

NUTRIENT FLOWS AND CYCLES

Earth receives very little mass from outer space and also loses little to that enormous void. Earth does gain small quantities of space materials through meteorite showers, but these extraterrestrial inputs are insignificant in comparison with Earth's mass. Earth, therefore, is essentially an isolated system in terms of matter. Because of this, at the global and biospheric levels, nutrients and other materials "cycle" within and between ecosystems. Energy (discussed in Chapter 4) always "flows through" ecosystems and the biosphere, while nutrients and other matter "cycle."

Nutrient cycling refers to the transfers, chemical transformations, and recycling of nutrients in ecosystems. A **nutrient budget** is a quantitative estimate of the rates of input and output of nutrients to and from some designated ecosystem, as well as the amounts present and transferred within the system.

The major elements of a nutrient cycle are shown in Figure 5.1. The outer boundary of the diagram defines the limits of an ecosystem. (It could represent the entire

FIGURE 5.1 | A REPRESENTATIVE NUTRIENT CYCLE

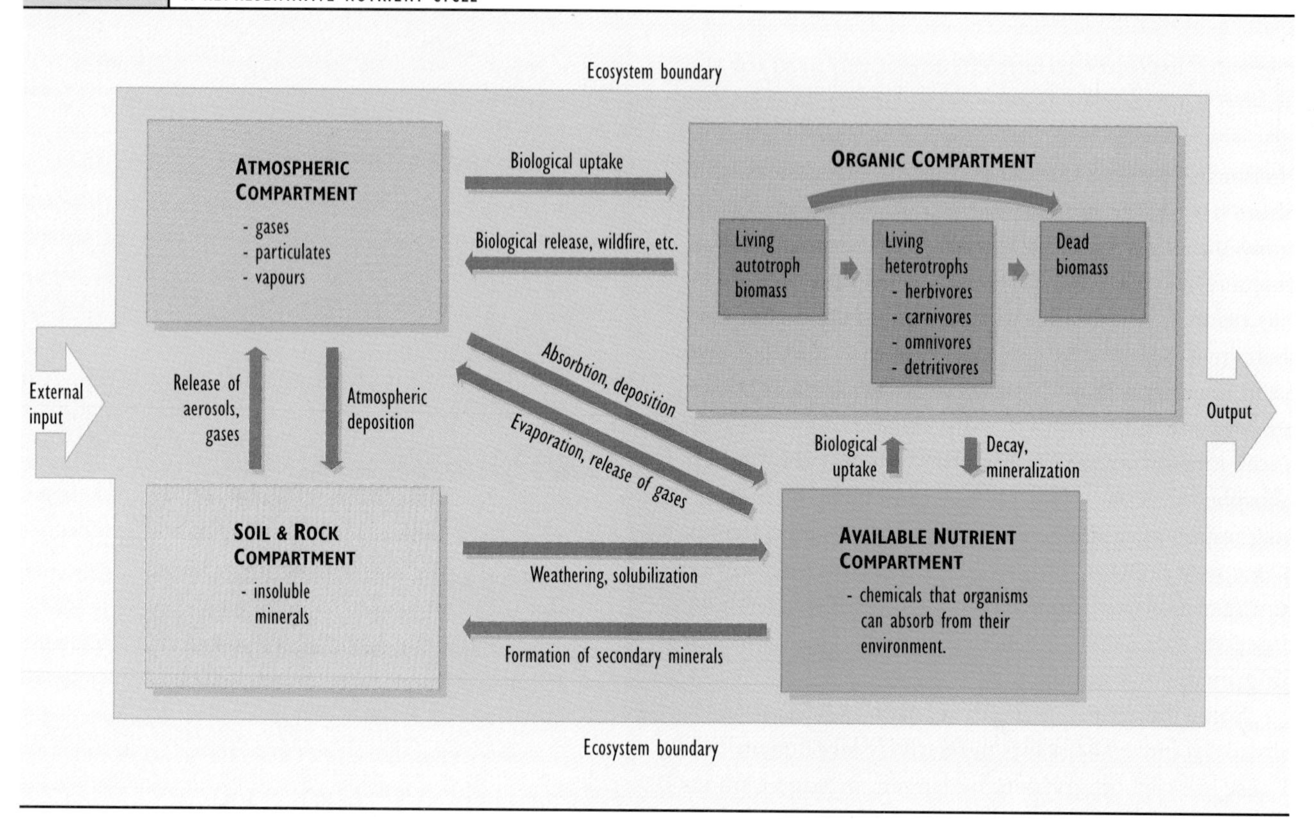

Major elements of a representative nutrient cycle for a particular, defined ecosystem, such as a watershed. The boxes represent compartments that "store" materials. The arrows represent fluxes, or transfers of materials between compartments.

Source: Modified from Likens *et al.* (1977)

biosphere, in which case there would be no external inputs to or outputs from the system.) More commonly in ecological studies, the system is defined as a particular landscape, lake, or *watershed* (i.e., the terrestrial basin from which water drains into a stream or lake). Each of these systems has inputs and outputs of nutrients, the rates of which can be measured.

The boxes within the boundary represent **compartments**, each of which stores quantities of materials. Compartment sizes are commonly expressed in units of mass per unit of surface area. Examples of such units are kilograms/hectare (kg/ha) or tonnes/ha (t/ha). In aquatic studies, compartment sizes may be expressed per unit of volume, as in g/m^3 of water. The arrows in the diagram represent **fluxes**, or transfers of materials between compartments. Fluxes are rate functions, measured in terms of mass per area per time, for example, kg/ha·y.

We can divide the system into four major compartments.

1. The atmosphere consists of gases and small concentrations of suspended particulates and water vapour.
2. Rocks and soil (the *lithosphere*) consist of various insoluble minerals that are not directly available for uptake by organisms.
3. Available nutrients, present in chemical forms that are water soluble to some degree, can be taken up by organisms from their environment and contribute to their mineral nutrition.
4. The organic compartment consists of nutrients present within living and dead organic matter. The organic compartment can be divided into three functional groups: living autotrophs such as plants, algae, and autotrophic bacteria; living heterotrophs including herbivores, carnivores, omnivores, and detritivores; and all forms of dead biomass.

The major transfers of materials between compartments, called fluxes, are also shown in Figure 5.1. These are the critically important transfer pathways within nutrient cycles. For instance, insoluble forms of nutrients in rocks and soil become available for uptake by organisms through various chemical transformations, such as *weathering* or *solubilization*, that render the nutrients soluble in water. This process is reversed by reactions that produce insoluble compounds from soluble ones. These reactions form *secondary minerals* such as clays, carbonates (e.g., $CaCO_3$, $MgCO_3$), oxides of iron and aluminum (e.g., Fe_2O_3, $Al(OH)_3$), sulphides (e.g., FeS_2), and other compounds that are not directly available for biological uptake.

Other fluxes in nutrient cycles include the biological uptake of nutrients from the atmosphere or from the available pool in soil. Plant foliage, for example, assimilates carbon dioxide (CO_2) from air, and roots absorb the nitrate (NO_3^-) and ammonium (NH_4^+) ions dissolved in soil water. These nutrients can be fixed by plants into their growing biomass. The organic nutrients then enter the ecological food web and may eventually be deposited as dead biomass. Organic nutrients in dead biomass are recycled through the processes of decay and mineralization to regenerate the supply of available inorganic nutrients.

We further investigate these concepts in the following sections. Initially, we examine the soil ecosystem, which is where most nutrient cycling occurs within terrestrial ecosystems. We will then examine important aspects of the cycling of carbon, nitrogen, phosphorus, and sulphur.

Soil in natural ecosystems often develops a pronounced vertical stratification, with organic-rich horizons on the surface and mineral-rich ones below. This soil "pit" was dug in an open aspen-grassland ecosystem in Saskatchewan known as "parkland." Beneath the dark, highly organic surface layer is a horizon moderately stained with organic matter leached from the surface, and then a mineral, lighter-coloured parent material.
Source: D. Gautier

The Soil Ecosystem

Soil is a complex and variable mixture of fragmented rock, organic matter, moisture, gases, and living organisms that covers almost all of Earth's terrestrial landscapes. It provides mechanical support for growing plants, even for trees as tall as 100 m. Soil also stores supplies of water and other critical nutrients for growing plants and other organisms and provides habitat for the many organisms that are active in the decomposition of organic matter and recycling of its nutrient content. Soil is a component of all terrestrial ecosystems, but it can also be studied as a dynamic ecosystem in and of itself.

Soils develop over very long periods of time toward a mature condition. Fundamentally, they are derived from so-called parent materials, which are rocks and minerals that occur with a metre or so of the surface. Parent materials in most of Canada were deposited as a result of glacial processes, often as a complex mixture known as till, which contains rock fragments of various sizes and mineralogies. In some areas, however, the parent materials were deposited beneath immense inland lakes, usually in post-glacial times. Such places are typically rather flat and have relatively uniform, fine-grained soils ranging in texture from clay to sand. (Clay particles have diameters less than 0.002 mm, while silt ranges from 0.002 to 0.02 mm, sand from 0.02 to 2 mm, and gravel from 2 to 20 mm. Coarse gravel and rubble are larger than 20 mm.)

In other regions, parent materials known as loess are derived from silts that were transported by the wind from other places. Because of their very small particle size, clays have an enormous surface area, giving them important chemical properties such as an ability to bind many nutrient ions.

The characteristics of the parent material have an important influence on the type of soil. Soil development is also, however, profoundly affected by biological processes and climatic factors such as precipitation and temperature. For example, water from precipitation can dissolve certain chemicals from parent materials and carry them downwards. This process, known as leaching, modifies the chemistry and mineralogy of both the surface and the deeper parts of the soil. In addition, inputs of litter and other debris from plants increase the content of organic matter in mature soils. Fresh litter is a food substrate for many species of soil-dwelling animals, fungi, and bacteria. These organisms progressively oxidize the organic debris into carbon dioxide, water, and inorganic nutrients such as ammonium, with some material remaining as complex organic matter known as humus.

As soils develop, they assume a vertical stratification known as a profile, which can have discrete, easily recognizable layers known as horizons. From the surface downwards, the major horizons of a well-developed soil profile are as follows:

HORIZON	DESCRIPTION
L	Litter layer contains organic matter that is readily identifiable as being plant litter.
F	Fermentation or duff layer contains partly decomposed organic matter with small fragments still visible.
H	Humus layer contains well-decomposed (or humified) organic matter with few readily identifiable fragments.
A_l	Transitional A horizon has a high organic concentration mixed with inorganic materials.
A_2 or A_e	Eluviated A horizon has a relatively light colour with low concentrations of organic matter and certain minerals (such as iron and aluminum) which have been leached downward (or eluviated) by percolating water.
B	Accumulation horizon has a darker colour because of the depositions of clay, iron, and organic matter leached from the A horizon.
C	Parent material, a layer in which the parent material has been little influenced by soil-forming processes.
R	Regolith or underlying rock.

Soils modified by human influences can be stratified quite differently from that just described. In agricultural lands, for example, a rather homogeneous plough layer (A_p) develops on the surface 15–20 cm or so of the soil. The plough layer is typically uniform in structure because it has been mixed up repeatedly for many years. In addition, the soil of agricultural lands is often highly deficient in organic matter and may also be degraded in structure, nutrient concentrations, and other qualities important to the potential ability of the soil to support plant productivity. These subjects are discussed in more detail in Chapter 14.

Broadly speaking, soils within a particular kind of ecosystem, such as tundra, conifer forest, hardwood forest, or a prairie, tend to develop in a similar way. Soils are classified by the ecological conditions under which they developed. The highest level of classification arranges soils into *orders*, which can be divided into more complex groups. The most important soil orders in Canada are the following:

SOIL ORDER	DESCRIPTION
Chernozem	Forms in cool temperate climates with sufficient rainfall to support tall-grass and mixed-grass prairie. Has a thick, blackish, organic-rich A horizon, rich in calcium carbonate at the surface. The B horizon is lighter coloured, and the C horizon is again rich in calcium carbonate. A fertile soil, also known as "black earth."
Podzol	Forms in cool, temperate, humid climates, especially under coniferous and mixed conifer and angiosperm (hardwood) forest. Has a thick, acidic, LFH layer, a highly leached A_e horizon, and often a reddish B horizon due to the deposition of iron oxides. Also known as spodosol.
Brunisol	Forms in temperate, humid climates, under angiosperm forest, and usually from calcium-rich parent material. Little accumulation of litter, with a dark-brown A horizon and a lighter coloured B horizon. Also known as brown forest soil.
Luvisol	Develops under a range of climatic conditions from boreal to temperate, and under a range of forest types from coniferous to angiosperm. Little accumulation of litter, with a slightly acidic A_e horizon and a neutral, clay-rich B horizon.
Regosol	Develops under various climatic conditions from poorly consolidated parent materials such as sand or silt. Has little profile development.
Gleysol	Develops in cool, temperate climates on sites subject to periodic waterlogging, usually because the C horizon is not fully permeable to downward movement of water. The surface waterlogged layers become anoxic (or oxygen-depleted), fostering the leaching of iron and manganese compounds which deposit lower down in grey-red mottled bands.
Solonetz	Develops in semi-desert to arid climates, under moderate drainage and somewhat saline conditions that support salt-tolerant plants. Has a thin surface layer over a darker, alkaline horizon.
Organic	Develops in a cool, humid climate in wetlands such as bogs and fens. It is characterized by surface peat deposits that can be up to 10 m thick.

The Importance of Soil

The soil ecosystem is extremely important ecologically. Terrestrial plants obtain their supplies of water and much of the nutrients they need for growth from the soil, absorbing these through their underground roots, rhizomes, and mycorrhizae. Soil also provides habitat for a great diversity of animals and microorganisms that play a crucial role in litter decomposition and nutrient cycling.

Soil is also economically important because it critically influences the kinds of agricultural crops that can be grown in a region (Chapter 14 discusses soil factors and crop productivity). Some of the most productive soils for agriculture are alluvial soils, typically found along rivers and their deltas, where periodic flooding and silt deposition bring in abundant supplies of nutrients. As long as they are not too stony, chernozems and brunisols are also fertile and favoured for agriculture. Much prairie agriculture is developed on chernozem soils of various types, while much of the fertile agricultural land of southern Quebec and Ontario contains brunisols.

The Carbon Cycle

Carbon is one of the basic building blocks of life and the most abundant element in organisms, accounting for 45–50% of typical dry biomass. Key aspects of the global carbon cycle are presented in Figure 5.2 (see also Chapter 17 and Figure 17.1). Gaseous carbon dioxide (CO_2) is the most important atmospheric form of carbon, occurring in a concentration of about 355 ppm, although methane (CH_4, about 1.5 ppm) is also significant.

Atmospheric CO_2 is a critical nutrient for photosynthetic organisms, such as plants, which absorb this gas through tiny pores in their foliage, fix it into organic matter, and then use the fixed energy to support their

FIGURE 5.2 | MODEL OF THE GLOBAL CARBON CYCLE

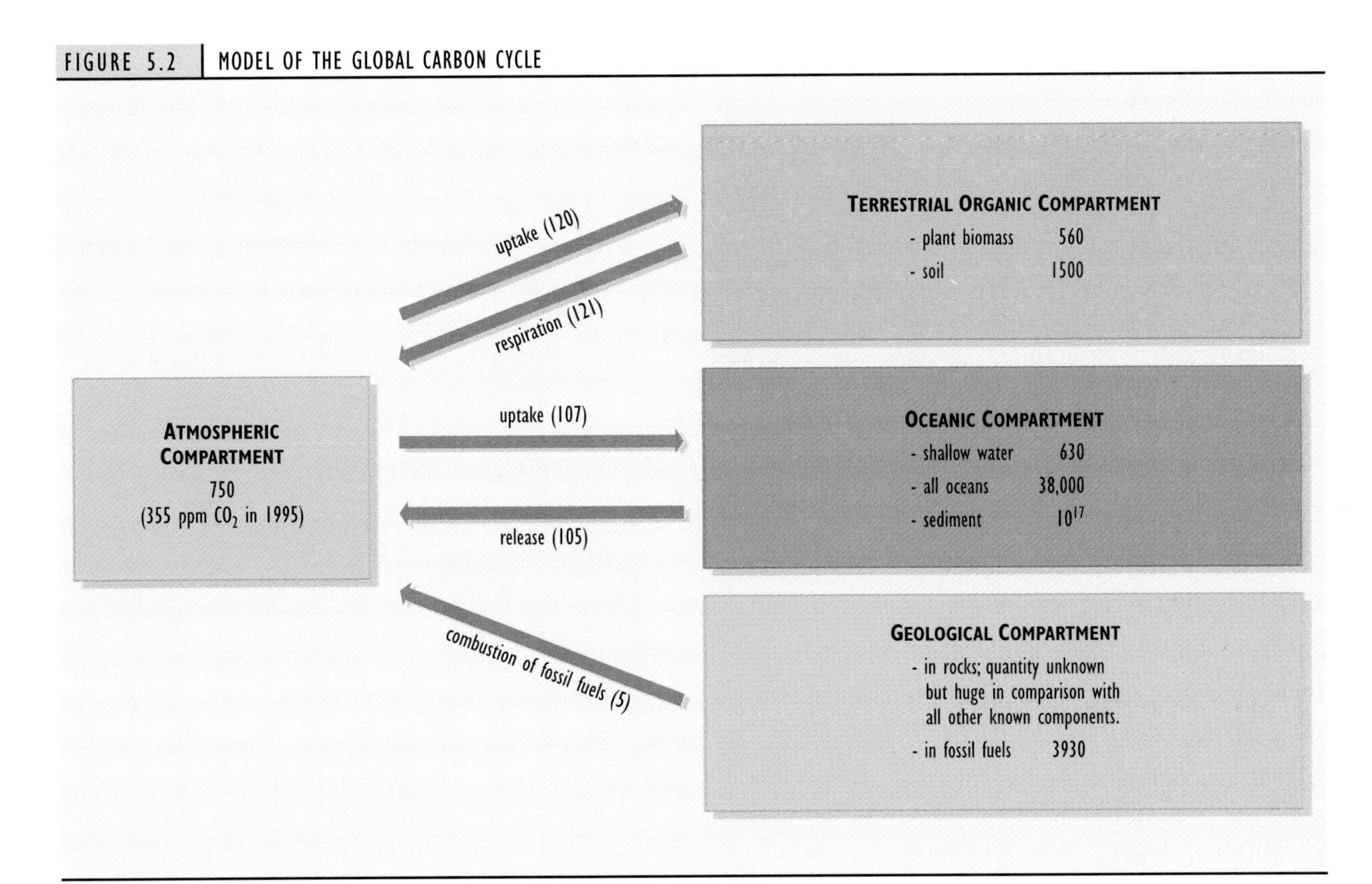

The amounts of carbon stored in the various compartments are expressed in units of billions of tonnes of carbon (10^9 t), while fluxes between compartments are in 10^9 t/y.

Sources: Blasing (1985), Solomon et *al.* (1985), and Freedman (1995)

respiration and to achieve growth and reproduction. Some of the organic matter of plants and other autotrophs is consumed by heterotrophs and passed through ecological food webs. All organisms release CO_2 to the atmosphere as a waste product of their respiratory metabolism.

CO_2 is also the most common emission associated with the decomposition of dead organic matter. However, if this process occurs under *anaerobic* conditions (i.e., oxygen (O_2) is not available), CH_4 and CO_2 are both emitted. Anaerobic decomposition is relatively inefficient, which explains why dead organic matter in wetlands, such as swamps and bogs, often accumulates, eventually forming peat. Under suitable geological conditions (i.e., anaerobic, deep burial, high pressure, high temperature) peat may be slowly transformed into carbon-rich fossil fuels such as coal, petroleum, and natural gas (see Chapter 13).

Atmospheric CO_2 also dissolves in oceanic waters, forming the bicarbonate ion (HCO_3^-), which can be taken up and fixed into organic matter by photosynthetic algae and bacteria. These autotrophic organisms are the base of the marine food web. Various marine organisms also use oceanic CO_2 and HCO_3^- to manufacture their shells of calcium carbonate ($CaCO_3$), an insoluble mineral that slowly accumulates in sediments and may eventually lithify into limestone (also $CaCO_3$).

Overall, the amount of CO_2 absorbed by the global biota from the atmosphere is quite similar to the amount released through respiration and decomposition. Consequently, the cycling of this nutrient can be viewed as a steady-state system. In modern times, however, anthropogenic emissions have significantly upset the atmospheric carbon balance. Global emissions of CO_2 and CH_4 are now larger than the uptake of these gases, an imbalance that has led to their increasing concentrations in the atmosphere. This phenomenon may be important in intensifying Earth's greenhouse effect (see Chapter 17).

The Nitrogen Cycle

Nitrogen is another critically important nutrient for organisms, being an integral component of many biochemicals, including amino acids, proteins, and nucleic acids. Like the carbon cycle, the nitrogen cycle has an important atmospheric phase; but unlike carbon, nitrogen is not a significant constituent of crustal rocks and minerals. Consequently, the atmospheric reservoir of nitrogen plays a paramount role in its environmental cycling (Figure 5.3).

Virtually all atmospheric nitrogen occurs in the form of nitrogen gas (N_2, sometimes referred to as dinitrogen), which is present in a concentration of 78%. Other gaseous forms of nitrogen include ammonia (NH_3), nitric oxide (NO), nitrogen dioxide (NO_2), and nitrous oxide (N_2O). These trace gases typically occur in the atmosphere in concentrations considerably less than 1 ppm, although there may be larger amounts in environments influenced by anthropogenic emissions (see Chapter 16). Nitrogen also occurs in trace atmospheric particulates containing nitrate (NO_3^-) and ammonium (NH_4^+), such as ammonium nitrate (NH_4NO_3) and ammonium sulphate ($(NH_4)_2SO_4$), both of which can be important pollutants (see Chapters 16 and 19).

Nitrogen occurs in many other chemical forms in terrestrial and aquatic environments. "Organic nitrogen" refers to the diverse variety of nitrogen-containing molecules in living and dead biomass. These chemicals range in character from simple amino acids, through proteins and nucleic acids, to very large and complex molecules (known as humic substances) which are important components of dead organic matter in ecosystems. Nitrogen in ecosystems also occurs in a small number of inorganic compounds, the most important of which are N_2 and NH_3 gases and the ions nitrate, nitrite (NO_2^-), and ammonium. The nitrogen cycle involves the transformation and cycling of the various organic and inorganic forms of nitrogen within ecosystems.

Nitrogen Fixation

Because the two nitrogen atoms in dinitrogen gas are held together by a strong triple bond, N_2 is a virtually inert compound. For this reason, N_2 can be directly utilized by only a few specialized organisms, even though it is extremely abundant in the environment. These so-called nitrogen-fixing species, all of which are microorganisms, have the ability to metabolize N_2 into NH_3 gas. The microorganisms can then use the "fixed" nitrogen for their own nutrition. More importantly, the NH_3 also becomes available to the great majority of autotrophic plants and microorganisms that cannot themselves fix N_2.

Biological nitrogen fixation is a critical process — most ecosystems depend on it to provide the nitrogen nutrition that sustains their primary productivity. In fact,

FIGURE 5.3 | MODEL OF THE GLOBAL NITROGEN CYCLE

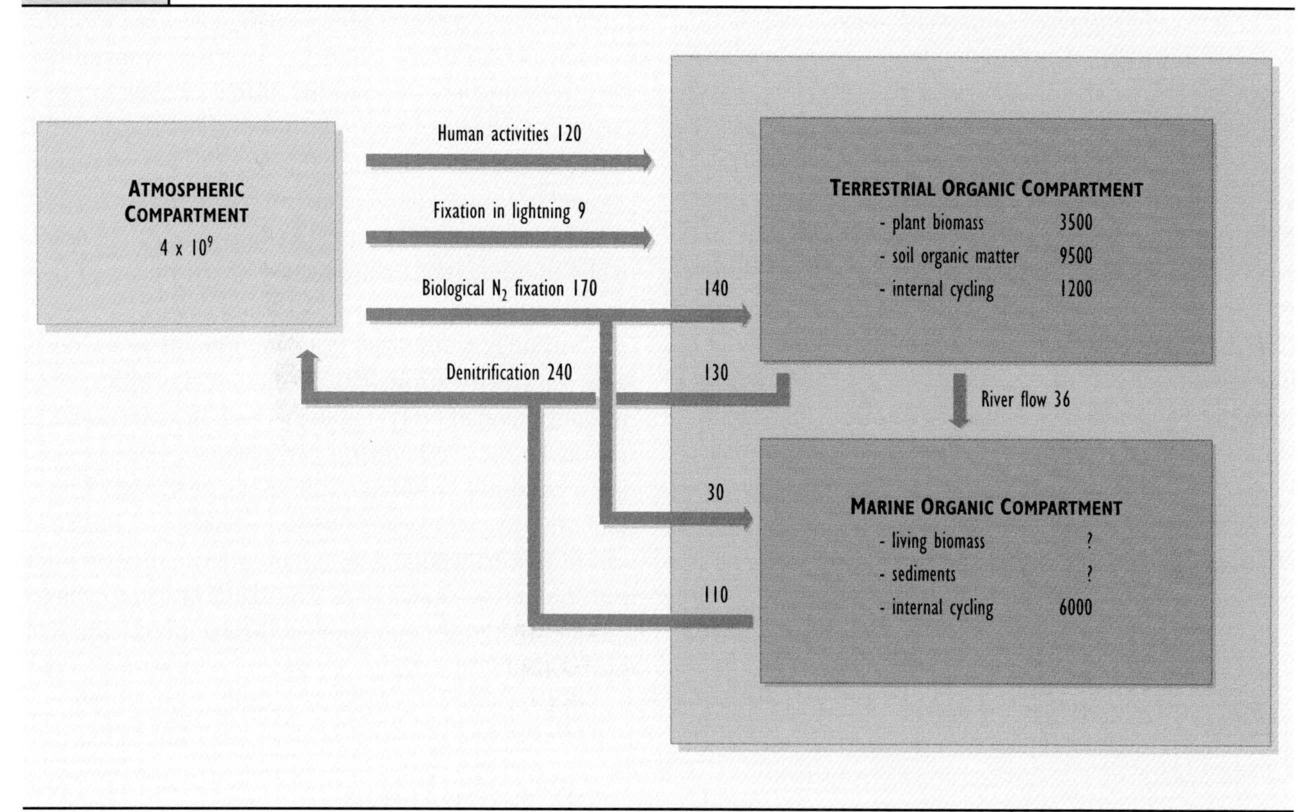

The amounts of nitrogen stored in compartments are in millions of tonnes (10^6 t of N), while fluxes between compartments are in 10^6 t/y.

Sources: Hutzinger (1982) and Freedman (1995)

because nitrogen is not an important constituent of rocks and soil minerals, N_2 fixation is ultimately responsible for almost all of the organic nitrogen in the biomass of the world's organisms and ecosystems. The only other significant sources of fixed nitrogen to ecosystems are atmospheric depositions of nitrate and ammonium through precipitation and dustfall, and the direct uptake of NO and NO_2 gases by plants. These are, however, minor sources in comparison with biological N_2 fixation.

The best known of the N_2-fixing microorganisms are bacteria called *Rhizobium* which live in specialized nodules in the roots of certain species of leguminous plants, such as peas and beans. Some other plants, such as alders, also live in a **mutualism** (i.e., a mutually beneficial *symbiosis*; see Chapter 9) with N_2–fixing microorganisms. So do most lichens, which are a mutualism between a fungus and an alga. Many other N_2–fixing microbes are free-living in soil or water — for example, cyanobacteria (or blue-green bacteria).

Most species in the pea family (Fabaceae), such as these soybeans, develop a mutualism with *Rhizobium* bacteria. The *Rhizobium* live in nodules on the roots and fix nitrogen gas (N_2) into ammonia (NH_3), which the plant can use as a nutrient.

Source: D. Patriquin

Non-biological nitrogen fixation also occurs naturally—for instance, during a lightning event, when atmospheric N_2 combines with O_2 under conditions of great heat and pressure. Humans also cause N_2 to be fixed in various ways. Nitrogen fertilizers are manufactured by combining N_2 with hydrogen gas (H_2, which is manufactured from CH_4, a fossil fuel) in the presence of iron catalysts to produce NH_3. In addition, the gas NO is formed in the internal combustion engine of automobiles, where N_2 combines with O_2 under conditions of high pressure and temperature. Large quantities of NO are emitted to the atmosphere with vehicle exhaust, contributing to air pollution (Chapter 16). Anthropogenic N_2 fixation now amounts to about 120 million tonnes per year, about 83% of which is associated with the manufacturing of nitrogen fertilizers. This is a globally important component of the modern nitrogen cycle, and is comparable in magnitude with non-human N_2 fixation (about 170 million t/y).

Ammonification and Nitrification

After organisms die, their organically bound nitrogen must be converted to inorganic forms; otherwise the recycling of the fixed nitrogen would not be possible (Figure 5.4). In **ammonification**, the initial stage of this process, the organic nitrogen of dead microbial, plant, and animal biomass is transformed to ammonia, which acquires a hydrogen ion (H^+) to form ammonium (NH_4^+). As such, ammonification represents a single component of the highly complex process of decay, one that is specific to the nitrogen cycle. Ammonification is carried out by a variety of microorganisms. Ammonium is a suitable source of nitrogen nutrition for many species of plants, particularly those that live in environments with acidic soils. Most plants, however, cannot utilize NH_4^+ effectively, and they require nitrate (NO_3^-) as their main nitrogen nutrient.

Nitrate is synthesized from ammonium by a process known as **nitrification**. The initial step in nitrification is the conversion of NH_4^+ to nitrite (NO_2^-), a function carried out by bacteria known as *Nitrosomonas*. Once formed, nitrite is rapidly oxidized to nitrate by *Nitrobacter* bacteria. Both *Nitrosomonas* and *Nitrobacter* are very sensitive to acidity, so nitrification does not occur at significant rates in acidic soil or water. This is why plants in acidic habitats must be able to use ammonium as their source of nitrogen nutrition.

Denitrification

In **denitrification**, also performed by a wide variety of microbial species, nitrates are converted to either of the gases N_2O or N_2, which are emitted to the atmosphere. Denitrification occurs under anaerobic (O_2-poor) conditions, and its rate is greatest when concentrations of ni-

FIGURE 5.4 | IMPORTANT TRANSFORMATIONS OF FIXED NITROGEN IN ECOSYSTEMS

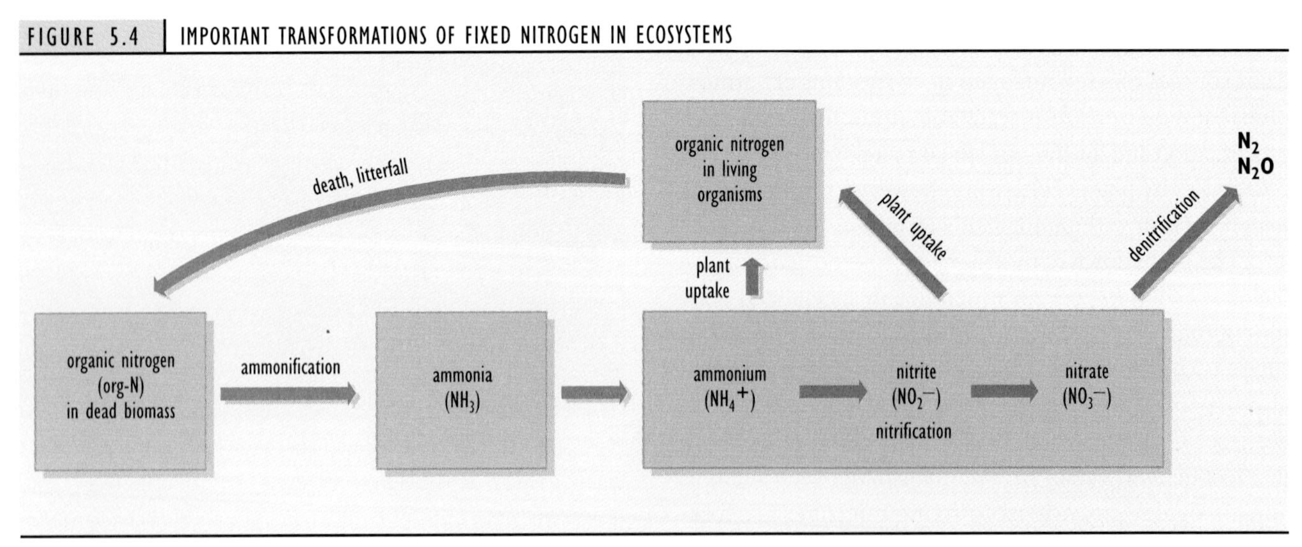

Source: Modified from Freedman (1995)

trate are large — for example, in fertilized agricultural lands that are temporarily flooded. In some respects, denitrification can be considered to be a counter-balancing process to nitrogen fixation. In fact, global rates of nitrogen fixation and denitrification are in a rough balance, so the total quantity of fixed nitrogen in Earth's ecosystems is not increasing or decreasing substantially over time.

The Phosphorus Cycle

Phosphorus is a key constituent of many biochemicals, including lipids, nucleic acids such as DNA and RNA, and energy-carrying molecules such as ATP. Phosphorus is, however, required by organisms in considerably smaller quantities than nitrogen or carbon. Compared with the rates of cycling of carbon and nitrogen, relatively small quantities of phosphorus cycle in ecosystems. Because phosphorus is commonly in short supply, it is a critical nutrient in many ecosystems, particularly in freshwater and agriculture.

Unlike the carbon and nitrogen cycles, the phosphorus cycle does not have a significant atmospheric phase. Phosphorus compounds do occur in the atmosphere, as particulates present in trace quantities. However, atmospheric inputs to ecosystems are typically very small compared with the amounts available from soil minerals or from artificial fertilization of agricultural lands.

Phosphorus tends to move from the terrestrial landscape into surface waters and eventually into the oceans, where it deposits to sediments which act as a long-term sink. Some phosphorus minerals in oceanic sediments are eventually recycled back to the land by geological uplift associated with mountain building. However, this is an extremely slow process, not meaningful in ecological time scales. Therefore, elements of the global phosphorus cycle represent a flow-through system.

Some ecological mechanisms do return some marine phosphorus to certain portions of the continental landscape. For example, some species of fish spend most of their lives at sea but migrate up rivers to breed. When they are abundant, fish such as salmon import substantial quantities of organic phosphorus to the higher reaches of rivers, where it becomes decomposed to phosphate after the fish spawn and die. Fish-eating marine birds are also locally important in returning oceanic phosphorus to land. Their phosphorus-rich excrement is so abundant on some islands that it is mined as a fertilizer.

Soil is the principal source of phosphorus nutrition for terrestrial vegetation. The phosphate ion (PO_4^{-3}) is the most important form of plant-available phosphorus. Although phosphate ions are typically present in small concentrations in soils, they are constantly produced from slowly dissolving minerals such as calcium, magnesium, and iron phosphates ($Ca_3(PO_4)_2$, $Mg_3(PO_4)_2$, and $FePO_4$, respectively). Phosphate is also produced by the microbial oxidation of organic phosphorus, a component of the more general decay process. Water-soluble phosphate is quickly absorbed by microorganisms and plant roots and used in their syntheses of biochemicals.

Aquatic autotrophs also utilize phosphate as their principal source of phosphorus nutrition. In fact, phosphate is commonly the most important **limiting factor** in freshwater ecological productivity. This means that the primary productivity of most freshwater ecosystems will increase if they are fertilized with phosphate, but not if they are treated with sources of nitrogen or carbon nutrition (unless they first have sufficient PO_4^{-3} added; see Chapter 20). Lakes and other aquatic ecosystems receive most of their phosphate supply by runoff from the terrestrial parts of their watershed and through recycling of phosphorus from sediment and organic phosphorus suspended in the water column.

Humans are greatly affecting the global cycling of phosphorus by mining it, manufacturing fertilizers, and applying these to agricultural lands to increase their productivity. For some time, the major source of phosphorus fertilizers was guano, the dried excrement of marine birds. Guano is mined on islands, such as those off coastal Chile and Peru, where colonies of breeding seabirds are abundant and the climate is dry, which allows the guano to accumulate over time. During the twentieth century, however, deposits of sedimentary phosphate minerals were discovered in several places such as southern Florida. Phosphorus became geologically concentrated in sedimentary deposits through the deposition of marine organisms over millions of years. These deposits are now being mined to supply mineral phosphorus used to manufacture agricultural fertilizers. When these easily exploitable mineral deposits become exhausted, however, phosphorus may turn out to be a limiting factor for agricultural production in the not-so-distant future.

About 50 million tonnes of phosphorus fertilizer are now manufactured each year. This is a highly significant input to the global phosphorus cycle, in view of the estimate that about 200 million tonnes of phosphorus per year are absorbed naturally from soils by vegetation.

THE SULPHUR CYCLE

Sulphur, another important nutrient, is a key constituent of certain amino acids, proteins, and other biochemicals. Sulphur is found relatively abundantly in many minerals and rocks and also has a significant presence in soils, waters, and the atmosphere.

Atmospheric sulphur occurs in various compounds, some of which are important air pollutants (see Chapter 16). Sulphur dioxide (SO_2), a gas, is naturally emitted by volcanic eruptions and forest fires. SO_2 is also emitted in large quantities by coal-fired power plants and when metal ores are roasted. Rarely occurring in the atmosphere in concentrations of more than several ppm, SO_2 is, however, toxic to many plants at concentrations lower than 1 ppm. Important ecological damage has been caused by this gas in some places, such as the Sudbury area (see Chapter 16).

In the atmosphere, SO_2 becomes oxidized to the *anion* (i.e, negatively charged ion) sulphate (SO_4^{-2}), which occurs as tiny particulates or dissolved in droplets of moisture. In these forms the negative charges of sulphate must be balanced by positive charges of *cations* such as ammonium (NH_4^+), calcium (Ca^{2+}), and hydrogen ion (H^+, an important element of "acid rain" — see Chapter 19).

Hydrogen sulphide (H_2S), which has a strong smell of rotten eggs, is emitted from volcanoes and deep-sea vents. It is also emitted from ecosystems in which organic sulphur compounds are decomposed under anaerobic conditions, and from anaerobic aquatic systems where SO_4^{-2} is reduced to H_2S. Dimethyl sulphide is another reduced-sulphur gas that is produced in the oceans and emitted to the atmosphere. In oxygen-rich environments, such as the atmosphere, H_2S is oxidized to sulphate — as is dimethyl sulphide, but more slowly.

Most emissions of SO_2 to the atmosphere are associated with human activities, but almost all H_2S emissions are natural. Overall, the global emissions of all sulphur-containing gases are equivalent to about 251 million tonnes of sulphur per year. About 41% of this emission is anthropogenic and the rest is natural (see Chapter 16 for more details).

Sulphur occurs in rocks and soils in a variety of mineral forms, the most important of which are sulphides, which occur as compounds with metals. Iron sulphides (such as FeS_2, called pyrites when they occur as cubic crystals) are the most common sulphide minerals, but all of the heavy metals (e.g., copper, lead, nickel, etc.) can also occur in this mineral form. Whenever metal sulphides become exposed to an oxygen-rich environment, certain bacteria begin to oxidize the sulphide, generating sulphate as a product. These autotrophic bacteria, known as *Thiobacillus thiooxidans*, obtain energy from this chemical transformation to sustain their growth and reproduction. This type of primary productivity is called **chemosynthesis** (in parallel with the photosynthesis of plants). In places where large quantities of sulphide are oxidized, enormous amounts of acidity are associated with the sulphate product, a phenomenon referred to as "acid-mine drainage" (see Chapter 19).

Sulphur also occurs in a variety of organically bound forms in soil and water. These compounds include proteins and other sulphur-containing chemicals from dead organic matter. Soil microorganisms oxidize organic sulphur to sulphate, an ion that plants can readily use in their nutrition.

Plants satisfy their nutritional requirements for sulphur by assimilating its simple mineral compounds from the environment, mostly by absorbing sulphate dissolved in soil water, which is taken up by roots. In environments where the atmosphere is somewhat contaminated by SO_2, plants can absorb this gas through their foliage (although too much absorption can be toxic — there is a fine line between SO_2 as a plant nutrient and as a poison).

As was already mentioned, human activities have significantly influenced the rates of certain fluxes of the sulphur cycle. Important environmental damage has been caused by SO_2 toxicity, acid rain, acid-mine drainage, and other sulphur-related problems. Sulphur is also, however, an important mineral commodity, with many industrial uses in manufacturing and as an agricultural fertilizer. Most commercial sulphur is obtained by cleaning "sour" natural gas (methane, CH_4) of its content of H_2S and by removing SO_2 from waste gases at metal smelters.

Key Terms

nutrients
nutrient cycle
nutrient budget
compartment
fluxes
soil
nitrogen fixation
mutualism
ammonification
decay
nitrification
denitrification
limiting factor
chemosynthesis

Questions for Discussion

1. Outline the basic elements of a nutrient cycle, including the roles of compartments and fluxes.
2. Compare and contrast key aspects among the cycles of carbon, nitrogen, phosphorus, and sulphur.
3. Describe how soils are formed from parent materials, including the influences of physical and biological processes.
4. Differentiate among the major types of soils.
5. Discuss how your daily activities affect key aspects of the carbon cycle.

References

Atlas, R.M. and R. Bartha. 1987. *Microbial Ecology.* Menlo Park, CA: Benjamin/Cummings.

Blasing, T.J. 1985. Background: Carbon Cycle, Climate, and Vegetation Responses. pp. 9–22 **In**: *Characterization of Information Requirements for Studies of CO_2 Effects: Water Resources, Agriculture, Fisheries, Forests, and Human Health.* Washington, D.C.: DOE/ER-0236, U.S. Department of Energy.

Botkin, D.B. and E.A. Keller. 1995. *Environmental Science: Earth as a Living Planet.* New York: J. Wiley & Sons.

Brady, N.C. 1989. *The Nature and Properties of Soils.* New York: Macmillan.

Foth, H.D. 1990. *Fundamentals of Soil Science.* New York: J. Wiley & Sons.

Freedman, B. 1995. *Environmental Ecology. 2nd Edition.* San Diego, CA: Academic Press.

Hutzinger, O. (ed.) 1982. *The Handbook of Environmental Chemistry.* New York: Springer-Verlag.

Likens, G.E., F.H. Bormann, R.J. Pierce, J.S. Eaton, and N.M. Johnson. 1977. *Biogeochemistry of a Forested Ecosystem.* New York: Springer-Verlag.

Margulis, L. and L. Olendzenski (eds.). 1992. *Environmental Evolution.* Cambridge, MA: MIT Press.

Post, W.M., T. Peng, W.R. Emanual, A.W. King, V.H. Dale, and D.L. DeAngelis. 1990. The global carbon cycle. *Amer. Scient.*, **78**: 310–326.

Schlesinger, W.H. 1992. *Biogeochemistry: An Analysis of Global Change.* San Diego, CA: Academic Press.

Solomon, A.M., J.R. Trabolka, D.E. Reichle, and L.D. Voorhees. 1985. The global cycle of carbon. pp. 1–13 **In**: *Atmospheric Carbon Dioxide and the Global Carbon Cycle*, Washington, D.C.: DOE/ER-0239, U.S. Department of Energy.

6

EVOLUTION

Chapter Outline

In the beginning...

"Progression" of life

Evolution

Relatedness and descent of species

Evolution observed

Religion and evolution

The theory of evolution by natural selection

The importance of genetics

Other mechanisms of evolution

Chapter Objectives

After completing this chapter, you will be able to:

1. Explain major differences in environmental conditions before and after the genesis of life.
2. Discuss the differences between creationism and evolution as explanations of the origin of life and species.
3. Describe the theory of evolution by natural selection.
4. Explain the role of genetics in understanding evolution as well as differences among individual organisms and species.

IN THE BEGINNING...

Based on geological and astronomical data, the planet Earth is believed to have originated by the condensation of interstellar dust about 4.5 billion years ago. The primordial, pre-life environments of Earth were quite different from what exists today. For example, it appears that Earth's initial atmosphere contained large concentrations of hydrogen sulphide (H_2S), methane (CH_4), ammonia (NH_3), carbon dioxide (CO_2), and other gases that today exist only in trace concentrations. In contrast, the modern atmosphere has large concentrations of oxygen (O_2) and nitrogen (N_2).

A major reason for these profound changes in atmospheric chemistry was the evolution of photosynthetic organisms, which emit O_2 as a waste product of their autotrophic metabolism. The presence of large amounts of O_2 changed Earth's atmosphere from one that favoured reducing reactions (in which the reaction products have a net gain in electrons) to one in which oxidizing reactions predominate. H_2S, CH_4, and NH_3 are all chemically reduced compounds, but in an O_2-rich atmosphere these become oxidized to sulphate (SO_4^{-2}), CO_2, and nitrate (NO_3^-), respectively. In addition, O_2 can react photochemically to produce small amounts of ozone (O_3). Ozone in the upper atmosphere absorbs solar ultraviolet radiation and thereby shields organisms from many of the damaging effects of this kind of electromagnetic energy.

The genesis of life on Earth is thought to have occurred in a primordial aquatic environment about 3.5 billion years ago, only about 1 billion years after the origin of the planet. It is not known exactly how life first began from inanimate matter, although most biologists believe that it occurred spontaneously. In other words, the origin of life happened naturally, as a consequence of the existence of appropriate conditions of chemistry, temperature, pressure, energy, and other environmental factors.

As such, the origin of life could have happened as a series of random events occurring under suitable conditions. Some biologists, however, believe that genesis could have occurred in a more purposeful manner, under the influence of autocatalytic reactions (self-catalyzed) that favoured the synthesis and persistence of particular organic chemicals. Under these selective influences, molecules and their inter-relationships became increasingly complicated, and they eventually developed the qualities that define the simplest forms of life: complex metabolism, growth, and reproduction.

Dinosaurs (order Dinosauria) were dominant animals on Earth for about 160 million years, but the last species became extinct about 65 million years ago. We known that dinosaurs used to exist because their fossilized bones have been discovered on each continent. Modern reptiles are relatives of dinosaurs, and birds are considered to be their surviving descendants.

Source: Victor Last/Geographical Visual Aids

The appropriate conditions for the genesis of life probably included the presence of many simple organic compounds in primordial waters. It is believed that simple organic compounds were naturally synthesized by inorganic (that is, non-living) reactions among the ammonia, methane, hydrogen sulphide, and other compounds occurring in Earth's atmosphere. These reactions were favoured because the atmosphere at that time was a high-energy environment associated with ultraviolet radiation and lightning strikes. The resulting organic compounds were deposited into the primordial oceans by rainfall, where they became progressively concentrated, especially in pools on the oceanic shores.

Modern scientists have performed simple laboratory experiments that simulate primordial conditions. In airtight flasks, mixtures of water and gaseous CH_4, NH_3, and H_2S are sparked by electric arcs. These experiments yielded various types of hydrocarbons, amino acids (precursors of proteins), nitrogenous bases (precursors of nucleic acids), and other organic chemicals. Scientists think something very similar happened prior to the origin of life on Earth.

It is still, however, an enormous step from the occurrence of appropriate environmental conditions to the spontaneous genesis of living microorganisms. Scientists do not yet understand how this momentous process occurred — the origin of the first organisms. In fact, the boundary between complex chemical systems and living organisms is thought to be somewhat arbitrary (viruses, for example, are near this boundary). Nevertheless, there is a broad consensus among scientists that microorganisms did appear in Earth's oceans about 3.5 billion years ago (Table 6.1). These first microorganisms were heterotrophic consumers of the rich soup of organic compounds that had accumulated in Earth's pre-biological oceans over hundreds of millions of years. The first autotrophic microorganisms evolved several hundred million years later, and the first photosynthetic ones about 2.5 billion years ago.

The earliest cellular life forms were **prokaryotic** microorganisms (that is, single-celled organisms without an organized nucleus containing the genetic material, which was likely DNA or RNA; see In Detail 6.1). Eventually, **eukaryotic** microorganisms (having an organized nucleus bounded by a membrane) evolved from simpler, prokaryotic predecessors.

More complex microorganisms, containing subcellular *organelles* such as mitochondria, plastids, and cilia, are thought to have evolved through symbiotic associations among different species. According to this theory, smaller microorganisms became encapsulated within larger ones in a mutually beneficial **symbiosis** (or **mutualism**; see Chapter 9). Some smaller microorganisms evolved into specialized energy-processing organelles, known as mitochondria. Other encapsulated microbes were specialized to capture light and to use that energy in photosynthesis; they became chloroplasts. Mitochondria and chloroplasts in living organisms contain small quantities of DNA that is distinctive in character, and is believed to be residual from ancient times when these organelles were independent microorganisms.

In Detail 6.1

Genetic Information, Genotype, and Phenotype

Every organism has an individual complement of genetic information, contained in the specific arrangement of nucleotides in the DNA of its chromosomes. The following is a brief outline on the storage and translation of genetic information in organisms.

DNA

Deoxyribo**n**ucleic **a**cid (or **DNA**) carries the genetic information in most species. In some viruses, the genetic information is contained in RNA (see below). DNA, a nucleic acid, is a linear sequence of four nucleotides (sometimes known as bases): adenine, cytosine, guanine, and thymine. The linear sequences are arranged as two strands, which coil as a double helix (or spiral). The two strands in the double helix are held together by hydrogen bonds, which develop between complementary pairs of nucleotides: adenine with thymine, and cytosine with guanine. Because these pairs of nucleotides exhibit this complementary binding, the two strands of DNA are mirror images. The genetic information is carried in the precise sequence of the nucleotides.

RNA

RNA (**R**ibo**n**ucleic **a**cid) is composed of a single strand of nucleotides. In RNA a fifth nucleotide, uracil, substitutes for the thymine of DNA. RNA occurs in all organisms, and it permits the translation of the genetic information of DNA into the structure of proteins (see below).

Chromosomes

Chromosomes are composed of DNA and protein and contain the genetic information of the cell. Chromosomes are self-duplicating structures, which means they create exact copies of themselves through the process of replication (see below). An exact copy is passed to each daughter cell when the cell divides. Chromosomes in the body (or somatic) cells of plants and animals occur as complementary pairs (known as homologous pairs). The number of pairs of chromosomes varies greatly among species, from one to hundreds.

Genes

A **gene** is a region of a chromosome that determines the development of a particular trait, coding for a specific protein during transcription (see below).

Because chromosomes occur in pairs, the genes also are paired. Genes commonly occur in more than one form, the different forms being called alleles. Frequently, one allele is dominant (D) and another recessive (r). The dominant allele is expressed when both alleles in the gene pair are of this type (i.e., DD), and also when both dominant and recessive alleles occur (i.e., rD or Dr). Recessive alleles are expressed only if both alleles are of this type (i.e., rr). The condition in which both alleles are the same (i.e., DD or rr) is referred to as homozygous, while the mixed condition (rD or Dr) is known as heterozygous.

Replication

Replication is the biochemical process during which the nucleotide sequence of each strand of DNA is copied, producing two identical DNA molecules. Replication is necessary for cellular division, because each new cell requires identical copies of the DNA from the parent. During replication, the double helix of the DNA "unzips," allowing free nucleotides to hydrogen-bond with those in each strand, producing new but identical DNA molecules. An error occurring during replication can result in a change in the genetic information, known as a mutation.

Transcription

In **transcription**, DNA unzips and a complementary strand of RNA is made on one of the DNA strands, in a manner similar to replication. Then the RNA floats free and the DNA zips up again. One of three types of RNA can be made: ribosomal RNA (rRNA), which forms small bodies in the cytoplasm called ribosomes; messenger RNA (mRNA), which transports the information from the DNA to the ribosome; and transfer RNA (tRNA), discussed under Translation (below).

Translation

In **translation**, the messenger RNA, which contains information from a certain portion of a DNA strand, attaches to a ribosome in the cytoplasm. There, transfer RNA molecules bind to specific amino acids and transport them to the mRNA in the correct sequence for protein synthesis. (Amino acids are the building blocks of proteins. Only about 20 amino acids are common, but these make up the extraordinary diversity of proteins found in organisms. Proteins are extremely important, mainly as structural chemicals and metabolism-regulating enzymes.) The information on the mRNA, copied from the information on the DNA, determines the sequence of amino acids in a protein. That amino acid sequence determines the function of the protein.

Meiosis

In sexual reproduction, two "sex" cells, one from each parent, combine to start a new life. If those cells were somatic cells, they would have the same number of chromosomes as the parent (the diploid number), and the progeny would then have double the number of the parent. This does not occur — the sex cells are not diploid. Instead, through a process called meiosis, the number of chromosomes in the sex cells is halved (to haploid) and the progeny has the same number of chromosomes as the parent.

During meiosis, the paired chromosomes separate, with one of each pair going randomly to each daughter "sex" cell. Just before they separate, exchanges of genetic material may occur between the paired chromosomes — a phenomenon known as crossing-over. Both of these processes increase the variability of genetic information in sex cells. When the haploid sex cells (one from each parent) combine, a diploid progeny results. Having chromosomes from each parent, the progeny is genetically different from them, but also rather similar. This is, essentially, how parents pass their genetic information to their progeny.

Genotype

Individual organisms have unique genetic information embodied in the nucleotide sequences of their DNA. This unique genetic information is known as the **genotype** of that individual, which is fixed and constant (except for very low rates of mutation). However, the genotypes of populations and species are quite variable, although this is restricted by the range of genetic variation among the constituent individuals.

PHENOTYPE

An individual's genotype contains all the information encoded in its DNA, but its **phenotype** shows the actual expression of the information in terms of anatomical development, behaviour, and biochemistry. For example, recessive alleles, unless in a homozygous condition, are not expressed, even though they appear in the genotype.

Also, the expression of genetic potential is affected by environmental conditions and other circumstances. For instance, a geranium plant, with a fixed complement of genetic information, may be relatively tall and robust if grown under well-watered, fertile, uncrowded conditions. However, if that same individual were grown under drier, less fertile, more competitive conditions, its productivity and form would be different. Such varying growth patterns of the same genotype represent a phenotypically "plastic" response to environmental conditions. In contrast, the flower colour of individual geraniums (which can be white, red, or pink) is fixed genetically, and is not affected by environmental conditions.

The ability of a species to exhibit phenotypically plastic responses to environmental variations is itself genetically determined to some extent. Therefore, **phenotypic plasticity** reflects both genetic capability and varying expression of that capability, depending on circumstances.

TABLE 6.1 ESTIMATED DATES OF THE ORIGINS OF IMPORTANT LIFE FORMS

The data represent the time of first appearance of each type of organism in the fossil record.

EVOLUTIONARY EVENT	MILLIONS OF YEARS AGO
Formation of planet Earth	4500
First life, anaerobic microorganisms	3500
Chemoautotrophic microorganisms	3100
Photosynthetic microorganisms	2500
First eukaryotes	1200
First multicellular organisms	600
Animals with external skeletons	570
Lampreys	550
Crustaceans and mollusks	500
Plants	435
Jawed fish	415
Land plants	410
Amphibians	355
Insects	310
Reptiles	300
Conifers	270
Dinosaurs	223
Mammals	214
Birds	150
Angiosperm plants	135
Anthropoid primates	43
Hominids	5.5
Genus Homo	2
Homo sapiens	0.5

Source: Modified from Raven and Johnson (1992)

Multicellular organisms were the next major category of life forms to appear, likely in late Precambrian times (see Tables 6.1 and 3.1). The evolution and radiation of these relatively complex organisms was driven by physiological and ecological adaptations associated with interactions of specialized cell types. The first multicellular organisms were relatively small and simple, but these eventually evolved into the larger, more complex organisms that are now so prominent on Earth, including vertebrates, the *phylum* of animals to which humans belong.

"PROGRESSION" OF LIFE

Modern biologists believe that all living organisms are similarly "advanced." The two reasons are: first, all organisms have had the same amount of time to evolve since the first organisms appeared; second, they are all well adapted to coping with the opportunities and constraints presented by the environments in which they live. Therefore, all organisms from the smallest and simplest, such as viruses tinier than 1 μm, to enormous blue whales exceeding 30 m in length, represent similarly advanced, well adapted, and marvellous examples of the diversity of living organisms.

This is not to say that organisms do not vary enormously in their complexity, which of course they do. We should, however, be careful when we use the terms "simple" and "complex," because these concepts are difficult to define precisely. All living organisms display a mixture of traits, some of which evolved in ancient times, while others are more recent adaptations. For example, most organisms (except some viruses) have DNA as their genetic material, so this is an ancient trait. Other attributes, such as flight in bats or intelligence in humans, represent specific adaptations that occurred relatively late in the particular evolutionary lineages.

The fossil record clearly demonstrates that, over time, a progression of life forms has existed on Earth. The first prokaryotic organisms were relatively tiny and simple, but through evolution these led to the development of larger, more complex eukaryotic microorganisms, and so on until large, exceedingly complex animals and plants evolved. This evolutionary pattern implies a clear temporal sequence. It is important to understand, however, that relatively complex, more recent species (including humans) do not represent the epitome of evolution, nor have they inherited the Earth and its opportunities. Rather, all living species share this bountiful planet, the only place in the universe known to sustain life.

EVOLUTION

Evolution can be simply defined as genetically based changes in populations of organisms, occurring over successive generations. Evolution is a critically important biological theory, because it accounts for the natural development of existing species from progenitors that may have been quite unlike their descendants in form and function. The reality of evolution is widely accepted by modern scientists, as much so as the theory of gravity, which explains how the Earth revolves around the sun.

Evolution occurs in response to various driving forces. The modern theory of evolution suggests that natural selection is an especially important cause of evolutionary change. In essence, **natural selection** predicts that individual organisms that are better adapted to coping with the opportunities and limitations of their environment will have an increased probability of leaving descendants. If the adaptive advantages are genetically determined, they will be passed to some of these descendants. This will result in evolutionary change.

Evolution can also occur in response to catastrophic influences on populations of organisms, such as a forest fire or flood. This may result in more haphazard (or random) changes in the genetic structure of the population. Small populations are particularly subject to *non-selective* evolutionary influences.

It is important to understand that individual organisms do not evolve. Evolution is a process of genetic change from generation to generation, occurring in populations or higher-order groupings of organisms (such as whole species). This is not to say that individual organisms cannot display variable responses to environmental conditions. These responses are, however, constrained by the degree of biochemical, developmental, and behavioural flexibility allowed by the genetic complement of each individual (its genotype; see In Detail 6.1). The variable expression of the genetic information of an individual is called phenotypic plasticity, but this kind of response to variations in environmental conditions is not evolutionary change. For evolution to occur, there must be change in the collective genetic information of a population or species.

Evolution can occur at various scales. For convenience, evolutionary biologists use the term **microevolution** to refer to relatively subtle changes within a population or species, often within only a few generations, that may lead to the evolution of a variety, race, or subspecies. In contrast, the term **macroevolution** describes the evolution of species or higher taxonomic groups, such as genera, families, or classes. Evolutionary biologists continue to debate and discuss the linkages of these scales of evolution. Are patterns of macroevolutionary change simply the cumulative effects of many microevolutionary changes over long periods of time? Or is macroevolution actually large changes occurring over a short time, each representing a great step (or saltation) of evolution? Or does macroevolution happen in both ways?

Despite debates regarding its details, the theory of evolution is a unifying theme in biology because it can be used successfully to understand so many phenomena in nature. Evolution is commonly used by scientists to explain both the origin of life and the extraordinary changes that have occurred in organisms over the billions of years of life on Earth.

RELATEDNESS AND DESCENT OF SPECIES

A biological definition of **species** is "a group of organisms that is reproductively isolated from other such groups."

Within a species, individual organisms tend to resemble each other; but more importantly, they can breed with each other and produce fertile offspring. An inability to successfully interbreed implies reproductive isolation (see In Detail 6.2).

That species have evolved from progenitors is a well-established biological theory, richly supported by a great deal of evidence. Two of the most compelling examples of evidence, showing evolutionary patterns of relatedness and descent, follow.

Patterns Inferred from the Fossil Record

One of the best-known examples of evolution as seen in the fossil record is that of the horse lineage. One of the earliest horse-like progenitors was *Eohippus*, a dog-sized creature that lived 50 million years ago. Its foot was characterized by two fused and three separate toes. Comparing the morphology (body structure) of fossil bones suggests that *Eohippus* was an ancestor of *Mesohippus*, a larger animal that lived some 35 million years ago. Its foot had three

IN DETAIL 6.2

THE NATURE OF SPECIES

If a species is a group of organisms that is reproductively isolated from other such groups, then only individuals of the same species can successfully mate with each other and have fertile offspring. In practice, however, it can be extremely difficult to demonstrate reproductive isolation. For example, two populations of small animals may be separated by a mountain range and may never interact. However, if they encountered one another in nature they might interbreed. This can be tested by attempting to breed them in captivity, but this might not be representative of their behaviour in nature. Limitations such as this have led some biologists to alter their definitions of species. However, definitions of the term "species" are not cast in stone; rather, the meaning is a human construct.

In all species, individual organisms share many physical, chemical, and behavioural attributes, and they look similar. However, there can also be some stunning differences among individuals within species. Such variations are illustrated by the following examples.

- *sex:* Male lions are larger than females, and have an impressive mane framing the head. And vive la différence in the sexual dimorphism of humans!
- *life-history stage:* Compare frogs of different ages: an egg, a tadpole, or the legged adult.
- *phenotypic plasticity:* If growing in environmentally stressful conditions beneath a closed forest canopy, a 50-year-old sugar maple plant may be only 10–20 cm tall and have fewer than ten leaves. Under free-to-grow conditions, the same individual might be 10–15 m tall.
- *differentiation of populations:* Compare the size and shape of Canada geese from different parts of their range. These vary from the duck-sized "cackling goose" (*Branta canadensis minima*) of the Aleutians, weighing only 1.3–1.5 kg, to the large "giant Canada goose" (*Branta canadensis maxima*) in many urban parks in Canada, weighing about 5 kg.
- *selection by humans:* Compare tiny Chihuahua dogs, some of which can fit into a soup bowl, with mighty Newfoundland dogs, which can weigh more than 100 kg. These and all other breeds of dog are of the same species, *Canis familiaris*.

In spite of the differences, each of these cases represents variations of individuals of the same species, because they can still interbreed. (At least potentially — in the case of the Chihuahua and Newfoundland dogs, there is obviously a physical incompatibility, although this could be overcome by artificial insemination.)

A classic example of reproductive incompatibility concerns two related species, the horse (*Equus caballus*) and donkey (*E. asinus*). Interbreeding these species produces live young: a mule is a cross between a male horse and a female donkey, while a hinney is a cross between a female horse and a male donkey. Mules are, in fact, rather common domestic animals and have many favourable attributes, including the ability to work hard as a beast of burden. Mules and hinneys are, however, biological and evolutionary dead ends, because both are sterile. Therefore, although horses can interbreed with

donkeys, the result is not evolutionarily "successful" and the two are considered separate species.

On the other hand, consider the case of the savannah sparrow, a common bird of grassy habitats throughout North America, with a streaky brown colour and a body length of 11–14 cm. For many years, the larger (15–16 cm) and paler Ipswich sparrow, which only breeds on Sable Island off Nova Scotia, was thought to be an isolated species and was named *Passerculus princeps*. However, ornithologists recently found a few cases of successful interbreeding of Ipswich and savannah sparrows on the Nova Scotia mainland. Consequently, these two kinds of savannah sparrows are now lumped as one species, *P. sandwichensis*. (The Ipswich retains subspecies rank, as *P. sandwichensis princeps*).

The discovery of hybrids does not always result in changes in taxonomy. For example, hybridization is quite common among species of willow (*Salix* spp.), particularly if there is a habitat intermediate to those typically used by the parent species. In such cases, plant taxonomists commonly retain the parents as separate species, because they are generally identifiable as unique and different, and hybridize only under unusual circumstances. The botanists are partly being pragmatic, recognizing that the taxonomic system must have practical value to scientists studying plants in the field.

Bird taxonomists are also starting to adopt this view of the nature of species, and are de-emphasizing the importance of occasional hybridizations in intermediate habitats. Consequently, some birds that were once considered separate species, such as the savannah and Ipswich sparrows, and were later "lumped" because of interbreeding, may soon be separated again as distinct species.

In the text, a species is defined as "a group of organisms that is reproductively isolated from other such groups." Perhaps a better, working definition is "species are groups of organisms that differ in one or more characteristics and do not successfully interbreed extensively if they occur together in nature."

fused central toes and two free outer ones. Subsequent evolution led to *Merychippus*, a somewhat larger animal living about 20 million years ago, also with three fused and two free toes. Next came *Pliohippus*, a pony-sized animal living some 10 million years ago, which had all five toes fused into a hoof. Modern horses evolved several million years ago, and have a hoof like that of *Pliohippus*. Modern horses include the horse (*Equus caballus*), ass or donkey (*E. asinus*), Mongolian wild horse (*E. przewalskii*), and zebra (*E. burchelli*).

Patterns Inferred from Studies of Modern Species

Modern species display many obvious similarities and dissimilarities which can be used to group them on the basis of inferred relatedness. Early studies of this sort mostly involved comparative anatomy. Studies of animals relied mostly on characteristics of bones, shells, skins, and other enduring structures, while studies of plants largely involved the anatomy of flowers and fruits. More recent studies gather a much wider range of comparative information to examine relatedness among groups of species, including information about behaviour, ecology, proteins, and — most recently — specific base sequences of DNA. For example, studies involving DNA and blood proteins have clearly demonstrated that humans are closely related to other great apes, such as chimpanzee, orangutan, and gorilla. Of these, humans are most closely related to chimpanzees — in fact, these species share about 99% of the information encoded in their DNA. These observations do not suggest that humans evolved from modern apes. Rather, the appropriate interpretation is that humans and living apes share common, ape-like ancestors.

Evolution Observed

As was just described, patterns of relatedness and descent can be inferred from comparative studies of the fossil record and of modern species. It is important to understand, however, that the evolution of a new species has never been directly observed. This is because it takes a very long time for populations of related organisms to diverge sufficiently to become new species — perhaps hundreds of thousands of years, and in many cases millions. In spite of this, biologists have no doubt that new species have been evolving for billions of years — in fact, throughout the history of life.

Although speciation has not been observed in nature, examples of microevolution are well known. These cases provide important evidence in support of the theory of evolution.

Industrial Melanism

One such example of microevolution is that of the peppered moth (*Biston betularia*) of western Europe. This moth's normal coloration is a mottled, whitish tan. During the day, the moth commonly rests on lichen-covered treebark where it is difficult to see against the bark. This camouflage is important to the moth's survival, because its predators, such as birds, hunt using vision.

About a century ago in England, it was observed that some populations of peppered moths had developed a black coloration, known as melanism. This had occurred in response to changes in the tree bark, which had lost its lichen cover because of air pollution and had become blackened by soot deposition. Under such habitat conditions in industrial areas, the normal light-coloured moths were highly visible to predators and were at a distinct disadvantage compared to the blacker moths. Studies showed that melanism was genetically based, and that melanistic moths occurred, but were rare, in unpolluted habitats. However, melanistic individuals became dominant in populations of peppered moths in polluted habitats, representing a population-level genetic change. This famous example of microevolution, which has also been demonstrated in other species of moths, is known as "industrial melanism."

Interestingly, air quality has greatly improved over most of western Europe in recent decades, largely due to clean-air legislation that has reduced the use of coal as a source of energy. As a result, lichens have again begun to grow on trees, and the bark surfaces have less soot on them. These ecological recoveries have been accompanied by the reappearance of light-coloured peppered moths in places where their populations had been dominated by melanistic moths — another evolutionary response to changing environmental conditions.

These are metal-tolerant ecotypes of the grass *Deschampsia caespitosa*, growing in metal-polluted soil close to a smelter near Sudbury, Ontario.

Source: B. Freedman

Metal-Tolerant Ecotypes

In another example of observed evolution, several widely distributed plant species were found growing on sites in England and Wales polluted by metal-rich mine wastes. Although the soil in these sites was toxic to most plants, local populations of several plant species were thriving. The most common species were grasses, such as bentgrass (*Agrostis tenuis*). Research showed that these populations had a genetically based, physiological tolerance of the toxic metals, and that they differed in this respect from other populations of the same species growing on nonpolluted sites. The locally adapted populations, referred to as "metal-tolerant ecotypes," were found to have evolved in as few as several years after their first exposure to the toxic soils. (This example was the first one to be documented and is famous for that reason. Canadian examples of metal-tolerant ecotypes, discovered later, are described in Chapter 18.)

Religion and Evolution

Some religious groups have long attacked the theory of evolution. This intensified after the publication of Charles

Darwin's ideas about the role of natural selection in evolution (see the next section). The Biblical description of creation is one of the oldest written explanations of the origin of life on Earth, the existence of so many species and ecosystems, and the roles and responsibilities of humans in their interactions with other organisms. But science and religion are not irreconcilable. Indeed for many people, physical concerns belong to the domain of science, whereas spiritual concerns belong to the domain of religion.

Nevertheless, a few religious groups have insisted upon a literal interpretation of the Bible as the ultimate authority for all knowledge. For example, **creationists** reject the theory of evolution in favour of a literal interpretation of Genesis, the first book of the Old Testament of the Bible. Creationists assert that the account given in Genesis means that God created the universe and all living organisms during a six-day period, culminating with the creation of humans. Humans were created in the physical image of God and were given powers to freely use the resources of Earth:

And God said, let us make Man in our image,
after our likeness; and let them have dominion over the fish of the sea, and over the fowl of the air, and over the cattle, and over all the Earth and over every creeping thing that creepeth over the Earth.

Humans were furthermore instructed to increase their populations and to exploit nature:

Be fruitful, and multiply, and replenish the Earth, and subdue it.

Based on their literal interpretation of Genesis and other passages in the Bible, creationists have drawn the following conclusions relevant to evolution:

1. Earth and its species are not ancient because creation occurred only some thousands of years ago.
2. Species are essentially immutable, having been created as entities which have not changed since their creation.
3. Because species were individually created, existing species did not descend from earlier ones through evolution.
4. Humans are particularly special, having been created in the Creator's image. Therefore, they are not related to nor descended from any other species.

However, these four points do not accord with scientific findings:

1. The geological record clearly demonstrates that Earth and the solar system are extremely old, having begun to develop at least 4.5 billion years ago. Life is also ancient, having originated about 3.5 billion years ago. Earth and organisms date back much further than a few thousand years.
2. The fossil record gives numerous signs of large changes in the characteristics of groups of species over time, as do some living species. Species are not immutable. Moreover, Earth's existing complement of species represents only a small sample of all species that have ever lived. The fossil record demonstrates that most species that evolved during Earth's long biological history are now extinct. Many of the extinct species, families, and even phyla have no descendants — their entire lineage has become extinct (see also Chapter 7).
3. The fossil record presents clear evidence of lineages among groups of organisms, indicating that living species have descended from earlier ones. In almost all cases, the progenitor species are now extinct. There are a few excellent examples in the fossil record of links between major groups. Perhaps the most famous of these is *Archaeopteryx*, a long-extinct, metre-long creature that lived about 150 million years ago. It had teeth and other typically dinosaurian characters, but it also had a feathered body and could fly. *Archaeopteryx* is considered to be a link between the extinct dinosaurs and the still-living birds. Quite recently, fossils were found of small dinosaurs with feather-like scales, possibly evolved for insulation, separate from the evolutionary development of flight.
4. Fossil and genetic information clearly indicate that humans are descended from earlier, now-extinct species. Fossil records show that the human species (*Homo sapiens*) is derived from an evolutionary lineage of anthropoid apes. There are a few other surviving species in that lineage, with chimpanzees, and to a lesser degree gorillas and orangutans, being the closest living relatives of humans. All surviving members of the ape family are descended from now-extinct progenitors.

People known as **scientific creationists** also insist that their interpretation of Genesis is the most reliable source of scientific knowledge about the origin and evolution of life. Scientific creationists have attempted to explain some of the discrepancies between their beliefs and current scientific understanding of evolution. For example, some acknowledge that geological and fossil evidence does suggest that Earth and life are ancient phenomena and that most species have become extinct. Most scientific creationists also acknowledge that species have changed over time, but only through microevolution — they do not agree that macroevolution has led to the development of new species from earlier ones. By extension, scientific creationists also do not believe that humans descended from previous species of hominids or are related to other ape-like creatures or other primates. Moreover, the theory of scientific creationism does not abandon the notion that, at one particular time in the past, God created all species that have ever lived on Earth.

Science proceeds by observation and hypothesis testing. But scientific creationism rests on a belief, not a hypothesis, concerning the literal interpretation of the Bible. Most predictions of scientific creationists cannot be tested by rigorous scientific methodology; and when they can be, they are refuted by the scientific evidence. In short, despite its name, scientific creationism is not science.

The Theory of Evolution by Natural Selection

Virtually all species are genetically variable: that is, individual organisms differ in the complements of information stored in their genetic material, DNA (see In Detail 6.1). All observable attributes of an individual organism (known as its phenotype) are influenced by its specific complement of genetic information (that is, its genotype), including morphology, physiology, behaviour, and other traits.

Not all variations of an organism's phenotype are due only to genetic differences. The phenotype is related to both genetic information and environmental influences on the expression of that genetic potential. Individual organisms display phenotypic plasticity, or differential growth, physiology, and behaviour that depends on environmental circumstances (In Detail 6.1).

Because organisms vary in character, they also differ in their abilities to deal successfully with stresses and opportunities in their environment. Under certain conditions an individual with a particular phenotype (which is substantially determined by its genotype) may be relatively successful compared with other individuals having different genotypes and phenotypes.

In the sense meant here, the "success" of an individual means successful reproduction: having progeny that themselves go on to reproduce successfully. This is also referred to as **fitness**, defined as the proportionate genetic contribution made by an individual to all of the progeny in its population. A central tenet of evolutionary theory is that individuals maximize their fitness by optimizing the degree to which their own genetic attributes will influence future generations of their species.

Biologists believe that evolution proceeds mainly by natural selection, which operates when genetically based variation exists among individuals within a population, but some individuals are better adapted to dealing with the prevailing environmental conditions than others. On average, the more fit organisms are more successful in reproducing, and consequently have a disproportionate influence on the evolution of subsequent generations.

The theory of evolution by natural selection is perhaps the greatest unifying concept in modern biology, giving context to virtually all aspects of the study of life. This theory was co-announced publicly in 1858 by the English naturalists Charles Darwin (1809–1882) and Alfred Russel Wallace (1823–1913). Darwin, however, had been working on aspects of the theory for about twenty years prior to its publication, and had collected detailed evidence in support of natural selection as a mechanism of evolution. This evidence was marshalled in Darwin's famous book, *On the Origin of Species by Means of Natural Selection*, first published in 1859. Because of this book, Darwin has become more closely linked than Wallace to the theory of evolution by natural selection. Darwin is also the more famous of the two scientists, largely because of his great contributions toward understanding the mechanisms of evolution. Perhaps the most influential biologist of all time, Darwin undertook an astonishingly broad research on a great variety of species and biological topics.

In his *Origin of Species*, Darwin summarized natural selection in the following way:

> *Can we doubt...that individuals having any advantage, however slight, over others, would have the best chance of*

surviving and of procreating their kind? On the other hand, we may feel sure that any variation in the least degree injurious would be rigidly destroyed. This preservation of favourable variations, I call Natural Selection.

In an essay that Wallace sent to Darwin for review, natural selection was expressed in a rather similar fashion:

The life of wild animals is a struggle for existence... in which the weakest...must always succumb...giving rise to successive variations departing further and further from the original type.

Darwin's and Wallace's theory was based on the following line of reasoning:

1. It is known that the fecundity of all species is high enough that they could easily overpopulate their habitats, yet this does not generally happen.
2. It is also known that the resources that species need to sustain themselves are limited, particularly in relatively stable habitats.
3. Therefore, in view of potential population growth and limited resources, there must be intense competition among individuals of each species for access to the requirements of life. Only some individuals can survive this struggle for existence.
4. Because individuals within a species are different from each other, and much of this variation is heritable, it is reasonable to suggest that survival in the struggle for existence is partly influenced by genetically determined differences in abilities.
5. Individuals that are more capable will have a better chance of surviving and reproducing, and their genetically based attributes will be disproportionately represented in future generations.
6. Over long periods of time, this process of natural selection will lead to evolutionary changes within populations and, eventually, to the evolution of new species.

When it was first presented publicly in 1858, the theory of evolution by natural selection created a sensation among scientists and also within society. The excitement and controversy occurred largely because the theory provided the first convincing body of evidence in support of the three notions that evolution occurs, that it proceeds under natural influences, and that it has resulted in the great diversity of living species. This was a radically different view from that of creationism, which was the prevailing explanation of the origin of life and species in the mid-19th century. According to the notion of creation widely believed at the time, all species had been directly created by God during a short period of time only several thousands of years previously, and those species had not substantially changed since then.

Interestingly, Charles Darwin's writings did not directly challenge the existence of a divine Creator. Darwin discussed mainly the causes of change in species over time: he did not suggest that their initial ancestors had not been created by God. Modern theories about the spontaneous genesis of life on Earth are based on relatively sophisticated science that was unknown to Darwin. Nor did he know of the mechanisms of genetics and the inheritance of traits.

Modern extensions of the theory of evolution by natural selection suggest that new species evolve from progenitors. This is thought to happen when populations are spatially isolated by an intervening barrier such as a mountain range, extensive glaciers, or other habitat discontinuities. Isolation is important in speciation, because it reduces or eliminates genetic interchange, allowing differentiation to proceed more effectively. Isolated populations that experience different environmental conditions are subject to different regimes of natural selection and can evolve in dissimilar ways. Eventually, there may be enough evolutionary change that the populations can no longer interbreed successfully, even if they become spatially re-united. At this point, the populations have achieved reproductive isolation, and therefore have become closely related but different species.

Speciation is also thought to occur in a more linear fashion, as when progenitor species gradually evolve over time in response to changes in environment. Eventually, the ancestral species may become extinct, but new species evolved from the progenitor may survive to continue the evolutionary chain.

The Importance of Genetics

Knowledge of genetics in Darwin's time was based on a highly incomplete understanding of how an organism's traits are passed to its offspring. One popular theory, the *inheritance of acquired traits*, was based on the observation that the morphology, behaviour, and/or biochemistry of individual organisms changed in response to environmental changes. According to the theory, these plastic responses to environmental conditions could be passed along to the individual's progeny. For example, during periods of drought or intense competition for food, individual short-necked ancestors of giraffes might have stretched their necks as far as they could to reach scarce foliage from higher up in trees, resulting in the development of a longer neck. The long-neckedness would have been passed to the giraffe's progeny, who developed it still further. Eventually, populations developed the familiar, long necks of modern giraffes.

Natural selection suggests a different mechanism: within populations of short-necked giraffes there existed a genetically determined variation in neck length among individuals. Because long-necked giraffes were better able to find food, they were more likely to survive and reproduce. This meant that more of the next generation had the long-necked trait, and this trait became increasingly prominent in the evolving population.

Humans can artificially select varieties of animals or plants for traits useful in agriculture, or for some other reason such as cultural preferences for breeds of dogs. Charles Darwin referred to this evolutionary process as "cultural selection."
Source: Mark Burnside/Tony Stone Images

Modern observations and experiments have shown that "acquired traits" are just a manifestation of phenotypic plasticity. There is no evidence that they can become genetically fixed in an individual and passed along to its offspring. In contrast, the science of genetics has provided convincing evidence in support of the theory of evolution by natural selection. The biochemical mechanisms that determine the genotype of an individual organism and how it is passed to progeny have been discovered. These topics are described briefly in In Detail 6.1, but the subject matter is rather complicated and cannot be dealt with in much detail here. It is, however, useful to examine the key experiments that first suggested the existence of genes in the genetic system.

This research was conducted by Gregor Johann Mendel (1822–1884), an Austrian scientist (and monk) who developed important ideas about inheritance through breeding experiments with the garden pea (*Pisum sativum*). Mendel was interested in producing pea hybrids, which involves crossing two parent plants, each having particular, distinctive traits. Prior research had shown that certain traits were fixed in cultivated varieties of peas, including flower colour (white or purple) and whether the seeds have a wrinkled or smooth coat. In total, Mendel worked with 32 traits of this sort. Pea flowers are bisexual, containing both female (pistil, containing the ovules) and male (anther, containing pollen) parts. These are compatible within the same individual, so that self-fertilization can occur. However, Mendel experimented by cross-fertilizing selected parents, producing known hybridizations.

In each experiment, Mendel cross-bred two inbred varieties in which certain traits "bred true" (were homozygous, e.g., white or purple flower colour; see In Detail 6.1 for definitions of genetic terms). The progeny (first generation) were all the same: all had purple flowers. Crosses between the first-generation plants, however, yielded a ratio of about 3:1 purple flowers to white flowers in the second generation. This fits the prediction for two-generation crosses between two homozygous lines, as follows:

1. Represent the original purple variety as AA. This trait is dominant over the white-flower trait (called recessive).
2. Represent the original white variety as aa.

3. When the two plants are crossed, the first-generation progeny all have purple flowers but are heterozygous (Aa).
4. A cross between the first-generation plants yields four possible outcomes: AA, Aa, aA, and aa. Each is equally probable among the progeny. Because A is dominant to a, the AA and Aa (which includes aA) progeny show purple flowers. Only aa has white flowers. Therefore, the expected ratio of purple to white among the progeny is 3:1.

The most important conclusion to emerge from Mendel's work was that the inheritance of genetic information occurred in a particulate form (which we now refer to as genes), often involving dominant and recessive alleles. Inheritance is not a blended condition — in the example previously described, a cross of purple- and white- flowered pea plants does not yield progeny with flowers of an intermediate colour. Therefore, flower colour and other traits are discrete units that remain intact during inheritance, and either are, or are not, expressed in progeny.

Mendel first published his results in 1865 in a relatively obscure journal. As a result, the work was unknown to the mainstream of science for many years. However, Mendel's work was re-discovered and re-published in 1900 and quickly became the basis of modern theories of genetic inheritance.

Mendel's work and the subsequent flourishing of the science of genetics have been extremely important in biology and in the development of the theory of evolution. This is because genetics allows a rational explanation of inheritance as a mechanism by which genetically fixed traits can be passed along to offspring. Subsequent research has found that new genotypes can arise through mechanisms such as hybridization, polyploidism (that is, a spontaneous increase in the number of chromosomes), and mutations. Genetic variation is, of course, the menu of possibilities from which natural selection can choose so that adaptive evolution can occur.

It is important to recognize that much genetic information in an individual does not appear to code for functional enzymes or other proteins, and hence does not code for traits that could be selected for or against. Because of its neutral response to natural selection, this type of genetic material is sometimes referred to as "junk DNA." However, we may be ignorant of other roles that so-called junk DNA may play in the functioning of the genome.

Other Mechanisms of Evolution

Although natural selection is the most important mechanism of evolution, it is not the only one. **Artificial selection**, for example, involves the deliberate breeding of plants, animals, and microorganisms to enhance certain traits that humans view as desirable. Artificial selection has obvious parallels to natural selection, in that individual organisms with particular, genetically based traits experience greater success in life and in breeding, so they become over-represented in subsequent generations. However, traits that are favoured in artificial selection may not be adaptive in the broader, natural world. In addition, evolutionary changes typically occur much more rapidly under artificial selection than under natural selection, because the breeding of desired genotypes can be controlled.

For example, maize or corn (*Zea mays*) is an important agricultural crop which, through artificial selection, now differs enormously from its closest wild progenitor, a Mexican grass known as teosinte (*Euchlaena mexicana*). Artificial selection has caused many specific evolutionary changes in maize. For example, the fruiting head (consisting of the seeds and cob) is much larger than in wild ancestors of corn; the seeds have different coloration; the seeds implant securely onto the cob so they do not scatter before harvesting; the ripe fruit is tightly wrapped within enclosing leaves known as bracts, again to prevent pre-harvest losses; and there are vigorous growth responses to fertilization, weed control, and other cultivation practices. Moreover, without the intervention of humans through cultivation, maize would likely become extinct within only a few generations. Artificial selection has rendered maize seeds virtually incapable of detaching from the cob, which in any event is tightly bound in leafy bracts. Unaided seed dispersal, therefore, is virtually impossible.

All domesticated plant, animal, and microbial species have undergone similar artificial selection for desirable traits. Sometimes, however, artificial selection proceeds in quite bizarre directions, with the fostering of genetic traits that are viewed as desirable for aesthetic reasons. For example, oriental breeders of pet fish have produced some truly amazing varieties of goldfish (*Carassius auratus*) and koi (a golden-coloured variety of common carp, *Cyprinus carpio*). These varieties, often with grotesque shapes and behaviours, would be rapidly eliminated in wild populations but are prized as unusual and valuable varieties by aficionados of these aquatic pets. Similar comments could be made about some unusual varieties of domestic pigeons, dogs, cats, and horticultural plants, among others.

Evolution can also occur through a process known as *genetic drift*, or random changes in the frequencies of genes occurring in small, isolated populations. Such populations often exist on small islands or may be created through a catastrophic reduction of a larger population as a result of disease, disturbance, or some other factor. The relatively small genetic base of small populations is sometimes called a genetic bottleneck. Subsequent evolution is based on the restricted genetic variation of only a few individuals, which may become further reduced through the effects of inbreeding (reproduction between closely related individuals, such as siblings). Given the small amount of genetic variation, the evolution of a small population may proceed very differently from the evolution of its larger, parent population.

Key Terms

prokaryotic
eukaryotic
DNA
RNA
chromosomes
genes
replication
transcription
translation
genotype
phenotype
phenotypic plasticity
symbiosis
mutualism
evolution
natural selection
microevolution
macroevolution
species
creationists
scientific creationists
fitness
artificial selection

Questions for Discussion

1. Explain how the evolution of living organisms, especially those capable of photosynthesis, resulted in important changes in the chemistry of the environment.
2. Outline some of the key evidence supporting the theory of evolution.
3. Explain natural selection as a mechanism of evolution and describe some other ways in which evolution can occur.
4. Explain why most biologists are reluctant to describe certain species as being "more advanced" or "more highly evolved" than others.
5. How have environmental conditions affected your own development? Relate that influence to phenotypic plasticity.
6. Explain why knowledge of genetics is important to understanding evolutionary processes.

References

Bengtson, S. (ed.). 1995. *Early Life on Earth.* New York: Columbia University Press.

Cowen, R. 1995. *History of Life.* London: Blackwell Scientific.

Darwin, C.D. 1859. *On the Origin of Species by Means of Natural Selection, or the Preservation of Favoured Races in the Struggle for Life.* London: Murray.

Dawkins, R. 1986. *The Blind Watchmaker.* New York: Penguin.

Gould, S.J. (ed.) 1993. *Book of Life.* London: Hutchinson.

Horgan, J. 1991. In the beginning... *Scientific American*, **264**: 116–125.

Lewin, R. 1982. *Thread of Life: The Smithsonian Looks at Evolution.* Washington, DC: Smithsonian Books.

Mayr, E. 1982. *The Growth of Biological Thought.* Cambridge, MA: Harvard University Press.

National Research Council. 1990. *The Search for Life's Origins.* Washington, DC: National Academy Press.

Numbers, R.L. 1993. *The Creationists: The Evolution of Scientific Creationism.* Berkeley, CA: University of California Press.

Raven, P.H. and G.B. Johnson. 1992. *Biology.* Toronto: Mosby Year Book.

Romaniuk, Roman B. 1997. *Roman's Notes on DNA.* Toronto: Trifolium.

Shapiro, R. 1986. *Origins: A Sceptic's Guide to the Creation of Life on Earth.* New York: Simon & Schuster.

Strickberger, M.W. 1995. *Evolution, 2nd ed.* Boston: Jones & Bartlett.

7

BIODIVERSITY AND THE SYSTEMATIC ORGANIZATION OF LIFE

CHAPTER OUTLINE

Biodiversity

The value and importance of biodiversity

Classification of organisms

The systematic organization of life

CHAPTER OBJECTIVES

After completing this chapter, you will be able to:

1. Outline the concept of biodiversity and explain its constituent elements.
2. Discuss the reasons why biodiversity is important and should be preserved.
3. Define the classification of life in terms of species, genus, family, order, class, phylum, and kingdom.
4. Describe the five kingdoms of life.

BIODIVERSITY

Biodiversity can be defined as the richness of biological variation. It is often considered to have three levels of organization: genetic variation within populations and species; numbers of species (also known as *species richness*); and the variety and dynamics of ecological communities on larger scales, such as landscapes and seascapes.

Genetic Variation within Populations and Species

In almost all species, individuals differ genetically, that is, in terms of the genetic information encoded in their DNA. This variation constitutes genetic biodiversity at the level of populations, and ultimately of the species.

There are, however, exceptions to this generalization. Some plants, for example, have little or no genetic variability within their populations, usually because the species relies on asexual (or vegetative) mechanisms of propagation. In such species, genetically uniform *clones* can develop, consisting of many plants that although discrete, nevertheless constitute the same genetic "individual." For example, clones of trembling aspen (*Populus tremuloides*) can develop naturally through vegetative propagation, in some cases covering more than 40 hectares and consisting of tens of thousands of trees. A single such aspen clone may be the world's largest "individual" organism (in terms of total biomass). Similarly, the tiny plant known as duckweed (*Lemna minor*), which grows on the surface of fertile waterbodies, propagates by developing small buds on the edge of its single leaf. These break off to produce "new" plants, resulting in a genetically uniform population. These interesting cases are exceptions, however, and most populations and species contain a great deal of genetic variation.

Species are one of the familiar elements of biodiversity. These yellow ladyslipper orchids (*Cypripedium calceolus*) are growing in boreal forest near Norman Wells in the Northwest Territories.
Source: B. Freedman

High levels of genetic diversity in populations are generally considered to be desirable. With greater genetic diversity, populations are more likely to survive, to have enhanced fecundity and resistance to diseases, and to be more adaptable to changes in environmental conditions. Very small populations with low genetic diversity are thought to be at extreme risk because of inbreeding and low adaptability. Examples of such populations include the several hundred beluga whales (*Delphinapterus leucas*) living in the estuary of the St. Lawrence River, and the population of only about 20 panthers (*Felis concolor*) in Florida.

Richness of Species

Another level of biodiversity concerns **species richness**, or the number of species in a particular ecological community or in some other defined area, such as a park, county, province, country, or ultimately the biosphere. Species richness is the aspect of biodiversity that most people can easily relate to and understand.

It is well known that many tropical countries support a much greater richness of species than do temperate countries (such as Canada). Tropical rainforests, in fact, support more species than any other kind of ecosystem. Unfortunately, species-rich rainforests in tropical countries are rapidly being destroyed, mostly through conversion into agriculture and other disturbances. These ecological changes are causing many species to become extinct or endangered, and are the overwhelming cause of the modern-day biodiversity crisis (see Chapter 24). The magnitude of this crisis is much smaller in Canada. Nevertheless, many of our native species have become **extinct** or **extirpated** because of over-exploitation or habitat losses, and hundreds of other species are considered to be at risk.

TABLE 7.1 | NUMBERS OF SPECIES IN MAJOR CLIMATIC ZONES

Estimates suggest that whatever the total number of species, most live in tropical rainforests. These data include only plants and animals.

ZONE	NUMBER OF IDENTIFIED SPECIES (× 10^6)	ESTIMATED TOTAL NUMBER OF SPECIES	
		ASSUME 5 × 10^6	ASSUME 10 × 10^6
Boreal	0.1	0.1	0.1
Temperate	1.0	1.2	1.3
Tropical	0.6	3.7	8.6
TOTAL	1.7	5.0	10.0

Source: Modified from World Resources Institute (1986)

In total, about 1.9 million species have been identified and given a scientific name. About 35% of these known species live in the tropics, 59% in the temperate zones, and the remaining 6% in boreal or polar latitudes (Table 7.1). It is important to recognize, however, that the identification of species on Earth is very incomplete. This is especially true of tropical ecosystems, which have not yet been thoroughly explored and characterized. According to some estimates, the total richness of species could range as high as 30–50 million species, with 90% of them living in the tropics, particularly in rainforests.

Most of the species that biologists have described and named are invertebrates, with insects making up the bulk of that total, and beetles (order Coleoptera) comprising most of the insects (Table 7.2). The scientist J.B.S. Haldane (1892–1964) was once asked by a theologian to succinctly describe, based on his knowledge of biology, what he could discern of God's purpose. Haldane reputedly said that God has "an inordinate fondness of beetles." This reflects the fact that, in any random sampling of all of the known species on Earth, there is a strong likelihood that a beetle would be the sampled specimen.

Furthermore, it is believed that many tropical insects have not yet been described by biologists — perhaps more than another 30 million species, with many of these being small beetles. This remarkable conclusion initially emerged from research by T.L. Erwin, an entomologist who studies tropical rainforests in South America. Erwin treated tropical-forest canopies with a fog of insecticide, which resulted in a "rain" of dead arthropods that was collected in sampling trays on the ground. In the trays were great numbers of previously undescribed species of insects, most of which have a very localized distribution, limited to only a single type of forest or even to a particular species of tree.

Clearly, biologists know remarkably little about the huge numbers of relatively small, unobtrusive species that occur in poorly explored, tropical habitats. However, even

TABLE 7.2 | NUMBERS OF SPECIES IN VARIOUS GROUPS OF ORGANISMS

GROUP	NUMBER OF IDENTIFIED SPECIES	ESTIMATED TOTAL NUMBER OF SPECIES
Bacteria	4 800	>10 000
Protists	40 000	>100 000
Fungi	77 000	>150 000
Non-vascular plants	150 000	200 000
Vascular plants	250 000	280 000
Invertebrates	1 300 000	4 400 000*
Fishes	21 000	23 000
Amphibians	3 125	3 500
Reptiles	5 115	6 000
Birds	8 715	9 000
Mammals	4 170	4 300
TOTAL	1 864 000	>5 000 000

**This is a conservative figure. Some recent estimates suggest that more than 30 million species of insects may live in tropical forests alone (see text).*

Source: Modified from World Resources Institute (1986) and Raven and Johnson (1992)

in a relatively well-prospected country like Canada, many indigenous species of invertebrates, lichens, microbes, and small organisms have not yet been discovered and studied. Of course, larger plants and animals are relatively well known, partly because for most people (including scientists), these have greater "charisma" than small beetles, microbes, and the like. Still, even in Canada and the U.S., new species of vascular plants and vertebrate animals are being discovered.

Compared with invertebrates and microbes, the species richness of other groups of tropical-forest organisms is better known. For example, a survey of rainforest in Sumatra, Indonesia, found 80 species of tree-sized plants (that is, with diameter larger than 20 cm) in an area of only 0.5 ha. A study in Sarawak, Malaysia, found 742 woody species in a 3-ha plot of rainforest, with 50% of the species being represented by only a single individual. A similar study in Amazonian Peru found 283 tree species in a 1-ha plot, with 63% of the species represented by only one individual and another 15% by only two. In marked contrast, temperate forests in North America typically have fewer than 9–12 tree species in plots of this size. The richest temperate forests in the world, in the Great Smokies of the eastern U.S., contain about 30–35 tree species, far fewer than the tropical forests. More northern boreal forests, which cover much of Canada, have only 1–4 species of trees present.

Communities are an element of biodiversity. They involve populations of different species that occur together at the same time and place. This is a subtidal marine community at a depth of 10 m near Campbell River, British Columbia. Some of the prominent species are tiger rockfish (*Sebastes nigrocinctus*), strawberry anemone (*Corynactis californica*), warty sponge (*Melanchora elliptica*), and glassy seasquirts (*Acidea paratrop*).
Source: C. Harvey-Clark

Only a few studies have been made of the richness of bird species in tropical rainforests. A study of Amazonian forest in Peru found 245 resident bird species, plus another 74 transients, in a 97-ha plot. Another study found 239 species of birds in a rainforest in French Guiana. In contrast, temperate forests in North America typically support about 30–40 species of birds.

Few systematic studies have been made of other types of biota in tropical ecosystems. In one study, a 108-km^2 area of dry tropical forest in Costa Rica was found to contain about 700 species of plants, 400 vertebrate species, and 13 000 species of insects, including 3140 species of moths and butterflies.

Richness of Communities

Biodiversity at the level of landscape (or seascape) is associated with the number of different communities occurring within a certain area, as well as their relative abundance, size, shape, connections, and spatial distribution. An area uniformly covered with a single type of community would be judged as having little biodiversity at the landscape level, compared with an area having a rich and dynamic mosaic of different communities.

Because natural landscapes contain many species and communities that have evolved together, it is as important to protect community- and landscape-level diversity as it is to protect genetic and species diversity. Natural communities, landscapes, and seascapes are being lost in all parts of the world, particularly through the destruction of tropical rainforests and coral reefs. Dramatic losses of this level of biodiversity are also occurring in Canada. For example:

- Only about 0.2% of the original area of tall-grass prairie remains, the rest having been converted to agricultural use during the past century.
- Virtually all of the Carolinian forest of southern Ontario has been destroyed, mostly through conversion to agricultural and urbanized landscapes.
- The survival of old-growth forests in coastal British Columbia is at risk, with the dry coastal Douglas-fir forest being especially depleted to only about 7% of its original extent. The losses

of old-growth forests are mostly due to harvesting of the trees, which converts the ecosystem into a younger, second-growth forest (see Chapter 23).

- Throughout southern Canada, wetlands of all kinds have been destroyed or degraded by pollution, in-filling, and other disturbances.
- Natural fish populations have been widely decimated, including mixed-species communities in the Great Lakes, communities of salmonid species (salmon and trout) in western Canada, and cod and redfish populations off the Atlantic Provinces.

In all of these Canadian examples, only *remnant* patches of endangered natural communities, landscapes, and seascapes remain. These are at great risk, because they are no longer components of robust, extensive, naturally organizing ecosystems.

Many elements of biodiversity provide products useful to humans as foods, materials, and medicines. In the 1980s, chemicals in the rosy periwinkle were found to be useful for treating certain kinds of cancers. The native habitats of this tropical plant, on the island of Madagascar, are threatened by human encroachments.
Source: World Wildlife Fund

The Value and Importance of Biodiversity

Biodiversity is valuable and important for many reasons. These can be categorized into several groups.

Direct Utilitarian Value

Humans are not isolated from the rest of the biosphere, mostly because we need the products of certain elements of biodiversity. Because of this requirement, humans exploit species and ecosystems as sources of food, biomaterials, and energy — in other words, for their **utilitarian value**. Virtually all human foods are ultimately derived from biodiversity. Moreover, about one-quarter of the prescription drugs dispensed in North America contain active ingredients extracted from plants. In addition, probably a huge wealth of other, as yet undiscovered, products of biodiversity are potentially useful to humans. Research on wild species of plants, animals, and microorganisms has frequently discovered bio-products useful to humans as food, as medicines, or for other purposes.

To illustrate the importance of medicinal plants, take the case of the rosy periwinkle (*Catharantus roseus*), a small plant native to the island of Madagascar. One method used in the search for anti-cancer drugs involves screening large numbers of wild plants for the presence of chemicals that slow tumour growth. During one study, an extract of rosy periwinkle was found to counteract the reproduction of cancer cells. Further research identified the active chemicals as several alkaloids, probably synthesized by the rosy periwinkle to deter herbivores. These natural biochemicals are now used to prepare the drugs vincristine and vinblastine, which are extremely useful in chemotherapy to treat childhood leukemia, a cancer of the lymph system known as Hodgkin's disease, and several other cancers. Children with leukemia now have a 94% chance of remission, compared with 5% prior to the discovery of the therapeutic properties of periwinkle extracts. Hodgkin's patients now have a 70% chance of survival, compared with less than 1% previously. Therefore, an obscure species of plant, until recently known only to a few tropical botanists, has proven to be of great medicinal benefit to humans.

Human exploitation of wild biodiversity can be conducted in ways that allow the renewal of harvestable stocks. Unfortunately, many potentially renewable, biodiversity resources are managed as if they were non-renewable resources (that is, they are "mined"; see Chapters 12 and 14), usually by excessive harvesting and inadequate fostering of post-harvest regeneration. This results in biological resources becoming degraded in quantity and quality.

Sometimes, over-exploited species are locally extirpated or even rendered globally extinct, and their unique resource values are no longer available for human use. The great auk, passenger pigeon, and sea mink are examples of Canadian species rendered extinct through over-harvesting. Local and regional extirpations have been more numerous, and include cougar, grizzly bear, timber wolf, and wild ginseng over most of their former ranges (see Chapters 14 and 24 for details).

Provision of Ecological Services

Examples of **ecological services** provided by biodiversity include nutrient cycling, biological productivity, cleansing of water and air, control of erosion, provision of atmospheric oxygen, removal of carbon dioxide, and other functions important to maintain the stability and integrity of ecosystems. Even though all of these services are critical to the welfare of humans and other species, they are not usually assigned economic value by society. In part, this is because humans do not yet have sufficient understanding and appreciation of the "importance" of ecological services, and of the particular species and communities that provide them. According to Peter Raven, a famous botanist and advocate of biodiversity, "In the aggregate, biodiversity keeps the planet habitable and ecosystems functional."

Landscapes and seascapes, other elements of biodiversity, are composed of mosaics of various communities occurring at relatively large scales. This rather flat landscape east of the Mackenzie River in the Northwest Territories is characterized by a meandering river, its old cut-offs known as ox-bow lakes, and various communities in their terrestrial watersheds. Because the river "wanders" over time, ox-bow lakes are periodically created and reconnected to the river, illustrating the dynamic nature of landscapes.

Source: B. Freedman

Intrinsic Value

Biodiversity has its own **intrinsic value**, regardless of any direct or indirect worth in terms of the needs or welfare of humans. Its intrinsic value suggests certain ethical questions about actions that threaten biodiversity. Do humans have the "right" to impoverish or exterminate unique and irretrievable elements of biodiversity, even if our species is technologically able to do so? Is the human existence somehow impoverished by extinctions caused by our actions? These are philosophical issues, and they cannot be resolved by science. However, enlightened people and societies do not facilitate the endangerment or extinction of species or natural communities.

People Believe Biodiversity Is Important

Many people firmly believe that wild biodiversity and natural ecosystems are worthwhile and important. They cite the above reasons but also mention less tangible opinions, such as the charisma of many species (lions, pandas, and baby harp seals among others) and the spirituality of natural places (such as towering old-growth forests and other kinds of wilderness). Because this belief is becoming increasingly widespread and popularized, it is having a major influence on politicians, who are now including biodiversity issues in their agendas. Threats to biodiversity have, therefore, become politically important.

Undoubtedly there is an undiscovered wealth of products of biodiversity that are potentially useful to humans. Many of these bio-products will be found in tropical species that have not yet been "discovered" by biologists. Clearly, the most important argument in favour of preserving biodiversity is the need to maintain natural ecosystems so they can continue to provide their vast inventory of useful products and their valuable ecological services. Biodiversity must also, of course, be preserved for its own, intrinsic value.

Classification of Organisms

Biologists classify species into higher-order groupings on the basis of relatedness and similarities. Similarity is judged using information about anatomy, development, biochemistry, behaviour, and habitat selection. These classifications are made by systematists (biologists who study the evolutionary relationships among groups of organisms) and taxonomists (biologists who focus on naming groups of organisms).

The systematics of life is traditionally organized hierarchically, with the levels ranging through sub-species, species, genus, family, order, class, phylum, and kingdom. This system is illustrated in Table 7.3.

Individual species are always described using two latinized words, known as a binomial. In cases where a subspecies is also recognized, the name has three latinized words, or a trinomial. Many species also have a scientifically recognized "common name," and they may have informal common names as well. For example, the "proper" (or scientifically recognized) common name of the widespread tree *Populus tremuloides* is trembling aspen, but this species is also known as aspen, golden aspen, mountain aspen, poplar, quaking asp, quaking aspen, trembling poplar, and that old-time favourite, "popple." Some of these common names have only a regional use, and are unknown in other parts of the range of the species. Common names may also overlap among species — for instance, balsam poplar (*Populus balsamifera*) and large-toothed aspen (*P. grandidentata*) are both often called "poplar."

To avoid these ambiguities associated with common names, all species are assigned a globally recognized scientific binomial and sometimes a "proper" common name. Therefore, biologists working in Canada, the U.S., Mexico, Germany, Turkey, Russia, China, and other countries where the animal *Ursus arctos* occurs all know it by its binomial. In English, this animal is known as the grizzly or brown bear, and in other languages, by other common names. But no one is confused by its scientific binomial name.

The Systematic Organization of Life

Most biologists divide all of Earth's species into five major groups, known as kingdoms. Although somewhat controversial and subject to ongoing refinement, this systematic organization is believed to reflect the evolutionary relationships among groups of organisms. The five kingdoms of living organisms and their major characteristics are briefly described below.

TABLE 7.3 | BIOLOGICAL CLASSIFICATION

The hierarchical, systematic classification of organisms, illustrated by three representative species.

GROUPING	DOUGLAS FIR (INTERIOR)	MONARCH BUTTERFLY	HUMANS
Kingdom	Plantae	Animalia	Animalia
Phylum (Division)	Coniferophyta	Arthropoda	Chordata
Class	Gymnospermae	Insecta	Mammalia
Order	Coniferales	Lepidoptera	Primates
Family	Pinaceae	Danaidae	Hominidae
Genus	*Pseudotsuga*	*Danaus*	*Homo*
Species	*menziesii*	*plexippus*	*sapiens*
Subspecies	*glauca*	*plexippus*	(not recognized)
Scientific name	*Pseudotsuga menziesii glauca*	*Danaus plexippus plexippus*	*Homo sapiens*

Monera

Monerans are the simplest of single-celled microorganisms and include bacteria and blue-green bacteria, the latter being photosynthetic. Monerans are **prokaryotic**, which means the genetic material is not contained within a membrane-bounded organelle (called a nucleus). Organisms in the other kingdoms have nuclei within their cells and are **eukaryotic**. Prokaryotic microorganisms also do not have other kinds of organelles, such as chloroplasts, mitochondria, or flagella, and they differ from eukaryotes in other respects as well. Prokaryotes were the first organisms to evolve, about 3.5 million years ago. It was not until 2 billion years later (that is, 1.5 billion years ago) that the first eukaryotes evolved.

About 4800 species of bacteria have been named, but many other species have not yet been described by microbiologists. The great diversity of bacteria includes species capable of exploiting a phenomenal range of ecological and metabolic opportunities. Many are decomposers, found in "rotting" biomass. Some species are photosynthetic; others are chemosynthetic; and still others can utilize virtually any organic substrate in their heterotrophic nutrition, either in the presence of oxygen or under anaerobic conditions. Some species of bacteria can tolerate extreme environments, living in hot springs at temperatures as torrid as 78°C, while others are active in sub-zero temperatures as deep as 400 m in glacial ice.

Many bacteria species live in *mutualistic symbioses* with more complex organisms. For example, some bacterial species live as a community in the rumens of cows and sheep, and others live in the human gut, in both cases aiding in the digestion of complex organic foods. Other bacteria, known as *Rhizobium*, live in a mutualism with the roots of leguminous plants (such as peas and clovers), fixing atmospheric nitrogen gas into a form (ammonia) that plants can utilize as a nutrient (see Chapter 5).

Many bacteria are parasites of other species, causing various diseases. For example, *Bacillus thuringiensis* is a pathogen of many species of moths, butterflies, and blackflies, and has been used as a biological insecticide against certain pests in agriculture and forestry. Species of bacteria also cause important diseases of humans, including various infections, bacterial pneumonia, cholera, diphtheria, gonorrhea, Legionnaire's disease, leprosy, scarlet fever, syphilis, tetanus, tooth decay, tuberculosis, whooping cough, most types of food poisoning, and the "flesh-eating disease" caused by a virulent strain of *Streptococcus*.

Protista

Protists include a wide range of simple, eukaryotic organisms, comprising both unicellular and multicellular species. Familiar protists include protozoans, foraminifera, slime moulds, single-celled algae, and multicellular algae.

In Detail 7.1

Some Non-Living Entities — Viruses and Prions

In addition to the living organisms described in this chapter, there are other entities that display some, but not all, of the characteristics of life. These "pseudoorganisms" are incapable of reproduction, or even of metabolism, unless they have first invaded and parasitized a living cell. The best known of these non-living entities are viruses, which consist of bits of nucleic acid (DNA or RNA) surrounded by a protein capsule. Viruses cannot reproduce themselves, but they can invade specific living cells, take over the genetic metabolism, and replicate more viruses of their kind.

Some viruses cause diseases, with specific strains likely infecting all kinds of living cells. Some viral diseases of humans include colds and flus and more serious maladies such as smallpox, yellow fever, rabies, herpes, polio, and human immunodeficiency syndrome (also known as AIDS). Some viral diseases can be controlled by a practice known as vaccination, which involves infecting hosts with nonvirulent, attenuated or killed viruses, which subsequently cause the host to develop resistance to virulent strains. This practice has allowed the recent eradication of smallpox, a debilitating disease of humans.

Prions are another kind of infectious particle, consisting only of bits of transmissible protein. Prions are known to cause certain diseases, including sheep scabies and bovine spongiform encephalopathy or "mad-cow disease."

The latter group includes the large seaweeds known as kelps, some of which are over 10 m long. The kingdom Protista consists of 14 phyla and about 40 000 named species, which vary enormously in their genetics, structure, and function. Systematists may eventually divide the Protista into several kingdoms, because many think that the other, relatively complex eukaryotic kingdoms (that is, fungi, plants, and animals) all evolved from different protistan ancestors.

Several phyla of protists, broadly known as algae, are photosynthetic. These groups include the diatoms (Bacillariophyta); green algae (Chlorophyta); dinoflagellates (Dinoflagellata); euglenoids (Euglenophyta); red algae (Rhodophyta); and brown algae, such as kelps (Phaeophyta). Algae are important primary producers in aquatic ecosystems, both marine and freshwater. Some seaweeds are harvested as a natural resource, mostly to extract chemicals known as alginates, important additives to many foods and cosmetics. Uncommon marine phenomena known as "red tides" are natural blooms of certain species of dinoflagellates that produce toxic chemicals.

Other phyla of protists are heterotrophic in their nutrition. These groups include the ciliates (Ciliophora), forams (Foraminifera), plasmodial slime moulds (Myxomycota), amoebae (Rhizopoda), and unicellular flagellates (Zoomastigina). Forams are unicellular microorganisms that form a shell of calcium carbonate, the remains of which can accumulate over geological time to form a mineral known as chalk. (The white cliffs of Dover in southern England are made of foram remains.) Trypanosomes are unicellular flagellates responsible for sleeping sickness, a debilitating disease of humans and other vertebrate animals. Certain species of *Amoeba* are parasites of animals, and one species causes amoebic dysentery in humans. A ciliate known as *Giardia* causes a water-borne disease known as hiker's diarrhea (or beaver fever), which is the reason why even the cleanest-looking natural waters should be boiled or otherwise disinfected before drinking.

Fungi

This kingdom consists of yeasts, which are single-celled microorganisms, and fungi, which are multicellular and filamentous in their growth form. Fungi evolved at least 400 million years ago, but may be much older, because their remains do not fossilize well. Fungal cells excrete enzymes into their surroundings (growth medium). These enzymes externally digest complex organic materials, and the fungus then ingests the resulting simple organic compounds. All fungi are heterotrophic, with most species being decomposers of dead organic matter and others parasitic on plants or animals.

There are three major divisions (or phyla) of fungi, distinguished mainly by their means of sexual reproduction. (Asexual reproduction is also common in most species of fungi.)

The zygomycetes (division Zygomycota) achieve sexual reproduction by the direct fusion of hyphae (the thread-like tissues of fungi), which form resting spores known as zygospores. There are about 600 named species of zygomycetes, the most familiar of which are the bread moulds, such as *Rhizopus*, with its fluffy mycelium (or loosely organized mass of hyphae).

The ascomycetes (division Ascomycota) include about 30 000 named species, some of which are commonly known as a cup fungus or morel. During their sexual reproduction, ascomycetes form many microscopic, cup-shaped bodies known as asci, located in specialized fleshy structures called ascocarps. Relatively familiar species of ascomycetes include yeasts, morels, and truffles, as well as pathogens such as the chestnut blight and Dutch elm disease fungi (see below).

The basidiomycetes (division Basiodiomycota) include about 16 000 named species. Sexual reproduction of basidiomycetes involves the development of a relatively complex spore-producing structure known as a basidium. Depending on its shape, a basidium may be commonly called a mushroom, toadstool, puffball, or shelf fungus. In Canada, the largest of these structures is developed by the giant puffball (*Calvatia* spp.), which can grow a ball-like basidium with a diameter of up to 50 cm.

Lichens are specific mutualisms between a fungus and either an alga or a blue-green bacterium. Most of the lichen biomass is fungal tissue, which provides habitat and inorganic nutrients for the photosynthetic partner, which provides organic nutrition to the fungus. Another type of mutualism, known as a mycorrhiza, involves an intimate relationship between plant roots and certain fungi. This relationship is highly beneficial to the plant because it allows more efficient absorption of inorganic nutrients from soil, particularly phosphate. About 80% of plant species develop mycorrhizae.

Fungi are extremely important ecologically because they are excellent decomposers, allowing nutrients to be recycled and reducing the accumulation of dead biomass.

Economically important fungi spoil stored grains and other foods, are serious parasites of agricultural or forestry plants, or cause diseases in humans and other animals. One example of a fungal pathogen is the chestnut blight fungus (*Endothia parasitica*): it was accidentally introduced to North America from Europe and wiped out the native chestnuts (*Castanea dentata*), which used to be prominent and valuable trees in many eastern forests. The Dutch elm disease fungus (*Ceratocystis ulmi*), also introduced from Europe, is now causing similar damage to native elm trees (especially white elm, *Ulmus americana*). Ringworm is a disease of the skin, usually the scalp, which can be caused by various species of fungi.

Economically useful fungi include a few species of yeast which have the ability to ferment sugars under anaerobic (O_2-deficient) conditions, yielding gaseous CO_2 and ethanol. The CO_2 produced during fermentation raises bread dough prior to baking, while brewers take advantage of the alcohol production to make beer, wine, and other intoxicating beverages. Other fungi are employed in the manufacture of cheese, tofu, soy sauce, food additives such as citric acid, and certain antibiotics such as penicillin.

Some mushroom-forming fungi are cultivated as a source of food for people, while other edible species are collected from natural ecosystems. The most commonly cultivated species is the meadow mushroom (*Agaricus campestris*), while the most prized wild mushroom is the extremely flavourful truffle (*Tuber melanosporum*). Some wild mushrooms contain chemicals that induce hallucinations, feelings of well-being, and other pleasurable mental states, and are sought by people for religious or recreational use. These include the fly agaric (*Amanita muscaria*), a species widespread in Canada and elsewhere, and psilocybin (*Psilocybe* spp.) of more southern regions of North America and Central America. Some wild mushrooms are deadly poisonous even when eaten in tiny quantities. The most toxic species in Canada are the destroying angel (*Amanita virosa*) and deathcap (*A. phalloides*).

Plantae

Plants are photosynthetic organisms, which manufacture their food by using the energy of sunlight to synthesize organic molecules from inorganic ones. Plants evolved from multicellular green algae about 430 million years ago, and the first tree-sized plants evolved about 300 million years ago. Plants are different from algae in that they are always multicellular; have cell walls rich in cellulose; synthesize a mixture of photosynthetic pigments, including chlorophylls *a* and *b* and carotenoids; and use starch as their principal means of storing energy. Plants are extremely important as photosynthetic fixers of CO_2 into organic carbon, and are dominant in terrestrial ecosystems, where algae and blue-green bacteria are relatively unimportant.

Plants are generally divided into 12 divisions, which themselves are aggregated into two functional groups.

Bryophytes

The relatively simple **bryophytes** lack vascular tissues and do not have a cuticle covering their foliage, a character that restricts these plants to moist habitats. The bryophytes consist of:

- liverworts (division Hepaticophyta), of which there are about 6500 species
- mosses (Bryophyta), including about 10 000 species, which are ecologically important in some types of wetlands, especially in bogs, where the dead biomass of peat-mosses (species of *Sphagnum*) accumulates as a partially decayed organic material known as peat. Peat is mined as a horticultural soil conditioner and as a source of energy
- hornworts (Anthocerophyta), with 100 species

Vascular Plants

The relatively complex **vascular plants** have specialized, elongated, tube-like, vascular tissues in their stems for conducting water and nutrients. The nine divisions of vascular plants are briefly described below:

- whisk ferns (division Psilophyta), containing several species
- club mosses and quillworts (Lycophyta), about 1000 species
- horsetails or scouring rushes (Sphenophyta), 15 species
- ferns (Pterophyta), 12 000 species
- cycads or sago palms (Cycadophyta), 100 species
- gnetums or mormon tea (Gnetophyta), 70 species
- ginkgo (Ginkgophyta), with many extinct species but only one relict, extant species (*Ginkgo biloba*)

- conifers (Coniferophyta), including about 550 species of firs, hemlocks, pines, redwoods, spruces, yews, and others
- flowering plants (Anthophyta), containing a diverse assemblage of about 235 000 species

The flowering plants are also known as **angiosperms**, because their ovules are enclosed within a specialized membrane and their seeds within a seedcoat. The conifers, ginkgo, and gnetums lack these structures and are referred to as **gymnosperms**. Together, the angiosperms and gymnosperms are known as *seed plants*. Their seeds develop from a fusion between specialized haploid cells known as pollen and ovules, in a process called pollination.

The seed plants are extremely diverse in their form and function. The tallest species are redwood trees (*Sequoia sempervirens*), which can exceed 100 m in height. The smallest is an aquatic plant known as watermeal (*Wolffia* spp.), only the size of a pinhead. Many seed plants live for less than one year (these are "annual" plants), while the age of others can exceed 4500 years, for example, the oldest bristlecone pines (*Pinus aristata*).

Many flowering plants grow as shrubs or trees. Rigid, woody tissues in their stems confer mechanical strength that allows these plants to grow tall against the forces of gravity and wind. Other species of angiosperms lack these rigid stem tissues and grow as herbaceous plants that die back to the ground at the end of the growing season.

Species of angiosperms are the most important crop plants in agriculture, while both conifer and angiosperm trees are important in forestry. Plants are also economically important as sources of biochemicals in industry and medicine, and because they provide the productivity and habitat required by so many other organisms, including many species of animals that are hunted by humans for food.

Animalia

All animals are multicellular organisms and most are mobile during at least some stage of their life history, having the ability to move about to search for food, to disperse, or to reproduce. All animals are heterotrophs: they must ingest their food, ultimately consuming the photosynthetic products of plants or algae.

Most animals (except the sponges) have their cells organized into specialized tissues that are further organized into organs. Almost all animals reproduce sexually, a process involving a transfer of haploid gametes between males and females to produce a fertile egg. Animals comprise the bulk of identified species of organisms, with insects being the most diverse group. Apart from these broad generalizations, animals are extremely diverse in their form and function. They range in size from the largest blue whales (*Balaenoptera musculus*), which can reach 32 m in length and 136 t in weight, to the smallest beetles and soil mites, which are less than 1 mm long and weigh a few milligrams.

The animal kingdom includes about 35 phyla. The majority of these are exclusive to marine habitats, with a smaller number occurring in freshwater and on land. All animals in all phyla except one are considered to be **invertebrates** (no backbone), while Phylum Chordata includes the **vertebrates** — animals with backbones.

The most prominent phyla of animals are described below.

Sponges

Sponges (phylum Porifera) are a marine group of about 5000 species plus about 150 freshwater species. Sponges are simple, sessile (non-mobile) animals, with no differentiation of their tissues into organs. Sponges filter-feed on organic matter suspended in water. The slow flow of water through sponges is driven by surface cells that use flagella, tiny whip-like structures, to move water over their surface.

Cnidarians

Cnidarians (phylum Cnidaria) include about 9000 species, almost all of which are marine. Familiar cnidarians include jellyfish, sea anemones, hydroids, and corals. Cnidarians have a simple, rather gelatinous body structure. They display radial symmetry, meaning a cross-section in any direction through their central axis yields two parts that are mirror images. Jellyfish are weakly swimming or floating animals, with a body form known as a medusa. Most other cnidarians are sessile as adults, being firmly attached to a bottom substrate. Cnidarians are carnivores, capturing their prey using tentacles that ring their mouth opening, often after subduing the prey by stinging it with specialized cells. Corals develop a protective casement of calcium carbonate, and are important reef-building organisms in shallow tropical seas.

Flatworms and Tapeworms

The flatworms and tapeworms (phylum Platyhelminthes) include about 12 000 species of soft-bodied, ribbon-shaped animals. Many flatworms are free-living scavengers or predators of small animals, while tapeworms and flukes are parasites in larger animals, including humans.

Nematodes and Roundworms

Nematodes (phylum Nematoda) include 12 000 species of small, worm-like creatures. These animals are round in cross-section, and are abundant in almost all habitats containing other forms of life, ranging from aquatic habitats to the driest deserts. Many species are parasites, living in or on their hosts. Virtually all plants and animals are parasitized by one or more species of nematodes, often specialized to live in a specific host. Species of hookworms, pinworms, and roundworms are important parasites of humans. The *Trichinella* roundworm causes a painful disease known as trichinosis, while *Filaria* causes filariasis, a tropical disease.

True Worms

The worms (phylum Annelida) include about 12 000 species of long, tubular, segmented, soft-bodied animals. Most worm species are marine, but others occur in freshwater and moist terrestrial habitats. Worms are divided into three major groups: bristleworms or polychaetes, typical worms or oligochaetes (including earthworms), and leeches or hirudineans. Most feed on dead organic matter, but leeches are blood-sucking parasites of larger animals. Earthworms provide an important ecological service by helping to recycle dead biomass in many terrestrial habitats, including some agricultural ecosystems.

Mollusks

Mollusks (Phylum Mollusca) comprise about 110 000 species of clams, cuttlefish, octopuses, oysters, scallops, slugs, snails, and squid. Many mollusks have a hard shell made of calcium carbonate which protects the soft body parts inside. Other mollusks, such as squid and octopuses, lack this hard shell. Mollusks are most abundant in marine and freshwater habitats, with far fewer terrestrial species. Most mollusks are herbivores or scavengers, and a few are predators of other animals. Various species of mollusks are an important resource used by humans as food. Several species produce pearls, valuable for making jewellery. Some species of slugs and snails are pests in agriculture, while others are alternate hosts for certain parasites such as the tropical fluke that causes schistosomiasis in humans.

Arthropods

Arthropods (phylum Arthropoda) comprise the largest phylum of organisms. There are about 900 000 named species of arthropods and probably millions of others that have not yet been described. Arthropods have an exterior skeleton (or exoskeleton) made of a polysaccharide known as chitin, with the body parts segmented to allow movement. They have at least three pairs of legs. The most important groups of arthropods are the spiders and mites (class Arachnida), crustaceans (Crustacea), centipedes (Chilopoda), millipedes (Diplopoda), and insects (Insecta). Insects alone make up more than half of all named species.

Arthropods are of great economic importance, with some species being used by humans as food (e.g., lobsters, crayfish), and others used to produce food (e.g., the honey of certain species of bees). Some other species are important pests. Termites damage buildings by eating wood, while various insects are important pests in agriculture. Species of mosquitoes, blackflies, fleas, and ticks spread diseases of humans and other animals, including malaria, yellow fever, encephalitis, and plague.

Echinoderms

Echinoderms (phylum Echinodermata) include about 6000 species of marine animals, such as brittle stars, sand dollars, sea stars, sea cucumbers, and sea urchins. Echinoderms have radial symmetry as adults. Most have an internal skeleton rich in calcium carbonate; some are covered with spiny projections; and some move about using large numbers of small, tube-like feet. Sea urchins and sea cucumbers are harvested as a minor source of food, popular in some Asian countries.

Chordates

Chordates (phylum Chordata) are the most familiar group of animals. Distinctive characters of chordates (in at least the embryonic phase) include a hollow nerve cord that runs along the dorsal (or top) surface and a flexible rod-like structure (the notochord) on the dorsal surface, which is replaced by the vertebral column in adult vertebrates.

IN DETAIL 7.2

WARM OR COOL BLOOD

Animals can be divided into two functional groups on the basis of whether they control their body temperature. So-called cold-blooded animals (or poikilotherms) do not expend metabolic energy to warm their bodies. Therefore, their activity tends to be strongly influenced by the ambient temperature (because muscles do not work well when they are cool). In contrast, warm-blooded animals (or homeotherms) allocate a significant part of their metabolism to warming their body, allowing these animals to keep active even in sub-zero temperatures.

The most familiar examples of homeothermic animals are mammals and birds. On average, these animals expend about 80% of their metabolic energy on keeping warm. However, smaller animals (such as shrews and hummingbirds) must spend even more energy keeping warm because they have a relatively large surface area in comparison with their mass. Larger animals (such as ostriches and whales) expend proportionately less energy. Some mammals and birds allow their body temperature to decrease greatly during winter hibernation, when the animals enter a temporary, quiescent state. Hibernation is relatively common in mammals that live in climates featuring a winter period, but only one bird (the poorwill, *Phalaenoptilus nuttallii*) is known to hibernate.

A few other species of animals are homeothermic to some degree. For example, the bluefin tuna (*Thunnus thynnus*) is a large, oceanic fish that is capable of swimming as fast as 90 km/hr. To optimize its muscle function, the body temperature of bluefin tuna is maintained at 24–35°C, even in water as cold as 6°C. Even some insects are somewhat warm-blooded: when disturbed under cool conditions, large moths such as the cecropia (*Hyalophora cecropia*) flutter their wings rapidly for several minutes, until sufficient heat is generated to allow the animal to fly.

There are about 42 500 species of chordates, divided among three subphyla. The tunicates (Urochordata) are composed of about 1000 species of marine animals, including sea grapes and sea peaches. Tunicates have a small notochord and are sessile filter feeders as adults. The lancets (Cephalochordata) consist of 23 species of filter-feeding marine animals, with long, laterally compressed bodies.

The vertebrates (Vertebrata) are a group of about 41 000 species, most of which have a vertebral column as adults. The major classes of living vertebrates are:

The jawless fishes (class Agnatha) include 63 species of lampreys and hagfishes, which first evolved 470 million years ago. These marine or freshwater animals have a notochord throughout their life and a skeleton made of cartilage (cartilaginous).

The cartilaginous jawed fishes (class Chondrichthyes) consist of 850 species of dogfish, rays, sharks, and skates, all of which occur in marine habitats. Cartilaginous fishes evolved more than 410 million years ago.

The bony fishes (class Osteichthyes) includes 18 000 species of typical fish, such as cod, salmon, tuna, and guppies. The first bony fishes evolved about 390 million years ago.

The amphibians (class Amphibia) consist of 4000 species of frogs, salamanders, toads, and legless animals known as caecilians. The first amphibians evolved about 330 million years ago. Early stages in the life history (egg and larva) are usually aquatic, but adult stages of many species can life in terrestrial habitats.

The reptiles (class Reptilia) include 6000 species of crocodilians, lizards, snakes, and turtles. Reptiles first evolved about 300 million years ago. Extinct groups of reptiles include the dinosaurs, plesiosaurs, and pterosaurs, the last of which became extinct about 65 million years ago. Reptiles were the first fully terrestrial animals, capable of completing all stages of their life history on land (although some species, such as turtles, are highly aquatic as adults). Reptiles have a dry skin and lay eggs on land. Their young are miniature versions of the adults.

The birds (class Aves) consist of 9000 living species which first evolved about 225 million years ago from small, dinosaurian ancestors. All birds are homeothermic (see In Detail 7.2), are covered in feathers, lay hard-shelled eggs, and have a specialized horny covering of the jaws known as a beak. Most species of birds can fly, the exceptions being the largest birds, penguins, and many of the species that evolved on islands lacking predators.

The mammals (class Mammalia) consist of 4500 living species, which first evolved about 220 million years ago (the earliest fossil mammals are difficult to distinguish from reptiles). Mammals became especially prominent in ecosystems after the extinction of the last dinosaurs, about 65 million years ago. Mammals are homeotherms, have at least some hair on their bodies, feed their young with milk, and have a double circulation of the blood (that is, a four-chambered heart and fully separate circulatory systems for oxygen-poor and oxygen-rich blood).

There are three major groups of mammals. The *monotremes* are a few species of egg-laying mammals that live in Australia and New Guinea — the duck-billed platypus (*Ornithorhynchus anatinus*) and several species of echidnas. *Marsupials* bear live young which are at an extremely early stage of development. After birth the tiny young migrate to a special pouch on the mother's belly (the marsupium), where they develop for a further time while feeding on milk. Examples of marsupials include kangaroos, koala, and wallabies, which live only in Australia, New Guinea, and nearby islands, and the opossum of North and South America. The *placental mammals* include all of the familiar species of the Americas, Africa, and Eurasia. Placental mammals give birth to live young which are suckled by the female. Humans are a species of placental mammal.

Key Terms

biodiversity
species richness
extinct
extirpated
utilitarian value
ecological services
intrinsic value
prokaryotic
eukaryotic
bryophyte
vascular plant
angiosperm
gymnosperm
invertebrate
vertebrate

Questions for Discussion

1. Identify the major components of biodiversity and give an example of each.
2. Pick a species and demonstrate the hierarchical classification of life by giving its species, genus, family, order, class, phylum, and kingdom.
3. Discuss the various reasons why biodiversity is considered to be important.
4. Identify each of the five kingdoms of life, and give examples of several of the groups in each kingdom.
5. Discuss the notion that all species are similarly "advanced" in the evolutionary sense, although they may vary greatly in their complexity.

REFERENCES

Begon, M., Harper, J.L., and Townsend, C.R. 1990. *Ecology: Individuals, Populations, and Communities, 2nd ed.* London: Blackwell Scientific.

Bolandrin, M.F., J.A. Klocke, E.S. Wurtele, and W.H. Bollinger. 1985. Natural plant chemicals: Sources of industrial and medicinal materials. *Science*, **228**: 1154–1160.

Boyd, R. 1988. *General Microbiology.* St. Louis, MO: Mosby Year Book.

Ehrlich, P.R., and A. Ehrlich. 1981. *Extinction: The Causes and Consequences of the Disappearance of Species.* New York: Ballantine.

Erwin, T.L. 1991. How many species are there? Revisited. *Conserv. Biol.*, **5**: 330–333.

Freedman, B. 1995. *Environmental Ecology, 2nd ed.* San Diego, CA: Academic Press.

Heywood, V.H. (ed.). 1995. *Global Biodiversity Assessment.* Cambridge: Cambridge University Press.

Janzen, D.H. 1987. Insect diversity in a Costa Rican dry forest: why keep it, and how. *Biol. J. Linn. Soc.*, **30**: 343–356.

Miller, K. and L. Tangley. 1991. *Trees of Life.* Boston: Beacon.

Myers, N. 1983. *A Wealth of Wild Species.* Boulder, CO: Westview.

Pough, F.H., J.B. Hirser, and W.N. McFarland. 1996. *Vertebrate Life, 4th ed.* Upper Saddle River, NJ: Prentice Hall.

Raven, P.H. 1990. The politics of preserving biodiversity. *BioScience*, **40**: 769–774.

Raven, P.H. and G.B. Johnson. 1992. *Biology.* Toronto: Mosby Year Book.

Terborgh, J., S.K. Robinson, T.A. Parker, C.A. Muna, and N. Pierpont. 1990. Structure and organization of an Amazonian forest bird community. *Ecol. Monogr.*, **60**: 213–238.

Wilson, E.O. (ed.). 1988. *Biodiversity.* Washington, D.C.: National Academy Press.

World Resources Institute (WRI). *World Resources 1986: An Assessment of the Resource Base That Supports the Global Economy.* New York: Basic Books.

8

BIOMES AND ECOZONES

CHAPTER OUTLINE

Biomes: global ecosystem types

Major biomes and their characteristics

Ecozones of Canada

Canadian anthropogenic habitats

CHAPTER OBJECTIVES

After completing this chapter, you will be able to:

1. Identify Earth's major biomes and outline their characteristics.
2. Identify Canada's ecozones and list their prominent species.
3. Outline the character of urban, agricultural, and industrial habitats in Canada.
4. Discuss the differences between natural and anthropogenic ecosystems.

Biomes: Global Ecosystem Types

A **biome** is a geographically extensive type of ecosystem, occurring wherever the environmental conditions are suitable for its development, throughout the world. Biomes are characterized by their dominant **life forms**, but not necessarily their particular species. Terrestrial biomes are generally identified on the basis of their mature or older-growth vegetation. Aquatic biomes, especially marine ones, are usually distinguished by their dominant animals. Earth's biomes are classified by a system that is used internationally, that is, by ecologists working in many countries.

Figure 8.1 shows a map of the distribution of the most extensive terrestrial biomes. The distribution of biomes is determined by environmental conditions, which must be appropriate to support their dominant species (Figure 8.2). Typically, the most important environmental factors influencing the distribution of terrestrial biomes are moisture, temperature, and soil types. The distribution of wetland types within terrestrial biomes is largely influenced by the amount and permanence of surface water and the availability of nutrients. Marine biomes are strongly influenced by water depth and upwellings, which affect the amounts of light and nutrients.

As long as environmental conditions are suitable for their development, biomes can occur in widely divergent regions, even on different continents. Widely separate regions of the same biome may be dominated by different species, but their life forms are typically ecologically convergent. In other words, the different species are similar in form and function, because the natural selection regimes of broadly similar environments elicit parallel, or convergent, evolutionary responses. Therefore, biomes are defined primarily by their ecological structure and function, not necessarily by their species composition.

We can illustrate this important point by referring to the boreal coniferous forest, an extensive biome that

FIGURE 8.1 | DISTRIBUTION OF EARTH'S MAJOR TERRESTRIAL BIOMES

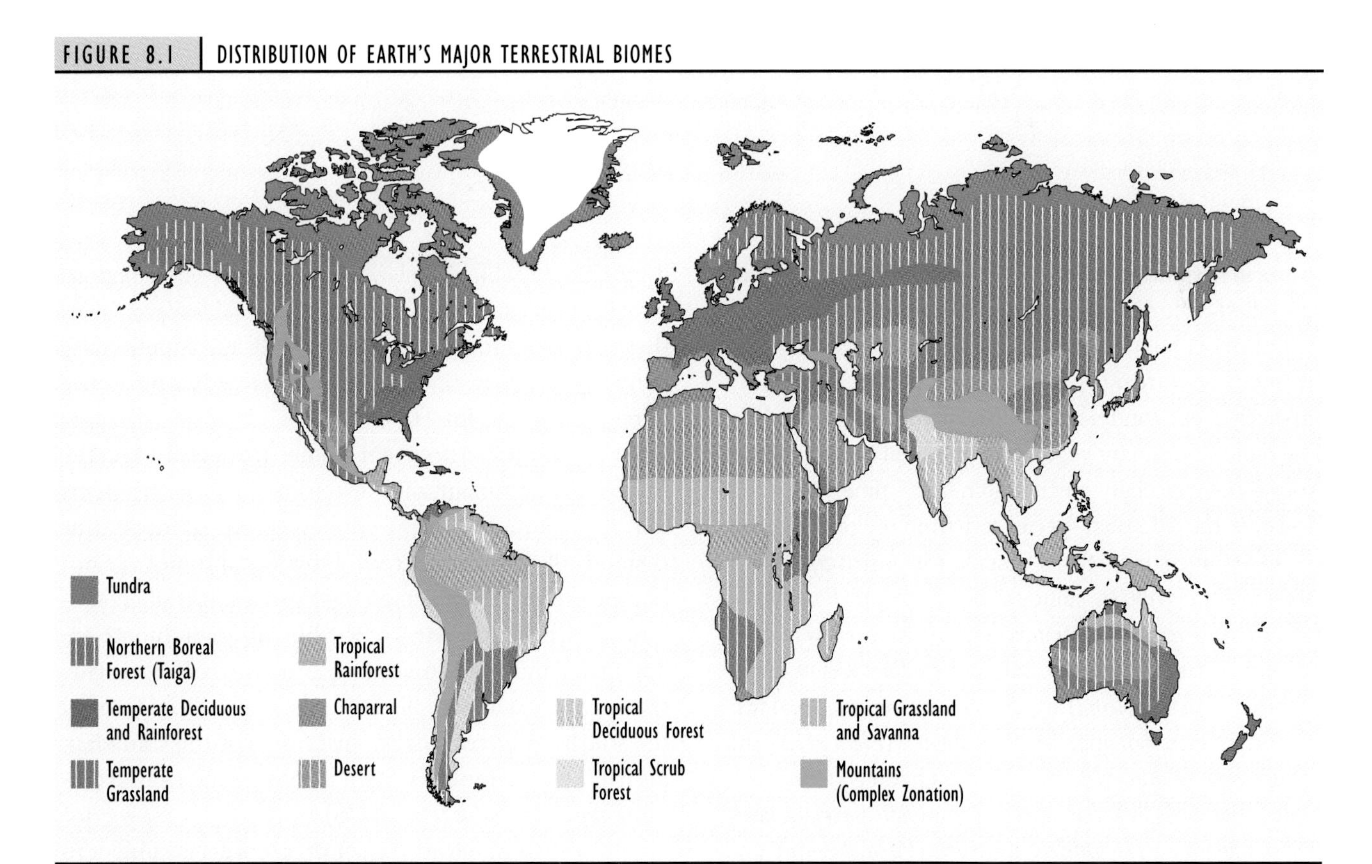

Note that the spatial complexity is greatest in regions with mountainous terrain, such as the western Americas and southern Asia.

Source: Modified from Odum (1983)

FIGURE 8.2 | ENVIRONMENTAL INFLUENCES ON THE DISTRIBUTION OF BIOMES

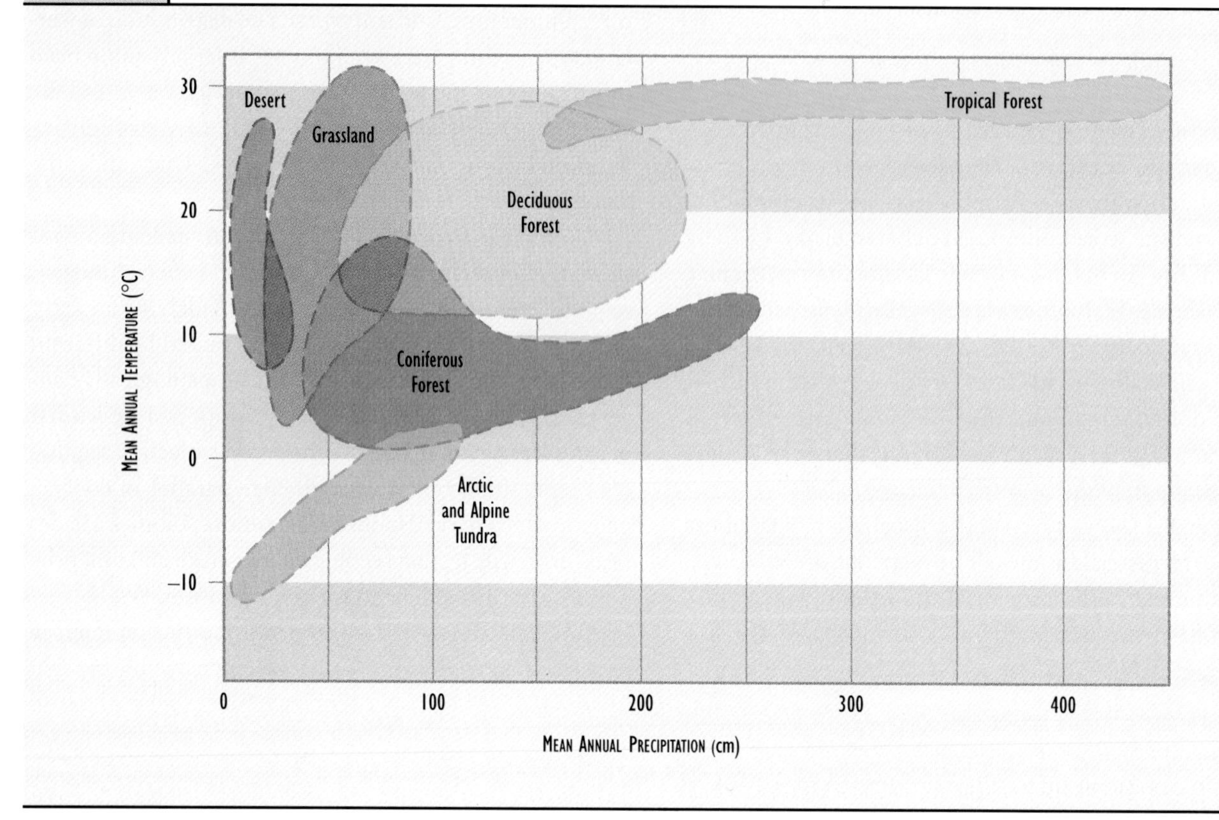

This figure suggests the reasons why temperature and moisture are believed to be the most important environmental factors affecting the distributions of terrestrial biomes.

Source: Modified from Odum (1983)

occurs in northern Canada, Alaska, and Eurasia. The boreal forest occurs at high latitudes, in regions with cold winters, short but warm summers, and moist growing conditions. The boreal forest is situated between the more northerly arctic tundra and temperate forests to the south. The typical dominant vegetation of boreal forest is coniferous trees, especially species of fir, larch, pine, or spruce. However, the particular species vary from region to region.

Over much of northern Canada, the boreal coniferous forest is dominated by stands of black spruce (*Picea mariana*). In some regions, however, white spruce (*Picea glauca*), jack pine (*Pinus banksiana*), balsam fir (*Abies balsamea*), or tamarack (*Larix laricina*) are dominant. In the boreal forest of northern Europe, Norway spruce (*Picea abies*) and Scotch pine (*Pinus sylvestris*) are the characteristic trees, while Siberia has different species of fir (*Abies sibirica*), larch (*Larix sibirica* and *L. gmelinii*), spruce (*Picea obovata*), and pine (*Pinus sibirica*). The boreal forests of northern Japan, Korea, and the Pacific coast of Russia occur in a more maritime climate, which is relatively moderate. These regions have yet other species of coniferous trees. However, all of these different forests are structurally and functionally convergent ecosystems within the same biome — the boreal coniferous forest.

We should also note that biomes are described on the basis of the dominant, most extensive kind of ecosystems that they contain. For the coniferous boreal forest, this is stands of coniferous trees. However, biomes are not homogeneous; they also contain other kinds of ecosystems. For instance, parts of the boreal-forest biome are dominated by broad-leaved (angiosperm) trees, especially species of alder, aspen, birch, poplar, and willow. Parts of the southern boreal forest, just north of the Canadian prairies, for example, contain extensive stands of trembling aspen (*Populus tremuloides*).

In addition, because the boreal forest is subject to periodic catastrophic **disturbances**, the landscape is composed of a number of stands in various stages of the post-disturbance process of ecological recovery, called **succession**. Disturbances of boreal forest are commonly caused by wildfire and sometimes by epidemics of insects, such as spruce budworm, that kill trees after several years of defoliation (see Chapter 21). The boreal forest biome also contains various kinds of wetland ecosystems, including bogs, fens, and marshes, as well as ponds, lakes, streams, and rivers.

The tundra is a biome of short vegetation growing in climatically stressed environments of the Arctic and Antarctic and on mountaintops. This is a view of arctic tundra at Alexandra Fiord on Ellesmere Island.
Source: B. Freedman

Major Biomes and Their Characteristics

Earth's natural biomes are characterized by their dominant natural ecosystems, which are themselves composed of communities of plants, animals, and microorganisms. Earth also has anthropogenic ecosystems that are strongly influenced by humans and their activities, such as cities and agricultural lands. In fact, all of Earth's modern biomes have been influenced by humans to some degree. At the very least, all organisms now contain trace contaminations of certain organochlorine chemicals that humans have manufactured and dispersed widely into the environment, such as DDT and PCBs (see Chapter 21).

Ecologists have used a number of systems to divide the biosphere into major biomes (Figure 8.1). The classification of global biomes described in this chapter is modified from a system proposed by the American ecologist, E.P. Odum. In the following sections, we describe the world's biomes within continental and global contexts. This is appropriate because biomes are widespread ecological units, and their boundaries and species do not respect political boundaries.

Terrestrial Biomes

Tundra

The **tundra** is a treeless biome that occurs in environments with long, cold winters and short, cool growing seasons. There are two broad types of tundra, alpine and arctic. Alpine tundra occurs at higher elevations on mountains, even in tropical latitudes. Arctic tundra occurs at high latitudes, that is, in the northern regions of the Northern Hemisphere and the southern parts of the Southern Hemisphere. Most tundras are meteorological deserts because they receive little precipitation (typically less than 25 cm/y). Nevertheless, tundra soil can be moist or even wet because the cold environment allows little evaporation to occur, and deep drainage of water is often prevented by frozen soil. The coldest, most northerly, high-Arctic tundras are extremely unproductive, and are dominated by long-lived small plants, generally growing less than 5–10 cm above the surface. In the less cold environments of the lower Arctic, well-drained tundras can be dominated by shrubs as tall as 1–2 m, while wetter habitats develop productive meadows of sedge, cottongrass, and grass.

Boreal Coniferous Forest

The **boreal coniferous forest**, sometimes known as taiga, is an extensive biome occurring in environments with cold winters, short but warm growing seasons, and moist soils. It is most extensive in the Northern Hemisphere. The boreal forest is dominated by coniferous trees, especially species of fir, larch, pine, and spruce. Some angiosperm trees may also be important, particularly aspen, birch, and poplar. Stands of boreal forest are poor in tree species, and may be dominated by only one or a few species. Most regions of boreal forest are subject to periodic disturbances, usually by wildfire, but sometimes by windstorms or insect epidemics.

The boreal coniferous forest (or taiga) is extensive in northern regions of Canada, Alaska, and Eurasia. This photo shows a stand of black spruce (*Picea mariana*) with a carpet of feather mosses, near Inuvik in the Northwest Territories.
Source: B. Freedman

Montane forest occurs at sub-alpine altitudes on mountains in temperate latitudes. It is similar in structure to high-latitude boreal forest and is also dominated by conifers.

Temperate Deciduous Forest

The **temperate deciduous forest** occurs in relatively moist, temperate climates with short, moderately cold winters and warm summers. This biome is dominated by a mixture of species of angiosperm tree. Most of the tree species have seasonally deciduous foliage, meaning their leaves are shed each autumn and then regrow in the springtime. This is an adaptation to surviving the drought and other stresses associated with winter. Common trees of the temperate deciduous forest in North America are species of ash, basswood, birch, cherry, chestnut, dogwood, elm, hickory, magnolia, maple, oak, sassafras, tulip-tree, and walnut. These tree species are found in distinctive communities based on their preferences for particular qualities of soil moisture and fertility, soil and air temperature, and other environmental factors.

Temperate Rainforest

Temperate rainforests develop in a climate in which winters are mild and precipitation is abundant year-round. Because this climate is too moist to allow frequent wildfires, the forest often develops into an old-growth condition. The old-growth forest is dominated by coniferous trees of mixed age and species composition. Some individual trees are extremely large and can be hundreds of years old. Common species of trees in temperate rainforests of the humid west coast of North America are Douglas fir, hemlock, red cedar, redwood, Sitka spruce, and yellow cypress.

Temperate Grassland

Temperate grassland ecosystems occur in temperate regions where the annual precipitation is 25–60 cm per year. Under these conditions, soil moisture is adequate to prevent desert from developing, but insufficient to support forest. Temperate grasslands are called prairie in North America and steppe in Eurasia, and this biome occupies vast regions in the interior of both continents. North American prairie is commonly divided into three types according to height of the dominant vegetation: tall-grass, mixed-grass, and short-grass. Tall-grass prairie is dominated by various species of grass and herbaceous angiosperm plants, such as blazing stars and sunflowers, some as tall as 2–3 metres. Fire is an important natural factor that prevents tall-grass prairie from developing into an open forest. Tall-grass prairie is a critically endangered ecosystem in North America, because almost all of it has been converted into agricultural lands. Mixed-grass prairie occurs where there is less rainfall, and is characterized by shorter species of grasses and herbaceous angiosperms. Short-grass prairie develops where precipitation is even less, and can be subject to unpredictable, severe drought.

Chaparral

Chaparral develops in south-temperate environments with a so-called Mediterranean climate, with winter rains and summer drought. Typical chaparral is characterized by dwarfed trees and shrubs with interspersed herbaceous vegetation. Periodic fires are characteristic of chaparral. In North America, chaparral is best developed in coastal southern California.

Desert

Deserts can be temperate or tropical, commonly occurring in continental interiors or in rain shadows of mountains. The distribution of deserts is determined by the amount of soil moisture, which in the temperate zones generally requires an annual precipitation of less than about 25 cm. The driest deserts support almost no plant productivity, while less-dry conditions may support communities of herbaceous and succulent plants, both annual and perennial. Occasional moist places with perennial springs of groundwater develop relatively lush vegetation of shrubs or trees and are known as desert oases.

Tropical Grassland and Savanna

Tropical grasslands and savannas occur in regions with as much as 120 cm of annual rainfall, but a pronounced dry season. Savannas are dominated by grasses and herbaceous angiosperms, with scattered shrubs and tree-sized plants that provide an open canopy. Some tropical grasslands and savannas support substantial populations of large animals, including seasonally abundant migratory mammals. This is particularly true of Africa, where this biome supports a diverse community of large mammals, including elephant, gazelle and other antelopes, hippopotamus, rhinoceros, water buffalo, and various predators of these herbivores, such as cheetah, hyena, leopard, lion, and wild dog.

Semi-Evergreen Tropical Forest

This type of tropical forest develops in a warm climate with pronounced wet and dry seasons. Most trees and shrubs of this biome are seasonally deciduous, shedding their foliage in anticipation of the drier season. This biome supports a great richness of biodiversity, though less than occurs in tropical rainforests.

Evergreen Tropical Rainforest

This biome occurs in tropical climates with copious precipitation throughout the year. Tropical rainforests commonly develop into an old-growth condition, because wildfire and other catastrophes are naturally uncommon in this wet climate. Old-growth tropical rainforests support a tremendous richness of tree species of many sizes

The temperate deciduous forest contains various species of angiosperm trees, which drop their leaves in the autumn, plus some species of coniferous trees. This forest type is widespread south of the boreal forest. This stand is dominated by sugar and red maple (*Acer saccharum* and *A. rubrum*), white and yellow birch (*Betula papyrifera* and *B. lutea*), beech (*Fagus grandifolia*), and red spruce (*Picea rubens*).

Source: Al Harvey/The Slide Farm

Temperate grasslands are widespread in the dry interior of North America and other continents, and are dominated by species of grasses and other herbaceous plants. This view is of mixed-grass prairie in Saskatchewan.

Source: D. Gautier

Desert is a very sparsely vegetated biome of extremely dry environments. This view is of a rare example of desert in Canada, located at Osoyoos Lake, South of Penticton, BC.
Source: S.B. Canning

and ages, most of which retain their foliage throughout the year. These forests also sustain an extraordinary diversity of other plants, animals, and microorganisms. Tropical rainforests represent the peak of development of terrestrial ecosystems. The biome supports huge biomass, great productivity, and enormous biodiversity under relatively benign climatic conditions.

Evergreen tropical forests occur in warm regions where rainfall is abundant throughout the year. Tropical rainforests, such as this one in Costa Rica, sustain more species than any other ecosystem.
Source: E. Greene

Freshwater Biomes

Lentic

Lentic ecosystems, such as lakes and ponds, contain standing or very slowly flowing water. The ecological character of lentic systems is greatly influenced by their water chemistry, particularly transparency and nutrient concentration. Waters well supplied with nutrients are highly productive (or *eutrophic*), while infertile waters are unproductive (or *oligotrophic*). In general, shallow waterbodies are much more productive than deeper waterbodies of a comparable surface area. However, waterbodies with poor transparency are much less productive than might be predicted on the basis of nutrient supply. Waters that are brown-coloured because of dissolved organic matter, and turbid waters with fine suspended particulates, have poor transparency.

Lentic ecosystems are often characterized by zonations in two dimensions. Horizontal zonation is due to changes in water depth, and is usually related to the slope and length of the shoreline. Vertical zonation occurs in deeper waters, and is related to the amount of light, changes in water temperature, and nutrient and oxygen concentrations. Lentic ecosystems often develop distinct communities along their shores (known as the littoral zone), in their deeper open waters (the pelagic zone), and on their sediments (the benthic zone).

Lotic

Lotic ecosystems are characterized by flowing water and include rivers, streams, and creeks. The quantity, velocity, and seasonal variations of water flow are important environmental factors in these ecosystems. Within streams or rivers, silt-sized particles are deposited in places with relatively calm water, leaving a fine-grained or muddy substrate. In contrast, the substrate of places with vigorous water flow is rocky in character, because fine particles are selectively eroded from the bottom. For similar reasons, the turbidity of the water is greatest during times of high water flows. Turbidity is an ecologically important factor because it interferes with light penetration and thereby restricts primary productivity.

Lotic ecosystems sustain some productivity of algae and aquatic plants, but their primary production is not usually large enough to support the higher trophic levels that may occur. Most of the productivity of aquatic invertebrates and fish in lotic ecosystems is typically sustained by inputs of organic matter from upstream lakes and from the terrestrial watershed in the form of plant debris.

Wetlands

Freshwater **wetlands** (also known as mires) occur in wet places on land. There are four major types of wetland: **marsh**, **swamp**, **bog**, and **fen**. Marshes are the most productive kind of wetlands, typically dominated by species of angiosperm plants that are rooted in sediment but grow as tall as several metres above the water surface, such as reed, cattail, and bulrush. Open-water areas of marshes have floating-leaved plants, such as water lily and lotus.

Swamps are forested wetlands, and can be flooded seasonally or permanently. North American swamps are often dominated by such tree species as silver maple, white elm, or bald cypress.

Bogs are acidic, relatively unproductive wetlands that develop in cool, wet climates. Their supply of nutrients is very sparse, because these ecosystems are fertilized only by atmospheric inputs in the form of dusts and chemicals dissolved in precipitation. Bogs are typically dominated by species of *Sphagnum* moss (also known as peat moss).

Fens also develop in cool and wet climates, but since they have a better nutrient supply than bogs, they are less acidic and more productive.

A swamp is a forested wetland. This swamp is located near Barrie, Ontario, in a flat area where the Nottawasaga River floods in the springtime, making it possible to canoe through the forest. The dominant trees are silver maple (*Acer saccharinum*), red ash (*Fraxinus pennsylvanica*), and elm (*Ulmus americana*).
Source: B. Freedman

Marine Biomes

The Open Ocean

The open ocean consists of pelagic and benthic ecosystems. The ecological character of the pelagic (or open-water) oceanic ecosystem is determined by physical and chemical factors, particularly waves, tides, currents, salinity, temperature, light intensity, and nutrient concentration. The rates of primary productivity of this ecosystem are small, comparable to those of deserts, the least productive of terrestrial biomes. The primary production is associated with phytoplankton, which range in size from extremely small photosynthetic bacteria to larger (but still microscopic) unicellular and colonial algae. Oceanic phytoplankton are grazed by tiny animals known as zooplankton (most of which are crustaceans), which are eaten in turn by larger zooplankton and small fish. Large predators such as bluefin tuna, sharks, squid, and whales are at the top of the pelagic food web.

The benthic ecosystems of the open-ocean biome are supported by a sparse rain of dead biomass from the surface waters. Benthic ecosystems of the deep oceans are not yet well described, but they appear to be generally rich in species, low in productivity, and extremely stable over time.

A marsh is a relatively fertile wetland dominated by taller herbaceous plants, such as bulrush (*Scirpus* spp.) and cattail (*Typha latifolia*). Marshes are often productive of animal wildlife, such as this great blue heron (*Ardea herodias*).
Source: B. Freedman

Continental-Shelf Waters

Oceanic waters near the shores of continents are relatively shallow because they overlie continental shelves, that is, underwater projections of the landmass. Compared with the open ocean, nearshore waters are relatively warm and well supplied with nutrients. The nutrients come from inputs from rivers, and from deeper, relatively fertile oceanic waters, occasionally stirred from the bottom to the surface by turbulence caused by severe windstorms. Because the nutrient supply of coastal waters is relatively high, phytoplankton are rather productive and support a higher productivity and biomass of animals than occurs in the open ocean. The world's most important oceanic fisheries are supported by the continental-shelf biome — for example those on the Grand Banks and other shallow waters of northeastern North America, on the nearshore waters of western North and South America, and in the Gulf of Mexico.

Regions with Persistent Upwellings

Oceanographic conditions in certain regions favour upwellings to the surface of relatively deep, nutrient-rich waters. The increased nutrient supply allows areas with persistent upwellings to sustain relatively high rates of primary productivity. This ecological foundation supports great populations of animals, including large fish, sharks, marine mammals, and seabirds. Some of Earth's most productive fisheries occur in upwelling areas, such as those off the west coast of South America, and extensive regions of the Antarctic Ocean.

Estuaries

Estuaries are a complex group of coastal ecosystems that are both open to the sea and semi-enclosed. Estuaries are transitional between marine and freshwater biomes, typically having large fluctuations of salinity associated with inflows of fresh water from the nearby land, twice-daily tidal cycles, and storms. Estuaries typically occur as coastal bays, river mouths, salt marshes, and tropical mangrove forests. They are highly productive ecosystems, largely because their semi-enclosed water circulation tends to retain much of the water-borne input of terrestrial nutrients. Estuaries provide critical habitats for juvenile stages of many commercially important species of fish, shellfish, and crustaceans.

Seashores

The interface of terrestrial and oceanic biomes supports a complex of ecosystems known as the seashore biome. This biome is locally influenced by physical environmental factors, especially bottom type, the intensity of wave action, and the frequency of major disturbances such as storms. Hard-rock and cobblestone bottoms in temperate regions usually develop ecosystems dominated by large species of seaweeds or kelp. These are productive ecosystems, and can maintain large quantities of algal biomass. Areas with softer bottoms of sand or mud develop ecosystems supported by the primary productivity of benthic algae and inputs of organic detritus from elsewhere. These soft-bottom ecosystems are usually dominated by benthic invertebrates, especially mollusks, echinoderms, crustaceans, and marine worms.

Coral Reefs

Coral reefs are a tropical marine biome, developing in shallow, relatively infertile places close to land. The physical structure of coral reefs is composed of the calcium carbonate shells of dead coral and mollusks. Coral reefs support a biodiverse veneer of crustose algae, living corals, other invertebrates, and fish. The biome is physically dominated by corals, cnidarian animals that live in a symbiosis with unicellular algae. Because this symbiosis is efficient in acquiring nutrients from water, coral reefs can sustain high rates of productivity even though they generally occur in infertile waters.

Human-Dominated Ecosystems

Urban-Industrial Techno-Ecosystems

This complex of anthropogenic ecosystems is characterized by urbanized regions and dominated by the dwellings, businesses, factories, and other infrastructures of human society. This biome supports many species in addition to humans, but these are mostly non-indigenous species that have been introduced from other places. Typically, the introduced species cannot live locally outside this biome (although they may occur in the foreign, natural biome to which they are indigenous).

Urban-industrial techno-ecosystems are dominated by the dwellings, businesses, factories, and other infrastructure of human society. These areas are created and maintained to support the economic activities of large numbers of people, and are sustained by enormous flows of resources from the surrounding landscape, and even from other countries. This aerial view of Halifax shows an area used entirely for roads, hospitals, homes, schools, and recreational parks.
Source: B. Freedman

Rural Techno-Ecosystems

These anthropogenic ecosystems occur outside urbanized areas, and consist of the extensive technological infrastructure of civilization. These ecosystems include rural transportation corridors (highways, railways, and electricity transmission corridors) as well as small towns that support industries involved in the extraction and processing of natural resources. Rural techno-ecosystems support a blend of introduced species, plus those native species that are tolerant of the disturbances and other stresses associated with human activities in rural areas.

Agroecosystems

An **agroecosystem** is a complex of agricultural ecosystems, and is managed to cultivate products for use by humans. The most intensively managed agroecosystems involve **monocultures** (single-species crops) of non-native species of plants or animals that are cultivated in agriculture, forestry, or aquaculture. These valuable and necessary crops are grown under conditions optimized for their productivity, although intensive management systems may cause many ecological problems. Some less intensive agroecosystems cultivate mixtures of species (or polycultures), which may also provide habitat for some indigenous species.

Ecozones of Canada

The terrestrial, freshwater, and marine ecosystems found in Canada have been described in various ways, including a hierarchical classification of distinctive types. The largest ecological zones in the national classification are referred to as **ecozones**. Ecozones are similar in many respects to biomes, particularly in that they represent extensive types of ecosystems. The classification of Canadian ecozones is, however, specifically related to the ecological conditions that occur in Canada, in terms of such biophysical factors as natural communities, species, soils, and landforms.

There are 15 terrestrial and five marine ecozones in Canada (Figure 8.3). These natural ecozones are characterized on the basis of key aspects of their physical environment, such as dominant landforms and climate, as well as their natural ecosystems and dominant species. Because so much of the Canadian landscape has been intensively modified through human activities, we will also examine three additional, anthropogenic ecosystems: urban, agricultural, and industrial.

It is important to recognize that each of the Canadian ecozones represents a hierarchical agglomeration of distinct ecosystems of more local character. *Ecoregions* are sub-ecozone units, largely characterized by distinctive regional factors related to climate and landform, and to some degree by soil, vegetation, fauna, and land use. There are 194 terrestrial ecoregions in Canada.

Ecodistricts are smaller elements within ecoregions that are themselves distinctive and characterized by more local assemblages of these same biophysical factors, particularly landform. There are 1020 terrestrial ecodistricts in Canada. Canada's marine ecozones and anthropogenic ecosystems have not yet been divided into ecoregions or ecodistricts.

For our purposes in this book, it is most useful to describe the national ecozones. These units allow us to broadly characterize the ecological qualities of Canada, and to gain an impression of how these vary over large regions. The following descriptions of Canadian ecozones are necessarily brief overviews. More detailed characterizations of the ecological zonations and communities of Canada can be obtained from some of the references listed for this chapter (especially Ecological Stratification Working Group, 1995; Scott, 1995; and Wiken *et al.*, 1996).

Terrestrial Ecozones of Canada

The Arctic Cordillera Ecozone is a tundra ecozone of the mountainous regions of northern Labrador and some of the eastern arctic islands, especially Baffin, Devon, and Ellesmere Islands. This ecozone contains the most spectacular mountains in eastern Canada, with elevations up to 2000 m, including extensive fields of glacial ice. The climate is cold and dry, with mean annual temperature ranging from –20°C to –6°C, average summer temperature from –2°C to 6°C, and precipitation from 10 to 60 cm/y. The growing season is short and cool, but is enhanced by the almost continuous daylight of summer in these high latitudes. About 75% of the terrain is covered by glacial ice or exposed bedrock, and the rest is sparsely vegetated. Permafrost, or permanently frozen soil, has a continuous distribution; that is, it is found throughout the ecozone. It occurs beneath a seasonally thawed *active layer* some 30 to 50 cm thick.

The climate at higher elevations is so extreme that the land is almost devoid of vegetation, with only a meagre cover of mosses, lichens, and a few hardy, low-growing species of vascular plants. Large mammals in this ecozone include sparse populations of caribou, muskox, and wolf, while smaller species include arctic fox, arctic hare, collared lemming, and ermine. Rock ptarmigan and raven are resident birds, while migratory species include gyrfalcon, hoary redpoll, peregrine falcon, ringed plover, and snow bunting. (The scientific names of Canadian species mentioned in this section are listed in Appendix 8A following this Chapter.)

The Northern Arctic Ecozone is a tundra ecozone in northern Quebec and the northeastern Northwest Territories, particularly on the arctic islands. The climate is cold and dry. Mean annual temperature ranges from –17°C to –11°C, average summer temperature ranges from –1.5°C to 4°C, and precipitation is only 10 to 20 cm/y. The growing season is short and cool, but is enhanced by almost continuous sunlight during the arctic summer. There is a continuous distribution of permafrost in this ecozone. Precipitation inputs are low enough for this ecozone to be characterized as a meteorological desert. However, rates of evaporation are small because of cool temperatures, and deep drainage is impeded by permafrost, so many sites are moist.

Because of the harsh climate, most of the terrain is very sparsely vegetated, and is referred to as arctic desert

FIGURE 8.3 | ECOZONES OF CANADA

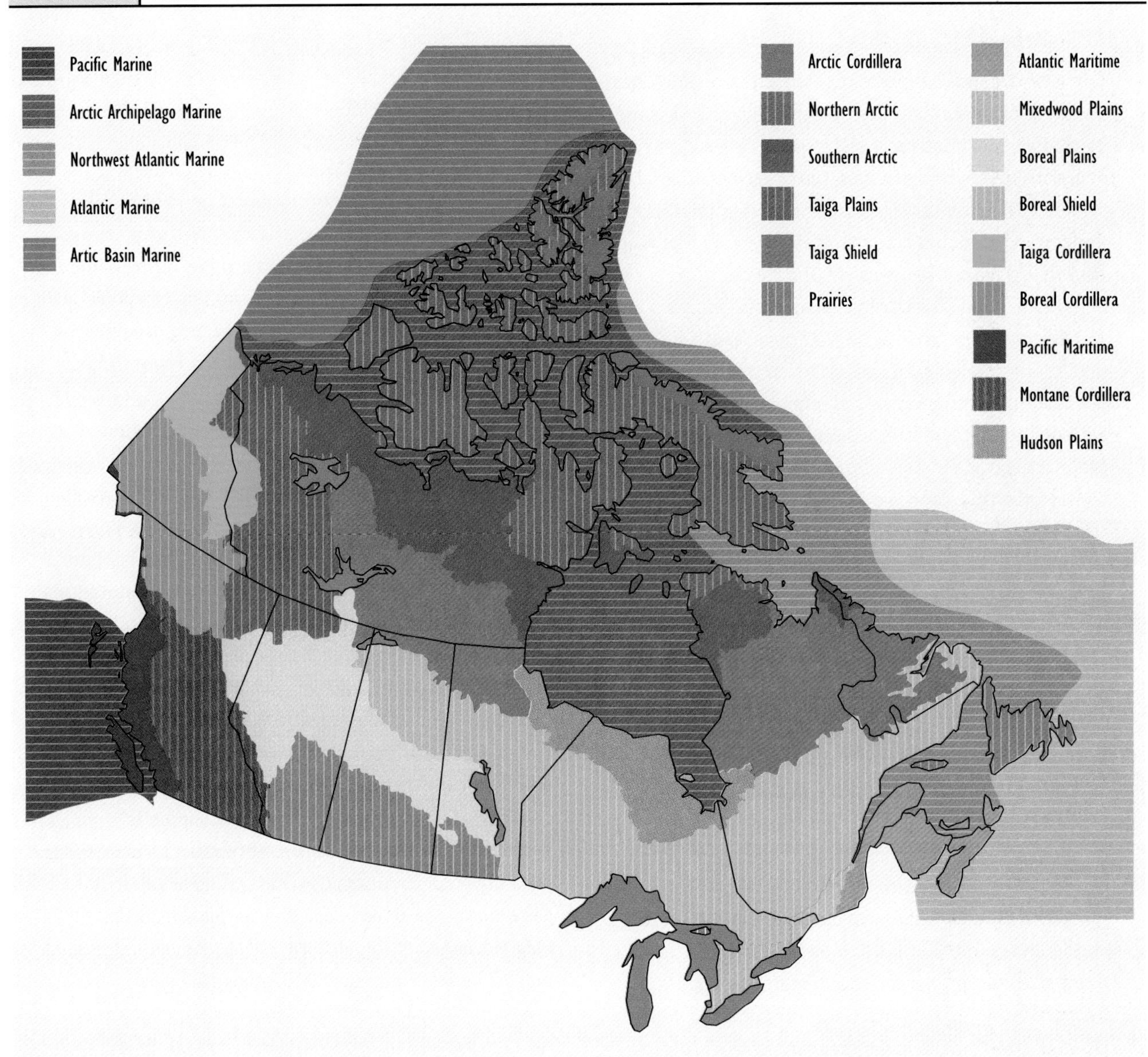

These ecological units are described by the nature of the dominant biota and by aspects of the physical environment, particularly climate, soil, geology, and other landscape-scale geographic features for terrestrial ecozones, and climate and ice cover for marine ecozones.

Source: Ecological Stratification Working Group (1995)

or semi-desert. However, protected lowlands and river valleys can be relatively warm, moist oases that sustain more productive vegetation and provide critical habitat for denser populations of animals (see Canadian Focus 8.1). These oases are uncommon, accounting for only a few percent of the land area.

The Northern Arctic Ecozone supports mammal and bird populations similar to those of the Arctic Cordillera Ecozone. Additional migratory bird species include Baird's sandpiper, Brant goose, Canada goose, greater snow goose, Lapland longspur, semipalmated sandpiper, and snowy owl. Sea ducks, such as common eider and oldsquaw, nest along the coast, but feed in nearshore environments of the Arctic Archipelago Marine Ecozone.

The Southern Arctic Ecozone is another tundra ecozone, with extensive rolling hills and lowlands. This low-arctic

Canadian Focus 8.1

Alexandra Fiord: A High Arctic Oasis

Some of the most important research in ecology has involved integrated studies of particular ecosystems, carried out by teams of ecologists, geologists, chemists, meteorologists, and other environmental scientists. From research of this kind, scientists can understand the physical, chemical, and biological factors that govern the structure and function of ecosystems and sustain their indigenous species.

Beginning in 1979, some 20 Canadian scientists undertook a team study of a particularly fine example of a high-arctic oasis, located in a coastal lowland adjacent to Alexandra Fiord on central Ellesmere Island, at about 79° N latitude. Their objectives were to describe the plant and animal communities of the oasis ecosystem and to determine the environmental factors that influence the ecological productivity and biodiversity. Specific research topics included studies of local and regional climate; geology; soils; species composition, distribution, and productivity of plant communities; the ecological relationships of prominent plant species; responses of vegetation to experimental changes in environmental conditions; and animal populations and habitat relationships.

This was a multidisciplinary research program, but because all the component studies were carried out in the same place, the results could be integrated to develop a larger picture of the structure and function of the oasis ecosystem. This kind of understanding is of great scientific importance because arctic ecosystems have not yet been well studied. This research also contributes toward the knowledge required to assess the many kinds of ecological damages that are potentially associated with increasing resource exploitation and ecotourism in the Canadian Arctic.

The team at Alexandra Fiord found that the climate of the lowland is, indeed, more moderate than that of the larger landscape. In general, air and soil temperatures are warmer, soil moisture is greater, and there is less wind. Dark-coloured cliffs on nearby uplands absorb solar radiation and then re-radiate long-wave infrared energy. This warms the oasis in a similar manner to an oven being heated by its enclosing walls. The lowland is also relatively sheltered, so energy-dispersing winds are less vigorous. In addition, snow meltwater from the surrounding uplands helps to keep local soils moist, so wet meadows and other communities that depend on adequate moisture can develop.

The moderate environmental conditions allow the lowland to support well-developed vegetation, including lush meadows dominated by sedges (*Carex membranacea* and *Carex stans*) and cottongrass (*Eriophorum angustifolium*). Vegetation in drier places is dominated by dwarf shrubs and cushion plants, which are woody, long-lived plants that grow no taller than 5 cm above the soil surface. These include arctic avens (*Dryas integrifolia*), arctic bilberry (*Vaccinium uliginosum*), arctic white heather (*Cassiope tetragona*), arctic willow (*Salix arctica*), and purple saxifrage (*Saxifraga oppositifolia*). Disturbed habitats beside rivers and streams or near human habitations (the lowland contains an abandoned Royal Canadian Mounted Police post) support profuse flowerings of herbaceous plants, such as arctic poppy (*Papaver lapponicum*) and willow-herb (*Epilobium latifolium*).

These plant communities are much more productive than those on the prevailing polar desert nearby, and consequently the oasis supports relatively large populations of breeding animals. Birds are especially abundant, including snow bunting (*Plectrophenax nivalis*), Baird's sandpiper (*Calidris bairdii*), hoary redpoll (*Carduelis hornemanni*), arctic tern (*Sterna paradisaea*), oldsquaw (*Clangula hyemalis*), greater snow goose (*Chen caerulescens atlantica*), rock ptarmigan (*Lagopus mutus*), parasitic jaeger (*Stercorarius parasiticus*), and another 19 species. Studies of the arctic skipper butterfly (*Gynaephora groenlandica*) discovered that its slow-growing larvae take 14 years to accumulate enough energy to undergo metamorphosis to the adult stage, resulting in a remarkably long life cycle.

Because of its relatively small area (only 8 km^2), this lowland oasis is not able to support a population of muskox (*Ovibos moschatus*), the most important large herbivore in the larger biome. However, small numbers of this impressive animal occasionally feed in the oasis while passing through on their way to larger oases nearby.

tundra is characterized by relatively taller, somewhat lusher vegetation than the two higher-arctic ecozones. This ecozone occurs in northern Quebec and across the northern part of the continental Northwest Territories. The climate is cold but more moist than in higher-arctic ecozones. Mean annual temperature ranges from –11°C to –7°C, average summer temperature from 4°C to 6°C, and precipitation from 20 to 40 cm/y. The growing season is short and somewhat warm, enhanced by long days. Permafrost is extensive and continuous, except beneath large lakes and rivers. Glacial deposits are common and highly evident, including long, sinuous mounds of mixed debris known as eskers. Ice in the soil (ground ice) has caused unusual surface features to develop. These include patterns on the ground created by underground ice wedges, and unusual hills, with a core of ice, up to 40–50 m tall. Known as pingos, they are common near Tuktoyaktuk, east of the Mackenzie Delta. Lakes, ponds, rivers, streams, and wetlands are common over most of this ecozone.

Compared with the higher Arctic, this ecozone is well vegetated. Shrubs can be as tall as several metres in protected sites, but are usually less than 1 m tall. Prominent shrubs include species of alder, birch, and willow. Drier, more exposed sites have shorter shrubs and herbaceous plants, and wet sites have meadows of sedges, cottongrass, and grasses. A few protected places support scattered, stunted trees of white spruce or balsam poplar.

The Southern Arctic Ecozone supports low densities of large mammals. However, large herds of caribou, which spend most of the year in or near the southern tree-line, migrate to low-arctic tundra to calve and feed during the short growing season. Other large mammals include grizzly bear, moose, muskox, and wolf. Smaller mammals include arctic ground squirrel, brown lemming, muskrat, red fox, snowshoe hare, and weasel. Willow ptarmigan and raven are resident birds, and migratory species include arctic and red-throated loon, Canada goose, common redpoll, golden eagle, gyrfalcon, Lapland longspur, lesser snow goose, sandhill crane, snow bunting, snowy owl, tundra swan, and various species of duck, such as greater and lesser scaup and oldsquaw.

The Taiga Plains Ecozone is a northern ecozone, characterized by rolling plains and uplands with a taiga (boreal) forest, mostly of relatively open stands of short trees. The Taiga Plains Ecozone is distributed in the western Northwest Territories, northern British Columbia, and northwestern Alberta, and essentially encompasses the taiga regions of the drainage of the MacKenzie River, Canada's largest watercourse. The overall climate is cool and winters are very cold. Mean annual temperature ranges from –10°C to –1°C, average summer temperature from 7°C to 14°C, and precipitation from 20 to 50 cm/y. The growing season is moderately long and warm, and is enhanced by long days. Permafrost is extensive but discontinuous. Although there is not much precipitation, soil moisture is generally adequate because evapotranspiration rates are low and permafrost keeps water from draining away in many places. Deposits of glacial debris are common and widespread. Over most of this ecozone, surface waters and wetlands are common features.

Pingos are ice-cored hills that slowly rise out of partially drained lakes in certain regions of permafrost. These unusual features are common near Tuktoyaktuk in the Southern Arctic Ecozone, where this 60-m-high pingo is located.
Source: B. Freedman

The dominant vegetation is an open, slow-growing forest of black spruce. Some upland areas develop forests of lodgepole pine, white spruce, white birch, trembling aspen, or balsam poplar, and large white spruce and balsam poplar can occur along rivers. These open forests have well-developed shrub components, including various species of willow, birch, and heaths such as Labrador tea and bearberry. Mosses and lichens are prominent in the ground vegetation, as are various species of herbaceous plants. The forest is periodically subject to wildfire, and after a fire, spectacular stands of purple-coloured fireweed appear during the initial stages of secondary succession.

Large mammals of this ecozone include black bear, moose, woodland caribou, wood bison, and wolf. Smaller mammals include lynx, pine marten, muskrat, snowshoe hare, and species of lemming and vole. Bird species include common redpoll, fox sparrow, gray jay, northern shrike, peregrine falcon, raven, red-throated loon, and sharp-tailed grouse. The numerous ponds and lakes support breeding waterfowl such as greater and lesser scaup, pintail, and Canada goose, as well as shorebirds such as greater and lesser yellowlegs and snipe. The Mackenzie corridor is an important migratory route for great numbers of waterfowl and shorebirds that breed in the tundra to the north.

The Taiga Shield Ecozone, a forested ecozone, overlies the hard, quartzitic rocks and thin soils of the Canadian Shield of central Quebec, most of Labrador, southeastern Northwest Territories, and northern Saskatchewan and Manitoba. The taiga forest is composed of open stands of relatively short, unproductive trees. The general climate is cool and winters are extremely cold. Mean annual temperature ranges from –8°C to 0°C, average summer temperature from 6°C to 11°C, and precipitation is 20 to 50 cm/y, although parts of the Labrador coast are wetter. The growing season is of moderate length and warmth, and is enhanced by long days. Permafrost is widespread but discontinuous. Deposits of glacial debris are common and widespread, including frequent eskers, and surface waters and wetlands are abundant.

The Taiga Shield Ecozone is an extensive area of open boreal forest, rocky outcroppings, and wetlands. This is a view of an open forest of black spruce (*Picea mariana*) near Nain in Labrador.
Source: B. Freedman

The dominant vegetation type is an open forest of black spruce, but in some areas, forests are dominated by balsam fir, balsam poplar, jack pine, tamarack, trembling aspen, white birch, and white spruce. Non-forest communities are also common, including tundra-like vegetation on open sites with shallow or no soil over bedrock, and many wetlands and open-water ecosystems.

Black and grizzly bear, caribou, moose, and wolf are prominent large mammals, while smaller species include arctic and red fox, beaver, lynx, and snowshoe hare. Some of the breeding birds include arctic and red-throated loons, gray-cheeked thrush, tree sparrow, white-crowned sparrow, willow ptarmigan, and yellow-rumped warbler.

The Boreal Shield Ecozone, another forested ecozone, largely occupies the quartzitic rocks and thin soils of more southern reaches of the Canadian Shield, including northern Saskatchewan, central Manitoba, much of northern Ontario, most of southern Quebec, and Newfoundland. The general terrain is a rolling mosaic of upland forests and lowland wetlands and surface waters. Most of this ecozone has a continental climate (that is, not influenced by proximity to an ocean), with long cold winters and warm summers. These conditions are moderated in areas closer to the Atlantic Ocean, where winters are less cold and summers less warm. Mean annual temperature ranges from –4°C in continental areas to 5.5°C in maritime Newfoundland, average summer temperature from 11°C to 15°C, and precipitation from 10 to 40 cm/y in continental places and from 90 to 160 cm/y in maritime areas. Deposits of glacial debris are widespread, and surface waters and wetlands are common features.

The ecozone is dominated by closed forests composed of various species of conifers, especially balsam fir, black and white spruce, and tamarack. More southern parts of the ecozone also have stands of jack, red, and white pine, and angiosperm species can be locally abundant, especially balsam poplar, trembling aspen, and white birch. Areas with shallow soil have more open communities, and bogs and fens are abundant in poorly drained places.

Prominent large mammals include black bear, caribou, moose, white-tailed deer, and wolf, while smaller species include bobcat, fisher, lynx, pine marten, raccoon, red squirrel, striped skunk, and eastern chipmunk. Some of the breeding birds of this ecozone include blue and grey jay, boreal owl, common loon, evening grosbeak, great horned owl, red-eyed vireo, white-throated sparrow, and numerous species of warblers.

The Atlantic Maritime Ecozone, a forested ecozone, includes New Brunswick, Nova Scotia, Prince Edward Island, and adjacent parts of the Gaspé and southeastern Quebec. Bedrock and soils in this region are complex, and include quartzitic rocks such as granite, gneiss, and greywacke, along with sedimentary limestone, sandstone, and shale. The terrain is rolling and extensively forested, with abundant wetlands and surface waters. The climate ranges from continental to maritime, with cold winters and warm summers, being more moderate closer to the Atlantic Ocean. Mean annual temperature ranges from 4°C to 7°C, average summer temperature from 13°C to 16°C, and precipitation from 90 to 150 cm/y. Glacial debris is widespread.

The forest is a relatively productive mosaic of mixed-species communities, with prominent trees being eastern hemlock, red spruce, red and sugar maple, red and white pine, and white and yellow birch. Boreal species are also prominent, such as black and white spruce, balsam fir, and jack pine. Understorey shrubs include alder, blueberry, cherry, spiraea, and viburnum.

Large mammals include black bear, coyote, moose, and white-tailed deer, while smaller species include bobcat, eastern chipmunk, mink, porcupine, raccoon, red squirrel, striped skunk, and various mice and voles. Breeding birds include barred owl, black duck, black-capped chickadee, broad-winged hawk, common crow, great horned owl, rose-breasted grosbeak, sharp-shinned hawk, and diverse warblers, vireos, and sparrows.

The Mixedwood Plains Ecozone, also forested, covers relatively southern regions of the Great Lakes-St. Lawrence River valley, and includes the most southern regions of Ontario and Quebec. Bedrock in this region is mostly limestone, and soils are relatively deep and rich. The terrain is gently rolling and was originally forested, although most of the natural vegetation has been cleared and the land converted into agricultural and urban uses (about half of Canada's population lives in this rather small ecozone). Climate is continental, with cold winters and warm summers, but more moderate conditions occur closer to the Great Lakes and St. Lawrence River. Mean annual temperature ranges from 5°C to 8°C, average summer temperature from 16°C to 18°C, and precipitation from 70 to 100 cm/y. Glacial debris is widespread in some areas, but the relatively flat, former beds of the post-glacial Great Lakes and St. Lawrence River have deep, fine-textured, fertile soils. Ponds, small lakes, and wetlands are less common than in more northern ecozones.

The natural forests of the Mixedwood Plains Ecozone are much richer in species than anywhere else in Canada. Forest cover is, however, now reduced to less than 10% of the area, and some communities and many species are endangered or extirpated. Forest stands vary greatly in species composition. Some stands are dominated by red or white pine, others by rich mixtures of basswood, beech, eastern hemlock, sugar and red maple, white elm, and yellow birch. Remnant stands of Carolinian forest in southern Ontario have rare species of trees that are more typical of the eastern United States, such as black walnut, butternut, cucumber-tree, sassafras, sycamore, tulip-tree, and various species of southern ash and hickory.

Black bear and white-tailed deer are the most prominent large mammals. Cougar used to occur, but this predator has been extirpated, and wolf has also been eliminated from most of the ecozone. Coyotes have invaded the region and are now abundant. Smaller mammals include cottontail rabbit, grey (or black) squirrel, groundhog, raccoon, and striped skunk. Many bird species breed in the ecozone, including blue jay, cardinal, red-shouldered hawk, and northern oriole. Some relatively southern birds also breed in southern parts of this ecozone, including bob-white, Carolina wren, green-backed heron, orchard oriole, prothonotary warbler, and scarlet tanager.

The Boreal Plains Ecozone occurs as a broad forested band extending from northeastern British Columbia, across northern Alberta, to central Saskatchewan and Manitoba. This ecozone is characterized by rolling terrain derived from extensive moraine deposits and by flatter landscapes with deep soils derived from post-glacial lake sediments. There are few exposures of bedrock. Climate is continental, with cold winters and warm summers. Mean annual temperature ranges from –2°C to 2°C, the average summer temperature from 13°C to 16°C, and precipitation from 30 to 63 cm/y. Lakes are less abundant than on the Canadian Shield, but wetlands are extensive, covering more than 25% of most landscapes.

Conifer-dominated stands in this ecozone contain black and white spruce, jack pine, and tamarack. Angiosperm-dominated stands are more common in southern areas, closer to the forest-prairie border, and include balsam poplar, trembling aspen, and white birch.

Prominent large mammals include black bear, caribou, elk, moose, white-tailed and mule deer, coyote, and wolf. Smaller mammals include fisher, lynx, pine marten, striped skunk, and eastern chipmunk. Representative breeding birds include blue jay, boreal owl, evening grosbeak, great horned owl, and white-crowned sparrow. Lakes and wetlands support Caspian tern, double-crested cormorant, great blue heron, white pelican, and many species of waterfowl. The endangered whooping crane breeds in extensive wetlands in Wood Buffalo National Park.

The Prairies Ecozone is an open, mostly non-forested ecozone which occurs in southern regions of Alberta, Saskatchewan, and Manitoba. This ecozone is characterized by rolling terrain derived from extensive glacial moraine and by flatter landscapes with deep, fertile soils derived from post-glacial lake sediments. The climate is continental, with cold winters and hot summers. Mean annual temperature ranges from 2°C to 4°C, average summer temperature from 14°C to 16°C, and precipitation from 25 to 70 cm/y. Soil moisture is often limiting to plant growth during the summer. This occurs because of the relatively sparse precipitation, coupled with hot, windy summers that increase evapotranspiration. Small lakes and ponds with fringing wetlands, known as potholes and sloughs, are a common feature, particularly in relatively rainy years.

The Prairies Ecozone is divided into categories based on the height of the dominant species. The richest, most productive, and most endangered type is tall-grass prairie. This ecosystem can have herbaceous plants as tall as several metres, such as this perennial sunflower (*Helianthus giganteus*) in a tall-grass prairie near Windsor, Ontario.
Source: B. Freedman

Grass and forb-dominated **prairie** is extensively developed throughout this ecozone (forb is a general term for broad-leaved herbaceous plants other than grasses). There are two types of prairie, identified by vegetation height and species composition. Tall-grass prairie has grasses 1–2 m tall such as big bluestem, plus many tall species of forbs, such as sunflowers. Mixed-grass prairie has a mixture of medium and short grasses and forbs. These natural prairie types have been largely converted into agricultural land uses. Tall-grass prairie is now an endangered natural ecosystem, as are the other kinds of natural prairie but to a lesser degree.

Northern parts of this ecozone support an open forest known as aspen parkland, dominated by groves of balsam poplar and trembling aspen, with intermittent, prairie-like glades. There are also some small areas of desert-like habitat in the southernmost parts of this ecozone, characterized by prickly pear cactus and other plants typical of dry habitats.

Large mammals of this ecozone include coyote, elk, white-tailed and mule deer, and pronghorn antelope, while smaller mammals include badger, black-tailed prairie dog, northern pocket gopher, Richardson's ground squirrel, and white-tailed jack rabbit. Buffalo were once important in this ecozone, but these large animals were extirpated from almost all of their natural range during the nineteenth century. Bird species include black-billed magpie, burrowing owl, ferruginous hawk, horned lark, northern oriole, prairie falcon, and Swainson's hawk. Large numbers of waterfowl and shorebirds breed in potholes and sloughs during wetter years, including American avocet, canvasback, coot, mallard, lesser scaup, pintail, and western grebe. Because so much of this ecozone has been converted into agriculture, the remnants of natural habitat tend to occur as small, isolated fragments. Consequently, some of the indigenous species of plants and animals are endangered.

The Taiga Cordillera Ecozone is a subarctic, largely forested ecozone occurring in northern Yukon and parts of the western Northwest Territories. The region covers the northernmost extension of the Rocky Mountains, and contains steep terrain with deep valleys and wild rivers as well as more gently sloping foothills. The climate is continental arctic, with long, cold, dark winters, and short, cool summers with long days. However, as in mountainous terrain everywhere, altitude, slope, and aspect are important influences on local climate. Mean annual temperature ranges from –10°C to –5°C, average summer temperature from 7°C to 10°C, and precipitation from 30 to 70 cm/y. Surface waters are mostly rivers and streams, with some mountain lakes. Ponds and lakes are more abundant on the flatlands around the Old Crow River and northern coastal plain. Most of the ecozone has continuous permafrost.

Vegetation of this complex ecozone ranges from alpine tundra at high elevation and arctic tundra on the coastal plain of Yukon, to open taiga at lower elevations in the south. The forest is dominated by white spruce and white birch, with a shrub understorey of dwarf birches and willows.

Large mammals include black and grizzly bear, caribou, Dall's sheep, moose, mountain goat, and wolf, while smaller mammals include arctic ground squirrel, hoary marmot, lynx, pika, and wolverine. Breeding birds include gyrfalcon, peregrine falcon, rock and willow ptarmigan, sandhill crane, white-crowned sparrow, yellow warbler, and many species of waterfowl and shorebirds. The Old Crow Flats are an important staging area for Canada goose, tundra swan, and other species of waterfowl during their southern migration.

The Boreal Cordillera Ecozone covers rugged mountainous terrain of northern British Columbia and southern Yukon. The region is characterized by steep terrain with deep, wide valleys, high-altitude plateaus, and gently sloping foothills and lowlands. The climate is continental, with long, cold winters and warm summers, although altitude and aspect greatly affect local climate. Mean annual temperature ranges from 1°C to 6°C, average summer temperature from 10°C to 12°C, and precipitation from 150 cm/y in high-elevation areas receiving orographic precipitation (see In Detail 3.1, page 34), to less than 30 cm/y in rain-shadow areas. Surface waters are largely streams and rivers, with some mountain lakes. Discontinuous permafrost is widespread in the northern part of the ecozone.

Southern areas of the ecozone have grasslands and open forests on the relatively warm and dry south-facing slopes, and boreal forest on the northerly exposures. Dominant tree species are balsam poplar, black and white spruce, trembling aspen, and white birch, joined by alpine fir and lodgepole pine in southern areas. Alpine tundra occurs at higher elevations.

Large mammals include black and grizzly bear, caribou, Dall's sheep, moose, mountain goat, and wolf, while smaller mammals include arctic ground squirrel, hoary marmot, lynx, marten, and pika. Breeding birds include peregrine falcon, rock and white-tailed ptarmigan, white-crowned sparrow, and many other migratory species.

The Pacific Maritime Ecozone includes mainly rugged mountainous terrain of the coastal mainland and islands of British Columbia, where the climate is greatly moderated by the Pacific Ocean. The climate is humid temperate, with short, cool winters and long, warm summers, although altitude and aspect have a great influence on local climate. Mean annual temperature ranges from 5°C to 9°C and average summer temperature from 10°C to 16°C. Precipitation ranges from 60 cm/y in some of the relatively dry Gulf Islands to as much as 400 cm/y in places strongly influenced by orographic precipitation. More typically, however, precipitation ranges from 150 to 300 cm/y. Surface waters are mostly streams and rivers, with some lakes.

The dominant natural vegetation is a mixed-species, conifer-dominated, temperate rainforest. Because the humid climate makes disturbance by wildfire an uncommon occurrence, old-growth forests were naturally extensive. These natural old-growth forests have, however, largely been replaced by younger, more productive, second-growth forests through logging and silviculture, and to some degree they have also been converted into agricultural and urban land uses. Prominent species of trees include amabalis and grand firs, coastal Douglas fir, mountain and western hemlock, red alder, Sitka spruce, western red cedar, and yellow cedar. Individual trees of some of these species can attain huge size and great age (sometimes older than five hundred years) in old-growth forests. Higher-altitude sites can develop boreal-like montane forest or alpine tundra. Because the Queen Charlotte Islands and parts of Vancouver Island were not glaciated, they have some endemic subspecies and species that occur nowhere else.

Prominent large mammals in this ecozone include black and grizzly bear, black-tailed deer, elk, and wolf, while smaller species include river otter and raccoon.

Although much less extensive than formerly, old-growth rainforests are characteristic of the Pacific Maritime Ecozone. Old-growth forests have a complex structure, with large, old trees of various species, as well as standing dead trees (or snags) and large logs lying on the ground. This is a view within an old-growth forest near Maugher's Bay on Vancouver Island.
Source: B. Freedman

Representative species of birds include bald eagle, black oystercatcher, blue grouse, California and mountain quail, chestnut-backed chickadee, dipper, northwestern crow, Steller's jay, and many migratory species.

The Montane Cordillera Ecozone is a largely mountainous ecozone that covers most of southern British Columbia and adjacent southwestern Alberta. The climate is temperate, with cold winters and warm summers. Altitude and aspect greatly influence local climate. Mean annual temperature ranges from 1°C to 8°C and average summer temperature from 11°C to 17°C. Precipitation is highly variable, ranging from only 30 cm/y in drier valleys and plateaus in the rain shadow of coastal mountains to 120 cm/y at higher elevations along the Alberta-British Columbia border. Surface waters are mostly streams and rivers, with some lakes.

The ecozone is largely forested, but community types are highly diverse, reflecting the heterogenous nature of the terrain and environmental conditions. Alpine tundra occurs at higher altitude with montane forest below, typically dominated by alpine fir, Engelmann spruce, and lodgepole pine. Forests at lower elevation often include these species along with interior Douglas fir, ponderosa pine, trembling aspen, western hemlock, western red cedar, and western white pine. Some drier sites support semi-desert, with shrubs such as antelope bush, rabbit bush, and sagebrush, and various species of grasses and forbs.

Prominent mammals of this ecozone include bighorn sheep, black and grizzly bear, coyote, woodland caribou, elk, moose, mountain goat, and mule and white-tailed deer, while smaller mammals include Columbian ground squirrel and hoary marmot. Prominent bird species include black-billed magpie, blue grouse, Clark's nutcracker, dipper, golden eagle, and Steller's jay.

The Hudson Plains Ecozone covers the broad lowlands of previously more extensive, post-glacial Hudson and James Bays. It extends from northeastern Manitoba through northern Ontario to adjacent Quebec. Because of the influence of the cold waters of Hudson and James Bays, the climate is more subarctic than might be expected on the basis of latitude, with long, cold winters, and short, warm summers. Mean annual temperature ranges from –4°C to –2°C, average summer temperature from 11°C to 12°C, and precipitation from 40 to 80 cm/y. Permafrost has a discontinuous distribution. Surface waters are abundant, with numerous streams, rivers, ponds, small lakes, and wetlands.

Tidal habitats can develop into extensive *salt marshes* of salt-tolerant grasses and forbs. Shrub-dominated, low-arctic tundra vegetation is extensive closer to the coasts of Hudson and James Bays. This tundra changes gradually to an open-canopied, short-treed taiga further inland, and then to a more closed-canopied boreal forest. The dominant tree species are black spruce and tamarack. Much of the terrain is characterized by series of old raised-beach ridges, representing former shorelines of Hudson and James Bays. The relatively dry tops of these ridges are generally forested, while the lower areas develop into wetlands such as fens, bogs, and marshes.

Species of large mammals include black bear, caribou, moose, white-tailed deer, and wolf, while smaller species include beaver, muskrat, river otter, and porcupine. Shorebirds and waterfowl, including Canada goose and snow goose, breed in large numbers in this ecozone.

Marine Ecozones of Canada

Canada's marine ecozones have not yet been fully designated. The descriptions that follow are somewhat tentative, but they illustrate the essential character of these distinctive ecological regions.

The Pacific Ecozone occurs in Pacific waters off the mainland and islands of western British Columbia. The climate is temperate, with long, warm summers and short, cool winters. There is no sea ice.

Various species of Pacific salmon, such as chum, coho, and sockeye, are important in this marine ecozone. These fish spend most of their adult lives at sea, but migrate up accessible rivers to spawn. Individual adult salmon return to their birth rivers for breeding. Pacific herring is another abundant species of fish. Prominent marine mammals in the Pacific Ecozone include grey whale, harbour seal, killer whale, northern fur seal, northern sea lion, sea otter, and sperm whale. Prominent seabirds include double-crested and pelagic cormorant, glaucous gull, marbled murrelet, pigeon guillemot, and tufted puffin. Many species of waterfowl winter in this ecozone.

The Arctic Archipelago Ecozone occurs in the waters between the arctic islands, the Beaufort Sea, and Hudson and James Bay. The climate is marked by long, cold winters and short, cool summers. The ice-free season is as long as several months, with the ocean surface covered with metres-thick ice for the rest of the year. In a few places, underwater seamounts force currents to the surface, creating ice-free areas known as *polynya*, which are a critical habitat for marine mammals and birds. The polynya are marine counterparts of the oases of the Arctic Cordillera and Northern Arctic Ecozones.

Arctic cod and char, a migratory species of salmonid fish, are abundant. Marine mammals in this ecozone include bearded seal, beluga, bowhead whale, narwhal, ringed seal, and walrus. Although polar bear usually have their young on land or den there when the sea ice has melted,

The Pacific Ecozone is rich in marine life. Pictured here is a "forest" of large seaweeds known as kelps (*Nereocystis* sp.), which provide critical elements of habitat for many species, such as black rockfish (*Sebastes melanops*).
Source: C. Harvey-Clark

they are essentially marine animals, hunting seals as their major food. On certain islands (such as King Christian Island in Lancaster Sound), thick-billed murres have huge colonies which can contain millions of breeding birds. Other important seabirds include arctic tern, black guillemot, dovekie, fulmar, and glaucous gull. The most abundant sea ducks are common eider and oldsquaw.

The Arctic Basin Ecozone is a northern, deep-water marine ecozone that occurs north and west of the arctic islands of Canada. The climate is characterized by long, cold winters and short, cool summers. Ice cover is continuous in the more northern reaches of this ecozone, and almost continuous further south. Some regions break up into discontinuous pack ice which shifts about in a huge, counterclockwise circle because of a current known as the arctic gyre, which is centred roughly on the North Pole.

Sparse populations of fish and marine mammals occur in this ecozone, living in regions of pack ice. Species of mammals include bearded and ringed seals, beluga, bowhead whale, narwhal, polar bear, and walrus. The ivory gull is a characteristic seabird.

The Northwest Atlantic Ecozone includes Atlantic waters of the continental shelf off eastern Baffin Island, northern Quebec, Labrador, Newfoundland, and the Gulf of St. Lawrence. The southeast-flowing Labrador Current is a major water flow and the Gulf of St. Lawrence is strongly influenced by the flow of the St. Lawrence River. The climate ranges from subarctic to boreal, with cold winters and warm summers. Sea ice forms extensively, but breaks up and melts completely in the springtime.

Populations of fish are abundant and widespread, including capelin, cod, herring, and redfish. Marine mammals in this ecozone include gray, harbour, harp, and hooded seals; harbour porpoise; white-sided dolphin; and beluga, blue, fin, humpback, minke, pilot, and sperm whales. Seabirds include black guillemot, common murre, common and arctic tern, double-crested and great cormorant, fulmar, gannet, herring and black-backed gull, and razorbill. The most abundant breeding sea duck is the common eider, but other species of ducks and loons winter in this ecozone.

The Atlantic Marine Ecozone extends offshore over Atlantic waters of the continental shelf off eastern Newfoundland and Nova Scotia. Some extensive shallower-water areas (less than about 150 m) are known as banks, but the ecozone also contains deep-water habitats. The climate is generally temperate, partly because of the influence of the Gulf Stream, which flows northward from the Caribbean Sea. Sea ice does not form, but icebergs drift south with the Labrador Current, and pack ice is imported from the Gulf of St. Lawrence.

Populations of fish are abundant and widespread on the offshore banks, including cod, haddock, redfish, silver hake, and turbot. Marine mammals in this ecozone include gray and harbour seals and blue, bottlenosed, fin, humpback, minke, pothead, right, and sperm whales. Seabirds include black guillemot, common and arctic tern, common eider, common murre, common puffin, double-crested cormorant, fulmar, gannet, great cormorant, herring and black-backed gull, kittiwake, razorbill, and Wilson's storm petrel.

Canadian Anthropogenic Habitats

Anthropogenic ecosystems, such as urban, agricultural, and industrial habitats, are not natural. Rather, they are dominated by humans and include their constructed and disturbed habitats, and associated species of plants and animals.

Urban Habitats

Urban habitats are characterized by the places where people live in metropolitan areas, cities, towns, and villages. Most of the natural ecozones have elements of urban habitats embedded in them, although to widely varying degrees. Urban habitats are developed most extensively in southern parts of Canada, particularly within several hundred kilometres of the border with the United States.

The vegetation of urban habitats is extremely complex. Mature residential city neighbourhoods commonly have an urban forest, but in general the most abundant trees are non-native species such as horse chestnut, linden, and Norway maple. These trees, and most of the shrubs and herbaceous plants cultivated in the urban ecosystem, were introduced to Canada from Europe, Asia, and elsewhere. Some native trees are also planted in urban habitats—for example, American elm, Manitoba maple, and silver maple—but these are usually less common than introduced species. Some additional native plants may survive in younger neighbourhoods that have been recently converted from natural habitats.

Along with humans, the most abundant mammals in urban habitats are domestic cat and dog, house mouse, and Norway rat, none of which are indigenous to Canada. Some native mammals, such as coyote, raccoon, and striped skunk, also live successfully in urban habitats. The most characteristic and abundant species of birds are house sparrow, rock dove (or pigeon), and starling, all of which were introduced from Eurasia. Some native species also occur in urban habitats, including American crow, American robin, cardinal, black-capped chickadee, blue jay, and song sparrow.

Agricultural Habitats

Agricultural habitats occur in the agroecosystems (i.e., agricultural ecosystems) of Canada, where plants and animals are grown for food, fibre, or energy for human consumption. These habitats were created through extensive conversions of natural ecosystems into agroecosystems. Agricultural habitats are most extensive in the Prairie Provinces, but they occur wherever soils and climate are suitable for the growth of crops, including much of southern Ontario and Quebec, the Maritime Provinces, and southern British Columbia.

Agroecosystems in which food plants are grown are mostly dominated by species of annual crops, such as barley, corn (or maize), potato, soybean, tomato, and wheat, or by perennial fruit-bearing crops such as apple, blueberry, cherry, and peach. Except for blueberry, all of the above species were introduced to Canada as non-indigenous crop plants. Often, these crops are managed as *monocultures*, containing as few other species as possible that could compete with the crop species.

Agroecosystems in which livestock are raised are either pastures dominated by introduced species of grasses, or semi-natural grasslands containing species adapted to being grazed by large animals. Grasslands are also harvested for their hay crop, which is supplemented by grains and other foods and fed to agricultural livestock. Increasingly, domestic livestock are cultivated indoors, in crowded, intensively managed, factory farms (see also Chapter 14).

The most common large animals in agricultural habitats are cow, goat, horse, pig, and sheep, while smaller animals include chicken, domestic rabbit, and turkey. All of these animals were introduced to Canada as agricultural livestock (except perhaps for the turkey, which is native to North America). Many indigenous animals also use agricultural habitats, particularly if the land is neither managed very intensively nor sprayed with poisonous insecticides and other pesticides. The most intensively managed agroecosystems support few native species.

Intensively managed forestry plantations can also be considered agricultural habitats. Such plantations may be managed as virtual monocultures in which trees are grown as a long-lived crop. Usually only one species of tree is planted, commonly in straight, evenly spaced rows, which optimizes growth rates and makes silvicultural management easier. Sometimes non-native species of trees are planted, such as hybrid poplars, Norway spruce, Scotch pine, or Eurasian larch. More commonly, however, native trees are planted, such as Douglas fir or species of pine or spruce. Some indigenous species of forest plants and animals can utilize these plantations as habitat, but many cannot.

Most urbanized habitats in Canada are extremely unnatural and support few indigenous species of plants and animals. However, a few urban habitats are used by native species, such as this pond in a Toronto park where people come to feed and view "wild" ducks and geese during the winter.

Source: B. Freedman

Industrial Habitats

Industrial habitats are associated with the technological infrastructure of Canadian civilization, and occur extensively in many rural landscapes. Transportation corridors such as highways and railroads are important industrial habitats, as are electricity transmission corridors. Also prominent are mine sites and other areas affected by the extraction, processing, and manufacturing of products from natural resources, such as metals, coal, petroleum, and natural gas.

Vegetation in this biome is typically highly disturbed, and may be actively managed to keep it in an early stage of ecological succession, usually by mowing or herbicide spraying. However, vegetation in industrial habitats is usually dominated by indigenous species, and useful habitat is available for many native species of animals. Like the plant species, these are mostly animal species of the early stages of succession.

Key Terms

biome
life form
disturbance
succession
tundra
boreal coniferous forest
montane forest
temperate deciduous forest
temperate rainforest
temperate grassland
chaparral
desert
lentic ecosystem
lotic ecosystem
wetland
marsh
swamp
bog
fen
estuary
agroecosystem
monoculture
ecozone
prairie

Questions for Discussion

1. Make a list of Earth's major biomes and describe the characteristics of each.
2. Identify the major ecozones occurring in the province or territory where you live and describe the characteristics and typical species of each.
3. Characterize the species occurring in urban, agricultural, and industrial habitats in the province or territory where you live.

References

Barbour, M.G. and W.D. Billings. 1988. *North American Terrestrial Vegetation.* New York: Cambridge University Press.

Begon, M., J.L. Harper, and C.R. Townsend. 1990. *Ecology: Individuals, Populations, and Communities, 2nd ed.* London: Blackwell Scientific.

Crabtree, P. (ed.). 1970. *The Illustrated Natural History of Canada (9 vol.).* Toronto: NSL Natural Science of Canada.

Ecological Stratification Working Group. 1995. *A National Ecological Framework for Canada.* Ottawa: Environment Canada.

Heywood, V.H. (ed.). 1995. *Global Biodiversity Assessment.* Cambridge: Cambridge University Press.

National Wetlands Working Group. 1988. *Wetlands of Canada.* Ecological Land Classification Series No. 24. Ottawa: Environment Canada.

Odum, E.P. 1983. *Basic Ecology.* New York: Saunders.

Odum, E.P. 1993. *Ecology and Our Endangered Life-Support Systems.* Sunderland, MA: Sinauer.

Phillips, D. 1990. *The Climates of Canada.* Ottawa: Environment Canada.

Rowe, J.S. 1972. *Forest Regions of Canada.* Ottawa: Forestry Canada.

Scott, G.A.J. 1995. *Canada's Vegetation: A World Perspective.* Montreal: McGill-Queen's University Press.

Shelford, V.E. 1974. *The Ecology of North America.* Urbana, IL: University of Illinois Press.

Walter, H. 1977. *Vegetation of the Earth.* New York: Springer.

Wiken, E., D. Gauthier, I. Marshall, K. Lawton, and H. Hirvonen. 1996. *A Perspective on Canada's Ecosystems: An Overview of the Terrestrial and Marine Ecozones.* Occ. Pap. No. 14. Ottawa: Canadian Council on Ecological Areas.

Appendix 8A

Scientific names of wild Canadian species mentioned in Chapter 8.

Plants

alpine fir, *Abies lasiocarpa*
amabalis fir, *Abies amabalis*
American elm, *Ulmus americana*
antelope bush, *Purshia tridentata*
balsam fir, *Abies balsamea*
balsam poplar, *Populus balsamifera*
basswood, *Tilia americana*
bearberry, *Arctostaphylos uva-ursi*
beech, *Fagus grandifolia*
big bluestem, *Andropogon gerardii*
black spruce, *Picea mariana*
black walnut, *Juglans nigra*
blueberry, *Vaccinium angustifolium*
butternut, *Juglans cinerea*
cucumber-tree, *Magnolia acuminata*
Douglas fir, *Pseudotsuga menziesii*
eastern hemlock, *Tsuga canadensis*
Engelmann spruce, *Picea engelmannii*
fireweed, *Epilobium angustifolium*
grand fir, *Abies grandis*
horse chestnut, *Aesculus hippocastanum*
jack pine, *Pinus banksiana*
Labrador tea, *Ledum groenlandicum*
linden, *Tilia cordata*
lodgepole pine, *Pinus contorta*
Manitoba maple, *Acer negundo*
mountain hemlock, *Tsuga mertensiana*
Norway maple, *Acer platanoides*
Norway spruce, *Picea abies*
ponderosa pine, *Pinus ponderosa*
prickly pear cactus, *Opuntia fragilis*
rabbit bush, *Chrysothamnus nauseosus*
red alder, *Alnus rubra*
red maple, *Acer rubrum*
red pine, *Pinus resinosa*
red spruce, *Picea rubens*
sagebrush, *Artemisia tridentata*
sassafras, *Sassafras albidum*
Scotch pine, *Pinus sylvestris*
silver maple, *Acer saccharinum*
Sitka spruce, *Picea sitchensis*
sugar maple, *Acer saccharum*
sycamore, *Platanus occidentalis*
tamarack, *Larix laricina*
trembling aspen, *Populus tremuloides*
tulip-tree, *Liriodendron tulipifera*
western hemlock, *Tsuga heterophylla*
western red cedar, *Thuja occidentalis*
western white pine, *Pinus monticola*
white birch, *Betula papyrifera*
white elm, *Ulmus americana*
white pine, *Pinus strobus*
white spruce, *Picea glauca*
yellow birch, *Betula lutea*
yellow cedar, *Chamaecyparis nootkatensis*

Fish

Arctic char, *Salvelinus alpinus*
Arctic cod, *Boreogadus saida*

capelin, *Mallotus villosus*
chum salmon, *Oncorhynchus gorbuscha*
cod, *Gadus morhua*
coho salmon, *Oncorhynchus kisutch*
haddock, *Melanogrammus aeglefinus*
herring, *Clupea harengus*
redfish, *Sebastes fasciatus*
silver hake, *Merluccius bilinearis*
sockeye salmon, *Oncorhynchus nerka*
turbot, *Reinhardtius hippoglossoides*

Mammals

arctic fox, *Alopex lagopus*
arctic ground squirrel, *Spermophilus parryi*
arctic hare, *Lepus arcticus*
badger, *Taxidea taxus*
bearded seal, *Erignathus barbatus*
beaver, *Castor canadensis*
beluga, *Delphinapterus leucas*
bighorn sheep, *Ovis canadensis*
black bear, *Ursus americanus*
black-tailed deer, *Odocoileus hemionus columbianus*
black-tailed prairie dog, *Cynomys ludovicianus*
blue whale, *Balaenoptera musculus*
bobcat, *Lynx rufus*
bottlenosed whale, *Hyperoodon ampullatus*
bowhead whale, *Balaena mysticetus*
brown lemming, *Lemmus sibiricus*
buffalo, *Bison bison*
caribou, *Rangifer tarandus*
collared lemming, *Dicrostonyx torquatus*
Columbian ground squirrel, *Spermophilus columbianus*
cottontail rabbit, *Sylvilagus floridanus*
cougar, *Felis concolor*
coyote, *Canis latrans*
Dall's sheep, *Ovis dalli*
domestic cat, *Felis catus*
domestic dog, *Canis familiaris*
eastern chipmunk, *Tamias striatus*
elk, *Cervus canadensis*
ermine, *Mustela erminea*
fin whale, *Balaenoptera physalus*
fisher, *Martes pennanti*
gray seal, *Halichoerus gryptus*
gray (or black) squirrel, *Sciurus carolinensis*
gray whale, *Eschrichtius robustus*
grizzly bear, *Ursus arctos*
groundhog, *Marmota monax*
harbour porpoise, *Phocoena phocoena*
harbour seal, *Phoca vitulina*
harp seal, *Phoca groenlandica*
hoary marmot, *Marmota caligata*
hooded seal, *Cystophora cristata*
house mouse, *Mus musculus*
humpback whale, *Megaptera novaeangliae*
killer whale, *Orcinus orca*
lynx, *Lynx lynx*
mink, *Mustela vison*
minke whale, *Balaenoptera acutorostrata*
moose, *Alces alces*
mountain goat, *Oreamnos americanus*
mule deer, *Odocoileus hemionus*
muskox, *Ovibos moschatus*
muskrat, *Ondatra zibethicus*
narwhal, *Monodon monoceros*
northern fur seal, *Callorhinus ursinus*
northern pocket gopher, *Thomomys talpoides*
northern sea lion, *Eumetopias jubata*
Norway rat, *Rattus norvegicus*
pika, *Ochotona princeps*
pilot or pothead whale, *Globicephala melaena*
pine marten, *Martes americana*
polar bear, *Ursus maritimus*
porcupine, *Erethizon dorsatum*
pronghorn antelope, *Antilocapra americana*
raccoon, *Procyon lotor*
red fox, *Vulpes vulpes*
red squirrel, *Tamiasciurus hudsonicus*
Richardson's ground squirrel, *Spermophilus richardsonii*
right whale, *Balaena glacialis*
ringed seal, *Phoca hispida*
river otter, *Lontra canadensis*
sea otter, *Enhydra lutris*
snowshoe hare, *Lepus americanus*
sperm whale, *Physeter catodon*
striped skunk, *Mephitis mephitis*
walrus, *Odobenus rosmarus*
Atlantic white-sided dolphin, *Lagenorhynchus acutus*
Pacific white-sided dolphin, *Lagenorhynchus obliquidens*
white-tailed deer, *Odocoileus virginianus*
white-tailed jack rabbit, *Lepus townsendii*

wolf, *Canis lupus*
wolverine, *Gulo gulo*
wood bison, *Bison bison*

Birds

American avocet, *Recurvirostra americana*
American robin, *Turdus migratorius*
arctic loon, *Gavia arctica*
arctic tern, *Sterna paradisaea*
Baird's sandpiper, *Calidris bairdii*
bald eagle, *Haliaeetus leucocephalus*
barred owl, *Strix varia*
black duck, *Anas rubripes*
black guillemot, *Cepphus grylle*
black oystercatcher, *Haematopus bachmani*
black-backed gull, *Larus marinus*
black-billed magpie, *Pica pica*
black-capped chickadee, *Parus atricapillus*
blue grouse, *Dendragapus obscurus*
blue jay, *Cyanocitta cristata*
bob-white, *Colinus virginianus*
boreal owl, *Aegolius funereus*
broad-winged hawk, *Buteo platypterus*
Brant goose, *Branta bernicla*
burrowing owl, *Athene cunicularia*
California quail, *Callipepla californica*
Canada goose, *Branta canadensis*
canvasback, *Aythya valisineria*
cardinal, *Cardinalis cardinalis*
Carolina wren, *Thryothorus ludovicianus*
Caspian tern, *Sterna caspia*
chestnut-backed chickadee, *Parus rufescens*
Clark's nutcracker, *Nucifraga columbiana*
common crow, *Corvus brachyrhynchos*
common eider, *Somateria mollissima*
common loon, *Gavia immer*
common murre, *Uria aalge*
common puffin, *Fratercula arctica*
common redpoll, *Carduelis flammea*
common tern, *Sterna hirundo*
coot, *Fulica americana*
dipper, *Cinclus mexicanus*
double-crested cormorant, *Phalacrocorax auritus*
dovekie, *Alle alle*
evening grosbeak, *Hesperiphona vespertina*
ferruginous hawk, *Buteo regalis*
fox sparrow, *Passerella iliaca*
fulmar, *Fulmarus glacialis*
gannet, *Sula bassanus*
glaucous gull, *Larus hyperboreus*
golden eagle, *Aquila chrysaetos*
grey-cheeked thrush, *Catharus minimus*
grey jay, *Perisoreus canadensis*
great blue heron, *Ardea herodias*
great cormorant, *Phalacrocorax carbo*
great horned owl, *Bubo virginianus*
greater yellowlegs, *Tringa melanoleuca*
greater scaup, *Aythya marila*
greater snow goose, *Anser caerulescens atlanticus*
green-backed heron, *Butorides striatus*
gyrfalcon, *Falco rusticolus*
herring gull, *Larus argentatus*
hoary redpoll, *Carduelis hornemanni*
horned lark, *Eremophila alpestris*
house sparrow, *Passer domesticus*
ivory gull, *Pagophila eburnea*
kittiwake, *Rissa tridactyla*
Lapland longspur, *Calcarius lapponicus*
lesser scaup, *Aythya affinis*
lesser snow goose, *Anser caerulescens caerulescens*
lesser yellowlegs, *Tringa flavipes*
mallard, *Anas platyrhynchos*
marbled murrelet, *Brachyramphus marmoratus*
mountain quail, *Oreortyx pictus*
northern oriole, *Icterus galbula*
northern shrike, *Lanius excubitor*
northwestern crow, *Corvus caurinus*
oldsquaw, *Clangula hyemalis*
orchard oriole, *Icterus spurius*
pelagic cormorant, *Phalacrocorax pelagicus*
peregrine falcon, *Falco peregrinus*
pigeon guillemot, *Cepphus columba*
pintail, *Anas acuta*
prairie falcon, *Falco mexicanus*
prothonotary warbler, *Protonotaria citrea*
raven, *Corvus corax*
razorbill, *Alca torda*
red-eyed vireo, *Vireo olivaceus*
red-shouldered hawk, *Buteo lineatus*
red-throated loon, *Gavia stellata*

ringed plover, *Charadrius hiaticula*
rock dove (or pigeon), *Columba livia*
rock ptarmigan, *Lagopus mutus*
rose-breasted grosbeak, *Pheucticus ludovicianus*
sandhill crane, *Grus canadensis*
scarlet tanager, *Piranga olivacea*
semipalmated sandpiper, *Calidris pusilla*
sharp-shinned hawk, *Accipiter striatus*
sharp-tailed grouse, *Tympanuchus phasianellus*
snipe, *Gallinago gallinago*
snow bunting, *Plectrophenax nivalis*
snow goose, *Chen caerulescens*
snowy owl, *Nyctea scandiaca*
song sparrow, *Melospiza melodia*
starling, *Sturnus vulgaris*
Steller's jay, *Cyanocitta stelleri*
Swainson's hawk, *Buteo swainsoni*
thick-billed murre, *Uria lomvia*
tree sparrow, *Spizella arborea*
tufted puffin, *Fratercula corniculata*
tundra swan, *Cygnus columbianus*
western grebe, *Aechmophorus occidentalis*
white pelican, *Pelecanus erythrorhynchos*
white-crowned sparrow, *Zonotrichia leucophrys*
white-tailed ptarmigan, *Lagopus leucurus*
white-throated sparrow, *Zonotrichia albicollis*
whooping crane, *Grus americana*
willow ptarmigan, *Lagopus lagopus*
Wilson's storm petrel, *Oceanites oceanicus*
yellow-rumped warbler, *Dendroica coronata*
yellow warbler, *Dendroica petechia*

9

ECOLOGY: FROM INDIVIDUALS TO THE BIOSPHERE

Chapter Outline

Chapter Objectives

After completing this chapter, you will be able to:

1. Outline how species are adapted to different levels of stress and disturbance in their habitat.
2. Explain the elements of population growth and the constraints on population size.
3. List major factors that influence the nature of ecological communities.
4. Describe an ecological landscape (or seascape) and the factors that influence its spatial and temporal dynamics.
5. Outline the Gaia hypothesis and discuss its applicability to the functioning of the biosphere.

INTRODUCTION

Ecology involves the study of the interrelationships among organisms and their environment. In the sense meant here, "environment" includes both non-living influences such as temperature, moisture, nutrients, and physical disturbances, as well as organisms, which exert biological influences through such interactions as competition, herbivory, predation, and disease, and by providing elements of habitat.

An "individual organism" is genetically unique, and is different from other individuals of its species. However, the development of an individual can vary depending on the environmental conditions it experiences, a phenomenon known as "phenotypic plasticity." This is a Turk's-cap lily (*Lilium superbum*) growing near London in southern Ontario.

Source: B. Freedman

This population of northern gannets (*Morus bassanus*) breeds at Cape St. Mary's in southern Newfoundland.

Source: B. Freedman

Some environmental influences are resources that organisms can exploit as opportunities, allowing them to gain the necessities of life and livelihood. Other environmental influences are **stressors** — constraints on productivity and reproductive success. Many environmental stressors operate in a relatively continuous fashion, as is often the case for soil and water pollution, climatic factors, and many biological interactions. Other stressors influence organisms and ecosystems as events of **disturbance**, which cause substantial damage during a short period of time. Disturbance is followed by an often-extended period of ecological recovery called **succession**. Disturbances may be caused by natural forces such as wildfire and windstorms, and by anthropogenic influences such as clear-cutting of forests or ploughing of fields.

Ecology considers the structures and functions of the web of life at the following hierarchical levels:

1. **Individual organisms** can be defined, in an evolutionary context, as genetically unique entities. Some species, however, reproduce by asexual mechanisms, and in these cases clones of genetically identical "individuals" may develop.
2. **Populations** are groups of individuals of the same species that are co-occurring in time and space, and can potentially interbreed with each other.
3. A **species** consists of one or more populations that can potentially interbreed with each other and are reproductively isolated from other such groups.
4. **Communities** consist of populations of various species that are co-existing at the same time and place and are interacting ecologically.
5. **Landscapes** and **seascapes** are spatial integrations of various communities over large areas. These comprise a dynamic mosaic known as a landscape in terrestrial environments and a seascape in marine environments.
6. The **biosphere** consists of all of Earth's life and all the environments where life occurs.

Each of these levels of ecology is meaningful and all are highly relevant to environmental science. These tiers of ecology are not, however, totally discrete — they are all interconnected, and each level influences every other level. This chapter discuss issues relevant to the various hierarchical levels of ecology.

Autecology: Individuals and Species

Autecology is the field within ecology that deals with the study of individuals, populations, and species. Important topics in autecology include:

- differences among species in their life-history characteristics and in adaptations to different kinds of environmental conditions
- influences of the environment on individual organisms, including effects on development and behaviour
- the dynamics and causes of changes in the size and makeup of populations

Life-History Characteristics

Each species is unique and can be described by its physical, biochemical, behavioural, and ecological attributes. These characteristics are ultimately determined by the genetic variation encompassed by the unique individuals that comprise the species.

Although species are unique, they can still be aggregated into groups on the basis of similarities in their attributes. These affinities may be due to ancestral relatedness — that is, many species are somewhat alike because they share aspects of their evolutionary history (i.e., they have a similar phylogenetic lineage). For example, all maple trees (genus *Acer*) look rather similar and occur in comparable, temperate-forest habitats. All members of the cat family (Felidae) also bear a certain resemblance and are ecologically comparable in that all are predators, although of different species of prey and in different habitats or ecosystems.

Unrelated species may also display strikingly similar attributes, usually because they have undergone parallel changes through a phenomenon known as *evolutionary convergence* (sometimes called parallel evolution). Convergence suggests that through natural selection, unrelated species in similar environments may evolve to resemble each other and play similar functional roles in their ecosystems.

Coral reefs are shallow-water marine communities in tropical seas, and are extremely rich in species. Prominent species in this nearshore coral-reef community near Grand Cayman, in the Caribbean Sea, are blue-striped grunts (*Haemulon sciurus*) and elkhorn coral (*Acropora palmata*).

Source: Comstock/Frank Biola

There are numerous examples of evolutionary convergence among unrelated groups of organisms. For instance, all perennial (that is, long-lived) plants growing in arid environments have a need to conserve moisture. This critical function is enhanced by growth forms which are adaptations for reducing water losses, such as relatively cylindrical trunks and branches; tissues protected by a thick, waxy outer cuticle; and a leafless condition. Thorniness is another useful trait in arid environments, because sharp spines deter herbivores from consuming plant biomass and stores of water. Desert-inhabiting plants in many different families have developed one or more of these adaptations, including species of cacti (family Cactaceae), euphorbs (Euphorbiaceae), and succulents (Crassulaceae). Although species in these families are not related in an evolutionary sense, they often resemble each other because of evolutionary convergence.

Examples of convergence among animals include the similarities of the timber wolf (*Canis lupus*) of Eurasia and the marsupial wolf (or thylacine, *Thylacinus cynocephalus*) of Australia. A comparable example is the groundhog (*Marmota monax*) of North America and the marsupial wombat (*Vombatus ursinus*) of Australia. Also, the penguins (family Spheniscidae) of the Southern Hemisphere are structurally and functionally similar to the guillemots, murres, puffins, and related auks (family Alcidae) of the Northern Hemisphere.

Plant ecologists often categorize plant species on the basis of their life (or growth) forms. For example, a system proposed by the Danish botanist C. Raunkiaer classifies biennial and perennial plants largely on where the shoots or buds are positioned (In Detail 9.1). Although this system is simplistic, it is a useful tool in research examining the relationships between growth forms and habitat.

Another categorization is based on plants' adaptations for coping with large differences in ecological conditions. The British ecologist Philip Grime has suggested that plant strategies can be divided into three categories, determined by life history and its relationship to habitat conditions. This system proposes that two groups of environmental factors, disturbance and stress, strongly influence the evolution of plant life-history strategies. Disturbance can be frequent or uncommon, and severe or mild in its intensity. Stress is a longer-term site condition, and it can be intense if associated with extreme shortages of moisture, light, or nutrients, or innocuous if these necessary factors are all available in adequate levels.

In Detail 9.1

A Life-Form Classification of Plants

This widely used system classifies plant species into functional groups based on the location of their perennating buds or shoots (next year's buds seen in autumn on biennial or perennial plants). This is largely a structural classification, and does not imply any relatedness among species within each group.

Phanerophytes

These are woody, perennial plants that bear their buds well above the ground surface (higher than 10 cm). Trees, shrubs, and erect vines are familiar examples of this growth form.

Chamaephytes

These plants bear their perennating buds and shoots close to the ground surface. Examples include low-growing and trailing plants, such as cushion plants and trailing vines.

Hemicryptophytes

The perennating buds of these plants occur on the soil surface. Most of these plants, such as primroses and hawkweeds, form basal rosettes of leaves. Flowering stalks may grow into the air, but these are temporary, herbaceous structures that die back at the end of the growing season.

Cryptophytes

These plants have perennating buds buried in the soil or in sediment under water. Familiar examples include species with bulbs, corms, rhizomes, and tubers, which are the perennating structures of such plants as dandelion, lilies, irises, potato, and most ferns.

Therophytes

These are annual plants, which complete their life cycle (from seed to seed) in one growing season and then die. In the next growing season, new individuals are established by seed germination.

Any particular environment can be characterized by the importance of these two factors. This results in four kinds of habitat conditions:

1. low stress and rare disturbance
2. low stress but frequent disturbance
3. intense stress but rare disturbance
4. intense stress and frequent disturbance

Grime suggests that plants exhibit only three primary life-history strategies, because plants cannot cope with an environment that is both stressful and frequently disturbed (case 4 above). According to Grime, the three primary strategies of plants result in competitors, ruderals, and stress-tolerators.

Competitor plants are dominant in habitats in which disturbance is rare and environmental stresses are unimportant. Under such conditions, competition is the major selective influence on plant evolution and on the organization of plant communities. Competitive plants effectively acquire resources, and use them to achieve a dominant position in their community by interfering with the productivity of other plants. Useful adaptations in competitors include rapid, tall growth; a spreading canopy; and a widely spreading root system, all of which effectively occupy space and appropriate resources. Seedlings of competitive plants can also establish themselves beneath a closed canopy of vegetation.

Ruderals are characteristic of frequently disturbed environments with abundant resources, so that stress is not great. Ruderal plants are, therefore, well adapted to utilizing rich but temporary habitats. They are typically short-lived and intolerant of stress and competition. Ruderals produce large numbers of seeds which usually have mechanisms for long-distance dispersal so that new habitats can be discovered and colonized.

Stress-tolerators are adapted to environments that are marginal in terms of climate, moisture, or nutrient supply, but are infrequently disturbed and therefore stable. Typical of arctic, desert, and other stressful environments, stress-tolerant plants are generally short, slow-growing, and intolerant of competition from more vigorous species.

Another system of categorizing organisms, more commonly applied to animals, involves two groups of life-history characteristics. One group consists of longer-lived organisms that produce relatively few progeny, but invest a great deal of energy in each to improve their chances of survival. These are known as *K-selected* species. The other group, referred to as *r-selected*, comprises short-lived species that produce large numbers of smaller offspring, each of which has a relatively small chance of survival. However, because of the enormous numbers of offspring, it is likely that some will manage to survive. K-selected species are dominant in relatively stable, mature habitats in which competition is the controlling influence on community structure, while r-selected species occur in younger, recently disturbed habitats in which resources are freely available and rapid population growth is possible. (The source of the "K" and "r" labels comes from the logistic equation, a fundamental element of population ecology that for simplicity is not described here.)

Species can also be considered in terms of other elements of their reproductive strategy—for example, how often they engage in reproduction. Some species only achieve one reproductive event during their lifetime, usually dying afterward. This type of reproduction, known as *semelparous*, is seen in annual and biennial plants, many insects and other invertebrates, and most species of salmon. Most semelparous species are short-lived, but some can live for many years, gradually accumulating enough energy to sustain one massive reproductive effort. Semelparous reproduction is favoured in relatively rich habitats that are frequently disturbed, and is common among ruderal and r-selected species.

Species that reproduce a number of times during their lives are known as *iteroparous*. These are typically long-lived species that live in relatively stable habitats. Iteroparous species may produce large numbers of small offspring (r-selected), or they may produce fewer, larger young, each of which receives a large investment of parental resources (K-selected).

Relationships of Individual Organisms with Their Environment

Autecology also deals with the lives of individual organisms and how they are influenced by their physical and biological environments.

As was discussed in Chapter 6, all individual organisms have a fixed complement of genetic information, known as their *genotype*. However, the expression of genetic information (the *phenotype*) is influenced by environmental conditions, a phenomenon known as *phenotypic plasticity*. If individuals experience difficult environmental conditions, their phenotypic expression of genetic potential may include sub-optimal growth rates and the production of few or no progeny. Other individuals that live in a more benign environment can achieve higher productivity and have many offspring. The latter, more prolific circumstance is highly desirable in terms of an individual achieving evolutionary "success." By definition, "successful" individuals have managed to maximize their *fitness*, that is, their genetic contribution to future generations of their population.

The success of an individual organism is also affected by unpredictable (or stochastic) events of disturbance, which may result in untimely injury or premature death. Even if it is growing in a relatively benign environment, with optimal access to resources and other necessities of life, an unlucky and "unsuccessful" individual may just happen to be scorched by a wildfire, injured during a hurricane, devoured by a predator, debilitated by a disease, or hit by a truck.

Population Ecology

The study of populations of organisms is another aspect of autecology. Natural populations of all species change in size over time in response to environmental factors that affect four population-related (or *demographic*) variables: birth rate (BR), immigration rate (IR), death rate (DR), and emigration rate (ER). The change in population size (ΔP) during a unit of time (say a year) is described by the following equation:

$$\Delta P = BR - DR + IR - ER$$

This demographic relationship is true of all species, including humans. In some cases, isolated (or closed) populations do not receive any immigration of new individuals, and do not lose any individuals to emigration. Under such conditions, ΔP is calculated as BR – DR, a value known as the *natural rate of population change*.

Often, ΔP is expressed as a percentage change, by dividing its value by the initial population size (for example, a population of 100 individuals that increased by 10 in one year had a 10%/y growth rate). If the percentage change in any population is constant, then there will be an accelerating rate of population increase or decrease, called *exponential change*.

Imagine a circumstance in which a fertile pair of individuals discovers a new habitat — one that is suitable but has not been previously occupied by their species. Under such conditions the founder individuals will breed and the population will grow over time. Initially, resources will be abundant and will not constrain growth of the population. Consequently, the percentage rate of population increase will be constant, being limited only by how quickly progeny can be produced and become fertile (that is, by birth and maturation rates), and countered only by longevity of individuals in the population. This maximum rate of population growth, limited only by the biology of the species in the given environment and not by competition for resources, is referred to as the *intrinsic rate of population increase*. Any population growing at the intrinsic rate of population increase (or indeed at any fixed percentage rate) will quickly explode in abundance (see In Detail 9.2).

Eventually, however, as the **carrying capacity** of the available habitat is approached, space and resources become limiting. (The carrying capacity is the size of population that can be sustained without degrading the habitat.) At or beyond the carrying capacity, opportunities are constrained by the limited availability of resources, and individuals in the population compete severely with each other. Intense competition produces physiological stress, which generally results in a decrease in the birth rate and an increase in the death rate.

In some cases, the rate of population increase may then decrease to zero (that is, birth rate equals death rate—a condition sometimes referred to as *zero population growth*, or ZPG). If ZPG is maintained, the population size will eventually level off, perhaps at an abundance appropriate to the carrying capacity of the habitat.

However, the earlier exponential population growth may have resulted in an abundance that exceeded what the habitat could support. Such an over-population would over-exploit the environment, resulting in its degradation and a consequent decrease in its carrying capacity. If this happens, the population size will decrease through a rapid increase in the mortality rate, or perhaps by a surge of emigration in search of new habitats. These may result in a rapid and uncontrolled crash in the numbers of individuals in the population. Usually, a crash takes the population to a level below the carrying capacity of the habitat, creating a circumstance for renewed population growth. In small habitats, however, the population crash can be massive enough to extirpate a local population.

IN DETAIL 9.2

EXPONENTIAL GROWTH

A constant rate of increase leads to extremely rapid growth in the sizes of populations of organisms. This happens for the same reason that money invested at a fixed rate of interest can quickly increase in quantity. This phenomenon, known in finance as compound interest, is illustrated below.

Consider, for example, an investment of $100 at an interest rate of 10% per year, locked in for a 10-year period. After one year, the initial investment grows to $110, representing the initial investment plus accumulated interest. In the second year, the 10% interest rate is applied to the $110, so the earned interest is larger ($11) than in the first year ($10). In the third year, the 10% interest is applied to the accumulated $121, so the earned interest is larger yet ($12.10), and the accumulated value of the investment is $133.10. At the end of the fourth year, the initial investment of $100 is worth $146.41; $161.05 at the end of the fifth year; and $259.37 at the end of the tenth year, representing an impressive 159% return on the initial investment. Clearly, the compound interest leads to extremely rapid increases in capital.

Exponential growth refers to the accelerating growth of an initial quantity due to a constant rate of increase. Sometimes an important parameter known as *doubling time* is calculated—that is, the time required for a twofold increase in capital. The doubling time can be roughly calculated as 70 divided by the rate of increase. In the example above, 70 divided by 10%/y yields 7 y. Therefore, the initial $100 would double in amount in only 7 y, and the accumulated $200 would again double (to $400) in another 7 y, and so on as long as the investment conditions do not change.

The calculations of compound interest can also be applied to the exponential growth of populations of organisms. One example will suffice: In 1996, the global human population was about 6 billion people, growing at about 1.5%/y. Therefore, in only 47 y (i.e., 70 divided by 1.5%/y), the human population will double to 12 billion (as long as the growth rate remains the same). The ecological implications of such a population increase are awesome (see Chapter 10).

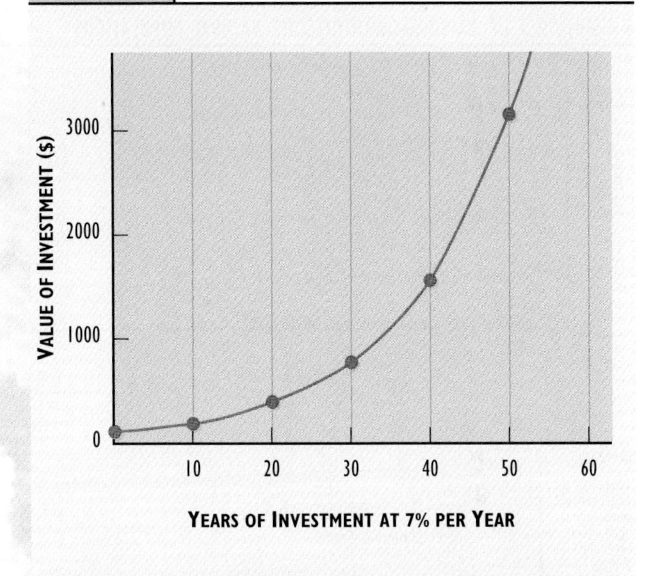

This curve shows the growth of an initial investment of $100 invested at a compound interest rate of 7% per year. Biological populations also grow at exponential rates if their rate of increase is constant, although when resources become limiting, the rate of increase decreases.

Population ecologists have developed mathematical models of population dynamics that account for the influences of such factors as the intrinsic rate of population increase, the carrying capacity of habitats, the effects of predation and disease, and even the effects of unpredictable disturbance events. These models are described in introductory textbooks of ecology and are not dealt with here in any detail. For present purposes, there are several important points to understand about population ecology:

- Populations of all species are dynamic. They change over time due to varying rates of birth, death, immigration, and emigration.
- Populations of all species can, potentially, increase very rapidly, under conditions in which resource availability and other factors are not constraining. Several examples of rapid population growth are illustrated in Figure 9.1. However, unlimited growth cannot be sustained forever — in all of the cases in

Figure 9.1 the population sizes eventually levelled off, decreased, or crashed.

- Ultimately, the abundance of a species that can be sustained is limited by the carrying capacity of the available habitat. Two examples of population growth that level off at the carrying capacity of the habitat are illustrated in Figure 9.2.
- Some populations are relatively stable. Usually these exist in environments in which resource availability is predictable, so that a balance can be achieved with the carrying capacity. For example, relatively little change occurs in the year-to-year populations of trees growing in old-growth forests, unless a rare, catastrophic disturbance occurs.
- Other populations are relatively dynamic, changing greatly over time and rarely achieving even a short-term balance with the carrying capacity of their habitat. This is commonly true of species living in habitats that are disturbed frequently or are in an early, dynamic stage of succession. Some populations are cyclic, achieving great abundances at regular intervals, interspersed by longer periods of lower abundance. Cyclic populations are obviously unstable over the short term, but they may be stable over the long term (Figure 9.3).

FIGURE 9.1 | RAPID GROWTH OF SOME NATURAL POPULATIONS

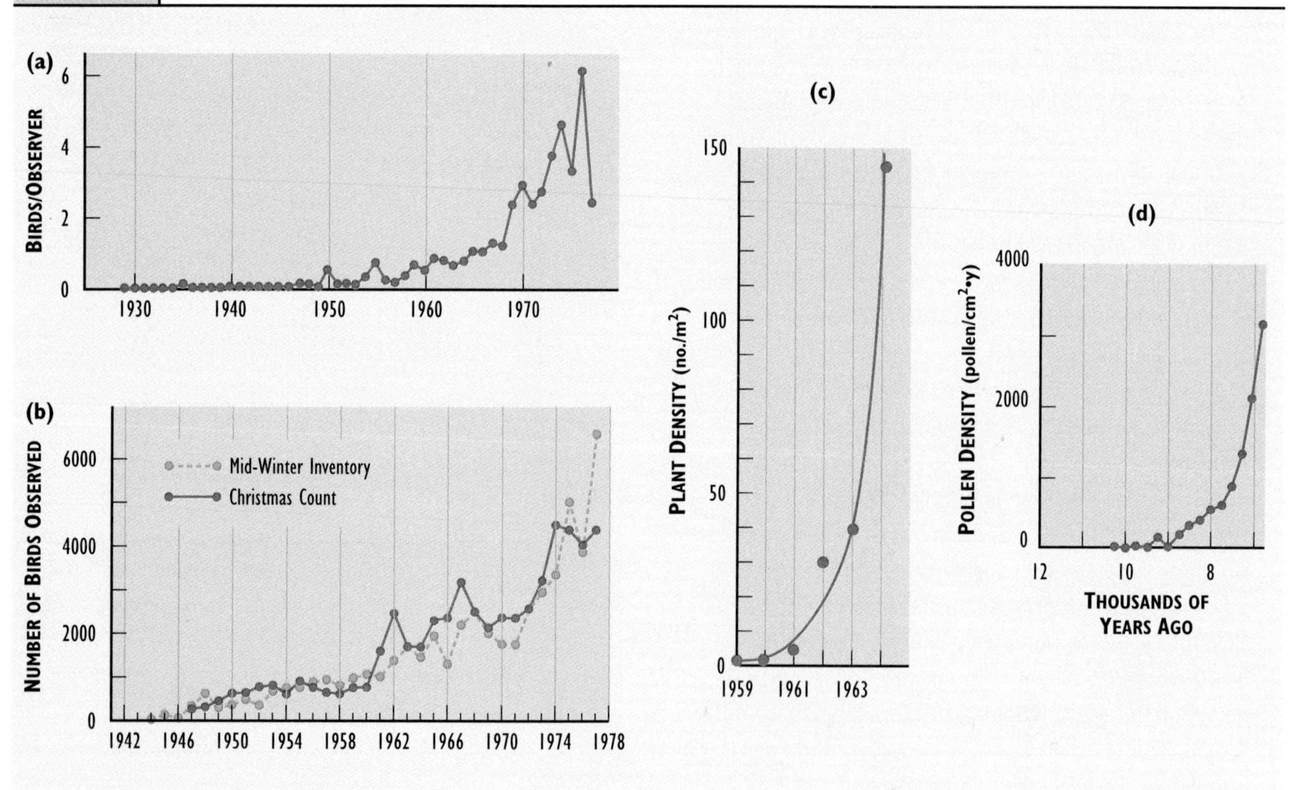

(a) The population of mourning doves (*Zenaida macroura*) wintering in southern Ontario over 48 years. This used to be a rare species, but it has apparently benefitted from a warming climate, improved suburban habitat, and winter feeding.

(b) The population of mallards (*Anas platyrhynchos*) wintering in southern Ontario over 35 years, illustrated with two independent sets of data. This duck has expanded its breeding and wintering ranges into eastern Canada, largely in response to habitat made available by the clearing of forested land.

(c) The population of wild oats (*Avena fatua*) in an annually ploughed field over six years. This grass is an invasive weed in agriculture.

(d) The population of lodgepole pine (*Pinus contorta*) near Snowshoe Lake, British Columbia, occurring during natural reforestation following deglaciation about 7000–9000 years ago. In this case, tree populations are indicated by the amounts of pollen recovered from dated layers of lake sediment.

Sources: (a) Freedman and Riley (1980); (b) Goodwin *et al.* (1977); (c) modified from Silverton (1987); (d) modified from MacDonald and Cwynar (1991)

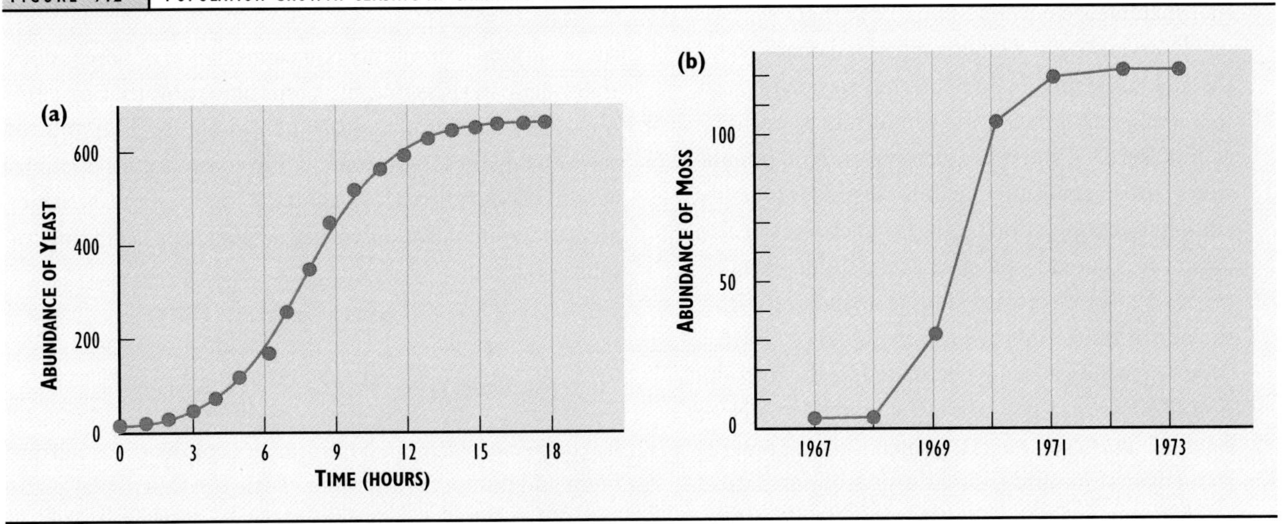

(a) The population growth of yeast cells cultivated in a flask is initially exponential, but then levels off at the carrying capacity of the habitat. Carrying capacity is determined by the volume of the flask, the quantity of nutrients available, and the increasing concentrations of toxic metabolites, including ethyl alcohol.

(b) The population of a moss colonizing a suitable, but initially bare, rock substrate in Iceland. The carrying capacity is limited by the amount of two-dimensional space available.

Source: (a) modified from Krebs (1985); (b) modified from Silverton (1987)

FIGURE 9.3 | CYCLIC POPULATIONS

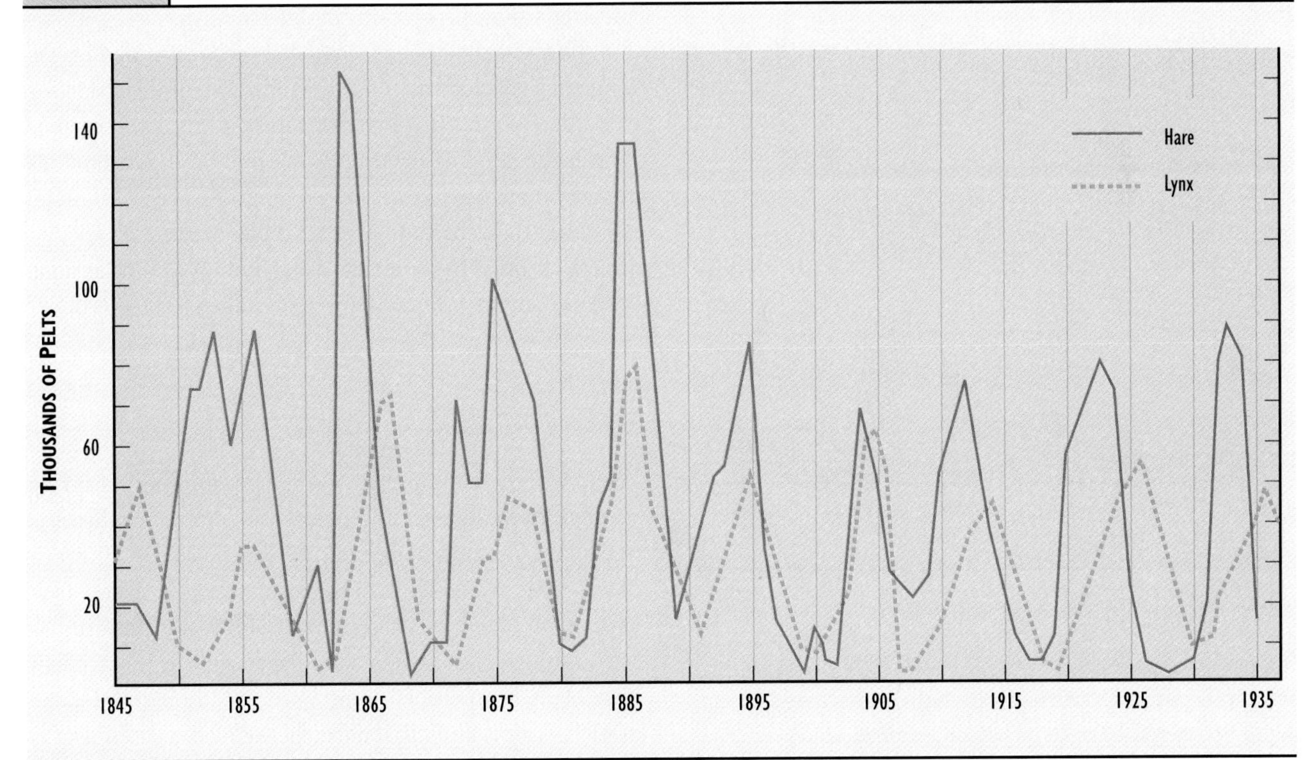

Populations of lynx (*Lynx canadensis*) and snowshoe hare (*Lepus americanus*) over much of northern Canada, as indicated by the numbers of pelts received from fur trappers by the Hudson Bay Company. Ecologists interpret these data as suggesting ten-year population cycles, with the predatory lynx responding to changes in abundance of snowshoe hare, its major prey.

Source: Modified from Odum (1983)

- Populations that exceed the carrying capacity of their habitat are never sustainable at that level, partly because of the environmental damage they cause. Unsustainable populations eventually crash to a smaller abundance and sometimes to extinction. Figure 9.4 shows an example of rapid population growth, resulting in habitat degradation and a subsequent population crash. Populations can also crash for other reasons, such as the sudden occurrence of a deadly disease. This is happening with the native white elm (*Ulmus americana*) of North America, which is being decimated by the introduction of a Eurasian pathogen (the Dutch elm disease fungus, *Ceratocystis ulmi*) to which this tree has virtually no immunity. Other causes of population crashes include unsustainable levels of predation, and extensive disturbances such as wildfire, windstorms, or clear-cutting.

FIGURE 9.4 | POPULATION GROWTH AND CRASH

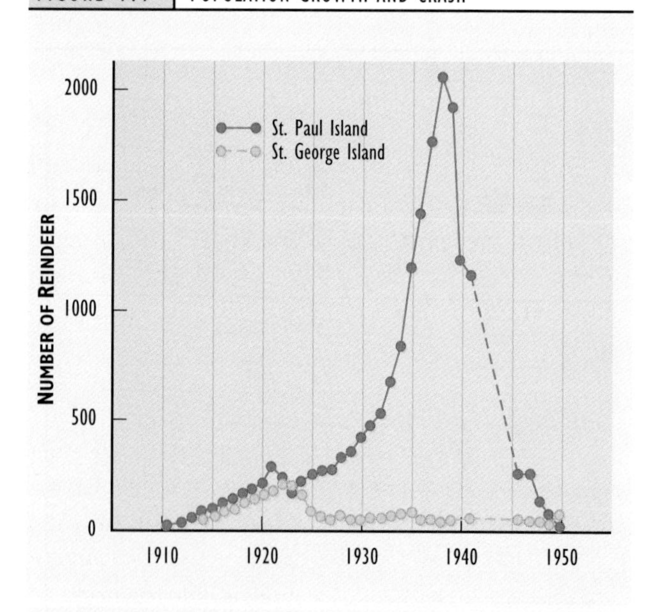

In 1910, reindeer (*Rangifer tarandus tarandus*; the Eurasian subspecies of caribou) were introduced to two islands in the Aleutian chain off western Alaska in an attempt to establish a new resource of meat and hides. On both islands, the reindeer population increased rapidly. They exceeded the carrying capacity of the habitat and caused severe damage through overgrazing. The populations then crashed.

Source: Modified from Krebs (1985)

Community Ecology

An ecological community is an aggregation of populations that occur in the same time and place and that interact physically, chemically, and/or behaviourally. The study of relationships among species within communities is known as **synecology**. Strictly speaking, a community consists of all plant, animal, and microbial populations occurring together on a site.

The Niche

Each species within a community exploits the environment and interacts with other species in a particular manner. This can be considered to its "occupation" or livelihood. Ecologists use the word **niche** to describe the role of a species in its community. Some niches are relatively narrow and specialized, as is the niche of bats that feed only on flying insects of a certain size, or of wasps that pollinate only one or a few species of plants. Other niches, however, are much broader, such as those of bears and humans, both of which forage over extremely broad ranges and affect their ecosystems in diverse ways.

The so-called *fundamental niche* is determined by the range of a species' tolerance to environmental factors (for example, to hot and cold temperatures). The *realized niche* reflects the range of environmental conditions that a species actually manages to exploit in nature. The realized niche is smaller than the fundamental niche because all species are constrained to some degree by biological interactions such as competition, predation, and disease.

Functional Communities

Because of their complexity, entire communities are rarely studied by ecologists. Ecological studies are usually limited by the amounts of funding, person-power, and breadth of expertise available. Instead, community-level research usually involves the examination of selected functional "communities" of similar organisms, such as insect, fish, bird, plant, or microbial communities. Although the scope of such work is limited, it nevertheless allows ecologists to investigate important aspects of community ecology.

Forest communities, for example, contain a wide range of organisms of various species and sizes, including plants, animals, and microorganisms. The populations of these diverse species interact with each other in myriad ways. Trees, for instance, contribute most of the physical structure of the habitat, provide food for many species of herbivorous animals, and drop leaf litter which is decomposed by animals and microorganisms who recycle the nutrients through the detrital food web. Other ecological interactions within a forest community include predatory, parasitic, and disease relationships among species, as well as many symbioses, such as pollination, seed dispersal, and root mycorrhizae. Because of the inherent complexity of forest communities, most ecological studies only investigate selected components.

This pragmatic approach to community-level forest research can be illustrated by studies of the ecological effects of forestry conducted in the Maritimes by my students and me. We divided the larger community into the following *functional groups*:

1. trees, which we define as woody plants with a stem diameter greater than 10 cm
2. shrubs, which have a stem diameter less than 10 cm, are taller than 1 m, and include shrub-sized, young individuals of tree species as well as "true" shrubs
3. ground vegetation, which includes all plants, mosses, and lichens that are growing within 1 m of the ground surface
4. epiphytes that grow on other plants, such as species of lichens, mosses, and liverworts that grow on the bark-covered surfaces of trees
5. small mammals such as mice, shrews, voles, squirrels, and marten
6. large mammals such as deer, bear, and coyote
7. birds
8. reptiles and amphibians
9. insects
10. fungi and other microorganisms living in the soil

During some of our relatively detailed studies of birds, specific work has been done with those species that nest in holes in trees. These comprise a "cavity-tree" element of the larger avian community. Similarly, studies of insects and other invertebrates have involved functional groups that live in soil, in rotting deadwood, or on foliage. Even with all of these (and other) functional groups, we have not yet managed to examine all of the important elements of the forest communities that we are studying.

Factors Influencing Ecological Communities

The nature of ecological communities is influenced by various environmental factors, particularly those described below.

Species Present

Obviously, only those species that are present in a habitat, or are capable of dispersing into it, can play a role in the community that develops. A species' ability to colonize an available habitat is influenced by its biology, intervening barriers such as a mountain range or ocean, the disturbance regime in the habitat, and other factors. Increasingly, humans are influencing the species composition of communities, often by introducing non-indigenous species beyond their natural ranges.

Appropriate Habitat

If a habitat is not suitable, then a particular species will not be able to utilize it even if it is capable of dispersing to the site. There are many aspects of habitat suitability, and all of these must be satisfied within the limits of tolerance of a species if it is to become a component of a local community.

Interactions with Other Species

Species interact through herbivory, predation, competition, disease, and symbiosis, the latter including mutualism, commensalism, and parasitism. All of these interactions can influence the presence and abundance of different species within communities. The following examples illustrate these influences.

Herbivory: Larvae of the hemlock looper (*Lambdina fiscellaria*) are voracious feeders on the foliage of spruce, fir, and other coniferous trees. When conditions are suitable, this moth can proliferate rapidly, causing severe damage over a large area of forest, as periodically happens in eastern Canada. Forests defoliated for several years have many dead trees, representing an important element of community

Species interact with each other in various ways, such as herbivory, predation, competition, disease, and symbiosis. This photo shows elk (*Cervus canadensis*) grazing on herbaceous plants in Jasper National Park, Alberta.

Source: B. Freedman

change. The loss of much of the forest canopy has many indirect effects, such as allowing understorey plants to grow more vigorously. The changes in vegetation affect the habitat available for species of insects, birds, and other animals. Microorganisms and other detritivores are also affected, because large quantities of dead tree biomass are available to be decomposed.

Predation: An extremely effective predator can greatly reduce the abundance of its prey, thereby changing the structure of the community. For instance, during the summer, most forest birds feed on insects, spiders, and other invertebrates, which are nutritious food for both adults and their rapidly growing nestlings. Avian predation can greatly change the invertebrate community, as has been demonstrated by studies in which small areas of forest were enclosed with netting. Avian predators were excluded, but invertebrates could move in or out. Under these conditions, the abundance of many species of insects and other avian prey increased, with species particularly vulnerable to avian predation benefitting the most.

Competition occurs when the biological demand for an ecological resource exceeds the supply, causing organisms to interfere with each other. Plants, for example, often compete for access to a limited supply of sunlight, water, nutrients, and space. Animals commonly compete for food, nesting sites, mates, and other resources. Intraspecific competition occurs when individuals of the same species vie for access to resources, while interspecific competition occurs between different species. If a species is particularly effective at co-opting resources to its own benefit, it may displace other species, a phenomenon known as competitive displacement (or in extreme cases, competitive exclusion). This affects the presence and relative abundance of species in its community. For example, sugar maple (*Acer saccharum*) is a very competitive tree in hardwood forests of eastern Canada. Where environmental conditions are well suited for this species, it can dominate mature stands. If large sugar maple trees are removed from a stand, perhaps by a selective timber harvest, other tree species (as well as small sugar maples) will benefit greatly from the release from competition, and will grow more vigorously.

Disease: The health of individuals and the population of a species are affected by their vulnerability to certain diseases. Virulent diseases can cause enormous changes in the composition of ecological communities. In the early 1900s, the American chestnut (*Castanea dentata*) was afflicted by chestnut blight (*Endothia parasitica*), an introduced fungal pathogen. Because chestnuts have little immunity to this disease, the species was virtually eliminated from the forests of eastern North America by the 1950s. This change released other tree species from competition with the previously dominant chestnut, and they quickly filled gaps in the canopy created by its demise.

Symbiosis refers to intimate relationships among species. Some symbioses involve obligate relationships in which the symbionts cannot live apart, but more commonly the association is somewhat flexible. Symbioses can greatly influence the success of species in particular environments by improving their competitive ability and decreasing their vulnerability to predation, disease, and other stresses.

The main types of **symbiosis** are known as **mutualism**, in which both partners benefit; **parasitism**, in which one organism benefits and the other is harmed; and **commensalism**, in which one organism benefits without harming the other. While symbioses are critical to one or both partners, they can also indirectly affect the habitat and the resources available to other members of the community.

A familiar example of a mutualism, lichens are an obligate association between a fungus and either an alga or a blue-green bacterium. The fungus benefits from the productivity of the photosynthetic partner, while the latter gains a relatively moist microhabitat and improved access to inorganic nutrients.

Another mutualism, called a mycorrhiza, is an intimate association between fungi and the roots of most vascular plants. The plant benefits through enhanced access to nutrients, especially phosphate, while the fungus receives nutritious exudates from the roots of the plant. This mutualism also provides a broad, community-level benefit through increased primary productivity. Some species of legumes live in a mutualism with the bacterium *Rhizobium japonicum*, which fixes nitrogen gas (N_2) into ammonia, a critical nutrient.

Some species of Dinoflagellates (single-celled algae) live within corals (small cnidarian animals), where they receive protection and access to inorganic nutrients. The corals benefit through access to the photosynthetic productivity of the algae.

Many animals eat plant biomass, but few are able, on their own, to digest complex polymeric biochemicals such as cellulose and lignin. Consequently, many herbivores live in a mutualism with microorganisms, which inhabit their gut and secrete specialized enzymes that digest cellulose and lignin, making those abundant sources of nutrition available to the animal. Cows, deer, and sheep host their digestion-aiding microorganisms in a specialized pouch of their fore-stomach, called the rumen. Humans also harbour a diverse community of microorganisms in their gut, many of which are important to nutrition.

Other mutualisms include the many species of flowers that are pollinated by specific kinds of insects. Pollination is crucial to the reproductive success of plants, while the insects benefit from abundant nectar or pollen as food. In addition, herbivores in the community benefit from the plant fruits that are produced because of pollination.

An example of commensalism is the epiphyte community of plants, lichens, and mosses that often grows on large trees. The epiphytes achieve an ecological benefit from the relationship, and the host trees are not affected to any meaningful degree. There are many familiar examples of parasitism, including fleas on a dog, and tapeworms in humans. The parasite benefits by taking nutrition from the host, and the host usually suffers, or may even die.

A mutualism is an intimate symbiosis in which both partners benefit from the relationship. Lichens, such as the light-coloured *Cetraria nivalis* in the centre of the photo, are an obligate mutualism between a fungus and an alga, meaning that the two species cannot live apart in nature. Thus, they are treated by taxonomists as a single "species."

Source: B. Freedman

Community Dynamics

All ecological communities are dynamic, changing over time in their species composition and functional attributes (such as productivity, decomposition, and nutrient cycling). The nature and rate of community change depend on the stability of environmental conditions. The most dynamic communities are associated with the younger stages of *succession* after *disturbance*. As we noted in Chapter 1, disturbance can occur on two spatial scales:

Stand-replacing catastrophes are caused by wildfire, a disease epidemic, glaciation, clear-cutting, and other cataclysmic events. This kind of disturbance is relatively extensive, and results in the immediate replacement of a community with a fundamentally different one. At this point, successional recovery begins. Over time, succession may regenerate a community similar to what existed before the disruption, or a completely different community may result.

The younger stages of a sere (or successional sequence) are especially dynamic in terms of the rate of community change. During the initial years of recovery, competition is not very intense. Ruderal, r-selected species predominate. Later stages of succession are much less dynamic in terms of the rate of community change.

Microdisturbances are local disruptions, affecting small areas within an otherwise intact community. A microdisturbance may, for instance, be associated with the death of an individual tree or of a small group of trees within a forest. Such a microdisturbance results in a gap in the canopy, within which community change is relatively dynamic as species compete to take advantage of the temporary resource opportunity of additional sunlight on the forest floor. Similarly, the deaths of individual coral heads initiates a microdisturbance within a coral-reef community. Although ecological changes are relatively dynamic within a gap created by a recent microdisturbance, at the stand level the community is relatively stable. Gap-phase community dynamics occur in all ecosystems, but are most important during later stages of succession—for example, in older-growth forests.

Spatial Variation of the Environment

Environmental conditions are always variable from place to place, sometimes extremely so. These spatial variations greatly influence the character of ecological communities, as described below.

Gradual spatial changes in environmental conditions are associated with varying altitude on a mountain, changing climatic regimes over large distances across continents, and other relatively continuous gradients. This type of spatial change is reflected in gradual variations of communities, because of individual species' different but overlapping tolerances and requirements of environmental conditions. This results in overlaps of the distributions of species, which can make it difficult for ecologists to determine the locations of boundaries (or ecotones) between some types of communities.

Rapid spatial changes in environmental conditions occur at sharp boundaries between distinctive types of soil or bedrock, at interfaces between aquatic and terrestrial habitats, and in places affected by different disturbance regimes. The latter influence can occur, for instance, between a burned and unburned area of forest, or between an eco-

Ecosystems are occasionally subjected to catastrophic disturbances, such as this forest fire in La Mauricie National Park, Quebec.
Source: L. Foisy

logical reserve and its surrounding area, which may be affected by agriculture or forestry. Relatively discrete changes in environmental conditions favour large differences in community types, with distinct, readily identifiable boundaries between them.

LANDSCAPE ECOLOGY

Landscapes (or seascapes in the marine context) are a mosaic of patches, each of which represents an ecological community. A landscape may contain various kinds of communities because:

- each reflects particular environmental conditions, such as different soil or bedrock types or variations of standing water (as in lakes, streams, or wetlands)
- they represent various stages in succession, such as communities of different ages after wildfire or insect damage
- they are related to the nature of land use, as when parts of landscapes are affected by urbanization, agriculture, forestry, transmission corridors, roads, or other human influences

Over time, the spatial patterns of communities on landscapes are highly dynamic. This largely reflects the influence of disturbances and successional recovery. A patch that today is a pasture, or a recent clear-cut or burn, may be a mature forest after fifty years of succession. Similarly, a pond may in-fill over the centuries and become a wetland, which with further time may transform into a forest. Ecologists use the term "shifting mosaic" to integrate the spatial and temporal variations of communities on landscapes. The following important factors affect the shifting mosaic of communities on landscapes.

Patch size relates to the area of particular stands of communities (a "stand" is a community in a particular place). All species need some minimal area of habitat to support their populations over the long term, and small patches of ecosystem may not be adequate for this purpose. Relatively small patches may, however, help to support a population living in several patches on the landscape (an extensive population of this sort is known as a metapopulation). This can happen if the patches are connected by corridors to other suitable habitat, or if the species is capable of dispersing through surrounding, inhospitable habitats (that is, the landscape matrix must be "permeable" to movements of the species).

Landscapes are subjected to patch dynamics associated with natural disturbances, such as wildfire, windstorms, and insect outbreaks. The patch dynamics of forested landscapes in Canada are being increasingly structured by management. In this aerial view of a forest in New Brunswick, the natural forest is being harvested by clear-cutting (the lighter patches are snow in the clear-cut areas), which initiates a succession that restores a forest for another harvest in 60–80 years. Unless areas are set aside for protection, this entire landscape may become used in this way.

Source: M. Sullivan

The amount of edge is also important, largely because it influences the amount of ecotone habitat associated with the patch. A circular patch has the smallest ratio of edge to area, and smaller patches have higher ratios than larger ones of the same shape. Ecotones between patch types are a habitat in themselves, and may be selectively utilized by so-called "edge species." However, the greater the ratio of edge to area, the less "interior" habitat there is (interior habitat is uninfluenced by ecological conditions associated with ecotones). Ecologists have identified "interior species" that are less successful if they try to use habitat close to an edge. Certain forest birds, for instance, experience greater rates of predation and nest parasitism in small remnants of mature forest (see Chapter 24).

Connectedness refers to the presence of links between otherwise discrete patches of similar habitat. These may be used by species as corridors to move among patches, allowing metapopulations to function on the landscape. As was noted previously, connectedness is also related to the ability of a species to disperse among habitable patches through the surrounding, inhospitable habitat.

Age-class adjacency is important in a landscape in which patches represent different stages of a successional sequence. This commonly occurs in landscapes affected by disturbances such as wildfire, insect epidemics, clear-cutting, or agriculture. In general, patches of a similar post-disturbance age will be comparable in many aspects of habitat quality, while patches of different age will be less similar. This can be an important consideration for movements of species among patches suitable as habitat.

Complex habitat requirements are characteristic of some larger species of animals, such as deer, bears, and wolf. These require different kinds of habitat patches for various purposes at different times of year. Because these animals participate in various kinds of communities, all the habitat patches they need must be present on the landscape if viable metapopulations are to be sustained.

Landscape-level biodiversity is related to the richness of community types over a large area (see Chapter 7). A landscape uniformly covered by a single community has less biodiversity than a landscape composed of a rich and dynamic mosaic of different communities.

Landscape-level functions operate over extensive areas, and integrate the influences of many kinds of communities. A **watershed**, for example, can be defined as the expanse of terrain from which water drains into a stream, lake, or some other kind of waterbody. Most watersheds contain various types of habitat patches, each with particular influences on hydrology and water chemistry. In general, well-vegetated watersheds covered with mature forest yield the cleanest flows of water. Other environmental services provided by well-vegetated landscapes include evapotranspiration, control of erosion, moderation of climatic extremes, and absorption of atmospheric carbon dioxide and release of oxygen.

Landscape ecology is an important subject area in environmental science. Humans commonly affect individual stands of particular kinds of communities, but many of the ecological effects of these actions must be managed at the scale of landscapes.

The Biosphere

The biosphere consists of all life and ecosystems on Earth. It is bounded by the presence of living organisms and it is the only place in the universe known to support life.

Ecological processes at the level of the biosphere include global climatic, oceanic, and atmospheric regimes (Chapter 3), Earth's energy budget (Chapter 4), and planetary nutrient cycles (Chapter 5). These biospheric processes influence all of Earth's life and ecosystems. It is important to recognize, however, that Earth's life and ecosystems also influence biosphere-level processes.

The Gaia Hypothesis

In fact, some scientists have suggested that there may be a degree of homeostatic control, or feedback, between the reciprocal influences of Earth's ecosystems and their biosphere-level environment. A popular notion describing these global biosphere-environment relationships is known as the **Gaia hypothesis**, a recent, highly controversial idea popularized by the British scientist James Lovelock. Lovelock suggests that Earth's organisms and ecosystems have caused substantial changes to occur in certain physical and chemical attributes of the environment, resulting in improvements in living conditions on the planet. The hypothesis envisions all of Earth's species and ecosystems as encompassed by a sort of "superorganism," or Gaia. Gaia attempts to optimize environmental conditions toward enhancing its own health and continuity, and uses homeostatic (or feedback) mechanisms that help maintain environmental conditions within the range that life can tolerate.

The ancient Greeks believed that Gaia (or Gaea) was the prolific ancestor of many of their most important gods. The Romans, who adopted many Greek gods and ideas as their own, also believed in Gaia, whom they knew as Terra. More recently, the Gaian myth has been personified as "mother Earth."

The Gaian idea is attractive and interesting, largely because it integrates so many ideas and large-scale observations into a single, consolidated belief and world-view. However, Earth is the only planet in the universe known to support life and ecosystems, and therefore is the only known replicate in the great experiment of life. Consequently, the Gaia hypothesis cannot be tested by rigorous, scientific experimentation, and for this reason many scientists reject many of its inferences. Except in the broadest of terms, Gaian ideas may not be very useful in helping humans to manage the detrimental impacts of their increasing population and agro-industrial activities on the biosphere.

Nevertheless, some intriguing lines of evidence can be marshalled in support of the Gaian notion. Two examples follow.

Atmospheric Oxygen

Earth's primordial atmosphere did not contain oxygen (O_2). This gas appeared only after the first photosynthetic organisms, blue-green bacteria, evolved. These, and the somewhat-later green algae, give off O_2 as a waste product of photosynthesis. The present concentration of O_2 in the atmosphere, 21%, has originated entirely with photosynthesis, and is a critically important environmental factor for most of Earth's species and many key ecological processes.

Atmospheric O_2 concentrations have probably been stable for as long as several billions of years. This suggests a long-term equilibrium between O_2 production by photoautotrophs and its consumption by respiration, including decomposition. Interestingly, if the concentration of oxygen were much higher than 21%, say about 25%, then biomass would be much more combustible. This condition would lead to more frequent and extensive wildfires, which would severely damage Earth's terrestrial ecosystems.

All this can be interpreted as suggesting the existence of a homeostatic control of the concentration of atmospheric O_2, operating at the biospheric scale. This control may achieve a balance between sufficient O_2 to sustain Earth's most abundant organisms (which have an aerobic metabolism) and larger O_2 concentrations which would result in extremely destructive conflagrations.

Earth's Greenhouse Effect

The concentration of atmospheric carbon dioxide (CO_2) is substantially regulated by a complex of physical and biological processes by which this gas is emitted and absorbed. Atmospheric CO_2 is important in Earth's greenhouse effect, which maintains the surface temperature within a range that organisms can tolerate (see Chapters 4 and 17). Earth's greenhouse effect helps maintain an average surface temperature of about 15°C, compared with the –18°C that would otherwise occur and would be too cold for organisms to tolerate. Advocates of the Gaia hypothesis suggest that these observations imply a homeostatic control of atmospheric CO_2 directly, and of the greenhouse effect and climate indirectly.

There is clear evidence that organisms and ecosystems cause substantial changes in their environment and that they are also affected by their environmental conditions. The scientific community does not, however, widely support the notion that Earth's species and ecosystems have somehow integrated into a mutually benevolent symbiosis, aimed at maintaining a comfortable range of environmental conditions.

The Gaia hypothesis is nevertheless useful in environmental science. Gaian ideas emphasize the diverse connections within and among ecosystems, as well as the damaging consequences of human actions that are increasingly causing environmental and ecological changes. If these changes exceed the biospheric limits of homeostatic tolerance and repair, the consequences for the planet's geophysiology and ecology could be catastrophic.

Key Terms

ecology
stressors
disturbance
succession
individual organism
population
species
community
landscape
seascape
the biosphere
autecology
competitor
ruderal
stress-tolerator
carrying capacity
synecology
niche
symbiosis
mutualism
parasitism
commensalism
watershed
Gaia hypothesis

Questions for Discussion

1. Distinguish between autecology and synecology, giving examples of each.
2. Define birth rate, death rate, immigration rate, and emigration rate and describe how these demographic factors influence population growth.
3. Choose an ecological community with which you are familiar and outline the environmental and biological influences that affect its structure and function.
4. What are the attributes of an ecological landscape (or seascape)? Illustrate your answer with reference to a landscape in the region where you live.
5. Outline the core elements of the Gaia hypothesis. Describe some of the evidence in support of this idea.

References

Begon, M., J.L. Harper, and C.R. Townsend. 1996. *Ecology: Individuals, Populations, and Communities*. London: Blackwell Scientific.

Freedman, B. and J. Riley. 1980. Population trends of various species of birds wintering in southern Ontario. *Ontario Field Biologist*, **34**: 49–79.

Goodwin, C. E., B. Freedman, and S. M. McKay. 1977. Population trends in waterfowl wintering in the Toronto region, 1929–1976. *Ontario Field Biologist*, **31**: 1–28.

Grime, J.P. 1979. *Plant Strategies and Vegetation Processes*. London: J. Wiley.

Kimmins, J.P. 1987. *Forest Ecology*. New York: Macmillan.

Krebs, C.J. 1985. *Ecology: The Experimental Analysis of Distribution and Abundance*. New York: Harper & Row.

MacDonald, G.M. and L.C. Cwynar. 1991. Post-glacial population growth rates of *Pinus contorta* ssp. *latifolia* in western Canada. *J. Ecol.*, **79**: 417–429.

Odum, E.P. 1983. *Basic Ecology*. New York: Saunders.

Silverton, J. 1987. *Introduction to Plant Population Ecology*. Harlow, UK: Longman Scientific and Technical.

Smith, L.E. 1991. *Gaia: The Growth of an Idea*. New York: St. Martin's.

GLOBAL POPULATIONS

CHAPTER OUTLINE

CHAPTER OBJECTIVES

After completing this chapter, you will be able to:

1. Outline the process of cultural evolution and describe how it has resulted in changes in the environment's carrying capacity for human populations.
2. Describe the growth of the global human population during the past 10 000 years.
3. Explain the causes of global population growth during the past several centuries.
4. Discuss the reasons for differences in population growth rates between developed and less-developed countries.
5. Explain the influences of the demographic transition and age-class structure on population growth.
6. List the major methods of birth control.
7. Discuss the reasons why certain methods of birth control are controversial.

INTRODUCTION

About 10 000 years ago, there were only a few million humans on Earth. Today there are more than six billion, and the numbers are climbing steadily (by about 90 million per year). In terms of consequences for the biosphere, the enormous growth of the human population is the most significant earthly event in the past 15 000–20 000 years (that is, since the most recent glaciation; see Chapter 3).

The global population of humans has been increasing for several millennia, but the growth has been particularly rapid during the past few centuries. Moreover, there is every indication that the present, extremely large population will continue to increase in the foreseeable future. We will examine some possible scenarios of future population growth later in this chapter.

The environmental consequence of the presence of a human population is a function of two interacting factors: the actual number of people and the per capita environmental impact. The per capita impact is related to both the lifestyles of individual people and the level of technological development of their society. Both of these influence resource use, pollution, and the degradation of ecosystems (see also Chapter 1).

The growth of the human population during the past several millennia is quite remarkable, and may be unprecedented in scale during the history of life on Earth. This inference is based on:

- the rather long period of time during which population growth has been sustained
- the extraordinary abundance that has been achieved

The human population is growing rapidly, and now numbers more than six billion. This scene shows a crush of urban people in Puskar, India.

Source: Philip Reeve/Tony Stone Images

- the similarly impressive collateral population growth of mutualist species (such as cows, pigs, chickens, and agricultural plants)
- the remarkable breadth of ecosystems and species that are being exploited as resources in support of the human enterprise

In large part, these phenomenal achievements of *Homo sapiens* have been realized through **cultural evolution**, the progression of adaptive discoveries of increasingly sophisticated tools and social systems. The capacity to learn from the experience of others, including the passage of information from one generation to the next, has allowed cultural evolution to proceed. In turn, cultural evolution has allowed humans to be more efficient in the capture of resources through the exploitation of other species, ecosystems, and nonrenewable natural resources (Chapter 12). Cultural evolution has allowed humans to achieve unparalleled success in their domestication of planet Earth.

Unfortunately, intense damage has been caused to the biosphere by the combined influences of the increases in population and per capita environmental impact. Some of this damage represents a substantial reduction in the **carrying capacity** of the biosphere for humans (i.e., the abundance that can be sustained without degrading the habitat). Moreover, the extent of natural ecosystems has been severely reduced through human actions, causing a great reduction in Earth's carrying capacity for innumerable other species (see Chapter 24). This ecological damage is so increasingly extensive, and of such a great intensity, that it has produced a global environmental crisis that is worsening with time. Regrettably, enhancing the human enterprise through cultural evolution has largely been achieved by reducing the ability of the biosphere to support wild species and natural ecosystems.

The increased size of the human population is not, on its own, the root cause of the environmental crisis. The rapid escalation of per capita resource usage is also critically important. However, we cannot achieve a sustainable resolution of the environmental crisis without knowing about, and ultimately dealing with, the explosive growth in the abundance of people on Earth.

In this chapter we examine the changes that have occurred in the human population during the past 10 000 or so years, and in the intensity of resource use during cultural evolution. We also look at predictions of change in the human population in the near future. Global patterns of change are emphasized in this chapter; we will examine the population of Canada in Chapter 11.

Cultural Evolution, Carrying Capacity, and Population Growth

The biological history of hominids, including *Australopithecus africanus*, extends to perhaps four million years. The genus *Homo*, of which *Homo sapiens* is the only surviving species, goes back about two million years. For almost all of the evolutionary history of our species, relatively small populations were engaged in subsistence lifestyles, foraging over extensive areas — these people hunted wild animals and gathered edible plants. These hunter-gatherers probably roamed the landscape in small family groups, searching for food and other resources and using simple weapons and tools made of wood, bone, stone, shells, and other natural materials. The foraging lifestyle characterized the first 99% or so of human history, and during that lengthy time the global population of our species was likely less than one million individuals.

By some 12 000–15 000 years ago, all of the major habitable landmasses had been discovered by these early humans, including the Americas. The latter were colonized relatively late, when small groups of people roved eastward across a broad (up to 1000 km wide) but temporary land-bridge that connected Siberia and Alaska, through what is now the Aleutian Islands. The land-bridge was present as recently as about 11 000 years ago, and existed because so much water was tied up in continental glaciers that sea level was about 100 m lower than it is now. (Note, however, that some archaeologists believe there may have been an earlier colonization of the Americas, occurring as early as 60 000 years ago.)

The wandering Siberians found a landscape with bountiful resources which had never before been exploited by humans. Descendants of the first human colonists of the Americas spread rather quickly, like an expanding wave, to occupy and exploit all habitable regions of North, Central, and South America. Coincident with the colonizing surge of humans was a mass extinction of many species of large mammals and birds. Probably naive to the lethal prowess of the novel two-legged predators that hunted in well-coordinated packs, these unfortune animals were unable to adapt to the onslaught (see Chapters 12 and 24).

During the protracted phase of foraging societies, cultural evolution was not static, and there were many significant advances of culture and technology. The adaptive innovations included:

- improvements of tools and weapons
- discoveries of edible and medicinal species
- development of better social organizations for more efficient exploitation of natural resources
- the mastery of fire
- the domestication of the dog, which allowed more efficient hunting, served as a pack animal, and helped keep encampments clean

Each of these prehistoric breakthroughs enhanced the ability of humans to exploit ecological resources. This increased the effective carrying capacity of the ecosystems they were utilizing, and allowed the population to increase. By the end of this period (about 9 000–10 000 years ago), during which most human societies employed foraging lifestyles, the global population was likely between one and five million individuals.

About 10 000 years ago, the first significant developments of primitive agriculture began, signalling the beginning of the neolithic revolution (sometimes known as the new stone age; Figure 10.1). The first agricultural innovations included the initial stages of domestication of a few edible species of plants and animals, such as barley and sheep, and the discoveries of simple ways of cultivat7ing these to achieve greater yields. Because crops must be tended and protected, adoption of agricultural technologies required a relatively sedentary lifestyle. The eventual achievement of agricultural food surpluses allowed some people to be supported as non-agricultural workers

FIGURE 10.1 HISTORY OF HUMAN POPULATION GROWTH

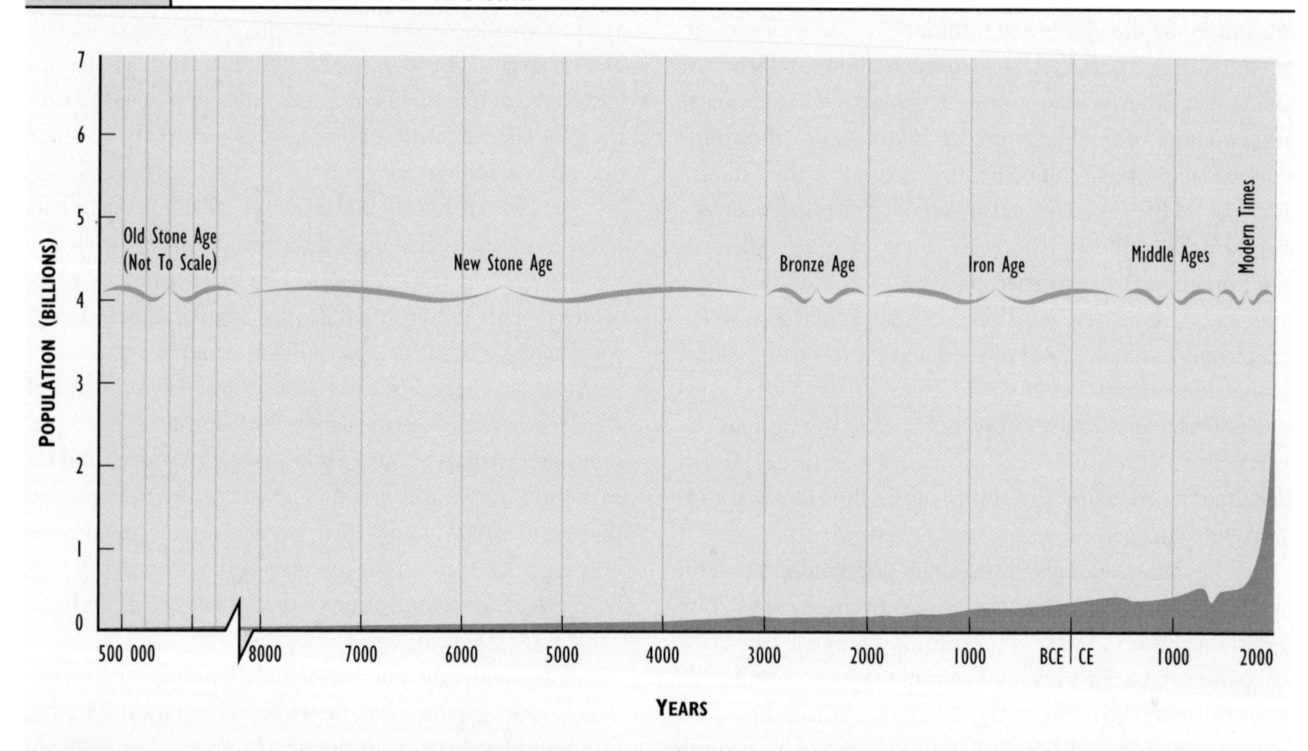

For most of the history of *Homo sapiens*, the global population remained stable at several millions or less. Successive innovations through cultural evolution allowed more efficient exploitation of environmental resources, so the effective carrying capacity for humans increased. This process has intensified greatly during the past several millennia and especially during the past several centuries. Human populations are now exhibiting explosive growth of a magnitude that is probably unprecedented by any large animal in the history of the biosphere.

Source: Modified from Ehrlich *et al.* (1977)

in towns. This eventually fostered the development of city-states and then nation-states, along with their relatively sophisticated cultures and technologies.

The development of agriculture and its associated sociocultural systems is one of the great leaps of human cultural evolution. The neolithic revolution provided an enormous increase in the carrying capacity of the environment for humans and their domesticated species. Steady population growth was one result of this change, because even primitive agricultural systems could support many more people than could subsistence lifestyles based on foraging for wild plants and animals.

The initial development of agriculture was followed by innovations that greatly increased agricultural yields. These improvements included the domestication of additional species of crop plants and animals, their genetic improvement through artificial selection (or selective breeding), and the discovery of better ways of managing the environment to increase crop productivity. In addition, there were many nonagricultural enhancements of the carrying capacity, including the discoveries of the properties of metals and their alloys, which allowed the manufacturing of tools and weapons that were superior to those made from bone, wood, or stone. Further, the domestication of beasts of burden and the invention of wheeled vehicles and ships made it easier to transport large quantities of valuable commodities, greatly stimulating trade.

Even this brief outline suggests that the cultural evolution of human sociotechnological systems has involved a long series of adaptive discoveries and innovations. Each of these increased the ability of people to exploit the resources of their environment. This increased the effective carrying capacity for people, which fostered growth of the population of humans and their mutualist species.

As a result of this adaptive progression, there were about five million people alive at the dawn of agriculture about 10 000 years ago, 200–300 million at the beginning of the Common Era 2000 years ago (0 CE), and 500 million in 1650.

The rate of population growth then began to increase markedly, a trend that has been maintained to the present. These recent, extremely rapid increases of human populations have occurred for several reasons. Of primary importance have been the discoveries of increasingly effective technologies in sanitation and medicine, which resulted in great decreases in death rates. Protection from communicable diseases has been especially important in this regard.

In addition, recently discovered technologies allow increasingly effective extraction of resources, manufacturing of improved products, and better methods and infrastructures for transportation and communications. These have been achieved as a result of the "industrial revolution," which began in the mid-18th century. Agricultural systems have also been enormously enhanced through the development of new crop varieties and improved cultivation methods, both of which have greatly increased yields. Again, all of these progressions of cultural evolution have further increased the carrying capacity of the environment for humans.

The global population of humans reached 1 billion in 1850, 2 billion in 1930, 4 billion in 1975, and 6 billion in 1997 (see In Detail 10.1). **Doubling time** is the length

IN DETAIL 10.1

THE HUMAN POPULATION IN CONTEXT

The global population of humans in 1997 was about 6 billion individuals. That enormous population was growing at about 1.5%/y, equivalent to an additional 90 million people each year. If that rate of growth is maintained, the human population will double in only 47 years, at which time 12 billion people would live on Earth. To put these gigantic numbers into perspective, it is useful to consider the abundances of other species of large animals ("large" is defined as weighing more than about 44 kg, or 100 pounds).

Some large animal species have been domesticated and live in a mutualism with humans. The most populous of these include sheep and goats (*Ovis aries* and *Capra hircus*), which have a combined population of about 1.7 billion. There are also about 1.3 billion cows (*Bos taurus* and *B. indica*), 0.85 billion pigs (*Sus scrofa*), 0.12 billion equines (mostly horses, *Equus caballus*), and 0.16 billion camels and water buffalo (*Camelus dromedarius* and *Bubalus bubalis*). Some smaller mutualists of humans are even more abundant, including an estimated 10–11 billion fowl, most of which are chickens (*Gallus gallus*).

It is doubtful that any wild large animal ever had such enormous populations as have humans and their domesticated mutualists. Within historical times, the most populous wild large animal was the American bison (*Bison bison*), which, prior to its near extermination by over-hunting, may have numbered as many as 60 million individuals. At the present time, the most populous large animals in the wild are the white-tailed deer (*Odocoileus virginianus*) of the Americas with 40–60 million individuals, and the crabeater seal (*Lobodon carcinophagus*) of the Antarctic with 15–30 million. The populations of these wild species are equivalent to only 1% or less of the present abundance of humans.

A few other wild species of large animals maintain populations in the millions, including:

- about 20 million grey and red kangaroos (*Macropus gigantea* and *M. rufus*) in Australia
- 6–7 million ringed seals (*Phoca hispida*) in the Arctic
- about 5 million harp seals (*Phoca groenlandica*) in the North Atlantic
- about 3 million caribou (or reindeer, *Rangifer tarandus*) in the Arctic and subarctic
- more than 2 million each of striped and spotted dolphins (*Stenella* spp. and *Lagenorhynchus* spp.) in the eastern Pacific
- 2 million northern fur seals (*Callorhinus ursinus*) in the North Pacific
- 1–2 million sperm whales (*Physeter macrocephalus*) in all oceans
- 1.4 million wildebeest (*Connochaetes taurinus*) in Africa
- some species of large fish may also have populations in the millions, but reliable data are not available

Clearly, humans and their large-animal mutualists are unusually, even unnaturally, abundant. The huge populations of domesticated large animals (humans included) can be maintained only by using an extremely large fraction of the productivity of Earth's ecosystems. Some ecologists have estimated that fraction to be as large as about 50% (Vitousek *et al.*, 1986).

of time required for a population to double in size. Between 8000 BCE and 1650 CE, the human population increased from 5 million to 500 million; the average doubling time was 1500 years, and the growth rate was about 0.01% per year (Table 10.1). During the late 1960s, the growth rate of the global population was at about its historical maximum, equal to about 2.1% per year. At this rate of increase, the human population is capable of doubling in only 33 years. Population growth rates have slowed significantly since then, to about 1.5% during the 1990s. If maintained, however, even this rate of increase will double the population in only 47 years. In fact, during the 1990s there was an annual net addition of about 90 million people to the global population. For perspective, this annual increase is equivalent to more than three times the population of Canada.

TABLE 10.1 GROWTH AND DOUBLING TIMES OF THE HUMAN POPULATION

DATE	GLOBAL POPULATION (MILLION)	DOUBLING TIME (YEARS)
8000 BCE	5	1500
1650 CE	500	200
1850	1000	80
1930	2000	45
1975	4000	36
1996	6000	41

Source: Modified from Ehrlich *et al.* (1977)

The cultural evolution of social, technological, and economic systems has allowed lifestyle improvements for many people. The main advancements are in food and health security, resulting from better access to sanitation, health care, food, shelter, and other elements of subsistence. (In the sense meant here, "security" is related to the likelihood of living to old age and of having access to the necessities and amenities of life.) The quality of life has also been improved through greater access to aesthetic resources and amenities, such as culture and recreation. Of course, these betterments of lifestyles are not shared equally among people, and are virtually unavailable to enormous numbers of poor people in the modern world.

This is not to say that hunter-gatherers did not enjoy aspects of their lifestyle. These people undoubtedly had a rich cultural life, and many were able to satisfy their subsistence requirements by "working" only a few hours each day, leaving much time for relaxing and socializing. In fact, the transitions to agricultural and then industrial societies may have involved greater workloads for average people and less time for relaxation. The additional work has, however, reaped benefits of the sort noted above, and for much larger numbers of people.

Nevertheless, the many improvements in human security noted above have involved a great intensification of per capita environmental impacts. We can easily understand this important change by examining patterns of energy usage, which is the best simple indicator of per capita impacts (Table 10.2; see also Chapter 1). Compared with hunter-gatherers, people living in an advanced technological society use at least 50 times more energy, and their environmental impact is greater by at least a similar degree.

The intensification of per capita energy usage has been especially great during the past century or so of accelerating technological development and innovation. In fact, during the present century, average per capita economic output and fuel consumption have both increased more rapidly than has the human population (Table 10.3). These changes have, of course, been most substantial in industrialized countries such as Canada. Per capita consumption of industrial energy in industrialized countries averages about 160 gigajoules/year, compared with 17 GJ/y in less-developed countries (1990 data; WRI, 1995). Although industrialized countries account for about 22% of the human population, they use about 72% of the global energy production.

TABLE 10.2 | CULTURAL EVOLUTION AND ENERGY USAGE

These data estimate the typical, per capita energy usage by people engaged in various kinds of lifestyles. Energy use is presented here as a simple indicator of environmental impacts.

STAGE OF CULTURAL EVOLUTION	PER CAPITA ENERGY USE (MJ/day)
Foraging	20
Primitive agriculture	48
Advanced agriculture	104
Industrial society	308
Advanced industrial society	1025

Source: Modified from Goldemberg (1992)

Regional Variations in Population Growth

The data cited to this point have been global in scale. It is important to recognize, however, that modern rates of population growth vary enormously among regions and countries. In recent decades, some countries have achieved natural rates of population growth as high as 4% per year, which if maintained would double their populations in only 18 years.

Populations are growing most quickly in Africa, with a rate of increase of 2.9% per year in 1995 and a projected rate of 2.7% up to 2025 (Table 10.4). Although countries vary considerably, as a whole, Africa is also the world's

TABLE 10.3 | A CENTURY OF CHANGE IN GLOBAL POPULATION AND PER CAPITA ENVIRONMENTAL IMPACT

Economic activity is given in 1980 U.S. dollars, so inflation does not account for the increases over time. Environmental impact is indicated by energy use, which is strongly related to the consumption of fossil fuels.

		ECONOMIC ACTIVITY		FOSSIL-FUEL CONSUMPTION	
YEAR	POPULATION (BILLIONS)	GROSS WORLD PRODUCT (GWP) (1980$ × 10^9)	PER CAPITA GWP (1980$ × 10^3)	TOTAL (10^9 t/y)	PER CAPITA (t/y)
1900	1.6	0.6	0.4	1	0.6
1950	2.5	2.9	1.2	3	1.2
1986	5.0	13.1	2.6	12	2.4

Source: Modified from Brown and Postel (1987)

TABLE 10.4 | REGIONAL POPULATION GROWTH

Data for 2025 are estimated from recent trends in demographic parameters (that is, rates of birth, immigration, death, and emigration).

	POPULATION (MILLIONS)				INTERVAL GROWTH RATE (%/y)		
COUNTRY	1950	1990	1995	2025*	1980–85	1990–95	1995–2025*
WORLD	2516	5295	5759	8472	1.75	1.68	1.43
Africa	222	642	744	1583	2.91	2.93	2.70
Asia	1377	3118	3407	4900	1.91	1.78	1.39
North America	166	277	292	360	0.92	1.06	0.80
Central America	50	141	157	243	2.27	2.10	1.71
South America	112	294	320	452	2.15	1.67	1.35
Europe	398	509	516	542	0.31	0.27	0.27
USSR (former)	175	281	289	345	0.89	0.51	0.65
Oceania	13	27	29	41	1.51	1.51	1.37

* *Estimated.*

Source: World Resources Institute (1995)

poorest region. Its condition is due partly to legacies of its colonial history, including national boundaries that often make little sense in view of the distributions of ethnic and tribal groups. In some countries, other factors also detract from development, especially endemic corruption in government and business and strife between tribes. The exploding populations of Africans are also making it extremely difficult to deal with the problem of chronic poverty. The population of Africa was about 222 million in 1950, 744 million in 1995, and an anticipated 1,583 million (or 1.6 billion) in 2025. With human populations doubling every 20–25 years, while the amount of agricultural land and other resources are remaining static or even declining because of over-exploitation, it will be a formidable challenge to avert social and ecological catastrophes in some African countries. Resolving these problems will likely require coordinated international assistance.

Populations are growing least quickly in Europe, where the present rate of increase is about 0.27%/y (doubling time of 250 years), a rate that is expected to hold for several decades. Populations are growing somewhat more quickly in North America, presently at 1.1%/y (doubling time of 64 years), and projected to be 0.8%/y (88 years) up to 2025.

Much of the population growth in Canada and the United States is due to relatively free immigration from other regions rather than to intrinsic growth (births minus deaths) (see Chapter 11). The major source regions for this immigration are Asia, Central and South America, and Africa, mostly from countries with growing populations and few prospects for most poor people to improve their lifestyles. Substantial immigration is also coming from parts of eastern Europe where there is little or no population growth but much unemployment and economic hardship.

Population Growth in Particular Countries

Data for recent population growth in selected countries are listed in Table 10.5. Countries with the most rapidly increasing populations are located in Africa and Asia, and to a somewhat lesser degree, Central and South America. Populations are increasing rapidly in almost all countries in those regions.

There are, however, some anomalous situations. Afghanistan, for instance, had a population decrease of 2.0%/y during 1980–85. This was largely because that country's devastating civil war killed so many of its citizens, while others emigrated to neighbouring countries, especially Pakistan. In contrast, the population of Afghanistan increased during 1990–95 by an extraordinary 6.7%/y. This happened because many displaced civilians returned after a sporadic cease-fire restored relative peace to the country. Overall, the intrinsic rate of increase of the Afghani population has been 3.0–3.5%/y. In 1950 the population of Afghanistan was 9 million, by 1995 it had increased 2.5 fold to 23 million, and it is projected to double to 46 million by 2025.

Some countries have had recent population growths even more rapid than that in Afghanistan. In Kenya, for example, the population was 6 million in 1950 and is projected to reach 64 million in 2025, a 10-fold increase. (Note that such predictions for Kenya and some other countries may prove to be inaccurate, because of huge mortality that could result from the accelerating AIDS epidemic or some other disaster; see the last section of this chapter.) In Côte d'Ivoire, the population increase over the same period is projected to be more than 13 fold. Try to imagine the ecological and resource stresses associated with population explosions like these! Contemplate being a politician or government bureaucrat who is charged with the responsibility of ensuring livelihoods and an acceptable quality of life for so many citizens, while also protecting the ecological heritage and environmental quality of the country! The challenges are extraordinarily daunting.

The countries with the most stable populations occur mainly in Europe (Table 10.5). The populations of most European countries are growing at less than 0.5%/y, and the doubling times are longer than 100 years. As was previously noted, the intrinsic rates of population increase in Canada and the United States are similar to these values, although because of substantial immigration their populations are growing at about 1%/y.

Some countries are showing rapid decreases in their rate of population increase (Table 10.5). China, for example, had population growth rates exceeding 3%/y earlier in this century, but this has decreased to 1.4%/y today, and is expected to be 0.8%/y in 2025. The slowdown is occurring because the national government of China has recognized the acute nature of that country's population problem, has developed policies to reduce the rate of growth, and is starting to implement them in an effective manner. China's government has imposed its population policy with a determination that has sparked debates over human rights. The government's controversial measures include coerced sterilization and the enforcement (particularly enthusiastic in some urban areas) of the one-child-per-family guideline. Nonetheless, China appears to be firmly on the road to rapidly decreasing its rate of population growth. We can only hope this necessary action has occurred in time. China's population in 1995 was 1.24 billion people, and even with its aggressive population policy, this will increase to 1.54 billion by 2025. These are immense numbers of people to accommodate within the bounds of China's landmass and natural resources, which are not increasing in area or quantity.

The situations in Brazil, India, Indonesia, Korea, Mexico, and Thailand are rather similar, although none of these countries is displaying declining rates of population growth as rapid as China. India, the world's second-most populous country, had 931 million people in 1995, and may somehow have to support 1.4 billion in 2025. Although all of these countries have started to develop population policies aimed at reducing growth rates, they are not being implemented as effectively as in China.

In general, countries that are experiencing rapid population growth are relatively undeveloped and poor, and tend to be tropical and subtropical in distribution. However, not all poor countries have high population growth rates. The population growth rates of Cuba, for example, have been consistently less than 1%/y since the 1950s. Although Cuba is relatively poor, almost all of its citizens are literate. Its social system also provides ready access to housing, food, social security, and health care, including effective means of birth control.

It can be broadly generalized that the wealth and state-of-development of nations correlate inversely with their population growth rates. Nevertheless, any country with an appropriate and effectively delivered population policy can achieve a measure of control of the rate at which its population is increasing. The case of Cuba demonstrates this fact.

The explosive rates of population increase in so many poor countries are straining the ecosystems that must somehow sustain the burgeoning numbers of people and their livelihoods. For instance, the population of central Sudan was 2.9 million in 1917, but it had irrupted to 18.4 million by 1977, an increase of 6.4 times (Olsson and Rapp, 1991) (an *irruption* is a rapid increase). Because most Sudanis are engaged in agricultural livelihoods, the populations of livestock have also increased tremendously during that period. The numbers of cattle increased 20 fold (to 16 million), camels 16 times (to 3.7 million), sheep 12.5 times (to 16 million), and goats by 8.5 times (to 10.4 million). Such enormous growth in the abundance of humans and their livestock has extensively degraded the carrying capacity of drylands in Sudan and other regions of Africa. Extensive damage has also been caused to natural ecosystems.

Another example concerns the numbers of people in the province of Rondonia in Amazonian Brazil. Between 1970 and 1988 this population increased 12 fold, mostly because of immigration of poor people from overpopulated cities in southern Brazil to this region of Amazonia,

TABLE 10.5 POPULATION GROWTH IN SELECTED COUNTRIES

COUNTRY	POPULATION (MILLIONS) 1950	1990	1995	2025	INTERVAL GROWTH RATE (%/y) 1980–85	1990–95	1995–2025
RAPIDLY GROWING POPULATION							
Afghanistan	9	17	23	46	−2.0	6.7	2.4
Bangladesh	42	114	128	223	2.7	2.4	2.2
Dem. Rep. of Congo	12	37	44	105	3.2	3.2	3.0
Egypt	20	52	59	94	2.6	2.2	1.9
Ethiopia	20	50	58	131	2.1	3.1	2.8
Iran	17	58	67	145	4.4	2.7	3.0
Iraq	5	18	21	46	3.3	3.2	2.9
Kenya	6	24	28	64	3.6	3.4	3.1
Madagascar	4	12	14	34	3.1	3.3	3.1
Nigeria	33	109	126	286	3.2	3.1	3.0
Pakistan	40	118	135	260	3.3	2.7	2.5
South Africa	14	38	43	73	2.6	2.4	2.1
Tanzania	8	26	31	74	3.3	3.4	3.0
DECLINING RATE OF POPULATION GROWTH							
Brazil	53	149	161	220	2.2	1.6	1.2
China	555	1153	1238	1540	1.4	1.4	0.8
India	358	846	931	1394	2.1	1.9	1.7
Indonesia	80	184	201	283	2.1	1.8	1.3
Korea (North)	10	22	24	33	1.7	1.9	1.2
Korea (South)	20	44	45	50	1.4	0.8	0.6
Mexico	27	85	94	138	2.4	2.1	1.6
Philippines	21	62	69	106	2.6	2.1	1.7
Sri Lanka	8	17	18	25	1.7	1.3	1.0
Thailand	20	55	58	72	1.8	1.3	0.9
RELATIVELY SLOW RATE OF POPULATION GROWTH							
Australia	8	17	18	25	1.4	1.4	1.2
Canada	**14**	**27**	**29**	**36**	**0.9**	**1.4**	**1.2**
Cuba	6	11	11	13	0.8	0.9	0.6
New Zealand	2	3	4	4	0.8	0.9	0.8
United States	152	250	263	322	0.9	1.0	0.8
RELATIVELY STABLE POPULATION							
France	42	57	58	61	0.5	0.4	0.3
Germany	68	80	81	84	−0.2	0.4	0.2
Italy	47	58	58	56	0.3	0.1	0.1
Japan	84	124	126	127	0.7	0.3	0.3
Sweden	7	9	9	10	0.1	0.5	0.4
United Kingdom	51	57	58	60	0.1	0.2	0.2
USSR (former)	174	281	289	344	0.9	0.5	0.7

Source: World Resources Institute (1995)

which is considered to be a development frontier. During the same period the population of cattle increased 30 times. Inevitably, these enormous population increases have been accompanied by intense ecological damage, as natural tropical forests are cleared to develop lands needed to sustain humans and their livelihoods. Commonly, the cleared land has proven unsuitable for agricultural use and has been abandoned in a highly degraded condition, while the farmers move on to clear additional forest elsewhere.

It must be remembered, however, that comparable economic growth occurred in Europe and North America in previous centuries. In Britain, for example, deforestation and other habitat losses resulted in the extirpation of numerous species of indigenous wildlife. In addition, deforestation and the grazing of hillsides in Scotland has virtually eliminated the native forest that once covered those hills. These ecologically destructive activities have largely been forgotten, and most inhabitants of Britain now regard the transformed landscape of their country as being "natural."

Birth and Death Rates

Human societies living in relatively primitive, undeveloped conditions have always tended to have high birth rates and death rates, typically about 40–50 per thousand. (Birth and death rates are measured as the average number per thousand individuals in the population per year.) These were the usual rates of natality and mortality throughout virtually all of human history. As long as both birth and death rates remained high and similar, population growth rates were small or zero. It is mostly during the past two centuries that explosive growth has occurred.

This has happened because death rates have decreased substantially in all countries. The death rate declined because of improved sanitation, medicine, immunization, and social welfare, and widespread access to education (which teaches awareness of the benefits of sanitation and medicine). The death rate is reduced until an increased life expectancy stabilizes within the population.

The benefits of sanitation, immunization, and medicine are particularly consequential for younger people, especially those less than five years old. This group tends to have the highest death rates under "primitive" conditions. Other relatively vulnerable groups that have benefitted significantly include the elderly and women during childbirth. As well, reductions in deadly infectious diseases, such as diphtheria, malaria, plague, smallpox, tuberculosis, and yellow fever, have been important in reducing death rates.

These medical and social benefits have not been shared equally among the world's countries or among disparate income groups within nations. For this reason, poorer, less-developed cultures or income groups have considerably higher death rates than do wealthier, more developed ones. This trend is readily apparent if data for death rates in poorer countries with increasing populations are compared with those of wealthier countries having more stable populations (Table 10.6).

Compared with the relatively large decreases in death rates in virtually all of the world's nations and cultures, decreases in birth rates have been less rapid and less uniform (Table 10.6). In general, the world's wealthiest, most-developed countries have relatively low birth rates, typically about 10 per thousand. Moreover, these are almost in balance with death rates, so the rate of natural population increase is low or zero. In large part, the relatively low birth rates of these countries result from the development of a cultural inclination to have small families. In addition, choices to have a small family can be achieved because effective methods of birth control are readily available.

Changes in cultural attitudes about family size appear to be a natural outcome of increasing affluence and health in developed societies. Such cultural changes are critically important for dealing with the potentially explosive population growth of modern times. It is not known exactly how these changes in attitudes came about. In less-developed societies, children are commonly viewed as sources of inexpensive labour and providers of material comfort for their parents in old age. In wealthier countries, however, all children are considered to be substantial economic and social responsibilities for their parents. As such, they become expensive consumers of space, education, energy, food, clothing, and other necessities. This is especially true in relatively affluent societies, in which each child is an expensive investment. This context provides an incentive for having smaller families.

Birth rates have remained quite high in most less-developed countries. Because death rates have fallen considerably in those countries, their populations are growing rapidly (Table 10.6). In general, birth rates have remained high because of cultural preferences for large families— a factor that may have been influenced by historically high death rates, particularly of young children. Fifty years ago, a family might have six children with only three surviving; today, all six might survive. Some religions also influence birth rates, because they favour large families or strongly disapprove of modern methods of birth control. These attitudes favour relatively high fertility and large families, resulting in a lag in the cultural adjustment of birth rates

TABLE 10.6 | BIRTH AND DEATH RATES IN SELECTED COUNTRIES

Note that birth rate minus death rate is equal to the intrinsic rate of population change. A difference of +10 units is equal to a 1% increase per year. Fertility rate is the number of children born to an average woman over her lifetime. Life expectancy is the average number of years lived, from birth.

	BIRTH RATE (BIRTHS/1000)		DEATH RATE (DEATHS/1000)		FERTILITY RATE (BIRTHS/WOMAN)		LIFE EXPECTANCY (YEARS)	
COUNTRY	1970–75	1990–95	1970–75	1990–95	1970–75	1990–95	1970–75	1990–95
RAPIDLY GROWING POPULATION								
Afghanistan	52	53	26	22	7.1	6.9	38	44
Bangladesh	49	39	21	14	7.0	4.7	45	53
Dem. Rep. of Congo	48	48	19	15	6.3	6.7	46	52
Ethiopia	48	49	23	19	6.8	7.0	41	47
Iran	44	40	15	7	6.5	6.0	56	67
Iraq	47	39	15	7	7.1	5.7	57	66
Kenya	53	44	17	10	8.1	6.3	51	59
Malawi	57	55	24	22	7.4	7.6	41	44
Nigeria	49	45	20	14	6.9	6.4	45	53
Pakistan	48	41	18	11	7.0	6.2	49	59
Tanzania	50	48	19	15	6.8	6.8	47	51
RELATIVELY SLOW RATE OF POPULATION GROWTH								
Australia	20	15	9	8	2.5	1.9	72	77
Brazil	34	23	10	7	4.7	2.8	60	66
Canada	**16**	**14**	**7**	**8**	**2.0**	**1.8**	**73**	**77**
China	31	21	9	7	4.8	2.2	63	71
Cuba	27	17	7	7	3.6	1.9	71	76
India	38	29	16	10	5.4	3.9	50	60
Indonesia	38	27	17	9	5.1	3.1	49	63
Korea (South)	29	16	5	2	4.1	1.8	62	71
Mexico	43	28	9	6	6.4	3.2	63	70
Philippines	38	30	10	7	5.5	3.9	58	65
Sri Lanka	29	21	8	6	4.0	2.5	65	72
Thailand	35	21	9	6	5.0	2.2	60	69
United States	16	16	9	9	2.0	2.1	71	76
RELATIVELY STABLE POPULATION								
France	16	14	11	10	2.3	1.8	72	77
Germany	11	11	12	11	1.6	1.5	71	76
Italy	16	10	10	10	2.3	1.3	72	77
Japan	19	11	7	8	2.1	1.7	73	79
Sweden	14	14	10	12	1.9	2.1	75	78
United Kingdom	15	14	12	12	2.0	1.9	72	76
USSR (former)	18	17	9	9	2.4	2.3	69	70

Source: World Resources Institute (1995)

to offset the rapid decline in mortality rates. The ensuing imbalance has resulted in the rapid rates of population growth seen in most less-developed countries.

Actually, the situation is not quite as simple as this. In many countries, fertility rates are maintained at a considerably higher level than many people, particularly women of child-bearing age, might freely choose. This happens because these women do not have sufficient access to effective means of birth control. Exceptions are countries such as China, and to a lesser degree Brazil, India, Indonesia, Korea, Sri Lanka, and Thailand, all of which are substantially reducing their population growth rates, mainly by ensuring that their citizens have access to effective means of controlling their fertility.

The Demographic Transition

During most of human history, characterized by relatively primitive living conditions, relatively stable or zero population growth (ZPG) occurred because high death rates were balanced by high birth rates. In modern times, ZPG occurs in relatively "developed" nations and cultures in which low death rates are balanced by low birth rates.

Unfortunately, it has typically taken a rather long time, usually several generations, for societies to make it through the so-called **demographic transition** from a condition of high birth and death rates to one of low birth and death rates (Figure 10.2). During this period of demographic transition, population growth rates typically sustain high values. This imbalance occurs because modern sanitation, immunization, medicine, and related measures all contribute to achieving rapid reductions of death rates. However, the declines in mortality occur without similar, offsetting decreases in birth rates. If, for example, annual birth rates remained at 55 per thousand while death rates declined to 22 per thousand, the population would grow at 3.3%/y. If this situation continued, the population would double in only 21 years. These numbers are, by the way, actual demographic parameters for the African country of

FIGURE 10.2 | THE DEMOGRAPHIC TRANSITION

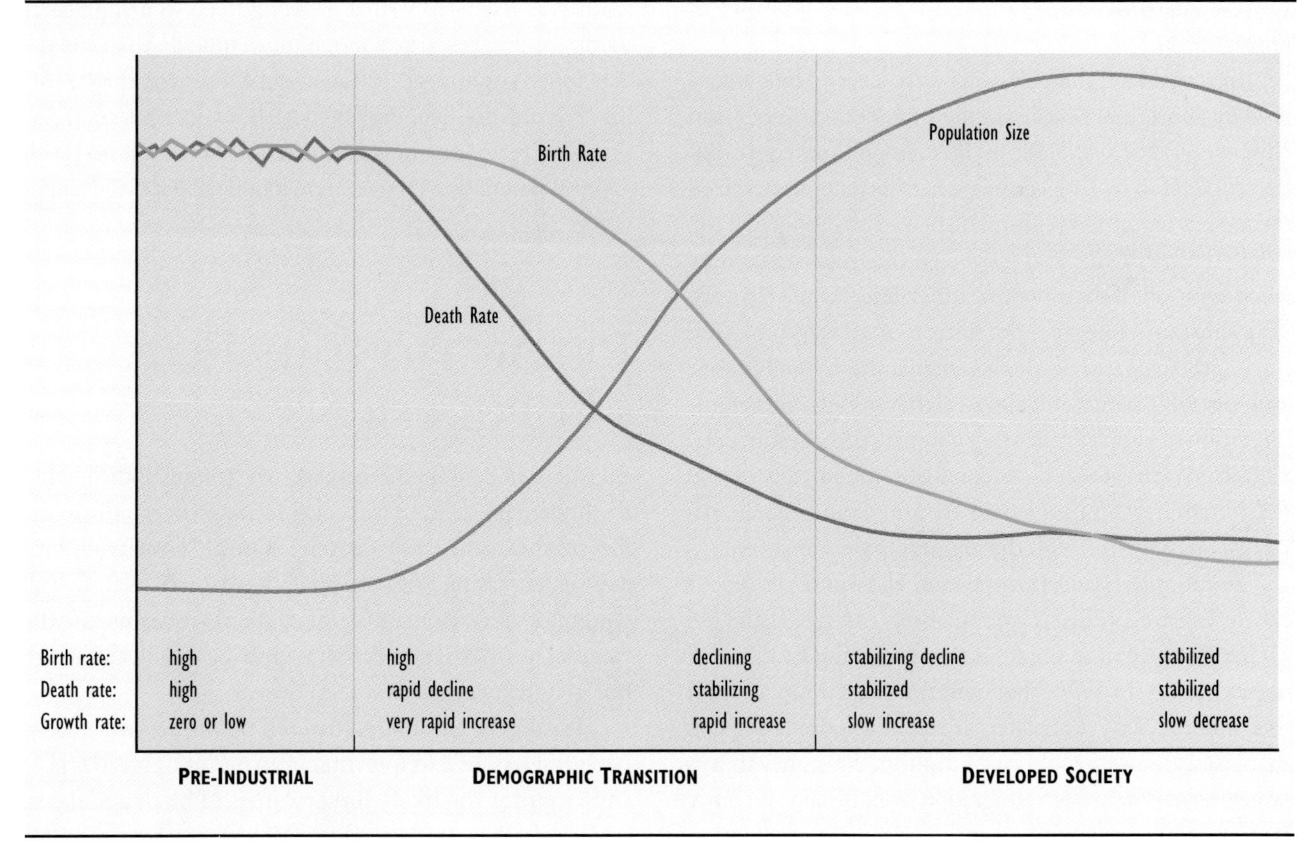

This illustration models the transition that must be made from a condition of high birth and death rates to one of low birth and death rates. Typically, death rates fall more quickly than birth rates, and the corresponding imbalance has an explosive influence on population growth rates. Zero population growth occurs when birth and death rates offset each other. If the birth rate falls below the death rate, the population size will decrease.

Malawi in 1995. These are not exceptional data for poorer countries, as can be seen by inspection of data for birth and death rates in Table 10.6.

No cultures prefer high death rates, but people often prefer large families. Recent history has shown that it takes two or more generations to overcome cultural inclinations toward having large families and for birth rates to decline to a level that is in balance with modern death rates.

Many of the world's developed countries had the great fortune of passing through their demographic transition during times when their populations were relatively small, and under circumstances in which their "surplus" people could be exported to other places. Many European countries, in particular, encouraged their surplus of poor, landless people to emigrate to colonial "frontiers" in the Americas, Australia, and elsewhere. Many countries, such as Argentina, Australia, Brazil, Canada, Mexico, New Zealand, South Africa, the United States, and Venezuela, were then colonies of European nations. These were considered to be "underpopulated" places with bountiful resources, capable of assimilating a large amount of immigration.

In actual fact, these regions were already fully occupied by aboriginal peoples at the time of their European "discovery." Nevertheless, in the sociopolitical context of that time (16th to 18th century), European powers seized ownership of foreign regions, displaced or subjugated the original inhabitants, and colonized the freed-up land by an emigration of their poor or otherwise mobile citizens. To a substantial degree, the notion of underpopulation lingers today, particularly in Canada, the United States, and Australia, which still allow relatively high rates of immigration of people from other countries. As a result, population growth rates in those countries substantially exceed their birth:death ratio, which are almost in balance, reflecting passage through the demographic transition.

Immigrants from Europe (and elsewhere) in recent centuries have swelled the populations of Canada and the United States. For example, at the time of the first United States census, in 1790, that country had a population of four million. Sixty years later, in 1850, the population had increased almost six fold to 23 million. This growth was largely achieved through emigration from Britain and other European countries. (The natural increase in population during that period was also vigorous, adding 4–8 million of those people.) Similar changes occurred in Canada (described in Chapter 11). The ability of many European countries to export so much of their surplus population was critical to their relatively smooth passage through the demographic transition.

Today only a few countries still allow and even encourage a substantial rate of immigration (notably Canada, the United States, and Australia). However, the actual number of people involved in trans-national immigration to developed countries, several million per year, is very small in comparison with global population growth (about 90 million per year during the first half of the 1990s). Moreover, there is no reason to expect that these and other host nations will continue to be willing, or able, to absorb population surpluses from other countries.

One of the bitter truths of modern times is that the relatively poor, less-developed countries of the world, with the most explosive population growth, have no significant outlets for their burgeoning surpluses of people. All nations today have substantial access to the mortality-reducing benefits of modern sanitation and medicine, but in many countries these are not being balanced by effective control of birth rates. Consequently, many less-developed countries are faced with a pressing need to bridge their demographic transition much more quickly than today's developed countries ever had to do. Moreover, this daunting feat must be accomplished without access to emigration. There are no underpopulated frontiers left on Earth — local population crises can no longer be exported somewhere else.

Predictions of Future Populations

All trends in demographic indicators strongly indicate that the human population will continue to grow rapidly in the foreseeable future. Fortunately, there are convincing signals of decreasing rates of population growth in almost all countries. This is encouraging, but it does not negate the fact that the global population is growing rapidly (although not as quickly as it was several decades ago).

It is never possible to foretell the future accurately. Nevertheless, by extrapolating from recent trends it is possible to infer the likely future values of birth and death rates and other demographic variables. Such predictions can then be modified according to anticipated changes in social policies that may influence birth or death rates.

There is also, of course, the possibility that some catastrophic event or environmental change could cause a massive increase in human death rates, resulting in a population crash. Such a potential calamity could be caused by a virulent disease, a collapse in resource availability, a nuclear holocaust, or some other emergency. Events such as these are extremely unpredictable and can never be forecast with accuracy, the most that can be suggested is that they might occur some time in the future.

Population scientists have developed sophisticated mathematical models to predict future populations of humans. Each of these models can be run using various demographic scenarios: for example, by changing the values of birth rates, death rates, population structure, or other variables. These models are not catastrophist — they assume that future population sizes will be determined by relatively small changes of birth and death rates and not by a huge increase in death rates (that is, a population crash).

In less-developed countries, governments will have to find livelihoods for increasingly larger numbers of young people, even as free space, resources, and environmental quality are rapidly diminishing. These students live in Nausori, Fiji.

Source: Victor Last/Geographical Visual Aids

FIGURE 10.3 | PREDICTIONS OF GLOBAL POPULATION

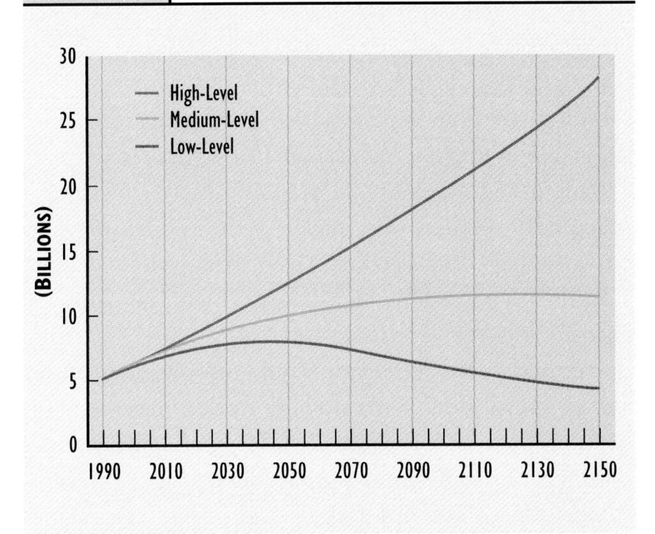

Three population scenarios are presented here, ranging from low-level to high-level in terms of the assumptions of demographic parameters. The low-level model makes optimistic and probably unrealistic assumptions about the development and implementation of population policies. Consequently it predicts relatively small populations in the future. The higher-level model is more conservative, and assumes that the imbalance between birth and death rates will be addressed rather slowly. The low-level model assumes that average fertility will stabilize at 1.7 children per woman; the medium-level, at 2.06; and the high-level, at 2.5. In 1995, the global average fertility was 3.3 children per woman.

Source: Modified from United Nations (1992) and World Resources Institute (1995)

The various population models are commonly divided into low-, medium-, and high-level projections (Figure 10.3). Low-level models use the most optimistic demographic predictions; they assume that effective population policies will be implemented rapidly and will allow stable populations to be achieved as quickly as can be hoped. The high-level models use relatively conservative parameters, suggesting that effective policies will not be implemented until considerable time has passed, so that demographic transitions will take a long time to occur in rapidly growing populations. Of course, high-level models forecast the largest future populations.

The medium-level predictions are perhaps most realistic, because they use the most likely scenarios of political and economic factors that affect population policies and their influences on demographic parameters. As such, the medium-level models portray the most reasonably expected outcome of modern trends of growth of the human population. The medium-level model shown in Figure 10.3 suggests that the global human population could increase from its present (1997) abundance of about 6 billion to 8 billion in 2025 and to 10 billion in 2050, and that abundance would stabilize at about 12 billion. It appears, therefore, that the global population will approximately double from its late-1990s level before it stabilizes. This assumes, of course, that there is no intervening catastrophe such as a collapse of the environmental carrying capacity for our species, an unprecedented pandemic, or a destructive war.

Moreover, it appears that the populations of almost all of the world's countries will increase. The population growth will not, however, be equitably shared between the less-developed and more-developed regions of the world. In 1950, about 31% of the world's population of 2.5 billion lived in developed countries. However, recent population growth has been much more rapid in less-developed countries, so that in 1996, only about 21% of the world's 5.9 billion people lived in developed regions. This disparate trend will intensify in the near future, and by 2050 only 14% of the world's 10 billion people may be living in developed regions (see Figure 10.4).

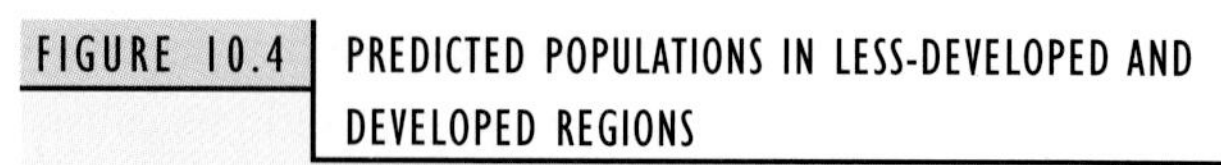

FIGURE 10.4 PREDICTED POPULATIONS IN LESS-DEVELOPED AND DEVELOPED REGIONS

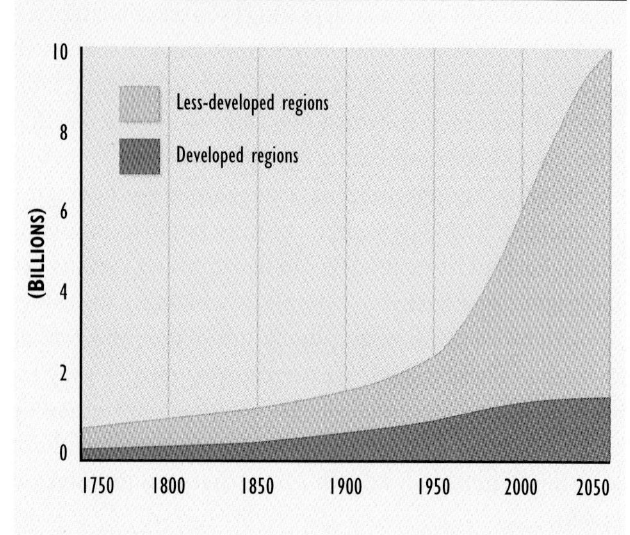

These predictions are based on medium-level population projections.

Source: Modified from United Nations (1992) and World Resources Institute (1995)

Age-Class Structure

A *population structure* shows the relative abundance of specified groups of people in a specific area, as well as their spatial patterns of distribution. The most important is the **age-class structure**, or the proportions of individuals in various age groups. This type of population structure differs greatly between growing and stable populations, and has important implications for their future growth.

Populations that have been stable for some time have similar proportions of people in various age classes (Sweden, Figure 10.5). In other words, there are roughly comparable numbers of people aged 5–10, 10–15, 15–20, 20–25 years, and so on. This equitable distribution of population structure holds for most age-classes, except for the elderly, which always have a higher risk of mortality.

In marked contrast, the age-class distribution of a rapidly growing population reflects the fact that there are more younger individuals than older people in these populations. Consequently, growing populations have a triangular age-class structure — that is, much wider at the bottom than at the top. In fact, almost one-half of the people in a rapidly growing population are typically less than fifteen years old (Table 10.7). This kind of population structure suggests an enormous potential for future growth, as increasingly larger numbers of young people mature to reproductive age.

The growth potential (or inertia) of populations with a triangular age structure is an extremely important demographic fact. Because of this inertia, it is difficult for populations to *stop* growing quickly, and it usually takes several generations to pass through the demographic transition. For example, a "young" population (that is, one with a triangular age-class structure) could rapidly achieve a **replacement fertility rate**, such that the number of progeny would only replace their parents (equivalent to about 2.1 children per family; slightly larger than 2 per family to account for the fact that some people are infertile). Nevertheless, the population would continue to grow for some time, although at a slowing rate of increase. This happens because for several decades, increasingly larger numbers of people come to reproductive age, a circumstance related to the initial, triangular age-class structure of the population. Eventually, replacement fertility rate would bring about a stable age-class structure, and zero population growth would finally occur.

The national population policy of China advocates one-child families, a fertility rate considerably smaller than that which would replace the numbers of parents. During the past decade or so, the one-child policy has been pursued with varying levels of official enthusiasm, generally more aggressively in urbanized areas than in rural ones. Most families of child-bearing age are subjected to intense social pressure to follow the one-child guideline, as well as to significant economic disincentives such as fines and lack of educational opportunities for sec-

FIGURE 10.5 | AGE-CLASS STRUCTURES OF HUMAN POPULATIONS

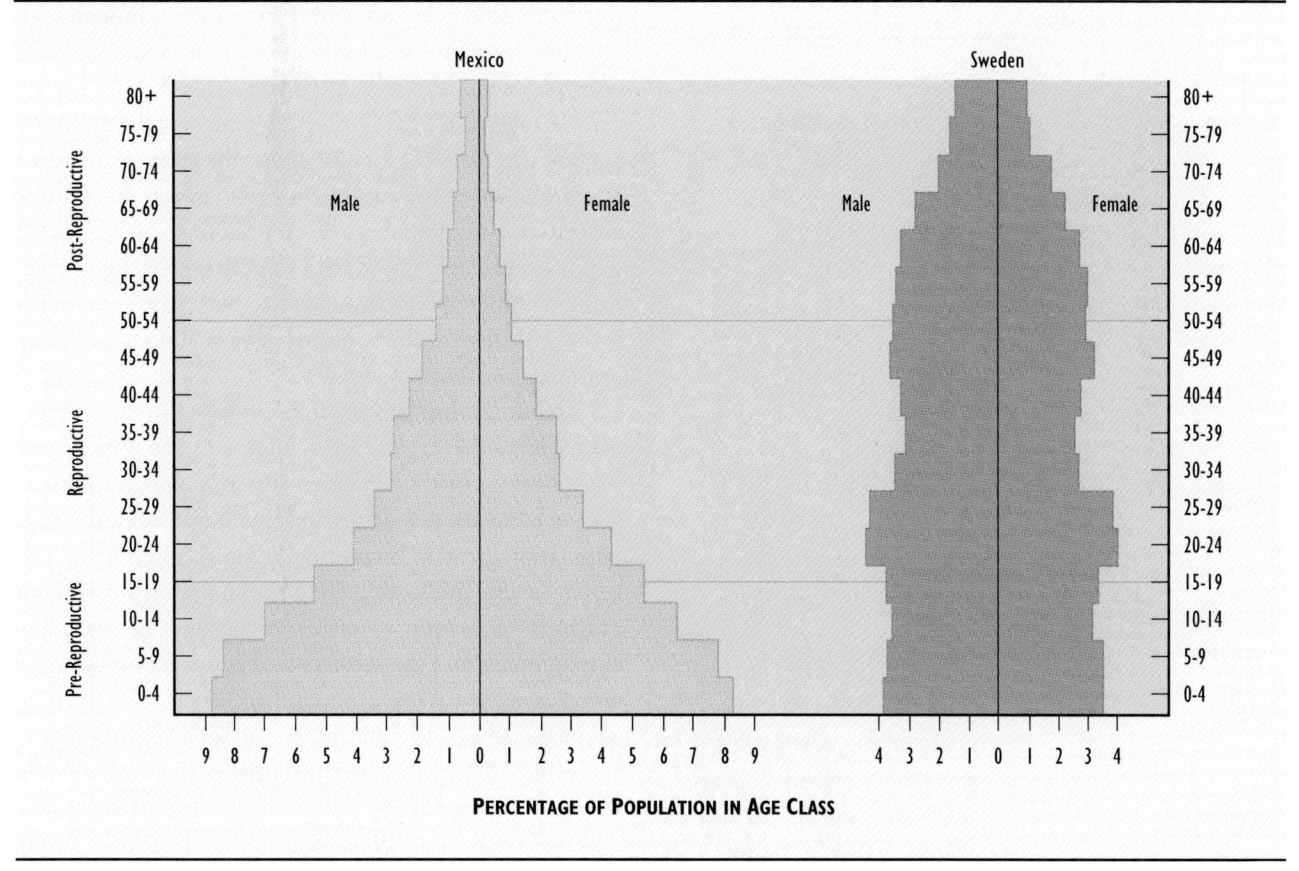

Compare the age-class structures of a relatively young, rapidly growing population (Mexico, 1970) with a stable population (Sweden, 1970).

Source: Modified from Ehrlich *et al.* (1977)

ond children. Many reluctant parents have been sterilized against their will.

In addition, there is a strong cultural preference in China for male children because of the prevailing system of inheritance of property and family lineage by the first-born son. This cultural attitude has led to widespread abortion of female fetuses, giving up female newborns for adoption, and even cases of female infanticide. These are not agreeable choices, but they are taken so that another attempt can be made to ensure that the single child allowed the family will be a male. If female abortion and infanticide become widespread, an imbalance in the female:male ratio in the population will develop. This could have important social implications when many men have difficulties finding a spouse and starting a family.

Some of the population-control measures that occur in China are regrettable. And from some points of view they abuse human rights. However, the Chinese federal government considers such measures necessary in view of their country's enormous population and the further growth implicit in its triangular age-class structure. In other words, in countries where population growth is causing a desperate situation, a vigorous implementation of aggressive population policies seems appropriate.

Distribution of Populations

Another important element of population structure is the spatial distribution of people and how this changes over time. This is a highly complex topic, because the distribution of people varies enormously within countries (compare, for example, urban and rural populations), and also among countries in different stages of development.

TABLE 10.7 AGE STRUCTURE OF THE POPULATIONS OF SELECTED COUNTRIES

Data are for 1995.

COUNTRY	PERCENTAGE OF POPULATION		
	<15	15–65	>65
RAPIDLY GROWING POPULATION			
Afghanistan	40.0	57.4	2.6
Bangladesh	40.3	56.7	3.0
Dem. Rep. of Congo	48.1	49.0	2.8
Ethiopia	46.5	50.7	2.8
Iran	45.9	50.3	3.8
Iraq	43.8	53.3	3.0
Kenya	47.4	49.7	2.9
Malawi	49.2	48.2	2.6
Nigeria	47.0	50.4	2.6
Pakistan	43.6	53.4	2.9
Tanzania	48.0	49.6	2.5
RELATIVELY SLOW RATE OF POPULATION GROWTH			
Australia	21.7	66.8	11.6
Brazil	32.2	62.7	5.2
Canada	**20.7**	**67.3**	**12.0**
China	27.3	66.4	6.3
Cuba	23.2	67.8	9.0
India	34.9	60.3	4.9
Indonesia	33.4	62.2	4.4
Korea (South)	23.3	71.1	5.4
Mexico	36.0	60.0	4.0
Philippines	38.4	58.3	3.3
Sri Lanka	30.3	63.8	5.9
Thailand	29.2	66.3	4.5
United States	21.9	65.5	12.6
RELATIVELY STABLE POPULATION			
France	19.8	65.3	14.9
Germany	17.1	68.1	14.8
Italy	15.4	69.0	15.6
Japan	16.8	69.2	13.9
Sweden	18.8	63.8	17.4
United Kingdom	19.7	64.7	15.6
USSR (former)	25.1	64.0	10.9

Source: World Resource Institute (1995)

As we previously noted, most of the world's people live in relatively poor, less-developed countries, largely in the tropics and subtropics. Because populations are growing most quickly in those areas, this pattern of global distribution will intensify further over time (see Figure 10.4). The exploding populations in those countries present enormous challenges to national and global governments, international aid agencies, and the security of human society more broadly. Even today, most of the great masses of poor people in less-developed countries do not have access to acceptable livelihoods, or even to minimally acceptable standards of food, shelter, education, health care, and other necessities of life, not to mention the cultural and entertainment amenities that help to make living a happy experience. Faced with rapidly increasing populations, will these poor countries be able to do better in the future? Will wealthier countries be willing to aid them to the degree that is necessary?

Urbanization (that is, the development of cities and towns) is another critical aspect of population distribution. Most of the world's countries are urbanizing rapidly. In fact, urbanization is occurring much more rapidly than population growth. Increasing urbanization is being driven by several interacting factors: population growth, migrations of people to cities and towns in search of employment, and the services and cultural attractions of urban areas. On average, about three-quarters of the population of developed countries lives in urban environments, and a third in the less-developed parts of the world (World Resources Institute, 1995). By 2005, however, it is expected that half of the world's population will be urban, and in 2025 that fraction may be two-thirds. Global urbanization represents an extraordinary change in the distribution of people, compared with the essentially agrarian societies of only one century ago, when more than 95% of people lived in rural areas.

Populations of the world's greatest metropolitan areas, or **megacities**, are shown in Table 10.8. Most of the world's largest urbanized areas are located in developing countries, a trend which will increase in the future.

Urban people live under densely crowded conditions, and their livelihoods tend to involve manufacturing, government administration, financial institutions, education, and services. In addition, many urban people depend on social assistance. No cities are self-sufficient in food, energy, or raw materials for building and manufacturing, and few have enough potable water. Urban areas depend on trade with the surrounding countryside and with foreign nations to provide these necessities of life and economy. The development and maintenance of the complex physical, economic, and social infrastructures required to care for enormous numbers of urban people is an extraordinary challenge for governments, particularly in relatively poor countries.

TABLE 10.8 | MEGACITIES

Growth of some of the world's great metropolitan areas, ranked by their anticipated population in 2000.

CITY	COUNTRY	POPULATION (MILLIONS) 1980	1990	2000	GROWTH RATE (%/y; 1980–90)
Tokyo	Japan	21.9	25.0	28.0	1.4
Sao Paulo	Brazil	12.1	18.1	22.6	4.1
Bombay	India	8.1	12.2	18.1	4.2
Shanghai	China	11.7	13.4	17.4	1.4
New York	United States	15.6	16.1	16.6	0.3
Mexico City	Mexico	13.9	15.1	16.2	0.8
Beijing	China	9.0	10.9	14.4	1.9
Lagos	Nigeria	4.4	7.7	13.5	5.8
Jakarta	Indonesia	6.0	9.2	13.4	4.4
Los Angeles	United States	9.5	11.5	13.2	1.9
Seoul	South Korea	8.3	11.0	13.0	2.9
Buenos Aires	Argentina	9.9	11.5	12.8	1.4
Calcutta	India	9.0	10.7	12.7	1.8
Manila	Philippines	6.0	8.9	12.6	4.1
Tianjin	China	7.3	9.2	12.5	2.4
Rio de Janeiro	Brazil	8.8	10.9	12.2	2.2
Karachi	Pakistan	5.0	7.9	11.9	4.7
Delhi	India	5.6	8.2	11.7	3.9
Dacca	Bangladesh	3.3	6.6	11.5	7.2
Cairo	Egypt	6.9	8.6	10.8	2.3
Osaka	Japan	10.0	10.5	10.6	0.5

Source: World Resources Institute (1995)

BIRTH CONTROL

Many individual people and families make conscious choices about their reproduction, including how many children to have. Historically, the available methods of controlling childbirth were few, unreliable, and sometimes unsafe.

Two of the best methods involve the avoidance of sexual intercourse, either before marriage (because society accepts that matrimony is a social institution involving a commitment by both parents to care for their children), or for some time after the birth of a child, which allows progeny to be well spaced. Both of these behaviours affect population-level birth rates by delaying and spacing reproduction. A complementary effect is gained by delaying marriage until relatively late in life, which if accompanied by pre-marital chastity also delays reproduction and decreases birth rates. Coitus interruptus, or withdrawal of the penis prior to ejaculation, likewise contributes to lower rates of impregnation associated with copulation. This practice has also been widely used through history. In addition, breast-feeding mothers have a lower probability of conceiving another child, so delaying the weaning of children helps to space births.

Some birth control practices have long been used, even by hunter-gatherer and early agricultural cultures, particularly when they had to deal with resource crises such as insufficient food. These include:

- the use of traditional medicines to prevent conception or to induce abortion
- infibulation, or the insertion of small pebbles or other objects into the uterus, where they can be retained for years, preventing implantation of fertilized ova
- the use of mechanical means of inducing abortion
- infanticide, or the killing of recently born infants
- mechanical means of raising the temperature within the scrotum, such as wearing a warm pouch, which inhibits sperm production and reduces fertility of the male

The Polynesian culture, for example, inhabits islands in the southwestern Pacific Ocean. Perhaps because of the obvious resource constraints associated with living on islands, the Polynesians were acutely aware of overpopulation and carrying capacity, and actively practised several methods of birth control. Prior to the modern era, Polynesians engaged in a subsistence economy, cultivating various species of crop plants (particularly coconut, sweet potato, taro, and yam); raising pigs and chickens; and hunting marine mammals, fish, mollusks, and other edible invertebrates in shallow, inshore waters. Their populations were closed to varying degrees, depending on the isolation of their island homes. On some islands, such as Easter Island, there was likely no exchange of people after the initial colonization.

The methods of birth control practised by the Polynesians included polyandry (in which one woman has several husbands at the same time), non-marriage of men who did not own land, abstention from sex for some time after birth of a child, coitus interruptus, abortion, and infanticide. However, these methods were not always sufficient to prevent population growth. Consequently, the Polynesians are well known for their voyages of colonization, undertaken during times of population pressure. They constructed sea-going outrigger sailing canoes, and provisioned them with stores of water, crop plants, pigs, and chickens. Family units then embarked and headed toward the vast Pacific horizon in search of habitable unpopulated islands. These courageous, extremely risky dispersals allowed some of the population surplus to discover new ecological opportunities, although many less lucky people must have perished at sea.

Today, there are many means of birth control which are safer and more effective than most of the methods available in the past (see In Detail 10.2). Consequently, individuals and families in many countries can now control their reproduction, choosing from a range of safe and readily available methods of birth control.

IN DETAIL 10.2

METHODS OF BIRTH CONTROL

The methods listed below vary greatly in their effectiveness in achieving birth control, and all have various physiological and psychological drawbacks.

Contraceptives

Contraceptives work by preventing ovulation, implantation of ova, or entry of sperm into the cervical canal. Oral contraceptives are most common, and have a typical failure rate of less than 1% if properly used (the failure rate is the number of pregnancies per year while using the birth-control method, compared with no birth control). A slow-release, intramuscular implant (usually into the upper arm) is also available, which can provide a failure rate of less than 1% for five years. These methods are reversible, so that conception is possible soon after stopping use of the contraceptive.

Spermicides

Spermicides are chemicals that kill sperm and are applied by women using sponges, foams, jellies, or creams that are inserted into the vagina. The failure rate is 3–20% (the higher value corresponds to failures associated with inappropriate use of the method; this also applies to the ranges cited for other methods).

Intrauterine device (IUD)

IUDs are inserted into the uterus where they prevent implantation of ova, probably by mechanically stimulating an inflammatory response. IUDs have a failure rate of 1–5%. This method is now being used less frequently because of risks of pelvic inflammation and infertility.

Cervical cap

Cervical caps are mechanical blocks, inserted by a woman to block access of sperm to the cervix. The failure rate is 3–10%.

Contraceptive diaphragm

A diaphragm blocks entrance to the cervix and is used with spermicidal jelly and/or foam. The failure rate is 3–14%.

Condom

Condoms provide a mechanical barrier that contains sperm after ejaculation, thereby preventing entry into the vagina. The failure rate is 2–10%.

Douche

Douching involves flushing of the vaginal area with water (often containing a spermicide) after intercourse. Douching is not very effective because it does not completely remove sperm, so the failure rate is about 40%.

Coitus interruptus (or withdrawal)

Withdrawal involves the male withdrawing his penis from the woman's vagina prior to ejaculation. This method is not reliable because it runs contrary to powerful sexual urges and because sperm are present in pre-ejaculation fluids emitted from the penis. The failure rate is 9–20%.

Rhythm

The rhythm method involves the timing of sexual intercourse to avoid the period during which the woman is fertile. This method has a high failure rate (13–20%), largely because it is difficult to determine the fertile period accurately.

Vasectomy

A vasectomy is one method of sterilization. It involves cutting and/or tying the vas deferens, two tubes that carry the sperm from the testes before mixing with seminal fluids. This method has a failure rate of less than 0.2%.

Tubal ligation

A tubal ligation is also a method of sterilization, in which the oviduct connecting the ovary and the uterus is tied, preventing the passage of ova. This method has a failure rate of less than 0.1%.

Abortion

Abortions involve the use of medication or an operation to terminate a pregnancy before the fetus is viable (that is, before it is capable of living unassisted outside the womb). This is an effective method of birth control and is relatively safe if carried out by qualified medical personnel.

Abstinence

Avoiding heterosexual sex avoids conception, but this method has a very high failure rate.

No birth control

On average, heterosexual intercourse without any birth control will result in fertilization of the ova, and pregnancy, about 10% of the time. If pregnancy is not desired, this would correspond to a failure rate of 90%.

Sources: Health and Welfare Canada (1980), Kane (1983), Solomon *et al.* (1990), and Leisinger and Schmitt (1994)

However, the use of birth control raises questions beyond the purely medical and scientific, including religious, ethical, and philosophical questions. For example, because of beliefs regarding the sanctity of human procreation, the Roman Catholic Church opposes almost all methods of birth control. So do most fundamentalist Christian, Jewish, and Muslim religions.

In Canada and the United States, one of the most contentious methods of birth control is abortion. On the one hand, anti-abortionists (known as "pro-life" activists) contend that abortion violates the sanctity of life and should never be permitted. On the other hand, pro-abortion activists (known as "pro-choice") argue that a fetus is not a viable life and that in a democratic society each woman should have the right to make decisions regarding her own body and reproduction. The acrimonious debate over abortion has erupted in both legal and illegal demonstrations. On a few occasions the picketing of clinics and hospitals providing abortion services has escalated to arson and even the murder of physicians and other medical personnel.

Population Policies

The topic of **population policies** is closely related to that of birth control. However, population policies do not include individual choices in reproduction, but refer to larger social strategies, designed and implemented by governments. Whether or not they have an official population policy, all nations have a population problem. From the perspectives of ecology and environmental science, these population problems can be divided into two groups, which differ significantly in their nature and in their prospects for effective action through the development and implementation of a population policy.

Less-Developed Countries

As previously discussed, less-developed countries are relatively poor and have not progressed far in industrial and socioeconomic development. Their populations are growing rapidly because of an excess of births over deaths. Under such conditions, unfettered population growth is a huge barrier to the development process: it is an enormous undertaking just to provide the minimal requirements of life to the rapidly increasing numbers of people, let alone to achieve the improved standards of living that result from socioeconomic development.

Although many people migrate from these countries to wealthier ones, there are not enough emigrants to make much of a difference in the overpopulated home countries. Wealthy countries control the numbers of immigrants they are willing to receive, and, compared with global population growth, those numbers are relatively small. When they were going through their own demographic transition some 100–200 years ago, wealthy nations exported much of their surplus population to colonies in the Americas and elsewhere. This option is not available to poorer countries today.

National population policies in less-developed countries must focus on reducing birth rates as quickly as possible, so that population growth can be arrested and hopefully even reversed (this is known as negative population growth, or NPG). Ideally, this would be achieved by educating the populace about the national importance of population issues, so that individuals and families would make appropriate free choices about family size and reproductive planning. Also ideally, there would be ready access to effective means of birth control in support of this enlightened, population-level initiative.

In addition, a more equitable distribution of wealth in less-developed countries would contribute greatly to achieving the necessary population goals, because better-off people are more inclined to have smaller families. For a similar reason, many people and agencies (such as the United Nations Population Fund) support the idea of a transfer of wealth from developed to less-developed countries. This could include both direct transfers and relief of some of the foreign-held debt loads of poorer countries.

A less desirable, but effective, alternative to these population policies involves coercive initiatives in family planning, such as social and/or economic disincentives to having more than one or two children per family. Forcing people to use family-planning practices infringes on their human rights, even if there are clear, long-term social and environmental benefits. Consequently, coercion presents a social and ethical dilemma for governments. Such actions may, however, be necessary if free choices do not result in sufficient reductions of population growth.

There is, of course, a clear alternative, which is very easily implemented by all societies and governments. Instead of designing and adopting effective population policies, governments could do nothing to reduce birth rates, thereby allowing populations to grow rapidly. This would intensify the severe environmental damage already caused in most less-developed countries, decreasing further the carrying capacity for humans, their economy, and other species. The population and resource crises would then resolve themselves in a natural, biological fashion, as would happen for any species. There would be catastrophic increase in death rates — a crash.

Developed Countries

Developed countries are relatively wealthy and their average citizens have access to desirable lifestyles, compared to typical conditions in less-developed countries. Moreover, most developed countries have passed through the demographic transition or have substantially done so. Consequently, their natural rate of population growth is relatively slow.

Progress through the demographic transition requires that new cultural attitudes be developed about appropriate family size, and that people have access to effective means of controlling their reproduction. As such, population policies in developed countries tend to ensure that people are sympathetic with small-family goals and that they have free access to safe and effective means of birth control.

A major issue concerning the population policies of developed countries is the allowable rate of immigration. There are limitless numbers of poor people living in less-developed countries who would happily migrate to wealthier countries in search of better economic, lifestyle, and social opportunities. In fact, some "conventional" economists believe that immigration should be encouraged. They believe that it is desirable for developed countries to have a growing population, which would be an increasing marketplace for saleable commodities while providing an abundant source of labour. In contrast, ecologically minded economists and environmental scientists argue that there are severe limits to such population and economic growth. This is because the relatively intense lifestyles of increasing numbers of people living in developed countries would require increasing amounts of resources.

The characteristically high per capita environmental impact of people living in wealthier societies integrates closely with population issues. As we saw in Chapter 1, people living a resource-rich lifestyle cause a greater intensity of environmental damage than do typical people in poorer countries. This fact must be recognized in the development of population policies in wealthier countries — the governments of those countries must acknowledge that there are limits to the numbers of resource-intensive people that can be sustained within their national boundaries.

Compared with people living in wealthy countries, such as Canada, inhabitants of less-developed countries make much less use of material and energy resources. Therefore, on a per capita basis, the environmental impacts of poorer people are relatively small. This family is engaged in subsistence agriculture in Sangeh, Bali.

Source: Victor Last/Geographical Visual Aids

Possible Causes of a Population Crash

Environmental scientists have suggested that uncontrolled growth of the human population could eventually lead to a population crash. If such a catastrophe occurred, it would probably be due to one of the following scenarios.

New, Virulent Diseases

A population crash could be caused by the emergence of one or more new communicable diseases to which human populations have little or no immunity. There are historical precedents for such a deadly phenomenon. The best example is bubonic plague or black death, caused by the bacterium *Pasteurella pestis*. Bubonic plague is thought to have originated in a species of rat. Under unsanitary conditions, humans and rats may live in close proximity, allowing rat fleas to bite people, spreading bubonic plague. The first major outbreak of bubonic plague occurred in the early 14th century, starting in central Asia and spreading through Europe. An extremely virulent disease with no known cure at the time, it killed as much as half of some human populations. Around 1320, the population of Europe was about 85 million, but this fell to about 60 million by 1400 as a result of bubonic deaths. Today, bubonic plague is treated with antibiotics, and rat populations are controlled by routine sanitation measures.

Another killer plague swept the world in 1918–1919, caused by an epidemic of a novel strain of influenza. About 20 million people died. Other virulent diseases will probably emerge in the future, perhaps transferred to humans by close contact with other species, most likely primates. A recent example is Ebola virus, which causes a rare but deadly hemorrhagic (or bleeding) fever. It was probably spread to humans from a species of rainforest monkey in central Africa. Another prospect is Creutzfeldt-Jakob disease, which is characterized by degeneration of the brain and other neural tissues. This affliction is caused by a transmissible protein known as a prion, and is apparently spread when humans eat beef contaminated by so-called mad-cow disease, or eat the brains of infected monkeys or apes. Some epidemiologists believe that auto-immune deficiency syndrome (or AIDS), a slowly developing but almost always lethal syndrome, also has the potential to cause catastrophic mortality in human populations. So could new, deadly, antibiotic-resistant strains of *Staphylococcus-B* and other pathogenic bacteria.

So far, medical science has managed to deal with most of these new, lethal, communicable diseases. However, science may not be able to cope with some of the new pestilences. If this is true, there could be catastrophic results for dense, vulnerable populations of modern humans.

Famine

The spectre of famine has been present throughout human history. Famines can arise from various factors, including insect outbreaks, insufficient rainfall causing drought, excessive precipitation causing flooding, and warfare and other sociopolitical upheavals. Ancient historical records are full of descriptions of deadly famines in many cultures. Some of the most catastrophic famines of pre–20th-century Asia and Europe killed hundreds of thousands of people.

Even more enormous famines have occurred in the 20th century. For example, as many as 5–10 million people may have starved in the then-Soviet Union during 1932–1934. This happened as a combined result of drought and social upheavals associated with the forced collectivization of private farms by the communist government. Another famine, in West Bengal, killed 2–4 million people in 1943. More recent famines have occurred in various parts of Africa and Asia, caused by crop failures due to drought and other weather extremes, and often aggravated by the chaos of war or revolution.

Overpopulation exacerbates most factors that contribute to famine. Regions or countries are vulnerable to developing famine conditions if:

- they have large and dense populations
- they have small reserves of stored food
- environmental conditions for agriculture are marginal, partly because population pressures have led to lands being cultivated in semi-arid regions susceptible to drought
- there is little foreign exchange to purchase food from elsewhere during times of shortage, so that people and governments must rely on goodwill and aid
- the economic and political systems do not foster the stable governments and social systems that are required for effectively dealing with crises

Decline of Carrying Capacity

Environmental catastrophists suggest that the ecological carrying capacity for the extremely large human population may collapse. If this happened, severe resource constraints and extensive mass starvation would follow. It is already abundantly clear that some of the most important kinds of potentially renewable resources are being severely overharvested, and that stocks of non-renewable resources are being rapidly depleted (see Chapters 12, 13, and 14). Collapsing fish stocks, declines in agricultural soil capability, desertification, deforestation, and depleted groundwater are all evidence of this pressing phenomenon. Declining resources decrease the carrying capacity of the biosphere for the human enterprise.

A Nuclear Holocaust

Enormous numbers of people have died prematurely through the direct and indirect consequences of warfare. The most lethal conflicts in history were the First World War, which killed as many as 20 million people, and the Second World War, during which at least 38 million died. Potentially, however, modern humans are capable of killing enormously larger numbers of people than these, through the unbridled use of nuclear weapons. The world's nuclear powers have enormous arsenals of sophisticated, extremely powerful weaponry, particularly the United States and the former Soviet Union. The explosive power of their thousands of nuclear weapons is unbelievably huge, and is capable of causing environmental damage so extreme as to return any surviving humans back to the stone age. Conventional military theory holds that nuclear arsenals are most useful as deterrents against other nuclear-power nations, and recent treaties have resulted in large reductions in arsenals. Nevertheless, the remaining stockpiles remain active and immense, and it is not difficult to imagine scenarios of political instability and conflict that could ultimately lead to a nuclear holocaust. Until all nuclear weapons are beaten into ploughshares, a global nuclear disaster cannot be ruled out.

A Natural Big Bang

Although extraordinarily unlikely, it is conceivable that Earth and its ecosystems could suffer a natural, unpredictable, environmental catastrophe such as a meteorite impact. There are precedents for such a rare event, with clear evidence from the geological record (see Chapter 3). For instance, it appears that about 65 million years ago Earth was struck by a meteorite, an accident that caused enormous environmental changes and likely resulted in a mass-extinction event (see Chapters 6 and 24). Fortunately, cosmic calamities of this tremendous intensity are extremely rare, occurring only every 25–30 million years or so. It is much more likely that any crash that might occur in the human population would be caused by a virulent disease or collapse of carrying capacity, rather than by a big-bang cataclysm from a meteorite strike.

Summary

The human population has been growing exponentially in recent centuries. Further growth will occur in the foreseeable future, but likely at decreasing rates. Demographic models suggest that the population will eventually stabilize, but this may happen at a level of approximately double the more than 6 billion people alive today.

Accompanying the growth of the human population has been an even more rapid increase in per capita environmental impact. In combination, these have changed the biosphere on a scale and intensity that is comparable to the effects of such enormous geological events as full-blown glaciation. The damage includes deforestation, depletion of virtually all kinds of resources, pollution, and mass extinction. It is clearly apparent that the cumulative, anthropogenic impacts on the environment will escalate even further with increases in the abundance of people.

The enormous growth of the human population must be kept in mind whenever environmental problems are considered. To some degree, the environmental effects of people and their economies can be avoided or mitigated by technological strategies such as pollution control and the conservation of natural resources. However, the size of the human population remains a root cause of the ecological damage caused by our species.

Key Terms

cultural evolution
carrying capacity
zero population growth (or ZPG)
intrinsic (or natural) population change
doubling time
demographic transition
age-class structure
replacement fertility rate
urbanization
birth control
population policy

Questions for Discussion

1. Outline the major stages in cultural evolution and relate these to changes in carrying capacity and growth of the human population.
2. Describe the recent pattern of growth of the human population and discuss the possible scenarios for future growth.
3. Compare demographic parameters and population growth rates in developed and less-developed countries. Discuss reasons for the differences.
4. Define "demographic transition" and compare its dynamics in developed and less-developed countries.
5. Outline the reasons why age-class structure is so important in future population growth.
6. List the major means of birth control and discuss controversies associated with their use.
7. Discuss the likelihood and potential causes of a human population crash.

References

Bates, D.G. 1996. *Cultural Anthropology*. New York: Simon & Schuster.

Brown, L.R. and S. Postel. 1987. Thresholds of change. In: *State of the World 1987*. Norton, NY: Worldwatch Institute. pp. 3–19.

Ehrlich, P.R. and A.H. Ehrlich. 1990. *The Population Explosion*. New York: Simon & Schuster.

Ehrlich, P.R., A.H. Ehrlich, and J.P. Holdren. 1977. *Ecoscience: Population, Resources, Environment*. San Francisco: W.H. Freeman.

Freedman, B. 1995. *Environmental Ecology, 2nd ed*. San Diego, CA: Academic Press.

Goldemberg, J. 1992. Energy, technology, development. *Ambio*, **21**: 14–17.

Health and Welfare Canada. 1980. *Facts and fancy about birth control, sex education, and family planning*. Ottawa: Health and Welfare Canada, Health Promotion Directorate.

Jennings, J.D. 1979. *The Prehistory of Polynesia*. Cambridge, MA: Harvard University Press.

Kane, P. 1983. *The Which? Guide to Birth Control*. London: Hodder & Stoughton.

Kessler, B.A. (ed.). 1992. Special issue on population. *Ambio*, **21**: 1–120.

Leisinger, K.M. and K. Schmitt. 1994. *All Our People: Population Policy with a Human Face*. Washington, DC: Island Press.

Olsson, H. and A. Rapp. 1991. Dryland degradation in central Sudan and conservation for survival. *Ambio*, **20**: 192–195.

Raven, P.H., L.R. Berg, and G.B. Johnson. 1993. *Environment*. New York: Saunders .

Solomon, E.P., R.F. Schmidt, and P. Adragana. 1990. *Human Anatomy and Physiology, 2nd ed*. New York: Saunders.

Tanner, J.T. 1975. Population limitation today and in ancient Polynesia. *BioScience*, **25**: 513–516.

United Nations. 1992. *Long-range World Population Projections: Two Centuries of World Population Growth, 1950-2150*. New York: United Nations, Population Division.

Vitousek, P.M., P.R. Ehrlich, A.H. Ehrlich, and P.A. Matson. 1986. Human appropriation of the products of photosynthesis. *Bioscience*, **36**: 368–373.

World Resources Institute (WRI). 1995. *World Resources 1994–95: A Guide to the Global Environment*. New York: Oxford University Press.

Wrigley, E.A. 1969. *Population and History*. New York: McGraw-Hill.

THE CANADIAN POPULATION

Chapter Outline

Introduction

Aboriginal populations

Early immigration from Europe

Growth of the Canadian population

Population structure

Regional differences

Birth control and population policy

Chapter Objectives

After completing this chapter, you will be able to:

1. Outline changes in the Canadian population over time.
2. Describe the modern populations of the provinces and territories of Canada and recent changes in their growth and distribution.
3. Describe the phenomenon of urbanization in the Canadian context.
4. Discuss the desirability and nature of a population policy for Canada.

INTRODUCTION

In Chapter 10 we examined the dynamics of human populations in different countries as well as globally. This information provides an international context for examining population issues in Canada.

Canada ranks among the top 20% of nations in terms of its human population (about 30 million in 1997). Canada also ranks among the wealthiest of nations in terms of per capita indicators of economic development, resource use, and anthropogenic impacts on environmental quality (see Chapter 1). Because its citizens have an environmentally intense lifestyle, Canada has a much greater impact on Earth and its resources than would be predicted on the basis of its population alone.

Because Canada has achieved a relatively high level of economic and social development, it has an opportunity and a responsibility to manage its environmental quality in a sustainable fashion. It also has a responsibility to control its population growth within sustainable limits. Moreover, because of Canada's privileged and wealthy status, it has an obligation to demonstrate a vision of sustainable development to other nations, including poorer countries that are hoping to emulate our national achievements. A central element of sustainable development is the implementation of a sensible population policy.

It is important that Canadians become knowledgeable about national and global population issues. If Canadians understand those matters, they will be sympathetic to population policies appropriate both within Canada and abroad.

ABORIGINAL POPULATIONS

Around 1000 CE, the Norse explorer, Leif Ericsson, made several landfalls along the northeast coast of North America. The Norse attempted several brief colonizations, including a settlement at l'Anse aux Meadows in Newfoundland, but these failed. About 500 years later,

Although population densities are high in Canadian cities (illustrated by this view of downtown Toronto), they are much lower in the country as a whole. Most of Canada is not suitable for supporting large human populations, mainly because of a difficult climate.

Source: Comstock/Malak

other European explorers encountered great regions in the Americas that were previously unknown to them. They did not, however, find unpopulated lands. In fact, all of the Americas were already fully occupied by various **indigenous cultures**. At the end of the 15th century (at the time of the voyages of Christopher Columbus and John Cabot), the First Nations of the Americas had an estimated population of 35 million people. About 30 million of these people lived in South and Central America, and 5 million in North America.

Some of these First Nations had developed advanced cultures, particularly the Aztecs and Maya of Central America and the Inca of South America. These people built elaborate cities which contained great pyramids and other magnificent buildings. Their nations were supported by complex physical and social infrastructures. Like cities everywhere, those of advanced Amerindian cultures relied on the surrounding agricultural landscape for supplies of food, water, and other resources. Furthermore, taxes were collected from the people living in the producing regions to support the rulers, administrators, soldiers, and artisans living in the urbanized centres.

The First Nation cultures in what is now Canada were diverse, comprising twelve distinct language groups and many additional dialects. Some of the aboriginal cultures, such as the Huron and Iroquois of the eastern temperate woodlands, were essentially agrarian societies. These people supplemented their agricultural livelihoods by foraging for useful wild plants and by hunting deer, birds, fish, and other animals. They lived in grand longhouses in stockaded villages, surrounded by fields in which they cultivated maize, pumpkin, squash, sunflower, and other indigenous crops.

Other First Nations of Canada subsisted largely through foraging lifestyles. The Bella Coola, Haida, Nootka, Tlingit, and related nations of the rainy west coast exploited a relatively abundant and predictable resource base, and consequently they lived in permanent settlements. These people were mostly fishers of salmon, mollusks, and other inshore resources. They supplemented their aquatic foods with wild plants, deer, and other terrestrial resources.

Most of the aboriginal cultures of the western plains, such as the Assiniboine, Blackfoot, and Piegan, were semi-nomadic hunters of the enormous herds of buffalo and other prairie animals that existed at the time. The more northern Athapaskans, Chipewyan, Cree, Dene, Innu, and Montagnais of the sweeping boreal forest hunted mostly caribou, moose, beaver, and waterfowl, and fished streams and lakes for grayling, trout, whitefish, pike, and other fishes. The Beothuk, Mi'kmaq, and Malacites of the Atlantic region also hunted moose, deer, and caribou, fished in freshwaters, and gathered shellfish in shallow, marine waters. The northernmost Inuit hunted caribou when those migratory animals were nearby, but they subsisted mostly on marine mammals, including ringed seal, walrus, beluga whale, narwhal, and even great bowhead whales. All of these peoples also gathered wild plant foods when they were seasonally abundant.

We know little about the population sizes of these First Nations of Canada. Reasonable estimates are based on assumptions about their lifestyles and the presumed carrying capacity of their habitats. At about the time when the first Europeans came to Canada, the total population of aboriginal Canadians is estimated to have been about 300 000.

The European colonization of the Americas began in the early 16th century, following the "discovery" of these lands in 1492 by Christopher Columbus, a Genoan sailing on behalf of the Spanish Crown. Columbus was searching for an oceanic passage to the rich spices and silks of China, India, Japan, and Southeast Asia. In 1497, John Cabot, another Genoan employed by the King of England, sighted Newfoundland and possibly Cape Breton Island.

Within a century of the arrival of the Europeans, the numbers of aboriginal people began to fall precipitously. By the end of the 19th century, the population of the First Nations in North America was only about 20% of their initial five million. Infectious diseases, particularly measles, smallpox, tuberculosis, and influenzas, were the most important causes of this calamitous mortality. Europeans were relatively tolerant of these diseases that they brought to the Americas, but the indigenous peoples were extremely vulnerable. Epidemiologists refer to populations that are hypersensitive to infectious diseases as "virgin fields." Such populations can suffer intense rates of mortality from introduced diseases (known as **virgin-field epidemics**).

In addition, huge numbers of aboriginal peoples died as a direct and indirect result of conflicts associated with the European conquest. Others died during intertribal wars, some of which were precipitated when competing European nations upset previous balances of power among indigenous nations, in part by providing their aboriginal allies with advanced weaponry. Many people starved when they were dispossessed of their resources and livelihoods by European colonists and governments. For example, the rapacious 19th-century slaughter of the great buffalo herds of North America was partly a stratagem to deprive the Plains First Nations of their critical resource base.

In about 1500 CE, there were about 300 000 indigenous people in Canada, a population that subsequently collapsed to perhaps 60 000. In 1996, the First Nations of Canada comprised about 677 000 people. This number refers to Canadians who declare themselves to be of aboriginal origin.

Early Immigration from Europe

The initial wave of European colonists coming to Canada consisted mostly of French and British adventurers seeking furs, fish, timber, agricultural land, and trade. Compared with their European homelands, which even then were relatively densely populated, Canada represented an enormous frontier to these colonists, containing boundless opportunities to make money and develop livelihoods. The fact that these lands were already occupied by indigenous cultures did not matter much to the European colonists. The dominant world views at that time were harsh, aggressive, and imperialist. These beliefs served to legitimize the displacement of the First Nations people by the technologically empowered Europeans.

Slowly for the first century or so, and then as a great flood of immigration, colonists came to Canada from France and Britain, and later from many other countries. Today, the population of Canada is an amalgam derived from a rich diversity of immigrants from virtually all parts of the world, plus descendants of the original First Nations.

Between about 1500 and 1700, the population of the North American continent increased to about six million people. This included about one million black slaves, almost all of whom had been brought unwillingly from Africa to the southeastern colonies to work on plantations. Under laws of the time, slaves were the human property of their "owners," having no personal freedom and few rights. People in the northern colonies had few slaves, but they had large numbers of indentured servants, mostly of European origin, who were bound to their employers by contracts and debts that in most cases were impossible to pay off.

Following this initial phase of colonization, the pace of immigration quickened markedly. Data are not available for the entire period, but between 1820 and 1930 at least 50 million Europeans migrated to colonies and former colonies around the world, particularly to the Americas. This immense dispersal involved about one-fifth of the population of Europe during that period. The mass migration was stimulated by a combination of factors: rapid population growth in Europe, a shortage of arable lands there, famine in Ireland and other countries, and rivalries among the imperial powers to develop empires and dominate world trade. In addition, religious and ethnic minorities in European countries were frequently persecuted, and many of these oppressed peoples emigrated to North America and elsewhere.

As was noted in Chapter 10, this great 19th- and 20th-century dispersal was a critical factor allowing the European countries a relatively easy passage through the early stages of their demographic transitions.

Growth of the Canadian Population

Reliable information is available describing early population growth in some regions of Canada, notably in New France (see Canadian Focus 11.1 and 11.2). The first credible estimate of the population of all of Canada is for 1851, when there were about 2.4 million people (see Figure 11.1). By 1867, the year of Confederation, there were 3.3 million people in Canada. By the turn of the century, the population had increased to 5.4 million. Much of the population growth resulted from a natural increase of 1.3–2.0%/y, with birth rates of 36–45 per 1000 people in the population, and death rates of 18–21 per 1000. In fact, because of a relatively depressed economy during the first several decades after Confederation, immigrants to Canada were fewer than those emigrating from the country.

During the 20th century, birth and death rates both declined steadily, although the natural rate of population increase remained greater than 1%/y until the mid-1970s. This natural growth, coupled with vigorous immigration, led to rapid increases in the Canadian population. Population growth rates were as high as 3%/y and averaged about 1.6%/y overall. By 1950 there were about 14 million Canadians, and in 1997 more than 30 million.

The natural rate of growth of the Canadian population (birth rate minus death rate) has slowed markedly during the past century or more (Tables 11.1 and 11.2). This has happened mainly because of rapid decreases in birth

CANADIAN FOCUS 11.1

THE LEGACY OF DANIEL LEBLANC AND FRANÇOISE GAUDET

In the springtime of 1650, Daniel LeBlanc emigrated from France to Acadia. He married Françoise Gaudet and settled into subsistence farming near what is now Annapolis Royal in Nova Scotia. Many families during that time were large, especially because children helped with the onerous labour of clearing the forest, tending crops and livestock, and taking care of the home and extended family. In fact, fecundity remained high among French Canadians for more than three centuries, until the 1950s and 1960s, when birth rates began to plummet.

Daniel and Françoise had seven children—six sons and a daughter. Five of the sons married, presenting Daniel and Françoise with 35 grandchildren. Today, the LeBlanc family has an enormous legacy of descendants from Daniel and Françoise. The family is estimated to now number more than 300 000 in Canada and the United States (many of whom have the anglicized surname, White). The LeBlanc extended clan is the largest of the several Acadian lineages that have persisted into modern times. Although the LeBlancs' case is rather extraordinary, it demonstrates the awesome power of human population growth.

rates, which now almost counterbalance the death rates (which had declined earlier).

An exception to the general decline in birth rates is a demographic anomaly known as the **baby boom** that occurred between 1945 and 1965. This period of relatively high fecundity was due largely to several decades of social optimism that followed the end of the Second World War. In addition, during the war many couples delayed marriage and child-bearing, as many young men went to fight overseas and women were employed in factories and other wartime occupations. After the war ended, men and women turned their attention back to raising families. During the baby boom, the birth rate in Canada averaged about 27 per 1000 and the natural growth rate of the population was 1.9%/y.

An important reason for the end of the baby-boom phenomenon was the growing affluence and urbanization of typical Canadians, which led to a general desire for smaller families. Also important was the increasingly easy access to, and social acceptance of, methods of birth control. By the mid-1990s the birth rate in Canada was 14 per 1000 and the natural rate of increase of the population was about 0.7%/y.

Immigration has always been an important factor in the population growth of Canada, as can be readily appreciated by examining the data in Table 11.2. During most of Canada's history, considerably more migrants have moved to this country than away from it. The major exception was during the latter decades of the nineteenth century, when many people emigrated from Canada to the United States. During the twentieth century, however, Canada has had consistently high rates of immigration.

Immigration has been especially vigorous since the 1960s, when the government of Canada loosened restrictions associated with national and ethnic origins of immigrants. These were replaced with criteria based on education, occupational skills, and wealth. Since the early 1970s there has been a substantial increase in the numbers of immigrants coming to Canada. Between 1972 and 1981, an annual average of 145 000 people immigrated to Canada,

FIGURE 11.1 THE POPULATION OF CANADA

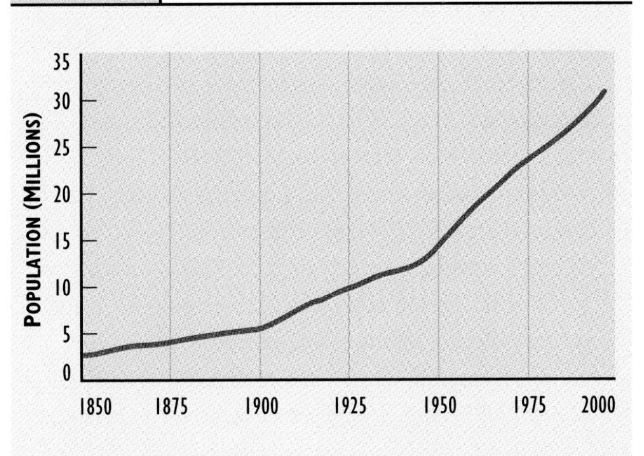

In 1851, the Canadian population was estimated at about 2.4 million people. This graph shows the steady growth of the population up to 1996, when the Canadian population was 29.8 million.

Source: Statistics Canada (1992, 1995b)

Canadian Focus 11.2

The Remarkable Growth of the Population of New France

The best early demographic data for any area of Canada are for New France. This region encompassed southern Quebec, particularly the valley of the St. Lawrence River, and the Acadian regions of what are now Nova Scotia, New Brunswick, and Prince Edward Island.

In the early 1600s, there were about 500 French colonists in New France. In 1666, after about 1 $\frac{1}{2}$ centuries of somewhat tentative colonization, a census found 3215 people of French origin in Quebec, and another in 1671 found about 400 in Acadia. Most of these people were single men who had journeyed to the Canadian frontier as soldiers, as priests hoping to convert indigenous people to Roman Catholicism, as government administrators, or as adventurers seeking their fortune through the fur trade.

In the following years, the pace of population growth quickened markedly. Complete families of settlers arrived from France, intent on developing agriculture in the fertile lowlands of Acadia and the valley of the St. Lawrence River. The immigration of single women was also actively encouraged, to offset the substantial deficit of females in the early population of New France. Many of these young women were recruited from Parisian orphanages and were known as *les filles du roi*.

By the 1670s, immigration slowed greatly because of dwindling prospects for finding work in New France, and because royal sponsorship of immigration had ended. French immigration to Acadia ceased after that area was ceded to Britain in 1713, and immigration to Isle Royale (Cape Breton Island) stopped after the fortress of Louisbourg was abandoned following a British siege in 1758. Immigration to Quebec ceased soon after, because of the British military victory at the Plains of Abraham in 1759, which resulted in the fall of Quebec City and the end of the French colonization of eastern Canada.

Birth rates were high in New France (and elsewhere) during the 18th century, typically about 50–60 per thousand people in the population. Anecdotal evidence suggests great fecundity in early colonial times. It was said that one soldier serving under the Marquis de Montcalm (1712–1759) had 250 descendants at the time of his death. Families of 15 to 20 children were not uncommon. Even though infant mortality was high, particularly from communicable diseases, populations grew quickly.

Because of earlier immigration and rapid natural growth, by 1770 the francophone population of Quebec had increased to 86 000. After 1759, all the growth of the French-Canadian population in Quebec was due to the natural excess of births over mortality, while much of the growth of the non-francophone population was due to immigration. By 1815 the francophone population of Quebec was 269 000 (there were also about 60 000 British colonists in Quebec at that time), and in 1885 there were 1.18 million (plus 250 000 non-francophones).

During the 19th century, the average number of births in Catholic families in Quebec was about 7 per family (this includes all Catholics, but the great majority of these were French). This rather high fecundity is typical of populations at the beginning of the demographic transition. It should be pointed out, however, that this high fecundity was not unique to Quebec — it was also typical of areas elsewhere in Canada, including Ontario.

In 1926, it was estimated that there were three million francophones in Canada and the United States. Almost all of these people were descendants of the original few hundred immigrants from France. At the present time, there are more than 6 million French Canadians. This includes about 5.6 million francophones living in Quebec, 0.4 million Acadians and smaller numbers in other provinces. There are also hundreds of thousands of Americans of French descent, many of whom live in Louisiana and New England.

---

and 143 000 during 1982–1991. This was followed by a substantial increase to 253 000 per year during 1992–1993.

Since the mid-1980s, net immigration has contributed about one-third of Canada's 1.2%/y population growth rate (the rest is due to natural population growth; see Table 11.2). If sustained, a 1.2%/y rate of increase would result in the Canadian population doubling in only 58 years. This rate of population growth is similar to that of the United States and Australia, also relatively wealthy, industrialized countries with small natural growth rates but substantial rates of immigration. These three countries are exceptional in these respects. Most other industrialized countries, particularly those of Europe, have much smaller rates of population increase, in part because they do not permit much immigration (see Chapter 10). This can be attributed mainly to the differences in population densities of European countries in comparison with Canada, the United States, and Australia.

Future Growth

Results of three models of future Canadian populations are summarized in Table 11.3. Scenario 1 is a low-growth model, involving lower rates of fertility and immigration than occur at present, and resulting in population stabilization by about 2030. Scenarios 2 and 3 are moderate-growth models. Scenario 2 assumes that population growth will

TABLE 11.1 COMPONENTS OF NATURAL GROWTH OF THE CANADIAN POPULATION

These data are standardized per thousand people in the population and are annual rates. Note that an annual growth rate of 10/1000 is equivalent to 1% per year.

DATE	BIRTH RATE	DEATH RATE	NATURAL GROWTH RATE
1921	29.3	10.6	18.7
1931	23.2	10.1	13.1
1941	22.4	10.0	12.4
1946	27.2	9.4	17.8
1951	27.2	9.0	18.2
1956	28.0	8.2	19.8
1961	26.1	7.7	18.4
1966	19.4	7.5	11.9
1971	16.8	7.3	9.5
1976	15.7	7.3	8.4
1981	15.3	7.0	8.3
1986	14.8	7.2	7.6
1991	14.3	7.0	7.3
1993	13.8	7.0	6.8

Source: Kalbach (1988) and Dumas and Belanger (1994)

TABLE 11.2 POPULATION GROWTH RATES IN CANADA

Natural growth rates are calculated as births minus deaths, while the actual growth rate also accounts for net immigration (that is, immigrants minus emigrants). The data are standardized per thousand people in the population, and are average annual rates.

TIME INTERVAL	NATURAL GROWTH RATE NO./1000	NET IMMIGRATION NO./1000	ACTUAL GROWTH RATE NO./1000
1851–1861	19.8	4.7	24.5
1861–1871	17.6	–5.2	12.4
1871–1881	16.7	–2.0	14.7
1881–1891	14.8	–4.3	10.5
1891–1901	13.4	–3.4	10.0
1901–1911	15.5	9.9	25.4
1911–1921	14.0	4.0	18.0
1921–1931	13.1	2.2	15.3
1931–1941	10.6	–0.8	9.8
1941–1951	14.0	1.2	15.2
1951–1961	17.3	5.9	23.2
1961–1971	12.5	3.4	15.9
1971–1981	8.1	3.5	11.6
1981–1986	7.6	1.6	9.2
1986–1991	7.3	5.4	12.7

Source: Kalbach (1988) and Dumas and Belanger (1994)

TABLE 11.3 | PROJECTED CANADIAN POPULATIONS

Population projections are based on sets of assumptions about fertility, mortality, and net immigration. Scenario 1 is a low-growth model, suggesting lower rates of fertility and immigration than occur at present. Scenarios 2 and 3 are moderate-growth models, with Scenario 3 assuming that current trends will continue into the future.

	PREDICTED POPULATION (MILLIONS)		
YEAR	SCENARIO 1	SCENARIO 2	SCENARIO 3
1994	28.8	28.8	28.8
1996	29.5	29.6	29.6
2000	30.8	31.1	31.4
2005	32.1	33.0	33.8
2010	33.1	34.7	36.3
2015	33.9	36.4	38.9
2020	34.6	38.1	41.4
2025	35.2	39.6	43.8
2030	35.5	40.8	46.0
2035	35.5	41.8	48.0
2040	35.3	42.6	49.9

Source: Modified from Statistics Canada (1994b)

stabilize around 2040. Scenario 3 assumes a continuation of current demographic parameters into the future, and population growth that continues beyond 2040.

All three scenarios are realistic to some degree because they involve plausible outcomes of reasonable anticipated changes in two groups of factors: government policy regarding immigration and population issues, and reproductive choices made by individual people and families. Overall, the models predict that the population of Canada will grow substantially from its 1997 value of about 30 million. The slow-growth projection suggests a population of about 35 million by 2040 (19% increase), compared with 50 million in the third scenario (65% increase), in which current trends are extrapolated.

Population Structure

The age structure of the modern Canadian population is illustrated in Figure 11.2. In general, the pattern is typical of a population that has progressed most of the way through the demographic transition. Note, however, the anomalous bulge of numbers corresponding to the "baby boomers." Fecundity dropped dramatically in Canada after the baby boom, and once this bulge has worked its way through the population structure, the age distribution will assume a more vertical shape. As the baby boomers age and retire from work, they are expected to exert significant strain on Canada's capacity for providing social and medical care for its elderly citizens.

Although Canada has a population structure characteristic of a country that has almost completed the demographic transition, its population growth rates remain relatively high, substantially exceeding 1%/y even in the 1990s. Much of the population growth is due to immigration. Although this factor is not closely related to age-class structure, it is notable that relatively young families and people of child-bearing age are prominent among immigrants to Canada. This adds further inertia to population growth.

Regional Differences

All regions of Canada have experienced substantial population growth during the past century or more (Table 11.4). The increases have been somewhat less substantial in the Atlantic Provinces and most rapid in the western provinces, particularly in British Columbia. In 1867, the Atlantic Provinces accounted for 21% of the Canadian population, Quebec 32%, Ontario 44%, and the rest of the country 3%. At that time the western regions of Canada were largely unsettled, but tremendous population growth has occurred there since. By 1996, a much larger fraction of the Canadian population lived in western regions of the country: the Atlantic Provinces now account for 8% of the Canadian population, Quebec 25%, Ontario 38%, the Prairie Provinces 17%, British Columbia 13%, and the two Territories 0.3% (Table 11.5).

There is little variation in birth and death rates among the provinces and territories of Canada (Table 11.6). The major exception is the Northwest Territories, which has by far the highest birth rate. First Nations communities, which tend to have relatively large families and a higher rate of population growth, make up a much larger proportion of the population in the Northwest Territories than of any other province or territory. The NWT also has a relatively low mortality rate, which is largely due to its comparatively young population. People younger than

FIGURE 11.2 | AGE STRUCTURE OF THE CANADIAN POPULATION

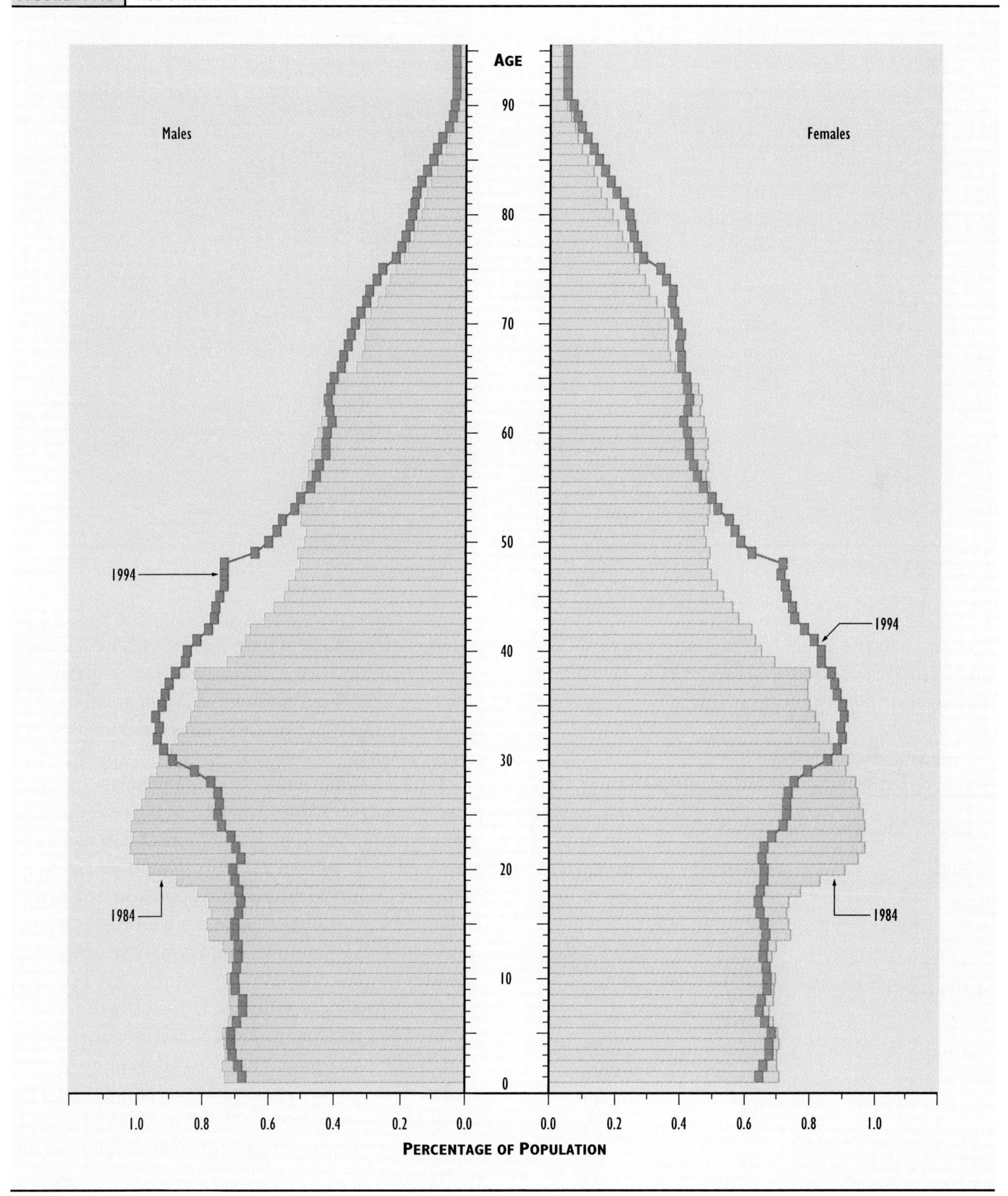

This diagram shows the relative numbers of people of various ages. The demographic "bump" of people who were 20–40 years old in 1984 and 30–50 years old in 1994 corresponds to a period of relatively high birth rates from the mid-1940s to mid-1960s, known as the "baby boom."

Source: Statistics Canada (1994a)

TABLE 11.4 REGIONAL DISTRIBUTION OF THE CANADIAN POPULATION

These data show the growth of the population of Canada and its regions since Confederation in 1867. Note that data for the Prairie Provinces prior to 1901 include only Manitoba. Data for Alberta, Saskatchewan, and Yukon were combined with the Northwest Territories until 1901, when they became separate political units. In 1949, Newfoundland joined Canada. Data are in thousands of people.

YEAR	CANADA	ATLANTIC PROVINCES	QUEBEC	ONTARIO	PRAIRIE PROVINCES	BRITISH COLUMBIA	YUKON + NWT
1867	3 466	726	1 123	1 525	15	32	45
1881	4 325	871	1 360	1 927	62	49	56
1891	4 833	880	1 489	2 114	153	98	99
1901	5 371	894	1 649	2 183	419	179	47
1911	7 207	938	2 006	2 527	1 327	393	16
1921	8 788	1 001	2 361	2 934	1 955	525	12
1931	10 375	1 008	2 874	3 431	2 354	694	14
1941	11 503	1 130	3 331	3 787	2 421	817	17
1951	14 001	1 618	4 056	4 598	2 549	1 165	25
1961	18 238	1 897	5 259	6 236	3 179	1 629	38
1971	21 557	2 058	6 027	7 703	3 542	2 184	43
1981	24 339	2 233	6 438	8 624	4 231	2 744	69
1991	27 001	2 332	6 846	9 914	4 609	3 219	81
1996	29 810	2 613	7 361	11 189	4 928	3 823	97

Source: Statistics Canada (1992, 1995a)

15 years comprise 32% of the population of the NWT, compared with a national average of 21%. Overall, the natural rate of population growth of the NWT is 2.1%/y, equivalent to a doubling time of 33 years. The age structure and growth rate are, in fact, typical of a region that is only beginning to pass through the demographic transition, and in that sense are anomalous compared with the rest of Canada.

TABLE 11.5 POPULATION BY PROVINCE AND TERRITORY IN 1996

Data are in thousands of people.

PROVINCE	POPULATION	PERCENT OF TOTAL
Canada	29 811	
Ontario	11 189	37.5 %
Quebec	7 361	24.7
British Columbia	3 823	12.8
Alberta	2 768	9.3
Manitoba	1 141	3.8
Saskatchewan	1 019	3.4
Nova Scotia	941	3.2
New Brunswick	761	2.6
Newfoundland	573	1.9
Prince Edward Island	137	0.5
Northwest Territories	66	0.2
Yukon	31	0.1

Source: Statistics Canada (1992, 1995a)

British Columbia has the fastest population growth rate in Canada, 2.7%/y, which is equivalent to a doubling time of only 26 years. This rapid rate of population increase occurs in spite of a 0.6%/y natural rate of population increase. Obviously, the population of British Columbia is growing because of high rates of immigration from other countries and from other regions of Canada.

Other provinces with population growth rates exceeding 1%/y are Ontario (1.6%/y) and Alberta (1.2%/y). The NWT has a population growth rate of 1.6%/y, smaller than its natural rate of increase of 2.1%/y, the result of a net migration of people from the NWT to other parts of Canada where there are greater economic opportunities.

The smallest population growth rates occur in Newfoundland (0.1%/y), Saskatchewan (0.3%/y), Manitoba (0.5%/y), Nova Scotia (0.6%/y), New Brunswick (0.6%/y), and the Yukon (0.6%/y). Natural population

TABLE 11.6 | DEMOGRAPHIC PARAMETERS BY PROVINCE AND TERRITORY IN 1993

Data are per thousand people in the population and are annual rates. Fertility rate is the average number of offspring per woman aged 15–49. Data for total increase include net immigration.

PROVINCE	BIRTH RATE	DEATH RATE	NATURAL INCREASE	TOTAL INCREASE	TOTAL FERTILITY RATE
	(NO./1000 IN POPULATION)				
Canada	13.8	7.0	6.8	13.2	1.7
Ontario	13.9	6.8	7.1	15.7	1.7
Quebec	13.2	7.1	6.0	8.4	1.7
British Columbia	13.0	7.0	6.0	27.4	1.7
Alberta	15.7	5.5	10.1	12.2	1.9
Manitoba	15.1	8.4	6.8	4.7	1.9
Saskatchewan	15.1	8.4	6.7	3.3	2.0
Nova Scotia	12.8	8.1	4.7	5.5	1.6
New Brunswick	12.4	7.5	4.9	5.5	1.6
Newfoundland	12.4	6.7	5.7	1.3	1.4
Prince Edward Island	14.1	9.4	4.6	7.7	1.9
Yukon	16.3	3.8	12.5	5.6	1.9
Northwest Territories	25.0	4.0	21.0	15.5	2.7

Source: Dumas and Belanger (1994)

growth in most of these regions is higher than the actual population increase, reflecting a net migration of people to other provinces or other countries.

In general, the population of Canada is much denser in the southern parts of the country. This pronounced spatial pattern reflects the distribution of economic opportunities in Canada, most of which are related to climate, suitability of the land for agriculture, and other factors.

Rural and Urban Populations

Until the latter half of the 20th century, most Canadians lived in rural environments, where they worked in agriculture and other country livelihoods. Since 1871, when the first data are available, the proportion of Canadians living in the countryside has steadily decreased, from 80% in 1871 to 23–24% since 1976 (see Table 11.7).

TABLE 11.7 | RURAL AND URBAN POPULATIONS IN CANADA

Data are in millions of people.

YEAR	TOTAL POPULATION	URBAN	RURAL
		(% OF TOTAL)	
1871	3.7	19.6	80.4
1881	4.3	25.7	74.3
1891	4.8	31.8	68.2
1901	5.4	37.5	62.5
1911	7.2	45.4	54.6
1921	8.8	49.5	50.5
1931	10.4	53.7	46.3
1941	11.5	54.3	45.7
1951	14.0	56.7	43.3
1961	18.2	60.7	39.3
1971	21.6	76.1	33.9
1981	24.3	75.7	24.3
1991	27.3	76.6	23.4

Source: Leacy (1983) and Statistics Canada (1994b)

People living in wealthy countries, such as Canada, use resources intensively, and therefore have a high per capita environmental impact.
Source: Karen Taylor

In large part, this mass migration of people from rural to urban environments has been caused by the mechanization of much of the routine labour in agriculture, forestry, mining, fishing, and other typically rural industries. In earlier times, most of this work was performed by human labour or draft animals, but this muscle power has largely been superseded by various kinds of mechanization. As a result, fewer Canadians are employed in rural economic sectors, even though the financial value of the outputs from these sectors has increased greatly over time. The displaced rural people have moved to the towns and cities of Canada, where they earn their livelihoods in manufacturing, financial and service industries, and government and education.

As time passes, urban Canadians are living in larger centres of population. These cities are increasing in size rapidly. In 1871 only 2.9% of Canadians lived in centres with a population greater than 100 000 people. In 1991, 61% of Canadians lived in the 25 largest cities in the country (Table 11.8), even though these metropolitan areas account for only 0.7% of the country's landmass. In fact, 32% of Canadians live in the three largest cities.

Birth Control and Population Policy

Most government planners and politicians do not consider Canada to be overpopulated. In fact, many of them think that Canada is underpopulated and fully capable of comfortably absorbing additional population growth. This attitude is debatable, in view of the relatively intense lifestyles of Canadians (typical of people living in developed countries), and their corresponding large per capita environmental impacts. Nevertheless, continuing the population-growth paradigm (or model) is the predominant way of thinking among decision makers in Canadian government. Consequently, Canada and its regions do not have well-developed **population policies**, other than those that establish targets and guidelines for the numbers and types of immigrants and refugees allowed into the country.

In addition, Canadian governments do not have a policy of encouraging other countries to develop their own population policies, particularly those that are less wealthy and have rapidly growing populations. Canadian governments also do not provide aid to those poorer countries to help them increase the availability of birth-control technologies. Our governments avoid high-profile controversies by not being directly involved in the population problems of poorer countries. This attitude contributes little to dealing with the global explosion of the human population.

Governments within Canada also lack policies to encourage their citizens to have small families as a means of slowing the growth rate of national or regional populations. In fact, some Canadian governments pursue policies that are distinctly pro-natalist. This is particularly true of the government of Quebec, which provides substantial cash payments to women based on the number of children they have — families with more than two children receive proportionately larger payments than those with one or two. In addition, all provinces and the federal government provide income-tax breaks to parents based on the number of children they are supporting. This can be interpreted as a pro-natalist aspect of the income-tax system.

Generally, governments in Canada permit their citizens to choose freely among a wide range of safe and effective birth-control options. However, this is not to say that all birth-control methods are freely available across the

TABLE 11.8 | CITIES OF CANADA

The populations of the 25 largest metropolitan areas of Canada, in decreasing order of population in 1991, as well as their population growth in recent decades.

		POPULATION (THOUSANDS)			% INCREASE
CITY	PROVINCE	1971	1981	1991	1971–1991
Toronto	Ontario	2628	2999	3893	48%
Montreal	Quebec	2743	2828	3127	14
Vancouver	British Columbia	1082	1268	1603	48
Ottawa-Hull	Ontario, Quebec	603	718	921	53
Edmonton	Alberta	496	657	840	69
Calgary	Alberta	403	593	754	87
Winnipeg	Manitoba	540	585	652	21
Quebec	Quebec	481	576	646	34
Hamilton	Ontario	499	542	600	20
London	Ontario	286	284	382	33
St. Catharines	Ontario	303	304	365	20
Kitchener	Ontario	227	288	356	57
Halifax	Nova Scotia	223	278	321	44
Victoria	British Columbia	196	234	288	47
Windsor	Ontario	259	246	262	1
Oshawa	Ontario	120	154	240	100
Saskatoon	Saskatchewan	126	154	210	66
Regina	Saskatchewan	140	164	192	36
St. John's	Newfoundland	132	155	172	30
Chicoutimi	Quebec	134	135	161	20
Sudbury	Ontario	155	150	158	1
Sherbrooke	Quebec	–	125	139	–
Trois-Rivières	Quebec	–	111	136	–
Saint John	New Brunswick	107	114	125	17
Thunder Bay	Ontario	112	121	124	11

Source: Statistics Canada (1994b)

country. For example, the governments of Prince Edward Island and Newfoundland severely restrict access to abortion as a means of terminating pregnancy. Consequently, women must travel to other provinces to have access to abortion services. Similarly, reproductive and family-planning education in schools varies considerably across Canada.

As discussed in Chapter 10, abortion is an extremely contentious issue in Canada and elsewhere. The controversy has resulted in public demonstrations and confrontations between pro-life and pro-choice groups. In a few instances, patrons and personnel of abortion clinics have been illegally harassed and physically assaulted, and clinics have been fire-bombed. A Vancouver doctor who had provided abortion services was shot in his home by an unidentified sniper.

Even though abortion and other means of birth control remain controversial issues, Canadians who desire to control the number and spacing of their children have relatively easy and inexpensive access to effective means of birth control. While most Canadians take advantage of this opportunity, it is not available to most of the world's peoples. Consequently, the natural rate of population increase in Canada is relatively small while it is high in most poorer countries.

Key Terms

indigenous cultures (or First Nations)
virgin-field epidemic
baby boom
population policy

Questions for Discussion

1. Describe the populations of aboriginal people in Canada before and after European colonization.
2. Outline the growth of the Canadian population over time. Discuss the factors influencing growth during the past several decades and in the immediate future.
3. Compare the populations and their recent growth among the provinces and territories of Canada.
4. Compare the age-class structure of the Canadian population with that of a less-developed country.
5. Outline the basic elements of the population policy of the government of Canada, and suggest how you think it should be changed.
6. Do you consider Canada to be underpopulated or overpopulated? Explain your reasons.

References

Bracq, J.C. 1926. *The Evolution of French Canada*. New York: Macmillan.

Burger, J. 1990. *The Gaia Atlas of First Peoples*. New York: Doubleday Dell.

Crosby, A.W. 1986. *Ecological Imperialism: The Biological Expansion of Europe, 900–1900*. London: Cambridge University Press.

Dumas, J. and A. Belanger. 1994. *Report on the Demographic Situation in Canada, 1994*. Ottawa: Statistics Canada.

Kalbach, W.E. 1988. Population. pp. 1719–1722 in: *The Canadian Encyclopedia, 2nd ed*. Edmonton: Hurtig Publishers.

Leacy, F.H. 1983. *Historical Statistics of Canada, 2nd ed*. Ottawa: Statistics Canada.

Ray, A. 1987. When two worlds meet. In: Brown, C. ed. *The Illustrated History of Canada*. Toronto: Lester. pp. 17–104.

Statistics Canada. 1992. *Postcensal Annual Estimates of Population by Marital Status, Age, Sex, and Components of Growth for Canada, Provinces, and Territories*. Ottawa: Statistics Canada.

Statistics Canada. 1994a. *Annual Demographic Statistics, 1994*. Ottawa: Statistics Canada.

Statistics Canada. 1994b. *Human Activity and the Environment 1994*. Ottawa: Statistics Canada.

Statistics Canada. 1995a. *Quarterly Demographic Statistics, 1995*. Ottawa: Statistics Canada.

Statistics Canada. 1995b. *Population Projections, 1993–2016*. Ottawa: Statistics Canada.

Young, B. and J.A. Dickinson. 1988. *A Short History of Quebec: A Socio-Economic Perspective*. Toronto: Copp Clark Pitman.

CBC CANADIAN CASE 2

WORLD POPULATION: CONTROLLING THE EXPLOSION

The population of humans on Earth has grown enormously during the past several hundred years. More than 6 billion people are alive today, with predictions of 12 billion by the middle of the next century. Humans and their mutualist species (such as domestic animals and crop plants) are now the dominant organisms on Earth, and are causing enormous ecological and environmental changes. Population growth, coupled with the industrialization of human economies, is widely considered to be the root cause of the environmental crisis.

The previous two chapters have examined the factors that have allowed rapid population increases to occur. Important ethical questions still remain, such as whether it is desirable to have increasingly large numbers of people living in abject poverty, as opposed to limiting population sizes to what can be sustained at an acceptable standard of living.

Birth control is the most controversial of the policies that could be implemented to manage the growth of the human population. Powerful interest groups, including several major religions, have steadfastly opposed the most workable methods of birth control. This dispute critically impedes the implementation of effective population policies.

Another crucial issue concerns the role of women in dealing with the population crisis. Many people believe that a key element of any successful population policy will be educating women about health and reproductive issues, while ensuring that they also have free access to safe and effective means of birth control, should they choose this option.

Population issues are volatile and uncomfortable, and hard for many people to discuss objectively. Yet they are too important to ignore, because sustainable economies cannot develop unless the population of humans on Earth is controlled.

Questions for Discussion

1. Do you think that there are too many humans on Earth? What about Canada — is our national population too large, in view of the relatively intense lifestyles of typical Canadians?
2. Should people be allowed to make free choices about their own reproduction, including having ready access to medically safe and effective methods of birth control?
3. What is meant by "the empowerment of women"? Why is this seen as an important component of an effective population policy?

Video Resource: "World Population: Controlling the Explosion," *News in Review* (November 1994)

12

RESOURCES AND SUSTAINABLE DEVELOPMENT

CHAPTER OUTLINE

CHAPTER OBJECTIVES

After completing this chapter, you will be able to:

1. Outline the difference between renewable and nonrenewable natural resources.
2. Discuss how appropriate management can increase the potential harvest of biological resources.
3. Describe at least two case studies of the degradation of potentially renewable resources and provide the reasons for those damages.
4. Distinguish between economic growth and economic development and outline the nature of a sustainable economy.

Introduction

For about four decades now, we have been able to examine photographs of Earth as viewed from space. Images from that perspective show that Earth is a spherical mass, with a blue oceanic surface, brownish-green landmasses, and a clear atmosphere except where visibility is obscured by whitish clouds. Such images also reveal that beyond Earth and its atmosphere is the immense, black void of space — an extremely dilute, universal matrix. If we divert our attention from the compelling image of "**spaceship Earth**" and focus instead on the unimaginably larger abyss of space, we cannot fail to be stirred by the utter isolation of our planet, the only place in the cosmos known to sustain life.

With such a lucid image of spaceship Earth in mind, it is not difficult to understand that the physical resources necessary to sustain life are limited to those already contained on the planet. That is, with one critical exception — the electromagnetic radiation that is continuously emitted by the sun. A tiny fraction of that solar energy irradiates Earth, warms the planet, and drives photosynthesis. With the exception of sunlight, Earth's resources are essentially self-contained and finite.

It is an undeniable biological reality that all organisms must have continuous access to resources obtained from their environment. Plants and algae, for example, require sunlight and inorganic nutrients, while heterotrophic animals and most microbes must feed on the living or dead biomass of other organisms. Ecological communities and ecosystems are similarly nourished by environmental capital. These resources must be available in at least the minimal amounts needed to sustain life, and in larger quantities in ecosystems that are increasing in biomass and complexity, as occurs during succession.

The same reality holds for individual humans, our societies, and our economic systems. All humans and their enterprises are subsidized by the harvesting of resources from the environment (including those harvested from ecosystems). These must be available in minimal quantities to sustain human life, and in much larger amounts in human economic systems that are growing over time. An obvious conclusion is that *economic and ecological systems are inextricably linked*.

The main connections between economic systems and ecological systems involve flows of resources from ecosystems and the environment to human economic systems, and flows of unused materials, by-products, and heat (all of which are sometimes referred to as "wastes") from the economy into the environment. Associated with these interchanges of materials and energy are many kinds of damage to natural and managed ecosystems. These may be caused by disturbances associated with the harvesting of natural resources, by emissions of pollutants, or by other stressors related to economic systems and industrialized societies.

An ultimate goal of environmental science is to understand how resource use and environmental quality contribute to a **sustainable economic system** and to the quality of human life. A sustainable economic system is one that operates without a net consumption of natural resources — the rates of resource use are equal to or smaller than the rates at which they are regenerated or recycled. This definition focuses on the resource-related aspects of sustainability. Also important are environmental damages that may be caused by the extraction and management of natural resources. The social context must also be considered, particularly how the wealth is shared among the people who are participating in the economy.

In this chapter, we discuss the broader issues related to the use of natural resources in economic systems. Initially, we examine the characteristics of nonrenewable and renewable natural resources. Renewable resources can be managed to maintain or increase their productivity, so we describe management practices that foster those goals. This is followed by a discussion of the reasons for a remarkably common, yet catastrophic, phenomenon — the degradation of potentially renewable resources through excessive use. Finally, we consider the notion of sustainability, a topic that is critically important to the long-term health of both economic *and* ecological systems (see In Detail 12.1). This chapter deals with natural resources in a conceptual manner; Chapters 13 and 14 give an overview of the actual use of resources in international and Canadian economies.

Natural Resources

All natural resources can be divided into two categories: nonrenewable and renewable natural resources.

IN DETAIL 12.1

EASTER ISLAND AS A METAPHOR FOR SPACESHIP EARTH

A prehistoric example of resource degradation occurred on Easter Island, a small (389 km^2), extremely isolated island in the southern Pacific Ocean (Jennings, 1979; Ponting, 1991). Easter Island was first discovered by wandering Polynesians in about the fifth century. The only useful crops these people brought with them were chicken and sweet potato (the climate of Easter Island is too temperate for the tropical crops known to prehistoric Polynesians, such as breadfruit, coconut, taro, and yam). Initially, the Easter Islanders could also hunt fish and porpoises in the rich, near-shore waters of their island, and on land they could hunt wild Polynesian rats, a species they had introduced.

By the 16th century, the Easter Islanders had developed a flourishing society, with a population estimated as larger than seven thousand. Because of food surpluses, the Easter Islanders had spare time to engage in a cultural activity that involved carving huge slabs of stone into human-faced monoliths, each erected on a great base of stone, at various places along the coast. The heavy monoliths (weighing as much as 75 t) and their similarly massive bases were carved at an inland quarry, and then moved with enormous human effort (there being no draught animals) to their coastal sites by rolling them on logs cut from the island's forests.

Easter Island was quickly deforested by the aggressive cutting of large numbers of trees to use as stone rollers and as timber for constructing buildings and fishing boats. Unfortunately, once the forest resource was gone, the central enterprises of the Easter Islanders also collapsed. Stone monoliths could no longer be moved, sturdy homes could not be built, and fishing and porpoise hunting became impossible because only wooden boats were strong enough to withstand the oceanic swells. It became difficult to cook food and keep warm because the only other fuels available were shrubs and herbaceous plants.

In other words, the deforestation of Easter Island caused the collapse of the economy of this prehistoric Polynesian society. The cultural and economic disintegrations were so great that, when Europeans first arrived at Easter Island in 1772, the indigenous people were not certain who had erected the stone monoliths or why. These people were living in squalid conditions in caves and reed huts, were engaged in a perpetual state of warfare among rival clans, and were cannibals, possibly to supplement the meagre food resources on their treeless island.

An obvious lesson of Easter Island is that even early human societies were capable of over-exploiting the vital ecological resources that they required for subsistence. Undoubtedly, some of the prehistoric Easter Islanders were keenly aware of their isolated and precarious circumstances — particularly the limited resources available to sustain their society on such a small and isolated island. As these resources diminished, they likely discussed the need to conserve their resource base. Any such deliberations obviously came to naught, and there was an irretrievable collapse of the economy and culture of these people.

Easter Island is an obvious metaphor for Earth as a planetary "island." Earth, too, has limited stocks of energy, minerals, and biological resources to sustain human economies and natural ecosystems. Any of these natural resources can be rapidly depleted by excessive usage. There was no alternative, resource-rich refuge to which the Easter Islanders could escape from their self-inflicted ecological catastrophe. Likewise, as far as we know, there is no alternative to planet Earth.

Nonrenewable Natural Resources

Nonrenewable natural resources are present in a finite quantity, and do not regenerate after they are harvested and used. (If they do regenerate, it is so slowly as to be insignificant compared with the rate of use; see the section on fossil fuels in Chapter 13.) Consequently, as nonrenewable resources are used, their remaining stocks in the environment become depleted. This means that nonrenewable resources can never be used in a sustainable fashion — they can only be "mined." Examples of nonrenewable resources include metal ores, petroleum, coal, and natural gas.

Although continuing exploration may discover "new" stocks of nonrenewable resources, this does not change the fact that there is a finite quantity of these resources on Earth. For example, the recent discovery of large quantities of nickel and copper ore at Voisey's Bay in Labrador substantially increased the known, exploitable reserves of those metals. The discovery did not, however, affect the quantities of these metals present on Earth.

To some degree, metals can be recovered after their initial uses and recycled back into the economy. Recycling is critical for extending the lifespan of metal reserves. However, due to the growth and increasing industrialization of the human population, the demand for metals is increasing over time. Because recycling cannot keep up with the increasing demands, additional metals must be mined from their known reserves in the environment.

Nonrenewable resources can only be mined. This scene shows an open-pit iron ore mine in Schefferville, Quebec.

Source: Victor Last/Geographical Visual Aids

Renewable Natural Resources

Renewable resources are capable of regenerating after harvesting, so potentially they can be utilized forever. Most renewable resources are biological, although some are nonbiological.

Biological Renewable Resources:

- wild animals that are hunted as food, such as deer, ducks, fish, lobster, and seals
- wild plants that are gathered as sources of food
- forest biomass that is harvested for lumber, fibre, or energy
- the capability of soil to sustain the productivity of agricultural crops

Nonbiological Renewable Resources:

- surface and ground waters, which are renewed through the hydrologic cycle
- sunlight, of which there are continuous inputs to Earth

Many renewable resources can be managed to increase their rates of recruitment and growth and to decrease rates of mortality. In the following section we explain how management practices can be used to increase the productivity of renewable resources.

Although a renewable natural resource can potentially regenerate after harvesting, it can also be degraded by excessive use or inappropriate management. These can damage a resource's ability to regenerate, and may ultimately cause a collapse of its stocks. If this happens, the renewable resource is, in effect, being "mined" as if it were a

Renewable resources, such as timber and fish, are capable of regenerating after they are harvested. Provided they are not over-harvested or managed inappropriately, renewable resources can potentially be harvested in a sustainable fashion. This photo shows a load of timber harvested on Vancouver Island.

Source: B. Freedman

nonrenewable resource. As such, it becomes depleted by excessive usage. For this reason, ecologists commonly use the qualified term **(potentially) renewable natural resources**.

Management of Renewable Resources

Potentially, populations of animals and plants and their communities can be harvested sustainably without depleting the stock sizes or their capability for renewal. In all species, potential fecundity and productivity are greater than the actual recruitment, growth, and maturation of new individuals and biomass. Therefore, part of the excess of potential productivity minus actual survival and growth can be harvested and used to sustain humans and their economies. Moreover, the potential harvest can often be enhanced considerably by management practices that increase the renewable yields of forest, fish, and agricultural ecosystems.

Ultimately, the potential productivity of individual organisms is limited by genetically determined factors that influence their fecundity, longevity, and growth rate. The same factors also constrain the potential productivity of populations. However, the expression of these genetic factors is influenced by environmental variables, many of which constrain productivity (Figure 12.1). Therefore, in the real world of ecosystems, actual productivity is considerably less than potential productivity.

If resource managers understand the nature of environmental constraints on the productivity of biological resources, and can devise systems to reduce those constraints, then the yield of harvested products can be increased. In any truly sustainable system of resource management, such increases in yield must be obtained without degrading the resource's capability for renewal (that is, they cannot be

FIGURE 12.1 | FACTORS AFFECTING THE YIELD OF A BIOLOGICAL RESOURCE

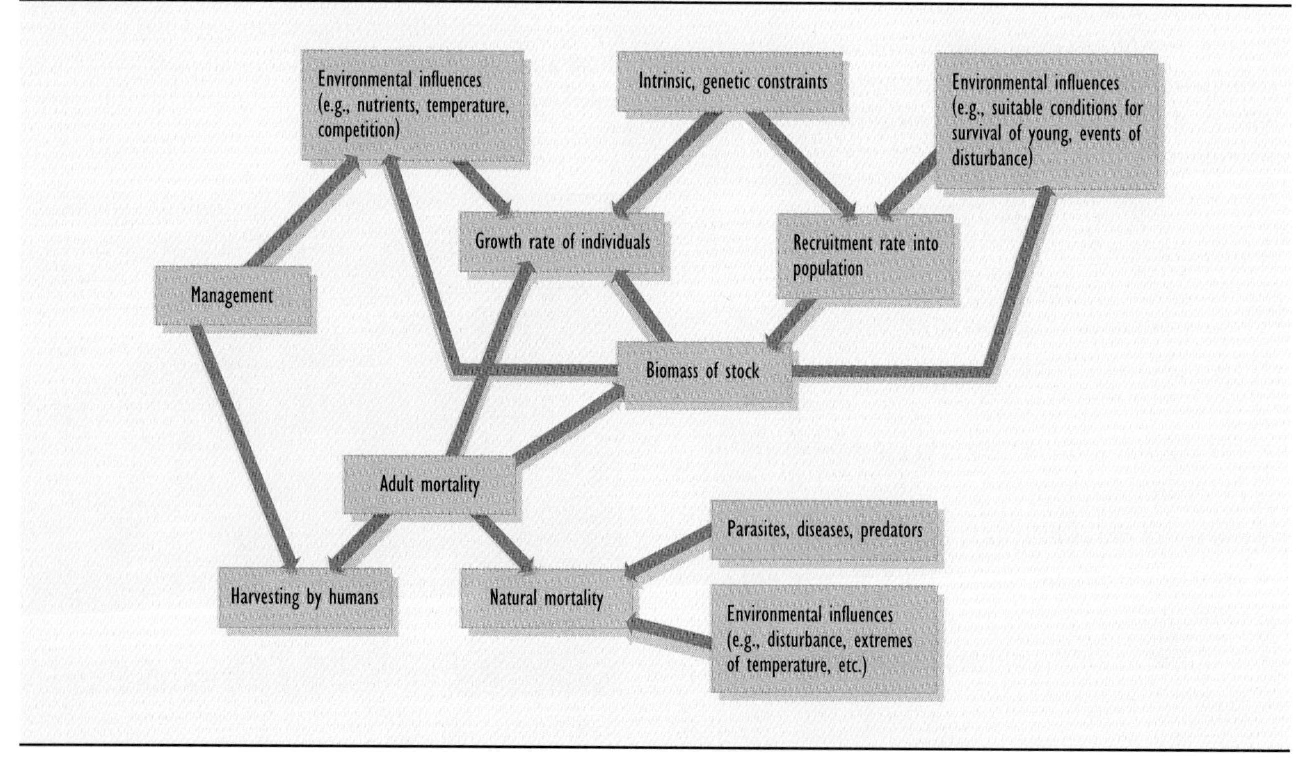

The biomass and productivity of a biological resource are determined by the recruitment of individuals into the population, by their growth rates, and by their rates of mortality, through either harvesting or natural means. These factors are affected by both genetically determined and environmental influences. Often, environmental factors can be managed, resulting in an increase in the productivity and size of the stock.

Source: Modified from Begon *et al.* (1990)

obtained by excessively harvesting the resource or by degrading the environment). Important management options for increasing the productivity of biological resources are briefly described below. (Note that this section illustrates commonly used management practices for increasing the productivity of renewable resources. However, all management options cause some degree of ecological damage, as is discussed in other chapters.)

Enhancement of Recruitment Rate

The rate of recruitment of new individuals into exploited populations can be increased in various ways. Some commonly used methods are described below.

Planting

In intensively managed agricultural, forestry, and aquacultural systems, managers attempt to achieve an optimally spaced monoculture of the crop species. This is done so that the crop productivity will not be limited by competition with non-crop species, or with individuals of the crop that are growing too closely together. Recruitment of agricultural plants is often managed by sowing seeds under conditions that favour their germination and establishment, while optimizing density to minimize competition among crop plants. Sometimes young plants are grown elsewhere and then out-planted—a practice commonly used to cultivate paddy rice, develop fruit-tree orchards, and establish plantations in forestry.

Regeneration of Perennial Crops

Some management systems encourage perennial crops to regenerate by re-sprouting from surviving roots, rhizomes, or stumps after the above-ground biomass is harvested. This kind of regeneration system is commonly used with sugar cane, and with stands of aspens, poplars, maples, and ashes in forestry. In some cases, the regenerating population may have to be thinned to optimize the density of recruitment.

Stock Enhancement

Recruitment of many wild fishes, particularly species of salmon and trout, is often enhanced by stripping wild animals or hatchery stock of their eggs and milt (sperm). The eggs are fertilized under controlled conditions and incubated to hatching. The larval and juvenile fish (called fry) are cultivated until they reach a fingerling size, when they are released to suitable habitats to supplement the natural recruitment of wild fish.

Site Preparation

Certain management practices in forestry favour the recruitment of economically desired tree species. For instance, some species of pine recruit well onto clear-cuts that have been site-prepared by burning, as long as an adequate supply of seed is available. Seedlings of other tree species establish readily onto exposed mineral soil, and are favoured by mechanical site preparation that exposes this substrate.

Managing Sex Ratios

Recruitment of some species of hunted animals can be maintained if only adult male animals are killed. For example, most species of deer are polygynous (that is, males will mate with more than one female). Consequently, the hunt can be restricted to male animals, on the assumption that the surviving bucks will still be capable of impregnating all the females in the population.

Harvest Season

Recruitment of most hunted animals can be managed by limiting the hunting season to particular times of the year. For example, restricting the hunt of waterfowl to the autumn allows ducks and geese to breed during the spring and summer, so that annual recruitment can occur. Spring hunting interferes with this reproduction, and is rarely permitted.

Enhancement of Growth Rate

As noted previously, the productivity of virtually all plants and animals is constrained by environmental influences, which include inorganic factors such as nutrient availability and temperature, and biological influences such as competition and disease. Often, management practices can be used to manipulate environmental conditions to reduce these limitations on growth rate and productivity, allowing an increased harvestable yield. Sometimes a **management system** is used, involving a variety of practices applied in a coordinated manner. Some examples follow.

Agricultural Systems

In intensive agricultural systems, high-yield varieties of crop species are grown and managed to optimize their productivity. Management practices typically combine some or all of the following: fertilization to enhance nutrient availability, irrigation to reduce the effects of drought, tillage (plowing and harrowing the land) or herbicide use to decrease competition from weeds, fungicide use and other practices to control diseases, and insecticide use and other practices to lessen damage caused by insects and other animal pests.

Forestry

The intensity of management used in forestry varies greatly, but crop-tree productivity can be increased through silvicultural practices such as thinning young stands to reduce competition among crop trees, using herbicides to control weeds, and using insecticides to cope with infestations of defoliating insects.

Aquaculture

In aquaculture, high-yield varieties of fish, crustaceans, or mollusks may be grown at high density in ponds or pens, where they are well fed and protected from diseases and parasites through antibiotics and other chemicals.

Management of Mortality Rate

Mortality of juveniles and adults can seriously affect the sizes of plant and animal stocks. By thinning out the stock, mortality also influences the intensity of competition within the population, and can increase the growth rates of the survivors. Mortality can be caused by natural influences such as predation, disease, or disturbance, or it can occur because humans have harvested some of the stock. Resource depletion occurs when the total rate of mortality (i.e., natural plus anthropogenic) exceeds the regenerative capability of the stock. Mortality associated with natural factors and harvesting can often be adjusted by management, as is described below.

Natural Mortality

Mortality associated with natural predators, parasites, diseases, and accidents can be decreased in various ways.

Diseases, Parasites, and Herbivores: Mortality of crop plants caused by herbivorous insects may be managed by using insecticides, or by changing the cultivation procedures to develop an ecosystem that is less favourable to the pest. Livestock are commonly affected by parasites, a problem that may also be reduced by using a pesticide. For example, sheep infested with ticks are dipped in chemical baths that kill the pests. Similarly, mortality caused by disease may be reduced by applying chemicals that manage the symptoms of disease, by administering antibiotics to deal with bacterial infections, or by changing cultivation systems to decrease vulnerability. All such practices allow diseases, parasites, and herbivores to be controlled over the short term, but none are long-term solutions to these causes of productivity losses.

Natural Predators: Coyote, wolf, cougar, and bears are rarely important predators of livestock, but many farmers consider any losses to these natural predators to be unacceptable. Many hunters feel the same way about the mortality these predators cause to some hunted wildlife, such as deer, moose, and caribou. Consequently, these large predators have been relentlessly persecuted by shooting, trapping, and poisoning. Access of predators to livestock may be restricted by fences, or by guard animals such as dogs and donkeys.

Harvesting Mortality

Mortality associated with harvesting must also be managed to ensure that the total mortality (natural plus anthropogenic) stays below the threshold for depleting the resource. For an ideal population, the **maximum sustainable yield** (MSY) is the largest amount of harvest mortality that can occur without degrading the productivity of the stock. Theoretically, harvest rates less than MSY would leave a "surplus" of the stock to natural sources of mortality, while harvest rates greater than MSY would impair regeneration. Note that a harvest rate equal to or less than MSY would sustain the resource.

Harvest-related mortality is influenced by many factors, including the quantity and kinds of harvesting equipment and personnel plus the amount of time that these units can spend harvesting. Resource managers can adjust the rates of mortality by controlling the total harvesting *effort*, which is a function of both the *means* (e.g., the kinds of fishing boats and their gear) and the *intensity* (e.g., the number of boats and the amount of time each spends fishing) of harvesting.

Technology has a great influence on harvesting rates. Consider, for example, the various methods of catching fish summarized in Figure 12.2. These technologies vary greatly in efficiency, which might be calculated as the quantity of fish caught per person fishing, per unit of energy expended, or per unit-value of investment in equipment. In general, much greater harvesting mortality is associated with the more intensive technologies, such as drift nets, trawls, and seines, compared with simpler methods such as hand lines and long lines. The more efficient methods also have a much greater *by-catch* of species that are not the target of the fishery and are often thrown away. Similarly, hunters armed with rifles are more efficient than those using bows and arrows, and trees can be harvested more quickly using a feller-buncher than with a chainsaw or an axe. (A feller-buncher is a large machine that cuts and de-limbs trees and stacks the logs into piles.)

Selection of Species and Sizes: The great variation in selectivity of harvesting methods, as regards both species and sizes, can be an important consideration in resource management. In a fishery, for example, a change in the netting mesh size directly influences the sizes of animals that are caught. Usually it is advantageous not to harvest smaller individuals, which may not yet have bred, and often have a smaller value-per-unit-weight than larger animals. In forestry, size- or species-selective cutting practices might be used in preference to clear-cutting, perhaps to encourage regeneration of the most desirable tree species. This also reduces environmental damage, by keeping the physical structure of the forest relatively intact.

Number of Harvesting Units: An obvious way to manage mortality rates associated with harvesting is to limit the number of units participating in the harvest. In a fishery, for example, the government could limit the number of

FIGURE 12.2 | FISHING TECHNOLOGY

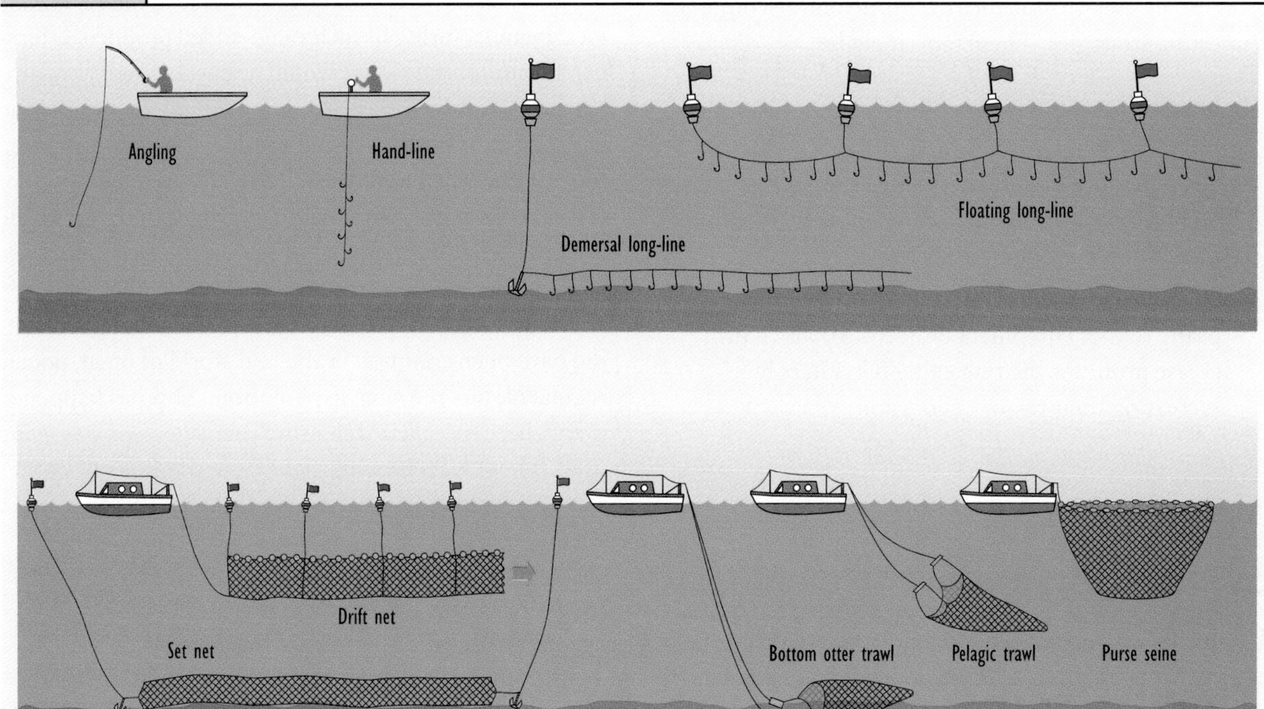

Methods of catching fish in open waters vary enormously in their efficiency and in the associated harvesting mortality.

(a) Line methods range from hand-lines with one or more hooks to floating or bottom lines running for kilometres and having thousands of hooks.

(b) Gill nets can be set on the bottom or attached to drifting buoys, and can range up to tens of kilometres in length, catching fish and other animals as they try to swim through the mesh.

(c) Trawls are open, broad-mouthed nets that are dragged along the bottom or through the water column, while purse seines are first positioned around a school of fish swimming near the surface, and are then pulled shut with a bottom draw-line.

fishers by issuing only a certain number of licences. Usually, the kind of technology that may be used by the harvesters is also specified—for example, the number of boats using particular fishing gear.

Time Spent Harvesting: Harvesting effort is also influenced by the amount of time that each harvesting unit works. Because of the economic value of investments in machinery and personnel, there is strong pressure on regulators to allow harvesting to occur as long as possible. Even so, harvesting time is in some cases very closely regulated. For example, certain herring fisheries off western North America operate for periods of time as short as several hours per year.

To achieve control over harvesting effort (and therefore over the mortality associated with exploitation), managers often use legal and administrative tools. Relatively direct controls include licenses that regulate the numbers of participants, the technology they may use, their resource quotas, and the times and places they may harvest. Indirect tools can be used to influence the profitability of different harvesting strategies. These tools include:

- fines for noncompliance, which decrease profit by raising costs
- taxes on more harmful harvesting methods or subsidies on less harmful ones, which influence profit by increasing or reducing costs respectively
- buyouts of inappropriate or excess harvesting capacity (either equipment or licenses), which increase profit for the remaining harvesters by increasing their relative allocation

Achieving Maximum Sustainable Yield

Potentially, all management options (that is, enhancement of growth and recruitment rates and management of mortality rate) can result in larger yields of biological resources for use by humans. However, the factors that influence the size and productivity of stocks of potentially renewable resources are imperfectly understood. Consequently, the management systems advocated by applied ecologists, such as foresters and fishery and agricultural scientists, are also imperfect. Despite this caveat about uncertainty, enough is usually known about ecological factors affecting biological resources to design sustainable harvesting and management systems that will not degrade the capability for renewal.

At the very least, harvesting intensities small enough to avoid *over-exploitation* of the resource can be predicted, even though the harvest might be smaller than the potential maximum sustainable yield. Harvests of natural resources need not be as large as are potentially attainable. If resource managers cannot predict accurate MSYs, then it is ecologically prudent to harvest at rates known to be smaller than the MSY, but clearly sustainable. Of course, such strategies result in smaller harvests and fewer short-term profits. These are, however, more than offset by the long-term economic and ecological benefits of adopting prudent strategies of resource harvesting and management.

Moreover, the regional economic benefits of smaller (but sustainable) harvests can be enhanced by ensuring that manufactured outputs of the resource-dependent industry focus more on "value-added" products. In forestry, for example, exports of raw logs would be prohibited, while local manufacturing of such value-added products as lumber, furniture, and violins would be encouraged. Similarly, a regional fishing industry might focus on the production and export of higher-valued products, such as prepared foods, rather than unprocessed fish. These integrations of resource harvesting and manufacturing can optimize the regional economic benefits of resource-based industries, while allowing smaller, sustainable harvests of the resource.

Regrettably, however, unsustainable rates of harvesting have been common in the real world of open, poorly regulated, bio-resource exploitation. This has happened even where so-called "scientific" management was being used. These facts become clear from the examples of resource degradation described in this chapter (and also in Chapter 24).

Unsustainable Use of (Potentially) Renewable Resources

Many potentially renewable natural resources have been used by humans in an unsustainable fashion. Either these resources have been exploited excessively (a condition known as **over-harvesting** or **over-exploitation**), or their post-harvest stocks have been managed inappropriately.

Either of these can result in depletion or exhaustion of the resource by so-called "mining."

There are numerous examples of the unsustainable use of potentially renewable natural resources. Some species have even been made extinct by excessive hunting—for instance, the dodo, the passenger pigeon, and the great auk (the latter two used to occur in Canada; see Chapter 24). In other cases, seemingly super-abundant species were rendered endangered by over-harvesting, including American ginseng, Eskimo curlew, northern fur seal, plains bison, right whale, trumpeter swan, whooping crane, and various other species that were once abundant in parts of Canada (see Chapter 24). In fact, it is remarkably difficult to cite examples of economically valuable, potentially renewable natural resources that have not been severely depleted at one time or another through excessive use or inappropriate management.

Additional examples of the mining of potentially renewable resources include:

- extensive deforestation of many parts of the world, resulting in losses of lumber and fuelwood resources as well as environmental damages such as erosion and regional changes in climate
- extensive degradation of the quality of agricultural soils, resulting in declining crop yields, and sometimes in the abandonment of previously arable land
- exhaustion of fisheries, such as those of cod and other groundfish off the Atlantic Provinces of Canada, and salmon and herring stocks off British Columbia
- depletion of many hunted resources—various species of Canadian fish, antelope, deer, furbearers, waterfowl, whales, and others (see Chapter 14)

Not all cases of the mining of potentially renewable resources have occurred in modern times. In Detail 12.1 (p. 184) and In Detail 12.2 describe examples that are ancient, even prehistoric. These well-known cases demonstrate that even unsophisticated human societies with few technological capabilities have caused enormous damage to their essential resource bases.

In some cases, early depletions of potentially renewable resources were followed by vigorous efforts at **conservation** or improved management, which subsequently restored stocks of the depleted resources (but not the extinct ones). (In the sense used here, "conservation" refers to the wise use of natural resources, including recycling and other means of efficient utilization, and also including ensuring that the harvest of renewable resources does not exceed the rate of regeneration of the stock.) For example, regulating the hunting of white-tailed deer across Canada has allowed that species to remain abundant in regions where habitat is suitable. Comparable successes have been achieved with other once-depleted populations of animals, such as certain sport fish, ducks, and geese. Some examples of conservation achievements are described as case studies at the end of Chapter 24.

Overall, however, there is more bad news than good about future stocks and regeneration of many potentially renewable resources. Although some renewable resources are being used in a fashion consistent with their future availability, many are not. If this situation does not change for the better in the near future, there will be grim consequences for human economies, and for natural ecosystems and biodiversity.

Patterns of Over-exploitation

In cases of single-species resources, in which only a particular species is harvested, overexploitation generally involves excessive harvesting rates, occurring without adequate attention to regeneration. Under such conditions the stocks are quickly mined, and they eventually collapse to economic or biological extinction. Sometimes, a "virgin" or previously unexploited resource is dominated by large, old-growth individuals, which are harvested selectively during the initial stages of resource "development." This changes the structure of the resource to one dominated by smaller, younger individuals. Because younger individuals are often relatively fast-growing, the productivity of the resource is not necessarily smaller than that of the original, old-growth stock, although the total biomass may be less. However, if this kind of resource degradation is taken too far, the population may collapse in both productivity and biomass. The collapse may be caused by inadequate recruitment into the harvested population, because the fecundity of younger individuals is not sufficient to offset the harvesting mortality.

Patterns of resource degradation are more complicated in the case of mixed-species resources, which are often overexploited in a sequential manner. At first, only certain species in the virgin, mixed-species resource may

IN DETAIL 12.2

PREHISTORIC EXTINCTIONS

Paleontologists around the world have found clear evidence of prehistoric mass extinctions of various species of animals, apparently caused by overhunting by stone-age humans (Martin, 1967, 1984; Diamond, 1982; Freedman, 1995). The extinction events occurred at different times and places, but each coincided with the discovery and colonization of a land mass previously uninhabited by humans. The extinct animals were likely naive to predation by efficient groups of human hunters, and were unable to adapt to the onslaught. The mass extinctions in these cases constituted instances of unsustainable mining of wild-animal populations, which were potentially renewable sources of food for the neolithic hunters.

In North America, the wave of prehistoric extinctions began about 11 000 years ago, soon after humans colonized the continent by migrating across a land bridge from Siberia. (The land bridge existed because the sea level was lower than today, with much water tied up in glacial ice.) Within a relatively short time, at least 56 species of large mammals (that is, weighing more than 44 kg), 21 species of smaller mammals, and several large birds had become extinct. The extinctions included 10 species of horses (genus *Equus*); the giant ground sloth (*Gryptotherium listai*); four species of camels (family Camelidae); two species of buffalo (genus *Bison*); a species of cow (genus *Bos*); the saiga antelope (*Saiga tatarica*); and four species in the elephant family, including the mastodon (*Mammut americanum*) and mammoth (*Mammuthus primigenius*). Predators and scavengers that depended on the large herbivores also became extinct, including the sabertooth tiger (*Smilodon fatalis*), the American lion (*Panthera leo atrox*), and a huge scavenging bird (*Terratornis merriami*). The best collections of fossil bones of many of the extinct animals have been excavated from natural tar pits at Rancho la Brea in southern California. However, bones of numerous species are quite widespread, and have been found in many places in Canada. As colonizing humans spread in waves into Central America, and then into South America, extinctions of other vulnerable prey species occurred in those regions.

Similar events of mass extinction have occurred elsewhere, also coinciding with initial colonizations by prehistoric, stone-age hunters. In New Guinea and Australia, waves of extinctions occurred about 50 000 years ago, following discoveries of those islands by Melanesians spreading south from Asia. These mass extinctions involved the losses of many large species of marsupials, flightless birds, and tortoises.

In New Zealand, an extinction wave occurred less than 1000 years ago, following the discovery of these Pacific islands by Polynesians. This event swept away numerous species of large, flightless birds, including a 250-kg, 3-m-tall giant moa (*Dinornis maximus*), 26 other species of moas, a goose (*Cnemiornis calcitrans*), a swan (*Cygnus sumnerensis*), a giant coot (*Fulica chathamensis*), a pelican (*Pelecanus novaezealandiae*), and an eagle (*Harpagornis moorei*), plus fur seals and species of large lizards and frogs from North Island. The extinctions of the moas progressed as a wave sweeping from North Island to South Island over a two-century period following the Polynesian colonization. Great quantities of bones were discovered at places where the moas had been herded and butchered. These deposits were mined by early European colonists and processed into phosphate fertilizer.

The human colonization of Madagascar occurred about 1500 years ago. This also resulted in many extinctions, including losses of 6–12 species of huge elephant birds, 14 species of lemurs, two giant tortoises, and various other large animals. Other prehistoric mass extinctions occurred in Hawaii, New Caledonia, Fiji, the West Indies, and other islands. All are believed to have resulted from overhunting by newly colonizing aboriginal humans.

Clearly, the unsustainable use of biological capital, resulting in irretrievable losses of species important to humans as resources, is not only a modern phenomenon. Prehistoric humans could be as rapacious as their modern descendants, given appropriate opportunities in the form of naive and edible species.

be considered desirable from the economic perspective. In addition, some individual organisms may be very large, especially in the case of old-growth resources. For instance, old-growth forests of coastal British Columbia are typically dominated by large individuals of valuable species of trees, coexisting with many smaller individuals (see Chapter 23). Many pre-exploitation communities of fish, whales, and other species are also typically dominated by large individuals of desirable species.

The exploitation of mixed-species resources usually involves a sequential harvest of commodities having increasingly smaller values (measured as value per individual, as well as per unit of harvested area). Initially, the largest individuals of the most valuable species are harvested selectively and are rapidly depleted. In an old-growth forest, for example, the largest logs of the most desirable species have the highest value per unit of biomass. These can be used to manufacture large-dimension lumber of precious species or valuable veneer products. The intention of post-harvesting management is not to re-create another old-growth forest, because this would take too much time and would also involve extended periods of relatively low productivity (see Chapter 23). Instead, the site is typically converted into a second-growth forest, such as one of the following stages of sequential exploitation.

Next, smaller individuals of the most desired species might be harvested selectively, along with the largest individuals of secondarily desired species. In a forest, the economic products might involve smaller-dimension lumber. If management of the regenerating stand is intended to produce timber for manufacturing into lumber, the subsequent harvests would be on a relatively long rotation, say 60–100 years, depending on growing conditions.

Area-harvesting methods might then be used to harvest virtually all individuals of all species for manufacturing into bulk commodities. In the case of forestry, biomass might be clear-cut for the production of pulp, industrial fuel, charcoal, or domestic fuelwood. Subsequent harvests for such purposes would be on a relatively short rotation, perhaps 30–50 years, depending on growing conditions.

Sometimes the area-harvesting system is followed by management that regenerates a productive resource, although it has a very different character from the original, natural community. In forestry, for example, natural mixed-species forests might be converted into single-species plantation forests, or perhaps into an agricultural ecosystem.

Intensive harvesting, sequential or otherwise, can also lead to a collapse of biological productivity useful for human purposes, and therefore to a huge loss of resource value. For instance, clear-cut forests sometimes regenerate into shrub-dominated barrens that resist the establishment of tree seedlings. This kind of severe resource degradation may require expensive management to restore another economically useful forest.

Ecologically rich old-growth forests in many tropical countries are being rapidly cleared to provide land for use in agriculture. This ecological conversion results in the destruction of the forest (the mining of a potentially renewable natural resource), often to develop agricultural lands that will probably not be productive for very long. This scene is from Sumatra, Indonesia.
Source: B. Freedman

Reasons for Resource Abuse

To function over the long term, economic systems depend on sustained inputs of natural resources. Given this context, it would appear to be economically self-destructive to degrade renewable resources by overharvesting or by inappropriate management. Nevertheless, this maladaptive behaviour has occurred frequently through human history — in fact, most uses of renewable resources have been decidedly nonsustainable, causing resources to become depleted quickly. The most important reasons for this seemingly foolish behaviour are outlined below.

The world's dominant cultures have developed an ethic which presumes that humans have the right to take whatever they want from nature for subsistence or economic benefits. This is an expression of the anthropocentric world view (Chapter 1). Particularly noteworthy is the so-called Judeo-Christian ethic (White, 1967). This ethic is based on the Biblical myth of creation, in which God directed humans to "be fruitful, and multiply, and replenish the earth, and subdue it," and to "have dominion over the fish of the sea, and over the fowl of the air, and over the cattle, and over all the earth and over every creeping thing that creepeth upon the earth" (Genesis 1:28). From the purely ecological perspective, this is an arrogant attitude, but it is typical of the world's dominant cultures and religions. Modern technological ethics have developed from this commanding kind of world view, and are now used to legitimize the mining of many potentially renewable resources, with the inflicting of associated ecological damage.

Individual people and their societies are intrinsically self-interested. This attitude is responsible for many cases of over-exploitation of resources in order to optimize short-term profits. At the same time, ecological damages associated with the resource degradation are discounted as being unimportant. This is readily done because in most economic systems, the consequences of most ecological damages are shared broadly across society, rather than being considered the responsibility of the persons or corporations who cause the damages.

Natural resources are perceived as being boundless. Many people think that nature and its resources are unlimited in their extent, quantity, and productivity. This is referred to as the "frontier" or "cornucopian" world view (a cornucopia is the mythical "horn of plenty," which yields food in boundless quantities). In actual fact, Earth has limited amounts of resources available for use by humans, and most of these are being rapidly depleted by overuse.

Investments of money in some sectors of the economy may accumulate profit faster than the growth rate of many renewable resources. Consequently, apparent profits can be maximized over the shorter term by liquidating natural resources through overexploitation, and then investing the money earned in a faster growing sector of the economy (see In Detail 12.3). The growth of many regional and national economies has been jump-started by economic "capital" gained through the unsustainable mining of ecological "capital."

Not all of the true costs of overexploitation are taken into account. The economic strategies suggested above only work if the ecological costs of overexploitation are not paid for. (In fact, some kinds of environmental damage can be interpreted as being "good" for the economy because they add to the Gross National Product (or GNP). For example, the wreck of the *Exxon Valdez* in Alaska and the cleanup of the spilled petroleum were responsible for billions of dollars of growth in the GNP of the United States over several years; see Chapter 22.) These environmental damages represent a depletion of ecological capital and are a sort of "natural debt." When apparent profits gained through overexploitation are being calculated, **conventional economics** (i.e. economics as it is usually practiced) does not account properly for costs associated with ecological damage and resource depletion. In theory at least, this damage would be fully costed in a system of **ecological economics** or full-cost accounting, which is advocated by many enlightened economists and environmentalists.

Within an economic context involving free access to **common-property resources** (resources owned by all of society), the above factors inevitably lead to an ecologically inappropriate overexploitation, overharvesting, and degradation of potentially renewable resources. In a highly influential essay, Hardin (1968) called this frequently observed phenomenon "the tragedy of the commons." He explained this economic misadventure using the analogy of a publicly owned pasture (the "commons"), to which all local farmers had free access for grazing their sheep. Individual, self-interested farmers believed that they would gain additional economic benefits by having as many of their own sheep as possible grazing the pasture. This led to an excessive aggregate use of the pasture, damaging and

IN DETAIL 12.3

INVESTMENTS IN RENEWABLE RESOURCES

Consider a simple case: tree biomass in a forest is increasing at a rate of 5%/y, and real interest rates (adjusted for inflation) on secure investments are 10%/y. Because the forest resource is growing at 5%/y, its value (i.e., its biomass) would double about every 14 years. If, however, the forest was harvested and the products sold, and the resulting money invested at an interest rate of 10%/y, the quantity of money would double in only 7 years, so that profit would be made twice as quickly.

Obviously, this kind of investment strategy only works if:

1. the objective is to gain short-term profit rather than achieve long-term resource sustainability;
2. the social perspective is that of individual people or corporations and not the welfare of society at large;
3. it is assumed that the natural resource has value only if it is harvested and converted into cash; and
4. only the costs of extraction are considered in the calculation of profit, while the costs of ecological damage and resource degradation are paid by society as a whole (that is, in economic terms, they are treated as externalities).

Over the long term, the liquidation of potentially renewable natural resources is clearly a losing strategy for society at large and for future generations (Clark, 1989; Burton, 1993). For individuals, firms, and local economies, however, this can be a highly profitable strategy because wealth can be accumulated during a short period of time. Therefore, this tactic is often advocated and pursued by influential people. Consider the following statement in 1986 by Bill Vander Zalm, a former Premier of British Columbia, one of the world's greatest exporters of forest products: "Let's cut down the trees and create jobs."[1] This is what has been happening, more or less, to many potentially renewable natural resources in most parts of the world.

[1] Luinenberg, O. and S. Osborne, compilers. 1990. *The Little Green Book: Quotations on the Environment.* Vancouver: Pulp Press Publishers.

degrading the forage resource. Hardin's major conclusion, that "freedom in a commons brings ruin to all," has generally been true of the way in which many renewable resources have been used.

Many nations are experiencing crises because of diminishing stocks of potentially renewable resources and the associated ecological damage caused by disturbance, pollution, and loss of biodiversity. Remarkably, many of these nations have not yet attempted to deal effectively with the resource degradation. With few exceptions, the design and implementation of intelligent strategies for using natural resources has proven to be beyond the capability of modern political and economic systems.

However, humans are capable of designing and implementing such systems, that would protect the flows of natural resources and the healthy ecosystems that are required to sustain economies. Solutions to resource-related predicaments require an integration of scientific knowledge and social change, along with the adoption of ecologically based economic policies that pursue true sustainability. Such solutions would be far preferable to unfettered economic growth based on the over-exploitation of resources.

GROWTH, DEVELOPMENT, AND SUSTAINABILITY

To economists, growth and development are very different phenomena. **Economic growth** is a feature of an economy that is increasing in size over time. It is usually associated with increases in both numbers of people and their per capita use of resources. Particularly in developed countries, economic growth is achieved by the rapid consumption of natural resources. Nonrenewable resources, such as metals and fossil fuels, are consumed in especially large quantities in growing economies. Potentially renewable resources are also frequently mined, rather than being harvested sustainably.

At the present time, almost all national economies are growing quite rapidly. Moreover, most economic planners and politicians hope for additional increases in economic activity in the foreseeable future. Economic growth is viewed as a means of generating more wealth for nations and providing a better life for their citizens.

Unfortunately, there are well-known limits to growth, related to the finite resources on planet Earth plus the laws of thermodynamics (Chapters 1 and 4). Consequently, in a resource-constrained world, unlimited economic growth can never be sustained over the long term. In the perspectives of ecologists and ecologically minded economists, growth is not necessarily desirable: "Economic growth as it now goes on is more a disease of civilization than a cure for its woes" (Ehrlich, 1989).

Economic development, fundamentally different from growth, implies an improving efficiency in the use of materials and energy and progress toward a sustainable economic system. Headway toward sustainable economic development involves the following actions:

- increasing the efficiency of use of nonrenewable resources—for example, by recycling and reusing metals and other materials; by minimizing the use of energy for industrial, transportation, and space-heating purposes; and by improving designs of other materials and products
- increasing the use of renewable materials in the economy, such as products manufactured from trees or agricultural biomass
- rapid increases in the use of renewable sources of energy, such as electricity generated using hydro, solar, wind, or biomass technologies (see Chapter 13)
- increasing social equity, ultimately to such a degree that all citizens (and not just the minority of relatively wealthy people) have access to the necessities and amenities of life

SUSTAINABLE DEVELOPMENT

The notion of **sustainable development** refers to progress toward an economic system which uses natural resources in a manner that does not deplete their capital or compromise their availability to future generations of humans. In this sense, the present human economy is obviously non-sustainable, because it involves rapid economic growth achieved by the vigorous mining of both nonrenewable and potentially renewable natural resources.

Many politicians, economists, resource managers, and corporate spokespersons have publicly stated that they support progress toward sustainable development. However, almost all of these people are confusing genuine sustainable development, as just defined, with "sustainable economic growth," which by definition is never possible.

The phrase "sustainable development" was first popularized by its use in the widely acclaimed report, *Our Common Future*, by the World Commission of Environment and Development, an agency of the United Nations. (The report, published in 1987, is sometimes referred to as the "Brundtland Report," after Gro Harlem Brundtland, the chairperson of the Commission at the time.) Even this report, however, obscured some important differences between economic growth and economic development. In fact, the Brundtland Report advocated a large expansion of the global economy: "It is ... essential that the stagnant or declining growth trends of this decade (i.e., the 1980s) be reversed." The report suggested that economic growth, coupled with a more equitable distribution of wealth, was required to improve the living standards of poorer peoples of the world. Once society had achieved the socioeconomic conditions required for stopping both population growth and rampant depletions of natural resources, real progress could be made toward a no-growth, equilibrium economy.

One of the recommendations of the Brundtland Report was that the global average per capita income should grow by 3%/y (if maintained, this rate of increase would double per capita income every 23 years). However, because the global population is increasing at about 2%/y, the economic growth rate would have to compensate. Therefore, achieving a 3%/y increase in per capita income would require a 5%/y increase in the total economy (which would result in a 14-year doubling time). Of course, in those regions where population growth is more rapid, such as most of Africa, south Asia, and Latin America (see Chapter 10), economic growth rates would have to be as high as 6%/y (doubling time of 12 years) or more in order to achieve a 3%/y increase in per capita income. Ultimately, the Brundtland Report estimated that a 5–10-fold expansion of the global human economy was required to set the stage for attaining a condition of sustainable development.

The authors of the Brundtland Report held that this growth would best be achieved through "policies that sustain and expand the environmental resource base." Such policies would include the advancement of efficient technologies that could help economic growth to be achieved, while consuming fewer material and energy resources than now occurs. In addition, a redistribution of wealth from richer people and regions to poorer ones would be central to achieving the growth of per capita income that is championed in the Brundtland Report.

It is important to understand that the Brundtland Report was developed through a consensus process, which involved wide-ranging consultations among diverse interested parties. Therefore, representatives of many nations and cultures had to agree on its content. Considering the diversity of the interests involved, it is not surprising that the report advocated enormous economic growth as a component of its "development" strategy. The growth-related aspects of the report make it easier for politicians and industry to support its recommendations.

From the ecological perspective, however, it is very doubtful that a 5–10-fold increase in the scale of the human economy could be sustained. Many have argued that it would be much more sensible to pursue solutions which aggressively attack both economic growth, as it is currently achieved, and population growth. Components of those solutions would include vigorous actions toward population control, a more equitable distribution of wealth, reduced use of resources by richer peoples of the world, more equitable access for women to education and social empowerment, development and use of more efficient technologies, and conservation of natural resources. These sustainable solutions are more difficult and unpopular than the policies advocated by most mainstream politicians and economists, but they appear to be necessary.

Sustainable and Nonsustainable Economies

The proper definition of a **sustainable economy** is one that can be maintained over time without resulting in any depletion of its capital of natural resources. Ultimately, a sustainable economy can be supported only by the wise use of renewable resources, which would be harvested at rates equal to or less than their productivity. Therefore, "economic development" should refer only to progress toward a sustainable economic system. Unfortunately, there have been few substantial gains in this direction. This has happened because most actions undertaken by politicians, economists, and planners have supported rapid economic growth, rather than sustainable economic development. In large part, they do this because they believe they are following the wishes of the public.

Because nonrenewable resources are always depleted by their use, they cannot provide the foundation of a sustainable economic system. However, nonrenewable resources still have an important role to play in a sustainable economy. Their use would be tied to improving the stocks of comparable renewable resources, so that a net depletion of resources would not occur. For example, if people wanted to use nonrenewable coal, they would act to provide compensating increases in forest area and biomass. This could result in no net consumption of potential energy (since tree biomass and coal are both fuels), and no net increase in the concentrations of atmospheric carbon dioxide or other pollutants (trees absorb CO_2 as they grow, and mature forests can store carbon for a long time as long as they are not disturbed). Of course, any nonrenewable materials that are already in use in the economy can continue to be used, and they should be recycled as efficiently as possible.

Symptoms of Nonsustainable Economies

As was previously mentioned, the dominant trends of local, national, and global economies are mostly toward vigorous (but unsustainable) economic growth, rather than toward sustainable development. The indicators of these trends of nonsustainable growth are summarized below.

Rapid Growth of Economies

Because of increases in both population and per capita consumption of materials and energy, virtually all economies are growing. This well-known fact is reflected by trends in many economic indicators, such as gross domestic product (GDP, or the value of all goods and services produced in a domestic economy) and stock-market and other indexes. For example, the Canadian economy grew by a factor of about 8 between 1947 and 1994, compared with a 2.3-fold increase in population during the same period (Figure 12.3).

Depletion of Nonrenewable Resources

All stocks of metal ores, petroleum, natural gas, coal, and other nonrenewable resources are finite, being limited to what is present on Earth. These resources are being consumed rapidly and their exploitable reserves will eventually become depleted. Discoveries of additional exploitable stocks extend the lifetimes of nonrenewable resources, as does efficient recycling. Nevertheless, global stocks of nonrenewable resources are being depleted at accelerating rates.

FIGURE 12.3 GROWTH OF THE CANADIAN ECONOMY AND POPULATION, 1943–1990

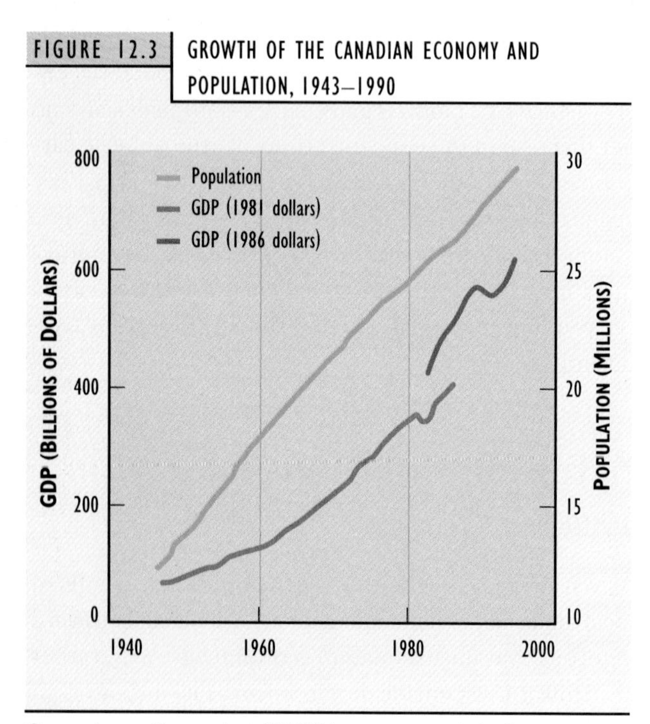

Gross domestic product (GDP) is an economic indicator related to the total size of a national economy. Because these data for Canadian GDP are standardized to either constant 1981 or constant 1986 dollars, the pattern of steady economic growth is not due to inflation. Compare the growth of GDP with the growth of the Canadian population.

Source: GDP data from Statistics Canada (1989, 1995), and population data from Figure 11.1 (p. 171)

Depletion of (Potentially) Renewable Resources

Around the globe there are crises of depletion of renewable resources. In many regions, for example, previously enormous fish stocks are collapsing, deforestation is proceeding rapidly, the fertility of agricultural soils is declining, supplies of surface and ground waters are being depleted and/or polluted, and hunted animals are becoming scarcer. Not all stocks of (potentially) renewable natural resources are severely depleted, but the declines are undeniably becoming more common and widespread.

Depletion of Noneconomic, Ecological Resources

Some resources that are necessary for the health of economic systems are not assigned value (or "traded") in the marketplace — that is, they are not valuated in dollars. Nevertheless, these resources are important to the health of the ecosystems that sustain economies. Examples of such non-valuated, ecological resources include the ecosystem's ability to cleanse the environment of toxic atmospheric pollutants associated with human activities, such as sulphur dioxide and ozone; ecological services such as the production of atmospheric O_2 and consumption of CO_2 (the latter being an important pollutant associated with human economies); and ecological functions that support the productivity of conventional resources, such as the plant and algal productivity that ultimately allows the growth of stocks of hunted deer, fish, and other animals.

Depletions of Other Ecological Values

Some ecological values are not directly or indirectly important in human economies, but still have intrinsic (or existence) worth. This makes these values significant, regardless of any real or perceived importance to human welfare (see also Chapter 1). The most important examples of non-economic, ecological values are associated with biodiversity, especially the many species and natural ecosystems that are indigenous to different regions. These biodiversity values are increasingly being threatened and lost in virtually all regions of the globe (see Chapter 24). Such losses would never be tolerated in an **ecologically sustainable economic system** (i.e., an economy in which renewable resources are used in ways that do not compromise their future availability, and do not endanger species or natural ecosystems).

Modern human economies deliver great benefits to those who are wealthy enough to purchase a happy and healthy lifestyle, with sufficient food, shelter, material goods, and recreational opportunities. For less wealthy people, however, current economic systems may allow only minimal access to the most fundamental basics of subsistence. If a fairer, more equitable delivery of economic benefits to the poorer people of the world is to be achieved, either unsustainable economic growth or a substantial redistribution of some of the existing wealth will be required.

Ultimately, the global scale and long-term sustainability of human economies will be limited by the ability of the biosphere to deliver renewable resources. The limits of many potentially renewable resources have already been reached or exceeded, resulting in stock declines or collapses. These well-documented damages should be regarded as warnings of the likely future of the human enterprise (assuming that resource-use systems do not change). If critical resources are no longer available to support economic activities, the unsustainable economic system will be forced to decrease in size, and may in fact collapse.

It must be recognized that ecologically sustainable economic systems might not be very popular among the public, politicians, government administrators, or industrial interests. All of these stakeholders would experience short-term pain (likely felt over decades) to achieve long-term, sustainable, societal gains. The pain would be associated with the less intensive use of natural resources, the abandonment of the economic-growth paradigm, and the rapid stabilization — and perhaps downsizing — of the human population. The gains would be associated with a sustainable economic system that could support human society for a very long time.

As we have repeatedly observed, humans undeniably rely on natural resources to sustain their enterprises. Throughout history, resources essential to economies have, technological capabilities permitting, been consumed to exhaustion. Clearly there are better, sustainable systems that can utilize renewable natural resources without depleting them and without degrading environmental quality. Human societies desperately require these sustainable systems. It remains to be seen, however, whether these essential systems will ever be designed and implemented.

Key Terms

spaceship Earth
sustainable economic system
nonrenewable natural resources
(potentially) renewable natural resources
management system
maximum sustainable yield (MSY)
overharvesting (or overexploitation)
conservation
conventional economics
ecological economics
common-property resource
economic growth
economic development
sustainable development
sustainable economy
ecologically sustainable economic system

Questions for Discussion

1. Distinguish between non-renewable and (potentially) renewable natural resources, giving examples of each.
2. Show how natural resources are important in your life by making a list of resources that you use daily for energy or food, or for materials in manufactured products.
3. Outline how productivity of biological resources can be increased through management.
4. Describe three cases of the "mining" of (potentially) renewable natural resources. Why did this over-exploitation occur?
5. Distinguish between economic growth and development, and relate these to the notion of sustainable development. Do you believe that the Canadian economy is making much progress toward sustainable development? Give reasons in support of your answer.
6. Can you think of any good examples of economically valuable, potentially renewable natural resources that have not been severely depleted through excessive use or inappropriate management? Explain your answer.

References

Begon, M., J.L. Harper, and C.R. Townsend. 1990. *Ecology: Individuals, Populations, and Communities*. London: Blackwell Scientific.

Burton, P.S. 1993. Intertemporal preferences and intergenerational equity considerations in optimal resource harvesting. *J. Environ. Economics & Manage.*, 24: 119–132.

Clark, W.C. 1989. Clear-cut economies: Should we harvest everything now? *The Sciences*, Jan./Feb.: 16–19.

Clark, W.C. and R.E. Munn (eds.). 1986. *Sustainable Development of the Biosphere*. New York: Cambridge University Press.

Costanza, R. 1991. *Ecological Economics: The Science and Management of Sustainability*. New York: Columbia University Press.

Costanza, R. and H.E. Daly. 1992. Natural capital and sustainable development. *Conserv. Biol.*, **6**: 37–46.

Cushing, D.H. 1988. *The Provident Sea*. Cambridge: Cambridge University Press.

Diamond, J.M. 1982. Man the exterminator. *Nature*, **298**: 787–789.

Freedman, B. 1995. *Environmental Ecology, 2nd ed*. San Diego, CA: Academic.

Hardin, G. 1968. The tragedy of the commons. *Science*, **162**: 1243–1248.

Martin, P.S. 1984. Catastrophic extinctions and late Pleistocene blitzkrieg: two radiocarbon tests. In: *Extinctions* (M.H. Nitecki, ed.). Chicago: University of Chicago Press. pp. 153-189.

Martin, P.S. and H.E. Wright (eds.). 1967. *Pleistocene Extinctions: The Search for a Cause*. New Haven, CT: Yale University Press.

Meadows, D.H., D.L. Meadows, and J. Randers. 1992. *Beyond the Limits: Global Collapse or a Sustainable Future*. 1992. London: Earthscan.

Mowat, F. 1984. *Sea of Slaughter*. Toronto: McClelland & Stewart.

Ponting, C. 1991. *A Green History of the World*. Middlesex, U.K.: Penguin.

Rees, W.E. 1990. The ecology of sustainable development. *Ecologist*, **20(1)**: 18–23.

Thirgood, J.V. 1981. *Man and the Mediterranean Forest: A History of Resource Depletion*. New York: Academic.

Tristram, H.B. 1873. *The Natural History of the Bible*. London: Society for Promoting Christian Knowledge.

White, L. 1967. The historical roots of our ecological crisis. *Science*, **155**: 1203–1207.

World Commission on Environment and Development (WCED). 1987. *Our Common Future*. Oxford: WCED, Oxford University Press.

13

NONRENEWABLE NATURAL RESOURCES

Chapter Outline

Chapter Objectives

After completing this chapter, you will be able to:

1. Describe the global and Canadian production and use of metals, fossil fuels, and other nonrenewable natural resources.
2. Discuss the reliance of industrialized economies on nonrenewable resources, and predict whether these essential sources of materials and energy will continue to be available into the foreseeable future.
3. Outline five alternative sources of energy available for use in industrialized countries and the potential roles of these in a sustainable economy.

INTRODUCTION

As we noted in Chapter 12, reserves of **nonrenewable natural resources** are inexorably diminished as they are used. This is because nonrenewable resources are finite in quantity and their reserves do not regenerate after they are mined from the environment. Note that the word "**reserve**" has a specific meaning here: to denote known quantities of materials that can be *economically* recovered from the environment (that is, while making a profit).

Of course, continuing exploration may discover exploitable but previously unknown deposits of nonrenewable resources, resulting in increases in the known reserves. For example, the world's known reserves of nickel and copper were increased recently because of the discovery of a rich deposit of those metals at Voisey's Bay in northern Labrador. There are, however, limits to the number of "new" discoveries of nonrenewable resources that can be made on planet Earth.

Changes in the value of nonrenewable commodities can also affect the sizes of their economically recoverable reserves. For example, an increase in the market value of gold can make it profitable to mine previously noneconomic ores, to engage in exploration in remote places, or to reprocess "waste" materials containing small quantities of this valuable metal. A change in technology can have the same effect — for instance, by making it profitable to recover and process lower-grade ores.

In addition, the useful lifetime of some nonrenewable resources, particularly metals, can be extended by **recycling**. This involves processing disused industrial and household products (sometimes referred to as "wastes") to recover reusable materials. However, thermodynamic and economic limits mean that recycling cannot be 100% efficient. Furthermore, the demand for nonrenewable resources is increasing over time because of population increases and industrialization. This growing demand must be satisfied by mining additional quantities from the environment.

The most important classes of non-renewable resources are metals, fossil fuels, and certain other minerals such as gypsum and potash. The production and uses of these important resources are described in the following sections.

METALS

Metals are materials with a wide range of useful physical and chemical properties. They can be used as pure elemental substances, as alloys (or mixtures) of various kinds of metals, and sometimes as compounds that also contain nonmetals. Metals are used to manufacture tools, machines, and electricity-conducting wires; to construct buildings and other structures; and for many other purposes.

The most prominent metals in industrial use are aluminum (Al), chromium (Cr), cobalt (Co), copper (Cu), iron (Fe), lead (Pb), manganese (Mn), mercury (Hg), nickel (Ni), tin (Sn), uranium (U), and zinc (Zn). The extremely valuable metals gold (Au), platinum (Pt), and silver (Ag) have some industrial uses (e.g., as conductors in computers and other electronics), but they are valued mostly for aesthetic reasons, as in the manufacture of jewellery. Some of the more common alloys are brass (containing at least 50% Cu, plus Zn), bronze (mostly Cu, plus Sn, and sometimes Zn and Pb), and steel (mostly Fe, but also containing carbon, Cr, Mn, and/or Ni).

Metals are mined from the environment, usually as minerals that also contain sulphur or oxygen. Deposits of metal-bearing minerals that are economically extractable contribute to the known reserves of the metal. Ores are mixtures of minerals, and are mined and processed to manufacture pure metals. The stages in metal processing, manufacturing, and recycling are summarized in Figure 13.1.

The initial step is ore extraction by mining, which may be conducted in surface pits or strip mines, or in shaft mines that can penetrate kilometres underground. In an industrial facility called a mill, mined ore is crushed to a fine powder by heavy steel balls or rods in huge, rotating tumblers. The ground ore is then separated into a metal-rich fraction and a waste known as tailings. Depending on local geography, the tailings may be discarded onto a contained area on land, into a waterbody, or into the ocean (see also Chapter 18).

If the metal-rich fraction contains sulphide minerals, it is concentrated in a smelter by roasting at high temperature in the presence of oxygen. This releases gaseous sulphur dioxide (SO_2) while leaving metals behind. The smelter concentrate is subsequently processed into pure metal in a refinery. The pure metal can then be used to manufacture industrial and consumer products.

After their useful lifetime, manufactured products can be recycled back into the refining and manufacturing processes, or they may be discarded into a landfill.

High-quality ores are geologically uncommon. Deposits that can be most economically mined are located fairly close to the surface, and the ores have relatively high

FIGURE 13.1 | METAL MINING AND USE

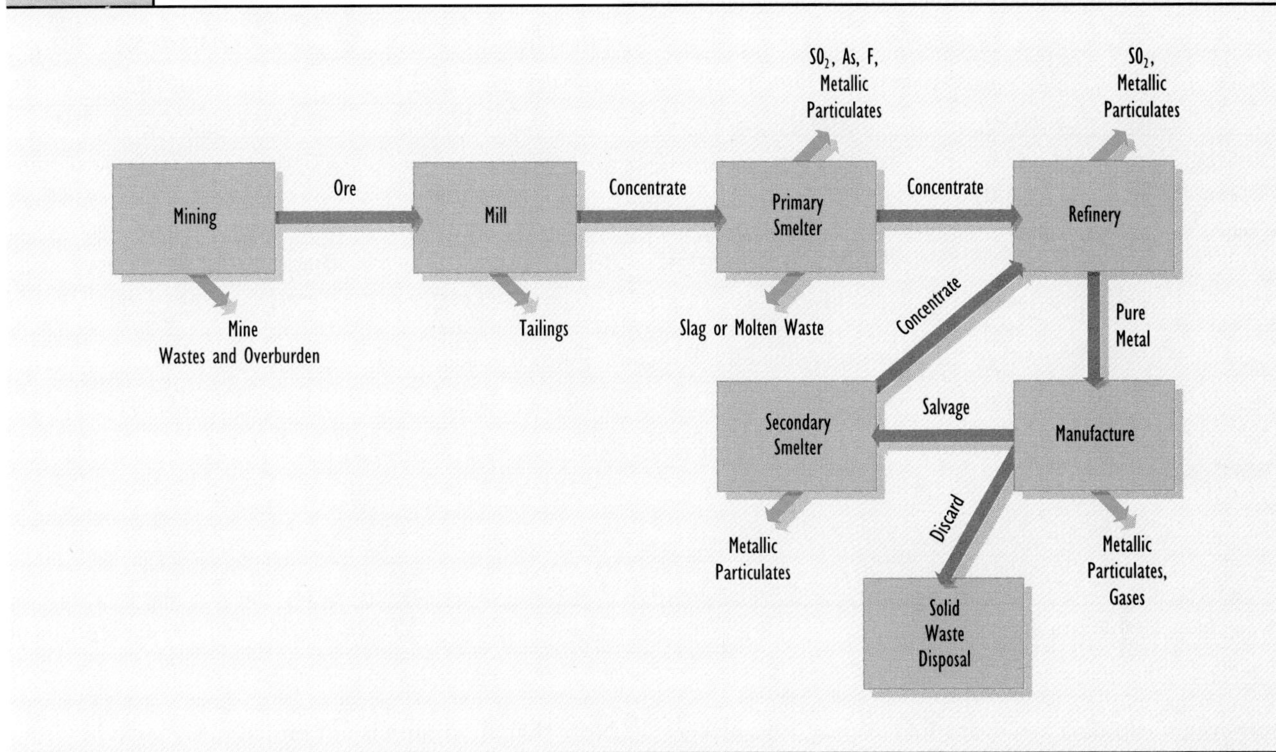

This diagram shows major stages of the mining, manufacturing, use, and reuse of metals, as well as the associated emissions of waste gases (particularly SO_2 or sulphur dioxide) and particulates to the environment. Overall, the system is a flow-through, with some recycling to extend the lifetime of metals within the system.

Source: Modified from Freedman (1995)

concentrations of metals. This varies, however, depending on the value of the metal being processed. Because gold and platinum are extremely valuable materials (on a unit-weight basis), ores with very small concentrations can be economically mined and processed. In contrast, less valuable metals such as aluminum and iron are mined as richer ores, in which the metals are present in high concentrations.

Data showing the global production and consumption of industrially important metals are given in Table 13.1. Note that for most metals the amounts consumed are larger than the annual production: this indicates that some of the consumption involves recycled metal that has been reclaimed from previous uses. Also note the rapidly increasing rates of production and consumption of most metals.

Iron and aluminum are the two metals produced and used in the largest quantities. The **life index** (or production life, calculated as the known reserves divided by the current rate of production) of aluminum is only 222 years, and for iron ore it is 178 years. Life indexes for most other metals listed in Table 13.2 are much less, suggesting that their known reserves are being depleted alarmingly quickly. It is important to remember, however, that known reserves are increased by new discoveries, changes in technology, and more favourable economics for the resource.

As depletion occurs, the specific value (that is, the per-tonne value) of the remaining reserves will increase in an open marketplace. The increasing value of diminishing supplies of metals stimulates more exploration, the exploitation of previously discovered deposits which may be of relatively low quality, more efficient recycling of metals from discarded products, and the substitution of less expensive alternative materials. For example, water-conducting pipes manufactured from polyvinyl chloride (a plastic) have largely replaced copper piping, because the metal has become relatively expensive.

The mining and processing of metals is big business in Canada, with a production value of $8.8 billion in 1993 (Natural Resources Canada, 1994a). In terms of quantities produced, the leading metals in Canada are iron, aluminum, zinc, copper, lead, and nickel (Table 13.2). In terms of value,

TABLE 13.1 | GLOBAL PRODUCTION, CONSUMPTION, AND RESERVES OF SELECTED METALS

METAL	PRODUCTION (10^6 tonne/year)		CONSUMPTION (10^6 t/y)		RESERVES 1992 (10^6 t)	
	1977	1992	1977	1991	KNOWN	LIKELY
Aluminum(1)	73.6	103.6	14.4	17.2	23 000	28 000
Cadmium	0.017	0.019	0.016	0.020	0.54	0.97
Copper	7.7	9.3	9.1	10.7	310	590
Iron(2)	831	930	891	960	150 000	230 000
Lead	3.4	3.4	4.4	5.3	63	130
Mercury	0.007	0.003	0.007	0.005	0.13	0.24
Nickel	0.77	0.92	0.64	0.88	47	110
Tin	0.23	0.18	0.23	0.22	8.0	10.0
Zinc	6.1	7.1	5.8	7.0	140	330
Steel (crude)	673	721	680	732	–	–

(1) Production data are for bauxite (aluminum ore).
(2) Data are for iron ore.

Source: World Resources Institute (1995)

however, the leading metals are gold, copper, zinc, nickel, and iron ore (Table 13.3). Regional Canadian production of these most valuable metals is also described in Table 13.3. Note that aluminum is not mined in Canada, but huge quantities of its bauxite ore are imported for processing into pure metal. Most aluminum smelters are in Quebec

TABLE 13.2 | RESERVES, PRODUCTION, AND CONSUMPTION OF SELECTED METALS IN CANADA IN 1992

Canadian consumption refers to the amount of Canadian production that is used in Canada.

METAL	LIFE INDEX (years)		CANADIAN RESERVES (% OF GLOBAL)	CANADIAN PRODUCTION (10^3 tonne)	CANADIAN CONSUMPTION	
	CANADIAN	GLOBAL			(10^3 t)	%
Aluminum(1)	–	222	0	1 972	502	25
Cadmium	53	27	5	1.4	0.030	2
Cobalt	21	161	1	2.2	0.20	9
Copper	14	35	4	762	176	23
Gold	8	20	4	0.153	0.042	27
Iron ore	171	178	8	31 582	11 774	37
Lead	20	20	10	337	91.7	27
Molybdenum	17	51	8	0.009	0.002	17
Nickel	30	51	13	178	12.1	7
Platinum group	23	190	<1	0.013	–	–
Silver	15	20	13	0.0012	–	–
Uranium	45	62	75	9.0	1.7	19
Zinc	15	19	15	1 196	115	10

(1) Aluminum is not mined in Canada, but bauxite ore is imported for processing.

Source: Natural Resources Canada (1994a)

TABLE 13.3 | PROVINCIAL PRODUCTION OF SELECTED METALS IN CANADA, 1993

REGION	GOLD (tonne/year)	COPPER (10^3 t/y)	ZINC (10^3 t/y)	NICKEL (10^3 t/y)	IRON ORE (10^6 t/y)
Newfoundland	–	0.4	–	–	17.5
Prince Edward Island	–	–	–	–	–
Nova Scotia	–	–	–	–	–
New Brunswick	0.49	10.5	308.6	–	–
Quebec	41.9	78.0	128.1	–	13.6
Ontario	72.0	268.9	182.9	124.3	0.5
Manitoba	3.0	61.6	95.5	56.5	–
Saskatchewan	–	–	–	–	–
Alberta	0.02	–	–	–	–
British Columbia	14.4	279.4	103.3	–	0.06
Yukon	3.4	–	33.9	–	–
Northwest Territories	13.0	–	146.0	–	–
NATIONAL TOTAL	152.6	698.8	998.2	180.8	31.7
Value ($\$10^6$)	2258	1760	1229	1216	1037

Source: Natural Resources Canada (1994a)

and British Columbia, because these facilities require relatively inexpensive and abundant hydroelectric energy.

Canada is one of the world's leading producers of metals, accounting for 21% of the global production of nickel in 1992, 18% of global zinc production, 10% of lead, 8.2% of copper, 8.0% of cadmium, and 3.7% of iron ore (World Resources Institute, 1995). In contrast, Canada accounts for only 0.50% of the global production of tin (in 1995; World Resources Institute, 1995). Most metal production in Canada is intended for export. Domestic consumption is less than 37% of the production of all metals (see Table 13.2).

The reserve life (or life index) of known Canadian reserves of metals are similar to or shorter than their global values (Table 13.2). Canadian reserves make up 75% of the global reserves of uranium and 10–15% of those of lead, nickel, silver and zinc.

Because petroleum and other fossil fuels are nonrenewable resources, their future reserves are diminished by mining. View of oil derricks in pumps in Taber, Alberta.

Source: Victor Last/Geographical Visual Aids

FOSSIL FUELS

Fossil fuels include coal, petroleum, natural gas, oil sands, and oil shales. All of these deposits are derived from the partially decomposed biomass of dead plants that lived many millions of years ago. The ancient biomass became entombed in sediments, which later were buried and lithified into sedimentary rocks such as shale and sandstone. Deep within the geological formations, under conditions of high pressure, high temperature, and low oxygen, the biomass transformed extremely slowly into hydrocarbon compounds composed entirely of carbon and hydrogen. In some respects, fossil fuels can be considered to be stored solar energy — fixed by plants into organic matter and then stored geologically.

Fossil fuels are still being produced today, also by geological processes involving biomass subjected to high pressure and temperature. Because the natural production of fossil fuels continues, it might be argued that these materials are renewable resources. However, fossil fuels are being used at a rate that is enormously faster than their extremely slow rate of natural regeneration. Under such circumstances, fossil fuels can only be regarded as nonrenewable resources.

The most abundant chemical compounds in fossil fuels are hydrocarbons. These fuels may also contain many other kinds of organic compounds, incorporating sulphur, nitrogen, and other elements. Coal is often contaminated with many inorganic minerals, such as shales and pyrites.

The most important use of fossil fuels is as a source of energy. They are combusted in vehicle engines, power plants, and other machines to produce the energy needed to perform work in industry, for transportation, and for household use. Fossil fuels are also combusted to produce energy to heat indoor spaces, a significant use in countries with a seasonally cold climate. Another important use of fossil fuels is for manufacturing synthetic substances, including most types of plastics. Asphalt materials obtained from fossil fuels are used in road construction and to manufacture roofing shingles and siding for buildings.

Continued exploration for non-renewable resources can discover new reserves. Because Earth is finite, however, there are limits to these discoveries, and these are being approached rapidly. This enormous off-shore production platform has been constructed to develop a petroleum deposit on the Grand Banks off Newfoundland.

Source: Greg Locke

Coal is a solid material that can vary greatly in character. The highest quality **coals** are anthracite and bituminous, which are hard, shiny, black minerals with a high energy density. Lignite, a poorer grade of coal, is a softer, flaky material with a lower energy density. Coal is mined in various ways. If coal deposits occur close to the surface, they are typically quarried using a strip-mining technique. This involves the use of huge shovels known as draglines to uncover and collect the coal-bearing strata. Deeper deposits of coal are mined from underground shafts and tunnels, which may follow seams far into the ground.

After mining, industrial coal may be washed to remove some of the impurities, and then ground into a powder. Most coal is combusted directly in large industrial facilities, such as coal-fired generating stations, a use which accounts for about half of the global use of coal. In addition, about 75% of the world's steel is manufactured using coal as an energy source, often as a concentrated material known as coke. Coal can also be used to manufacture synthetic petroleum, which can be refined into fuels or used as a feedstock to manufacture many kinds of useful chemicals, such as liquid and gaseous hydrocarbon fuels, plastics, and pigments.

Petroleum (or **crude oil**) is a fluid mixture of hydrocarbons with some impurities, including organic compounds containing sulphur, nitrogen, and vanadium. **Petroleums** from different places vary greatly in their physical and chemical qualities, from heavy, tarry materials that must be heated before they will flow, to extremely light liquids that quickly volatilize into the atmosphere. Petroleum is mined using wells drilled to various depths, from which the liquid mineral is forced to the surface by geological pressure. Often, this force is supplemented by pumping. Petroleum is also produced by mining and refining oil sands, which are strip-mined in large quantities in northern Alberta.

Mined petroleum is transported by overland pipelines, trucks, and ships to industrial facilities known as refineries, where the crude material is separated into various constituents. These include:

1. a relatively light hydrocarbon fraction known as gasoline, which is used to fuel automobiles

2. slightly heavier fractions, such as diesel fuel for trucks and trains, and a home-heating fuel
3. kerosene, used as a heating and cooking fuel and as a fuel for airplanes
4. dense residual oils, used in oil-fired power plants and as a ship fuel
5. tarry, semi-solid asphalts used to pave roads and manufacture roofing materials

Natural gas is likewise mined with drilled wells. Methane is the dominant hydrocarbon in **natural gas**. Other gaseous hydrocarbons such as ethane, propane, and butane are also present, as are sulphurous gases such as hydrogen sulphide. Most natural gas is transported in pipelines from the drilling sites to distant markets. Sometimes it is liquefied under pressure for transportation, particularly by ships. In Canada, however, it is distributed mostly through an extensive network of steel pipelines. Natural gas is used to generate electricity, to heat buildings, as a cooking fuel, as a fuel for light vehicles, and to manufacture nitrogen fertilizers.

Production, Reserves, and Consumption

Table 13.4 shows the global production and consumption of fossil fuels. The production of coal and natural gas increased by about 40% between 1971 and 1991, while production of petroleum increased by 9%. There is active exploration for all of these fuels, and additional reserves are being discovered in various regions of the world. Fossil fuels are, however, being used extremely rapidly, particularly in industrial economies. Consequently, the expected lifetimes of the known reserves are alarmingly short, equivalent to several hundred years for coal and only a few decades for natural gas and petroleum. These specific numbers must not be interpreted too literally, however, because continuing exploration for fossil fuels is discovering additional deposits, which add to the known reserves. Nevertheless, the discoveries will be limited by the finite amount present on Earth, so the fact remains that the stocks of these nonrenewable resources are being depleted rapidly.

At the present time, petroleum is the world's most important hydrocarbon resource, largely because it can easily be refined into portable, liquid fuels which are readily used as a source of energy for industrial and domestic purposes. In addition, petroleum is the major feedstock used to manufacture plastics and other synthetic materials.

TABLE 13.4 GLOBAL PRODUCTION (1991) AND RESERVES (1990) OF FOSSIL FUELS

"Proven" reserves are the total amounts of the commodity known to exist on Earth; "recoverable" reserves are economically recoverable at the present time. Production data are standardized as petajoules (10^{15} joules). Percent increases compare production levels in 1991 with those in 1971.

	RESERVES (10^9 t)		PRODUCTION	
FOSSIL FUEL	PROVEN	RECOVERABLE	(1×10^{15} joule)	% INCREASE
Hard coals	1213	521.4	–	
Soft coals	743.2	517.8	–	
Total coals	1956	1039	93 689	+42
Crude oil	–	134.8	132 992	+ 9
Natural gas[1]	–	128.9	76 275	+39

[1] *Reserves of natural gas are given in 10^{12} m^3.*

Source: World Resources Institute (1995)

About 58% of the world's proven recoverable reserves of petroleum occur in the Middle East (Table 13.5). This fact underscores the strategic importance of that region to the world's industrial economies. Saudi Arabia alone has 26% of the world's petroleum reserves, followed by Iraq, Kuwait, and Iran, each with about 10%. The world's most industrialized countries are in Europe, North America, and eastern Asia, and these depend heavily on petroleum imports from the Middle East, Russia, and Venezuela to maintain their consumption levels. The world's best-endowed country in terms of total hydrocarbon resources is Russia, which has the bulk of the proven reserves of natural gas, as well as enormous reserves of coal and petroleum (Table 13.5).

The production lives of proven recoverable Canadian reserves of fossil fuels are shown in Table 13.6. It must be remembered, however, that new reserves are still being discovered, which extend these production lives. Most reserves of fossil fuels in Canada occur in the western provinces, as does most production (Table 13.7). In the western provinces, coal is generally extracted using open-pit and strip mines, while coal in the eastern provinces is mined underground. In addition to crude oil, Canada also has a large resource of oil sands, which are strip-mined and used to manufacture petroleum. There are about 7.4 million

TABLE 13.5 | RESERVES OF FOSSIL FUELS IN SELECTED COUNTRIES

Countries are listed in order of decreasing reserves of petroleum (crude oil). Data are proved recoverable reserves.

COUNTRY	PETROLEUM (10^9 tonne)	HARD COAL (10^9 t)	SOFT COAL (10^9 t)	NATURAL GAS (10^{12} metre^3)
Saudi Arabia	35.65	–	–	5.50
Iraq	13.60	–	–	2.69
Kuwait	12.79	–	–	1.36
Iran	12.70	0.19	2.30	17.00
Venezuela	8.60	0.42	–	3.43
USSR (former)	8.00	104.00	137.00	54.53
Mexico	6.08	1.25	0.47	2.03
United States	3.56	112.67	127.89	4.74
China	3.26	62.20	52.30	1.13
Norway	1.03	0.01	0.04	1.23
India	0.81	60.65	1.90	0.73
Indonesia	0.73	0.96	31.10	1.80
Canada	*0.72*	*4.51*	*4.11*	*3.11*
United Kingdom	0.54	3.30	0.50	0.55
Brazil	0.40	–	2.36	0.12
Australia	0.16	45.34	45.60	0.49
Germany	0.06	23.92	56.15	0.35
Japan	<0.01	0.83	0.02	<0.04

Source: World Resources Institutes (1995)

hectares of oil sand deposits in northern Alberta, which can potentially yield about 30 billion m^3 of synthetic oil.

About half of the Canadian production of natural gas is consumed domestically, the rest being exported to the United States (Table 13.6). In contrast, about 80% of Canada's net production of coal and petroleum is used domestically. (Actually, a large fraction of the petroleum mined in western Canada is exported to the United States, but this is offset by a substantial import of foreign petroleum to the eastern provinces. The 80% figure is a net value.)

The mining, refining, export, and consumption of fossil fuels is an extremely important economic sector in Canada, with a production value of $23 billion in 1993 (Table 13.7) (Natural Resources Canada, 1994a). Canada produces about 5.7% of the global production of natural gas, 2.7% of the petroleum, and 1.7% of the coal. These

TABLE 13.6 | PRODUCTION, CONSUMPTION, AND RESERVES OF FOSSIL FUELS IN CANADA

Annual production and consumption data are for 1992, while reserves data are for 1990. Percent consumption refers to the fraction of production that is used in Canada. Reserve life is 1990 reserves divided by 1992 production.

FOSSIL FUEL	CANADIAN PRODUCTION	CANADIAN CONSUMPTION	%	RECOVERABLE RESERVES	RESERVE LIFE (years)
Coal (10^6 tonne)	65.6	51.0	78	8623	131
Petroleum (10^6 metre^3)	93.3	81.4	81	758	8
Natural gas (10^9 m^3)	116.7	50.7	51	3116	27

Source: Natural Resources Canada (1994a) and World Resources Institute (1995)

TABLE 13.7 PROVINCIAL PRODUCTION OF FOSSIL FUELS IN CANADA, 1993

Data for natural gas include byproducts.

REGION	PETROLEUM (10^6 metre^3/year)	NATURAL GAS (10^9 m^3/y)	COAL (10^6 tonne/year)
Newfoundland	–	–	–
Prince Edward Island	–	–	–
Nova Scotia	1.08	–	3.50
New Brunswick	–	–	0.39
Quebec	–	–	–
Ontario	0.25	0.41	–
Manitoba	0.63	0.01	–
Saskatchewan	14.75	6.49	9.95
Alberta	76.71	132.9	34.21
British Columbia	1.97	17.33	20.55
Yukon	–	0.40	–
Northwest Territories	1.87	0.24	–
NATIONAL TOTAL	97.25	157.7	68.60
Value ($\$10^6$)	11 155	10 042	1 783

Source: Natural Resources Canada (1994a)

are much larger than the 0.5% of global population that lives in Canada.

Other Mined Minerals

Other materials that are mined in large quantities in Canada include asbestos, gypsum, limestone, potash, salt, sulphur, aggregates, and peat. In general, these materials have a relatively small commodity value (i.e., value per tonne) compared with metals and fossil fuels. Global or Canadian shortages of these materials are not imminent.

Asbestos is a group of tough, fibrous, incombustible, silicate minerals, used for manufacturing fireproof insulation, cement, brake linings, and many other products. Certain kinds of asbestos minerals have been linked to human health problems, particularly the development of lung diseases. These health hazards have greatly reduced the market for this otherwise extremely useful mineral. Still, about 0.51 million tonnes of asbestos were mined in Canada in 1993, with an economic value of $215 million (Natural Resources Canada, 1994a). Almost all of the asbestos mining in Canada occurs in Quebec.

Gypsum, a mineral composed of calcium sulphate, is used mainly to manufacture plaster and wallboard for the construction industry. About 7.8 million tonnes of gypsum were mined in Canada in 1993, with an economic value of $83 million. About 78% of gypsum mining occurs in Nova Scotia, with the rest in Ontario, British Columbia, and Newfoundland.

Limestone is a rock composed of calcium carbonate. Limestone is used to manufacture cement, as well as lime for making plaster. In addition, some limestone, and the related metamorphic rock known as marble, is quarried for use as building stone and facings. About 9.8 million tonnes of cement were manufactured in Canada in 1993, with an economic value of $765 million. Another 89 000 tonnes of lime were manufactured, with a value of $201 million. Ontario, Quebec, and British Columbia have the largest cement manufacturing industries, and Ontario has the largest lime-making capacity.

Potash is a rock formed from the mineral potash feldspar, and it is mined to manufacture potassium-containing fertilizers. About 7.0 million tonnes of potash (expressed as K_2O) were mined in Canada in 1993, with an economic value of $900 million. Potash is mined in Saskatchewan and New Brunswick.

Salt, or sodium chloride, is used in the chemical manufacturing industry, for de-icing roads, as "table salt," and as a food additive and flavouring. About 11.4 million tonnes of salt were mined in Canada in 1993, with an economic value of $280 million. The largest salt mines are in Ontario, Alberta, Saskatchewan, and Nova Scotia.

Sulphur is manufactured from hydrogen sulphide obtained from sour-gas wells (i.e., gas wells rich in H_2S), from pollution-control "scrubbers" (for sulphur dioxide) at metal smelters, and from deposits of "native sulphur." Sulphur is used mostly in the chemical manufacturing industries and to make fertilizers. About 0.80 million tonnes of sulphur were produced in Canadian smelters in 1993, mostly in Ontario, British Columbia, Quebec, and New Brunswick, with an economic value of $95 million. Large quantities of sulphur are also produced as a byproduct of natural-gas refining in Alberta and Saskatchewan.

Aggregates include sand, gravel, and other materials that are mined for use for road construction and as fillers for concrete in the construction industry. Aggregates are a low-grade resource, having relatively little value per tonne. Close to large cities, however, these materials may be available only in small quantities, leading to local shortages of inexpensive aggregates. About 241 million tonnes of sand and gravel were quarried in Canada in 1993, with an economic value of $737 million. These materials are mined in all provinces and territories, at rates more or less determined by the local construction activity.

Peat, another mined commodity, is a subfossil material developed from dead plant biomass that is hundreds to thousands of years old. It accumulates in wetland bogs where it becomes partially decomposed (or humified). Peat is sometimes dried and burned as a source of energy, an important use in Ireland, parts of northern Europe, and Russia. In Canada, however, most peat is mined for use as a horticultural material, and to produce highly absorbent hygienic products such as diapers and sanitary napkins. About 0.82 million tonnes of peat were mined in Canada in 1993, with an economic value of $119 million. About 37% of peat mining occurs in Quebec, 33% in New Brunswick, and the rest in Alberta, Ontario, and other provinces.

Energy Use

Having ready access to relatively inexpensive and accessible energy is critical to any economy. The use of large quantities of energy is particularly characteristic of highly industrialized nations such as Canada. As has been discussed previously, relatively wealthy, developed countries use much more energy (on a per capita basis) than do poorer, less-developed countries.

Ever since the mastery of fire, people have used fuels for subsistence purposes — that is, to cook food and to keep warm. Initially, the fuels used for those purposes were locally collected wood and other plant biomass. When fire was domesticated, perhaps only one million people were alive, and their per capita energy usage was small. Consequently, biomass fuels were a renewable source of energy, because the rate at which they were being used was much smaller than the rate at which new biomass was being produced by vegetation.

The human population in modern times, however, is much larger than it was when fire was first put to work. Moreover, many countries now have intensely industrialized economies, in which per capita energy usage is extremely high. The combination of population growth and increased per capita energy use means that enormous quantities of energy must be expended in these developed countries. The energy is needed to fuel industrial processes, to manufacture and run machines, to provide warmth in winter and to cool in summer, and to prepare foods.

Most industrial energy supplies are based on the use of nonrenewable resources, although some renewable

Electricity generated by burning coal, oil, or natural gas is a nonrenewable source of energy. View of a generating station at Shawinigan, Québec.

Source: Victor Last/Geographical Visual Aids

sources of energy are also important. For comprehensiveness, renewable energy sources are discussed in this section, together with nonrenewable sources, rather than in the chapter on renewable natural resources (Chapter 14).

Sources of Energy

The world's major sources of industrial energy are fossil fuels and nuclear fuels, both of which are nonrenewable. Hydroelectric power, generated using the renewable energy of flowing water, is also important in some regions, including much of Canada. Minor energy sources, often called "alternative energy sources," include solar power, geothermal power, wind, waves, and biomass, all of which are potentially renewable.

All of the above sources can be used to drive a turbine, which spins an electrical generator that converts the kinetic energy of motion into electrical energy. Solar energy can also be used to generate electricity more directly, through photovoltaic technology (see below). One of the most important kinds of energy used in industrial societies, electricity is widely distributed to industries and homes through a network of transmission lines.

The following sections briefly describe how these various energy sources are used.

Fossil Fuels

Coal, petroleum, natural gas, and their various refined products can be combusted in power plants, where the potential energy of these fuels is harnessed to generate electricity. Fossil fuels are also used to power machines directly, particularly in transportation, in which gasoline, diesel, liquefied natural gas, and other "portable" fuels power automobiles, trucks, airplanes, trains, and ships. Fossil fuels are also combusted in the furnaces of many homes and larger buildings to provide warmth during colder times of the year. The burning of fossil fuels has many environmental drawbacks, including large emissions of greenhouse gases, sulphur dioxide, and other pollutants into the atmosphere.

Nuclear Fuels

Nuclear fuels contain unstable isotopes of the heavy elements uranium and plutonium (^{235}U and ^{239}Pu, respectively). These can decay through a process known as *fission*, which produces lighter elements while releasing two or three neutrons and enormous quantities of energy. The emitted neutrons may be absorbed by other atoms of ^{235}U or ^{239}Pu, causing them to become unstable and undergo fission themselves in a process known as a *chain reaction*. Uncontrolled chain reactions can result in a nuclear explosion. In a nuclear reactor, however, the flux of neutrons is carefully regulated, allowing electricity to be produced safely and continuously.

Nuclear reactions are fundamentally different from conventional chemical reactions, in which atoms recombine into different compounds without changing their internal structure. In nuclear fission reactions, atomic structure is fundamentally altered, and small amounts of matter are transformed into immense quantities of energy.

Most of the energy liberated by nuclear fission is released as heat. In a nuclear power plant, the heat is used to boil water. The resulting steam drives a turbine, which generates electricity. Most nuclear-fuelled power plants are huge commercial reactors that produce electricity for industrial and residential use in large urban areas. Smaller reactors are sometimes used to power ships and submarines, mostly for military use, although there are several Russian nuclear-powered icebreakers.

^{235}Uranium is the element used in conventional nuclear reactors, such as the CANDU system developed and used in Canada (see Canadian Focus 13.1). ^{235}Uranium is obtained from uranium ore, which is mined in various places in the world. (Canada is a major player in uranium mining, producing 9×10^3 t/y of uranium, of which 81% is exported; see Table 13.1.) Uranium metal produced by refining ore typically consists of about 99.3% non-fissile ^{238}U, and only 0.7% ^{235}U. Most commercial reactors require a fuel that has been further refined to enrich the ^{235}U concentration to about 3%. However, the Canadian-designed CANDU reactors can use non-enriched uranium as a fuel.

Various elements, most of which are also radioactive, are produced during fission and other nuclear reactions. One of these, ^{239}Pu, can also be used as a component of nuclear fuel in power plants. To obtain ^{239}Pu for this purpose (or for use in manufacturing nuclear weapons), spent fuel from nuclear generating stations is reprocessed. Other transuranium elements and any remaining ^{235}U (as well as nonfissile ^{238}U) can also be recovered; these can be used to manufacture new fuel for reactors. So-called breeder reactors are designed to optimize the production of ^{239}Pu (which occurs when an atom of ^{238}U absorbs a neutron to produce ^{239}U, which then forms ^{239}Pu by the emission of

CANADIAN FOCUS 13.1

THE CANDU SYSTEM FOR NUCLEAR GENERATION OF ELECTRICITY

The Canadian system for generating electricity using controlled nuclear fission, called CANDU (an acronym for **Can**ada **D**euterium **U**ranium), was developed by Atomic Energy of Canada Limited (AECL), a federal crown corporation. The CANDU reactor design uses unique technology, particularly in its reliance on pressurized heavy water as the *moderator* to slow down the neutrons emitted during the radioactive decay of ^{235}U. This is an important function, because slower-moving neutrons have a better chance of inducing fission in other ^{235}U atoms, as is required to sustain the chain reaction. Heavy water is also used in the CANDU system as a matrix (surrounding the nuclear fuel) to absorb much of the thermal energy produced by the controlled fission of uranium. (Heavy water contains a relatively large concentration of the deuterium isotope of hydrogen.)

Uranium fuel for CANDU reactors is manufactured as ceramic-like, cylindrical pellets of non-enriched uranium oxide (each about 2.0 cm long and 1.4 cm in diameter). These are stacked into a 50-cm-long tube of zirconium alloy, which is welded shut at each end. One fuel *bundle* is made up of 37 of these zirconium-sheathed elements and weighs about 22 kg. Within the reactor core, 12 fuel bundles are arranged end-to-end in a *pressure tube*, through which heavy-water coolant is pumped at about 100 atmospheres pressure and a temperature of 300°C. The pressure tube, plus its contained fuel, heavy-water bath, and associated fittings, is known as a *fuel channel*, of which there are several hundred in the *core* of a typical CANDU reactor. The fuel channels are arranged as a horizontal grid within a large tank containing heavy water as a moderator. The specific arrangement of the fuel channels, and the presence of heavy water, controls the rate of the fission reaction. If these conditions are significantly upset, the nuclear chain reactions terminate automatically. This is an important safety feature in the CANDU design.

The rate of the nuclear reaction is also controlled using a series of rods made of materials that absorb neutrons efficiently (such as the metal cadmium). The control rods can be selectively inserted between fuel channels in the reactor core, allowing the rate of the chain reaction to be precisely controlled.

Each nuclear reactor is heavily shielded with reinforced concrete about 1 m thick. The shield prevents the emission of extremely energetic gamma radiation, which would be a hazard to people working in the station. In addition, the reactor and its heavy-water coolant system are encased within a massive containment building made of concrete. This outermost structure is intended to prevent emissions of radioactivity into the environment in the event of a catastrophic reactor failure.

The pressurized heavy-water coolant circulates through the fuel channels as a closed system, carrying absorbed thermal energy to a heat exchanger. Ordinary water is heated in the exchanger, and the steam generated is used to drive a turbine, which spins a generator that produces electricity (see Figure on next page). The technology used to generate electricity from this steam is not unique to CANDU or to any other nuclear-fuelled power plant — essentially the same technology is used in fossil-fuelled generating stations.

A single fuel bundle, when used in a CANDU reactor, produces as much energy as 364 t of coal or 318 000 L of petroleum. A total of 1 619 t of uranium fuel was used in Canadian reactors in 1993. Canada has a total of 15 857 MW of nuclear generating capacity, about 14% of the total electricity generating capacity in the country (in 1993; NRC, 1994b). Ontario has 20 large reactors, each of which exceeds 500 MW in capacity. These account for about 91% of Canada's nuclear generating capacity. Quebec and New Brunswick each have one reactor, and together account for the other 9%.

The design of CANDU reactors allows them to be re-fuelled while still operating, a factor that minimizes shut-downs and contributes to the efficiency of operating the system. A fuel bundle is used for about 1.5 years, after which it is removed from the fuel channel. It is then stored in a water-filled tank for several years, where the heat of ongoing nuclear fission is dissipated and emitted radiation is absorbed.

Only about 1% of the initial content of ^{235}U of the fuel bundle is consumed during its time in a reactor.

When used fuel bundles are sufficiently cooled, they can be reprocessed to produce new fuel as well as other products such as radioisotopes for medicine and research, and — potentially — plutonium for use in nuclear weaponry. In fact, however, CANDU wastes are not being reprocessed on a commercial scale, so the uranium resource is being used very inefficiently.

Used fuel bundles contain substantial quantities of high-level radioactive wastes. High-level nuclear wastes also come from other sources, such as obsolete medical and research nuclear technology and old reactor parts. These wastes are potentially extremely toxic, and can remain so for tens of thousands of years. To avoid environmental damage, such high-level wastes must be permanently managed in a safe, fully contained manner. Canada has not yet implemented a system for the long-term disposal of high-level nuclear wastes. The adopted system will probably involve permanent placement into chambers mined into extremely stable geological masses known as plutons. Because these are located in granitic rocks of the Canadian Shield, the disposal site will probably be in northwestern Ontario. The disposal system must be capable of safely containing the nuclear wastes for at least tens of thousands of years.

In recent years, premature corrosion of heat-exchange piping and other components of CANDU reactors has necessitated shutting reactors down for repairs and modifications. If these technical problems are not readily solvable, some CANDU reactors may be permanently closed earlier than was anticipated when they were first planned and constructed. Early shutdowns would undermine part of the economic justification of CANDU technology, compared with some alternatives. This, along with environmental controversies associated with the production and disposal of high-level wastes, means that the future of nuclear power in Canada is somewhat uncertain.

THE CANDU SYSTEM OF GENERATING ELECTRICITY

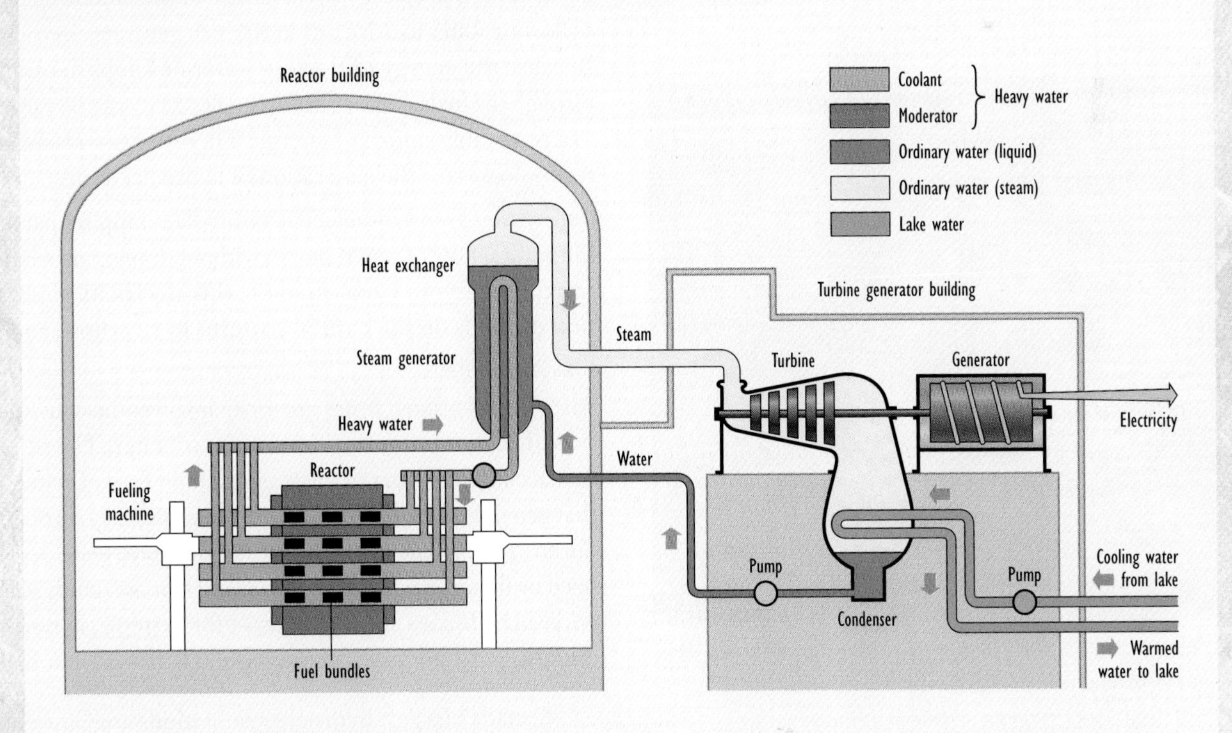

two beta electrons). Although breeder reactors have been demonstrated, they have not been commercially developed to any great extent. Even though breeder reactors produce "new" nuclear fuel (by producing plutonium) and thereby help to optimize use of the uranium resource, there are limits to the process, because the original quantity of ^{238}U is eventually depleted. Therefore, both ^{235}U and ^{239}Pu should be considered non-renewable resources.

A number of important environmental problems are associated with nuclear power. These include the small but real possibility of catastrophic accidents such as a meltdown of the reactor core, which can result in the emission of large amounts of radioactive materials into the environment (as happened at the Chernobyl reactor in Ukraine in 1986). Nuclear reactions also produce extremely toxic, long-lived radioactive byproducts (such as plutonium), which must be safely managed for very long periods of time (i.e., tens of thousands of years). Enormous quantities of these "high-level" nuclear wastes are stockpiled in Canada and in other countries that use nuclear power, but so far there are no solutions to the problem of their long-term management. Another problem is the emission of toxic radon gas and radioactivity from "low-level" wastes associated with uranium mines, structural elements of nuclear power plants, and other sources.

Hydroelectricity is a renewable source of energy, although significant environmental damage can be caused when large reservoirs are developed behind a dam such as this.

Source: Victor Last/Geographical Visual Aids

Another kind of energy-producing nuclear reaction is known as *fusion*. This process occurs when light nuclei are forced to combine under conditions of extremely high temperature (millions of degrees) and pressure, resulting in an enormous release of energy. Fusion usually involves the combining of hydrogen isotopes. The most common fusion reaction involves two protons (that is, two hydrogen nuclei, ^{1}H) fusing to form a deuterium nucleus (composed of one proton and one neutron, ^{2}H), while also emitting a beta electron and an extremely large amount of energy.

Fusion reactions occur naturally in the interior of the sun and other stars, and can be initiated by exposing hydrogen to the enormous heat and pressure generated by a fission nuclear explosion, as occurs in so-called hydrogen bombs. Nuclear technologists have not, however, designed a system that can control fusion reactions enough to generate electricity. If this technology is ever developed, it would be an enormous benefit to industrial society. Essentially unlimited supplies of hydrogen fuel could be extracted from Earth's oceans, virtually eliminating the constraints on energy use by humans. So far, however, controlled fusion reactions remain the stuff of science fiction.

Hydroelectric Energy

Hydroelectric energy involves use of the kinetic energy of flowing water to drive turbines which generate electricity. Because the energy of flowing water develops naturally through the hydrologic cycle, hydroelectricity can be viewed as a renewable source of energy. There are two classes of technologies for the generation of hydroelectricity.

Run-of-the-river hydroelectricity involves tapping part of the natural flow of watercourses without developing a large storage reservoir. Consequently, this electricity generation depends on the natural patterns of riverflow, and is highly seasonal.

Reservoir-generated hydroelectricity involves the construction of dams to store huge quantities of water. The reservoir accumulates part of the seasonal high flow of water so that generation can occur relatively continuously throughout the year. Some enormous reservoirs have been developed by flooding extensive areas of land that previously were covered by forests and wetlands — for example, in northern Quebec, Labrador, Manitoba, and British Columbia.

Canada's largest hydroelectric stations are Churchill Falls in Labrador, with a capacity of 5429 MW; La Grande

2 in northern Quebec, with 5328 MW; G.M. Shrum in British Columbia, with 2730 MW; and La Grande 4 and 3 in Quebec, with 2651 and 2304 MW respectively.

Although hydroelectric energy is renewable, important environmental impacts are associated with the technology. Changes in the amount and timing of water flow in rivers cause important ecological changes, as does the extensive flooding involved in the development of a reservoir.

Solar Energy

Solar energy is continuously available on Earth, although only during the day at any particular location. Solar energy can be tapped in various ways as a renewable source of energy. For example, it is stored by growing plants, the biomass of which can be harvested and combusted to release its potential energy (see Biomass Energy, below).

Solar energy can also be trapped within a glass-enclosed space. This happens because glass is transparent to visible wavelengths of electromagnetic radiation, but not to most of the infrared. This allows the use of passive solar or "greenhouse" designs to heat buildings. Solar energy can also be captured using black, highly absorptive surfaces to heat enclosed water, which can then be distributed through piping to warm the interior of a building.

Solar energy can also be used to generate electricity through photovoltaic technology (or solar cells), converting electromagnetic energy directly into electric energy. In another technology, large, extremely reflective, parabolic mirrors are used to focus sunlight onto a small, enclosed space containing water or some other fluid, which becomes heated and generates steam, driving a turbine to generate electricity.

Geothermal Energy

Geothermal energy can be tapped in only the few places on Earth where magma from Earth's mantle occurs relatively close to the surface and heats water. Boiling-hot water can be piped to the surface, where its heat content can be used to warm buildings or generate electricity. The energy of less hot water can be utilized using heat-pump technology, mostly for space heating or to provide warm water for a manufacturing process. Geothermal energy can be a renewable source as long as the cooled water is reinjected into the ground, so that the supply of groundwater available to be geothermally heated is not depleted.

Because there is a continuous input of solar radiation to Earth, sunlight is a renewable source of energy. This photo shows an array of solar cells in Odeillo, France, used to convert solar electromagnetic energy into electricity.

Source: Victor Last/Geographic Visual Aids

Wind Energy

Wind energy, the kinetic energy of moving air masses, can be tapped and used in various ways. Sailboats use wind energy to travel through the water, and windmills can be

A wind farm generating electricity in California. This technology for generating renewable energy could also be used in windy places in Canada, such as parts of Newfoundland, or the vicinity of Pincher Creek, Alberta.

Source: Al Harvey/The Slide Farm

designed to lift water in wells or to generate electricity. Extensive wind farms, consisting of arrays of many highly efficient wind-driven turbines, have been constructed to generate electricity in a few consistently windy places, for example, in northern California.

Tidal Energy

Tidal cycles develop because of the gravitational attraction between the Earth and the Moon. In a few coastal places, **tidal energy**, the kinetic energy of tidal flows, can be harnessed to drive turbines and generate electricity. The Bay of Fundy in eastern Canada has enormous tides, which can exceed 16 m at the head of the bay. A medium-scale (20 MW) tidal-power facility has been developed at Annapolis Royal in Nova Scotia. There is potential for much more tidal power development within the Bay of Fundy, although the technological challenges are great. Also, the construction and operation of the required dams would cause enormous environmental impacts.

Wave Energy

Waves on the ocean surface are another manifestation of kinetic energy. **Wave energy** can be harnessed using specially designed buoys that generate electricity as they bob up and down, although this has not yet been done commercially.

Biomass

The biomass of trees and other plants contains chemical potential energy. This **biomass energy** is actually solar energy that has been fixed through photosynthesis. Peat, mined from bogs, is a kind of subfossil biomass.

Like hydrocarbon fuels, biomass can be combusted to provide thermal energy for industrial purposes and to heat homes and other buildings. Biomass can also be combusted in industrial-scale generating stations, usually to generate steam, which drives a turbine that generates electricity. It is also used to manufacture methanol, which can be used as a liquid fuel in vehicles or for other purposes.

If the ecosystems from which biomass is harvested are managed to allow post-harvest regeneration of the vegetation, this source of energy can be considered a renewable resource. Peat, however, is always mined faster than the slow rate at which it accumulates in bogs and other wetlands, so it is not a renewable source of biomass energy.

Energy Consumption

Energy consumption varies greatly among the world's countries, largely depending on differences in their populations and their degree of development and industrialization (Table 13.8). In general, per capita rates of energy consumption are relatively small in less-developed countries, typically less than about 20 gigajoules per person per year (i.e., 20×10^9 J/person-year). The poorest, least-developed countries have a per capita energy consumption of less than 10/GJ/person-y (Table 13.8). Most energy consumption in the least-developed countries involves "traditional" fuels, such as wood, charcoal, dried animal manure, or food-processing residues such as coconut shells or bagasse (a residue of sugarcane pressing).

Countries that are developing rapidly are intermediate in their per capita energy consumption. Their rates of energy usage are, however, increasing rapidly due to their accelerating industrialization. In Thailand, for example, the national consumption of commercial energy increased by 425% between 1971 and 1991, while in Malaysia it increased by 344%, in Indonesia by 337%, in Brazil by 135%, in India by 107%, and in China by 66%. While use of commercial energy has grown in these countries, reliance on traditional fuels has dropped. This happens because traditional fuels are relatively bulky, smoky, and less convenient to use than electricity or hydrocarbon fuels, particularly in the urban environments where people are living in increasingly large numbers. In addition, supplies of wood, charcoal, and other traditional fuels have become severely depleted in most rapidly developing countries, particularly in urban regions and other densely populated areas.

Relatively industrialized countries have very high per capita consumptions of energy (Table 13.8). Their energy use is typically more than about 100 GJ/person-y, and almost entirely involves "commercial" sources such as electricity and hydrocarbon fuels. The world's most energy-intense economies are those of Canada and the United States (328 and 324 GJ/person-y respectively in 1991).

In terms of the total amounts, the world's largest consumers of energy are the United States and the former USSR, using 82×10^{18} J and 56×10^{18} J respectively (in 1991). Canada is also a highly industrialized country, but because of its relatively small population and moderate-sized economy, it uses much less energy in total: about 8.9×10^{18} J. However, if Canada's energy consumption is calculated on a per capita basis, it turns out that Canadians

TABLE 13.8 | ENERGY CONSUMPTION IN SELECTED COUNTRIES IN 1991

"Commercial energy" is produced by public and/or private utilities and in large industrial facilities, or is used as hydrocarbon fuels. "Traditional fuels" are used mainly in homes and small factories, particularly in rural areas. National energy consumption reflects the size of a country's economy and its population, while per capita data allow a comparison of the lifestyles of average citizens. Countries are arranged in order of total per capita energy use.

COUNTRY	TOTAL ENERGY USE		USE OF COMMERCIAL ENERGY		USE OF TRADITIONAL FUEL		
	NATIONAL (10^{15} joule)	PER CAPITA (10^9 J)	NATIONAL (10^{15} J)	PER CAPITA (10^9 J)	NATIONAL (10^{15} J)	PER CAPITA (10^9 J)	% OF TOTAL ENERGY USE
RELATIVELY UNDEVELOPED COUNTRIES							
Bangladesh	539	4	262	2	277	2.4	51
Afghanistan	136	8	85	5	51	2.8	37
Somalia	74	8	3	<1	71	8.0	96
Ethiopia	454	9	40	1	414	8.1	91
Sudan	267	11	47	2	220	8.5	82
Tanzania	359	13	29	1	330	12.3	92
Nigeria	1 701	15	691	6	1 010	9.0	59
Kenya	422	17	78	3	344	14.2	81
Dem. Rep. Congo	181	22	51	6	130	15.5	76
RAPIDLY DEVELOPING COUNTRIES							
Pakistan	1 328	10	1 032	8	296	2.4	22
India	10 835	12	8 011	9	2 824	3.3	26
Philippines	1 139	18	757	12	382	6.0	34
Indonesia	3 369	18	1 914	10	1 455	7.8	43
China	29 363	25	27 345	23	2 018	1.7	7
Thailand	1 807	33	1 281	23	526	9.5	29
Brazil	5 572	36	3 551	23	2 021	13.3	36
Iran	2 935	49	2 906	48	29	0.5	1
Malaysia	915	50	825	45	90	4.9	10
Mexico	5 082	59	4 834	56	248	2.9	5
DEVELOPED COUNTRIES							
Italy	6 816	118	6 768	117	48	0.8	1
Japan	17 394	140	17 384	140	10	0.08	<1
United Kingdom	9 069	157	9 065	157	4	0.07	<1
France	9 224	162	9 123	160	101	1.8	1
Germany	14 974	188	14 928	187	46	0.6	<1
Russia	55 522	196	54 730	193	792	2.8	1
Sweden	1 863	214	1 741	202	122	14.2	7
Australia	3 881	221	3 772	215	109	6.3	3
United States	81 755	324	80 839	320	916	3.6	1
Canada	*8 866*	*328*	*8 799*	*325*	*67*	*2.5*	*1*

Source: World Resources Institute (1995)

are the world's most intensive consumers of energy (328 GJ/person-y). This is more than 30 times the per capita usage by people living in the world's least-developed economies. Also, about 97.5% of Canada's total energy consumption comes from commercial sources, such as electricity and fossil fuels.

Canada's energy consumption increased by 307% between 1958 and 1992, and by 58% between 1970 and 1992, while per capita consumption increased by 84% and 18% during the same periods, respectively (Table 13.9). Because per capita energy consumption increased less quickly than national usage, the data suggest that Canadians became somewhat more efficient in their use of energy during that time. More small automobiles, improved gas economy of vehicles, improved insulation of residences and commercial buildings, and the use of more efficient industrial processes have all contributed to this increased efficiency. Although substantial, these gains in efficiency have been more than offset by growth in the per capita ownership and use of automobiles, consumer electronics, and other energy-demanding products and technologies.

The intense energy usage by Canadians reflects the high degree of industrialization of their national economy. Also significant is the relative affluence of average Canadians (compared to the global average). Wealth allows people to lead comparative luxurious lifestyles, with ready access to energy-consuming amenities such as automobiles, home appliances, space heating, and air conditioning. Canada is also a large country, so energy expenditures for travel are relatively high. In addition, Canada's cold climate means that people use a great deal of energy to keep warm.

TABLE 13.9 RECENT TRENDS IN CONSUMPTION OF COMMERCIAL ENERGY IN CANADA

YEAR	NATIONAL CONSUMPTION[1] (10^{15} joules)	PER CAPITA CONSUMPTION (10^9 J)
1958	2852	167
1960	3134	175
1965	4131	210
1970	5545	260
1975	7081	305
1980	7929	322
1985	7876	304
1990	8738	314
1992	8748	308

(1) In this case, energy commodities include hydrocarbon feedstock used for the manufacturing of plastics and other petrochemicals.

Source: Statistics Canada (1994b)

Energy Production

As was discussed in Chapter 12, truly sustainable enterprises are not supported by the mining of nonrenewable sources of energy or other resources. Therefore, a sustainable economy must be based on the use of renewable sources of energy.

However, most energy production in industrialized countries is based on nonrenewable sources. Averaged across the 13 developed countries shown in Table 13.10, natural gas accounts for 32% of commercial energy production, coal and liquid hydrocarbons for 27% each, and nuclear reactors for 11%. In total, these nonrenewable sources of energy account for 97% of the total production of commercial energy in those countries. Renewable sources, such as hydroelectric, geothermal, and wind power, account for only about 3%. These are sobering data. Considering the rapid rate at which reserves of nonrenewable energy resources are being depleted by mining and use, one wonders how long the energy-intensive economies of developed nations can be maintained.

Of Canada's production of commercial energy, 37% comes from natural gas, 32% from liquid hydrocarbons, 14% from coal, and 8% from nuclear energy. These nonrenewable energy sources account for 91% of the total production of commercial energy in Canada. Virtually all of the remaining production comes from hydroelectricity, which is renewable (although it can cause substantial environmental damage through flooding to create reservoirs, and can require large quantities of nonrenewable resources for the construction of dams, transmission lines, and related infrastructures).

Electricity produced by public or private utilities accounts for much of the energy used by industry, institutions, and residences in Canada. In 1992, 61% of the electricity produced in Canada was generated from hydroelectric sources, 23% from fossil fuels, and 16% by nuclear technology (Statistics Canada, 1994).

In summary, 91% of commercial energy production in Canada is based on nonrenewable sources, as is 39% of electricity generation. Reserves of all of the nonrenewable energy resources are being depleted rapidly, both in Canada and globally. Consequently, the long-term sustainability of the energy-intense economies of developed countries such as Canada, and of their citizens' lifestyles, is highly doubtful.

TABLE 13.10 | COMMERCIAL ENERGY PRODUCTION IN SELECTED INDUSTRIALIZED COUNTRIES IN 1991

Data are in units of 10^{18} joules.

COUNTRY	TOTAL PRODUCTION	COAL	PETROLEUM	GAS	GEOTHERMAL AND WIND	HYDRO	NUCLEAR
Belgium	0.49	0.02	–	–	–	<0.01	0.47
Sweden	1.07	<0.01	–	–	–	0.23	0.84
Italy	1.17	0.01	0.18	0.70	0.12	0.16	–
Spain	1.31	0.49	0.06	0.06	–	0.10	0.61
Netherlands	3.04	–	0.16	2.85	–	–	0.04
Japan	3.10	0.21	0.03	0.09	0.06	0.38	2.33
France	4.37	0.31	0.14	0.08	–	0.22	3.62
Australia	6.40	4.35	1.15	0.84	–	0.06	–
Germany	8.00	5.38	0.14	0.63	–	0.08	1.78
United Kingdom	8.77	2.37	3.84	1.78	–	0.02	0.77
Canada	*11.85*	*1.62*	*3.82*	*4.37*	*<0.01*	*1.11*	*0.93*
Russia	64.99	12.72	21.71	27.41	<0.01	0.85	2.31
United States	67.94	22.08	18.06	19.51	0.57	1.04	6.68
Average distribution	100%	27.2%	27.0%	31.9%	0.4%	2.3%	11.2%

Source: World Resources Institute (1995)

Key Terms

nonrenewable natural resources
reserve
recycling
metals
life index
fossil fuels
coal
petroleum (or crude oil)
natural gas
nuclear fuels
hydroelectric energy
solar energy
geothermal energy
wind energy
tidal energy
wave energy
biomass energy

Questions for Discussion

1. Using information from this chapter, describe the Canadian and global production and use of nonrenewable natural resources.
2. Show how industrialized economies rely mostly on non-renewable resources to sustain their economies. Will this kind of resource use be able to continue for very long? Why or why not?
3. List the various nonrenewable and renewable sources of energy available for use in industrialized countries. Discuss the future prospects for increasing the use of re-newable sources of energy.
4. Outline the ways in which you use energy, both directly and indirectly. For each of your major uses, consider how you could decrease your energy consumption, and how this would affect your lifestyle.

References

Craig, J.R., D.J. Vaughan, and B.J. Skinner. 1996. *Resources of the Earth: Origin, Use, and Environmental Impact, 2nd ed.* Englewood Heights, NJ: Prentice Hall.

Ehrlich, P.R., A.H. Ehrlich, and J.P. Holdren. 1977. *Ecoscience: Population, Resources, Environment.* San Francisco: W.H. Freeman.

Freedman, B. 1995. *Environmental Ecology, 2nd ed.* San Diego: Academic.

Kesler, S.E. 1994. *Mineral Resources, Economics, and Environment.* New York: Macmillan.

Miller, G.T. 1990. *Resource Conservation and Management.* Belmont, CA: Wadsworth.

Mitchell, B. (ed.). 1991. *Resource Management and Development.* Toronto: University of Toronto Press.

Natural Resources Canada (NRC). 1994a. *Canadian Minerals Handbook.* Ottawa: Natural Resources Canada.

Natural Resources Canada (NRC). 1994b. *Electric Power in Canada, 1993.* Ottawa: Natural Resources Canada.

Owen, O.S. and D.D. Chiras. 1995. *Natural Resource Conservation: Management for a Sustainable Future.* Englewood Heights, NJ: Prentice Hall.

Priest, J. 1991. *Energy.* New York: Addison-Wesley.

Ripley, E.A., R.E. Redmann, and A.A. Crowder. 1996. *Environmental Effects of Mining.* Delray Beach, FL: St. Lucie.

Robertson, J.A.L. 1988. Nuclear energy and nuclear power plants. In: *The Canadian Encyclopedia, 2nd ed.* Edmonton: Hurtig. pp. 1543–1546

Statistics Canada. 1994. *Human Activity and the Environment 1994.* Ottawa: Statistics Canada.

World Resources Institute (WRI). 1995. *World Resources 1994–95: A Guide to the C 'al Environment.* New York: Oxford University Press.

14

RENEWABLE NATURAL RESOURCES

Chapter Outline

Introduction
Fresh waters
Agricultural resources
Forest resources
Fish resources
Other hunted animals

Chapter Objectives

After completing this chapter, you will be able to:

1. List the major classes of renewable resources and outline the character of each.
2. Identify the ways in which renewable resources can be degraded by excessive harvesting or inappropriate management.
3. Describe the renewable resource base of Canada, and discuss whether those resources are being used in a sustainable fashion.
4. Show how the commercial hunts of cod and whales have represented the mining of potentially renewable resources.

INTRODUCTION

Renewable resources are capable of regenerating after harvesting, so that this use can potentially be sustained forever. For this to happen, however, the rate of usage must be equal to or less than the rate of regeneration: otherwise, (potentially) renewable resources are "mined"; that is, they are used as if they were nonrenewable resources.

The most important classes of renewable resources are surface and ground waters, agricultural site capability, forests, and hunted animals such as fish, deer, and waterfowl. In the following sections we examine the use and abuse of these potentially renewable resources. (Renewable sources of energy, such as hydro, solar power, wind, and biomass, were described in Chapter 13.)

FRESH WATERS

Surface Waters

Surface waters comprise lakes, ponds, streams, and rivers. They can be used as sources of drinking water, for irrigation, to generate hydroelectricity, for industrial purposes such as cooling, and for recreation. Surface waters are abundant in regions which have a climate characterized by more precipitation than evapotranspiration, which allows the resource to be recharged (see Chapter 3). In drier regions, this resource may be uncommon or rare, and its lack can present a natural constraint on ecological and economic development.

In regions with low rates of recharge, the excessive use of surface waters for irrigation or municipal requirements (such as drinking, washing, and toilet flushing) can greatly deplete the quantity of the resource. Shortages of surface waters in arid places can lead to conflicts between local regions, and even between countries. Each region wants as much water as possible to conduct its agricultural and industrial activities and to service its urban areas. Severe competition for surface waters is a chronic problem in various parts of the world, including the Middle East, much of Africa, and parts of southwestern North America.

In the Middle East, for example, the watershed of the Jordan River is shared by Israel, Jordan, Lebanon, and Syria. All of these countries have a semiarid climate, and all demand a share of the critical water resource. Also in the Middle East, the watersheds of the Tigris and Euphrates Rivers originate in Turkey, while Iraq and, to a lesser degree, Syria are highly dependent *downstream* users that are threatened by hydroelectric and other diversion schemes in Turkey. In northern Africa, the watershed of the Nile River encompasses huge territories in Egypt, Ethiopia, Sudan, and four other nations.

In North America, the most contentious water conflicts involve the Colorado River and the Rio Grande, which are shared by the United States and Mexico. Use of water from these rivers is extremely intensive, particularly for irrigated agriculture. In fact, virtually all the flow of the Colorado River can potentially be used in the United States before it even reaches northern Mexico, where there is also significant demand. In addition, water quality is degraded by inputs of dissolved salts from agricultural use in the United States, a factor that severely reduces the potential use of any remaining river-water flow in Mexico. Because this binational problem is important, the two federal governments have negotiated a treaty that guarantees minimum quantities of flow and water quality of the Colorado River where it crosses the international boundary. The Rio Grande has a similar problem, which has likewise been dealt with by a treaty between the United States and Mexico.

Even where surface waters are relatively abundant, their quality can be degraded by various kinds of pollution. They can be contaminated by chemicals such as nutrients, hydrocarbons, pesticides, metals, or oxygen-consuming organic matter. Excessive nutrients can increase the primary productivity of surface waters, causing a problem known as *eutrophication* (Chapter 20). Biological contamination from bacteria and viruses, and parasites from faecal matter of humans, pets, or livestock, can render the water unfit for drinking, or even for recreation. Thermal pollution, due to releases of excess heat from power plants or factories, can cause ecological damage in the receiving waterbodies. (Water pollution is discussed in greater detail in Chapters 18, 19, 20, and 23.)

Groundwater

Groundwater consists of underground reservoirs of water. Groundwater resources are known as **aquifers**, and they occur in the interstitial spaces and cracks of surface overburden and porous bedrock, such as sandstone, limestone, and other sedimentary rocks. Groundwater can be an extremely valuable natural resource, especially in regions where lakes and rivers are not abundant. It is typically accessed by drilling and pumping.

Groundwater stores are recharged through the hydrologic cycle. This happens as water from precipitation slowly percolates downward through surface overburden and bedrock, in a sometimes extensive area known as a recharge zone. In humid regions, where the quantities of precipitation are larger than the amounts dissipated by evapotranspiration and surface flows, the excess serves to recharge groundwater. Such rapidly recharging aquifers can sustain high rates of pumping of their groundwater and can be managed as a renewable resource.

In drier environments, however, the quantities of precipitation available to recharge groundwater are much smaller. Aquifers that recharge extremely slowly are essentially filled with old, "fossil" water that has accumulated over thousands of years or longer. These aquifers have little capability of recharging if their groundwater is used rapidly, so their stores are easily depleted. Slowly recharging aquifers are essentially non-renewable resources, whose reserves are mined by excessive use.

The world's largest aquifer, known as the Ogalalla, occurs beneath about 572 000 km^2 of arid lands in the southwestern United States. The Ogalalla aquifer is recharged very slowly by underground seepage that originates from the precipitation on distant mountains. Much of the groundwater in the Ogalalla is fossil water, accumulated during tens of thousands of years of slow infiltration. Although the Ogalalla aquifer is enormous (containing about 2.5 billion L), it is being depleted rapidly by pumping at more than 150 000 drilled wells. Most are withdrawing water for irrigation, and a few for drinking and other household purposes. The level of the Ogalalla aquifer is decreasing by as much as a metre per year in zones of intensive use, while the annual recharge rate is only about 1 mm/y. Clearly, the Ogalalla aquifer is being mined rapidly. Once it is effectively drained — as is likely to occur within several decades — irrigated agriculture in much of the southwestern United States may fail.

Groundwater resources are also threatened by pollution, caused when chemicals are accidentally or deliberately discarded into the ground. For instance, groundwater is often polluted by gasoline leaking from underground storage tanks at service stations, degraded by agricultural fertilizers and pesticides, and contaminated by bacteria and nutrients seeping from septic fields. Badly contaminated groundwater may not be usable as a source of drinking water, and may not even be suitable for crop irrigation.

Groundwater can also be degraded through intrusions of salt water, which can render the resource unfit for drinking or irrigation. In areas close to an ocean, saline deeper groundwater is overlain by a surficial layer of fresh water (which is less dense than salt water and therefore "floats" above it). If fresh groundwater is withdrawn at a rate faster than the recharge capability of the aquifer, the deeper salt water will rise to a higher level. Once this happens, it is extremely difficult to displace the salt-water intrusion, and the ability of the aquifer to supply fresh water is essentially destroyed.

Use of Water Resources

Water Supply

The regions of the world with the smallest per capita water resources are Asia, Africa, and Europe — a pattern that reflects the high population densities of those continents (Table 14.1). In general, Canada has abundant supplies of both surface and ground waters. Water is scarce only in relatively arid regions of the country, such as southern parts of the Prairie Provinces and southeastern British Columbia. There, the shortage of water for irrigation is a constraint on agricultural development, a problem that may intensify in the future.

Groundwater is an important resource in some regions of Canada, particularly where surface waters are not abundant or their chemical quality is poor. Often, groundwater is naturally cleaner than surface waters, and is therefore better suited for many purposes, especially household use. Sometimes, however, groundwater quality for certain purposes is impaired through naturally high concentrations of calcium, iron, and other chemicals.

Water Use

North America has the world's highest per capita water usage, mostly to meet the demands for irrigation and industry. The world's least-developed regions have the smallest per capita water usages, as seen in most countries of Africa, South America, and Asia (see Table 14.1).

National patterns of water use are influenced by the degree of development of the national economy, the population size, and the humidity of the climate (Table 14.2). Among the arid, less-developed nations, several depend on

TABLE 14.1 | WATER AVAILABILITY AND USE IN MAJOR REGIONS OF THE WORLD

The per capita resources are standardized to 1992 populations. The number in parentheses following total annual use refers to percent of the total resource that is used annually.

REGION	RENEWABLE RESOURCE		ANNUAL USE (1987)		SECTORAL USE (%)		
	TOTAL (kilometre^3)	PER CAPITA (10^3 metre^3)	TOTAL (km^3/y)	PER CAPITA (10^3 m^3/y)	DOMESTIC	INDUSTRY	AGRICULTURE
Asia	10 485	3.24	1531 (15)	0.52	6	8	86
South America	10 377	34.08	133 (1)	0.48	18	23	59
North America	6 945	17.31	697 (10)	1.86	9	42	49
Africa	4 184	6.14	144 (3)	0.25	7	5	88
Europe	2 321	5.43	359 (15)	0.71	13	54	33
USSR (former)	4 413	15.51	358 (8)	1.28	7	27	65
Oceania	2 011	73.05	23 (1)	0.91	64	2	34
World	40 673	7.42	3240 (8)	0.64	8	23	69

Source: World Resources Institute (1995)

irrigation for much of their agricultural production. Their per capita water use is relatively heavy, with agriculture accounting for more than 85% of the total. Other relatively poor, arid countries, such as Ethiopia, Kenya, Somalia, and Tanzania, would benefit greatly from more irrigated agriculture. However, these countries do not have access to sufficient water for this kind of management, so per capita water use, while largerly agriculture, remains small.

The patterns of water use in humid, less-developed countries are also complex. Countries such as those in southeast Asia cultivate rice extensively in irrigated fields, so their water use is relatively high. Other countries do not grow paddy rice or other irrigated crops, so their per capita water use is smaller, although most of their national water use still goes to agriculture. For example, Haiti would benefit from a more extensive development of irrigated agriculture. Unfortunately, Haiti is too impoverished to make the necessary investments to develop the infrastructure for irrigation.

Some developed countries have rather high per capita water use. Canada and the United States, for example, use much of their surface water for generating hydroelectricity. In addition, the western United States has invested heavily in irrigated agriculture. Some western European countries, such as France, Germany, and the United Kingdom, utilize much of their surface water for industrial purposes, such as cooling fossil-fuelled and nuclear power plants (see Table 14.2).

Canadians use about 5.4 billion m^3/y of water from both surface and groundwater sources. Only about 4% of that water is groundwater, used mainly in the rural sector.

Patterns of use of surface waters in the regions of Canada are summarized in Table 14.3. (The table does not include data on the use of surface waters for hydroelectricity.) Nationally, about 63% of surface-water withdrawals are used to cool fossil-fuelled and nuclear generating stations, while 16% is used as a coolant and process material in manufacturing, 11% in municipal distribution systems, 9% in agriculture, and 1% in mining. Often, as discussed earlier, the water returned after use is degraded in quality.

On average, 12% of the water used is *consumed* during the process — that is, it evaporates to the atmosphere, so that the discharges of used water are smaller than the quantity initially withdrawn. This is particularly true of heavy agricultural water use, as seen by the 58% of water use consumed in the Prairie Provinces (Table 14.3). Ontario only consumes about 2% of its surface water use; most of the rest is for flow-through cooling in industry.

Canada manages the flow of tremendous quantities of surface waters using dams and reservoirs. This is done to control flooding, to accumulate water for irrigation, to develop municipal reservoirs, and to generate hydroelectricity. Canada has 622 large dams (1991 data), of which 71% were used to provide hydroelectricity, 9% for irrigation, and 7% for storing municipal water (Table 14.4).

TABLE 14.2 | WATER USE IN SELECTED COUNTRIES

COUNTRY	PER CAPITA USE	SECTORAL USE (%)		
	(metre^3/year; 1987)	DOMESTIC	INDUSTRY	AGRICULTURE
ARID, LESS-DEVELOPED COUNTRIES				
Tanzania	35	21	5	74
Ethiopia	49	11	3	86
Kenya	51	27	11	62
Somalia	99	3	<1	97
India	612	3	4	93
Egypt	1028	7	5	88
Sudan	1093	1	<1	99
Iran	1362	4	9	87
Afghanistan	1775	1	<1	99
Pakistan	2053	1	1	98
HUMID, LESS-DEVELOPED COUNTRIES				
Republic of Congo	21	58	25	17
Nigeria	37	31	15	54
Indonesia	95	13	11	76
Bangladesh	212	3	1	96
Brazil	245	22	19	59
Viet Nam	416	13	9	78
China	462	6	7	87
Philippines	686	18	21	61
Malaysia	768	23	30	47
Mexico	921	6	8	86
DEVELOPED COUNTRIES				
United Kingdom	253	20	77	3
Germany	687	11	69	20
Japan	732	17	33	50
France	778	16	69	15
Russia	787	17	60	23
Italy	996	14	27	59
Australia	1306	65	2	33
Canada	*1688*	*18*	*70*	*12*
United States	1868	13	45	42

Source: World Resources Institute (1995)

The most important use of groundwater in Canada is as a source of domestic water in rural areas. Many rural homeowners tap into the resource using drilled wells or shallow dug wells. About 82% of rural domestic water use in Canada comes from groundwater accessed through wells. (Compare Table 14.3 with Table 14.5.) Groundwater is also an important source of irrigation water in some areas, and is sometimes used for industrial purposes.

AGRICULTURAL RESOURCES

The production of an agricultural crop is related to several factors, including the amount of land under cultivation and the productivity of the crop. Productivity is related to the management system and a quality of the land known as *site capability*. (Recall that in Chapter 4 ecological **production** was defined as the total yield of biomass, while **productivity** is standardized per unit area and unit time. For example, the global *production* of a crop such as wheat is measured in millions of tonnes, while the average *productivity* of a wheat field is expressed as tonnes per hectare per year.)

Ultimately, the agricultural land in any region, and on Earth, is a limited resource with a constrained area. To some degree, the area of land suitable for cultivating crops can be increased by clearing forests and grasslands growing on fertile sites and by draining certain kinds of wetlands. There are, however, finite areas of natural ecosystems suitable for *conversion* into arable lands. Only about 30% of Earth's surface is terrestrial, and most of that land is too cold, hot, dry, wet, rocky, or infertile to be converted into an agricultural land use.

Agricultural production relies on rainfall and soil capability, which supply moisture, nutrients, and other factors crucial to plant growth. This is a view of a landscape intensively used for agricultural production in the Provost Plain of Saskatchewan.
Source: D. Gauthier

TABLE 14.3 USE OF SURFACE WATER IN THE REGIONS OF CANADA, 1991

Data on use of surface waters for hydroelectricity are not included. Data are in 10^6 metre^3/year; information for B.C. incorporates the Yukon and Northwest Territories.

SURFACE WATER USE	ATLANTIC PROVINCES	QUEBEC	ONTARIO	PRAIRIE PROVINCES	BRITISH COLUMBIA	CANADA	% OF TOTAL
SECTOR							
Agriculture	15	100	186	3 014	676	3 991	8.9
Mining	77	74	87	50	75	363	0.8
Manufacturing	601	1 616	3 457	447	1 161	7 287	16.2
Thermal cooling	2 126	1 005	23 095	2 025	106	28 357	62.9
Municipal	365	1 703	1 660	685	698	5 102	11.3
Total	3 175	4 498	28 485	6 221	2 716	45 095	100
% increase 1981–1991	+10.2%	+7.5	+35.2	+16.4	–28.1	+21.1	
Net water consumption	118	383	512	3 680	724	5 367	
Consumption as % of withdrawal	3.7	8.5	1.8	58.4	26.7	11.9	

Source: Statistics Canada (1994)

TABLE 14.4 NUMBERS OF LARGE DAMS IN CANADA

A "large dam" is defined as being higher than 15 m, with a capacity greater than 1 million metre^3, and maximum discharge rate greater than 2000 metre^3/second.

PROVINCE	1969	1991
Newfoundland	15	80
Prince Edward Island	–	–
Nova Scotia	27	35
New Brunswick	16	16
Quebec	103	189
Ontario	74	80
Manitoba	24	34
Saskatchewan	34	40
Alberta	44	53
British Columbia	80	89
Yukon	3	3
Northwest Territories	3	3
Canada	423	622
Hydroelectric	267	444
Irrigation	48	55
Municipal	38	43
Flood control	18	22
Other	52	58

Source: Statistics Canada (1994)

Some countries still have substantial areas of natural ecosystems that are suitable for conversion. In most cases, this means of economic development is being actively pursued, particularly in countries that also have rapidly growing populations. Prominent examples include parts of Brazil, Indonesia, Kenya, Nigeria, Sudan, Thailand, and Viet Nam. These countries have been substantially increasing their areas of cultivated land, mostly by clearing natural tropical rainforests, savannas, and wetlands (see Table 14.6 for percent change between 1980 and 1990). Other less-developed countries have few remaining areas of natural lands that are suitable for agricultural development. In spite of their rapidly growing populations, these countries have not managed to create much additional cropland since 1980. This is the situation of Argentina, Bangladesh, Ethiopia, Haiti, India, Iraq, Iran, Malaysia, Mexico, Pakistan, and Tanzania (Table 14.6).

Most of the world's wealthier countries are not developing much additional arable land, largely because their areas with good site capability for agriculture are already being used. In fact, many developed countries have taken a great deal of land out of agricultural use since 1980 (Table 14.6). The land withdrawals have occurred for two major reasons: (1) to reduce the production of certain crops, which keeps prices relatively high; and (2) to conserve environmental quality through less-intensive use of marginal

TABLE 14.5 | USE OF GROUNDWATER IN THE REGIONS OF CANADA, 1981

The municipal sector refers to places with central systems for water treatment and distribution, while rural water use generally involves wells at individual homes. Water-use data are in 10^6 metre^3/year, and include groundwater only. The percentage of the total water use in each sector that is provided by groundwater is shown as "% of total," the rest being supplied by surface waters. Information for British Columbia incorporates the Yukon and Northwest Territories.

GROUNDWATER USE	ATLANTIC PROVINCES	QUEBEC	ONTARIO	PRAIRIE PROVINCES	BRITISH COLUMBIA	CANADA	% OF TOTAL
SECTOR							
Agriculture	0.01	0.08	0.15	2.32	3.11	3.11	13
Industrial	2.42	2.50	19.18	2.36	2.76	29.22	1
Municipal	0.37	1.60	1.77	0.60	0.52	4.87	9
Rural	0.07	0.06	0.08	0.08	0.06	0.34	82
Total	2.87	4.24	21.18	5.36	3.89	37.54	4

Source: Statistics Canada (1994)

lands that are prone to erosion and other kinds of degradation. In addition, some high-quality agricultural lands have been lost to urbanization in many developed countries, including Canada.

Agricultural Site Capability

Agricultural site capability (or site quality) can be defined as the potential of an area of land to sustain the productivity of agricultural crops. Site capability is a complex ecological quality, which depends on the maintenance of soil fertility, organic matter, drainage, and other factors that influence crop productivity. These factors are themselves influenced by climate and drainage, by the vigour of ecological functions such as decomposition and nutrient cycling, and by the nature of the plant and microbial communities.

Site capability is critically important to the productivity of agricultural systems, and therefore to the production and availability of food. Because the beneficial qualities of cultivated lands can be maintained and even improved by appropriate management practices, agricultural site capability represents a potentially renewable resource. It can also, however, be degraded by some agricultural practices.

Ultimately, site capability for agriculture depends on seven interrelated factors: soil fertility, soil organic matter, bulk density of soil, resistance to erosion, moisture status, salinization, and prevalence of weeds. We now examine each of these in turn.

Soil Fertility

Soil fertility is related to the ability of the ecosystem to supply the nutrients required to sustain crop productivity. Especially important are sources of inorganic nitrogen, particularly ammonium and nitrate, as well as phosphate, potassium, calcium, magnesium, and sulphur. Soil fertility is influenced not only by the quantities of these nutrients, but also by factors such as the following, which affect their availability to plants:

- cation exchange capacity, or the degree to which positively charged ions of such nutrients as ammonium (NH_4^+), potassium (K^+), and calcium (Ca^{2+}) are bound by soil
- anion exchange capacity, which is related to the binding of negatively charged ions such as nitrate (NO_3^-), phosphate (PO_4^{3-}), and sulphate (SO_4^{2-})
- soil acidity, or pH, which affects the solubility of many nutrients and many aspects of microbial activity
- rates of oxidation of organically bound nutrients into inorganic compounds that plants can take up and use more effectively
- nutrient addition to the land by agricultural fertilization

For example, consider nitrification (see Chapter 5), an important process performed by certain species of bacteria that transforms ammonium into nitrate. The rate of nitrification is decreased in acidic or waterlogged soils, resulting in decreased nitrate availability to crops.

Soil fertility can also be degraded by removing excessive quantities of nutrients with the harvested crops, as well as by compaction, depletion of soil organic matter, waterlogging, and acidification.

TABLE 14.6 AGRICULTURAL LAND IN SELECTED COUNTRIES

Land areas are averaged over 1989–1991; % refers to percent change of 1989–91 relative to 1979–81. (A positive value indicates an increase.)

COUNTRY	CROPLAND		PASTURE	
	(10^6 hectare)	%	(10^6 ha)	%
RELATIVELY UNDEVELOPED COUNTRIES				
Bangladesh	9.4	2.1	0.6	0.0
Dem. Rep. Congo	7.9	3.5	15.0	0.0
Ethiopia	13.9	0.4	44.9	−1.1
Kenya	2.4	6.5	38.1	0.0
Nigeria	32.2	6.2	40.0	0.0
Sudan	12.9	3.9	110.0	12.2
Tanzania	3.4	1.0	35.0	0.0
Viet Nam	6.4	−2.9	0.3	18.3
RAPIDLY DEVELOPING COUNTRIES				
Argentina	27.2	0.0	142.0	−0.7
Brazil	59.9	23.1	184.2	7.5
China	96.5	−4.0	400.0	19.9
India	169.6	0.7	11.8	−2.6
Indonesia	22.0	12.3	11.8	−1.5
Iran	15.1	2.8	44.0	0.0
Iraq	5.5	0.2	4.0	0.0
Malaysia	4.9	1.5	<0.1	–
Mexico	24.7	0.7	74.5	0.0
Pakistan	21.1	4.0	5.0	0.0
Philippines	8.0	2.5	1.3	24.0
Thailand	23.0	25.5	0.8	29.7
DEVELOPED COUNTRIES				
Australia	48.3	10.2	417.2	−4.8
Canada	*45.9*	*0.5*	*28.1*	*0.9*
France	19.2	1.5	11.4	−11.4
Germany	12.0	−4.2	5.3	−11.1
Italy	12.0	−3.7	4.9	−5.0
Japan	4.6	−5.8	0.6	11.8
Russia	132.1	−1.3	81.0	−3.2
United Kingdom	6.7	−4.3	11.2	−2.3
United States	187.8	−1.5	239.2	0.7

Source: World Resources Institute (1995)

Soil Organic Matter

Soil organic matter consists of plant debris and humified organic materials. Organic matter contributes to the ability of soil to form a loose, crumbly structure, called tilth. Soil with good tilth is well aerated, allows plant roots to grow freely, and retains moisture — all of which are factors critical to crop growth.

Some nutrients are components of soil organic matter. Organically bound nutrients can be slowly released for plant uptake through the process of decomposition, which in this respect can be viewed as a natural, slow-release, organic fertilization. Organic matter also helps the soil to retain ionic forms of nutrients through cation and anion exchange capacity. Intensive cropping with insufficient return of crop residues commonly leads to losses of soil organic matter and degradation of the valuable services it provides.

Bulk Density of Soil

A low bulk density of soil (that is, its weight per unit of volume) is also preferred for soil tilth and drainage. Bulk density can be degraded by losses of soil organic matter and by excessive compaction under heavy machinery, especially when fields are wet. Soils degraded by compaction may be wet, may be poor in oxygen, and may have impaired nutrient cycling and poor crop growth.

Resistance to Erosion

Soil mass can be lost by erosion, caused by particles being carried away by the forces of wind or running water. Any agricultural practices that increase rates of erosion should be viewed as a mining of soil capital. In severe cases, erosion can strip away the relatively fertile, surface horizons of soil. Even naked bedrock may be exposed in the worst cases.

Erosion is encouraged when soils are left without a cover of vegetation or crop residues during the winter, when contour ploughing is not practised (that is, when cultivation runs down a slope rather than along it), and when steep terrain is tilled. Sites are relatively resistant to erosion if their soil has good tilth, if they are relatively flat, and if the climate is not excessively wet or windy.

Moisture Status

Moisture status is another important aspect of site capability. In general, an intermediate moisture status (referred to as *mesic*) is preferred for the growth of most crops. Excessively dry sites will produce small yields, and crops may even die from extreme drought. In contrast, excessively wet sites tend to have cold soils with little or no oxygen present, conditions that are stressful to almost all crops.

The moisture status of most sites is largely affected by climatic factors, especially the rates of precipitation and evapotranspiration. Soil moisture is also affected by drainage characteristics — coarse-grained soils may drain too rapidly and have little ability to hold moisture, while heavy clay soils may not drain well enough, retaining water close to the surface. Soils with good tilth tend to have a degree of drainage midway between these extremes.

Salinization

Salinization refers to the accumulation of salts in soil, particularly excessive concentrations of salts of sodium, potassium, chloride, or sulphate. These and other salts are present in irrigation water and in certain fertilizers, and remain behind when water evaporates to the atmosphere. Salinization is a common problem in sites that are irrigated but do not have enough drainage to carry the salts away, causing them to accumulate in the surface soil.

Prevalence of Weeds

Weeds can be broadly defined as plants that are judged to be interfering with some human purpose (see Chapter 21). Weeds are considered a problem in agriculture if they present crop species with undue competition. Continuous cultivation of the same species of crop can result in an increased abundance of weeds, which may interfere with crop productivity. This problem is commonly managed by the use of herbicides. Often, however, the build-up of weed populations can be avoided by rotating crops, and by using other management practices that provide less favourable conditions for these unwanted plants.

Degradation of Site Capability

In the long term, continuous intensive agricultural management can degrade site capability. When this happens, the productivity of crops decreases, and in severe cases the land may no longer be suitable for agricultural use. Such damage can often be avoided or mitigated by changing the management system. For example, inorganic fertilizers may be applied to the soil in an attempt to compensate for declining fertility. Organic soil conditioners, such as compost and manure, can also be added to mitigate losses of organic matter, thereby helping to maintain the fertility and tilth of soil. In other cases, pesticides may be used to try to manage weeds and other pests. These management options are, however, intensive in their use of material and energy resources, and they may cause additional damage to the site and nearby ecosystems. Ultimately, truly sustainable agricultural systems involve management strategies that *conserve* site capability, while minimizing the use of nonrenewable sources of energy, nutrients, and pesticides.

Agricultural Production and Management

The number of people and domestic animals that can be supported depends on the production of agricultural crops. In addition to the more than 6 billion humans that had to be fed in the mid-1990s, there were about 1.3 billion cows, 1.8 billion sheep and goats, 0.86 billion pigs, 0.16 billion camels and water buffaloes, 0.20 billion horses and asses, and 17 billion chickens (World Resources Institute,

1995). Most of these domestic livestock forage extensively on wild plants, but many are fed crops grown in agriculture. In fact, about 37% of the global production of grains is fed to livestock. Eventually, food products derived from the livestock are eaten by humans, who are secondary consumers (and top predators) in the agricultural food chain.

Countries with an excess of agricultural production over domestic consumption have a surplus available for export, while nations with a deficit must import some of their food (Table 14.7). In general, the world's greatest food-exporting nations have relatively developed economies. Many poor countries export such foods as sugar, coffee, tea, palm oil, and other cash crops. Overall, however, most less-developed countries have food deficits or are only marginally self-sufficient. The food deficits must be made up by expensive purchases of food grown elsewhere, or by food aid from wealthier nations (Table 14.7).

TABLE 14.7 | INTERNATIONAL TRADE OF AGRICULTURAL PRODUCE

Net trade is calculated as exports minus imports. Positive numbers are net exports or donations, while negative numbers are net imports. Cereals include maize, rice, sorghum, and wheat. Oils are manufactured mainly from olive, oil palm, peanut, rapeseed (canola), and soybean. Data are averaged over 1989–1991.

	ANNUAL NET TRADE (10^6 tonne)		ANNUAL DONATIONS (10^6 t)	
	CEREALS	OILS	CEREALS	OILS
REGION				
Africa	–25.59	–2.14	–5.17	–0.22
Asia	–78.79	1.01	–3.25	–0.27
North America	115.64	0.75	7.48	0.54
Central America	–11.82	–1.55	1.53	–0.09
South America	1.70	2.59	–0.56	–0.02
Europe	23.48	–1.06	2.40	0.09
USSR (former)	–35.50	–0.72	–	–
Oceania	13.60	0.36	0.34	<0.01
MAJOR EXPORTERS				
United States	93.54	1.24	6.42	0.48
France	28.83	–0.01	0.22	<0.01
Canada	*22.10*	*–0.49*	*1.06*	*0.06*
Australia	14.28	0.09	0.33	<0.01
Argentina	9.53	2.40	0.02	–
Thailand	5.75	–0.02	–0.09	<–0.01
United Kingdom	3.55	–0.83	0.14	<–0.01
MAJOR IMPORTERS				
USSR (former)	–35.50	–0.72	–	–
Japan	–26.86	–0.55	0.48	<0.01
China	–14.70	–2.01	0.00	<–0.01
Korea (South)	–9.89	–0.39	0.00	0.00
Egypt	–8.11	–0.78	–1.43	–0.05
Mexico	–6.64	–0.26	–0.22	<–0.01
Algeria	–6.27	–0.46	–0.01	<–0.01
Iran	–6.17	–0.69	–0.02	<–0.01
Italy	–4.60	–0.49	0.13	<0.01
Brazil	–3.86	0.72	–0.02	0.01

Source: World Resources Institute (1995)

The world's most important crops are cereals, such as wheat, rice, maize, sorghum, and barley. Of secondary importance are tuber crops, such as potato, cassava, sweet potato, and turnip. In any country, the total production of cereals and other crops is a function of the amount of land devoted to the cultivation of those species, multiplied by the average productivity.

Productivity reflects the combined influences of site capability and the intensity of management. In agriculture, *management* is intended largely to mitigate some of the constraints associated with site quality, as well as the yield-reducing influences of inclement weather, insect infestations, weeds, and crop diseases. The productivity of cereals and other crops varies among countries, being relatively low in less developed countries and higher in more developed ones (see Table 14.8). In general, this difference reflects the use of highly mechanized agricultural systems

TABLE 14.8 | INDEXES OF AGRICULTURAL PRODUCTION IN SELECTED COUNTRIES

Data are averaged over 1990–1992; % refers to percent change between 1980–82 and 1990–92. The 1979–1981 value of the index of food production was assigned a value of 100, and 1990–92 average is shown relative to that number.

COUNTRY	CEREALS		ROOTS & TUBERS		INDEX OF FOOD PRODUCTION	
	PRODUCTIVITY (tonne/hectare·year)	%	PRODUCTIVITY (t/h·y)	%	TOTAL	PER CAPITA
RELATIVELY UNDEVELOPED COUNTRIES						
Bangladesh	2.6	30	9.9	–2	126	96
Dem. Rep. Congo	0.8	–7	7.6	9	131	92
Kenya	1.6	2	8.1	8	145	99
Nigeria	1.2	–14	10.5	31	174	122
Sudan	0.7	6	2.7	–19	113	81
Tanzania	1.3	12	8.3	–13	120	83
Viet Nam	3.1	40	7.7	14	156	123
RAPIDLY DEVELOPING COUNTRIES						
Argentina	2.6	15	18.9	29	113	97
Brazil	1.9	22	12.4	8	138	111
China	4.3	37	14.9	6	161	137
India	1.9	42	16.0	23	153	122
Indonesia	3.9	26	11.7	27	168	135
Iran	1.6	44	18.0	18	181	119
Malaysia	2.8	0	8.9	–1	227	171
Mexico	2.4	7	16.6	20	128	100
Pakistan	1.8	11	10.7	2	159	122
Philippines	2.0	20	6.6	5	118	90
Thailand	2.1	6	14.1	–6	129	109
DEVELOPED COUNTRIES						
Australia	1.7	52	27.7	14	113	96
Canada	*2.5*	*9*	*25.4*	*6*	*125*	*111*
France	6.4	31	32.9	8	106	100
Germany	5.7	–	30.0	–	–	–
Italy	4.3	20	20.5	13	100	98
Japan	5.7	11	25.3	9	99	93
Russia	1.7	35	10.6	9	–	–
United Kingdom	6.3	24	38.6	13	115	112
United States	4.9	18	33.1	13	108	97

Source: World Resources Institute (1995)

in wealthier countries, including intensive applications of fertilizers and pesticides.

The index of food production shown in Table 14.8 is a composite indicator that takes into account the production of all important crop species. The 1979–1981 value of the index of food production was assigned a value of 100, and other years are relative to that number. The index shows whether agricultural production has increased (>100) or decreased (<100) during the decade, both in total and on a per capita basis.

It is important to understand that the high yields from intensive agricultural systems are heavily subsidized by large inputs of nonrenewable resources. For example, the most important agricultural fertilizers are inorganic compounds of nitrogen, usually urea or ammonium nitrate, both of which are manufactured using natural gas. The second- and third-most-important fertilizer nutrients are compounds of phosphate and potassium, which are manufactured from mined minerals. In addition, most pesticides are manufactured from petrochemicals, using energy-intensive technologies. Moreover, the mechanization of agricultural systems in developed countries involves the use of large tractors pulling heavy equipment for tilling, harvesting, and other purposes. The manufacture of these machines requires large quantities of nonrenewable energy and materials, such as metals and plastics. Furthermore, the machines run on nonrenewable fuels.

Therefore, the high productivities of intensively managed agricultural systems used in developed countries are achieved mainly through the use of large amounts of nonrenewable resources. This fact is indicated by data that show much larger rates of fertilizer use and numbers of tractors in these countries, compared with less-wealthy nations (Table 14.9).

Some agricultural systems used in less-developed countries are also quite intensive and result in high yields.

TABLE 14.9 | INTENSITY OF AGRICULTURAL MANAGEMENT IN SELECTED COUNTRIES

COUNTRY	CROPLAND				
	AREA (10^6 hectare)	PER CAPITA (ha/person)	% IRRIGATED	FERTILIZER USE (kilogram/hectare)	NUMBER OF TRACTORS ($10^3/10^6$ ha)
RELATIVELY UNDEVELOPED COUNTRIES					
Bangladesh	9.1	0.08	31	101	0.6
Dem. Rep. Congo	7.9	0.20	0	1	0.3
Kenya	2.4	0.10	2	45	5.7
Nigeria	32.3	0.29	3	12	0.4
Tanzania	3.4	0.13	4	15	2.0
Viet Nam	6.4	0.09	29	96	4.2
RAPIDLY DEVELOPING COUNTRIES					
Brazil	61.4	0.40	4	54	11.7
China	96.6	0.08	49	284	8.6
India	169.7	0.20	27	73	5.7
Indonesia	22.2	0.12	37	109	1.2
Iran	15.1	0.25	38	77	7.6
Malaysia	4.9	0.27	7	188	2.5
Mexico	24.7	0.29	21	69	6.9
Pakistan	21.1	0.17	80	89	12.3
Philippines	8.0	0.13	20	65	1.3
Thailand	23.2	0.42	19	39	6.8
DEVELOPED COUNTRIES					
Australia	46.9	2.70	4	26	6.8
Canada	**45.9**	**1.70**	**2**	**46**	**16.9**
France	19.2	0.34	6	301	76.2
Germany	12.0	0.15	4	520	130.3
Italy	12.0	0.21	26	160	119.4
Japan	4.6	0.04	62	402	451.0
Russia	212.8	1.44	3	52	6.4
United Kingdom	6.6	0.11	2	350	76.5
United States	187.8	0.74	10	99	25.3

Source: World Resources Institute (1995)

For example, in many humid tropical countries, rice is cultivated using a system known (in parts of Asia) as paddy. Water buffalo are used to plough and till the dyked, flooded fields. People then hand-transplant the young rice plants, hand-weed the crop with hoes, and eventually harvest by hand-scything. In places with evenly spaced precipitation and naturally fertile soils, such as parts of Java, Sumatra, and the Philippines, as many as three rice crops can be grown each year. The paddy system, typically used by families working small plots of land, can achieve high yields with relatively small inputs of inorganic fertilizer or pesticides.

Other agricultural systems used in less-developed countries are much less productive than paddy rice, generally because of suboptimal rainfall and less-fertile soils. The least productive systems are used in semiarid regions. Under such conditions it is not possible to cultivate many plant crops. However, livestock such as cows, camels, goats, and sheep can forage extensively over the landscape, harvesting the sparse production of native forage. The dispersed plant biomass of semi-arid ecosystems is too small in quantity and too poor in nutritional quality for direct harvesting and use by humans. Grazing livestock can, however, convert the poor-quality forage into a form (such as meat or milk) that humans can utilize as food.

Increasingly, the agricultural systems used in less-developed countries are becoming more intensive in their management. In this sense they are proceeding toward the kinds of systems used in developed countries. Indicators of this change include increasing use of fertilizers, pesticides, and mechanization. This has resulted in the increasing yields seen in many less-developed countries since about 1980 (see Table 14.8). For example, the productivity of cereal crops in Bangladesh, China, India, Iran, and Viet Nam increased by more than 30% between about 1980 and 1990. This was largely due to the adoption of more intensive agricultural practices, particularly fertilization and pesticide use.

The industrialization of agricultural production in these countries also results in important social changes. Of particular importance is the rapid amalgamation of small family farms into larger industrial units. This results in the displacement of many poor people from agricultural livelihoods. These economic refugees then migrate to towns and cities, resulting in an increase of the urbanization rate far beyond that expected from population growth alone.

Agricultural Production and Management in Canada

Plant Crops

Canada is one of the world's great agricultural nations and a major contributor to the international trade in foods. Canada's 46 million hectares of cropland equals 3.2% of the global total. Canada's average grain production of 53 million t/y during 1990–1992 ranked sixth in the world (see Table 14.8). During the same period, Canada exported 23.2×10^6 t/y of grain (as trade plus aid shipments), which ranked third in the world after the United States (99.9×10^6 t/y) and France (29.0×10^6 t/y) (World Resources Institute, 1995).

Canadian exports of farm products in 1994 had a value of $11.3 billion, including $1.4 billion of livestock and $9.9 billion of all other commodities, mostly grains (Natural Resources Canada, 1996). These exports were largely offset by imports of $10.6 billion, for a net trade surplus of $0.64 billion for the agricultural sector.

At the beginning of the twentieth century, more than 80% of the Canadian labour force was employed in agriculture. Farming relied on animal and human labour as sources of energy for cultivation and harvesting. Most farms were relatively small, family-operated enterprises, run mainly as subsistence operations to produce foods and other crops for use by the family. Any surplus production was traded in local markets for cash or manufactured goods. The agricultural surplus was eventually sold in Canadian cities or exchanged internationally by traders. Much of today's agricultural activity in less-developed countries still has this socioeconomic character.

Today, Canadian agriculture involves highly mechanized, industrial operations. Only 3% of the national workforce is employed in farming (about 431 000 people in 1995). Virtually all cultivation, harvesting, and processing is accomplished by large, fossil-fuelled machines. Tractors haul cultivating, seeding, and spraying machines, and self-contained harvesters harvest and process crops. Canadian farmers used 775 000 tractors and 160 000 harvesters in 1991.

Canada has an immense land base, some 9.215 million km^2, making it the second-largest country in the world after Russia. However, the area suitable for agriculture is quite limited, and is largely restricted to the southern parts of the country. The ability of land to support agricultural

Canadian Focus 14.1

Canada Land Inventory

The ability of land in Canada to support agricultural uses is classified according to a system known as the *Canada Land Inventory*. Lands in classes 1, 2, and 3 have the highest capability, accounting for almost all of the crop production in Canada. Lands in Classes 4 and 5 are generally unsuited for plant crops, but may be suitable as rough pasture for livestock.

Class 1

The best agricultural lands, Class 1, have no significant limitations for the growth of crops. These lands have deep soil, gentle or no slope, and good drainage, and have sufficient organic matter to retain water and nutrients well. These lands are fertile, do not erode easily, and sustain a high productivity of crop plants suited to the local climate.

Class 2

Class 2 lands either have moderate limitations that restrict the productivity of at least some species of crop plants or require moderately intense management for conservation purposes. Limitations of these sites are generally associated with a mildly adverse local climate, moderate erosion, poor soil structure, imperfect drainage, acidic soils, moderate slope, or occasional wetness. Class 2 soils can, however, sustain a good productivity of a wide range of crop species.

Class 3

These lands have moderately severe limitations for crop productivity, or require special conservation practices to maintain productivity and site quality. Limitations of these lands include the same factors that restrict Class 2 lands: that is, climate, relatively strong slopes with high erosion potential, poor fertility, poor drainage or poor water-holding capacity, and sometimes salinity. If properly managed for carefully selected crops, Class 3 lands can achieve a moderate-to-fair agricultural productivity.

TABLE 14.10 | DISTRIBUTION OF CLASS 1–3 LAND IN CANADA

Soil capability classes 1–3 are suitable for the cultivation of crop species. Classes 4 and 5, used for pasture or rough grazing, are not reported here. Not all of the high-quality land is used for agriculture — some has been converted to urban and suburban land uses. Data are expressed in 10^3 kilometre^2.

PROVINCE	TOTAL LAND	CAPABILITY CLASS		
		1	2	3
Newfoundland	371.7	–	–	0.02
Prince Edward Island	5.7	–	2.5	1.4
Nova Scotia	52.8	–	1.7	9.8
New Brunswick	72.1	–	1.6	11.5
Quebec	1356.8	0.2	9.1	12.8
Ontario	891.2	21.6	22.2	29.1
Manitoba	548.4	1.6	25.3	24.4
Saskatchewan	570.7	10.0	58.7	94.2
Alberta	644.4	7.9	38.4	61.1
British Columbia	929.7	0.2	2.4	6.9
Yukon	478.9	–	–	–
Northwest Territories	3293.0	–	–	–
Canada total	9215.4	41.5	161.8	251.3

Source: Statistics Canada (1994)

uses is categorized by a system known as the Canada Land Inventory (Canadian Focus 14.1). The distribution of the most productive lands for agriculture in Canada is shown in Table 14.10.

Most of the highest-capability lands are located in southern parts of Canada, in regions where the growing season is relatively long and moist, and on sites with relatively fertile soil and flat terrain. Southern Ontario, for example, has 52% of the Class 1 lands in Canada, while Saskatchewan has 24%, Alberta 19%, and Manitoba 4%. Unfortunately, some of the best lands in southern Ontario are being lost rapidly through conversion to residential and industrial land uses, particularly in the greater Toronto area and the Niagara Peninsula. Excellent agricultural land is likewise being lost to urbanization in the delta of the Fraser River and the Okanagan Valley in southern British Columbia. These depletions of high-quality land directly affect the overall production of foodstuffs in Canada.

As we previously noted, the quality of land for agriculture is strongly influenced by such factors as soil fertility, organic-matter concentration, drainage, and weed populations. All of these qualities can be degraded through inappropriate management of the land. It is important to monitor changes in these site factors over time, in order to track changes in the sustainability of Canadian agriculture. Unfortunately, suitable monitoring data do not yet exist in most areas, although programs are being designed.

There are, however, some general indications that soil fertility and other site factors may be declining in quality over much of the agricultural land base of Canada. For example, in order to maintain the productivity of many Canadian agroecosystems, large quantities of fertilizers and soil conditioners must be added to the system. Similarly, herbicides, insecticides, fungicides, and other pesticides must be used to manage pest problems (see Chapter 21). The pressing need to use intensive management practices to maintain productivity could, in itself, be considered a symptom of unsustainable stress on the agroecosystem. Moreover, most fertilizers, pesticides, and their mechanized application systems are based on the mining and use of nonrenewable resources, representing another element of nonsustainability.

Huge quantities of fertilizers are used to increase crop productivity in Canada. In 1991, fertilizers were applied to 21.6 million ha of farmland, a 3-fold increase over 1971 (Table 14.11). Increases in the quantities of fertilizers used have been similarly great. In 1990, agricultural lands were fertilized with 1.11 million tonnes of nitrogen and 0.57 million t of phosphate (Statistics Canada, 1994). Compared with rates of fertilizer use in 1970, these represented increases of 348% and 118%, respectively.

The use of pesticides to deal with insect pests, weeds, and fungal pathogens is also intensive in Canadian agriculture. In 1991, herbicides were applied to 21.6 million ha of farmland, representing a 2.5-fold increase over 1971 (Table 14.11). Insecticides and fungicides were applied to 2.8 million ha in 1991, a 3-fold increase over 1971.

Canadian agricultural systems also utilize crop varieties that have been selectively bred to increase their potential productivity and resistance to important pests and pathogens, to respond vigorously to fertilization and other intensive management practices, and to grow well under Canadian climatic regimes. This is not to say that these varieties are optimally adapted to intensively managed agroecosystems. New pests and diseases often develop, so that the crop-breeding industry must continuously respond to changing environmental and ecological conditions.

The intensification of industrial agriculture in Canada has allowed great increases in crop production (see Figures 14.1 and 14.2). Similarly large increases in agricultural production have been accomplished in other countries that

TABLE 14.11 | AGRICULTURAL LAND USE AND CROPS IN CANADA

LAND USE	1971 (10^6 hectare)	1981 (10^6 ha)	1991 (10^6 ha)
Farmed area	68.66	65.89	67.75
Land in crops	27.83	30.97	33.51
Summerfallow	10.82	9.70	7.92
Improved pasture	4.14	4.40	4.14
Rough pasture	25.87	20.82	22.18
Grains: total	33.61	42.23	44.38
Wheat	19.41	30.77	35.00
Oats	6.70	3.81	3.05
Barley	5.66	5.46	4.52
Corn	0.57	1.14	1.11
Fodder crops	6.08	6.25	6.47
Oilseeds	3.13	2.28	4.32
Other field crops	0.35	0.42	0.58
Horticultural crops	0.18	0.20	0.21
Fertilizer application	6.93	18.51	21.56
Herbicide application	8.57	15.22	21.60
Insecticide or fungicide	0.91	1.65	2.77

Source: Statistics Canada (1992)

FIGURE 14.1 | INCREASES IN GRAIN PRODUCTION IN CANADA

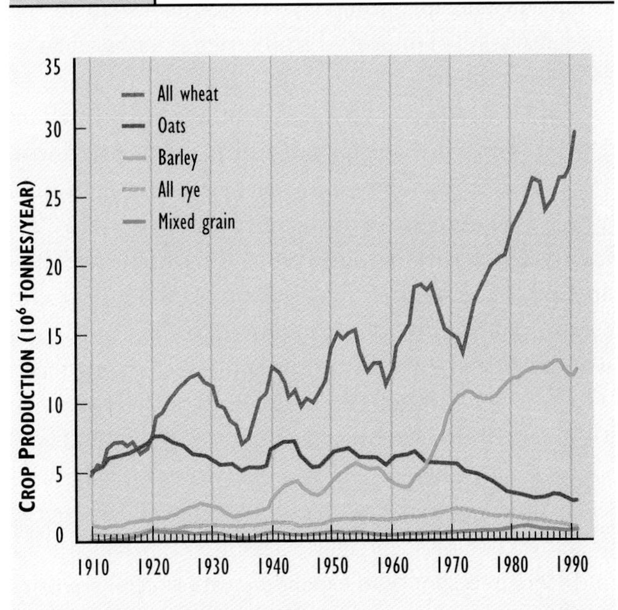

Source: Modified from Statistics Canada (1994)

FIGURE 14.2 INCREASES IN CORN AND PULSE PRODUCTION IN CANADA

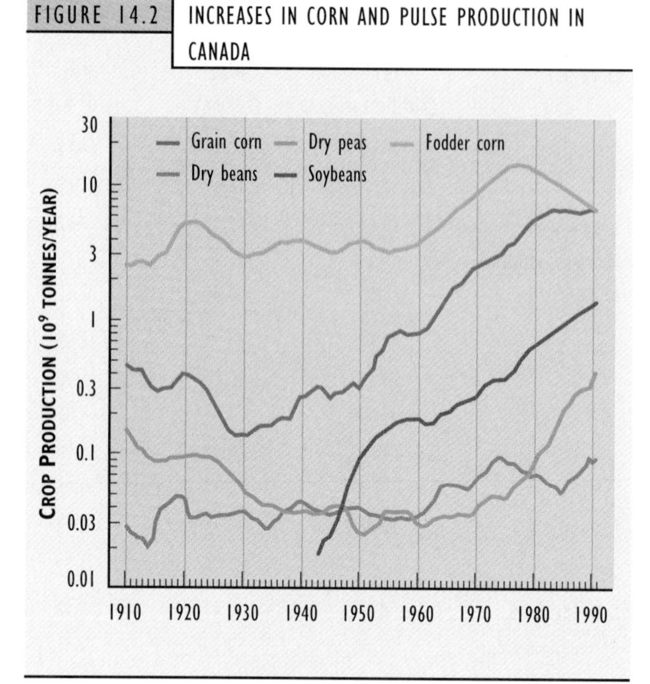

Pulses are dried legumes, such as peas, beans, and soybeans. Note that the data are presented as $\log_{10}$.

Source: Modified from Statistics Canada (1994)

have adopted intensive and mechanized agricultural systems. This includes the United States, most countries of western Europe, and — increasingly — China, India, Russia, Ukraine, and other rapidly developing countries. The increases in agricultural production have been accomplished mainly through intensified management and the cultivation of improved crop varieties, rather than by increasing the areas of cultivated land.

In 1991, 67.75×10^6 ha of land were cultivated on 290 000 farms in Canada (Table 14.11). Compared with 1971, this represented a 1.3% decrease in the cultivated area, a 24% decrease in the numbers of farms, and a 29% increase in the average farm size. Of the total cultivated area, 49.5% was actively managed for crops, 11.7% was in summer fallow, 6.1% was in improved pasture, and 32.7% was natural grazing lands, mostly in the western provinces. The area of cultivated crops increased by 20% between 1971 and 1991, while the area of other agricultural land uses decreased by 14%. Wheat is by far the most extensively grown crop in Canada, accounting for 79% of the area sown to grains, and increasing by 80% between 1971 and 1991 in response to expanding export opportunities (Table 14.11).

Livestock

Animal husbandry has also become intensive in Canada. Most production of chickens, cows, and pigs now occurs on so-called "factory farms." This is an industrial technology that involves raising livestock indoors under densely crowded conditions. The livestock are fed to satiation with nutritionally optimized diets, while diseases are managed with antibiotics and other medicines. Productivity may be enhanced with growth hormones.

In 1991, 94.9 million chickens were raised for their meat and eggs on Canadian farms, most of them in operations of an industrial scale and intensity. This represented a 32% increase in chickens since 1971. As well, 8.1 million turkeys were raised, again mostly on factory farms.

Larger livestock in 1991 included 13.0 million cows (of which 1.3 million were mature milk cows), 10.2 million pigs, 936 000 sheep, and 356 000 horses. Dairy cows and pigs are raised mostly on factory farms. Most beef cows spend part of their lives grazing outdoors, in pastures or on seminatural prairies. However, prior to slaughter, most are rounded up and kept in densely crowded feedlots, where they are well fed so that they can gain weight rapidly. Sheep, goats, and horses are raised under less-intensive conditions.

Overview

The 20th century has witnessed enormous increases in agricultural production. This has fed the similarly rapid increases in the global populations of humans and domestic animals. The rapid intensification of agriculture is, however, substantially dependent on nonrenewable sources of energy and materials, a fact that makes the sustainability of the overall production systems questionable. Moreover, intensive agricultural systems cause important damages to the environment, many of which are described in later chapters.

FOREST RESOURCES

Forests of various kinds are extremely important terrestrial ecosystems. They cover extensive areas of Earth's surface, fixing and storing large quantities of biomass. In 1990, Earth's cover of closed forests was about 3.4 billion ha, of which 48% occurred in temperate and boreal regions and 52% in tropical regions (World Resources Institute,

1995). Although temperate and boreal forests cover an area comparable to the tropical forests, their production is only about one-half as large, and they store only 61% as much biomass. There are also another 2 billion ha of more open woodlands and savannas. The most heavily forested regions are in North and South America, Europe, and Russia, all of which have more than 30% forest cover.

World-wide, an immense area of about 25 million ha of forest is cleared or harvested each year. Tree biomass is harvested for three major reasons:

1. as a fuel for subsistence purposes — that is, to burn as a source of energy for cooking and warmth
2. as an industrial fuel, used to generate electricity or to produce steam or heat for a manufacturing process
3. as a raw material for the manufacture of lumber; paper; composite materials such as plywood, masonite, or waferboards; and other products such as the synthetic fabric known as rayon

Clear-cutting is the most common method of harvesting forests in Canada. Mechanized harvesting systems are used in most areas, such as this machinery that cuts and de-limbs trees, cuts them into convenient lengths, and hauls the wood to a road.
Source: B. Freedman

In addition, forests may be cleared not so much for their biomass, but to create new agricultural or urbanized lands. These longer-term **ecological conversions** result in an essentially permanent loss of forest cover (or **deforestation**).

The global, net primary production of forests has been estimated to be about 48.7 billion t/y, of which an extraordinary 28% is used by humans (Vitousek *et al.*, 1986). This human use can in turn be subdivided into the following categories:

- short-term clearing of forests for shifting cultivation in less-developed countries (45%)
- more permanent conversion of forests to agricultural land uses (18%)
- harvesting of tree biomass (16%)
- productivity of trees in plantations (12%)
- loss during harvest (9%)

Changes in Forest Cover

Forest resources in many countries are being depleted rapidly by excessively high rates of clearing. This is particularly true in many tropical countries, where deforestation is largely driven by increasing populations and the resulting need for more agricultural land and wood fuels. Also important are the economic and industrial demands for tree biomass to manufacture into charcoal and into products for international trade.

Rates of deforestation of some less-developed and rapidly developing countries are listed in Table 14.12. Recently, some of these countries have been losing their forests at extraordinary rates. For instance, deforestation is occurring at 2.9%/y in the Philippines and Thailand, for a half-life of only about 24 years. Moreover, many tropical countries have been increasing their rates of deforestation, for example, by 61% in the Congo between 1979–81 and 1989–91, by 49% in Zambia, by 48% in Cameroon, and by 46% in Tanzania and the Democratic Republic of Congo (Table 14.12).

The rapid deforestation in most developing countries represents the mining of a potentially renewable natural resource. In addition to the losses of timber resources, deforestation in the tropics and subtropics has terrible ecological costs such as extinctions of indigenous biodiversity. These topics are discussed in Chapters 23 and 24.

In contrast to the rapid deforestation in most less-developed countries, the forest cover of many developed countries is relatively stable or even increasing (Table

14.12). This is happening in spite of industrial harvesting of forest resources in many of those countries, largely for the manufacture of lumber and paper. In the industrial forestry that is typically pursued in Canada, the United States, and western Europe, another forest is allowed or encouraged to regenerate on harvested sites. Consequently, deforestation does not occur — appropriate management can ensure that the forest resource does not become depleted by harvesting.

This does not, however, mean that forestry is ecolgically benign in developed countries. Industrial forestry results in substantial changes in the character of the ecosystem, with important consequences for biodiversity and other ecological values (see Chapter 23).

TABLE 14.12 | FOREST RESOURCES AND FORESTRY PRODUCTION IN SELECTED COUNTRIES

Forest cover data are for 1990; deforestation is the annual average between 1981 and 1990; harvesting data are averages for 1989–1991; percent increase refers to the increase of total harvest between 1979–1981 and 1989–1991.

COUNTRY	FORESTS (10^6 hectare)	DEFORESTATION (%/year)	FOREST HARVESTING TOTAL (10^6 metre^3/y)	FUEL (10^6 m^3/y)	INDUSTRIAL (10^6 m^3/y)	PERCENT INCREASE
RELATIVELY UNDEVELOPED COUNTRIES						
Angola	23.1	0.7	6.4	5.5	0.9	25
Dem. Rep. Congo	113.3	0.6	38.9	35.0	3.9	46
Cameroon	20.4	0.6	14.2	11.2	3.0	48
Congo	19.1	0.2	3.7	2.1	1.6	61
Myanmar	28.9	1.2	22.9	17.8	5.1	26
Nigeria	15.6	0.7	107.8	99.9	7.9	36
Sudan	43.0	1.0	22.8	20.7	2.1	35
Tanzania	33.6	1.2	34.3	32.3	2.0	46
Venezuela	45.7	1.2	1.3	0.8	0.6	9
Zambia	32.3	1.0	13.2	12.5	0.7	49
RAPIDLY DEVELOPING COUNTRIES						
Brazil	561.1	0.6	262.4	186.5	76.0	25
China	142.8	–	281.4	188.5	92.9	21
India	51.8	0.6	274.5	250.1	24.4	24
Indonesia	109.5	1.0	167.8	141.0	26.8	16
Malaysia	17.6	1.8	49.9	8.7	41.2	35
Mexico	48.6	1.2	23.4	15.5	8.0	27
Philippines	7.8	2.9	38.5	33.5	5.0	10
Thailand	12.7	2.9	37.7	34.6	3.2	12
DEVELOPED COUNTRIES						
Australia	145.6	–0.1	19.6	2.9	16.8	18
Canada	*453.3*	*0.1*	*179.0*	*6.8*	*172.2*	*15*
France	13.1	–0.1	44.9	10.4	34.5	16
Germany	10.5	–0.5	59.2	4.4	54.8	43
Italy	6.8	–	8.4	4.1	4.4	–3
Japan	24.2	0.1	29.6	0.3	29.3	–12
Russia	925.0	–0.2	375.4	81.1	294.2	5
Sweden	24.4	–	53.7	4.4	49.3	6
United Kingdom	2.2	–1.1	6.5	0.3	6.2	56
United States	296.0	0.1	508.2	90.3	417.9	22

Source: World Resources Institute (1995)

Although most developed countries presently have a stable forest cover, this has not always been the case. Many of these countries were being actively deforested as recently as the beginning of the 20th century. Much early deforestation occurred in order to develop land for agriculture. For instance, most of western Europe was still forested as recently as the Middle Ages (up until about 1500), as was eastern North America up until one to three centuries ago. Parts of these regions are now largely devoid of forest cover, which has been replaced by agroecosystems and urbanized lands.

This process of deforestation largely stopped around 1920 or 1930. At that time, forested areas began to increase in many developed countries. This happened because small farms of marginal agricultural capability were abandoned, and their inhabitants migrated to urban areas to seek work. Over time, the land reverted to forest. For example, because of these socioeconomic and ecological dynamics, the area of forest in the Maritime Provinces has approximately doubled since the beginning of the 20th century. Similar changes have occurred in other developed regions of the world.

Harvesting and Managing Forests

Globally, however, the net trend is one of accelerating deforestation. During the past decade, about 25 million ha/y of forest were cleared around the world, compared with 20 million ha in 1978, 16 million ha in 1950, 10 million ha in 1900, and 6 million ha in 1800 (Freedman, 1995). Most of this aggresive deforestation is associated with conversions of tropical forests into agricultural lands, but the harvesting of forest products is also important in some regions.

During 1989–1991, the global consumption of wood averaged 3.46 billion m^3/y, representing a 19% increase from a decade earlier (1979–1981) (World Resources Institute, 1995). The 1989–1991 wood consumption included:

- 0.49 billion m^3 of sawn timber (a 15% increase from 1979–1981)
- 0.13 billion m^3 of wood panels such as plywood (a 22% increase)
- 1.7 billion m^3 of industrial roundwood (a 15% increase; mostly used to manufacture 215 million t of paper)
- 1.8 billion m^3 of fuelwood (a 23% increase; about 85% of the fuelwood was consumed in less-developed countries)

In Canada, an enormous industrial complex depends on the harvesting of forest biomass, largely for manufacture into lumber, composite materials such as plywood and waferboard, and pulp and paper. The total value of Canadian forest products in 1994 was $49.3 billion.

Most of Canada's production of forest products is intended for export, providing foreign earnings that are crucial to maintaining Canada's positive balance of trade. In 1994, Canadian exports of forest products had a value of $32.4 billion, and contributed $27.8 billion to the country's international balance of trade (calculated as $32.4 billion of exports, minus $4.6 billion of forest-product imports, for a net value of $27.8 billion; Natural Resources Canada, 1996). In terms of dollar value, Canada is the world's leading exporter of forest products.

In fact, the net earnings from the export of forest products approximated the total of the net earnings for all other economic sectors in Canada, underscoring the importance of forest-product exports to the national economy. (The net earnings from trade of energy commodities in 1994 were $15.0 billion, while metals and minerals earned $11.7 billion, fish products $1.8 billion, and farm products $0.64 billion. Other sectors had a negative balance of trade: chemicals and fertilizers –$3.0 billion and manufactured goods –$28.3 billion.)

Of course, to achieve the great economic benefits of forestry, large areas of mature forest must be harvested each year. In 1993, 969 000 ha of mature forest were harvested in Canada, of which 87% was due to clear-cutting (Table 14.13). The total volume of the industrial harvest was equivalent to 161 million m^3 of stem biomass, of which 87% was from conifer trees (or "softwoods") and 13% from broad-leafed trees (or "hardwoods"). The area of forest harvested in 1993 was equivalent to 0.2% of the total area of Canadian forests and 0.4% of the area of "productive" forests. Economically "productive" forests are relatively productive and well stocked, and are most widespread in more southern regions of Canada.

Almost all of the industrially harvested area in Canada is allowed to regenerate to forest. Conversions to agricultural or urbanized land uses are relatively uncommon. The rate of net deforestation in Canada is consequntly at most 0.1% (Table 14.12), in spite of the annual harvesting of almost one million ha of forest.

Moreover, regeneration on most of the harvested area is actively encouraged by planting and other aspects of **silvicultural management**, such as thinning, fertilization, and herbicide and insecticide applications. In 1993, 418 000 ha (or 43% of the total) of harvested forest were planted with 656 million tree seedlings. Some of the planted areas are managed quite intensively, developing tree **plantations**. These tree-farms are generally more productive than natural forests, but they lack many elements of indigenous biodiversity and other ecological and aesthetic values (Chapter 23).

TABLE 14.13 | FOREST RESOURCES AND FORESTRY PRODUCTION IN CANADA

Land classified as "productive" of timber has a sufficiently high productivity and stocking of trees to be economically exploitable, while "non-productive" forests are considered uneconomic. Harvest data are for 1993, with percent increases calculated from 1975.

PROVINCE	FOREST LAND (10^6 hectare)			HARVEST	
	TOTAL	PRODUCTIVE	NON-PROD	(10^3 ha)	% INCREASE
Newfoundland	22.5	11.3	11.3	20.6	31
Prince Edward Island	0.3	<0.1	0.3	3.0	86
Nova Scotia	3.9	3.8	0.2	42.8	57
New Brunswick	6.1	6.0	0.2	100.7	7
Quebec	83.9	54.0	29.9	311.6	16
Ontario	58.0	42.2	15.8	206.0	5
Manitoba	26.3	15.2	11.0	11.0	−9
Saskatchewan	28.8	12.6	16.2	19.5	11
Alberta	38.2	25.7	12.5	44.6	108
British Columbia	60.6	51.7	8.8	207.7	32
Yukon	27.5	7.5	20.0	0.6	2
Northwest Territories	61.4	14.3	47.1	0.5	−26
Canada total	417.6	244.6	170.0	968.6	42

Source: Canadian Council of Forest Ministers (1994)

Virtually all the non-planted part of the harvested area in Canada (57% of the total) also reverts to forest. However, this occurs through a relatively "natural" regeneration of tree species, in a process which involves seedlings that existed on the site prior to harvesting and survived the disturbance (this is known as "advance regeneration"), seedlings that established from seeds dispersed onto the harvested area from nearby forest, and/or seedlings that established from seeds dispersed by "seed-trees" left on the site at the time of harvesting.

In general, industrial forestry as it is practised in Canada appears to be sustainable in the primary economic resource — that is, the productivity of tree biomass. Supporting this bold statement are three facts: (1) the rate of net deforestation is small in Canada; (2) almost all harvested sites regenerate back to forest, which will be available for harvesting again once the trees grow to an appropriate size; and (3) except in some local areas, for short periods of time, the amount of harvesting of forest biomass does not exceed the forest productivity.

There are, however, additional environmental considerations that must be weighed before Canadian industrial forestry can be considered ecologically sustainable in the sense explained in Chapter 12. These additional considerations, to be examined in Chapter 23, include the following:

- the long-term effects of harvesting and management on site capability, which may become degraded by nutrient losses and erosion
- the effects of forestry on populations of fish, deer, and other hunted species, which also represent an economic "resource"
- effects on indigenous biodiversity, including all native species and naturally occurring ecosystems
- effects on hydrology and aquatic ecosystems
- implications of management for carbon storage (this is important in the light of anthropogenic influences on the greenhouse effect; see Chapter 17)

FISH RESOURCES

Wild populations of fish have been long been exploited as food. In recent decades, the rate of harvesting of wild fish has increased enormously, as has the cultivation of certain species in semi-domestication — a practice known as **aquaculture**. Like crop plants, livestock, and forests, populations of fish can be harvested in a sustainable fashion, allowing yields to be maintained over time. Fish stocks can also, however, be overharvested to the degree that regeneration is impaired. When this happens, productivity declines and the resource can collapse disastrously. Regrettably, the recent history of almost all of the world's major fisheries provides abundant examples of overexploitation causing rapid declines in resources.

The global harvest of fish and shellfish during 1989–1991 was about 114 million t/y. This included 84 million t/y of marine fish (representing a 29% increase over 1979–1981), 14 million t of freshwater fish (a 91% increase), and 16 million t of biomass grown in aquaculture (Table 14.14).

Canada is a major fishing nation, with an annual marine harvest of 1.5 million t during 1989–1991 (Table 14.14). Recently, less than this has been harvested because of the closure of major Atlantic and Pacific fisheries. Canadian exports of raw and processed fish products in 1994 had a value of $3.0 billion (Natural Resources Canada, 1996). These exports were partly offset by fish imports of $1.2 billion, for a net trade balance of $1.8 billion in this economic sector.

The most important marine species harvested in Canada are summarized in Table 14.15. Note that these data are for Canadian landings only. Some foreign nations also fish waters within Canada's 320-km management jurisdiction. Their fish landings are not reflected in Table 14.15.

In 1992, the total catch of cod in Atlantic Canada was 239 000 t, of which 80% was landed by Canadian vessels and 20% by the foreign fleet working within the 320-km management zone (Statistics Canada, 1994). The 1992 catch was, however, much smaller than had been attained in recent decades, which averaged as much as 598 000 t during 1982–1986 (81% of this was Canadian landings). In fact, the declining harvest reflects a collapse of the cod stocks over most eastern Canadian waters, a resource calamity that resulted in closure of virtually all of the fishery in 1992 and 1993. The cod stocks were still closed to commercial exploitation in 1997, and will largely remain so for several years. The devastation of cod stocks in the northwestern Atlantic Ocean, mostly caused by Canadian overfishing, is a clear example of the mining of a potentially renewable natural resource (see Canadian Focus 14.2).

Trawling is a technology used to harvest fish from large volumes of water, and is the marine equivalent of clear-cutting a forest. This scene illustrates the catch from a bottom-trawl.

Source: Department of Fisheries and Oceans

Other Hunted Animals

Marine Mammals

Marine mammals have also been subjected to intense commercial hunting in many parts of the world. These animals were initially hunted as a source of oil, which in

TABLE 14.14 | FISH CATCHES AND AQUACULTURE IN SELECTED COUNTRIES

Data are in 10^6 t/y, averaged for 1989–1991, with percent increase since 1979–1981 given in parentheses. Countries are listed in order of decreasing catches of marine fish.

COUNTRY	MARINE FISH	FRESHWATER FISH	AQUACULTURE		
			FISH	SHELLFISH	OTHER
Global	83.62 (29)	14.42 (91)	8.29	3.74	3.51
Japan	10.07 (–1)	0.20 (–8)	0.35	1.22	0.58
USSR (former)	9.29 (9)	0.95 (22)	0.39	<0.01	0.01
China	6.94 (133)	5.21 (319)	4.17	1.31	1.90
Peru	6.86 (128)	0.03 (119)	0.01	0.01	–
Chile	5.88 (100)	<0.01 (–)	0.03	0.04	–
United States	5.43 (52)	0.27 (212)	0.22	0.13	–
South Korea	2.70 (24)	0.03 (–21)	0.02	0.68	0.47
Thailand	2.61 (48)	0.24 (60)	0.09	0.20	–
India	2.27 (52)	1.55 (70)	1.05	0.03	–
Indonesia	2.27 (66)	0.78 (70)	0.38	0.11	0.01
Norway	1.90 (–25)	<0.01 (–)	0.12	–	–
Denmark	1.71 (–8)	0.03 (76)	0.04	–	–
Philippines	1.62 (40)	0.58 (40)	0.30	0.08	0.28
North Korea	1.61 (21)	0.10 (44)	0.01	0.06	0.13
Canada	*1.53 (14)*	*0.05 (–2)*	*0.02*	*0.01*	–
Iceland	1.35 (–12)	<0.01 (–)	<0.01	–	–
Mexico	1.25 (2)	0.18 (1400)	0.01	0.05	–
France	0.83 (9)	0.05 (88)	0.04	0.22	–
United Kingdom	0.80 (–8)	0.02 (449)	0.05	<0.01	–

Source: World Resources Institute (1995)

CANADIAN FOCUS 14.2

MINING THE COD IN THE NORTHWEST ATLANTIC OCEAN

In 1497, John Cabot explored the waters around Newfoundland on behalf of the English crown. On his return, he wrote enthusiastically that the Grand Banks were so "swarming with fish [that they] could be taken not only with a net but in baskets let down [and weighted] with a stone."

Initially, the most important resource was the cod (*Gadus morhua*) that teemed on the Grand Banks, a marine ecosystem of about 25 million ha. Important cod stocks also occurred off Labrador, Nova Scotia, the Gulf of St. Lawrence, and New England.

By 1550, hundreds of ships were sailing from western Europe, catching cod and preserving it by drying or salting, to be sold later in the hungry marketplaces of Europe. Around 1600, about 650 ships were fishing for cod off Newfoundland, and by 1800, about 1600 vessels were sailing from Europe, Newfoundland, Canada, and New England (Mowat, 1984; Hutchings and Myers, 1995). Between 1750 and 1800 the average landings were about 190 000 t/y, increasing to about 400 000–460 000 t/y during 1800–1900, and almost 1 million t/y between 1899 and 1904 (Mowat, 1984; Cushing, 1988).

The cod were harvested using hand-lines, long lines, traps, and seines. Many men fished from small dories, often launched from a larger mother ship, such as one of the celebrated fishing schooners of

TABLE 14.15 | CATCHES OF SELECTED SPECIES OF MARINE FISH IN CANADA

Data are for 1992. Catch biomass is in 10^3 tonne/year, and economic value of the catch is in millions of dollars.

SPECIES	ATLANTIC COAST		PACIFIC COAST		CANADA	
	QUANTITY	VALUE	QUANTITY	VALUE	QUANTITY	VALUE
GROUNDFISH						
Cod	186.5	152.0	10.1	5.5	196.6	157.5
Hake	38.3	17.7	97.2	16.0	135.5	33.8
Redfish	98.0	27.7	24.8	18.3	122.7	46.0
Flatfishes	48.5	29.2	7.8	5.7	56.3	35.0
Pollock	34.1	24.0	3.0	0.9	37.1	24.9
Turbot	22.5	19.3	3.5	0.9	26.0	20.1
Haddock	21.9	30.2	–	–	21.9	30.2
Total	461.3	313.8	159.7	93.8	621.0	407.6
PELAGIC AND OTHER FINFISH						
Herring	215.4	27.8	34.5	46.4	249.9	74.2
Salmon	0.3	1.1	64.9	161.3	65.1	162.4
Capelin	31.0	4.8	–	–	31.0	4.8
Mackerel	25.9	7.1	–	–	25.9	7.1
Total	282.6	67.3	103.3	210.3	385.9	277.6
SHELLFISH						
Scallop	91.3	99.6	–	–	91.3	99.6
Shrimp	39.2	81.2	3.5	10.9	42.7	92.1
Lobster	41.6	314.0	–	–	41.6	314.0
Crab	37.9	58.8	2.5	9.3	40.4	68.2
Clams	16.8	15.4	4.0	18.6	20.8	34.0
Oysters	0.6	1.0	5.0	4.0	5.6	5.0
Total	233.1	575.0	28.9	57.0	261.9	631.9
Total marine	977.0	966.5	291.8	369.3	1269	1336
Total inland	–	–	–	–	64.9	63.0

Source: Statistics Canada (1994)

Newfoundland and Nova Scotia. Although this fishing technology was inefficient, the fishing efforts, and therefore the catches, were large. Consequently, some near-shore cod stocks became depleted, although not those of the offshore banks.

The fishery for cod greatly intensified during the 20th century, largely because of such technological innovations as:

- the development of extremely efficient netting technologies, particularly trawls and monofilament gill nets
- the use of sonar equipment to locate schools of fish
- great increases in ship-borne capacity for storing and processing fish, allowing vessels to stay at sea for a long time

The improved technology allowed enormous catches to be made in the northwest Atlantic, particularly during the 1960s when the fishery was essentially an unregulated, open-access enterprise. By this time, unsustainably high catches in the region were causing cod stocks to collapse rapidly (see Figure).

In response to the economic crisis caused by the declining stocks of cod, the federal government of Canada declared a 320-km-wide fisheries-management zone in 1977, and began to control and allocate quotas of fish. The conservation actions resulted initially in short-lived increases in cod stocks and landings (see Figure). However, exploitation levels were still too high, and the fishery experienced an even more serious collapse. In 1992, government regulators declared an outright moratorium on commercial fishing for cod, a ban that was still in force in early 1997 (when this item was written). Because only small populations of adult cod are available for spawning, the recovery of the stocks has been slow. However, if allowed, the cod population should eventually recover to the point where it can again be harvested as a natural resource.

There are many explanations of the collapse of cod stocks in the northwest Atlantic, each based on more or less convincing logic and information. The most important of these are discussed below (Freedman, 1995; Hutchings and Myers, 1995).

The hypothesis of *overexploitation* suggests that the cod resource was exploited faster than it could regenerate, causing a decline of stocks that became especially acute during the 1970s and 1980s. The excessive harvesting was caused by several factors. Over the years, scientists have estimated the size and productivity of cod stocks and the maximum sustainable yields (MSY). The *scientific information* is, however, imperfect. Firstly, it is extremely difficult to estimate the abundance of fish in the open ocean. Then, in the late 1980s, an error was discovered in a cod-population model being used to determine

RECENT HISTORY OF LANDINGS OF COD OFF EASTERN CANADA

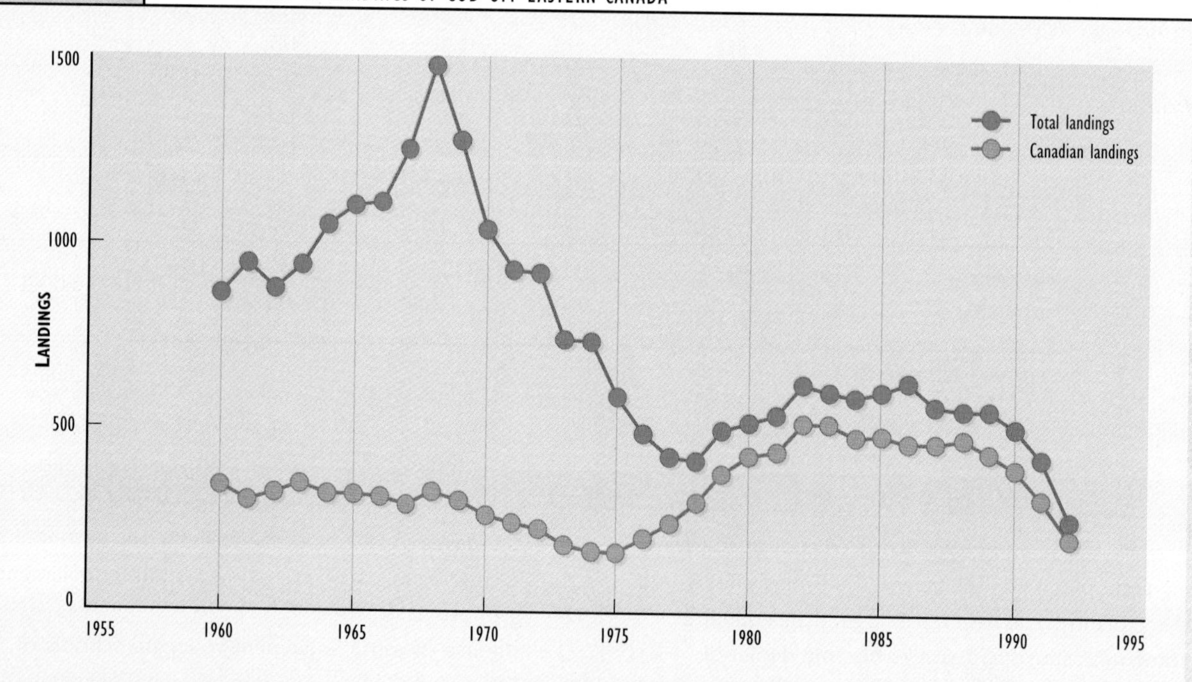

Note the large decrease in landings overall, and the increasing proportion of Canadian landings, after the declaration of the 320-km management zone in 1977. A moratorium on cod fishing was declared in 1992. Data are in thousands of tonnes.

Source: Statistics Canada (1994)

stock size and to set quotas. This error caused overestimations of cod biomass and MSY, resulting in the recommendation of excessive fishing quotas.

Moreover, politicians and decision makers in Canada (and everywhere else) are influenced by various *socioeconomic considerations* in addition to the advice of scientists. Often, these pressures come from individual fishers, their associations, and fish companies. All of these interest groups need cash flows and livelihoods, in a context where there are few alternatives to fishing for employment and revenue generation. These powerful socioeconomic influences can lead to decisions to set quotas that are larger than those recommended by fishery scientists, a factor that has contributed significantly to the mismanagement of the cod stocks and other resources.

Most of the Grand Banks falls within Canada's 320-km management zone. Some parts, however, extend into international waters, where until 1995 there was an unregulated multinational fishery. Because cod and most other marine species are highly mobile and do not recognize the boundaries of management zones, *foreign overfishing* in international waters compromised efforts being made to conserve the cod stocks. However, between 1977 and 1991, Canadians landed about 85% of the cod caught in the northwest Atlantic, and their fishery was more or less regulated.

Humans are not the only predators of marine resources. The *harp seal* (*Phoca groenlandica*) is the most abundant marine mammal in the northwest Atlantic (4–5 million). The seal population consumes about 1 million t/y of food. However, this seal's prey consists of a wide variety of species, especially crustaceans and small fish such as capelin (*Mallotus villosus*) and Arctic cod (*Boreogadus saida*). Even though the cod stocks collapsed at the same time that the population of harp seals was increasing, the minor role of cod in their diet makes it unlikely that these seals were an important cause of the demise of cod.

Before the stocks of cod were heavily exploited, individual fish were much larger than they are today. Immense codfish, such as these "mother cod," are now exceedingly rare, which is unfortunate because they have much greater spawning capacity than the smaller cod that occur today. This photo, from Battle Harbour, Labrador, was taken in the 1890s.

Source: National Archives of Canada

Finally, some people believe that the recruitment of cod may have been impaired by *environmental changes*, including several years of cold surface-water temperatures in parts of the Northwest Atlantic. However, there is no direct evidence to support this environment-related hypothesis.

The simplest and most compelling hypothesis offered to explain the collapse of cod stocks is this: the valuable resource was exploited at an intensity that exceeded its capability for renewal. That is, the cod stocks off eastern Canada, one of the world's greatest potentially renewable resources, were fished to *commercial extinction*.

pre-petroleum times was an extremely valuable commodity, used as a fuel in lamps and for cooking. In a few cases, species of marine mammals, including Steller's sea cow, the Caribbean monk seal, and the Atlantic population of grey whales (see Chapter 24), became extinct from overhunting. In addition, populations of other species of marine mammals became endangered by overharvesting. The best known commercial hunts of marine mammals involve the great whales of all oceans of the world (see below), and the harp seal of eastern Canada (see Canadian Focus 14.3).

Canadian Focus 14.3

Hunting Harp Seals in Eastern Canada

Because seals breed on land or on sea ice in dense populations, they can be easily killed with clubs, lances, guns, or nets. Consequently, over the years, huge numbers of seals have been commercially slaughtered for their skins, which can be manufactured into fur or leather; for their blubber, which can be rendered into oil; and for their meat, teeth, and other valuable products.

Until the middle of this century, seal hunting was an unregulated, open-access enterprise that depleted the populations of most species. There were numerous regional and local extirpations, and several species with initially small populations were rendered extinct (see Chapter 24). More recently, effective conservation measures have protected most seal populations. This has allowed some of the most severely depleted stocks, such as the northern elephant seal (*Mirounga angustirostris*) of the western United States and Mexico, and the northern fur seal (*Callorhinus ursinus*) of the Bering Sea, to increase in abundance.

One of the most famous commercial seal hunts focused on the harp seal (*Phoca groenlandica*), an abundant species that lives in the northern Atlantic Ocean. The seals breed on drifting pack ice in the Gulf of St. Lawrence and around Newfoundland and Labrador, and summer in the eastern Arctic.

Harp seals are especially vulnerable to hunters in March and April. During that time, large numbers of recently born pups, called "white-coats" because of the colour of their birth fur, lie about on the pack ice. Since they are not yet aquatic, the pups can easily be approached and clubbed to death. Adult animals are also concentrated during that time, and can be caught in nets, shot on the ice, or clubbed if they try to defend their young. The skins of harp seals are a valuable commodity, and many people enjoy eating their meat.

The largest hunts of harp seals were mounted by Newfoundlanders, mostly working from ships that sailed from St. John's. The number of seals harvested in any year largely depended on ice conditions, which determined how close sealers could get to the denser aggregations of breeding animals. During the heyday of this enterprise, more than 600 000 animals could be killed per year, as happened in 1831, 1840, 1843, and 1844 (Busch, 1985). Overall, between 1800 and 1914, about 21 million harp seals were harvested, mostly by sealers from Newfoundland. This huge slaughter of a large wild animal has only a few scattered parallels — among them the killing of American bison during the 19th century (Chapter 24), the modern hunt of kangaroos in Australia, and the hunt of deer species in the Americas.

In the 20th century, about 12 million animals were harvested between 1915 and 1982, as many as 380 000 of them in one year (in 1956; Busch, 1985). Since 1983, however, the harvests of harp seals have been smaller, mainly because of poor markets for seal products, and because of controversies about both conservation and the ethics of a commercial harvest of the babies of wild animals. Traditionally, the most profitable market for harp seal fur was in Europe, but this was greatly reduced in 1984 when the European Economic Community (EEC; now European Union, or EU) legislated a ban on the importing of whitecoat pelts. The loss of this market resulted in a temporary reduction in the harvesting of harp seals in Canada, from 190 000 animals during 1981–1982 to 19 000–80 000 during 1983–1990. (Young harp seals are not called whitecoats after about 9–10 days, when they begin to shed their white fur. Older young can still be killed and legally exported to the European Union. The most lucrative market was, however, for whitecoat pelts.)

In recent decades, some elements of the popular media, along with some animal rights and conservation activists, have engaged in sensationalized reporting of the hunting of harp seals in Canada. Consequently, many people in North America, Europe, and elsewhere developed an image of seal hunting in Atlantic Canada as a cruel and barbaric enterprise. In part, this happened because baby seals, which are extremely attractive animals, were the prime objects of the hunt. The young seals are typically killed by clubbing, which is a humane method of slaughter because their thin skulls are easily crushed

by a heavy, well-aimed blow with a truncheon or gaffe. Unfortunately, although most sealers are competent in this method of slaughter, some are not. Videos of seal hunting have shown that during the rush to kill whitecoat harp seals, some animals might only be stunned by an inadequate clubbing, and then be skinned while still apparently "alive" (or at least still twitching — the animals may be brain-dead). Video scenes such as these are extremely upsetting to most viewers, and have been widely publicized by well-organized opponents of the Canadian hunt of harp seals.

Many people, however, do not agree with the portrayal of the seal hunt as being unusually "cruel and brutal." These people contend that the commercial harvesting of wild seals is no more brutal than the slaughter of domestic livestock. (Hundreds of millions of animals are slaughtered each year in Canada, often under cruel conditions, to provide meat and other products for human use.) Consider, for example, this statement by W.T. Grenfell, a missionary and doctor in Newfoundland and Labrador during the late 19th and early 20th centuries, who observed a harvest of harp seals on the pack ice in 1896 (from Busch, 1985):

> Now the killing of young seals has been frequently described as brutal and brutalizing, and the seal hunters depicted as inconceivable savages, and this not only by shrieking faddists or afternoon tea drinkers. But, to my mind, the work is not nearly so brutalizing as the ordinary killing of sheep, pigs, or oxen, [brought] terror-stricken to the shambles [slaughter house],...already reeking with their fellows' blood. Here [in the seal hunt] the animal is too young to feel fear, and evinces no sign of it; no animal is wounded and left to die in vain.

Although Grenfell was commenting on a 19th-century controversy, his words are still relevant to the present debate about commercial harvests of wild animals. The respective slaughters of seals and livestock may in many respects be judged equal in cruelty. An analysis of the ethics of killing animals should also, however, recognize that seals are wild creatures while livestock are specifically bred, raised, and killed for consumption by humans. It is up to philosophers, and to individual consumers of meat and other animal products, to determine which of these slaughters of animals for food, if either, is the greater moral outrage.

The harp seal is an abundant marine mammal that breeds on pack ice off Newfoundland and in the Gulf of Saint Lawrence. "Whitecoat" harp seals, such as the one pictured here, were once hunted for their fur. Harp seals are still harvested in large numbers, but the hunt is now restricted to adults and young that have moulted their white birth coat.

Source: S. Iverson

Although the intense hunting of harp seals in eastern Canada caused this species to decrease in abundance, its populations were never depleted to the point of biological or commercial endangerment. This did not result from any particular conservation ethic on the part of the sealers or their industry. In fact, sealers typically killed as many as possible of any harp seals that were encountered, particularly before 1970. (In that year, the Canadian government began to regulate the hunt through a quota system.) However, the difficulties of hunting in the treacherous pack ice limited the numbers of seals that could be found and killed, and so prevented a severe depletion of the harp seal population.

When the commercial hunt was reduced in the late 1980s, the global abundance of harp seals was about 3 million animals, including about 2 million in Canadian waters. The harp seal was thus among the world's most populous species of large wild animals. In 1997, its abundance in Canadian waters was about 4–5 million.

In fact, the rapidly increasing harp seal populations are alarming some people, who are concerned that the seals are "eating too many fish" (although there is little evidence to support this idea; see Canadian Focus 14.2). In any event, harp seals are again being harvested in large numbers. This harvest culls the populations, while also providing economic benefits through the sale of meat, hides, and penises (the latter being sold to a recently emerged market in eastern Asia). In 1996, about 280 000 harp seals were harvested in eastern Canada, including adults, subadults, and recently moulted young seals, rather than whitecoats.

Whaling: Mining Marine Megamammals

Humans have been hunting whales for centuries. The first species of whale to be commercially hunted was the northern right whale (*Balaena glacialis*), considered the "right" whale to kill because it swims slowly and floats when dead. Early records tell of hunts in the Bay of Biscay. Men would row or sail near a right whale, harpoon it, allow it to tow their boat until exhausted, and then repeatedly lance the animal until it bled to death. The carcass was towed to shore and butchered, and its blubber boiled and rendered into valuable oil. This crude hunt eliminated right whales from European waters.

The development of steam ships made it possible to hunt swifter whales, such as the blue, fin, sei, and minke whales. The invention of the harpoon gun in 1873, and later the exploding-head harpoon, made it easy to kill even the largest whales. By 1925, giant factory ships could spend months or years in remote waters, processing whales killed by a small fleet of catcher-killer boats, sometimes guided to their prey by spotter aircraft. Whales of all species and sizes could be efficiently located, killed, and processed. This onslaught resulted in a rapid, and profitable, depletion of whale stocks.

With only a few exceptions, whale populations were not overharvested to extirpation, but only to **commercial extinction**, a small population that was no longer profitable to find and kill. The sequential exploitation of a whale community is best illustrated by the hunt in Antarctic waters, where six species initially co-existed in great abundance (Figure 14.3).

In response to concerns about declining populations of whales, the International Whaling Commission (IWC) was established in 1949. The IWC was given a mandate to develop and implement conservation-related controls over the multinational, fiercely competitive, highly capitalized, and profitable whaling industry. Unfortunately, the initial efforts of the IWC were not successful, mainly because it was extremely difficult to estimate whale stock sizes and recruitment rates, and so to determine accurate maximum sustainable yields. In addition, the whaling nations were not very cooperative, and the IWC was not aggressive in setting and enforcing quotas small enough to ensure that whale populations would not be depleted. These problems are to be expected whenever a for-profit enterprise is allowed to regulate and police itself. According to J.L.McHugh, a former commissioner and chairperson of the IWC, "From the time of the first meeting of the Commission...almost all major actions or failures to act were governed by short-range economic considerations rather than by the requirements of conservation" (cited in Ellis, 1991).

Because of its enormous size, reaching 32 m and 136 t for the largest male animals, the blue whale (*Balaenoptera musculus*) was initially the most profitable species in the Antarctic seas. The original population was about 180 000 blue whales, and as many as 29 000 animals were killed in a single year (1931; Figure 14.3; note that during the Second World War, as few as 59 animals were harvested in a year). Between 1955 and 1962, declining stocks of blue whales meant an annual harvest of only 1000–2000. After 1965, killing this species was prohibited by the IWC. In total, about 331 000 blue whales were killed in Antarctic waters between 1920 and 1965. The present Antarctic population of blue whales is fewer than 2000, only about 1%

FIGURE 14.3 | WHALING IN ANTARCTIC WATERS

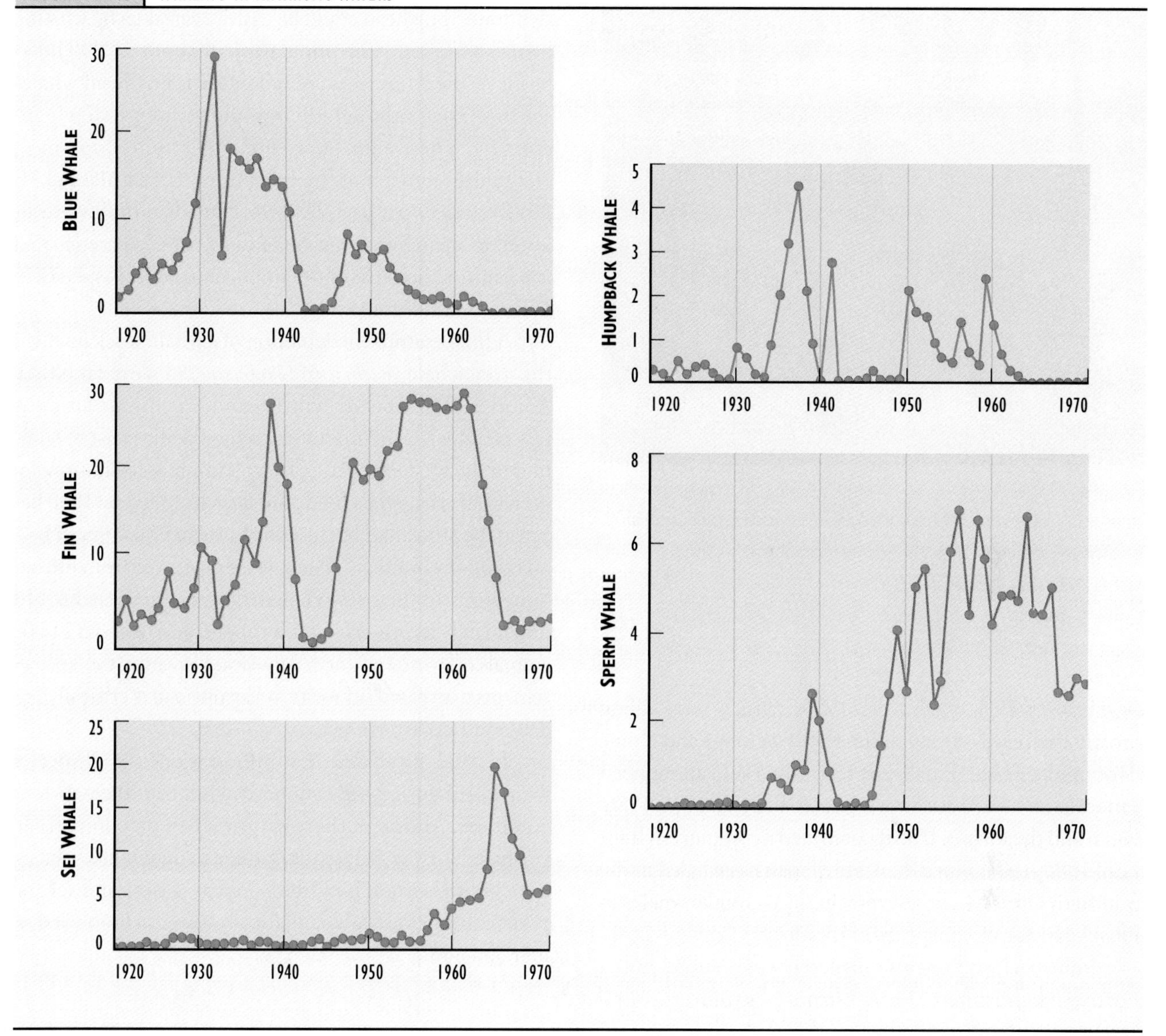

Annual catches in thousands of whales.

Source: Ellis (1991)

of their initial abundance. The global population is less than 10 000 individuals.

As blue whales became depleted, the fin whale (*B. physalus*) became the favoured species of the Antarctic hunt. This is the second-largest species, up to 21 m long. As many as 29 000 animals were harvested in a year, causing this species to decline, though not to commercial extinction. More than 704 000 fin whales were killed in this region. The present population is less than 85 000 animals, about 21% of the original abundance.

As the largest species became difficult to harvest because of their increasing rarity, initially "less desirable" species were hunted. These were the sei whale (*B. borealis*), humpback whale (*Megaptera novaeangliae*), sperm whale (*Physeter macrocephalus*), and minke whale (*B. acutorostrata*). These smaller species were also overharvested, and their populations declined markedly (Figure 14.3).

Toward the end of the hunt in the Antarctic Ocean, the population of blue whales had been reduced by about 99%, humpback whales by 97%, sei whales by 82%, and

About 250 000 white-tailed and mule deer (*Odocoileus virginianus* and *O. hemionus* respectively) are harvested by hunters each year in Canada. This mule deer was photographed in Jasper National Park, Alberta.

Source: B. Freedman

fin whales by 79%. By the early 1980s, whalers were killing mostly the relatively small (up to 9.1 m long) and abundant minke whale. Finally, in 1982, the IWC announced a moratorium on Antarctic whaling, to begin in 1985–1986. Japan and the former USSR continued a commercial hunt until 1986–1987. Since then, only Japan has whaled in the Southern Ocean, killing several hundred minke whales in most years for the purposes of "research."

Industrial whaling also depleted whale stocks in the northern hemisphere. Early European explorers found large populations of northern right whales in waters off Atlantic Canada, and these valuable animals were soon hunted by Basque whalers. The Basque hunt of 1530–1610 killed about 25 000–40 000 right whales, but very few afterward because of the severely depleted stocks. The right whale survives today in the western Atlantic in an endangered population of about a few hundred animals, only 3–4% of its original abundance. Although this species has been protected from hunting for more than 50 years, its abundance is not increasing. This is probably because of mortality caused by collisions with ships and entanglement in fishing gear.

Soon after right whales were depleted off eastern North America, northern populations of bowhead whales (*Balaena mysticetus*) were discovered in arctic Canada, Alaska, and eastern Siberia. Like right whales, the slow-swimming bowhead could be easily overtaken by whaling boats and killed. The population of about 55 000 bowheads in the Arctic was soon depleted. Bowhead whales are now rare, although their populations are slowly increasing. These animals are no longer hunted commercially, although a hunt by Inuit in northern Alaska kills 20–40 animals per year. In 1996, Canadian Inuit were allowed to again hunt a few bowhead whales, a practice that is permitted because of the importance of this species in Inuit food traditions.

A final example of depletion of a whale stock involves the grey whale (*Eschrichtius robustus*) of western North America. This species winters and breeds in warm waters off Mexico and migrates up the Pacific coast to summer in the western Arctic Ocean. Commercial hunting of grey whales began in 1845, and largely ended by 1900 because the stock had been reduced to an endangered several thousand animals. These were protected from further hunting, and the grey whale has since increased to approximately its pre-exploitation abundance of about 21 000 animals off western North America. However, the species remains extirpated off western Europe and is critically endangered in eastern Asia.

In total, more than 2.5 million whales of all species were killed during the commercial hunts of the past four centuries. Although there is now a ban on commercial whaling, Norway and Japan continue to hunt minke whales, each killing several hundred per year. These and several other countries are lobbying for a return to a limited commercial hunt.

Terrestrial Hunting

Many terrestrial animals are also hunted in large numbers, including large mammals such as deer, antelope, bears, pigs, and kangaroos. Many species of birds are also hunted, particularly waterfowl, grouse, pheasants, and shorebirds. Most hunting of wild terrestrial animals is undertaken for subsistence purposes, but sport hunting is also important in some regions.

Hunting is a popular activity in Canada. Many Canadians hunt, whether for subsistence, as a blood sport, or for both reasons. The most commonly hunted large mammals are species in the deer family, but other animals are also taken. The most important big-game species in

Canada (harvested during the 1994 hunt; these are rounded numbers) are:

white-tailed deer (*Odocoileus virginianus*)	208 000 harvested
moose (*Alces alces*)	69 300
mule deer (*Odocoileus hemionus*)	47 600
black bear (*Ursus americanus*)	20 700
caribou (*Rangifer tarandus*)	18 800
elk (*Cervus canadensis*)	7 200
pronghorn antelope (*Antilocapra americanus*)	5 100
mountain sheep (*Ovis canadensis*)	1 200
mountain goat (*Oreamnos americanus*)	1 000
wolf (*Canis lupus*)	600
grizzly bear (*Ursus arctos*)	400
cougar (*Felis concolor*)	400

(These data were compiled from information provided by provincial and territorial governments. However, the data are incomplete in that they are based on reports by hunters, with kills due to poaching (i.e., illegal hunting) not included. In addition, subsistence hunting by many aboriginal Canadians is not reported, or is considered proprietary information. This is particularly true in northern Canada.)

Waterfowl are also hunted in large numbers in Canada. In 1991, about 300 000 people purchased permits to hunt migratory birds in Canada, and about 68% of those people bagged at least one bird (Levesque *et al.*, 1993). In total about 1.64 million ducks were harvested by hunters in Canada in 1991, plus another 665 000 geese. On average, each successful hunter bagged 10.4 ducks and 6.9 geese. The most commonly hunted ducks in Canada in 1991 were:

mallard (*Anas platyrhynchos*)	607 000 harvested
black duck (*Anas rubripes*)	230 000
wood duck (*Aix sponsa*)	131 000
green-winged teal (*Anas crecca*)	115 000
ring-necked duck (*Aythya collaris*)	98 000
lesser scaup (*Aythya affinis*)	65 000
another 22 species	394 000

More than 600 000 geese are harvested by hunters each year in Canada. This snow goose (*Chen caerulescens*) was photographed on Ellesmere Island.

Source: B. Freedman

The most commonly hunted geese are:

Canada goose (*Branta vcanadensis*)	466 000 harvested
snow and blue goose (*Chen caerulescens*)	146 000
white-fronted goose (*Anser albifrons*)	46 000

Less-commonly hunted migratory gamebirds include:

woodcock (*Scolopax minor*)	100 000 harvested
snipe (*Gallinago gallinago*)	45 000
coot (*Fulica americana*)	11 000

A large, essentially unregulated, subsistence and market hunt (the latter being illegal) is made of murres (species of sea birds) in Newfoundland and Labrador. About 750 000 murres are killed annually in this hunt, of which 90% are thick-billed murres (*Uria lomvia*) and 10% common murres (*U. aalge*) (Elliot *et al.*, 1991).

Key Terms

renewable resources
surface waters
groundwater
aquifers
production (and roductivity)
agricultural site capability (site quality)
ecological conversion
deforestation
silvicultural management
plantation
aquaculture
commercial extinction

Questions for Discussion

1. Explain what is meant by a renewable natural resource, and illustrate the principle by referring to one of the following: surface waters and groundwater, agricultural site capability, timber, a hunted wild animal.
2. Identify a potentially renewable natural resource that has been overharvested and depleted in your region. Discuss the reasons for the unsustainable use.
3. What are the most important renewable resources in Canada? Indicate, giving reasons, whether you think those resources are being used in a sustainable manner.
4. Should relatively abundant whale species (such as minke whales) or harp seals be hunted? Your answer should consider whether the species can be harvested in a sustainable manner, and should also address the ethics of hunting.
5. Discuss the political and economic problems of water resources shared between countries or regions.

References

Bolen, E.G. and W.L. Robinson. 1995. *Wildlife Ecology and Management*. Englewood Heights, NJ: Prentice Hall.

Busch, B.C. 1985. *The War Against the Seals: A History of the North American Seal Fishery*. Kingston, ON: McGill-Queen's University Press.

Canadian Council of Forest Ministers. 1995. *Compendium of Canadian Forestry Statistics, 1994*. Ottawa: Canadian Council of Forest Ministers.

Cushing, D.H. 1988. *The Provident Sea*. Cambridge: Cambridge University Press.

Ehrlich, P..R., A.H. Ehrlich, and J.P. Holdren. 1977. *Ecoscience: Population, Resources, Environment*. San Francisco: W.H. Freeman.

Elliot, R.D., B.T. Collins, E.G. Hayakawa, and L. Metras. 1991. The harvest of murres in Newfoundland from 1977 to 1988. In: *Studies of High-Latitude Seabirds*. Ottawa: Canadian Wildlife Service. Occas. Pap. No. 69; pp. 36–44.

Ellis, R. 1991. *Men and Whales*. New York: Knopf.

Freedman, B. 1995. *Environmental Ecology, 2nd ed*. San Diego, CA: Academic.

Hutchings, J.A. and R.A. Myers. 1993. What can be learned from the collapse of a "renewable" resource? *Atlantic cod,* Gadus morhua, *of Newfoundland and Labrador.* St. John's NF: Department of Fisheries and Oceans, Science Branch.

Hutchings, J.A. and R.A. Myers. 1995. The biological collapse of Atlantic cod off Newfoundland and Labrador: An exploration of historical changes in exploitation, harvesting technology, and management. In: *The North Atlantic Fisheries: Successes, Failures, and Challenges*. Charlottetown, PE: Institute of Island Studies. pp. 39–93.

Levesque, H., B. Collins, and A.M. Legris. 1993. *Migratory game birds harvested in Canada during the 1991 hunting season*. Ottawa: Canadian Wildlife Service. Progress Note 204.

Miller, G.T. 1990. *Resource Conservation and Management*. Belmont, CA: Wadsworth.

Mowat, F. 1984. *Sea of Slaughter*. Toronto: McClelland & Stewart.

National Marine Fisheries Service (NMFS). 1991. *Endangered Whales: Status Update*. Silver Spring, MD: U.S. Dept. of Commerce, National Oceanic and Atmospheric Administration, NMFS.

Natural Resources Canada. 1996. *Selected Forestry Statistics, Canada, 1995*. Ottawa: Natural Resources Canada, Canadian Forestry Service; Industry, Economics, and Programs Branch.

Owen, O.S. and D.D. Chiras. 1995. *Natural Resource Conservation: Management for a Sustainable Future*. Englewood Heights, NJ: Prentice Hall.

Statistics Canada. 1992. *Agricultural Profile of Canada*. Ottawa: Statistics Canada.

Statistics Canada. 1994. *Human Activity and the Environment 1994*. Ottawa: Statistics Canada.

Vitousek, P.M., P.R. Ehrlich, A.H. Ehrlich, and P.A. Matson. 1986. Human appropriation of the products of photosynthesis. *Bioscience*, **36**: 368–373.

World Resources Institute (WRI). 1995. *World Resources 1994–95: A Guide to the Global Environment*. New York: Oxford University Press.

CBC CANADIAN CASE 3

HYDROELECTRIC POWER: ENVIRONMENTAL FRIEND OR FOE?

Because hydroelectric power is generated using flowing water, it is commonly regarded as a renewable source of energy. The water flows are regenerated through environmental services provided by the hydrologic cycle. The renewability of hydroelectric power makes it a desirable option when compared with nonrenewable energy sources such as fossil fuels and nuclear power.

It is important to recognize, however, that the generation of hydroelectric power can result in significant environmental damage. Enormous areas of land may be flooded to develop reservoirs, the natural flow of rivers may be changed, mercury may be released into the environment, and people and/or their livelihoods may be displaced. These actions can cause conflict between hydro companies and the people affected by the environmental changes.

Media reports of the plan to develop additional hydroelectric generating capacity in northern Quebec illustrate the controversies that can arise. The developers planned to dam and divert several large rivers running into James Bay, which would create enormous reservoirs to store water until it was needed to generate electricity.

Such industrial megaprojects are considered important to the economic development of Quebec. Huge quantities of relatively inexpensive power would be provided to fuel industrial activities in the province, while also generating revenue through sales of energy to other places, especially New York State. The enormous hydroelectric development would also, however, disrupt aboriginal communities living in the affected areas and cause important ecological damage.

Here as with all megaprojects and development policies, the benefits of large-scale industrial activities must be weighed against the inevitable socioeconomic and ecological damages.

Questions for Discussion

1. Do you think that governments should be allowed to force people out of their traditional land-use areas, in order to undertake development projects that may achieve a larger social benefit?
2. List the ecological damages that could result from the proposed hydroelectric developments in the James Bay region. Which of these damages are relatively certain to occur, and which are more difficult to predict?
3. What are the benefits of hydroelectric power compared with other sources of energy? To what degree do you think these benefits are offset by environmental damage associated with hydroelectric development?

Video Resource: "Power Struggle at James Bay." *News in Review* (September, 1991)

CBC CANADIAN CASE 4

THE SEAL HUNT

Marine mammals, such as seals and whales, have long been hunted as natural resources — as sources of meat, oil, and other useful products. Although marine mammals can potentially be utilized as renewable resources, until recently most species were overhunted, being harvested at rates that exceeded their rate of regeneration. The populations of some species of seals and whales were endangered by this overexploitation, and several became globally extinct. This is similar to the "mining" of a nonrenewable natural resource.

Extirpation is not, however, the present fate of the harp seal of the northwestern Atlantic Ocean. Although the populations of this species have been somewhat depleted by harvesting, the species was never endangered by the commercial hunts for its fur and meat. Today, in fact, the harp seal is one of the world's most populous large wild animals, numbering more than five million.

The intense controversy about seal hunting in eastern Canada has focused mainly on ethical concerns, rather than on issues associated with resource depletion. The most important economic commodities gained from the hunt of harp seals in Atlantic Canada have been furs, especially those of baby seals known as "whitecoats," and more recently penises, which are valued in traditional medicine in eastern Asia. The meat of hunted seals is also used, but this is not the primary objective of the modern commercial hunt — the pelts and penises are.

Although the products of the harp-seal harvest are valuable, they are not particularly crucial to the workings of modern economies. In fact, many people believe that wild animals should not be subjected to commercial hunts for relatively frivolous commodities such as furs, ivory, horns, or penises. Moreover, many people are outraged by the brutality of seal hunting, and by the fact that baby seals are killed in large numbers.

In contrast, proponents of the seal hunt in eastern Canada point out that they are attempting to engage in a sustainable exploitation of a renewable natural resource. They also contend that seals eat a great deal of fish (some of which could be caught and eaten by people), and that the slaughter of domestic animals in agriculture is also a brutal enterprise. In view of these considerations, why, they ask, pick on the sealers?

Questions for Discussion

1. Do you think that seals represent a renewable natural resource that can be commercially exploited in a sustainable fashion? What about other marine mammals, such as whales? Or terrestrial animals, such as deer and furbearing mammals?
2. Many people believe that the cultivation and slaughter of domestic animals in modern, industrial agriculture is a brutal and, from the biocentric perspective, dubiously ethical enterprise. What do you think about this issue? Is it appropriate to justify the hunting of wild animals on the basis of the methods that we use in industrial agriculture?
3. Do you think that it is ethical to kill wild animals to obtain non-essential materials, such as furs, ivory, or horns? Make a list of some plausible reasons for and against this kind of resource harvesting.

Video Resource: "Seal Hunt." *News in Review* (April, 1996)

15 POLLUTION AND DISTURBANCE AS ENVIRONMENTAL STRESSORS

Chapter Outline

Environmental stressors
Contamination and pollution
Disturbance
Ecotoxicology
Environmental risks

Chapter Objectives

After completing this chapter, you will be able to:

1. Describe the nature and causes of environmental stress and explain how ecosystems respond to changes in its intensity.
2. Compare and contrast contamination and pollution.
3. Identify examples of naturally occurring pollution and disturbance, and discuss how knowledge of these phenomena can contribute to understanding the effects of anthropogenic stressors.
4. Outline the differences between toxicology and ecotoxicology.
5. Compare voluntary and involuntary risks.
6. Identify the typical steps in a risk assessment of a predicted exposure to a toxic chemical.

Environmental Stressors

Environmental stressors are factors that can constrain productivity, reproductive success, and/or ecological development (see Chapter 9). To some degree, environmental stressors affect all organisms as well as their populations, communities, and ecosystems. Many stressors are natural in origin, being associated with such environmental factors as:

- competition, predation, disease, and other interactions among organisms
- constraints related to climate or to inadequate or excessive nutrients, moisture, or space
- disturbances such as wildfire and windstorms

The effects of natural stressors are not always negative. Some individuals, populations, and communities may benefit from the ecological effects of natural stress, even while others suffer.

Increasingly, however, stressors associated with human activities are critically influencing Earth's species and ecosystems. In too many cases these anthropogenic stressors are causing important damage to resources needed to sustain humans and their societies, and to Earth's biodiversity and ecosystems more generally.

Wildfire, windstorms, and insect outbreaks can result in extensive disturbances affecting entire communities. This photo shows an area of maple-birch forest in the Muskoka region of Ontario that has been completely defoliated by a natural epidemic of forest tent caterpillar (*Malacosoma disstria*). An area of non-affected pine forest (darker colour) occurs beside the defoliated forest.

Source: B. Freedman

Environmental stressors can occur as intense, short-lived events, also known as **disturbances**. Alternatively, stressors may exert their influence over an extended period of time, that is, in a *chronic* manner. The interaction of organisms with an environmental stressor at a particular place and time is called **exposure**. Exposure can be an instantaneous factor or it may accumulate over time. If an exposure is intense enough, it will cause some sort of biological or ecological change, called a **response**. It is important to understand, however, that individuals, populations, communities, and ecosystems are capable of tolerating a range of intensities of environmental stressors without suffering any significant damage. In other words, *thresholds* of biological or ecological *tolerance* must be exceeded before damage is caused.

Environmental damage occurs when stressors elicit responses that result in a degradation of environmental quality. Such responses include, for example, illness or death caused by the presence of pesticide residues in wild animals, reduced productivity of ecosystems, or the endangerment of elements of biodiversity. In this chapter we examine a conceptual framework for the study of damage caused by environmental stressors. In the following nine chapters, we will deal with specific stressors, and examine many examples and case studies of environmental and ecological damages.

Kinds of Environmental Stressors

The diverse kinds of environmental stressors can be grouped into several classes, although the classes are not all mutually exclusive.

Physical stress is a type of disturbance that involves an intense exposure to kinetic energy, which physically damages habitats and ecosystems. Physical stress includes such disruptive events as a hurricane or tornado, a seismic sea wave (or tsunami), the blast of a volcanic eruption, an explosion, or trampling by heavy machinery or hikers.

Wildfire is another disturbance, involving the uncontrolled combustion of the biomass of an ecosystem. Wildfires can be ignited by humans or naturally by lightning or lava flows. Severe fires consume much of the biomass of an ecosystem, particularly of trees; but even less-severe wildfires can kill many organisms by scorching and toxic-gas poisoning.

Chemical pollution occurs when one or more substances occur in concentrations high enough to elicit physiological responses by organisms, potentially resulting in ecological changes. Chemical stressors include pesticides, gases such as sulphur dioxide and ozone, and toxic elements such as mercury, lead, and arsenic. Pollution can also result from excessive nutrients, which can distort productivity and other ecological functions. Note that the presence of a potentially toxic agent in the environment does not necessarily cause pollution. (We examine the distinction between contamination and pollution later in this chapter.)

Thermal pollution is caused by releases of heat (thermal energy) into the environment, resulting in ecological changes because species vary in their tolerance of temperature extremes. Thermal stress may occur at natural springs and submarine vents where geologically heated water is emitted. It is also associated with discharges of hot water from power plants and other industrial facilities.

Radiation stress is caused by excessive exposure to ionizing energy. The radiation can be what is emitted by nuclear wastes or explosions, or it can be solar ultraviolet radiation.

Climatic stress is associated with insufficient or excessive regimes of temperature, moisture, solar radiation, wind, or combinations of these.

Biological stressors are associated with interactions among organisms, such as competition, herbivory, predation, parasitism, and disease (see Chapter 9). For example, individuals of the same or different species may compete for essential resources that are limited in supply. Herbivory, predation, parasitism, and disease are *trophic* interactions, in which one species exploits another. Exploitation can be anthropogenic, as when wild animals or trees are harvested, or it can be natural, perhaps associated with defoliating insects or disease-causing pathogens.

Biological pollution can occur when humans change ecological conditions by releasing organisms beyond their natural ranges. This might involve intentional or unintentional introductions of non-native species that invade and alter natural ecosystems, or the release of pathogens into the environment through discharges of raw sewage.

"Biological pollution" is caused when humans introduce species beyond their natural range, where they may cause ecological damage. Horses (*Equus caballus*) were introduced to Sable Island several centuries ago to provide food for shipwrecked mariners on that remote island off Nova Scotia. The horses continue to maintain a population of several hundred animals, and may have contributed to the extinction of several taxa of plants that previously grew only on Sable Island.

Source: B. Freedman

Ecological Responses to Changes in Environmental Stress

An ecosystem that is disrupted by a disturbance (i.e., an intense, short-lived event) typically suffers mortality among its species, along with various damages to its structural (e.g., species composition, biomass distribution) and functional properties (e.g., decomposition and nutrient cycling). Once the disturbance is over, a process of recovery through succession begins. If succession proceeds for a long enough time, it will eventually restore another mature ecosystem, perhaps one similar to the ecosystem existing before the disturbance.

Chronic stressors, which operate over time, include ionizing radiation and many types of chemical and thermal pollution. Depending on the intensity of exposure to the chronic stressor, organisms may suffer acute toxicity resulting in tissue damage or death, or chronic toxicity resulting in decreased productivity.

Increased exposures to environmental stressors can result in evolutionary changes within populations, especially if individual organisms vary in their tolerance to the stressors and those differences are genetically based. Under such conditions, natural selection for tolerant individuals will eventually result in increased tolerance at the population level. At the community level, relatively vulnerable species will be reduced in abundance or eliminated from sites if the intensity of stress increases markedly. The niches of those species may then be occupied by more tolerant members of the community, or by invading species that are capable of exploiting a stressful but weakly competitive habitat.

Long-term ecological change will result from a prolonged intensification of stress. Consider, for example, a new smelter which causes chronic pollution in a forested landscape. The intense toxic stresses can damage the tree-sized plants of the original forest, and may eventually make them give way to shrub-sized and herbaceous vegetation. If the long-term toxic stress is extremely severe, the landscape can lose its vegetation almost entirely. This type of ecological damage has actually occurred around some Canadian smelters, such as those near Sudbury, Ontario (Chapter 16).

These kinds of ecological damage involve changes in the composition and dominance of species in communities, in the spatial distribution of biomass, and in functions such as productivity, litter decomposition, and nutrient cycling. Because a smelter is a discrete, point source of environmental stress, the ecological responses will eventually stabilize as *gradients* of community change that radiate outward, in a downstream or downwind direction, from the source of pollution.

Occasionally, the intensity of environmental stress decreases in time and space. When this happens, the ecological responses are, in many respects, the reverse of the degradation seen when stress intensifies. These changes represent a process of recovery through succession. In the case of the Sudbury smelters, emissions of pollutants have decreased since the installation of pollution-control technology. This resulted in decreased toxic stress in the surrounding environment, which allowed some ecological recovery to occur (Chapters 16 and 18).

Ecologists have described the general attributes of ecosystems that have been subjected to severe stress for different periods of time. As environmental stress intensifies significantly (for example, by increasing pollution), the following changes are commonly observed:

1. Rates of mortality increase.
2. Species richness is decreased.
3. Nutrients and biomass are depleted.
4. Rates of community respiration exceed production, so net production becomes negative.
5. Sensitive species become replaced by more tolerant ones.
6. Top predators and large-bodied species are lost from the ecosystem.

Ecosystems that are chronically subject to intense stress (for example, climate-stressed tundra) eventually stabilize. Typically, the stable ecosystems are low in species richness, simple in structure and function, and dominated by relatively small, long-lived species. As well, they have low rates of productivity, decomposition, and nutrient cycling.

Contamination and Pollution

Pollution is caused by exposure to chemicals or energy that exceeds the tolerance of organisms. As such, pollution is judged to occur when toxicity to organisms, or other ecological damage, can be demonstrated. Pollution can affect humans and other species as well as communities and ecosystems.

Pollution is often caused by the exposure to toxic chemicals in concentrations large enough to poison at least some organisms. Pollution can also be caused by non-toxic exposures to environmental stressors — for example, the excessive fertilization of water (which causes eutrophication; see Chapter 20), the release of waste heat into the environment, and the discharge of raw sewage containing pathogenic microorganisms.

Contamination refers to those much more common situations in which potentially damaging stressors are present in the environment, but their intensities are too low to cause detrimental effects. For example, a certain chemical may occur in a higher concentration than normally encountered in the environment. However, if its concentration is too low to cause measurable toxicity to organisms, or to affect other ecological components or processes, the chemical is contaminating, not polluting.

For example, metals such as aluminum, cadmium, copper, lead, mercury, nickel, and zinc are present in all parts of the environment, including all organisms, in at least trace concentrations. If the detection limits of the available analytical chemistry are sensitive enough, this universal contamination by metals can be demonstrated. Potentially, all metals are toxic. However, metals (and any other chemicals) must be present in a high enough concentration for a long enough period of time to actually poison organisms and cause ecological damage. In other words, the exposure must exceed biological tolerances before damage is caused, and pollution can be said to occur.

Pollution and contamination are often judged from a human-focused bias. People decide whether pollution is causing damage at some place and time, and how important the damage might be. This anthropocentric bias tends, quite naturally, to favour humans themselves. It also favours those species, communities, and ecological functions that are recognized as being required to support human welfare, or that may be appreciated for other reasons, such as aesthetics.

Interestingly, some species, communities, and ecological processes actually benefit from many types of pollution. For example, certain species may take advantage of ecological opportunities made available when pollution reduces the numbers of a previously dominant species. Many of the case studies described in following chapters involve situations in which *opportunistic* species of plants, animals, and microorganisms have benefitted from ecological changes caused by pollution.

Pollution Can Be a Natural Phenomenon

Pollution is not only caused by human activities — in some cases it is a wholly natural phenomenon. Natural sources of pollution include emissions of particulates and gases, such as sulphur dioxide, from volcanoes; oil seeps on the ocean floor; metal concentrations in some soils and rocks; and the heat of geothermal springs. "Natural" pollution can cause severe ecological changes (which humans may view as damages). These can be as intense as those resulting from anthropogenic pollution, although natural pollution damage is usually more localized. The fact of natural pollution is interesting and well recognized, but it does not justify human activities that cause similar kinds of pollution.

Studies of the ecological effects of natural pollution can yield useful insights into the potential long-term effects of anthropogenic pollution. There are two reasons for this usefulness: many examples of natural pollution are ancient, and their patterns of ecological damage may be similar to those caused by anthropogenic pollution.

One interesting example occurs at the Smoking Hills in the Northwest Territories. This is a remote wilderness, little influenced by humans. At several places along the seacoast, erosion has exposed deposits of bituminous shale. These carbon-rich deposits have spontaneously ignited and have been smouldering for centuries, fumigating the nearby tundra with sulphur dioxide. The SO_2 is toxic to vegetation and causes severe acidification of soil and water. The natural pollution at the Smoking Hills has severely damaged both terrestrial and aquatic communities of the tundra (see Chapter 16).

Another kind of natural pollution occurs when metal-rich minerals occur close to the ground surface, resulting in toxic conditions for vegetation. For example, plant ecologists have studied soils containing "serpentine" minerals, which are rich in nickel and cobalt. In high concentrations, these metals are toxic to most plants. Habitats containing serpentine minerals develop a distinctive plant community, dominated by low-growing species that can tolerate the toxic stresses of the metal-rich soils (see Chapter 18).

An additional case of natural pollution involves certain species of marine phytoplankton that occasionally become abundant and cause ecological damage. In events called "toxic blooms," these algae release biochemicals that are poisonous to a broad range of animals exposed to them through the food web. In 1985, 14 humpback whales died at sea off Massachusetts after feeding on fish polluted with saxitoxin, a potent neurotoxin synthesized by dinoflagellate algae. Likewise, in 1987, some Canadians became ill after eating mussels containing domoic acid, which had originated in diatoms in Atlantic coastal waters.

Research and discussion of naturally occurring pollution is useful and informative in environmental science. However, in this book we emphasize pollution caused by human activities and its resulting ecological damage. This emphasis is sensible, because anthropogenic pollution is increasing rapidly in many countries, including Canada. There is a pressing need to avoid or manage the damage that affects both people and natural ecosystems.

Pollution Is Also Caused by Humans

In the modern world, an enormous amount of pollution is associated with human activities. This has caused important damage to ecosystems and sometimes to human health. Humans cause pollution in diverse ways, and we examine these in detail in following chapters of this book. Most commonly, anthropogenic pollution is associated with the following activities:

Natural disturbances such as wildfire initiate a process of ecological recovery known as succession. This photo shows a seven-year-old burn of boreal forest near Inuvik in the Northwest Territories. The community at this early stage of succession is dominated by a herbaceous plant known as fireweed (*Epilobium angustifolium*).

Source: B. Freedman

- accidental or deliberate emissions of chemicals, such as sulphur dioxide, metals, pesticides, and petroleum, into the environment
- releases of substances which become synthesized, as "secondary pollution," into chemicals of greater potential toxicity (as when ozone is synthesized by photochemical reactions in the atmosphere)
- emissions of chemicals that degrade stratospheric ozone, such as chlorofluorocarbons
- releases of waste industrial heat, as when a power plant discharges hot water into a river or lake
- discharges of nutrient-laden sewage or fertilizers into waterbodies

Disturbance

A disturbance is an episodic but intense influence that may cause severe biological and ecological damage (see also Chapter 9). Events of disturbance are followed by a sometimes protracted period of ecological recovery through the process known as succession. There are two broad types of disturbances, **community-replacing disturbances** and **microdisturbances**.

Community-replacing disturbances are relatively extensive in scale, resulting in a catastrophic destruction of the original community. Natural examples of community-replacing disturbances include wildfires, severe windstorms, avalanches, and glaciation, while anthropogenic ones include clear-cutting, ploughing, and conversions of natural ecosystems into agroecosystems. A community-replacing disturbance is followed by successional recovery that may eventually regenerate a community similar to what was destroyed. Younger communities in the successional sequence (or *sere*) have relatively dynamic rates of change in their structural and functional properties. Younger seres are typically dominated by species that are abundant only during the initial period of recovery, when competition is not intense. Community change in later seres is much less dynamic.

Microdisturbance involves local disruptions that affect small areas within an otherwise intact community. Anthropogenic microdisturbances include the selective harvesting of individual large trees or animals, while leaving the community otherwise intact. Ecological changes are relatively rapid within the habitat patch affected by a recent microdisturbance, but at the stand level the community is stable. So-called patch- or gap-phase successional dynamics occur in all natural forests, but are particularly important during later stages of succession. This is particularly the case in older-growth forests, where individual trees might die from disease, insect attack, or a lightning strike, creating a gap in the canopy.

Disturbance Is a Common, Natural Phenomenon

Disturbance is a natural force that affects all ecosystems. Examples of catastrophic natural disturbances include wildfires, which may kill mature trees over large areas, and are followed by regeneration through secondary succession. Fire is common in boreal forests and in seasonally drought-prone ecosystems, such as prairie and savanna. About 3 million ha of forest burns each year in Canada, mostly in fires started by lightning. Wildfire transforms habitat conditions and causes pollution through emissions of particulates, and of gases such as carbon dioxide and oxides of nitrogen, to the atmosphere.

Hurricanes and tornadoes, flooding, and glaciation (over geologically long periods of time) also cause extensive ecological damage, which is followed by successional recovery. After glaciation, which involves prolonged burial and abrasion of the land by an enormous mass of ice, the postglacial recovery requires recolonization of the landscape by organisms (i.e., primary succession).

Volcanic eruptions and earthquakes can generate devastating oceanic waves or tsunamis. In 1883, the cataclysmic eruption of Krakatau in Indonesia initiated a 30-m–high sea wave that killed about 36 000 people.

The blast and heat of a volcanic eruption can also damage ecosystems, as occurred in 1980 when Mount St. Helens in Washington erupted more or less sideways. The blast blew down 21 000 ha of conifer forest, killed another 10 000 ha by heat injury, and otherwise damaged an additional 30 000 ha. Devastating mudslides also occurred, and a huge area was covered in particulate ejecta (known as tephra) that settled from the atmosphere up to 50 cm in depth.

Volcanic eruptions can also emit huge quantities of sulphur dioxide, particulates, and other pollutants into the atmosphere. Quantities of SO_2 amounting to about 2–5 million t (expressed as sulphur, or SO_2-S), are typically emitted by volcanoes each year, and an individual eruption can emit more than 1 million t. This natural SO_2 contributes to the acidification of precipitation and to other environmental damage (Chapter 19).

Natural population outbreaks (or irruptions) of herbivores, predators, or pathogens can result in intense damage to natural habitats. For instance, spruce budworm (*Choristoneura fumiferana*) periodically defoliates huge areas of conifer forest in eastern Canada (more than 55 million ha in 1975). This causes extensive mortality of fir and spruce trees and other ecological changes (see Chapter 21). A marine example involves the green sea urchin (*Strongylocentrotus droebachiensis*), which occasionally irrupts in rocky, subtidal habitats off Nova Scotia. These invertebrates can overgraze mature "forests" of the kelps *Laminaria* and *Agarum*, resulting in a "barren ground" with much less productivity and biomass. After the population of sea urchins collapses, the kelp forest reestablishes rapidly.

In addition to community-replacing disturbances, microdisturbances also occur in natural ecosystems. Examples of local disturbances include the deaths of individual large trees or small groups of trees within an otherwise intact forest, perhaps caused by disease or an accident (such as a lightning strike). This creates a natural gap in the forest canopy, beneath which a microsuccession occurs as plants compete to take advantage of temporary resource opportunities such as extra light. The gap eventually becomes filled by the foliage of mature trees. Similarly, the deaths of individual coral heads within an otherwise intact coral reef initiate a microsuccession within that stable and biodiverse marine ecosystem.

Ecologists try to understand the effects of natural disturbances. They use this knowledge to help them design management systems that allow resources to be harvested or otherwise used, while controlling the ecological damage as much as possible. For example, understanding the characteristics of natural gap-phase disturbances in an old-growth forest can aid in the design of a selective harvesting system. Using such a system would leave the physical and ecological integrity of the forest substantially intact, even while individual trees are periodically harvested for industrial use. These individuals would be replaced by natural regeneration. In view of the natural, gap-phase disturbance dynamics of old-growth forests, clear-cutting followed by planting of tree seedlings might be considered a less "natural" management system. However, clear-cutting might be an appropriate practice to use when harvesting forests that are adapted to community-replacing disturbances, such as wildfire or insect outbreaks (see Chapters 21 and 23).

Disturbance Is Also Caused by Humans

Humans also disturb ecosystems in diverse ways, many examples of which are described in the following chapters. Anthropogenic disturbances are associated with many human activities: harvesting, conversion, species introductions, and war.

The harvesting of both renewable and nonrenewable natural resources often causes disturbances to ecosystems, as does the post-harvest management of renewable resources. For instance, intense disturbance is caused by strip-mining the surface for coal or metals. Similarly, harvesting a forest by clear-cutting represents a community-replacing disturbance, which is followed by regeneration through succession. Additional disturbances may be associated with silvicultural management, such as scarification of the site to prepare it for the planting of tree seedlings, or herbicide spraying to decrease the abundance of weeds. Harvesting may also be selective, as when particular species or sizes of trees or fish are targeted for harvesting. This represents a kind of gap-phase disturbance.

Conversions of natural ecosystems to agricultural or urbanized land uses also represent severe disturbances. In these cases, the successional recovery is intensively managed to foster the development of anthropogenic ecosystems and to prevent the regeneration of natural ecosystems. Usually, the anthropogenic ecosystems are dominated by introduced species of plants and animals, and sometimes by the bricks and concrete of the built environment. The conversion consequently displaces or eliminates indigenous species and natural ecosystems.

Humans have deliberately and accidentally introduced many species beyond their natural ranges. Often, the introduced species become invasive of natural communities, displacing native species and causing other ecological damages. North American examples include the introduction of zebra mussels (*Dreissena polymorpha*) into the Great Lakes; of purple loosetrife (*Lythrum salicaria*) into wetlands; and of starlings (*Sturnus vulgaris*), house sparrows (*Passer domesticus*), and domestic pigeons (or rock dove; *Columba livia*) into urban areas.

Warfare also produces a wide range of community-replacing and microdisturbances, through explosions, movements of heavy machinery over the landscape, spills of fuel and other toxic chemicals, hunting to provide food for large numbers of soldiers, and even — as occurred during the Viet Nam War — extensive spraying of herbicides over forests.

Anthropogenic Pollution and Disturbance in Context

To summarize: Pollution and disturbance can be natural phenomena. Since life began, both of these classes of environmental stress have affected the structure and function of natural ecosystems. In modern times, however, pollution and disturbance associated with human activities are becoming increasingly important sources of ecological damage. The prevention of anthropogenic pollution and disturbance, and management of the damages already caused, are among the most important challenges of the global environmental crisis.

Ecotoxicology

Toxicology is the science of the study of poisons. It examines their chemical nature and their effects on the physiology of organisms. If the dose (or exposure) is large enough, any chemical, even water, is potentially toxic (see In Detail 15.1).

Environmental toxicology is a somewhat broader field than conventional toxicology. It also examines the environmental factors that can influence the exposure of organisms to potentially toxic levels of chemicals. Important topics in environmental toxicology are:

1. the cycling and transport of potentially toxic chemicals
2. their transformations into other substances (which may be more or less poisonous than their precursors)
3. the determination of "sinks" where chemicals may accumulate in especially high concentrations, including within the bodies of organisms

IN DETAIL 15.1

WHAT IS TOXICITY?

In the biological sense, a chemical can poison an organism if it detrimentally affects some aspect of its metabolism. This effect is called *toxicity*. The toxic chemical may, for example, disrupt the function of an enzyme system or interfere with cellular division. The legal definition of *toxic substance*, as stated by the Canadian Environmental Protection Act, is as follows:

> A substance is defined as toxic if it enters or may enter the environment in a quantity or concentration or under conditions that: (1) have or may have an immediate or long-term harmful effect on the environment; (2) constitute or may constitute a danger to the environment on which human life depends; or (3) constitute or may constitute a danger in Canada to human life or health.

This definition has legal standing in Canada, and is used in the management and regulation of certain chemicals.

However, this definition is inadequate in some important respects, particularly because it deals only with extremely toxic chemicals, under conditions in which they occur in high concentrations. Substances of less acute toxicity may cause subtle, long-term damage to people, to other species, or to important ecological values. These kinds of exposures are not dealt with by this definition.

Ecotoxicology has an even broader domain, because it studies both the directly poisonous effects of chemicals and their indirect effects. Examples of indirect ecological effects of chemicals include changes in habitat or in the abundance of food. For instance, the use of a herbicide in forestry or agriculture will change the biomass and species composition of the vegetation. These are important changes in the habitat of animals. Even if the herbicide does not poison the animals directly, they may be affected by the changes in their habitat.

A complex of factors influences the ecotoxicological risks associated with exposures to chemicals in the environment. The most important factors are biological sensitivity, the intrinsic toxicity of the chemical being considered, the intensity of the exposure, and any indirect effects that might be caused.

Biological Sensitivity

Sensitivity to chemical exposures varies greatly among individual organisms and species. Toxicology studies, conducted under controlled laboratory conditions, compare the susceptibilities of organisms to toxic substances. **Acute toxicity** is defined as a short-term exposure to a chemical in a high enough concentration to cause measurable biochemical or anatomical damages, or even death (a common acute *endpoint*). **Chronic toxicity** involves long-term exposure, or low to moderate concentrations of the offending chemical, or both. Over time, chronic exposures may cause biochemical or anatomical damage, or perhaps a lethal condition such as cancer.

Data in Table 15.1 illustrate the sensitivities of several species to various routes of exposure to the extremely toxic chemical TCDD (2,3,7,8-tetrachlorodibenzo-*p*-dioxin). (TCDD has no industrial or medicinal use. It is, however, incidentally synthesized during high-temperature combustions in incinerators and fossil-fuelled power plants, during forest fires, in the whitening process for wood pulp using chlorine bleaches, and in the manufacture of certain industrial chemicals (particularly trichlorophenol, used to produce the herbicide 2,4,5-T and the antibacterial agent hexachlorophene). These various syntheses can result in the emission of TCDD into the environment, where humans and other organisms may be exposed.)

The data in Table 15.1 suggest that species vary greatly in their sensitivity to TCDD. Guinea pigs are extremely vulnerable, while hamsters and frogs are much less so. Sensitivity to toxic chemicals also varies with the route of exposure and with the sex of the animals.

Data showing acute and chronic toxicities are presented in Table 15.2. The chemical illustrated here is glyphosate, a herbicide widely used in agriculture, forestry, and horticulture (see Chapter 21). The data suggest that if the concentration of glyphosate is large enough, it will

TABLE 15.1 | ACUTE TOXICITY OF TCDD TO VARIOUS ANIMALS

Animals were exposed to TCDD in controlled tests under laboratory conditions. Oral exposure occurs by ingestion into the stomach; dermal exposure involves absorption through the skin; and intraperitoneal exposure involves injection into the abdominal body cavity. LD_{50} (lethal dose for 50%) is the dose of chemical required to kill one-half of a population of experimental animals. LD_{50} is measured in units of amount of chemical per unit of body weight of the animals (microgram TCDD/kilogram body weight).

SPECIES	ROUTE OF EXPOSURE	LD_{50} (μg/kg)
Guinea pig (male)	oral	0.8
Guinea pig (female)	oral	2.1
Rabbit (male and female)	oral	115
Rabbit (male and female)	dermal	275
Rabbit (male and female)	intraperitoneal	252–500
Monkey (female)	oral	<70
Rat (male)	oral	22
Rat (female)	oral	45–500
Mouse (male)	oral	<150
Mouse (male)	intraperitoneal	120
Dog (male)	oral	30–300
Dog (female)	oral	>100
Hamster (male and female)	oral	1157
Hamster (male and female)	intraperitoneal	3000
Frog	oral	1000

Source: Tschirley (1986)

TABLE 15.2 | ACUTE AND CHRONIC TOXICITY OF GLYPHOSATE

Toxicity to rats and mice is indicated by data from controlled tests under laboratory conditions. Acute toxicity is measured by oral LD_{50}, while chronic exposures are from long-term feeding experiments. The data given for chronic exposure are no-effect levels — that is, doses at (and below) which no observable effect is observed.

SPECIES	TYPE OF EXPOSURE	
Rat	acute:	oral LD_{50}; 5600 milligram/kilogram
	chronic:	fed for 90 days with food containing 2000 mg/kg; no observable effects
	chronic:	fed for 2 years with food containing 100 mg/kg; no observable effects
	chronic:	3 generations fed food containing 300 mg/kg; no observable effects
Mouse	acute:	oral LD_{50}; 1570 mg/kg
	chronic:	fed for 18 months with food containing 300 mg/kg; no observable effects

Source: Modified from Freedman (1991)

cause acute toxicity. However, long-term tests of chronic toxicity did not demonstrate any observable effects at the examined levels of exposure. (Note that the doses required to cause acute toxicity, and those tested for chronic toxicity, are much higher than exposures that would be encountered during the routine use of glyphosate as a herbicide.)

Intrinsic Toxicity of the Chemical

Chemicals vary enormously in their intrinsic, or relative, toxicity. In other words, some chemicals are extremely toxic even in minute doses, while others will only cause poisoning at much higher doses (i.e., exposures). This is illustrated by Table 15.3, which compares the acute toxicity (to rats) of a wide range of chemicals. There are two central messages in Table 15.3:

1. Chemicals vary enormously in their relative toxicity
2. At large enough doses, any chemical can be toxic.

Exposure

Exposure has a fundamental influence on toxicity. It is defined as the dose of chemical that any individual or group of organisms receives per unit time. An exposure to any potentially toxic chemical is affected by many factors, including environmental influences. For example, the exposure of, say, a mouse in an agricultural field sprayed with an insecticide could be affected by such factors as the spray rate, the types of equipment being used, the weather, the persistence of the chemical (how long it remains active), and the mouse's behaviour and choices of food and habitat. If toxicologists were evaluating exposures of humans to potentially toxic chemicals, they would consider the amounts ingested with solid and liquid food, the intake while breathing, and the amounts found in both working and ambient (i.e., non-occupational) environments.

TABLE 15.3 | ACUTE TOXICITIES OF VARIOUS CHEMICALS

Toxicity to rats is indicated by oral LD_{50} data from controlled laboratory tests.

CHEMICAL	ORAL LD_{50} (milligram/kilogram)
TCDD (dioxin isomer)	0.01
tetrodotoxin (globe fish toxin)	0.01
saxitoxin (paralytic shellfish neurotoxin)	0.3
carbofuran (insecticide)	10
strychnine (rodenticide)	30
nicotine (alkaloid in tobacco)	50
caffeine (alkaloid in coffee, tea)	200
DDT (insecticide)	200
fenitrothion (insecticide)	250
2,4-D (herbicide)	370
2,4,5-T (herbicide)	500
acetylsalicylic acid (aspirin)	1 700
sodium hypochlorite (household bleach)	2 000
sodium chloride (table salt)	3 750
glyphosate (herbicide)	5 600
ethanol (drinking alcohol)	13 700
sucrose (table sugar)	30 000
distilled water	44 000[1]
isotonic saline	68 000[1]

[1] *These data are for the house mouse, with the substance given intravenously rather than by oral ingestion. From Balazs (1970).*

Source: Freedman (1995)

Indirect Effects

Also important in ecotoxicology are the indirect effects of toxic chemicals — that is, effects other than the direct poisoning of organisms. Indirect effects are most commonly associated with changes in the habitat or in the condition of the organism's immune system. In some cases, indirect damage is worse than the direct, toxic effects of chemicals. For instance, the use of a herbicide in forestry causes changes in vegetation, affecting the animals that live in the habitat, even if the herbicide itself is not directly toxic to them.

Potentially, All Chemicals Are Poisonous

The above discussion suggests that if an exposure is intense enough, even routinely encountered chemicals can be poisonous. For example, poisoning can even be caused by water, if a person drinks enough in a short period of time. The physiological capacity for regulating the concentration of dissolved salts in the blood plasma can be overwhelmed by drinking too much water, too quickly. Depending on body weight, the lethal dose for an average adult is about 5 L, ingested over about an hour. Similarly, if the dose is large enough, carbon dioxide, table sugar, table salt, aspirin, ethanol (drinking alcohol), and other routinely ingested chemicals can cause poisoning (Table 15.3).

This fundamental rule of toxicology was first emphasized by Paracelsus (1493–1541), a Swiss physician and alchemist who is considered to be the father of "modern" toxicology. One of Paracelsus's most famous conclusions can be paraphrased, "Dosage determines poisoning."

In perhaps all cases there are thresholds of tolerance to potentially toxic chemicals. Tolerance occurs because organisms have physiological mechanisms to excrete toxins from the body, to metabolize them into less toxic chemicals, or to sequester (store) them in body tissues where they will not cause damage. Organisms also have mechanisms to repair damage to tissues or biochemical systems, providing the chemical exposures are not too high and damaging. For the chemical to cause toxicity, the capacities of these physiological systems must be overwhelmed.

Interpretation of Chemical Exposures and Damage

The notion of physiological thresholds of tolerance helps define the difference between contamination and pollution, which we examined previously. The notion of thresholds also indicates why it is best to couch the discussion in terms of *potentially* toxic levels — particularly when the actual environmental exposures to chemicals are not known, and when the biological risks of extremely small exposures are not sufficiently understood.

However, the notion of biological thresholds of tolerance is controversial, and some toxicologists do not agree with the explanation just given. These scientists believe that exposures to even one or a few molecules of certain chemicals may be of toxicological importance. This is particularly true in the case of chemicals believed to be carcinogenic at extremely small exposures, and also in the case of exposures to radionuclides and highly energetic forms of ionizing energy, such as X-rays and gamma radiation.

Often, the risks borne by humans exposed to chemicals are interpreted differently from those borne by other species, particularly wild animals and plants. This is because prevailing cultural attitudes place much greater value on the life and health of individual people than on the life of individuals of other species. There is a special reluctance, both social and regulatory, to permit exposures of humans to potentially toxic chemicals.

However, regulations and guidelines tend to be considerably less strict for human exposures in the workplace, compared with non-occupational exposures. This recognizes the fact that considerable risks are inherent in the normal activities and environmental conditions of many occupations. Particularly significant hazards confront firefighters, police officers, members of the armed forces, operators of heavy equipment in construction and agriculture, and workers in chemical industries. Within limits, chemical exposures associated with earning a living are generally interpreted as a "cost of doing business," and are, therefore, judged acceptable.

Such attitudes can, however, change markedly over time. Certain occupational hazards that were once considered routine and tolerable are now viewed as unacceptable. For instance, when synthetic organic insecticides, such as DDT, were first introduced in the mid-1940s and 1950s, people were remarkably cavalier about using them in agriculture, forestry, and public sanitation. Workers often applied these insecticides with only minimal attention to avoiding exposure to themselves or others. Such poorly controlled usages would be unthinkable today, at least in relatively well-regulated countries such as Canada.

Many people willingly choose to expose themselves to toxicologically significant doses of certain chemicals. These choices include taking up hazardous occupations, smoking cigarettes, and ingesting medicines and recreational drugs. The consequences of "voluntary" chemical exposures are interpreted using criteria different from those applied to "involuntary" exposures.

If chemicals cause toxicity to species other than humans, their importance is interpreted on the basis of the following considerations:

Are measurable changes seen in the populations of affected species? From the ecological perspective, population-level damage is the most important consideration (even while it is acknowledged that the deaths of individual organisms are regrettable). Populations of all species have a certain degree of resilience, and they can tolerate some mortality caused by toxic chemicals without exhibiting an overall decline.

Are affected species important in maintaining the integrity of their community? Ecological philosophies suggest that all species have a similar intrinsic value. Nevertheless, species vary greatly in their contributions to the functioning and structure of their community. So-called "keystone" species have a dominant influence on their community (Chapter 9). Substantial changes in their abundance should be judged as relatively important, compared with damage inflicted on more minor species.

Is the damage of economic importance? This consideration involves damage to resources which are needed by humans, and therefore have economic value. In this sense, damage is judged relatively important if it is caused to hunted animals such as deer or trout, to trees that can be harvested to manufacture pulp or lumber, or to important ecological services such as the provision of clean water and air. From the purely utilitarian perspective, damage caused to non-economic values, both species and services, is viewed as being less important.

Other considerations, less tangible than those just mentioned, involve appraising damage in aesthetic or ethical terms. These considerations are important. They are, however, difficult to interpret in terms of risks or benefits to human welfare, or in economic terms. As a result, aesthetic or ethical considerations are rarely reflected in regulatory criteria or in the management of chemicals in the environment.

Environmental Risks

Broadly interpreted, **environmental risks** are hazards — the probabilities of suffering damage or misfortune as a result of exposure to some environmental circumstance. Risks are associated with driving an automobile, flying in an airplane, participating in sports, hiking in the wilderness, exposing oneself to toxic chemicals, and getting out of bed in the morning. Environmental risks interact with biological factors to determine the likelihood of experiencing some damage, such as developing a cancer or suffering an accident.

Statisticians assign reasonable probability values to many kinds of risks, using data based on previous experience, such as the frequency of automobile accidents or cases of poisoning with a specific chemical such as a medicine. The procedure is illustrated in Table 15.4, which summarizes the causes of mortality in Canada. These data suggest that the average Canadian has an annual risk of dying of about 0.7% (calculated as total annual mortality divided by the national population).

Data concerning less common environmental risks are more difficult to acquire. Usually they must be developed from models based on knowledge about medical science and likely exposures to environmental influences. However, both of these kinds of information are imperfect because they are based on an incomplete understanding of interactions between environmental factors and biological responses. Consequently, the calculated risk factors are inaccurate and often controversial. These issues are particularly important for diseases, such as cancers, that have an extended latency period between exposure and development.

Cancers are a leading cause of mortality in Canada and many other relatively wealthy countries. Remarkably little is known, however, about the specific environmental and biological factors that predispose organisms to developing specific types of cancers. Table 15.5 summarizes data from a recent United States study that estimated the risks of dying from different cancers associated with several potentially contributing factors. Of the approximately 485 000 cancer deaths that occur each year in the United States, dietary factors are believed to be the most important predisposing factor, accounting for about 35% of the cancer mortality, followed by tobacco smoking (30%), infections (10%), and reproductive and sexual behaviour (7%).

The population of Canada is 10.8% that of the United States, while the number of cancer-related mortalities in

TABLE 15.4 | CAUSES OF MORTALITY IN CANADA

These data for 1991 summarize the most important attributed causes of deaths among Canadians.

RISK FACTOR	NUMBER OF DEATHS	% OF TOTAL
All causes of mortality	196 535	100%
MEDICAL CAUSES		
Diseases of circulatory system	76 211	38.8
Cancers (all malignancies)	54 840	27.9
Respiratory diseases	16 663	8.5
Diseases of nervous system	5 268	2.7
Diabetes	4 296	2.2
Infectious and parasitic diseases	2 744	1.4
Liver diseases	2 122	1.1
Congenital abnormalities	1 216	0.7
Perinatal (just before or after birth)	981	0.5
ACCIDENTS AND RELATED CAUSES		
Suicide	3 709	1.9
Motor vehicle accidents	3 456	1.8
Other accidental deaths	3 216	1.6
Accidental falls	2 138	1.1
Homicide	597	0.3
All other causes of death	18 901	9.6

Source: Statistics Canada (1993)

TABLE 15.5 | ESTIMATED RISKS OF CANCER MORTALITY

Cancers are grouped by their possible causes, in terms of environmental exposures. The data are the best estimates for the U.S. population, with the range of estimates in brackets.

RISK FACTOR		RELATIVE CAUSE OF U.S. CANCER MORTALITY (% OF TOTAL)
Diet factors		35 (10 – 70)
Smoking tobacco		30 (25 – 40)
Infections		10 (1 – ?)
Reproductive and sexual behaviour		7 (1 – 13)
Occupational exposures		4 (<2 – 8)
radiation in workplace	0.01	
pesticide application	<0.02	
chemicals in workplace	<0.06	
Geophysical factors		3 (2 – 4)
current sunlight exposures	2	
indoor radon	2 (1 – 4)	
Alcohol consumption		3 (2 – 4)
Exposure to pollution		2 (<1 – 5)
secondary tobacco smoke	1.0	
indoor organic chemicals	0.3	
pesticides on food	0.9 (0.6 – 1.2)	
hazardous toxic air pollutants	0.3 (0.2 – 0.4)	
chemicals in drinking water	<0.1 (0.04 – 0.09)	
other pesticide exposures	0.02 (0.02 – 0.03)	
Medicines and medical procedures		1 (<1 – 3)
Exposures to food additives		<1 (-5 – 2)
Exposures from consumer products		<1

Source: Modified from Gough (1989)

Canada is 11.3% that of the United States. These similar proportions, along with the comparable lifestyles of Canadians and Americans, suggest that the estimated risks in Table 15.5 are also relevant to Canadians.

In spite of excellent data (and common sense) about the known risks of many activities, many people choose to expose themselves to obvious high risks of injury or disease. Examples of such activities include skiing down steep slopes, bungee jumping, smoking cigarettes, and drinking alcohol. Moreover, people are exposed to hazards over which they have little control — that is, to involuntary risks, such as crime, polluted outdoor air, and pesticides in food.

Perceptions of risk are an important consideration. A recent survey of Canadians indicated that people are aware of, and concerned about, a wide range of risks to their health and well-being (Table 15.6). People were especially concerned about health-related risks associated with lifestyle choices, such as smoking cigarettes, using recreational drugs or alcohol, and behaviour involving exposure to the virus that causes AIDS (or **ac**quired **i**mmune **d**eficiency **s**yndrome). People are also concerned about exposures to potentially toxic levels of chemicals in the atmosphere, drinking water, and foods.

TABLE 15.6 | PUBLIC PERCEPTION OF RISKS OF VARIOUS ENVIRONMENTAL AND MEDICAL HAZARDS

The data, based on a national survey of 1500 Canadians in 1992, indicate the percentage of the survey group that chose the designated category. The totals do not add to 100% because some respondents indicated that they "didn't know."

RISK FACTOR	PERCEIVED RISK (% RESPONSE)			
	HIGH	MODERATE	SLIGHT	NONE
Smoking cigarettes	61	31	7	1
Use of silicon breast implants	61	25	10	6
Depletion of stratospheric ozone	58	28	9	1
Recreational drugs	57	32	9	1
Emotional and physical stress	54	29	9	2
Pollution by chemicals	54	35	10	1
Crime and civil violence	53	35	10	1
Suntanning	52	33	12	2
AIDS virus	49	39	10	2
Motor-vehicle accidents	48	30	10	2
Nuclear wastes	48	22	16	7
Use of alcohol during pregnancy	41	41	14	2
PCBs and dioxins	39	33	17	4
Pesticides in foods	37	38	19	3
Additives in foods	36	39	19	4
Consumption of alcohol	34	48	15	2
Nuclear power plants	31	34	23	10
Effects of climate change	28	39	22	7
Use of non-prescription medicines	26	36	28	9
Exposure to asbestos	26	32	28	7
Municipal incinerators	23	37	23	11
Effects of malnutrition	23	43	25	6
High-voltage power lines	21	30	30	15
Consumption of irradiated foods	21	32	31	10
Use of prescription medicines	19	43	27	10
Genetically engineered bacteria	18	34	25	6
Outdoor air quality	18	47	27	7
Pathogenic bacteria in food	18	38	35	8
Fungi and moulds in food	16	27	39	12
Mercury in dental fillings	13	30	29	16
Consumption of tap water	12	38	35	13
Exposure to medical X-rays	12	40	39	13
Indoor air quality	11	43	32	11
Video display monitors	10	36	26	8
Use of contraceptives	10	31	40	20
Use of a heart pacemaker	6	26	39	26
Consumption of bottled water	5	17	35	39
Use of contact lenses	4	17	45	34

Source: Krewski *et al.* (1994)

Obviously, people understand that environmental factors pose risks to human health. Often, however, they have little understanding of the *actual* risks, as opposed to the *perceived* risks. Sometimes, people view high risks to be inconsequential, while considering much smaller risks as unduly important. Nevertheless, public perceptions of risks have an extremely important influence on politicians, policy makers, and bureaucrats in government and industry, and on their decisions concerning the management and regulation of environmental and health hazards.

Environmental Risk Assessment

Environmental risk assessments are quantitative evaluations of the risks associated with some sort of environmental hazard. Risk assessments quantify these threats to humans, as well as to other species and to broader ecological values. A risk assessment requires knowledge of three factors:

1. the likelihood that the hazard will be encountered
2. the likely intensity of the hazard
3. the biological damage that is likely to result from the predicted exposure

Meteorologists, for example, may predict the probability that a particular place will be struck by lightning under various weather conditions. Such probabilities are much greater during a thunderstorm than during sunny conditions, and higher beneath a large tree in an open field than beside a shrub in a ditch. The energy content of a typical lightning strike is also known, as is the biological damage to a human struck by lightning. With this information, it is relatively straightforward to model the risks of a lightning-caused injury associated with standing in the middle of an open field or under a tree in that same field, on a sunny day or during an active thunderstorm. This is a simple example of an environmental risk assessment.

Risk assessments for potentially toxic levels of chemicals can be conducted for individual organisms, for populations, or for broader ecological functions such as productivity, decomposition, and nutrient cycling. To assess the risks associated with exposures to chemicals, one requires knowledge of two factors: the degree of exposure (i.e., the anticipated dose), and the biological damage that is likely to be caused by the predicted exposure. The integration of these two types of information is known as a **dose-response relationship** (Figure 15.1).

Dose-response relationships can be determined by conducting experiments in which, for example, populations of organisms are exposed to various doses of the chemical

FIGURE 15.1 CONCEPTUAL MODELS OF DOSE-RESPONSE RELATIONSHIPS

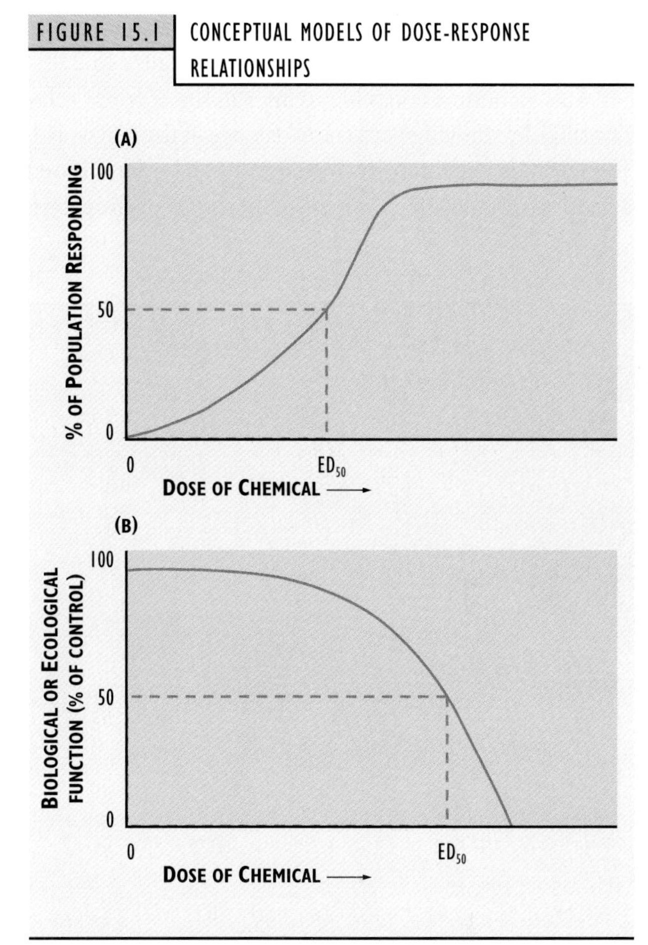

Model (a) suggests that the larger the dose encountered, the greater the proportion of the population that is affected. ED_{50} represents the dose that affects 50% of the test population (**E**ffective **D**ose). If the biological response being measured is death, the term LD_{50} is used, or the dose killing 1/2 of the population (**L**ethal **D**ose).

Model (b) suggests that larger doses have a more pronounced effect on physiology (or on an ecological function). In this case, the rate of a biological function is plotted versus the chemical exposure, and the data are expressed as a percentage of the control rate (that is, the rate occurring in the absence of the chemical). In this curve, ED_{50} represents the dose needed to decrease the rate of the function by 50%.

in question. Results of simple dose-response experiments involving several herbicides are shown in Figure 15.2.

It is sometimes possible to infer dose-response relationships by studying patterns of damage in the real world. For instance, the intensity of pollution can be determined at various distances from a large point source of emissions,

Many human activities result in emissions of pollutants into the atmospheric, aquatic, and soil environments. This image shows a 380-m–tall smokestack, known as the "superstack," at a metal smelter near Sudbury, Ontario.

Source: B. Freedman

FIGURE 15.2 | EXAMPLES OF DOSE-RESPONSE CURVES

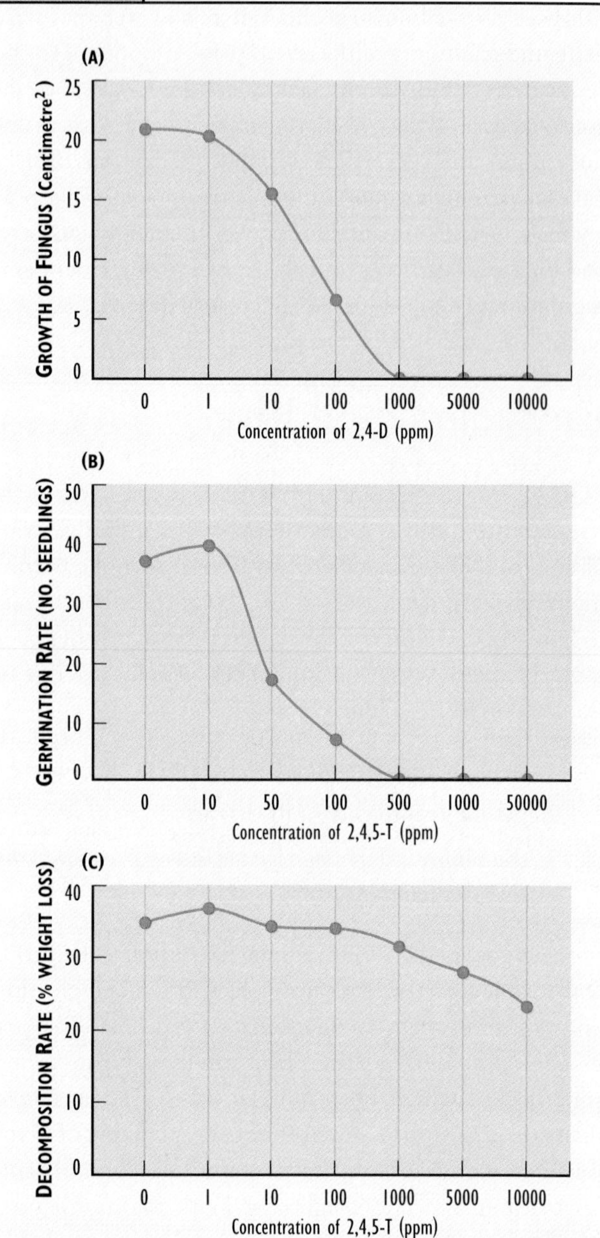

Note the extremely wide ranges of doses that were examined in these experiments. Each experiment includes a control treatment involving a zero dose of the chemical being tested.

Graph (a) describes the effects of the herbicide 2,4-D on the growth rate of a mycorrhizal fungus, *Hebeloma longicaudum*.

Graph (b) illustrates the effects of the herbicide 2,4,5-T on the rate of germination of seeds from the surface organic mat of a young clear-cut.

Graph (c) shows the effects of 2,4,5-T on the decomposition of leaf litter, an ecological function.

Source: Estok *et al.* (1989), Fletcher and Freedman (1989), and Morash and Freedman (1989)

such as a smelter. The exposure to pollution can then be related to the patterns of ecological damage that occur along the gradient of toxic stress. Patterns of pollution and ecological damage around a large smelter near Sudbury are one example of such a relationship (see Chapters 16 and 18).

An **exposure assessment** investigates all of the ways in which organisms may encounter a potentially toxic level of a chemical. For example, humans may be exposed to mercury through various pathways, each of which can be quantified (calculated). The principal avenues of mercury exposure include inhaling mercury vapour or particulates in the atmosphere; ingesting mercury dissolved in drinking water; and ingesting mercury in foodstuffs, especially in certain kinds of fish and animal organs. Also included among the principal avenues of exposure are such miscellaneous sources as mercury-amalgam dental fillings, and certain pigments used in ceramics and paints.

The assimilation rate of a chemical into the bloodstream and other organs varies greatly among the exposure pathways. Assimilation depends on several factors, including the specific metabolic characteristics of the organ into which the chemical is being absorbed — for example, into the lungs, into the gastrointestinal tract, or through the skin. The physical-chemical form of the substance also greatly affects uptake dynamics. For instance, mercury can occur as an elemental vapour or liquid, as inorganic compounds such as mercuric chloride, and as organo-mercurial complexes such as methylmercury (an extremely bioavailable and poisonous compound). The total exposure for an individual person is the weighted sum of the chemical assimilated through all pathways, which vary greatly in their effect on the person.

The relative importance of various sources of a chemical such as mercury depends to some extent on a person's lifestyle and occupation. These influence how often and to what degree the various sources are encountered. Dental workers, for example, commonly come into contact with mercury vapours, because this metal is used to make fillings. In addition, a diet rich in certain species of large oceanic fish, such as halibut, shark, swordfish, or tuna, is relatively rich in mercury (see Chapter 18). Therefore, both dental workers and fish consumers may have a higher risk of mercury exposure.

Once an exposure assessment has been performed, the biological hazards can be predicted on the basis of known dose-response relationships. Unfortunately, dose-response information is often incomplete, or even totally lacking.

Smoking entails a voluntary exposure to a wide range of chemicals that are known to be extremely toxic. In addition, nonsmokers are involuntarily exposed to "sidestream smoke" as a result of sharing space with smokers in restaurants, taverns, or their homes.

Source: Dick Hemingway

For instance, most hypotheses about potential dose-response relationships in humans are actually inferred from research conducted in laboratories using other species of mammals, such as mice, rats, dogs, pigs, and monkeys. These species have physiological, anatomical, and behavioural characteristics that are broadly similar to those of humans, but are also different in many important respects. Consequently, most assessments of human exposure to trace levels of environmental chemicals are inaccurate.

In addition, the information about dose-response relationships is almost nonexistent for wild species, as well as for community-level ecological functions such as productivity and nutrient cycling. As with human-focused assessments, it is common to use data for "surrogate" species, which are believed to be typical in their dose-responses.

Consider, for example, an attempt to predict the effects of chemical contamination in a particular lake. It is highly unlikely that relevant dose-response data will be available for the species of fish in the ecosystem. Consequently, predictions will typically be made using information for surrogate species, such as rainbow trout (*Salvelinus gairdneri*) and fathead minnow (*Pimephales promelas*). These two species have been well studied in toxicological laboratories and are widely used as indicators. Similarly, the potential effects on the community of zooplankton might be predicted using information available for well-studied species, such as the water fleas *Daphnia magna* and *Ceriodaphnia dubia*, while the risk assessment for phytoplankton might use data for the unicellular algae *Selenastrum capricornutum* and *Chlorella vulgaris*.

The results of risk assessments for a community, based on laboratory studies of surrogate species, are always uncertain. This is especially true if the potential effects of chemical exposures in natural, ecological contexts are being predicted. Such risk assessments are, however, the best that can be done under most circumstances, because there is rarely enough funding or time to do better. Nevertheless, because these methods deliberately overestimate the potential risks, they provide conservative guidance for management purposes.

Key Terms

environmental stressor
disturbance
exposure
response
pollution
contamination
community-replacing disturbance
microdisturbance
toxicology
environmental toxicology
ecotoxicology
acute toxicity
chronic toxicity
environmental risks
environmental risk assessment
dose-response relationship
exposure assessment

Questions for Discussion

1. Identify the various kinds of environmental stressors, giving examples of each.
2. What is the difference between pollution and contamination?
3. Describe several examples of naturally occurring pollution and disturbance.
4. Distinguish the following: toxicology, environmental toxicology, ecotoxicology.
5. Compare the ecological effects of community-replacing disturbances and gap-phase microdisturbances.
6. List what you consider to be the most important risks to your health, and compare these with the data in Tables 15.4 and 15.5.

REFERENCES

Barrett, G.W. and R. Rosenberg (eds.). 1981. *Stress Effects on Natural Ecosystems*. New York: Wiley.

Cote, R.P. and P.G. Wells. 1991. *Controlling Chemical Hazards: Fundamentals of the Management of Toxic Chemicals*. London: Unwin Hyman.

Estok, D., B. Freedman, and D. Boyle. 1989. Effects of the herbicides 2,4-D, glyphosate, hexazinone, and triclopyr on the growth of three species of ectomycorrhizal fungi. *Bull. Environm. Contam. & Toxicol.*, **42**: 835–839.

Fletcher, K. and B. Freedman. 1986. Effects of several herbicides used in forestry on litter decomposition. *Can. J. For. Res.*, **16**: 6–10.

Forbes, V.E. and T.L. Forbes. 1994. *Ecotoxicology in Theory and Practice*. New York: Chapman & Hall.

Freedman, B. 1991. Controversy over the use of herbicides in forestry, with particular reference to glyphosate usage. *J. Envir. Sci. & Health*, **C8**: 277–286.

Freedman, B. 1995. *Environmental Ecology, 2nd ed.* San Diego, CA: Academic.

Gough, M. 1989. Estimating cancer mortality. *Environ. Sci. & Technol.*, **23**: 925–930.

Hodge, R.A., P.R. West, and R. Gregory-Eaves. 1996. Toxic substances. Chapter in: *1996 State-of-the-Environment Report for Canada*. Ottawa: Environment Canada.

Krewski, D., P. Slovic, S. Bartlett, J. Flynn, and C. Mertz. 1994. *Health Risk Perception in Canada*. Calgary: University of Calgary. Environmental Risk Management Working Paper ERC 94–4, EcoResearch Chair.

Levin, S.A, M.A. Harwell, J.R. Kelly, and K.D. Kimball (eds.). 1989. *Ecotoxicology: Problems and Approaches*. New York: Springer.

Morash, R. and B. Freedman. 1989. Effects of several herbicides on the germination of seeds in the forest floor. *Can. J. For. Res.*, **19**: 347–350.

Odum, E.P. 1985. Trends expected in stressed ecosystems. *BioScience*, **35**: 419–422.

Schindler, D.W. 1987. Detecting ecosystem responses to anthropogenic stress. *Can. J. Fish. Aquat. Sci.*, **44**: 6–25.

Smith, W.H. 1984. Ecosystem pathology: A new perspective for phytopathology. *For. Ecol. Manage.*, **9**: 193–219.

Statistics Canada. 1993. *Mortality, Summary List of Causes, 1992*. Ottawa: Statistics Canada. Cat. 84-209.

Suter, G. 1993. *Ecological Risk Assessment*. Boca Raton, FL: CRC.

Tschirley, F.H. 1986. Dioxin. *Sci. Amer.*, **254** (2): 29–35.

GASEOUS AIR POLLUTION

CHAPTER OUTLINE

CHAPTER OBJECTIVES

After completing this chapter, you will be able to:

1. Outline the major sources of emission of air pollutants associated with sulphur, nitrogen, and hydrocarbons.
2. Explain the difference between primary and secondary air pollutants.
3. Compare the environmental problems respectively associated with stratospheric and tropospheric ozone.
4. Discuss the importance of air pollutants to human health.
5. Outline the lessons to be learned from air pollution at the Smoking Hills.
6. Describe the patterns of air pollution and ecological damage near Sudbury and other point sources of air pollution.

Introduction

Gaseous air pollutants are emitted from various natural sources, such as volcanoes, forest fires, and wetlands. In fact, the natural emissions of some gases exceed those due to human activities. However, anthropogenic emissions of other gases are larger than natural emissions, and are increasing because of the growth of the human population and its degree of industrialization.

In ancient times, humans were responsible for very little atmospheric pollution. Even then, however, there were local problems — smoky wood fires, commonly used for cooking and warmth, likely resulted in poor quality indoor air in poorly ventilated dwellings. Hunting cultures often used fires to drive game animals, and to improve forage availability by increasing the local abundance of prey species. These burns would have resulted in substantial emissions of particulate carbon (soot), carbon dioxide, and other gases, and would have temporarily impaired air quality. Still, these effects would have been rather minor.

As humans became more numerous and industrialized, air pollution developed into a much more extensive problem. Since coal was the major fuel used to generate heat and energy for machines, severe air pollution became especially common at the beginning of the Industrial Revolution in the early 1800s. The widespread use of coal led to severe pollution by sulphur dioxide (SO_2) and soot in the industrial cities and towns of Europe and the Americas. Since 1900, new technologies such as power plants and automobiles have further increased the emissions of pollutants.

Air pollution is especially severe when the lower atmosphere is unusually stable and calm. These conditions often occur beneath **atmospheric inversions**, which are characterized by cool air being trapped beneath a layer of warmer air. This stable condition prevents the mixing of polluted, ground-level air with cleaner air from higher altitudes (see In Detail 16.1). If an atmospheric inversion is accompanied by fog, the air pollution is known as "**smog**" (a composite word derived from "**sm**oke" and "**fog**"). As recently as the 1950s, "killer smogs," rich in sulphur dioxide and soot, caused the deaths of thousands of people, and many more suffered from acute respiratory distress. Smogs rich in sulphur dioxide are also called "reducing smogs." The most famous killer smogs occurred in London, Glasgow, some other industrial centres of Europe, and near Pittsburgh in the United States (these are described later).

Once the severe damage inflicted by air pollution was recognized, governments passed laws to decrease the emissions. Pollution control became particularly vigorous after medical researchers discovered convincing evidence of linkages between air pollution and increases in human diseases and deaths. Particular attention was paid to air qual-

In Detail 16.1

Atmospheric Inversions

Normally, the temperature of the atmosphere decreases with increasing altitude above the ground surface. Under certain conditions, however, a layer of relatively cool air may become trapped under a layer of warmer air, a phenomenon known as **atmospheric inversion** (or temperature inversion).

Atmospheric inversions can develop on clear, cloudless nights. Under such conditions, the ground surface cools quickly as it radiates the heat that was absorbed during the day. This can result in a layer of cool air occurring beneath a layer of warmer air. Hilly terrain is particularly vulnerable to developing atmospheric inversions, because cool air can drain from slopes and hilltops and accumulate in valleys.

Atmospheric inversions are relatively stable. Until they are dispersed by vigorous winds, they can trap and accumulate air pollutants emitted during the inversion event. Severe episodes of air pollution can occur when stable inversions develop and are maintained for several days.

Some places regularly develop less persistent inversions. This occurs in the Los Angeles basin, around Mexico City, and in the Greater Vancouver area. In these places, the inversions develop during the morning but are typically dispersed during the afternoon. In the meantime, however, large concentrations of oxidizing air pollutants such as ozone can accumulate into a photochemical smog.

ity in urban environments, where the killer smogs were most frequent. The most important control actions have typically included the following:

- switching from coal, which is a relatively "dirty" fossil fuel, to "cleaner" fuels such as natural gas or oil, or to alternative technologies such as nuclear power
- constructing tall smokestacks to spread emissions over a wider area, so that ground-level exposures become less common and less intense — a tactic known as the "dilution solution to pollution"
- centralizing energy production in large power plants, to replace much of the relatively dirty coal burning in home fireplaces and furnaces, and so allowing better control of emissions and less overall energy use.
- treating waste gases to remove some of their pollutant content, thereby reducing emissions into the atmosphere

Because of these measures, so-called "oxidizing smogs" have supplanted the SO_2-rich reducing smogs in many regions. Oxidizing smogs develop in the atmosphere through complex photochemical reactions in which hydrocarbons and nitrogen oxides are transformed into other gases, most notably ozone (O_3). Ozone and other oxidizing gases harm vegetation and irritate the respiratory systems and eyes of exposed people. Oxidizing smogs develop under sunny conditions, if hydrocarbons and nitrogen oxides are present from automobile and industrial emissions, and particularly if atmospheric temperature inversions reduce dispersion.

In this chapter we examine the most important types of gaseous air pollutants. Their emission sources, chemical transformations, and toxicity are described, and Canadian case studies are used to demonstrate the ecological damages that can be caused.[1]

Sulphur Gases

Emissions and Transformations

Sulphur dioxide (SO_2) is one of the most important of the gaseous air pollutants. SO_2 is a colourless but pungent gas. Humans can detect the bitter taste of SO_2 at a concentration of only 0.3–1 ppm (parts per million). Hydrogen sulphide (H_2S), another sulphur gas, can be detected as a foul odour, reminiscent of rotten eggs, at concentrations lower than 1 ppb (parts per billion).

After they are emitted to the atmosphere, SO_2 and H_2S become oxidized to other compounds, and ultimately to sulphate (SO_4^{2-}; In Detail 16.2). The sulphate ion carries negative charges, and is, therefore, an anion (see In Detail 16.3). Since H_2S is rather quickly oxidized to SO_2, its atmospheric residence time is less than one day. SO_2 oxidizes more slowly, at a rate of <1% to 5% per hour, depending on sunlight, humidity, the concentration of strong oxidants such as ozone, and the presence of metal-containing particulates that can serve as catalysts. A typical residence time of SO_2 in the atmosphere is about four days. Consequently, SO_2 may disperse a long distance from its point of emission before it becomes oxidized or is about deposited to a terrestrial or aquatic surface. This dispersal is referred to as a **l**ong-**r**ange **t**ransport of **a**ir **p**ollution, or LRTAP.

Atmospheric sulphate, formed by the oxidation of SO_2 and H_2S, can combine with positively charged ions (or cations) to form various compounds. Most atmospheric sulphates occur as tiny particulates, especially as ammonium sulphate ($(NH_4)_2SO_4$). Ammonium sulphate is the most prominent component (along with ammonium nitrate (NH_4NO_3) of the particulate haze that often impairs visibility in cities. Haze is also seen in some rural areas where air pollutants have been transported from emission sources elsewhere (i.e., by LRTAP). Other cations that combine with sulphate include calcium (Ca^{2+}), magnesium (Mg^{2+}), and sodium (Na^+). Often, however, there are not enough of these cations (that is, NH_4^+, Ca^{2+}, Mg^{2+}, or Na^+) to balance all of the negative charges of the sulphate (SO_4^{2-}) present. Under such conditions, hydrogen ions (H^+) serve to balance some of the negative charges of SO_4^{2-}. This results in an aerosol containing sulphuric acid (H_2SO_4), the most important component of acidic precipitation (see Chapter 19).

Sources of SO_2 Emissions

Volcanoes are natural sources of emissions of atmospheric sulphur gases. On average, volcanoes emit about 12 million t/y of sulphur gases, of which 90% is SO_2 and 10% H_2S. However, the enormous 1991 eruption of Mount Pinatubo in the Philippines emitted about 7–10 million t of SO_2 (expressed as sulphur, or SO_2–S). The much smaller

[1] Unless otherwise indicated, specific data cited in this and other chapters in this section on environmental damages are from Freedman (1995).

IN DETAIL 16.2

SOME AIR-POLLUTION CHEMISTRY

Air pollutants are commonly emitted as particular chemicals, which become transformed into other compounds through chemical reactions occurring in the atmosphere. Several examples follow:

Oxidation of SO_2

$SO_2 + OH \longrightarrow HO.SO_2$ (1)
$HO{\cdot}SO_2 + O_2 \longrightarrow HO_2 + SO_3$ (2)
$SO_3 + H_2O \longrightarrow H_2SO_4$ (3)
$H_2SO_4 \longrightarrow 2H^+ + SO_4^{2-}$ (in aqueous solution) (4)

Note that the SO_4^{2-} produced is important as a constituent of acid rain and as particulate ammonium sulphate, both of which are prominent air pollutants.

Oxidation of NO

$NO + HO_2 \longrightarrow NO_2 + OH$ (5)
$NO_2 + OH \longrightarrow HNO_3$ (6)
$HNO_3 \longrightarrow H^+ + NO_3^-$ (in aqueous solution) (7)

Similarly, the NO_3^- produced is important as a constituent of acid rain and as particulate ammonium nitrate.

Formation and destruction of stratospheric O_3

O_2 + UV radiation $\longrightarrow O + O$ (8)
$O + O \longrightarrow O_2$ (9)
$O + O_2 \longrightarrow O_3$ (10)
O_3 + UV radiation $\longrightarrow O_2 + O$ (11)

Reaction 11 is an ultraviolet photodissociation of O_3. Ozone can also be consumed by reactions with NO, NO_2, or N_2O, or with ions or simple molecules of chlorine, bromine, and fluorine. These latter reactions are too complex to describe here (see Freedman, 1995).

Formation and destruction of ground-level O_3

$O + O_2 \longrightarrow O_3$ (10)
NO_2 + UV radiation $\longrightarrow NO + O$ (12)
$NO + O_3 \longrightarrow NO_2 + O_2$ (13)
$NO + RO_2 \longrightarrow NO_2 + RO$ (14)

The formation of O_3 (reaction 10) requires atomic O, formed by the photodissociation of NO_2 (reaction 12). Ozone can be consumed by reaction with NO (an emitted gas), which regenerates NO_2 (reaction 13). Atmospheric O_3 can, however, accumulate if other reactions (such as reaction 14) convert NO to NO_2, because these operate in competition with reaction 13 for NO. (The species RO_2 in reaction 14 includes various chemicals known as peroxy radicals, formed by the degradation of organic molecules through reaction with hydroxyl radicals, followed by the addition of molecular O_2. RO is the chemically reduced form of RO_2.)

eruption of Mount St. Helens in Washington State in 1980 vented about 0.2 million t of SO_2–S. Natural emissions of SO_2 also occur when biomass burns during wildfires.

The global anthropogenic emissions of SO_2 are about 63–72 million t/y, about double the natural emissions (Table 16.1). Fossil-fuel combustion accounts for 54% of the anthropogenic emissions. Coal and petroleum contain mineral compounds of sulphur, such as pyrite (FeS_2), as well as many organic sulphur compounds. When these fuels are burned, the sulphur is oxidized to gaseous SO_2, which

IN DETAIL 16.3

CATIONS AND ANIONS

Cations and anions are atoms or molecules that carry an electrical charge. Cations have one or more positive charges, and anions have one or more negative charges. Common examples of cations include Na^+ (sodium ion), Ca^{2+} (calcium ion), and Al^{3+} (aluminum ion), having one, two, and three positive charges respectively. Some anions include NO_3^- (nitrate), SO_4^{2-} (sulphate), and PO_4^{3-} (phosphate), with one, two, and three negative charges respectively. See also In Detail 19.2 in Chapter 19 for a description of the *conservation of electrochemical neutrality*.

TABLE 16.1 GLOBAL EMISSIONS AND OTHER CHARACTERISTICS OF IMPORTANT AIR POLLUTANTS

Data are from the late 1980s.

POLLUTANT	NATURAL EMISSIONS (10^6 tonne/year)	ANTHROPOGENIC EMISSIONS (10^6 t/y)	TYPICAL CONCENTRATION		ATMOSPHERIC RESIDENCE TIME
			BACKGROUND (ppm)	POLLUTED AIR (ppm)	
SO_2	40	70	0.0002	0.2	4 days
H_2S	100	3	0.0002	–	<1 day
CO	33	304	0.1	40–70	<3 years
NO_X (as NO)	430	35	<0.002	–	5 days
NO_X (as NO_2)	658	53	<0.004	0.2	–
NH_3	1160	4	0.01	0.02	7 days
N_2O	18	6	0.3	–	4 years
Hydrocarbons	200	88	<0.001	–	?
CH_4	1600	–	1.5	2.5	4 years
Particulates	3700	3900	–	–	–
O_3	–	–	0.03	0.5	–

Source: Modified from Freedman (1995)

is usually vented to the atmosphere. Hard coals mined in eastern North America contain 1–12% sulphur, while softer coals from western regions have <0.3–1.5%; crude oil has 0.8–1.0%; residual fuel oils such as bunker-C have 0.3–0.4%; and motor fuels such as diesel and gasoline have 0.04–0.05%. Other important sources of SO_2 emissions are manufacturing processes (accounting for 23% of global emissions) and the smelting of metal ores (accounting for 7%).

Anthropogenic emissions of SO_2 have increased enormously since the beginning of the Industrial Revolution. Emissions in 1860 were about 5 million t, compared with about 130 million t/y in the late 1980s. Since then, some wealthier countries have invested in clean-air technologies for power plants and for other industrial applications of coal and oil, in order to reduce emissions of damaging SO_2 to the atmosphere. The clean air actions include:

- installation of technologies to capture SO_2 from post-combustion waste gases (also known as flue-gas desulphurization, or "scrubbing")
- removal of some of the sulphur content of fuels (also known as fuel desulphurization or "coal washing")
- installation of particulate-control devices (such as electrostatic precipitators) to greatly reduce emissions of particulates (although this has little effect on SO_2 emissions)
- switching to no-sulphur fuels such as natural gas, or to no-sulphur energy sources such as hydroelectricity and nuclear power
- energy conservation to reduce the overall demands for fuels and the associated emissions of air pollutants
- building taller smokestacks to disperse emissions of SO_2 more widely, helping to decrease local, ground-level pollution

However, future global emissions are bound to increase. China, India, and other rapidly industrializing countries supply much of their burgeoning energy needs by burning coal and petroleum fuels. In China, for example, coal is the major source of industrial energy, accounting for 76% of the energy supply in 1986. Due to increasing industrialization, emissions of SO_2 in China increased from 10 million t in 1980 to 14 million t in 1985, a 40% increase in only six years.

The quantities of SO_2 emissions differ greatly among nations, depending on their populations, the nature of their industrialization, and the types of fuels they use. For instance, Canadian emissions of SO_2 are about 16% of those of the United States (Table 16.2). However, the population of Canada is only about 11% that of the United States (Chapter 10). Therefore, per capita emissions of SO_2 are about 48% larger in Canada.

TABLE 16.2 COMPARISON OF EMISSIONS OF AIR POLLUTANTS IN CANADA AND THE UNITED STATES

Data are for 1991, in units of million tonne/year. "%" refers to percentage of total emissions of each stated type by country.

SOURCE OF EMISSIONS	SO_2		NO_X (AS NO_2)		CO	
	(10^6 t/y)	%	(10^6 t/y)	%	(10^6 t/y)	%
CANADA						
Transportation	–	–	1.22	63	7.5	70
Industrial processes	2.51	76	0.30	16	0.9	9
Stationary sources	0.62	19	0.30	16	0.2	1
Miscellaneous	0.18	5	0.10	5	2.1	20
Total	3.31		1.92		10.7	
UNITED STATES						
Transportation	–	–	8.03	43	48.3	78
Industrial processes	5.0	24	3.81	20	8.3	13
Stationary sources	14.4	70	6.02	32	0.6	1
Miscellaneous	1.3	6	0.90	5	4.9	8
Total	20.7		18.76		62.1	

Source: World Resources Institute (1994) and Freedman (1995)

About 76% of Canadian emissions of SO_2 are from large industrial sources, particularly metal smelters, while 19% are from "stationary sources" such as fossil-fuelled power plants. In comparison, 70% of United States emissions are from power plants, and 24% from industrial sources. Most of this difference is due to two factors: a relatively large proportion of electricity generation in Canada is from nuclear and hydro technologies, which do not emit SO_2; and SO_2-emitting metal smelting is a major Canadian industry.

Because of human-health and environmental damages associated with SO_2 and other air pollutants, most industrialized nations have taken measures to reduce their emissions. In Canada, for example, total emissions of SO_2 were reduced by about 20% between 1980 and 1990, compared with a 9% reduction in the United States during the same period. In both countries the reductions have been achieved by:

- switching to low- or no-sulphur fuels for some major uses, especially for electricity generation
- removing sulphur from some fuels prior to combustion, mostly by coal washing
- reducing energy demands by conservation
- installing scrubbers to remove SO_2 from post-combustion waste gases before they are vented to the atmosphere

Sources of Sulphide Emissions

In contrast to SO_2, most global emissions of sulphide gases are from natural sources (Table 16.1). The most important sources are H_2S emitted from oxygen-poor sediments in shallow marine and inland waters, and dimethyl sulphide $(CH_3)_2S$ produced by marine phytoplankton and out-gassed from oceanic waters. The natural emission of H_2S is about 100 million t/y, and of dimethyl sulphide about 15 million t/y (both expressed as "sulphur equivalent"; that is, as tonnes of S). Anthropogenic emissions of H_2S are smaller, about 3 million t/y, mostly associated with chemical industries, sewage-treatment plants, and agricultural livestock manure.

The global total emissions of all sulphur-containing gases is equivalent to about 251 million t/y of sulphur. About 41% of this global emission is from anthropogenic sources and 59% from natural sources.

Clean air typically contains less than 0.2 ppb of SO_2 or H_2S. Concentrations of SO_2 and H_2S in air polluted by emissions are highly variable. However, they are typically about 0.2 ppm in urban atmospheres, and can exceed 3 ppm close to large emission sources.

Toxicity of Sulphur Gases

Concentrations of H_2S in the environment are usually too low to be toxic to plants. However, concentrations of SO_2 in cities and near industrial sources are often high enough to injure wild and cultivated plants. Near smelters, vegetation has been severely damaged, as we examine later.

An exposure to 0.7 ppm SO_2 for one hour will result in acute injury to most plant species, as will an exposure to 0.2 ppm over an 8-hour period. It is important to note, however, that some species of plants are particularly sensitive to SO_2 exposures (that is, they are hypersensitive), and can suffer acute injuries at SO_2 concentrations smaller than those just noted.

In addition, plants often exhibit reductions in yield when exposed to concentrations of SO_2 smaller than those required to cause acute injuries. This type of response, occurring without symptoms of acute tissue damages, is referred to as "**hidden injury**." Hidden injuries to wild and agricultural vegetation have been documented by enclosing plants in chambers and exposing them either to ambient concentrations of SO_2 or to air that has been filtered through activated charcoal, which removes this gas. If the plant productivity is greater in the filtered air, it follows that ambient concentrations of SO_2 are sufficient to cause hidden injuries. Studies of pasture grasses in England, for example, have found that exposure to SO_2 concentrations averaging only 0.04 ppm causes hidden injuries, manifested by reductions in yield.

Interestingly, the Canadian air-quality guideline for ambient SO_2 in the atmosphere is for a maximum of 0.34 ppm over a 1-hour exposure. This guideline is based on the concentration of SO_2 that is required to cause acute foliar (leaf) injuries to most agricultural plants. Although regions meeting this guideline would not have much acute damage to vegetation, relatively sensitive species might be damaged through hidden injuries, resulting in significant losses of yield.

Humans and most other animals are much less sensitive to SO_2 than plants are. Guidelines for allowable exposures of humans to SO_2 and other potentially toxic gases accommodate the fact that, in terms of dose received, longer-term exposures to low concentrations can be as important as higher acute exposures (Table 16.3).

Guidelines for occupational exposures to air pollutants are frequently greater than those for continuous exposures. In North America, it is recommended that occupational exposures to SO_2 be no higher than 2 ppm over the long term, and no higher than 5 ppm for shorter exposures. However, some people are relatively sensitive to SO_2, and concentrations less than 1 ppm can cause them to develop acute asthmatic or other distresses related to impaired pulmonary function. In addition, some studies have suggested that long-term exposures of large human

TABLE 16.3 | CANADIAN GUIDELINES FOR PERMISSIBLE EXPOSURES TO SEVERAL AIR POLLUTANTS

Note that in terms of the dose received by an organism, longer-term exposures to lower concentrations of chemicals can be equivalent to shorter-term exposures to higher concentrations. Data are in ppm (except O_3, in ppb).

POLLUTANT	AVERAGING TIME	MAXIMUM DESIRABLE CONCENTRATION	MAXIMUM ACCEPTABLE CONCENTRATION	MAXIMUM TOLERABLE CONCENTRATION
Sulphur dioxide	annual	0.01	0.02	–
	24 h	0.06	0.12	0.31
	1 h	0.17	0.34	–
Nitrogen oxide	annual	0.03	0.05	–
	24 h	–	0.11	0.16
	1 h	–	0.21	0.53
Carbon monoxide	8 h	5	13	17
	1 h	13	31	–
Ozone (ppb)	annual	–	15	–
	1 h	50	82	153

Source: Furmanczyk (1994)

populations to sulphate particulate aerosols (which are ultimately derived from gaseous SO_2) in cities may negatively affect health. The studies noted small increases in the incidence of diseases of the respiratory and blood-circulation systems, most probably in hypersensitive people.

NITROGEN GASES

Emissions and Transformations

The most important of the nitrogen-containing gases are nitric oxide (NO), nitrogen dioxide (NO_2), nitrous oxide (N_2O), and ammonia (NH_3). NO and NO_2 are often considered together as a complex, referred to as NO_x.

Ammonia, a colourless gas, is emitted mostly from wetlands, where it is produced during anaerobic decomposition of dead biomass. Natural emissions of NH_3 are about 1 billion t/y (Table 16.1). Sources of anthropogenic emissions include fossil-fuel combustion (4 million t/y) and animal husbandry (0.2 million t/y). The residence time of NH_3 in the atmosphere is about 7 days (the NH_3 eventually being oxidized to nitrate).

Nitrous oxide (N_2O) is a colourless, non-toxic gas that produces a mild euphoria when inhaled. This gas is also known as "laughing gas," and is used as a mild anesthetic in medicine, and sometimes as a recreational drug. Because N_2O is a rather unreactive compound, it has a long residence time in the atmosphere, about four years. Most N_2O emissions are associated with microbial denitrification in soil and water. These are equivalent to about 18 million t/y, while industrial emissions amount to about 6 million t/y. Agricultural soils fertilized with nitrate can have quite high rates of N_2O emission, and modern agricultural practices are thought to have increased global emissions of N_2O by about 40%.

Nitric oxide is a colourless and odourless gas, while nitrogen dioxide is reddish, pungent, and irritating to respiratory and eye membranes. Natural emissions of NO_x are about 430 million t/ (expressed as NO; the same emissions expressed as NO_2 are 658 million t/y). The most important natural emissions of NO_x are due to bacterial denitrification of nitrate in soils, fixation of atmospheric nitrogen gas (N_2) by lightning, and oxidation of biomass-nitrogen during fires.

Anthropogenic emissions of NO_x, about 35 million t/y (expressed as NO), result mostly from the combustion of fossil fuels, especially in automobiles and power plants (Table 16.1). These emissions are mostly NO, which is secondarily oxidized to NO_2 by reactions in the atmosphere. Ultimately, most atmospheric NO_x gases become oxidized to nitrate (NO_3^-), an ion that is important in the acidification of precipitation and ecosystems (see Chapter 18).

Toxicity of Nitrogen Gases

It is rare that concentrations of NH_3 or NO_x gases are high enough to injure vegetation. The environmental damages associated with NO_x gases include the photochemical reactions by which ozone, a much more toxic chemical, is produced (see below), and acidification of precipitation and ecosystems.

Ambient concentrations of NH_3 and NO_x are rarely high enough to bother humans. Guidelines for long-term exposures in an occupational setting are 25 ppm for NO and 5 ppm for NO_2. Occupational guidelines for short-term exposures are 35 ppm and 5 ppm, respectively. Intense occupational exposures to NO_x can cause impaired pulmonary function in humans.

ORGANIC GASES AND VAPOURS

Emissions and Transformations

Hydrocarbons are a diverse group of chemicals with molecular structures containing various combinations of hydrogen and carbon. The simplest hydrocarbon is methane (CH_4), a gas. Hydrocarbons with larger weights and more complex structures commonly occur as vapours, liquids, or solids (see Chapter 22). Other volatile organic compounds contain oxygen, nitrogen, and other light elements in addition to carbon and hydrogen, and include such groups as aldehydes, alcohols, and phenols.

The background concentration of methane in the atmosphere is about 1.5 ppm, while all other hydrocarbons and volatile organics together amount to less than 1 ppb (Table 16.1). Most emissions of CH_4 are natural, and are associated with the fermentation of organic matter by microbes in anaerobic wetlands. Smaller amounts of CH_4 are outgassed from fossil-fuel deposits and during wildfires, and to some degree from ruminant animals (such as cows and sheep) and termites, which produce CH_4 as they digest their plant foods. The global emissions of CH_4 are estimated to be about 1.6 billion t/y.

Atmospheric hydrocarbons other than CH_4 are referred to as *non-methane hydrocarbons*. Natural emissions of these and many organics occur mainly as gases and vapours evaporated from living vegetation, along with smaller quantities outgassed from deposits of fossil fuels. The highest emissions from forests typically occur during hot, sunny days. Natural emissions of non-methane hydrocarbons are estimated to be about 200 million t/y, compared with anthropogenic emissions of 88 million t/y (Table 16.1). The most important anthropogenic sources involve unburned fuel emitted from vehicles and aircraft, releases during fossil-fuel mining and refining, and evaporation of solvents and oil-based paints.

Toxicity of Organic Gases and Vapours

Organic gases and vapours can be toxic, but rarely are atmospheric concentrations high enough to damage vegetation or animals. The environmental importance of these gases and vapours lies mainly in their role in the photochemical reactions that produce toxic ozone. In addition, CH_4 is an important greenhouse gas (Chapter 17). In some workplaces, however, relatively toxic organics such as benzene and formaldehyde may be important pollutants.

Ozone and Other Photochemical Pollutants

There are two quite different ozone-related environmental issues: O_3 in the troposphere (ground-level ozone) and O_3 in the stratosphere. High concentrations of ozone are naturally present in the upper-atmospheric layer known as the stratosphere, located about 8–17 km above the Earth's surface, depending on latitude and season. Stratospheric O_3 causes no damage and is not an air pollutant. Rather, by absorbing solar ultraviolet radiation, stratospheric O_3 helps to protect organisms on Earth's surface from many of the damaging effects of exposure to this harmful part of the electromagnetic spectrum (see below). In contrast, O_3 in the lower atmosphere (i.e., the troposphere) is an important air pollutant, known to cause extensive damage to vegetation, materials, and human health. Tropospheric O_3 is known as ground-level ozone.

Ground-Level Ozone

Ground-level ozone (O_3) is the most damaging of the so-called **photochemical air pollutants**. Less important are peroxyacetyl nitrate (PAN), hydrogen peroxide (H_2O_2), and other oxidant gases. **Oxidizing smogs** are rich in O_3 and these other oxidant gases. These chemicals are all called **secondary pollutants**, meaning that they are not actually emitted to the atmosphere (as are **primary pollutants** such as SO_2 and NO_x). Instead, they are synthesized within the atmosphere by photochemical reactions (chemical reactions requiring light). These proceed at faster rates (resulting in a build-up of O_3 in the lower atmosphere) if NO_x and hydrocarbons are present in high concentrations — a condition typically due to anthropogenic emissions.

Some localities tend to develop weak atmospheric inversions in the morning (see In Detail 16.1). Because inversions are relatively stable and resist the in-mixing of cleaner, ambient air, they encourage the development of oxidizing smogs during the morning and early afternoon. Later in the day, the inversion is broken up by stronger winds and the air pollution is dispersed. Such inversions and their associated ozone-rich smog are common phenomena around Vancouver, the Los Angeles basin, Mexico City, and some other places.

The concentrations of ambient, ground-level O_3 vary greatly among different regions of North America. Average concentrations in the southwestern United States are relatively high, at about 100 ppb, and generally range from 40 to 60 ppb in other regions of the United States and in Canada.

Canada, the United States, and other countries have developed air-quality standards for O_3. These are intended to reflect concentrations that would prevent severe damage to agricultural and wild vegetation. For some time, the United States O_3 standard was 80 ppb (for an average

1-hour exposure), but in 1979 this was relaxed to 120 ppb. The authorities made the change because the original standard (i.e., 80 ppb) was frequently exceeded over large regions, and was therefore essentially unenforceable. In fact, even the 120 ppb criterion is commonly exceeded in some regions, particularly in the southwestern United States.

The Los Angeles basin suffers especially intense photochemical air pollution. Concentrations of O_3 can exceed 500 ppb (1-hour average), and they typically exceed 100 ppb for more than 15 days during the summer. Maximum O_3 concentrations are lower in other cities of North America, typically reaching up to 150–250 ppb (1-hour average). These concentrations are well within the range at which O_3 can cause acute injuries to plants, which is why O_3 is considered such an important air pollutant. The emissions of ozone precursors (NO_x and hydrocarbons) occur mainly in cities, but extensive ecological damage is caused when polluted air masses are transported over rural areas with agricultural or natural vegetation.

In Canada, the regions that most often experience ozone pollution are the lower Fraser Valley in southwestern British Columbia, the Windsor–Quebec City corridor, and the southern Maritimes (Environment Canada, 1996). The lower Fraser Valley, a coastal lowland bounded by mountains to the east, is frequently subject to atmospheric inversions and receives large emissions of ozone precursors from the Greater Vancouver area to the west. Some places in the lower Fraser Valley have maximum O_3 concentrations of about 100 ppb.

The Windsor–Quebec City corridor likewise has local emissions associated with people and industries (about 60% of the Canadian population lives in the region), but this area also receives ozone and its precursors in airmasses blowing from industrial regions in the northeastern United States. Many places in the corridor have maximum O_3 concentrations of about 110–160 ppb, and values of 190 ppb have been measured.

The southern Maritime area has a moderately large population, but is affected mainly by ozone-rich airmasses blowing into the region from industrial parts of southern Ontario, Quebec, and the northeastern United States. Some places in southern New Brunswick and western Nova Scotia have maximum O_3 concentrations of about 90–110 ppb.

Toxicity of Ozone

Humans and some animals are rather sensitive to O_3 exposures, which can irritate and damage membranes of the eyes and respiratory system and cause some loss of pulmonary function. The guideline for long-term exposure to O_3 in an occupational setting is 100 ppb, with 300 ppb for short-term exposures. Sensitive people can, however, be severely affected by O_3-related symptoms at lower concentrations. Exposure to O_3 can result in asthmatic attacks and can exacerbate bronchitis and emphysema.

Ozone causes important damage to wild and agricultural plants over widespread areas. Distinctive foliar injuries damage the photosynthetic abilities of the plants and therefore reduce productivity. Acute foliar injuries are caused to most species by 2–4-hour exposures to 200–300 ppb O_3, while long-term exposures to only 40–100 ppb may cause hidden injuries (and reduced yield). However, many species of plants are more sensitive to O_3 and suffer acute and hidden injuries at lower concentrations. Some varieties of tobacco (*Nicotiana tobacco*), for example, can suffer acute foliar injuries from a 2–3-hour exposure to only 50–60 ppb, and spinach (*Spinacea oleracea*) from 1–2 hours at 60–80 ppb. Sensitive species of conifer trees may suffer acute injuries from exposures to 80 ppb O_3 over 12 hours.

Researchers have grown agricultural plants in chambers receiving ambient air, or in chambers with air filtered through charcoal, which removes O_3. These studies have been extremely useful in defining the extent of the damage caused to agricultural crops by exposure to ambient O_3. One series of field experiments demonstrated that crop yields were reduced in all regions of the United States. The worst damage occurred in the southwest, where sunny conditions and large emissions of NO_x and hydrocarbons result in relatively high O_3 concentrations. That study estimated that crop damages due to O_3 were equivalent to 2–4% of the total agricultural yield in the United States, with economic losses equivalent to more than $5 billion/y. Because O_3-related damages to vegetation occur over extensive areas of North America, O_3 is by far the most important air pollutant in agriculture, and probably the leading air pollutant in forests and other natural ecosystems as well.

Stratospheric Ozone

In contrast, ozone in the stratosphere protects life on Earth. The fact that it is being destroyed by anthropogenic air pollutants is cause for alarm.

Ozone is produced in the stratosphere by natural reactions. These involve the absorption of solar ultraviolet radiation by oxygen molecules (O_2), forming highly reactive oxygen atoms (O) which join with other O_2 molecules to form O_3 (see In Detail 16.2). These reactions proceed relatively quickly in the stratosphere because high-energy ultraviolet radiation is abundant there. As a result, O_3 concentrations are typically 200–300 ppb in the stratosphere, about ten times higher than in the ambient troposphere.

Stratospheric O_3 provides a critical environmental service. It efficiently absorbs high-energy ultraviolet (UV) radiation, which, if intense, can be extremely damaging to organisms. In particular, the genetic material DNA (deoxyribonucleic acid) is a strong absorber of UV and can be damaged by UV radiation. This can increase the risk of developing skin cancers, including melanoma, an often-fatal disease. Other health risks from UV exposure include the development of cataracts in the eyes and suppression of the immune system. UV radiation also damages plants, in part because chlorophyll (the major photosynthetic pigment) can be degraded by UV absorption, leading to decreases in productivity. The waxy covering of the cuticle of plants can also be damaged by UV radiation.

Stratospheric O_3 can be destroyed by various processes, including reactions with the trace gases NO_x and N_2O, and with chlorine, bromine, and fluorine. Because of anthropogenic emissions, the concentrations of some of these O_3-consuming chemicals have been increasing in the stratosphere, leading to concerns about possible depletions of stratospheric O_3. It is widely believed that emissions of chlorofluorocarbons (CFCs), particularly the industrial gases known as "freons," have been especially important in this regard. Because CFCs are extremely unreactive in the troposphere, they eventually migrate to the stratosphere, where they are bombarded with ultraviolet radiation and slowly degrade (photodissociate) to release free chlorine. The Cl efficiently reacts with and destroys O_3.

The O_3-destroying reactions proceed most effectively under extremely cold and stagnant conditions in the stratosphere, such as those occurring above polar latitudes at the end of the antarctic or arctic winters. These polar-focused O_3 depletions result in the development of so-called "ozone holes" during the early springtime. These phenomena have been observed regularly since the late 1970s, although they may have occurred before then without being recognized. The O_3 holes over Antarctica are particularly extensive, and typically involve decreases of total O_3 concentration of 30–50% during the springtime. Smaller depletions of O_3 occur above the Arctic, including northern Canada. The affected areas are less extensive than their counterparts in Antarctica.

The sizes of the ozone holes vary from year to year. In 1996, Antarctica had its largest-ever ozone hole, covering an area of about 20 million km^2. Although the seasonal depletions of stratospheric O_3 occur only over polar regions, lower latitudes may be affected when O_3-depleted polar air becomes dispersed during the breakup of the ozone holes, temporarily reducing stratospheric O_3 concentrations throughout the hemisphere.

In summary: stratospheric O_3 provides a critical UV shield that helps protect organisms on Earth's surface from damages caused by ultraviolet radiation. This function is impaired by the development of O_3 holes above the polar regions, a phenomenon believed to be caused largely by emissions of CFCs and other pollutants such as NO_x and N_2O into the atmosphere.

Air Pollution and Human Health

An extraordinary case of a natural emission of gas causing human deaths involved the release of a large volume of CO_2 from a lake in Cameroon, West Africa. Lake Nyos is a 200-m–deep volcanic lake in which the deep waters are supersaturated naturally with CO_2, similar to bottled soda water. One night in 1986, a large slump of sediment apparently eroded into the steep-sided lake, causing bottom waters to churn to the surface. The water de-gassed its CO_2 content into a dense, ground-level airmass, which then flowed into the low areas in the surrounding landscape. The CO_2-rich airmass asphyxiated almost 2000 sleeping people, plus thousands of cattle, and uncounted numbers of wild animals. Plants are much less vulnerable to CO_2 toxicity, so no vegetation was damaged by this extreme natural event.

Anthropogenic emissions of gaseous pollutants have frequently caused increases in human mortality and diseases. Some people, especially those with chronic respiratory or heart diseases, are particularly vulnerable to the effects of air pollution. Exposures of people to toxic gases can occur within several contexts, including the following:

The ambient environment: The urban atmosphere typically contains relatively high concentrations of potentially toxic chemicals. This is true in general; but air quality is especially bad during smog events, often caused by poor dispersion during an atmospheric inversion. Consequently, city people living their normal urban lives are routinely exposed to higher concentrations of air pollutants than those living in cleaner, rural environments.

The working environment: Many people are exposed to high concentrations of pollutants as a consequence of their occupation. Of course, the specific exposures depend on the job — workers in metal smelters may be exposed to sulphur dioxide and metallic particulates, while auto mechanics may be affected by exhaust fumes containing carbon monoxide and hydrocarbons, and dental workers may inhale mercury vapours.

The indoor environment is often contaminated by various gases and fumes. For example, space heaters, furnaces, and fireplaces burning wood, kerosene, or fuel oil may emit carbon monoxide into the indoor environment. All high-temperature combustions emit nitric oxide, and many synthetic materials and fabrics vent formaldehyde and other organic vapours. These chemicals can accumulate if indoor air is not exchanged frequently with cleaner, outdoor air.

The smoking of tobacco is the leading source of easily avoidable air pollution. Smoking is also the most important cause of preventable diseases, particularly lung cancer and heart ailments (see Chapter 15). People directly inhale a great variety of toxic gases and fumes when they smoke tobacco in cigarettes, cigars, or pipes. In addition, non-smokers are indirectly exposed to lower concentrations of all of those chemicals, because of the lingering residues of so-called "sidestream" and "secondhand" smoke that can occur in indoor atmospheres.

All of these exposures to air pollutants have important implications for human health. However, only the pollution of the ambient, urban environment is discussed in the following paragraphs.

Since the beginning of the Industrial Revolution in western Europe in the early 19th century, people living in cities and working in certain types of factories have been exposed to high concentrations of air pollutants. Especially important have been sulphur dioxide, soot, and other emissions associated with the combustion of coal and other fossil fuels. The most severe exposures to pollutants in urban environments typically occur during prolonged temperature inversions, which prevent the dispersion of emissions and result in smogs rich in SO_2 and sooty particulates.

Coal has long been used around the world to heat homes and other buildings. Its associated emissions have been regarded as a problem in cities and towns in Europe since at least 1500. With the beginning of the Industrial Revolution, which initially used coal as its principal energy source, air pollution worsened markedly. The first convincing link between air pollution and a substantial increase in the death rate of an exposed human population was made in relation to Glasgow, Scotland, in 1909, where about 1000 deaths were attributed to a noxious smog that developed during an episode of atmospheric inversion.

The most infamous "killer smog" in North America occurred in 1948 in Donora, Pennsylvania. A temperature inversion and fog persisted in the Donora Valley for four days, but emissions from several large factories continued. High concentrations of SO_2 and particulates built up in the local atmosphere. The smog resulted in increased mortality in the local population (20 deaths in a population of only 14 100). An additional 43% of the population became ill, 10% severely so. The most common symptoms were irritation of the eyes and respiratory tract, sometimes accompanied by coughing, headache, and vomiting.

The world's most notorious "killer smog" afflicted London, England, in 1952, when an extensive temperature inversion and fog stabilized over most of southern England. In London, emissions of pollutants, mostly associated with coal combustion, transformed the natural "white fog" into a noxious "black fog." Visibility was terrible — people lost their way while walking or driving, even falling off wharves into the Thames River. Airplanes became lost while trying to taxi at the airport. The air pollution episode lasted for four days, but was followed by another 14 days with a higher-than-usual death rate. Overall, about 3900 deaths were attributed to this episode of noxious air pollution. Most affected were people who were elderly or very young, or had severe, pre-existing respiratory or heart diseases.

Until about the early 1960s, severe episodes of urban air pollution were rather common in the industrialized cities of North America and western Europe. Most of these smogs were caused by the widespread burning of coal in fireplaces and furnaces in homes, in electrical utilities, and in factories. The poor-quality urban air affected the health of humans and animals, and also damaged vegetation. In many cities, only certain species of plants that can tolerate air pollution could grow. Examples of pollution-tolerant trees that are commonly grown in Canada include Norway maple (*Acer platanoides*), silver maple (*Acer saccharinum*), linden (*Tilia europaea*), tree-of-heaven (*Ailanthus altissima*), and ginkgo (*Ginkgo biloba*).

More recently, governments have brought in legislation that has required large reductions in the emissions of air pollutants, particularly in cities. In Canada, for example, the enactment of various federal, provincial, and municipal laws related to air pollutants has substantially improved urban air quality. Air quality has been similarly improved under legislation enacted in the United States, Britain, and other countries since the 1960s.

Of course, the killer smogs were extreme air-pollution events. More typically, the urban atmosphere is contaminated by smaller concentrations of SO_2, NO_x, O_3, volatile organic compounds, and particulates. Many studies have investigated the effects of chronic exposures to urban air contaminants on human health. The results of some studies suggests that modern urban air quality is sufficiently degraded to cause chronic damages to human health (especially by increasing the incidence of lung disease, asthma, and eye irritation), although other studies have not found this to be the case. In any event, effective actions have been taken in Canada and other relatively wealthy nations. Visibly threatening, even lethal, episodes of air pollution like those described above are now, it may be hoped, things of the past in those countries.

Unfortunately, in the cities of some rapidly developing countries, such as China, India, Indonesia, and Mexico, poorly regulated industrial growth is producing drastic declines in air quality. Although not yet well studied, these will be the modern tragedies of urban air pollution.

Case Studies of Air Pollution Damage

In this section we examine two case studies of ecological damage caused by air pollution. The first example describes natural air pollution at the Smoking Hills, a remote locality in the Northwest Territories. The second case examines the ecological effects of emissions from large smelters at Sudbury, Ontario.

Natural SO_2 Pollution at the Smoking Hills

The Smoking Hills are located in the Northwest Territories, on the mainland shoreline of the Beaufort Sea in the western Arctic. At various places along the coast, bituminous shales occur as seams in the seacliffs. The shales contain pyritic sulphur, which becomes oxidized to sulphate when exposed to atmospheric oxygen through erosion of the cliffs. The oxidation produces heat (i.e., the reaction is exothermic), which under insulating conditions can increase the temperature enough to spontaneously ignite the bituminous materials. These smoulder, releasing SO_2 to the atmosphere. The nearby tundra is fumigated with this toxic gas. The first recorded sighting of the Smoking Hills was in 1826 by John Richardson, an explorer. However, the burns were long known to Inuit in the area, and are probably thousands of years old.

Winds at the Smoking Hills often blow the SO_2-laden air masses (or "plumes") inland at ground level, so the tundra ecosystem is directly fumigated. The air pollution is most intense at the edge of the seacliff, where the plumes begin to spread inland. Concentrations of SO_2 at the cliff-edge are as high as 2 ppm, and rapidly decrease farther inland in a more-or-less *exponential* manner (Table 16.4). This *spatial gradient* of air pollution occurs because the gases become progressively diluted in the ambient atmosphere with increasing distance from the points of emission.

The pollution by SO_2 has severely acidified the soil. Acidic conditions in freshwater ponds reach pH 2 or less, compared with pH 8 or greater outside the fumigation area (see In Detail 19.1 for an explanation of pH). The ex-

TABLE 16.4 SULPHUR DIOXIDE CONCENTRATION AT THE SMOKING HILLS, NWT

The data are averages of 8- and 14-day sampling periods, respectively. The averages include times when SO_2 concentrations were high, as well as times when the sampling sites were not fumigated because the plumes were blowing out to sea.

DISTANCE FROM EDGE OF SEACLIFF (metres)	CONCENTRATION OF SO_2 (ppm) 1975	1977
20	0.27	0.61
40	0.20	0.53
80	0.17	0.43
160	0.10	0.25
320	0.06	0.16
640	0.02	0.11
1280	0.02	0.06
2560	0.01	0.04

Source: Gizyn (1980)

Natural air pollution at the Smoking Hills, at the edge of the Beaufort Sea in the Northwest Territories. Seams of bituminous shale have spontaneously ignited, and the sulphurous plumes blow inland and fumigate the nearby tundra.

Source: B. Freedman

treme acidification causes metals to become solubilized from minerals in the soil and aquatic sediments (Table 16.5). High concentrations of dissolved metals are toxic to terrestrial and aquatic organisms. In addition, sulphate occurs in high concentrations in both soil and water at the Smoking Hills. This is mostly a result of the dry deposition of SO_2 from the atmosphere, and its subsequent oxidation to SO_4^{2-} within the ecosystem (see Chapter 19).

The high concentrations of SO_2 in the air, and the acidity and soluble metals in the soil and water, render fumigated habitats toxic to most species of plants, animals, and microorganisms. Close to the edge of the seacliff in fumigated areas, where the pollution is most intense, no vegetation occurs at all — there is total ecological degradation. Farther inland, the pollution becomes less severe, and a few pollution-tolerant species of plants grow. The most notable of these species are arctic wormwood (*Artemisia tilesii*), polargrass (*Arctagrostis latifolia*), a lichen (*Cladonia bellidiflora*), and a moss (*Pohlia nutans*). These few pollution-tolerant species have replaced the much greater richness of species of the unpolluted tundra, which includes arctic willow (*Salix arctica*), mountain avens (*Dryas integrifolia*), and more than 70 other species (Freedman *et al.*, 1990).

Pollution-tolerant communities also occur in acidic ponds at the Smoking Hills. Even the most acidic ponds, which have a pH as low as 1.8, support at least six species of algae. These algae are extremely tolerant of acidity and dissolved metals, and are not found in non-acidic waterbodies. In contrast to the acidic ponds, unpolluted ponds are alkaline, with pHs greater than 8, and they support rich algal communities of more than 90 species. A few acid-tolerant invertebrates also occur in acidic ponds (but only at pHs greater than 2.8), including a crustacean (*Brachionus urceolaris*) and an insect midge (*Chironomus riparius*). The invertebrate fauna of non-acidic ponds is much richer in species and more productive (Havas and Hutchinson, 1983).

TABLE 16.5 | CHEMISTRY OF TUNDRA PONDS AT THE SMOKING HILLS, NWT

The data are in ppm, and are averages for the indicated sample sizes.

pH RANGE	NUMBER OF PONDS	ALUMINUM	IRON	MANGANESE	NICKEL	SULPHATE
1.8 – 2.5	4	270	500	61	6.3	8200
2.6 – 3.5	14	5.5	18	15	0.21	890
3.6 – 4.5	9	1.1	1.2	3.6	0.04	156
4.6 – 5.5	1	<0.6	0.5	2.3	0.04	813
5.6 – 6.5	1	<0.2	0.2	1.8	0.06	713
6.6 – 7.5	4	<0.8	<0.04	0.7	0.02	360
7.6 – 8.5	8	<0.7	0.1	<0.5	0.004	106
8.6 – 9.7	5	<0.2	0.1	0.5	0.01	31

Source: Havas and Hutchinson (1983)

The most important lesson to be learned from the Smoking Hills is that "natural" pollution can cause ecological damages as intense as those associated with anthropogenic pollution. Clearly, SO_2 can damage ecosystems regardless of the source of emissions. In addition, the natural pollution at the Smoking Hills has stressed ecosystems for a long time, at least thousands of years, and the ecological effects have likely reached an equilibrium. The study of the Smoking Hills gives some understanding of the long-term effects of severe air pollution, which include:

- simplification of biodiversity and ecosystems
- disruption of productivity and nutrient cycling
- the development of unusual communities of pollution-tolerant species

Soil and surface waters at the Smoking Hills have been severely acidified by the deposition of sulphur dioxide. The acidity causes metals to go into solution, exacerbating the toxic conditions. This pond has been affected by atmospheric SO_2, as well as by acidic, metal-laden drainage water that has passed through roasted shales.

Source: B. Freedman

Pollution from Metal Smelters near Sudbury

In 1883, while blasting bedrock during construction of the Canadian Pacific Railroad, a worker with some knowledge of prospecting discovered a rich body of metal-bearing ore in the vicinity of Sudbury, Ontario. The principal metals in the ores are nickel and copper. However, valuable quantities of iron, cobalt, gold, silver, and other metals are also produced from the Sudbury mines, as are sulphur and selenium.

One of the world's largest industrial complexes has been developed to mine and process the rich ore bodies near Sudbury. The facilities include underground mines, an open-pit mine, two ore-processing mills with tailings-disposal areas, several smelters, metal refineries, sulphuric-acid plants, and various secondary installations. The industrial activities around Sudbury provide the primary economic base for a regional population of more than 150 000 people.

The metals in the Sudbury ores occur as sulphide minerals. Consequently, an important step in processing is high-temperature roasting of the minerals in the presence of oxygen, which converts the sulphide-sulphur into gaseous SO_2. The roasting increases the concentration of valuable metals in the residual material, which can then be smelted and refined into pure metals (see also Figure 13.1).

The large-scale roasting and smelting have resulted in huge emissions of SO_2 and metal-containing particulates to the atmosphere in the Sudbury area, causing severe pollution and ecological damage. Until 1928, the ore roasting was conducted in huge open pits known as roast beds. These consisted of a layer of locally harvested cordwood covered with heaps of sulphide ore. The wood was ignited, and the resulting heat ignited the metal sulphides, releasing additional heat because the oxidation reaction is exothermic. Eventually, the roast bed was hot enough to start a self-sustaining combustion of the ore, which continued to burn and smoulder for several months. After the roasting was completed, the combustion was quenched with water. When the nickel and copper concentrates had cooled, they were collected and shipped to a smelter, and then to a refinery, for further processing.

As is evident in the accompanying photographs, this crude roasting process resulted in intense, ground-level fumigations of the landscape with toxic SO_2, acidic mists, and metal-containing particulates. The pollution devastated ecosystems near the roast beds. The denuding of the terrestrial ecosystems resulted in massive erosion of soil from slopes, exposing the bedrock, which became pitted and blackened by reaction with the sulphurous fumes.

About 30 roast beds operated in the Sudbury region. These emitted an estimated 270 000 t/y of SO_2, plus huge but undocumented amounts of metallic particulates. These early, extremely toxic, ground-level emissions caused the worst of the ecological damage in the Sudbury area.

In 1928, the government of Ontario prohibited further use of roast beds. All roasting was then conducted at smelters located at Coniston, Copper Cliff, and Falconbridge, all in the vicinity of Sudbury. Smelters are huge industrial facilities that contain their roasting chambers within buildings. Most of their emissions of waste gases and particulates are vented high into the atmosphere through smokestacks, which greatly reduces the severity of ground-level pollution.

The largest smelter was built at Copper Cliff in 1929. Initially it had a single smokestack. Two others were added in 1936. In 1972, the three stacks were replaced with a single, 381-m–tall "superstack" (the world's tallest chimney). At the same time, the Coniston smelter was closed and its

A view of a roast bed near Sudbury, around 1925. The roast bed had a bottom layer of wood (in the foreground), upon which heaps of sulphide ore were piled (background), using the track-mounted gantry with its continuous-feed apparatus.

Source: International Nickel Company Archives

Heat from the burning wood ignited the sulphide ore, which then smouldered for several months, giving off dense plumes of SO_2 and metal-laden particulates. After the sulphur was oxidized and driven from the ore into the atmosphere, the fires were quenched and the metal concentrates collected and taken away for further processing. The toxic plumes killed the local vegetation, caused local soil and surface waters to become toxic because of acidity and metals, and resulted in severe erosion and exposure of naked bedrock.

Source: International Nickel Company Archives

A hillside damaged by emissions of pollutants close to the Copper Cliff smelter near Sudbury. The worst damage was caused by roast beds, but fumigations from the smelter have also been important. Following the devastation of the forest that once grew here, soils eroded and collected in nearby basins. The exposed, naked bedrock became blackened and pitted through reactions with the acidic fumigations.
Source: B. Freedman

production shifted to Copper Cliff. The Falconbridge smelter, with smokestacks of 93 and 140 m, is owned by another company and continues to operate.

The commissioning of the superstack in 1972 allowed pollutants to be vented high enough into the atmosphere to make ground-level fumigations infrequent. This resulted in a great improvement of air quality in the Sudbury area. Tall smokestacks facilitate the dispersion and mixing of emissions into ambient air. This is known as the "dilution solution to pollution."

Because the superstack is so tall, its emissions are well dispersed into the atmosphere. Little of the vented SO_2 is deposited locally, a fact reflected by the relatively good air quality in the region since 1972. In fact, studies have indicated that only about 1% of the SO_2 emissions are deposited within 40 km of the superstack. This means that 99% of the SO_2 is exported over longer distances, contributing to the acidification of precipitation over a large region (see Chapter 19).

In addition to the use of tall smokestacks, other technologies have been installed to reduce the emissions of toxic chemicals. As a result, the environment in the Sudbury region has been improved. Devices such as electrostatic precipitators recover metal-containing dusts from the smelter flue-gases, while SO_2 emissions have been reduced by installing wet scrubbers, building sulphuric-acid plants, and constructing a facility to separate iron sulphides from the more valuable sulphides of nickel and copper.

Emissions of SO_2 in the Sudbury area peaked during 1965–1970, when annual discharges from the three smelters averaged 2.7 million tonnes. At that time Sudbury was the world's largest source of anthropogenic SO_2 emissions, responsible for about 4% of global releases through human activities. Emissions of SO_2 from the Sudbury smelters have decreased greatly since that time, to about 1.4 million t/y during 1973–1977, 0.39–0.94 million t/y during 1980–1989, and 0.24–0.64 million t/y during 1990–1994. The decreased emissions are due mainly to expensive investments in pollution-abatement technologies, including equipment for flue-gas desulphurization. Still, the emissions of SO_2 from the Sudbury smelters remain very large.

Clearly, the post-superstack air quality in the Sudbury region represents a great improvement over the sulphurous past. Toxic fumigations with SO_2, acidic mists, and metal particulates were much more frequent and intense when roast beds were in use, as well as prior to 1972 when the smelters had shorter stacks and little pollution-control technology had been installed. In large degree, the extensive ecological damage in the Sudbury region resulted from the earlier emissions of pollutants. Other disturbances added to that damage, however, including the clear-cutting of forests to provide fuel for roast beds, and the starting of fires both by prospectors and by sparks from steam-powered railroad engines. The modern emissions of SO_2 and other pollutants in the Sudbury region, while still large, are well dispersed and only infrequently cause acute ecological damages. In fact, a substantial ecological recovery has occurred since the superstack was commissioned in 1972.

Large areas of land and surface waters in the Sudbury region were severely damaged by pollution from the roast-beds and smelters. Before 1972, the smelters were large point sources of emissions, so the severity of the ecological damages decreased rapidly with distance. Over the years, ecologists have documented vegetation damages in the Sudbury area. Around 1970, about 100 km^2 of land around the smelters were characterized as "severely barren," and another 360 km^2 had "impoverished" vegetation, including a lack of conifers in the forest (Watson and Richardson, 1972). White pine (*Pinus strobus*), an economically important tree in the region, is sensitive to air pollution. This species showed diagnostic SO_2-injuries over an area of about 6400 km^2.

Pollution was especially damaging around Copper Cliff, the location of the largest smelter. The most degraded terrestrial habitats occur within several kilometres of this facility, and support almost no vegetation. Hills and slopes in this zone are denuded of vegetation, their soils are eroded, and the exposed bedrock has been blackened by reaction with acidic fumigations. Only a few plant species that have evolved pollution-tolerant populations grow in this area, including several species of grasses, other herbaceous plants, and stunted, shrub-sized plants of red maple (*Acer rubrum*).

The intensity of pollution-related stresses rapidly lessens at greater distances from the smelters (that is, a spatial gradient exists), and damages to vegetation are correspondingly fewer. About 3 to 8 km from the Copper Cliff smelter, remnants of forest survive in places where the local topography provided some protection from fumigations. However, denuded and blackened hilltops are common in this patchily vegetated zone. Trees in the remnant stands are stunted, with many dead branches and other injuries. These trees include such pollution-tolerant species as red maple, white birch (*Betula papyrifera*), red oak (*Quercus rubra*), trembling aspen (*Populus tremuloides*), and large-toothed aspen (*P. grandidentata*). The forest cover beyond 8 km is almost continuous, but the biomass, productivity, and biodiversity of the stands are impoverished. Beyond about 20–30 km from the smelters, the forests are little affected by pollution, and stands consist of a mixture of conifer and hardwood trees, as is typical of the region (Freedman and Hutchinson, 1980).

Lakes close to the Sudbury smelters have also been severely degraded by atmospheric pollution. The lakes have been acidified by the deposition of SO_2, and contain high concentrations of toxic nickel, copper, and other metals. The toxified water bodies contain species-poor communities of pollution-tolerant species of algae, plants, and zooplankton. They lack fish, mostly because of their acidity.

Improvements in ground-level air quality since the building of the superstack in 1972 have resulted in dramatic ecological recoveries in the Sudbury area. Where eroded soils collected in flat, moist basins, wet meadows have developed. These are dominated by hairgrass (*Deschampsia caespitosa*), the local populations of which are genetically tolerant to toxic nickel and copper in the soil (see Chapter 18). Other plant communities have also benefitted greatly from the reductions in air pollution, although their ecological recovery is still impeded by residual soil toxicity associated with acidification and metals.

Lakes have also begun to recover. For example, the pH of a small lake near the closed Coniston smelter increased from 4.1 to 5.8 between 1972 and 1984, while the aqueous concentrations of copper, nickel, and sulphate declined by 60–90% (Hutchinson and Havas, 1985). The reduced toxicity of waterbodies has resulted in increased aquatic productivity, and species previously excluded by toxicity have invaded the lakes.

The case of Sudbury is perhaps the world's best documented example of ecological damage caused by toxic gases and metals emitted from smelters. There are, however, additional examples of ecological damage around smelters in Canada, affecting smaller areas. These examples include smelters processing lead and zinc at Trail, British Columbia; gold at Yellowknife, Northwest Territories; nickel at Thompson, Manitoba; copper at Flin Flon, Manitoba; iron at Wawa, Ontario; and copper at Rouyn-Noranda, Quebec.

All of these smelters are large, point sources of emissions of pollutants into the atmosphere. All have developed spatial gradients in the intensity of pollution, which decreases exponentially with increasing distance from the source of emissions until the ambient condition is reached. The patterns of ecological damages track these spatial gradients of toxic stress.

Key Terms

atmospheric inversion
smog
hidden injury
photochemical air pollutants
secondary pollutants
primary pollutants
oxidizing smogs

Questions for Discussion

1. Compare the natural and anthropogenic emissions of sulphur and nitrogen compounds. Explain why the sources of emissions vary between regions and countries.
2. Why is too much ozone in the lower atmosphere an environmental problem? Contrast this with the stratosphere, where too little ozone is considered a problem.
3. Explain the difference between primary and secondary air pollutants, giving examples of each.
4. Compare, in broad terms, the patterns of ecological damage caused by natural air pollution at the Smoking Hills with those caused by smelters near Sudbury.
5. Existing clean-air technologies could be used to greatly reduce the emissions of air pollutants. Considering the damage they cause to human health, ecosystems, and other values, why are these technologies not being used more extensively? Your explanation should consider factors associated with economics, politics, scientific uncertainty about air pollution damage, and the benefits of cleaner air.

References

Barker, J.R. and D.T. Tingey. 1992. *Air Pollution Effects on Biodiversity*. New York: Van Nostrand Reinhold.

Environment Canada. 1996. *NO_x/VOC Science Program. Data Analysis Work Group Report*. Ottawa: Environment Canada. Draft Report.

Freedman, B. 1995. *Environmental Ecology, 2nd ed*. San Diego, CA: Academic.

Freedman, B. and T.C. Hutchinson. 1980. Long-term effects of smelter pollution at Sudbury, Ontario, on forest community composition. *Can. J. Bot.*, **58**: 2123–2140.

Freedman, B., V. Zobens, and T.C. Hutchinson. 1990. Intense, natural pollution affects arctic tundra vegetation at the Smoking Hills, Canada. *Ecology*, **71**: 492–503.

Furmanczyk, T. 1994. *National Urban Air Quality Trends, 1981–1990*. Ottawa: Environmental Protection Service. Environment Canada, EPS 7/UP/4.

Gizyn, W. 1980. *The chemistry and environmental impact of the bituminous shale fires at the Smoking Hills, NWT*. M.Sc. Thesis, University of Toronto.

Havas, M. and T.C. Hutchinson. 1983. The Smoking Hills: Natural acidification of an aquatic ecosystem. *Nature*, **301**: 23–27.

Hemond, H.F. and E.J. Fechner. 1994. *Chemical Fate and Transport in the Environment*. San Diego, CA: Academic.

Hutchinson, T.C. and M. Havas. 1985. Recovery of previously acidified lakes near Coniston, Canada following reductions in atmospheric sulphur and metal emissions. *Water, Air, & Soil Pollution*, **20**: 20–32.

Roberts, T.M. 1984. Effects of air pollutants in agriculture and forestry. *Atmospheric Environment*, **18**: 629–652.

Stern, A.C. 1976. *Air Pollution, 3rd ed*. San Diego, CA: Academic.

Shriner, D.S. 1990. Responses of vegetation to atmospheric deposition and air pollution. In: *Acidic Deposition: State of Science and Technology. Vol. III. Terrestrial, Materials, Health, and Visibility Effects*. Washington, DC: Superintendent of Documents, U.S. Government Printing Office.

Watson, W.Y. and D.H. Richardson. 1972. Appreciating the potential of a devastated land. *Forestry Chronicle*, **48**: 312–315.

Winterhalder, E.K. 1978. A historical perspective of mining and reclamation in Sudbury. *Proceedings, 3rd Annual Meeting, Canadian Land Reclamation Association*. Guelph, ON.

Wise, W. 1968. *Killer Smog*. New York: Ballantyne.

World Resources Institute (WRI). 1994. *World Resources, 1994–95. A Guide to the Global Environment*. New York: Oxford University Press.

17

ATMOSPHERIC GASES AND CLIMATE CHANGE

CHAPTER OUTLINE

Introduction

Earth's energy budget and the greenhouse effect

Carbon dioxide in the atmosphere

Climate change

Direct effects of CO_2 on plants

Reducing atmospheric carbon dioxide

CHAPTER OBJECTIVES

After completing this chapter, you will be able to:

1. Outline the physical basis of Earth's greenhouse effect, and describe how human influences may be causing it to intensify.
2. Explain the term "radiatively active gas" (or RAG).
3. Describe how the various RAGs vary in their effectiveness and their relative importance in Earth's greenhouse effect.
4. Identify which RAGs have increased in concentration in the atmosphere during the past century or so, and give the reasons for these changes.
5. Explain the probable climatic consequences of an intensification of the greenhouse effect, and describe the resulting ecological effects.
6. Discuss different strategies for reducing the intensity of the greenhouse effect.

Introduction

In this chapter we examine how Earth's naturally occurring "*greenhouse effect*" keeps the planet's surface relatively warm. We also describe how certain atmospheric constituents influence this phenomenon. These constituents are known as *radiatively active gases* (or *RAGs*; see below). It is well documented that the concentrations of some of these RAGs, particularly carbon dioxide, are increasing because of emissions associated with various human activities. Potentially, these increased emissions could intensify Earth's greenhouse effect, which could result in global warming. This would be an extremely important environmental change, with potentially devastating consequences for both natural ecosystems and human economic systems.

Earth's Energy Budget and the Greenhouse Effect

Earth's greenhouse effect is a well-understood physical phenomenon, thought to be critical in maintaining the planet's average surface temperature at about 15°C. Without this influence, Earth's average surface temperature would be about −18°C, or 33° cooler than it actually is. This would be frostier than organisms could tolerate over the long term, mainly because at −18°C, water is in a solid state. Liquid water is crucial to the proper functioning of organisms and ecosystems. At Earth's actual average temperature of 15°C, water is unfrozen for much or all of the year (depending on location). This means that enzymes can function and physiology can proceed efficiently, as can the many important ecological processes that involve liquid water.

The combustion of fossil fuels for transportation and industrial energy is the leading anthropogenic source of emissions of carbon dioxide to the atmosphere.

Source: Al Harvey/The Slide Farm

To understand the nature of Earth's greenhouse effect it is necessary to comprehend the planet's **energy budget**. As described in Chapter 4, an energy budget is a physical analysis that deals with:

1. all of the energy coming into a system
2. all of the energy going out
3. any difference that might be internally transformed or stored

Solar electromagnetic radiation is the major input of energy to Earth. On average, this energy arrives at a rate of about 8.4 joules/cm^2·minute. Much of the incoming solar radiation penetrates Earth's atmosphere and is absorbed by the surface of the planet. However, temperatures do not increase excessively, because Earth dissipates its absorbed solar energy by emitting long-wave infrared radiation. Earth's surface temperature is determined by the equilibrium rates at which solar energy is absorbed by the surface, and the absorbed energy is reradiated in a longer-wavelength form (see Figure 4.2, p. 47).

If Earth's atmosphere was transparent to the long-wave infrared radiated by the surface, then that radiation would travel unobstructed to outer space. However, this is not the case, because so-called **radiatively active gases** (or **RAGs**; sometimes known as "greenhouse gases") are present in the atmosphere. RAGs efficiently absorb infrared radiation, becoming heated as a consequence. They then dissipate some of this thermal energy through yet another reradiation. (This reradiated energy has a longer wavelength than the electromagnetic energy that was originally absorbed. This is necessary to satisfy the second law of thermodynamics.) The reradiated energy of the RAGs is emitted in all directions, including back toward Earth's surface. The net effect of the various energy transformations and reradiations involving atmospheric RAGs is a reduction in Earth's rate of cooling. Thus, the equilibrium temperature of the planet's surface is warmer than it would be if the RAGs were not present in the atmosphere.

The process just described is known as the **greenhouse effect** because its physical mechanism is similar to the warming of a glass-encased space by solar radiation. The encasing glass of a greenhouse is transparent to incoming solar radiation. The solar energy is absorbed by, and therefore heats, internal surfaces of the greenhouse, such as plants, soil, and other materials. These warmed surfaces then dissipate the absorbed energy by reradiating longer-wave infrared radiation. However, much of the infrared energy is absorbed by the greenhouse's glass and humid atmosphere, which are somewhat opaque to these wavelengths of electromagnetic radiation. The absorption of some reradiated infrared slows the rate of cooling of the greenhouse, causing it to heat up rapidly on sunny days. (In addition, a greenhouse is an enclosed space, so it traps heat because its warmed interior air cannot be dissipated by convection higher into the atmosphere, with cooler air drawn in below.)

Radiatively Active Gases

Water vapour (H_2O) is the most important of the radiatively active constituents of Earth's atmosphere, followed by carbon dioxide (CO_2). More minor roles are played by trace atmospheric concentrations of methane (CH_4), nitrous oxide (N_2O), ozone (O_3), carbon tetrachloride (CCl_4), and chlorofluorocarbons (CFCs). These latter compounds are, however, much stronger absorbers of infrared wavelengths than is CO_2; i.e., on a per molecule basis they are more efficient RAGs. A molecule of CH_4 is 11–25 times more effective than CO_2 at absorbing infrared radiation, while N_2O is 200–270 times more effective, O_3 2000 times, and CFCs 3000–15 000 times.

There is no evidence that the concentration of water vapour in the atmosphere has increased recently. However, concentrations of all the other RAGs have increased markedly during the past several centuries, because of emissions associated with human activities. Prior to 1850, the atmospheric concentration of CO_2 was about 280 ppm, while in 1997 it was about 358 ppm (this change is discussed in more detail in the next section). During the same period, CH_4 increased from 0.7 ppm to 1.7 ppm, N_2O from 0.285 ppm to 0.304 ppm, CCl_4 from essentially zero to 0.12 ppb, and CFCs from zero to 0.7 ppb. These concentration increases have been especially rapid since the middle of the 20th century, coinciding with enormous increases in human populations and the industrialization of their economies.

Because the various RAGs are known to help control Earth's greenhouse effect, it is reasonable to hypothesize that their increasing concentrations may intensify that process. A stronger greenhouse effect may lead to global warming. Such an environmental change could be considered an anthropogenic enhancement of Earth's naturally occurring greenhouse effect. Overall, the increased concentration of CO_2 has been estimated to account for about 60% of this possible enhancement of the greenhouse effect, while CH_4 is responsible for 15%, CFCs for 12%, O_3 for 8%, and N_2O for 5%.

Carbon Dioxide in the Atmosphere

Concentrations of CO_2 in the atmosphere have been increasing steadily for at least the past century. The data record supporting this change is excellent, and in fact demonstrates one of the most convincing examples of long-term changes in environmental chemistry. Atmospheric CO_2 has been monitored continuously since 1958 at an observatory located on Mauna Loa, a mountain on the island of Hawaii (Figure 17.1). These data clearly show steadily increasing concentrations of CO_2 in the atmosphere during the past four decades.

FIGURE 17.1 | INCREASES IN ATMOSPHERIC CO_2 CONCENTRATION

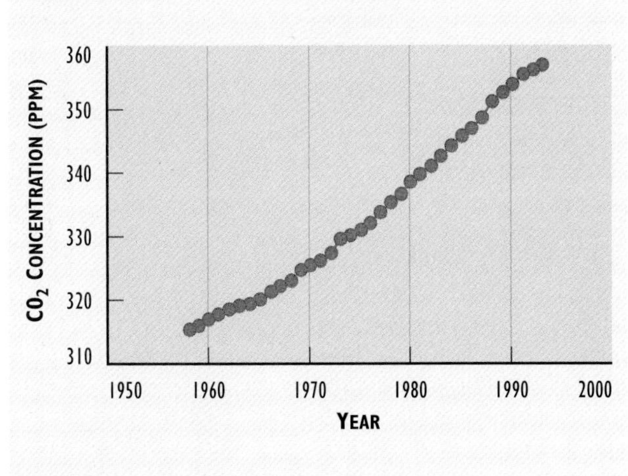

These data are from measurements made on Mauna Loa, Hawaii. Each datum represents an annual average.

Source: Boden *et al.* (1994)

A seasonal cycle of CO_2 concentration is illustrated in Figure 17.2, using data from a site in the Canadian Arctic. The annual periodicity is caused by high rates of CO_2 uptake by vegetation of the northern hemisphere during the growing season. This seasonal CO_2 fixation occurs at high enough rates to depress the concentration of this gas in the global atmosphere. The obvious long-term pattern, however, is one of a gradual net increase of CO_2 concentration over the years, as is clearly shown in Figure 17.1.

The increased concentrations of atmospheric CO_2 are due to emissions associated with various human activities. The two most important causes of anthropogenic CO_2 emissions, discussed in more detail in the following sections, are:

- the combustion of fossil fuels — a process which results in most of the carbon content of the fuel being oxidized to CO2, which is emitted into the atmosphere
- deforestation — an ecological change in which mature forests storing large amounts of organic carbon in their biomass are converted into ecosystems that contain much less, with the dif-

FIGURE 17.2 SEASONAL CHANGES IN ATMOSPHERIC CO_2 CONCENTRATION

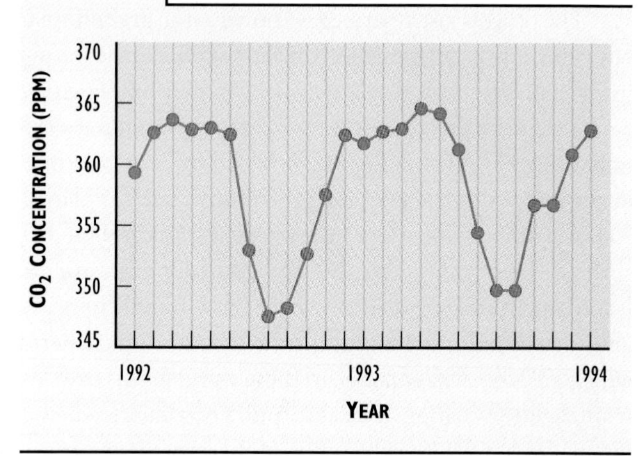

These data are based on measurements made at Alert, Ellesmere Island, Northwest Territories. A similar pattern is seen at Mauna Loa.
Source: Boden *et al.* (1994)

TABLE 17.1 EMISSIONS OF IMPORTANT GREENHOUSE GASES

Data are for 1991.

SOURCE OF EMISSIONS	EMISSIONS (10^6 tonne/year)		
	UNITED STATES	CANADA	GLOBAL
EMISSIONS OF CARBON DIOXIDE			
Industrial:			
Combustion of fossil fuels			
Coal	1804	94.7	8581
Liquid fuels	2034	178.1	9493
Natural gas	1052	127.9	3752
Natural gas flaring	7.5	4.5	256
Cement manufacturing	34.2	5.5	594
Land-use changes	22.0	25.0	3400
EMISSIONS OF METHANE			
Solid-waste disposal	8.4	1.20	40
Coal mining	9.1	0.63	40
Oil and gas production	5.3	0.80	26
Wet rice agriculture	0.75	0.00	72
Livestock	6.0	0.74	76
EMISSIONS OF CHLOROFLUOROCARBONS			
Various sources	0.090	0.008	0.40

Source: World Resources Institute (1994)

ference in carbon storage balanced by CO_2 releases to the atmosphere

CO2 Emissions from Burning Fossil Fuels

Fossil fuels are the most important source of energy in industrialized countries, followed by hydroelectricity, nuclear power, and relatively minor sources such as wood, solar, and wind energies (Chapter 13). The rates of utilization of coal, petroleum, and natural gas have increased enormously during the past century, mostly to satisfy surging energy demands for industry, transportation, and space heating.

Between 1860 and 1869, during the early part of the Industrial Revolution, the combustion of fossil fuels, mainly coal, resulted in the global emission of about 118 million t of CO_2–C/y (i.e., of carbon occurring within CO_2). By the late 1980s, global emissions from fossil fuel combustion had increased by a factor of more than 40, to 5–6 billion t/y. About 97% of the current industrial emissions of CO_2 is due to the combustion of fossil fuels, of which 42% is from liquid hydrocarbons, 38% from coal, and 17% from natural gas. The remaining 3% is associated with cement manufacturing.

The current global industrial emissions are equivalent to about 1.15 tC/person·year (in 1991; Table 17.2). Of course, per capita use of fossil fuels differs greatly among countries, depending on their degree of industrialization, types of energy sources, climate, and other factors. The per capita emissions of CO_2 are greatest in relatively wealthy, fuel-intensive, industrialized countries (Table 17.2). Several petroleum-producing countries have particularly large per capita emissions of CO_2 because huge quantities of natural gas are flared at their wellheads and refineries. For example, the United Arab Emirates has emissions of 9.95 tC/person·year.

Not surprisingly, people living in relatively poor, less-developed countries are responsible for much smaller emissions of CO_2 from fossil-fuel combustion. For example, Burundi, Cambodia, Chad, Nepal, and Uganda emit only 0.01 tC/person·year.

Future emissions of CO_2 from fossil-fuel combustion are predicted to be substantially larger than those occurring today, mainly because of the anticipated industrialization of less-developed countries as they develop economically. One prediction suggests that global emissions during the mid-21st century will be in the range of 8–15 billion t/y of CO_2–C, exceeding current levels by a factor as large as three.

TABLE 17.2 PER CAPITA EMISSIONS OF CARBON DIOXIDE BY SELECTED COUNTRIES

Data are for industrial sources of emission (mostly fossil fuels), in units of tonnes of CO_2–carbon per person·year in 1991 (1 t of CO_2–C is equivalent to 3.67 t of CO_2).

COUNTRY	EMISSIONS (t/y PER CAPITA)
GLOBAL AVERAGE	1.15
United Arab Emirates	9.95
United States	5.33
Canada	*4.15*
Australia	4.12
Saudi Arabia	3.81
Norway	3.75
Russia	3.36
Czechoslovakia (former)	3.33
Germany	3.31
United Kingdom	2.73
Japan	2.40
Poland	2.20
Italy	1.90
France	1.79
Sweden	1.70
Venezuela	1.68
South Korea	1.65
Mexico	1.07
Iran	1.01
China	0.60
Thailand	0.50
Brazil	0.39
Indonesia	0.25
India	0.22
Nigeria	0.22
Philippines	0.19
Guatemala	0.12
Viet Nam	0.08
Kenya	0.05
Bangladesh	0.04
Sudan	0.04
Haiti	0.03
Rwanda	0.02
Cambodia	0.01
Uganda	0.01

Source: World Resources Institute (1994)

CO2 Emissions from Clearing Mature Forests

Mature forests store large quantities of organic carbon in the living and dead biomass of their vegetation, on the forest floor, and in the soil. All other kinds of ecosystems, including relatively young forests that are regenerating from disturbance, store much less organic carbon. This observation suggests that whenever an area of mature forest is disturbed by harvesting its trees, or is cleared to provide new land for agricultural use, less organic carbon will be stored on the land. The depletion of stored carbon may be a relatively short-term phenomenon, as occurs when a harvested stand is allowed to regenerate back to another mature forest. However, where a forest is converted into an agricultural ecosystem, there is a long-term decrease in the amount of carbon stored on the land. In either case, the difference in the average quantity of organic carbon stored in the ecosystems is balanced by an emission of CO_2 into the atmosphere. The CO_2 release occurs by decomposition of the biomass of the mature forest or by fire.

It is well known that humans have caused enormous reductions in the area of mature forests in most regions of the world (Chapters 12 and 14). These changes began slowly, with the domestication of fire and its widespread use in improving the habitat of hunted animals. Deforestation proceeded more rapidly when it was discovered that fertile agricultural lands could be developed by removing the natural forest cover. (The harvested trees were also valuable commodities.) Deforestation has proceeded especially quickly during the past several centuries because of population growth and industrialization.

Prior to any substantial clearing of Earth's natural forests, the global terrestrial vegetation stored an estimated 900 billion t of organic carbon (Figure 17.3). About 90% of that carbon was stored in forests, of which 50% was in tropical forests. Now, only about 560 billion t of carbon are stored in terrestrial vegetation, a 38% decrease overall. Moreover, the quantities of global vegetation biomass are diminishing further, as additional areas of forest are converted into ecosystems, especially agroecosystems, that store less carbon.

During the 130-year period between 1850 and 1980, anthropogenic changes in land use (mostly conversions of forests into agricultural lands) resulted in the total emission of an estimated 108–120 billion t of CO_2–C. This quantity is about the same as the emissions due to fossil-fuel combustion during that same period (about 170–190 billion t of CO_2–C). More recently, in 1991, combustions of fossil fuels emitted about 6.0 billion t of CO_2–C into the atmosphere each year, while deforestation accounted for another 1.0 billion t/y.

Global emissions of CO_2 resulting from land-use practices in recent times are summarized in Table 17.3. The most important source of CO_2 emissions is conversions of mature forests into lands used for growing crops or grazing livestock. Because the forests previously stored large quantities of organic carbon, and the agricultural lands store much less, the result is a large emission of CO_2 for every hectare of forested land that is converted. Harvesting a forest's trees as a resource also results in a large emission of CO_2 into the atmosphere — a release which occurs relatively slowly if the wood is used to manufacture lumber, since this product has a rather long lifespan in furniture or buildings. However, the data in Table 17.3 suggest that the forestry-related emissions are largely balanced by subsequent regeneration of other forests.

Table 17.4 shows large differences between regions in their emissions of CO_2 from changes in land use. In North America, extensive forest clearing began when the continent was colonized by Europeans, and continued until the 1920s. Since then, however, large areas of economi-

TABLE 17.3 ANTHROPOGENIC EMISSIONS OF CO_2 TO THE ATMOSPHERE

These estimated global emissions are associated with changes in land use and the combustion of fossil fuels. A negative value indicates a net release of CO_2 to the atmosphere, while a positive number indicates a net withdrawal. Data are for the period 1958–1980.

CATEGORY	TOTAL EMISSIONS DURING 1958–1980 (10^9 tonne CO_2–C)
EMISSIONS FROM CHANGES IN LAND USE	
Forest conversion to cultivated land	−29.1
Forest conversion to grazed pasture	−11.1
Decay of wood in disturbed forests	−5.9
Harvesting of forests	−58.3
Regeneration of harvested forests	+45.8
Afforestation of agricultural lands	+1.3
Total terrestrial net flux	−57.3
COMBUSTION OF FOSSIL FUELS	−85.5

Source: Houghton (1991)

The conversion of carbon-dense ecosystems, such as forests, into agricultural and urban ecosystems which store much less carbon is an important source of CO_2 emissions. This site on the island of Sumatra has just had its tree cover felled and the resulting woody debris burned. The land will be planted with a variety of crops. Deforestation is proceeding rapidly in this region of Indonesia, and in most tropical countries.

Source: B. Freedman

cally marginal agricultural lands have been returned to forest. Overall, the net emission of CO_2 by changes in forest area has recently been approximately zero — that is, agricultural lands are regenerating back to forest about as quickly as forests elsewhere in North America are being converted into agriculture. The European situation is similar, because forest biomass (and carbon storage) has increased since the 1920s. In tropical countries, however, forests are being cleared rapidly, mostly to develop agricultural lands to provide livelihoods and grow food for increasing numbers of people.

TABLE 17.4 NET EMISSIONS OF CO_2 TO THE ATMOSPHERE AS A RESULT OF LAND-USE CHANGES IN SELECTED REGIONS

REGION	EMISSION OF CO_2-C (10^6 tonne/year) 1920	1980
North America	290	10
Europe	100	−50
Latin America	140	800
Tropical Africa	100	700

Source: Houghton *et al.* (1983)

Overall, in modern times, most CO_2 emissions associated with deforestation have been occurring in relatively poor, less-developed, tropical countries of Africa, Asia, and Latin America. In contrast, most CO_2 emissions from the combustion of fossil fuels have been occurring in relatively wealthy, industrialized, higher-latitude countries, of which Canada is a leading example.

Global Carbon Geochemistry

The anthropogenic influences on Earth's carbon budget are summarized in Figure 17.3, which shows the major compartments in which carbon is stored as well as the transfers between those compartments. Figure 17.3 greatly

FIGURE 17.3 | KEY COMPARTMENTS AND FLUXES OF THE GLOBAL CARBON CYCLE

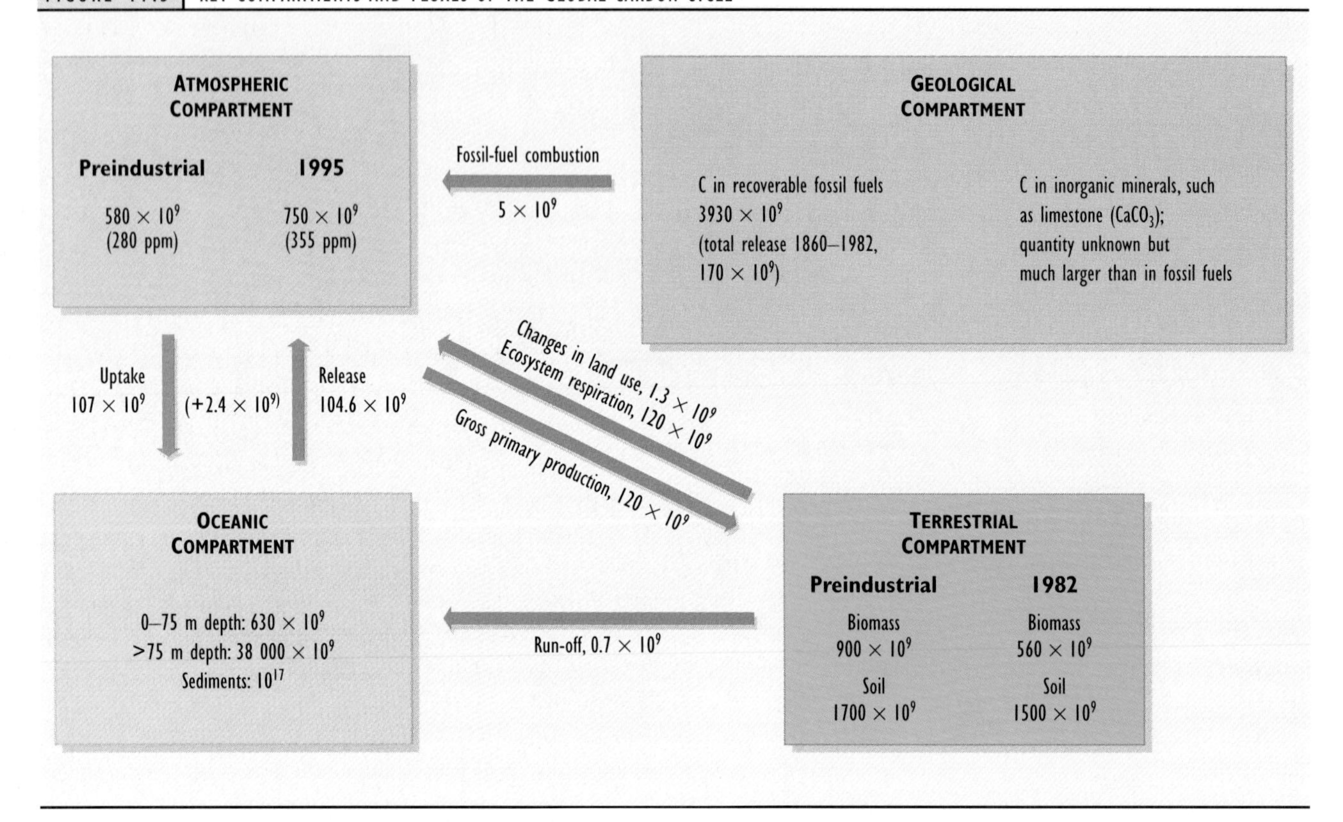

The amounts of carbon stored in the various compartments are in units of tonnes of carbon, while the transfers between compartments are in t/y of carbon.

Source: Blasing (1985), Solomon *et al.* (1985), and Freedman (1995).

simplifies the complex nature of the global carbon cycle. Nevertheless, some important conclusions relevant to the greenhouse effect can be drawn, as is described below.

Anthropogenic emissions have caused a 29% increase in the amount of CO_2 stored in the atmosphere, from about 580×10^9 t of CO_2–C in preindustrial times to 750×10^9 t in 1995. The atmospheric concentration of CO_2 has increased accordingly, from about 280 ppm to 355 ppm during the same period.

Before humans began to modify the character of Earth's ecosystems, particularly through deforestation, the global emission and fixation of atmospheric CO_2 were probably in balance. In other words, on a global basis, gross primary production (GPP) was approximately equal to ecosystem respiration (ER), and biologically fixed carbon was not changing over time. However, deforestation is now resulting in huge emissions of CO_2, amounting to about 1.3×10^9 t/y of CO_2–C. Overall, modern terrestrial ecosystems are storing about 38% less carbon in their vegetation and 12% less in soils, compared with preindustrial times.

Ultimately, the oceans are the most important sink for CO_2 emitted through human activities. The oceans have a net absorption of about 2.4×10^9 t/y of CO_2–C from the atmosphere. However, this is less than the anthropogenic emissions of 6.3×10^9 t/y of CO_2–C, and therefore the amounts of CO_2 stored in the atmosphere are increasing. The oceans have an enormous capacity for absorbing atmospheric CO_2, which is ultimately stored as calcium carbonate ($CaCO_3$), a mineral that accumulates in sediments (mostly as the shells of mollusks, forams, and other invertebrates). However, the rate of formation of $CaCO_3$ is affected by various factors, including the concentration of inorganic carbon in seawater. This concentration is determined by the rate at which CO_2 enters the oceans from the atmosphere, minus its biological uptake (mostly by phytoplankton during photosynthesis). Although anthropogenic CO_2 eventually ends up as $CaCO_3$ in oceanic sediments, there is a substantial time-lag in the response of oceanic sinks to increasing concentrations of CO_2 in the atmosphere. This lag allows atmospheric CO_2 concentrations to increase with increased anthropogenic emissions.

Climate Change

As previously described, Earth has a naturally occurring greenhouse effect, the physical mechanism of which is relatively simple and well understood by scientists (Chapter 4). Moreover, the greenhouse effect helps to maintain Earth's surface temperature within a range that is comfortable for organisms. That temperature averages about 15°C, or 33° warmer than it would be under a non-greenhouse atmosphere. It is also well documented that the concentrations of CO_2 and other radiatively active gases are increasing in Earth's atmosphere. It has been hypothesized that this increase will intensify the planet's natural greenhouse effect.

Although this potential intensification of the greenhouse effect remains a hypothesis, it is an extremely important one. If this environmental change did happen, it would cause many secondary climatic and ecological changes, some of which could be catastrophic for both exploited and natural ecosystems.

One of the most important indicators of **climate change** is the temperature of the surface atmosphere. Air temperatures are measured routinely in many parts of the world. These data can be used to calculate estimates of the average surface temperature of Earth and to detect changes over time. However, the air-temperature records suffer from several important problems:

- Air temperature is extremely variable over time and space. The unfavourable ratio of signal to noise in such data makes it difficult to detect long-term trends.
- Most of the older data are less accurate than modern records. (Detailed and accurate recordings of surface air temperatures began around 1880.)
- Weather-monitoring stations are commonly located in urbanized areas, and their data are influenced by the so-called urban "heat island." This effect is characterized by typically warmer conditions in urban areas than in surrounding, rural places. Some initially rural weather stations have become surrounded by urban land uses, resulting in a "contamination" of their air-temperature records.
- Global temperatures can respond to influences other than changes in the greenhouse effect, such as the cooling effects of volcanic eruptions that inject reflective aerosols into the upper atmosphere.

In spite of the difficulties with the data used to estimate Earth's average surface temperature, recent analyses suggest that there has been a warming trend since the mid-19th century. The average global surface temperature has increased by about 0.5°C over the past 150 years. This warming reflects the end of a 400-year period of climate cooling, known as the "Little Ice Age," which lasted until the mid-1800s (Figure 17.4). However, there appears to have been a particular intensification of warming during the most recent several decades.

Moreover, paleoclimatic studies of long-term changes have provided rather convincing evidence of a link between concentrations of atmospheric CO_2 and climatic warming. Especially valuable data came from a core of glacial ice taken in Antarctica, representing a time record of 160 000 y (Figure 17.5). Results of this important study suggest a strong correlation between CO_2 concentration and air temperature, implying a possible causal relationship. It is not clear, however, whether increased concentrations of CO_2 caused warming via an intensified greenhouse effect, or the opposite. An increase in CO_2 emissions from ecosystems could have been a result of climatic warming, perhaps because the rate of biomass decomposition increased, or because frozen soils warmed in polar latitudes (which would release biomass in permafrost for decomposition). Although Figure 17.5 suggests a relationship between CO_2 and temperature changes, the possible interpretations are ambiguous because of "chicken or egg" considerations. (That is, it is unclear which development came first.)

Other valuable insights have been obtained by running sophisticated mathematical models of global climate processes on high-powered supercomputers. These "virtual experiments" examine the potential climatic responses to increases in atmospheric CO_2. The computer simulations are known as "three-dimensional general circulation models" (or GCMs). The GCMs simulate the complex movements of energy and mass in the global circulation of the atmosphere. They also examine the interactions of these processes with other physical variables that are

FIGURE 17.4 PATTERNS OF DEVIATION OF GLOBAL AVERAGE SURFACE TEMPERATURE FROM PRESENT CONDITIONS

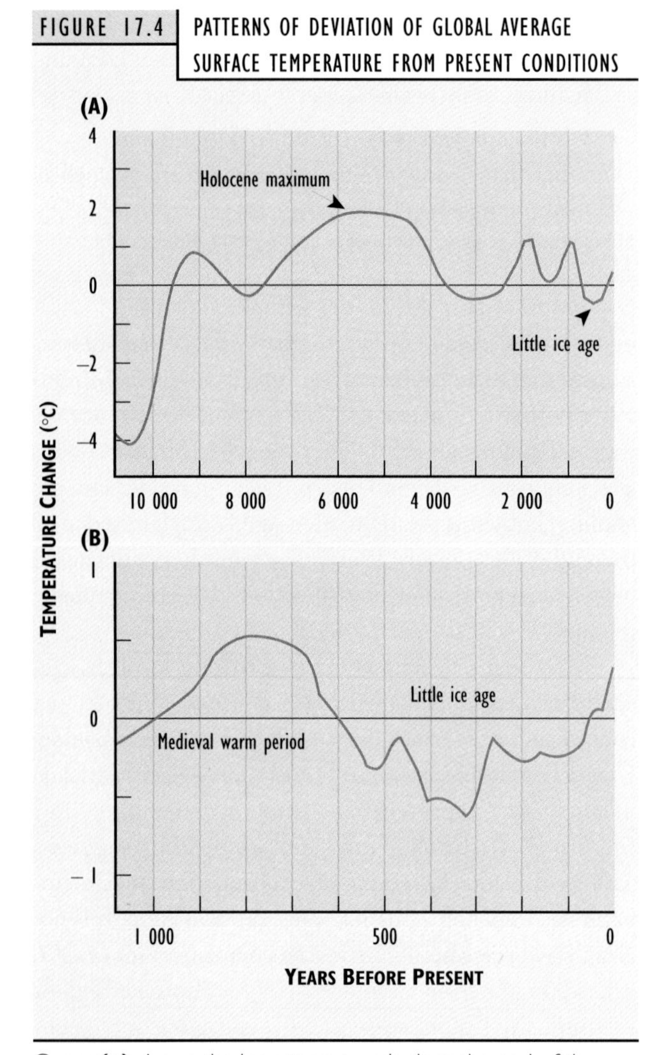

Curve **(a)** shows the long-term trends since the end of the most recent ice age.
Curve **(b)** shows the trends during the past millennium.
Note that a value of "zero" means that no temperature change (deviation) has occurred.

Source: Modified from Intergovernmental Panel on Climate Change (1990) and Environment Canada (1995)

FIGURE 17.5 VARIATIONS IN ATMOSPHERIC CO_2 AND SURFACE-TEMPERATURE DEVIATIONS

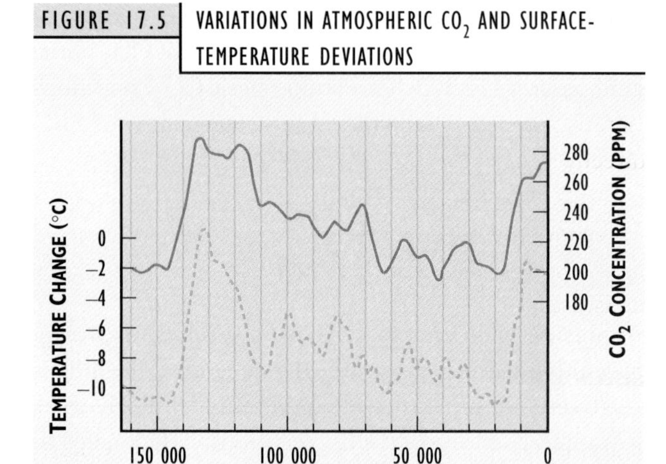

These data were obtained by studying a 160 000-year glacial-core record from Vostok, Antarctica. The lower line shows the temperature deviation and the upper line shows the CO_2 concentration.

Source: Modified from Barnola *et al.* (1987)

important in climate, such as temperature and precipitation. Many simulation experiments have been run using various GCMs, and the results are somewhat variable. However, a central tendency emerging from these experiments suggests that global warming and associated climate changes are a likely consequence of increased concentrations of CO_2 and other RAGs in Earth's atmosphere.

Numerous experiments have simulated the scenario of an eventual doubling of CO_2 concentration from its present concentration of about 357 ppm. These experiments suggest that such a doubling could result in an increase of between 1° and 4°C in the average temperature of Earth's surface atmosphere. The intensity of warming is predicted to be greatest in high-latitude regions, where the temperature increases might be two to three times greater than in the tropics.

Warming of the lower atmosphere will be one change likely caused by an increased intensity of the greenhouse effect. However, other changes in climate will also occur indirectly, in response to an increase in air temperature. Among the most important of the indirect changes will be large-scale shifts in the patterns of atmospheric circulation. Such shifts would likely result in changes in the quantities, spatial distribution, and seasonality of precipitation. Changes in precipitation regimes would influence soil moisture, which would greatly affect the distribution and productivity of vegetation, both natural and managed.

Ecological Effects of Climatic Changes

In terrestrial ecosystems, the direct, biological effects of global warming and associated climatic changes would be restricted mainly to plants. Animals and microorganisms would also be affected, but largely through secondary responses to changes in their habitat caused by changes in vegetation.

The predicted increases in air temperature might not directly affect plants, because these changes would probably not be sufficient to increase heat-related stresses. Much more important would be any substantial changes in the amounts and seasonal patterns of precipitation. Soil moisture is often a critical environmental factor influencing the distribution and productivity of vegetation. For instance, a decrease in the quantities of precipitation and soil moisture in the Canadian prairies might cause the natural tall-grass prairie to change into short-grass prairie, or even to semi-desert. Decreased soil moisture would also affect the types and productivity of crops that could be grown in many regions, making present agricultural systems more difficult or even impossible, unless irrigation was practised.

About 10 000–12 000 years ago, the continental glaciers melted. During the warming climates that followed, the vegetation in various regions in Canada changed substantially. Paleoecological studies of these vegetation changes involve the examination of fossil pollen grains found in dated sections of cores of lake sediment (these studies are known as *palynology*). This kind of analysis has provided a record of vegetation changes extending to early deglaciation. Research in Canada and elsewhere suggests that plants responded to post-glacial climatic warming in a species-specific manner. This occurred because of the different abilities of species to migrate to and colonize newly available habitats released by glacial meltback. As a result, the species composition of early post-glacial plant communities was different from the composition occurring today under similar climatic regimes. We can expect the responses of natural vegetation to future climate changes to be similarly species-specific. This will result in the development of plant communities that are different from those occurring now.

If climatic changes result in substantial changes in the character of plant communities, there will probably be changes in the species of animals, microbes, and other organisms that can be supported on the landscape. Challenges to indigenous biodiversity will be an important consequence of climate change around the world.

Climate changes in tropical countries, which support much larger numbers of species than Canada does, will also have great ecological consequences. For example, most of northern and central South America now possesses a warm, humid climate. However, this region is thought to have been considerably drier during the past glacial period, which ended 12 000–15 000 years ago. During that time, much of the tropical region was covered by an open-canopied savanna vegetation, while rainforests occurred only in isolated regions of relatively greater rainfall, known as *refugia*. On the larger landscape, the isolated refugia of tropical forest apparently occurred as "islands" within a more extensive matrix of savanna, which is an inhospitable habitat for species of moist forests. The restructuring of tropical ecosystems during the Pleistocene Ice Age, driven by climate change, must have had enormous effects on the multitudes of rare species in the rainforests. Many of these species probably became extinct as a result of the habitat changes. If an anthropogenic intensification of the greenhouse effect were to cause substantial changes in the character of tropical habitats over large areas, similar ecological calamities would again occur.

It is important that we acknowledge that scientists do not fully understand the likely dynamics of climatic change. Therefore, they are not able to make accurate predictions about changes in precipitation, air temperature, evapotranspiration, and other climatic factors that may occur in particular regions of Canada or elsewhere. It can, however, be reasonably suggested that any large changes in climate, and especially in precipitation, would result in fundamental changes in the structure and productivity of natural and agricultural vegetation. These ecological changes would have important consequences for the flows of resources that are required by humans, as well as for the habitats of other species.

As was just noted, changes in climate would influence the ability of landscapes to support agriculture. In Canada, this would be especially true of the great expanses of agricultural land in the Prairie Provinces. Much of this land is already ecologically marginal from a rainfall perspective, and is vulnerable to years of severe drought. Wheat, for example, is an economically important crop that is grown extensively in areas that were originally short-grass prairie. In North America, an estimated 40% of this 400-million-hectare, semi-arid region has already been desertified as a result of changes associated with agricultural use. Sporadic crop-threatening droughts occur widely. The critical limitations of precipitation in this region can be alleviated by irrigating the land. However, insufficient water is available for this purpose, and secondary problems, such as salinization, can be caused by this practice. Therefore, any further losses of soil moisture in this important agricultural region of North America would be extremely damaging to agricultural production as well as to food security.

The extent and severity of forest fires would also likely change in response to decreases in the amounts and distributions of precipitation and evapotranspiration, and to changes in their secondary effects, such as soil moisture. In a typical year, about 1–2 million ha of forest burns in wildfires in Canada. This is highly variable — in some years more than 6 million ha are burned. Modelling experiments have suggested that an increased intensity of the greenhouse effect would cause drier climates over much of the boreal region of Canada. This would likely result in about a 50% increase in the annual burned area (Flannigan and Van Wagner, 1991).

In marine ecosystems, increases in seawater temperature would adversely affect some marine biota. Prolonged increases in water temperature can cause corals to lose their symbiotic algae (known as zooxanthellae), sometimes resulting in the death of the coral. This syndrome of damage, known as coral bleaching, can be induced by unusually high or low water temperatures, changes in salinity, and other environmental stresses. Coral reefs are the world's most biodiverse marine ecosystems, and they are already threatened by many environmental stressors associated with human activities, including coastal pollution, mining of the coral, and overly intensive fisheries.

An "island" of trees in the midst of tundra near Tuktoyaktuk in the Northwest Territories. These short individuals of white spruce (*Picea glauca*) are remnants of a more widespread population that established during a period of warmer climate several centuries ago. If an anthropogenic enhancement of Earth's greenhouse effect results in a warming climate, as has been predicted, then these isolated tree-islands may be focal points from which trees can colonize the tundra.

Source: B. Freedman

Sea level has also been predicted to increase as a result of global warming. This would be caused by two factors: an expansion of seawater volume caused directly by warming, and a possible increase in the rate of melting of polar glaciers. A commonly cited model suggests that global sea level might be about 30–50 cm higher in 2100 than today. A change of that magnitude would cause serious problems for low-lying coastal cities, agricultural areas, and natural ecosystems, particularly if the rate of sea-level rise were rapid.

Most of the climate-modelling studies have suggested that the intensity of warming will be greatest at high latitudes. This means that climate changes in countries like Canada, where the climate ranges from temperate to polar, will be much greater than in subtropical and tropical countries. Therefore, relatively wealthy, well-developed countries like Canada and the United States may be exposed to most of the damage associated with greenhouse-related climate changes. Less-developed, equatorial countries may be less directly affected by these changes. These predictions are, however, highly uncertain.

Direct Effects of CO_2 on Plants

Increased concentrations of atmospheric CO_2 can directly stimulate the productivity of some species of plants, especially if they are growing under conditions in which moisture and nutrients (particularly nitrogen, phosphorus, and potassium) are relatively abundant. Under such conditions, plant productivity can be constrained by the rate at which CO_2 can be acquired from the atmosphere during photosynthesis.

Numerous laboratory experiments have demonstrated that agricultural plants can be more productive when fertilized by CO_2. Some commercial greenhouses increase the productivity of crops such as cucumbers, tomatoes, and some ornamental plants by fertilizing with CO_2 at concentrations of 600–2000 ppm.

However, the productivity of most crops grown under field conditions is usually constrained by an inadequate supply of nutrients other than CO_2. The limiting nutrients are most commonly nitrogen, phosphorus, or potassium, and often the availability of water is also a constraint. Under these kinds of field conditions, the responses of plants to CO_2 fertilization are relatively small and short-term, or non-existent.

Increased concentrations of CO_2 can also affect many plants by decreasing their rate of water loss by transpiration. Most water loss by plants occurs through tiny pores, known as stomata, on their leaf surfaces. The size of the stomatal opening is controlled by specialized guard cells. Activity of the guard cells is influenced by CO_2, and stomata tend to close partially or entirely when atmospheric concentrations of CO_2 are high. The availability of moisture is an important factor influencing plant productivity in many agricultural and forestry ecosystems. Consequently, decreased water losses because of lessened transpiration could be viewed as a beneficial change.

It appears that some benefits might be realized from CO_2 fertilization and decreased transpiration, particularly in intensively managed agricultural systems. It is important to recognize, however, that these gains are likely to be relatively minor. Moreover, the possible benefits would likely be overwhelmed by the negative consequences of climate changes that might result from an intensification of Earth's greenhouse effect. The distribution and composition of natural and managed ecosystems could change greatly in response to changes in precipitation and other climatic factors. Anthropogenic climate changes could have enormous consequences for economic resources in agriculture, forestry, and fisheries, as well as for natural biodiversity.

Reducing Atmospheric Carbon Dioxide

Because of the potential consequences of anthropogenic climate change, governments are considering actions that would reduce the concentrations of CO_2 and other RAGs in the atmosphere. This goal could be achieved by reducing the emissions of RAGs or by increasing the rates at which RAGs are removed from the atmosphere. The latter tactic is especially relevant to CO_2, the most important of the anthropogenic greenhouse gases.

Ultimately, substantial decreases in the emissions of RAGs, particularly CO_2, will be the major component of any program designed to deal with an intensification of the greenhouse effect. Emissions of CO_2 are, however, associated with many economically important activities, making it extremely difficult to reduce them rapidly. As was previously described, the major CO_2-emitting activities include the use of fossil fuels in industry, transportation, and space heating; the manufacturing of cement; and ecological conversions, particularly of forests to agriculture.

Planting trees is another option that would contribute to a net reduction of CO_2 concentration in the atmosphere. As trees and other plants grow, they fix atmospheric CO_2 into the organic carbon of their accumulating biomass. Depending on the species and growing conditions, that biomass can eventually reach several tonnes of dry weight (per large tree), about half of which is carbon.

Ecological studies have shown that substantial "**carbon credits**" can be achieved by planting large numbers of trees in urban or rural environments. The carbon credits are especially large if the tree-planting involves the **afforestation** of agricultural areas. (Afforestation converts land into a forest, while reforestation ensures that another forest regrows on a clear-cut site.) Agroecosystems typically store small quantities of carbon in biomass, while forests store much more per unit area. The carbon-storage function would be optimized if mature or old-growth forests were established as "carbon reserves," and if these ecosystems were maintained in their high-carbon condition for as long as possible. (Harvesting of the mature forests would detract from the carbon-storage function.) Moreover, afforestation of extensive areas would achieve many additional, non-carbon environmental benefits, such as the enhancement of biodiversity.

Although tree-planting and reforestation are attractive options toward reducing CO_2 in the atmosphere, these tactics cannot offset more than a portion of the CO_2 emitted by fossil-fuel combustion and deforestation. An enormous area of land would have to be reforested to achieve greater offsets. For example, to fully offset the CO_2 emissions from one 200-megawatt, coal-fired generating station (which would emit about 0.34 million t/y of CO_2–C) would require the carbon-fixing services of about 500 000 ha of natural forest of the kind typical of eastern Canada. If the forest productivity were increased by intensive management on a fertile site, as little as 1/10 of that area might be required; but that would still be a huge area (Freedman *et al.*, 1992). Only a limited amount of land is available, in Canada or elsewhere, for reforestation to provide carbon offsets. The use of larger areas would withdraw too much land from economically productive uses, especially for agriculture.

Dealing effectively with an anthropogenic enhancement of the greenhouse effect will require a comprehensive, integrated strategy. Reduced emissions of RAGs must be the major component of that strategy. Carbon offsets such as tree planting will be a useful element of an integrated strategy, but will not be sufficient in themselves.

The most important means of reducing CO_2 emissions would potentially involve the following:

- conservation of energy through more efficient use, which would result in a decreased demand for fossil fuels
- increased use of non-carbon energy technologies (such as solar, wind, tidal, hydro, and nuclear), to displace some uses of fossil fuels
- prevention of further conversions of mature forests into agricultural and other land uses, to avoid the CO_2 emissions that are associated with deforestation
- afforestation, which would increase carbon stored in ecosystems

Implementation of an integrated strategy involving these actions would be politically and economically difficult. Industrialized nations rely heavily on fossil fuels. Changes in this reliance would have huge implications for economic systems, industrial capitalization, resource use, and citizens' expectations for life style. Similarly, deforestation in tropical countries is a primary means for poor people to gain access to opportunities and livelihoods, and harvested timber helps to earn the foreign exchange that is necessary to fund development activities.

In summary: The societal changes that would be necessary to deal with the intensified greenhouse effect are revolutionary in their nature and magnitude. Designing the required economic and energy systems will be a tremendous challenge, and implementing them will need enlightened and forceful leadership. Unfortunately, there are no easy solutions to an environmental problem as potentially damaging as an anthropogenic enhancement of the greenhouse effect. Moreover, it is crucial that effective actions to avoid the damages be implemented as soon as possible, even before it is definitely known that the damages are occurring.

Key Terms

energy budget
radiatively active gases (or RAGs)
greenhouse effect
climate change
carbon credits
afforestation

Questions for Discussion

1. Explain how human influences may be making Earth's greenhouse effect more intense.
2. What is a "radiatively active gas" (or RAG)? What are the most important RAGs in Earth's atmosphere, and how are human actions affecting their concentrations?
3. What are the likely climatic and ecological consequences of an intensification of Earth's greenhouse effect?
4. How might the Canadian economy and the lifestyles of typical Canadians be affected if serious actions are taken to deal with the consequences of an intensified greenhouse effect?

REFERENCES

Barnola, J.M., D. Raynaud, Y.K. Korotkevich, and C. Lorius. 1987. Vostok ice core provides 160,000-year record of atmospheric CO_2. *Nature*, **329**: 408–414.

Blasing, T.J. 1985. Background: Carbon Cycle, Climate, and Vegetation Responses. In: *Characterization of Information Requirements for Studies of CO_2 Effects: Water Resources, Agriculture, Fisheries, Forests, and Human Health*. Washington, DC: U.S. Department of Energy. DOE/ER-0236. pp. 9–22.

Boden, T.A., D.P. Kaiser, R.J. Sepanski, and F.W. Stoss (eds.). 1994. *Trends '93: A Compendium of Data on Global Change*. Oak Ridge, TN: Oak Ridge National Laboratory. Pub. ORNL/CDIAC-65, Carbon Dioxide Information Centre.

Bolin, B., B.R. Doos, J. Jager, and R.A. Warrick. 1986. *The Greenhouse Effect, Climatic Change, and Ecosystems*. Chichester, UK: Wiley & Sons. SCOPE Rep. 29.

Charlson, R.J. and T.M.L. Wigley. 1994. Sulphate aerosol and climatic change. *Sci. Amer.*, **270** (2): 48–57.

Dale, V.H., R.A. Houghton, and C.A.S. Hall. 1991. Estimating the effects of land-use change on global atmospheric CO_2 concentrations. *Can. J. For. Res.*, **21**: 87–90.

Detwiler, R.P. and C.A.S. Hall. 1988. Tropical forests and the global carbon cycle. *Science*, **239**: 42–47.

Environment Canada. 1995. *The State of Canada's Climate: Monitoring Variability and Change*. Ottawa: Environment Canada. SOE Rep. No. 95-1.

Flanigan, M.D. and C.E. Van Wagner. 1991. Climate change and wildfire in Canada. *Can. J. For. Res.*, **21**: 66–72.

Flavin, C. 1996. Facing up to the risks of climate change. In: *State of the World 1996*. Washington, DC: Worldwatch Institute. pp. 21–39.

Freedman, B. 1995. *Environmental Ecology, 2nd ed.* San Diego, CA: Academic.

Freedman, B. and T. Keith. 1996. Planting trees for carbon credits: a discussion of context, issues, feasibility, and environmental benefits, with particular attention to Canada. *Environ. Rev.*, **4**: 100–111.

Freedman, B., F. Meth, and C. Hickman. 1992. Temperate forest as a carbon-storage reservoir for carbon dioxide emitted by coal-fired generating stations: a case study for New Brunswick, Canada. *For. Ecol. & Manage.*, **15**: 103–127.

Gates, D.M. 1985. *Energy and Ecology*. New York: Sinauer.

Harrington, J.B. 1987. Climatic change: a review of causes. *Can. J. For. Res.*, **17**: 1313–1339.

Houghton, R.A. 1991. The role of forests in affecting the greenhouse gas composition of the atmosphere. In: *Global Climate Change and Life on Earth*. (R.C. Wyman, ed.). New York: Routledge, Chapman, and Hall. pp. 43–56.

Houghton, R.A., J.E. Hobbie, J.M. Melillo, B. Moore, B.J. Peterson, G.R. Shaver, and G.M. Woodwell. 1983. Changes in the carbon content of terrestrial biota and soils between 1860 and 1980: a net release of CO_2 to the atmosphere. *Ecol. Monogr.*, **53**: 235–262.

Houghton, R.A., B.A. Callander, and S.K. Varney (eds.); Intergovernmental Panel on Climate Change. 1992. *Climate Change 1992: The Supplementary Report to the IPCC Scientific Assessment*. Cambridge: Cambridge University Press.

Houghton, R.A., G.J. Jenkins, and J.J. Ephraums (eds.); Intergovernmental Panel on Climate Change. 1990. *Climate Change: The IPCC Scientific Assessment*. Cambridge: Cambridge University Press.

Luther, F.M. and R.G. Ellingson. 1985. Carbon dioxide and the radiation budget. In: *Projecting the Climatic Effects of Increasing Carbon Dioxide*. Washington, DC: U.S. Department of Energy. DOE/ER-0237. pp. 25–56.

Marland, G. and R.M. Rotty. 1985. Greenhouse gases in the atmosphere: what do we know? *J. Air Pollut. Control Assoc.*, **35**: 1033–1038.

Mooney, H.A., B.G. Drake, R.J. Luxmoore, W.C. Oechel, and L.F. Pitelka. 1991. Predicting ecosystem responses to elevated CO_2 concentrations. *BioScience*, **41**: 96–104.

Peters, R.L., and Darling, J.D.S. 1985. The greenhouse effect and nature reserves. *Bioscience*, **35:** 707–717.

Ramanathan, V. 1988. The greenhouse theory of climate changes: A test by an inadvertent global experiment. *Science*, **240**: 293–299.

Rodhe, H. 1990. A comparison of the contributions of various gases to the greenhouse effect. *Science*, **248**: 1217–1219.

Rowland, F.S. 1988. Chlorofluorocarbons, stratospheric ozone, and the Antarctic "ozone hole." *Environ. Conserv.*, **15**: 101–116.

Schneider, S.H. 1989. The changing climate. *Sci. Amer.*, **261** (3): 70–79.

Solomon, A.M., J.R. Trabolka, D.E. Reichle, and L.D. Voorhees. 1985. The global cycle of carbon. In: *Atmospheric Carbon Dioxide and the Global Carbon Cycle*. Washington, DC: U.S. Department of Energy. DOE/ER-0239. pp. 1–13.

Solomon, A.M. and D.C. West. 1985. Potential responses of forests to CO_2-induced climate change. In: *Characterization of Information Requirements for Studies of CO_2 Effects: Water Resources, Agriculture, Fisheries, Forests, and Human Health*. Washington, DC: U.S. Department of Energy. DOE/ER-0236. pp. 145–169.

Trabalka, J. 1985. *Atmospheric Carbon Dioxide and the Global Carbon Cycle*. Washington, DC: U.S. Department of Energy. DOE/ER-0239.

Wigley, T.M.L. and S.C.B. Raper. 1992. Implications for climate and sea level of revised IPCC emissions scenarios. *Nature*, **357**: 293–300.

World Resources Institute (WRI). 1994. *World Resources, 1994–95. A Guide to the Global Environment*. New York: Oxford University Press.

Wyman, R.L. (ed.). 1991. *Global Climate Change and Life on Earth*. New York: Routledge, Chapman, and Hall.

TOXIC ELEMENTS

Chapter Outline

Chapter Objectives

After completing this chapter, you will be able to:

1. Describe the ubiquitous distribution of elements in the environment and discuss this distribution in terms of the difference between pollution and contamination.
2. Outline cases of natural pollution by toxic elements and discuss how these provide useful insights into the effects of anthropogenic pollution.
3. Describe cases of anthropogenic pollution of the environment with metals and outline the resulting ecological damages.

INTRODUCTION

All of the naturally occurring chemical elements are ubiquitous (found everywhere) in at least trace concentrations in soil, water, air, and organisms. As long as the detection limits of the available analytical chemistry are low enough, this universal contamination can always be demonstrated.

Organisms require some trace elements as essential micronutrients, including copper, iron, molybdenum, zinc, and in some cases aluminum, nickel, and selenium. Under certain conditions, however, these same elements can accumulate to high concentrations in organisms (this is known as **bioconcentration** and **food-web magnification**) and cause ecological damage (see In Detail 18.1). The trace elements most often associated with environmental toxicity are the heavy metals cadmium, chromium, cobalt, copper, iron, lead, mercury, nickel, silver, tin, and zinc, and lighter elements such as aluminum, arsenic, and selenium.

Some cases of elemental pollution are natural in origin. This usually happens when metal-rich minerals are exposed at the Earth's surface and cause local ecological changes. However, human activities have caused many additional examples of pollution by toxic elements, particularly in the vicinity of large industrial sources such as smelters. In addition, emissions of lead and mercury from power plants and automobiles have caused widespread contamination of remote environments, although it is not yet known whether this is causing ecological damage.

There are also cases of people being poisoned by exposure to toxic elements in their environment. Some historians believe that the decline of the Roman empire may have been hastened by neurotoxicity caused by chronic lead poisoning. The Romans had significant exposure to lead because they stored acidic beverages (such as wine) in pottery containing lead-based pigments and glazes. As well, their water piping was made of lead (the word "plumbing" is based on the Latin word for lead — *plumbum*). In 19th-century Britain, people who made felt top hats developed neurological damage because of their occupational exposure to mercury compounds, used to give a shiny finish to the hats. Hence Lewis Carroll's character in *Alice in Wonderland*, the "Mad Hatter," and the common expression, "mad as a hatter."

More recently, thousands of people suffered acute mercury poisoning during the 1960s after they ate seed grain treated with mercuric fungicides. In one particularly disastrous case, more than 6500 people were poisoned (about 500 died) in Iraq when they ate food prepared from mercury-treated grain. The grain had been donated by a foreign aid program, and was intended only for planting. Although the bags of grain were labelled to show the poisonous nature of the seed, many of the victims were illiterate or did not fully understand the implications of the message and ignored it. About the same time, similar poisonings were caused when people ate mercury-treated seed-grain in Iran, Pakistan, and Guatemala.

Mercury poisoning also caused thousands of cases of poisoning (including hundreds of deaths) at Minamata, Japan. A factory had discharged elemental mercury into Minamata Bay. Elemental mercury is not very poisonous, but microbes in anaerobic sediments metabolized the metal into methylmercury, which is extremely toxic and bioccumulating. The methylmercury entered the aquatic food web and caused extensive poisoning of fish-eating birds, domestic cats, and humans.

In this chapter we examine natural and anthropogenic pollution of the environment with toxic elements, and the resulting ecological consequences.

CONCENTRATION AND AVAILABILITY IN THE ENVIRONMENT

All of the naturally occurring elements occur in all samples of water, soil and rocks, air, and organisms in at least trace concentrations. The term "background concentration" describes the concentration of an element that is not significantly influenced by either anthropogenic emissions or unusual natural exposures. Background concentrations in soil and rock are usually much higher than in water, and most are also higher than in the tissues of organisms (Table 18.1).

However, elements dissolved in water occur in chemical forms (such as ions) that are much more easily absorbed by organisms. For this reason, even trace aqueous concentrations may be toxic. In contrast, the much higher concentrations that commonly occur in soil and rocks are mostly insoluble, and therefore are not particularly bioavailable. Scientists determine the **"total" concentrations** of metals in a component of the environment (such as soil, sediment, or rocks) by digesting samples in hot mixtures of strong acid. In contrast, **"available" concentrations**

IN DETAIL 18.1

BIOACCUMULATION AND FOOD-WEB MAGNIFICATION

Certain metals, or their compounds such as methylmercury, tend to occur in much higher concentrations in organisms than in the ambient, nonliving environment. This phenomenon is known as **bioaccumulation** (also called **bioconcentration** or **biomagnification**). Similar tendencies are shown by chlorinated hydrocarbons, such as DDT, PCBs, and dioxins (see Chapter 21). Bioaccumulation occurs because certain chemicals have a powerful affinity for organisms, and therefore concentrate within organisms in strong preference to the nonliving environment. Many of these chemicals dissolve in biological fluids and tissues, such as lipids (fats), in preference to the water or soil in the ambient environment.

Another phenomenon, known as **food-web magnification** (or **food-web concentration**), is the tendency for top predators to have the highest concentrations of these chemicals or their residues. Organisms are very efficient at assimilating methylmercury and organochlorines from their food. Therefore, these chemicals in an organism's food become stored within the organism, rather than being eliminated or excreted. This means that predators at the top of the food web develop the highest concentrations of residues of these chemicals. Usually, bioaccumulation and food-web magnification are progressive with age, so that the oldest individuals in any population are most contaminated.

FOOD-WEB MAGNIFICATION

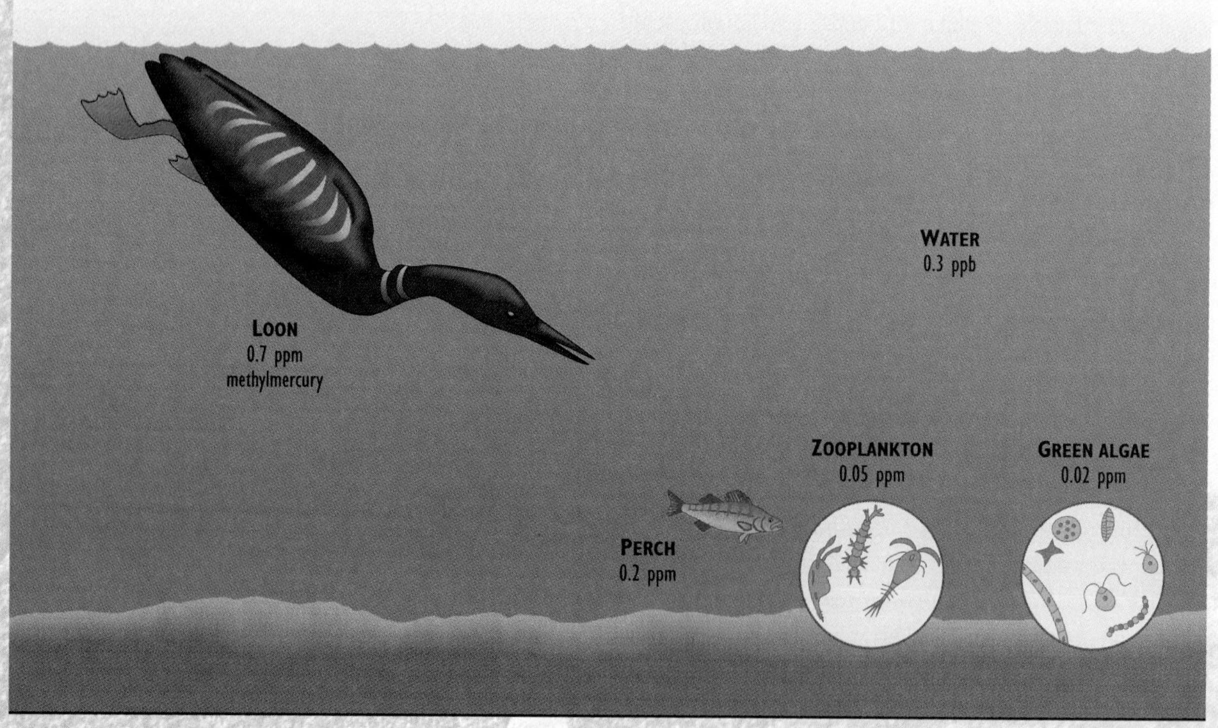

Food-web magnification leads to progressively higher concentrations of methylmercury and chlorinated hydrocarbons in organisms occupying higher positions in the food web. Common loons (*Gavia immer*) are a top predator in many lakes. In some regions of Canada, the bodies of these birds can harbour concentrations of methymercury so high as to impair their reproduction. The source of the environmental mercury is not yet known, but is probably associated with anthropogenic emissions, likely from power plants and smelters.

TABLE 18.1 | BACKGROUND CONCENTRATIONS OF TOXIC ELEMENTS IN SELECTED COMPONENTS OF THE ENVIRONMENT

All data are in ppm, except those for water, which are in ppb.

	ROCKS (ppm)				WATER (ppb)				
ELEMENT	GRANITE	BASALT	LIMESTONE	SOIL (ppm)	OCEANIC	FRESH	TERRESTRIAL PLANTS (ppm)	MAMMAL MUSCLE (ppm)	MARINE FISH (ppm)
Aluminum	77 000	87 600	9 000	71 000	2.0	300	90–530	0.7–28	20
Arsenic	1.5	1.5	1	6	3.7	0.5	0.2–7	0.007–0.09	0.2–10
Cadmium	0.1	0.13	0.03	0.35	0.1	0.1	0.1–2.4	0.1–3.2	0.1–3
Chromium	4	90	11	70	0.3	1.0	0.03–10	<0.002–0.84	0.03–2
Cobalt	1	35	0.1	8	0.02	0.2	0.005–1	0.005–1	0.006—0.05
Copper	13	90	5.5	30	0.3	3.0	5–15	10	0.7–15
Fluoride	1 400	510	220	200	1 300	100	0.02–24	0.05	1 400
Iron	27 000	56 000	17 000	40 000	2.0	500	70–700	180	9–98
Lead	24	3	5.7	35	0.03	3.0	1–13	0.2–3.3	0.001–15
Manganese	400	1 500	620	1 000	0.2	8.0	20–700	0.2–2.3	0.3–4.6
Mercury	0.1	0.01	0.18	0.06	0.3	0.1	0.005–0.02	0.02–0.7	0.4
Molybdenum	2	1	0.16	1.2	10.0	0.5	0.06–3	0.02–0.07	1
Nickel	0.5	150	7	50	0.6	0.5	1–5	1.2	0.1–4
Selenium	0.05	0.05	0.03	0.4	0.2	0.2	0.03	0.4–1.9	0.2
Silver	0.04	0.1	0.12	0.05	0.04	0.3	0.01–0.8	0.009–0.28	0.04–0.1
Tin	3.5	1	0.5	4	0.004	0.01	0.2–2	0.01–2	0
Uranium	4.4	0.43	2.2	2	3.2	0.4	0.005–0.04	0.001–0.003	0.04–0.08
Vanadium	72	250	45	90	2.5	0.5	0.001–0.5	0.002–0.02	0.3
Zinc	52	100	20	90	5.0	1.5	20–400	240	9–80

Source: Bowen (1979)

are determined from an aqueous (water) extract of the sample. Data on available concentrations are more relevant to potential toxicity than are data on total concentrations. In general, available concentrations of toxic elements in soil are much smaller than the total concentrations (generally less than 1% of the total value).

Most elements are found in only trace concentrations in the environment (see Table 18.1). In contrast, a few elements typically occur in much higher concentrations, particularly aluminium and iron. These are prominent constituents of rocks and soil, with aluminum concentration averaging about 8% and iron 3–4%. However, almost all of the aluminum and iron in soil and rocks occurs as insoluble minerals that are not readily available for uptake by organisms.

For example, virtually all soil aluminum occurs as insoluble silicate and clay minerals. Although aluminum in these forms comprises about 8% of the soil mass, it is not available for uptake by organisms and is, therefore, nontoxic. However, much smaller concentrations of aluminum, typically only a few parts per million (ppm), are found in soil as ions, either bound to organic matter and clay surfaces or freely dissolved in soil water. The ionic forms of aluminum are readily available for biological uptake, and may cause toxicity to some species.

Much higher concentrations of soluble, available aluminum occur in strongly acidic environments, especially when the pH is less than about 5.5. (In fact, most metals are much more soluble under acidic conditions.) Aluminum solubility is also greater in strongly alkaline environments, with pHs higher than about 8. Moreover, different ionic species of aluminum occur at different pHs. Al^{3+} is dominant in strongly acidic environments with a pH of less than 5.0, while $AlOH^{2+}$ and $Al(OH)_2^+$ are important under less acidic conditions of pH 4.5–5.5, $Al(OH)_3$ from pH 5.2 to 9, and $Al(OH)_4^-$ in alkaline environments with a pH greater than about 8.5. Aluminum toxicity is a common problem for organisms living in highly acidic or alkaline environments. This is because of the combined influences of greater solubility and the presence of relatively toxic ions under those conditions.

Toxicity

The toxicity of elements and other chemicals is related to two factors: the exposure or dose and the vulnerability of an organism to the specific poison. The dose received by an organism is influenced by the available concentration of the poison in the environment and the period of exposure. Therefore, a long-term exposure to a small available concentration may cause toxicity, particularly in cases in which the element can bioaccumulate and food-web magnify until a threshold of tolerance is exceeded.

Organisms vary greatly in their **tolerance** of exposures to toxic elements (as well as to other poisons). Consequently, an intense exposure to a potentially toxic chemical may result in some species being poisoned, while tolerant ones may not be damaged and may even benefit from the demise of sensitive species. In addition, genetically based variation for tolerance usually exists within species. This can lead to the evolution of populations, known as *ecotypes*, that are relatively tolerant of toxic exposures (we examine this topic in more detail in the next section).

The most common mechanism of poisoning by toxic elements is damage to an enzyme system. (Organisms have a huge diversity of enzymes, which are proteins that catalyze specific biochemical reactions and are critical to healthy metabolism.) The poisoning is caused because the metal ions bind to specific enzymes, changing the shape of the protein, which results in loss of its unique catalytic function. Toxic elements may also poison by binding to DNA or RNA, disrupting transcription and translation, the processes by which genetic information is used to synthesize specific proteins (particularly enzymes). Toxic metals can also disrupt DNA replication, and hence cell division.

Typical symptoms of acute toxicity caused by toxic elements in plants include abnormal patterns of growth and development, decreased productivity, impaired reproduction, disease, and ultimately death. Symptoms of chronic toxicity are more difficult to detect, and may include "hidden injuries" such as decreases in productivity. Animals can show a variety of symptoms associated with enzyme disruption, including neurotoxicity and impaired functioning of the kidneys and other organs.

Natural Pollution by Toxic Elements

Localized natural pollution is sometimes seen when metal-rich minerals occur at the surface and affect the chemistry of local soil, surface waters, and vegetation. These conditions can sometimes be identified by the presence of particular plant species, or by a distinctive, often stunted, growth form of the vegetation. In combination with chemical analyses, these biological indicators can be used to explore for metal-rich mineralizations, in a technique known as biogeochemical prospecting.

In some cases, natural pollution by metals can be very intense. For example, soil containing up to 3% lead and zinc was found at a site on Baffin Island. In another case, peat that was filtering metal-rich ground water in New Brunswick accumulated as much as 10% copper. Such high concentrations of metals in soil are also reflected in the chemistry of plants, particularly in certain genetically adapted, **hyperaccumulator species** that often occur in metal-rich habitats. For example, nickel concentrations as high as 10% have been measured in plants in the genus *Alyssum* growing in Russia, and up to 25% in the blue-coloured latex of the plant *Sebertia acuminata* from New Caledonia in the south Pacific. These hyperaccumulator plants grow on naturally metal-polluted sites.

Serpentine Soil and Vegetation

Some of the best-studied cases of natural metal pollution involve soils influenced by *serpentine* minerals, which are rich in nickel, chromium, and cobalt and are associated with asbestos deposits. Soils containing serpentine minerals are toxic to nonadapted plants because of the high concentrations of these metals, in combination with a severe imbalance of the nutrients calcium and magnesium. Serpentine soils typically contain several thousand ppm of nickel, but some can have as much as 25 000 ppm (or 2.5%) of this metal.

Natural vegetation on serpentine soils is often distinctively stunted. Extensive serpentine "barrens" occur on plateaus in eastern Quebec and western Newfoundland.

An extensive area of serpentine-rich rock in Gros Morne National Park in western Newfoundland. Soils rich in serpentine have high concentrations of toxic nickel and cobalt, and are poor in nutrients. These conditions are stressful to plants, and can result in the development of an atypically stunted vegetation of limited species diversity, as in this scene. The typical vegetation on nonserpentine soil in this region is a conifer-dominated forest.

Source: T. Keith

Those habitats support tundra-like ecosystems in a landscape otherwise covered by boreal coniferous forest.

In some areas, serpentine habitats support many plant species that occur only in those kinds of habitats, a narrow distribution that ecologists refer to as **endemic**. In other cases, widespread plant species have evolved locally adapted populations that can cope with the toxic and nutritional stresses of serpentine soils — these are known as **ecotypes**. On nonserpentine sites, the specifically adapted endemics and ecotypes are quickly eliminated by competition with plants that are better adapted to living in less stressful habitat conditions.

Serpentine sites in northern California support a relatively ancient vegetation. These habitats contain 215 endemic species or subspecies of plants. Some of the endemic plants occur only on particular serpentine sites in California and nowhere else in the world. In contrast, the serpentine barrens in eastern Canada are relatively young, being released from glaciation only about 8000 or fewer years ago. Consequently, not enough time has passed to allow many serpentine endemics or ecotypes to evolve. These specifically adapted plants are more common in places that have supported vegetation for longer periods of time.

Seleniferous Soil and Vegetation

Semiarid regions in various parts of the world often have areas of soil that naturally contain high concentrations of selenium, called seleniferous soils. These habitats can support plants that hyperaccumulate selenium, such as species in the genus *Astragalus*, or locoweeds. About 25 of the 500 North American species of *Astragalus* are hyperaccumulators of selenium. These plants may contain as much as 15 000 ppm of selenium in their tissues, storing it in unique amino-acid-like biochemicals, such as selenomethionine. The *Astragalus* species also emit dimethyl selenide and dimethyl diselenide to the atmosphere, giving the plants a distinctive, unpleasant odour. Livestock that feed on these plants are poisoned by a toxic syndrome known as "alkali disease" or "blind staggers."

Mercury in Aquatic Environments

Even in remote oceanic habitats, mercury often accumulates in high concentrations (as methylmercury, CH_3Hg) in fish, birds, and sea mammals. Offshore eastern and western Canada, for example, some large fish may have mercury concentrations in their flesh that exceed what is considered acceptable in fish intended for human consumption (that is, 0.5 ppm mercury on a fresh-weight basis; Table 18.2). Analysis of old specimens of fish and seabirds in museums has revealed levels of mercury contamination similar to those occurring in modern samples, suggesting that the phenomenon is natural. This contamination of marine animals represents a substantial increase from ambient seawater, which has a trace concentration of mercury of less than 0.1 ppb.

TABLE 18.2 MERCURY CONTAMINATION OF FISH CAPTURED OFFSHORE NORTH AMERICA

These data show the average mercury concentration in muscle tissue of commercially important species of marine fish, some of which are harvested in Canadian waters.The data are in ppm, measured on a fresh-weight basis. Some data refer to specific size ranges of fish.

SPECIES	MERCURY CONCENTRATION
Swordfish (*Xiphias gladius*) >45 kilogram	1.08
Bluefin tuna (*Thunnus thynnus*) >14 kg	0.89
Yellowfin tuna (*T. albacares*) >32 kg	0.62
Skipjack tuna (*Euthynnus pelamis*) >4 kg	0.21
Atlantic dogfish (*Squalus acanthius*)	0.41
Pacific dogfish (*S. acanthius*)	0.70
Pacific halibut (*Hippoglossus stenolepis*) >45 kg	0.42
Atlantic halibut (*H. hippoglossus*) > 45 kg	0.80

Source: Armstrong (1979)

The **bioconcentration** and **food-web magnification** of mercury occurs because of the progressive accumulation of this metal up the trophic web. Algae initially absorb mercury from the water (as methylmercury), while zooplankton accumulate even larger residues as they graze on the algae. Zooplankton-eating fish accumulate larger quantities still, but the highest residues occur in long-lived top predators, such as large fish and marine mammals (see In Detail 18.1, p. 310).

Within particular species of marine fish, larger (or older) fish generally have higher mercury concentrations than smaller (or younger) ones. A study of 224 swordfish caught off eastern Canada found that animals heavier than 45 kg had an average mercury concentration of 1.1 ppm. Those weighing 23–45 kg averaged 0.86 ppm, and animals smaller than 23 kg had 0.55 ppm (Armstrong, 1979). It appears that mercury biomagnification becomes more intense as these animals age and grow larger.

High concentrations of mercury also occur in fish-eating marine mammals and birds, which are top predators in their ecosystem. Studies of adult harp seals (*Phoca groenlandica*) in eastern Canada found an average mercury concentration of 0.34 ppm in muscle and 5.1 ppm in the liver (Armstrong, 1979). High mercury residues also occur in North Atlantic seabirds, with an average of 7 ppm found in feather samples of northern skua (*Catharacta skua*); 5 ppm in puffins (*Fratercula arctica*); and 1–2 ppm in fulmar (*Fulmarus glacialis*), kittiwake (*Rissa tridactyla*), razorbill (*Alca torda*), and common murre (*Uria aalge*) (Thompson *et al.*, 1991).

Mercury contamination of fish has also been observed in many remote lakes. For example, about three-quarters of 1500 lakes monitored in Ontario have at least some fish with mercury residues exceeding 0.5 ppm fresh weight (f.w.) in their flesh. In a remote lake in northern Manitoba, the average mercury concentration in muscle samples of 53 northern pike (*Esox lucius*) was 2 ppm f.w., and one animal had 5 ppm (McKay, 1985). In general, fresh-water fish that are top predators develop the highest contaminations of mercury, and larger or older individuals tend to be the most contaminated.

Some governments warn people against eating fish taken from particular lakes or regions where mercury contamination is known to be severe. For instance, fish consumption advisories have been issued for about 1200 lakes in Ontario. In Sweden, about 250 lakes have been "blacklisted," and that dubious status is being considered for another 9400 lakes (Spry and Wiener, 1991).

The causes of mercury contamination of lakes are not known for certain. It seems likely that the phenomenon is natural in regions that are remote from sources of emission. However, anthropogenic mercury is probably contributing to the problem closer to large emissions sources, such as coal-fired generating stations, municipal incinerators, and smelters. For example, Harp Lake in Ontario is located relatively close to municipal and industrial sources of mercury emissions. Studies found that atmospheric inputs accounted for 57% of the total mercury inputs to that lake, suggesting a significant anthropogenic influence (Mierle, 1990).

The above discussion of mercury in lakes refers to the many situations in which there are no direct anthropogenic sources of the metal. However, cases of pollution caused directly by industrial emissions are well known. For example, large industrial facilities such as chlor-alkali and acetaldehyde factories and some older pulp mills have caused local mercury pollution, resulting in high residues of methylmercury in fish and other animals (the case of Minamata Bay involved an acetaldehyde plant; a less-severe case in Canada, affecting parts of the English and Wabigoon Rivers in northwestern Ontario, involved a pulp mill). Significant bioaccumulation of mercury also occurs when hydroelectric reservoirs are developed. This happens because flooding leaches naturally occurring soil mercury into the reservoir, where it is metabolized into methylmercury by bacteria in oxygen-poor sediments, and is then bioaccumulated by fish. Bioaccumulation of mercury occurs especially rapidly in acidic lakes, because such conditions favour the production of methylmercury in the sediments, compared with less-available dimethylmercury in nonacidic waterbodies.

Anthropogenic Sources of Toxic Elements

In this section we examine examples of damages caused by emissions of toxic elements from agricultural practices, metal mining, and ore processing; by the use of lead shot in hunting; and by emissions of lead from automobiles.

Pollution from Metal Mining and Processing

During the mining and processing of metals, the various industrial processes can result in significant pollution of air, water, and land (see Figure 13.1, page 203).

Mining Residues

Areas near mine sites are sometimes degraded by the dumping of metal-rich overburden and excavation wastes. Because these wastes may be toxic to plants, vegetation development may be restricted to early-successional communities, such as sparse grasslands. In some cases, soil toxicity is so severe that few plants manage to establish even after hundreds of years. This can be seen around mine wastes from 2000-year-old Roman lead workings in England and Wales.

Ecologists studying British sites polluted by mine wastes have found that these places often support populations of plants that are genetically tolerant of the polluting metals. These locally adapted ecotypes can establish themselves and grow in metal-polluted environments, whereas nontolerant plants of the same species are quickly eliminated by the toxic stress. Conversely, metal-tolerant ecotypes are poor competitors in nonpolluted environments, and are therefore rare in habitats that are not affected by metal toxicity.

Research into metal-tolerant ecotypes has provided important insights into the process of evolution (see Chapter 6). Metal-tolerant individuals are present in populations growing on nonpolluted sites, but they are rare, generally comprising less than 1% of the population. However, the frequency of metal-tolerant genotypes can increase quickly after metal pollution occurs. In places where sharp boundaries exist between polluted and nonpolluted soils, a tolerant plant population can maintain itself over a distance of only a few metres. This is possible because the intense toxic stress strongly favours the survival and reproduction of tolerant individuals on the polluted soil. Such a population-level change in genetically based characters, occurring in response to an agent of natural selection (i.e., metal pollution), demonstrates evolution (more specifically, microevolution).

TABLE 18.3 TOLERANCE OF THE GRASS *DESCHAMPSIA CAESPITOSA* TO METALS

Populations of the grass *Deschampsia caespitosa* were collected from metal-polluted places near Sudbury and from reference sites where metal contamination was not a problem. The index of metal tolerance is based on the amount of root growth that occurs when plants are grown in solutions containing metals, compared with growth in a control solution. The smaller the index number, the greater the degree of metal tolerance. In all of the comparisons presented here, the two populations demonstrated statistically significant differences in tolerance to the metal tested, with a probability level of <0.001, except for aluminum, which had a probability level of <0.05.

METAL	METAL-POLLUTED SUDBURY SITE	NONPOLLUTED REFERENCE SITE
Nickel	0.96	1.97
Copper	1.06	2.01
Aluminum	1.06	1.21
Lead	1.37	1.90
Zinc	1.55	1.98

Source: Index of metal tolerance modified from Cox and Hutchinson (1979)

Metal-tolerant plant ecotypes have been studied near Sudbury, Ontario, where severe pollution by nickel and copper has been caused by emissions from smelters and roast beds (see Chapter 16). Plant communities of metal-polluted sites are dominated by metal-tolerant ecotypes of grass species, particularly *Agrostis gigantea* and *Deschampsia caespitosa*. Meadows of these grasses developed rapidly after the extremely tall "superstack" was commissioned in 1972. By dispersing emissions widely, the superstack reduced ground-level SO_2 pollution substantially. However, soils in the Sudbury area remained acidic and polluted with nickel, copper, and other metals. The local ecotypes of these grasses can tolerate toxic stresses associated with acidity and metals, but are intolerant of SO_2, which is why the grasslands did not develop until after the superstack began to operate.

The metal tolerance of the grass *D. caespitosa* from the Sudbury area has been well studied. Plants were grown in solutions containing the metals of interest, and the growth was compared with that occurring in solutions without metals (Table 18.3). The data show that the Sudbury population is markedly more tolerant of nickel and copper, which occur in their native soil environment at concentrations of about 400 ppm, compared with only 20 ppm at nonpolluted sites. The Sudbury population of the grass is also more tolerant of aluminum. This is a response to the greater solubility and toxicity of aluminum in acidic soils near the smelters (which have a pH of 3.5–3.9, compared with pH 6.8–7.2 at the reference sites).

Metal-Containing Tailings and Their Reclamation

During the milling process, ore is ground to a fine powder. The powder is separated into a metal-rich fraction, which is roasted and then smelted, plus large quantities of waste "tailings" (see Figure 13.1, page 203). Although the tailings are a waste, they still contain rather high concentrations of metals. Consequently, it can be difficult to revegetate tailings-disposal areas once they become filled. Sulphide minerals, if present, may also cause toxicity, be-

TABLE 18.4 CHEMICAL ANALYSES OF SOME METAL-CONTAMINATED TAILINGS

Samples taken from various sites in the Yukon and northern Ontario. Metal data are in ppm, sulphur in %.

SITE	pH	ARSENIC	CADMIUM	COPPER	NICKEL	LEAD	ZINC	SULPHUR
Gold mine A	1.9	5 200	18	140	13	952	2 400	1.1
Gold mine B	3.2	400	15	613	14	130	9 600	44.4
Gold mine C	6.0	1 350	102	172	15	3 600	6 200	5.7
Gold mine D	6.9	50 000	114	330	20	2 300	18 000	13.5
Gold mine E	7.1	7	3	33	21	1 130	1 060	4.0
Tungsten mine	7.0	–	<1	1 420	22	12	288	6.5
Copper mine	9.0	15	1	1 710	21	9	178	0.1
Nickel mine	3.2	–	<1	392	290	18	291	0.9

Source: Kuja (1980)

cause sulphides generate acidity when oxidized by bacteria. Chemical analyses of tailings from several Canadian mines are shown in Table 18.4. The tailings contain high concentrations of various metals, depending on the quality of the processed ore. The acidic tailings are especially toxic, because most metals are much more soluble and bioavailable under acidic conditions.

Canadian regulators require that tailings disposal areas be covered with vegetation once they are full of waste or after their associated mine closes. This is done because tailings dumps are not only aesthetically undesirable, but can be sources of wind-borne dusts. In addition, if their associated dams and berms are not structurally sound, tailings-disposal areas are potential sources of water pollution. These environmental problems can be substantially avoided if abandoned tailings dumps are covered with a stable cover of vegetation.

To be successfully vegetated, most tailings have to be treated. If the tailings are acidic, a liming treatment is needed to raise the pH to a neutral condition and reduce the availability of metals. This may be followed by fertilization to alleviate nutrient deficiencies, by the addition of organic matter to improve soil structure and water-holding capacity, and by the sowing of a mixture of perennial plants. Sometimes, novel techniques are used — for example, the use of acid- or metal-tolerant plant ecotypes in the planting mixture. If the tailings are extremely toxic or acid-generating, they may have to be covered with a locally available overburden, such as glacial till, which is then vegetated. Canadian Focus 18.2 describes the reclamation of tailings disposal areas in the vicinity of Sudbury.

Tailings are fine wastes that remain after ore is ground and processed to remove the metal-rich minerals. Tailings may contain high concentrations of toxic metals, and can generate large amounts of acidity when exposed to the atmosphere. These conditions make it difficult to establish vegetation after disposal sites are filled.

This is a view of a reclaimed area of tailings near Sudbury, Ontario. Most of the vegetation was sown onto the site, but some native shrubs and trees are also becoming established. The pond in the background is used by migratory birds.

Source: B. Freedman

Smelters

Smelters are large industrial facilities in which ores are roasted. This process oxidizes sulphide minerals and

CANADIAN FOCUS 18.II

TAILINGS RECLAMATION AT COPPER CLIFF

The large smelter at Copper Cliff, near Sudbury, Ontario, is serviced by a mill that produces about 54 000 t of tailings per day (Peters, 1984). The tailings are mixed with water and piped as a slurry to be disposed in low areas surrounded by earthen dikes. At present, tailings dumps cover more than about 1100 ha. Once a tailings dump is full, it is stabilized with a cover of vegetation, which prevents the fine dusts from eroding into the atmosphere and improves aesthetics and environmental quality. The revegetated tailings dumps have a central pond, which is surrounded by gradually sloping grasslands.

The tailings are a finely ground substrate, composed mainly of minerals that are not particularly toxic. However, the tailings contain pyrites that oxidize when exposed to atmospheric oxygen. This generates acidity that can result in a pH lower than 3.7 in the tailings. Under these extremely acidic conditions, metals in the tailings become available for plant uptake, increasing the toxicity of the substrate. Plant-available metal concentrations have been analyzed

by extracting samples of tailings using a weak solution of acetic acid. Such analyses have found very high levels of available metals, with concentrations of nickel up to 87 ppm, of copper up to 81 ppm, and of iron up to 440 ppm.

The reclamation techniques attempt to establish a stable grassland, which can then be naturally invaded by native species of shrubs, trees, and other plants. The methods used include the following:

1. application of about 900 kg/ha of limestone ($CaCO_3$), which raises the pH of the tailings to 4.5–5.5 and thereby reduces metal availability
2. several fertilization treatments during the initial stages of grassland establishment, particularly with nitrogen-containing fertilizer
3. sowing with a mixture of pasture grasses and legumes

The seed mixture includes some annual rye (*Secale cereale*), which serves as a short-lived nurse crop that provides a less stressful microclimate for the tender seedlings of perennial grasses and legumes, thereby enhancing their establishment.

As vegetation develops on the reclaimed tailing-disposal areas, animals begin to invade the habitat. Birds that breed regularly in the grassy habitat and its central pond include mallard and black ducks (*Anas platyrhynchos* and *A. rubripes*), American kestrel (*Falco sparverius*), killdeer (*Charadrius vociferus*), and savannah sparrow (*Passerculus sandwichensis*). A larger number of bird species, about 90, have been observed to use the reclaimed tailings dump and its pond during migrations.

produces large quantities of waste SO_2 and metallic particulates. In some cases, installed pollution-control technologies recover much of the SO_2 and particulates before the flue gases are vented to the atmosphere. In other cases, however, these wastes are emitted into the environment, where intense pollution and ecological damage may be caused. As recently as several decades ago this was common practice, and it still is for some older smelters. Newer smelters operate more cleanly.

Smelters are **point sources** of toxic stress to surrounding ecosystems. Emissions from smelters result in well-defined spatial gradients of both pollution and the resulting ecological damages; these diminish with increasing distance from the facilities. Studies of damages near smelters indicate that, in general:

- Close to the point source, the pollution by atmospheric SO_2 and metals in soil is most severe.
- The intensity of pollution decreases rapidly with increasing distance from the smelter.
- Damage to vegetation varies with the intensity of toxic stress. The damages include decreases in ecosystem biomass, productivity, and species diversity, with only a few low-growing species occurring in the most polluted habitats.
- Ecological functions such as nutrient cycling and decomposition are disrupted by toxic metals, gases, and acidity.

The pattern of metal pollution around a point source can be illustrated by the Copper Cliff smelter near Sudbury. Figure 18.1 shows that metal concentrations in the environment decline rapidly with increasing distance from the smelter. These data specifically refer to the forest floor, but similar observations would result for soil, vegetation, lakewater, and other components of the ecosystem.

As we saw in Chapter 16, SO_2 has also been an important pollutant in the Sudbury area. Consequently, it is difficult to determine the specific role of toxic metals in causing ecological damage near the Sudbury smelters. One way of investigating the influence of metals is to grow plants in polluted soils in a greenhouse, where SO_2 is not present. Such *bioassay* experiments have demonstrated that soils collected near the Sudbury smelters are toxic, mainly because of their high concentrations of metals. To a substantial degree, the toxicity persists even after the soil acidity is neutralized by adding lime.

Not all smelters emit both SO_2 and metals. The ecological damages that result from those that emit only metal particulates have consequently been caused by metal pollution. One well-studied smelter, at Gusum, Sweden, has

FIGURE 18.1 | METAL POLLUTION NEAR SUDBURY

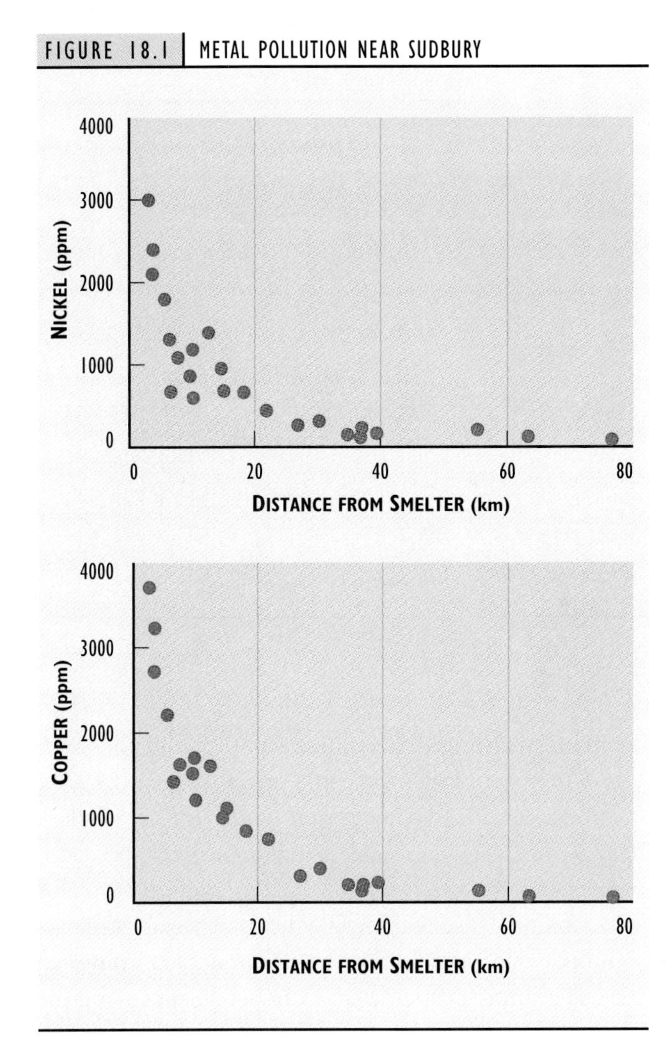

Decades of emissions of metals from the Copper Cliff smelter have caused an accumulation of nickel and copper in various components of the environment. The most intense pollution occurs close to the point source. These data are for metals in the forest floor, that is, the organic-rich layer that overlies the mineral soil. This component of the ecosystem effectively binds metals in organic complexes, and accumulates much higher residues than the underlying mineral soil. The forest-floor samples were collected along a transect running south of the smelter.

Source: Freedman and Hutchinson (1980)

been operating since 1661 (Tyler, 1984). Zinc is an important pollutant emitted at Gusum, reaching concentrations as high as 2% (or 20,000 ppm) in surface organic matter close to the point source, compared with less than 200 ppm farther than 6 km away. Levels of copper pollution are similar, reaching 1.7% within 0.3 km, compared with 20 ppm beyond 6 km. The zinc and copper pollution has caused local ecological damage. Pine and birch trees have died or declined close to the source, and understorey plants, mosses, lichens, and soil-dwelling invertebrates have been damaged. Rates of decomposition and nutrient cycling are also impaired in the most polluted sites. Some species, however, are quite tolerant of the metal pollution at Gusum. These include the grass *Deschampsia flexuosa* and the moss *Pohlia nutans*, which do relatively well in sites that are toxic to other plants.

Use of Inorganic Pesticides in Agriculture

Until the 1970s, inorganic chemicals were widely used to combat pests in agriculture (see also Chapter 21). This was especially true in fruit orchards, where pesticides based on lead arsenate, calcium arsenate, copper sulphate, and related compounds were used to control fungal diseases and arthropod pests. These compounds are now largely displaced by synthetic organic pesticides. Until the mid-1970s, annual spray rates in southern Ontario orchards were as high as 8.7 kg/ha of lead, while arsenic treatments reached 2.7 kg/ha·y, zinc 7.5 kg/ha·y, and copper 3.0 kg/ha·y (Frank *et al.*, 1976). The spray rates depended on the crop being grown, the pest being managed, and the pesticide used; but in some cases all of these toxic elements were applied in the same orchards.

Residues of these chemicals tended to accumulate in the soil of pesticide-treated orchards. Studies of apple orchards in Ontario found residues as high as 890 ppm of lead and 126 ppm of arsenic in surface soil, compared with background levels of <25 ppm Pb and <10 ppm As (Figure 18.2). These accumulations were caused up to 70 years of spraying lead arsenate as an insecticide, mostly against the codling moth (*Laspeyresia pomonella*), a pest that causes "wormy" apples.

Agricultural soils can also be contaminated by the use of mercury-containing fungicides, especially those that protect newly germinated seedlings from a fungal infection known as "damping-off." This pathogen attacks seedlings at the soil-air interface and causes the weakened plant to fall over and die. Mercury-containing pesticides are also used to control turfgrass diseases on lawns and golf-course putting greens. Mercury residues ranging from 24 to 120 ppm have been measured in the surface soil of putting greens in Ontario, while concentrations up to 9 ppm were found in golf courses in Nova Scotia.

FIGURE 18.2 ACCUMULATION OF ARSENIC AND LEAD IN SPRAYED ORCHARDS

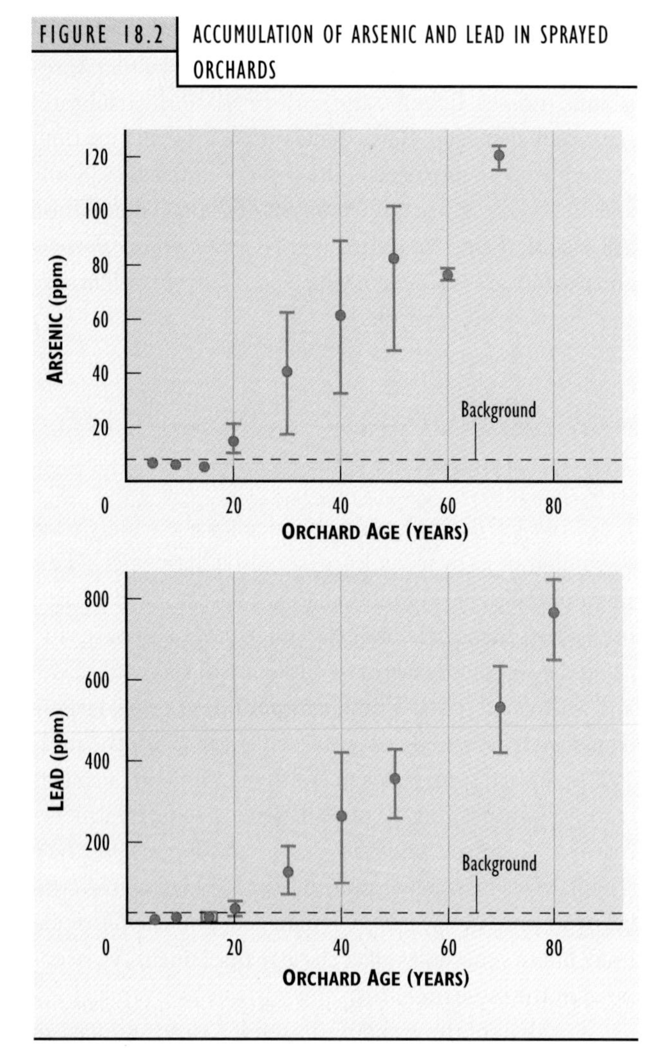

Lead arsenate has been used as an insecticide to combat infestations of apple orchards with codling moth. These data show the progressive accumulation of arsenic and lead in soils of orchards in southern Ontario. The largest residues occurred in the oldest orchards, which had been sprayed for many years. The error bars indicate the range of values about the average value for each orchard.

Source: Modified from Frank *et al.* (1976)

The sowing of seed coated with mercuric fungicides has frequently caused contamination and poisoning of wild animals that consumed the planted grains or scavenged dead herbivores. Alkyl-mercury compounds such as methylmercury are especially hazardous in this respect, because this form of mercury is extremely toxic and readily assimilated by animals from their food. Table 18.5 shows the significant mercury contamination of seed-eating wildlife in regions of Alberta where treated seed was used,

TABLE 18.5 MERCURY CONCENTRATIONS IN ANIMALS FEEDING ON TREATED SEED IN ALBERTA

Seed-eating rodents and birds became contaminated by feeding on seed treated with alkyl-mercury fungicide in agricultural areas. For comparison, data are presented for an area where mercury-treated seed was not used. Data are analyses of liver, and are expressed in ppm dry weight, as average ± standard deviation, with the sample size (n) given in parentheses.

ORGANISM	MERCURY CONCENTRATION	
	TREATED AREA	UNTREATED AREA
Rodents	1.25 ± 0.68 (n=6)	0.18 ± 0.15 (n=5)
Songbirds	1.63 ± 1.00 (n=10)	0.03 ± 0.01 (n=3)
Upland game birds	1.88 ± 0.44 (n=19)	0.35 ± 0.22 (n=12)
All seed eaters	1.70 ± 0.38 (n=35)	0.26 ± 0.14 (n=20)

Source: After Fimreite *et al.* (1970)

compared with areas where these exposures did not occur. Use of these fungicides was common until the early 1970s.

Most developed countries prohibited the use of alkyl-mercury fungicides as seed dressings in the late 1960s. This ban resulted from the recognition of ecological problems associated with use of these chemicals, especially the poisoning of wild animals. Sweden, for example, prohibited the use of these pesticides in 1966, while approving the use of alkoxyl-alkyl-mercury compounds, which are much less toxic, as replacements. This action rapidly led to decreased mercury contamination of wildlife, such as predatory birds (Figure 18.3). Canada took similar actions, although several years later.

As was noted in the introduction to this chapter, humans have also been poisoned by eating mercury-treated seed grains.

Metals in Sewage Sludge

Sewage sludge is produced by composting the organic-rich materials that settle out of waste waters during the treatment of municipal sewage (see Chapter 20). Because sewage sludge is composed mainly of well-humified organic matter, it is an extremely useful material that can be worked into soil to enhance its tilth and its water- and nutrient-holding capacities (these terms are defined in Chapter 14). Sewage sludge also contains important nu-

FIGURE 18.3 MERCURY CONTAMINATION OF SWEDISH HAWKS CAUSED BY ALKYL-MERCURY FUNGICIDES

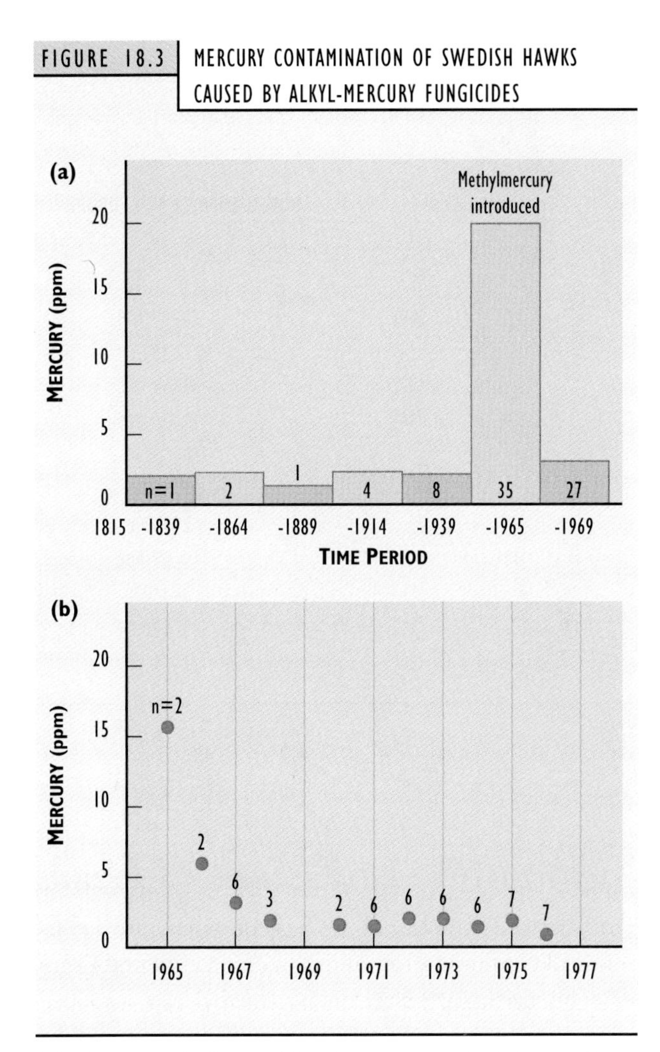

(a) Mercury in feathers of female goshawks (*Accipiter gentilis*), collected from nesting birds during various time periods.
(b) Mercury in feathers of marsh harriers (*Circus aeruginosus*). Note the large increase in mercury contamination caused by the use of alkyl-mercury fungicides and the rapid decrease that followed the banning of these chemicals in 1966. Sample sizes (n) are indicated.

Source: After Johnels *et al.* (1979)

trients such as nitrogen and phosphorus, and is useful as a slow-release organic fertilizer. Because of these favourable properties, sewage sludge is often applied to agricultural land.

Unfortunately, sewage sludge can contain considerable amounts of metals, particularly if there are significant inputs of industrial wastes to the sewerage. Typical concentrations of metals in sludge from several industrialized regions are summarized in Table 18.6. Obviously, metal concentrations vary among sewage sludges from different places, reflecting the kinds of industrial inputs to the wastewater systems. In general, cadmium, copper, nickel, and zinc are the metallic contaminants most likely to cause toxicity to plants when sewage sludges are applied to agricultural lands. By eating contaminated produce, humans can also be exposed to metals assimilated by crops from sewage sludges.

The problem of metals in sewage sludges can be largely avoided if industrial and sanitary sewage are treated separately. This is not always possible, however. Whatever the case, it is important to monitor the quantities of metals applied to agricultural land with sewage sludges, as well as any resulting contamination of crops. Because of their beneficial qualities when applied to agricultural or forestry lands, sewage sludges are potentially useful resources. If they are unacceptably contaminated with metals, however, sludges become wastes that must be discarded in landfills or the ocean or incinerated, causing environmental problems instead of providing economic benefits.

Birds and Lead

Millions of birds suffer lead poisoning in North America each year because they inadvertently ingest spent shotgun pellets. Most of the spent shot is associated with hunting. Although more localized, skeet shooting is also a problem because of the large amount of shot deposited in the vicinity of shooting ranges. It is not unusual for tonnes of lead shot to be spent each year at a single skeet shooting range.

Lead shot can be retained in the gizzard, the muscular forepouch of the stomach of seed-eating birds, after being ingested. Hard grit is normally retained in the gizzard and used to grind hard-coated seeds, aiding in their digestion. Unfortunately, shotgun pellets are similiar in size and weight to the grit that many species of birds pick up for this purpose. The lead shot becomes abraded in the gizzard, and the bits are swallowed and dissolved by acidic stomach fluids. The lead can then be absorbed into the bloodstream, allowing it to poison the nervous system of the bird, commonly leading to death.

Waterfowl are especially widely affected, with an estimated 2–3 million individuals, or 2–3% of the North American population, dying each year from lead-shot

TABLE 18.6 | METALS IN SEWAGE SLUDGE

Elemental concentrations in sewage sludge from a variety of locations. Data are in ppm dry weight, with average and/or ranges of values given.

ELEMENT	ONTARIO(1)	MICHIGAN(2)	NORTH AMERICA AND EUROPE(3)
Arsenic	–	7.8 (1.6–18)	1–18
Cadmium	29 (2–147)	74 (2–1 100)	1–1 500
Chromium	4 200 (16–16 000)	2 030 (22–30 000)	20–40 000
Cobalt	–	–	2–260
Copper	1 100 (162–3 000)	1 020 (84–10 400)	52–11 700
Lead	1 200 (85–4 000)	1 380 (80–26 000)	15–26 000
Manganese	310 (60–500)	–	60–3 860
Mercury	9 (1–24)	5.5 (0.1–56)	0.1–56
Molybdenum	–	–	2–1 000
Nickel	390 (7–1 500)	371 (12–2 800)	10–53 000
Silver	25 (4–60)	–	5–150
Vanadium	–	–	40–700
Zinc	4 500 (610–19 000)	3 320 (72–16 400)	72–49 000

Sources: (1) 10 sewage works, after Van Loon (1974);
(2) 57 sewage works, after Blakeslee (1973);
(3) 300 sewage works, after Page (1974)

poisoning. The gizzard retention of just one or two pellets can poison a duck, causing a wasting away of 30–50% of the body weight, neurological toxicity, and ultimately death. Typically, about 10% of the waterfowl in North America have one or more shotgun pellets in their gizzard.

Larger aquatic birds, such as swans, are known to retain lead fishing weights in their gizzard. Lead sinkers or shot have been cited as the cause of 20–50% of the known mortality of trumpeter swans (*Cygnus buccinator*) in parts of their range in western North America. Lead sinkers are also known to poison tundra swans (*C. columbianus*) wintering in the eastern United States, mute swans (*C. olor*) in Europe, and common loons (*Gavia immer*) in Canada and the United States.

A related syndrome, caused by ingesting lead shot and bullets, afflicts birds that scavenge dead carcasses. Although the numbers are not well documented, this kind of poisoning is known to kill vultures, eagles, and other scavenging birds. The critically endangered California condor (*Gymnogyps californianus*) has been relatively well studied — about 60% of the known deaths of this species in the wild between 1980 and 1986 were caused by lead toxicity from ingested bullets in carrion.

Because of the widespread poisoning of birds by lead shot, regulators have begun to restrict its use. Lead shot is now banned over most of the United States, and in Canada since 1997. The use of lead shot for hunting is being replaced mostly by steel shot, and to a lesser degree by bismuth shot. The restricted use of lead shot has caused some controversy, because many hunters believe that the alternative shot types might cause more crippling deaths than lead shot. Field tests have, however, shown this effect to be marginal, as long as the inferior ballistic qualities of alternative shot are compensated for by shooting at closer distances or by using a larger size of shot.

Automobile Emissions of Lead

Lead emitted by automobiles has contributed to a general lead contamination of urban environments. From 1923, but particularly after 1945, tetraethyl lead was added to gasoline as a so-called "anti-knock" compound. The lead increases mechanical efficiency and gasoline economy, while decreasing engine wear. In 1975, about 95% of North American gasoline was leaded at concentrations as high as 770 mg/L. In 1987, only 35% of the gasoline was leaded, the maximum permitted concentration being 290 mg/L. The decreased use of lead between 1975 and 1987 was largely due to the increased use of catalytic converters to reduce emissions of other automobile pollutants, especially carbon monoxide and hydrocarbons. Automobiles equipped with catalytic converters can use only unleaded gasoline, because the converter catalysts, usually platinum, are rendered inactive by lead. The increasing use of unleaded fuels between 1977 and 1989 resulted in a 93% decrease in lead particulates in the air of Canadian cities.

After 1990, the use of leaded gasoline was banned for most purposes in Canada and the U.S. (except for low-lead fuels that may still be used in some farm vehicles, marine engines, and large trucks). Consequently, emissions of lead from automobiles in Canada decreased from about 9500 t in 1978 to less than 100 t in 1995 (Environment Canada, 1996). However, many other countries, particularly in the less-developed world, continue to allow the use of leaded fuels.

Almost all of the lead in gasoline is emitted as particulates through the vehicle tailpipe. The heavier particulates settle out close to the roadway. This results in the build-up of well-defined gradients of lead pollution, the intensity of which is related to traffic volume. This pattern of roadside lead pollution is illustrated in Table 18.7 for urban roads of different traffic densities in Halifax (note that this study was made prior to the banning of leaded fuels in Canada). Finer lead particulates are more widely dispersed in the atmosphere and contribute to the general lead contamination that occurs in cities. Not surprisingly, studies have shown some effects of lead on urban wildlife. For example, pigeons (*Columba livia*) living in cities may contain significant residues of lead and may exhibit symptoms of acute lead poisoning.

TABLE 18.7 | LEAD AND AUTOMOBILES

Samples of soil, collected near roads of different traffic density in Halifax, Nova Scotia, were analyzed for lead content. Metal data are in ppm dry weight, while average daily traffic (ADT) is in vehicles per day. The background concentration in soil is 14 ppm.

DISTANCE FROM ROAD EDGE (m)		LEAD CONCENTRATION			
		ROAD A	ROAD B	ROAD C	ROAD D
	ADT:	50 000	18 500	16 000	3 000
0		3045	858	1075	465
1		2813	402	457	118
5		342	177	136	32
15		223	75	163	26
30		–	45	63	26
50		223	45	95	38
100		–	–	60	21

Source: Modified from Dale and Freedman (1982)

KEY TERMS

bioconcentration (bioaccumulation or biomagnification)
food-web magnification (food-web concentration)
"total" concentration
"available" concentration
tolerance
hyperaccumulator species
endemic species
ecotypes
point source

QUESTIONS FOR DISCUSSION

1. Considering the fact that metals and other elements are ubiquitous in the environment, explain how to distinguish normal levels, contamination, and pollution by these substances.
2. Do you think that damages similar to those seen near Sudbury are likely to be caused when a new smelter is constructed to process the ore mined at the newly discovered mineral deposit at Voisey's Bay, Labrador? (Note that the ores in both cases are similar, containing sulphide minerals of nickel and copper.)
3. Important environmental benefits have been gained by banning the use of leaded gasoline in Canada. Explain why there have been long delays in taking similarly vigorous actions against the use of lead shot in hunting and skeet shooting, and lead weights in fishing.
4. Pick an element discussed in this chapter and research its benefits, toxicity, effects on the environment, control, and amelioration.

REFERENCES

Armstrong, F.A.J. 1979. Mercury in the aquatic environment. In: *Effects of Mercury in the Canadian Environment*. Associate Committee on Scientific Criteria for Environmental Quality. Ottawa: National Research Council of Canada. NRCC No. 16739. pp. 84–100.

Blakeslee, P.A. 1973. *Monitoring considerations for municipal wastewater effluent and sludge application to the land*. Urbana, IL: U.S. Environmental Protection Agency, U.S. Department of Agriculture, Universities Workshop, July 9–13, 1973.

Bowen, H.J.M. 1979. *Environmental Chemistry of the Elements*. New York: Academic.

Bradshaw, A.D. and M.J. Chadwick. 1981. *The Restoration of Land*. Oxford: Blackwell.

Cox, R.M. and T.C. Hutchinson. 1979. Metal co-tolerances in the grass *Deschampsia caespitosa*. *Nature*, **279**: 231–233.

Dale, J.M. and B. Freedman. 1982. Lead and zinc contamination of roadside soil and vegetation in Halifax, Nova Scotia. *Proc. N.S. Inst. Sci.*, **32**: 327–336.

Environment Canada. 1996. *Urban Air Quality*. State of the Environment Reporting Organization. Ottawa: Environment Canada. SOE Bull. No. 96-1.

Fimreite, N., R.W. Feif, and J.A. Keith. 1970. Mercury contamination of Canadian prairie seed eaters and their avian predators. *Can. Field-Nat.*, **84**: 269–276.

Frank, R., H.E. Braun, K. Ishida, and P. Suda. 1976. Persistent organic and inorganic pesticide residues in orchard soils and vineyards of southern Ontario. *Can. J. Soil Sci.*, **56**: 463–484.

Freedman, B. 1995. *Environmental Ecology, 2nd ed.* San Diego, CA: Academic.

Freedman, B. and T.C. Hutchinson. 1980. Pollutant inputs from the atmosphere and accumulations in soils and vegetation near a nickel-copper smelter at Sudbury, Ontario, Canada. *Can. J. Bot.*, **58**: 108–132.

Gilmour, C.C. and E.A. Henry. 1991. Mercury methylation in aquatic systems affected by acid deposition. *Environ. Pollut.*, **71**: 131–169.

Johnels, A., G. Tyler, and T. Westermark. 1979. A history of mercury levels in Swedish fauna. *Ambio*, **8**: 160–168.

Kruckeberg, A.R. 1984. *California Serpentine: Flora, Vegetation, Geology, Soils, and Management Problems*. Los Angeles: Univ. of California Press.

Kuja, A.L. 1980. *Revegetation of mine tailings using native species from disturbed sites in northern Canada*. M.Sc. Thesis, Department of Botany, University of Toronto.

McKay, C. 1985. *Freshwater Fish Contamination in Canadian Waters*. Ottawa: Department of Fisheries and Oceans, Fish Habitat Management Branch, Chemical Hazards Division.

Mierle, G. 1990. Aqueous inputs of mercury to Precambrian Shield lakes in Ontario. *Environ. Contam. & Chem.*, **9**: 843–851.

Page, A.L. 1974. *Fate and Effects of Trace Elements in Sewage Sludge Applied to Agricultural Lands*. Cincinnati, OH: U.S. Environmental Protection Agency, Office of Research and Development. Pub. EPA-670/2-74-005.

Peters, T.H. 1984. Rehabilitation of mine tailings: a case of complete ecosystem reconstruction and revegetation of industrially stressed lands in the Sudbury area, Ontario, Canada. In: *Effects of Pollutants at the Ecosystem Level*. (P.J. Sheehan, D.R. Miller, and P. Bourdeau, eds.), New York: Wiley. pp. 403–421.

Ripley, E.A., R.E. Redmann, and A.A. Crowder. 1996. *Environmental Effects of Mining*. Delray Beach, FL: St. Lucie.

Sanderson, G.C. and F.C. Bellrose. 1986. *Lead Poisoning in Waterfowl*. Urbana, IL: Illinois Natural History Society. Spec. Pub. 4.

Scheuhammer, A.M. and S.L. Norris. 1996. The ecotoxicology of lead shot and lead fishing weights. *Ecotoxicology*, **5**: 279–295.

Spry, D.J. and J.G. Weiner. 1991. Metal bioavailability and toxicity to fish in low-alkalinity lakes: A critical review. *Environ. Pollut.*, **71**: 243–304.

Thompson, D.R., K.C. Hamer, and R.W. Furness. 1991. Mercury accumulation in great skuas (*Catharacta skua*) of known age and sex, and its effect on breeding and survival. *J. Appl. Ecol.*, **28**: 672–684.

Tyler, G. 1984. The impact of heavy metal pollution on forests: a case study of Gusum, Sweden. *Ambio*, **13**: 18–24.

Van Loon, J.C. 1974. *Analysis of Heavy Metals in Sewage Sludge and Liquids Associated with Sludges*. Presented at Canada/Ontario Sludge Handling and Disposal Seminar, Toronto, Sept. 18–19.

Wiemeyer, S.N., J.M. Scott, M.P. Anderson, P.H. Bloom, and C.J. Stafford. 1988. Environmental contaminants in California condors. *J. Wildl. Manage.*, **52**: 238–247.

19

ACIDIFICATION

CHAPTER OUTLINE

CHAPTER OBJECTIVES

After completing this chapter, you will be able to:

1. Describe the most important chemical constituents of precipitation and explain which of these can cause acidity to develop and how.
2. Outline the spatial patterns of acidic precipitation in North America and identify the factors that influence this distribution.
3. Explain the difference between wet and dry depositions of acidifying substances and how their rates vary depending on air quality and other factors.
4. Describe how water chemistry changes as precipitation interacts with plant canopies and soil, and discuss the implications of these changes for surface waters.
5. Identify the factors that make freshwaters vulnerable to acidification.
6. Describe the effects of acidification on major groups of freshwater organisms.
7. Discuss the roles of liming and fertilization in the reclamation of acidified lakes.

INTRODUCTION

Acidification is characterized by increasing concentrations of hydrogen ions (H^+) in soil or water. It can cause stable metals and metal compounds to ionize, producing metal ions (such as Al^{3+}) in concentrations high enough be toxic to plants, animals, and microorganisms. Consequently, increasing acidification is generally interpreted as a degradation of environmental quality. Acidification is caused by many influences, both natural and anthropogenic, but the most widespread problems are associated with a phenomenon commonly known as **acid rain**.

Acid rain has been an important problem in parts of North America since at least the 1950s, but it did not become a high-profile issue until the early 1970s. This rather sudden attention resulted from the discovery that acid rain was an important environmental problem in western Europe, and from the realization that the same conditions were likely to have developed in North America. This awareness stimulated much research in Canada and the United States — research which demonstrated that acid

IN DETAIL 19.1

ACIDS AND BASES

Acids are defined as substances that donate protons (H^+) in chemical reactions. An aqueous solution is acidic if its concentration of hydrogen ions is more than 1×10^{-7} moles/L. In contrast, bases (or alkalis) are substances that donate hydroxyl ions (OH^-) in chemical reactions. An aqueous solution is basic if it has a concentration of hydroxyl ions of more than 1×10^{-7} moles/L. (A mole is a fundamental unit used to measure the amount of a substance, and is equal to 6.02×10^{23} molecules, atoms, or ions. This number is known as Avogadro's constant, and is derived from the number of atoms of carbon contained in 12 grams (1 mole) of carbon-12.)

Acids and bases react together to form water and a neutral salt. If equal numbers of moles of each are present, the solution has both zero acidity and zero alkalinity: that is, the concentrations of both H^+ and OH^- happen to be exactly 1×10^{-7} moles/L. Such a solution is neutral.

Because extremely wide ranges of H^+ and OH^- concentrations can be encountered in nature and in laboratories, acidity is measured in logarithmic units, which are referred to as pH (an abbreviation for **p**otential of **H**ydrogen). pH is defined as $-\log_{10}[H^+]$ (that is, the negative logarithm to base 10 of the aqueous concentration of hydrogen ion, expressed in units of moles per litre). Acidic solutions have a pH less than 7, while alkaline solutions have a pH greater than 7. Note that a one-unit difference in pH implies a 10-fold difference in the concentration of hydrogen or hydroxyl ions. The pH scale illustrated here shows the pHs of some commonly encountered substances.

THE pH SCALE

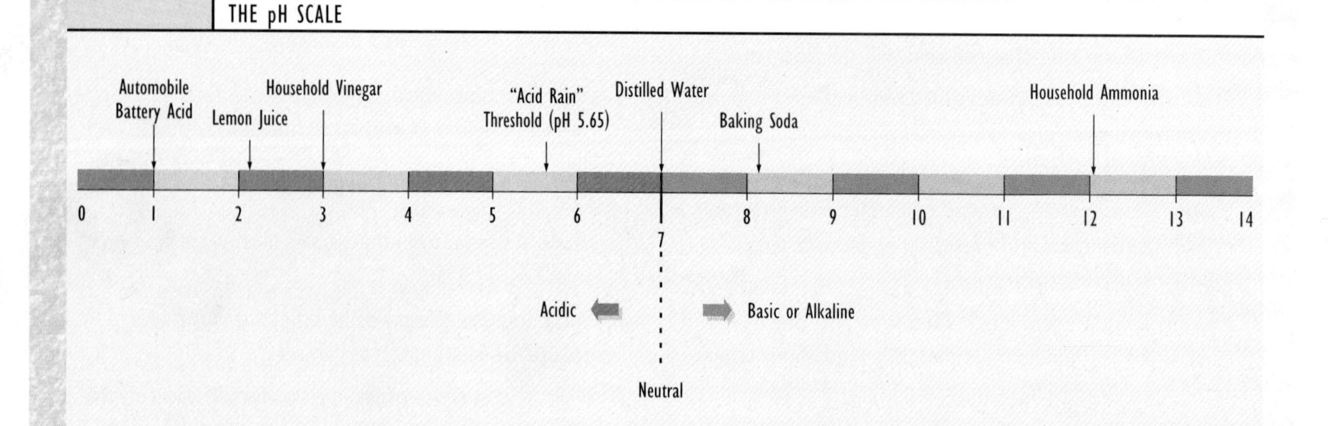

rain was causing extensive acidification in lakes and streams, and possibly in soils. The acidification of aquatic ecosystems was shown to be causing important ecological damages, including the extirpation of many fish populations. Buildings and other materials are also being damaged by acid rain, because metals, paint, bricks, and some kinds of quarried stone are eroded by acidic solutions.

Strictly speaking, the phrase "acid rain" refers only to acidic rainfall, which along with snowfall accounts for **wet deposition**. However, acidifying chemicals are also deposited from the atmosphere even when it is not precipitating, through the **dry deposition** of certain gases and particulates. A suitable phrase to define this complex of processes is **the deposition of acidifying substances from the atmosphere**, or more simply, **acidifying deposition**.

In this chapter we examine various natural and anthropogenic causes of the acidification of ecosystems. We will focus on the chemical qualities of acidic precipitation and dry deposition, their effects on terrestrial and aquatic ecosystems, and how acidification can be avoided or mitigated.

Chemistry of Precipitation

Scientists have adopted a functional definition of acidic precipitation as having a pH less than 5.65. This was chosen as the cutoff because at pH 5.65, carbonic acid (H_2CO_3) is in equilibrium with atmospheric CO_2, as follows:

$$CO_2 + H_2O \longleftrightarrow H_2CO_3 \longleftrightarrow H^+ + HCO_3^- \longleftrightarrow 2H^+ + CO_3^{2-}$$

This definition assumes that "non-acidic" precipitation is essentially distilled water, in which acidity is determined only by the atmospheric concentration of CO_2 and the amount of carbonic acid that subsequently develops. This is why the threshold below which precipitation is deemed "acidic" is set at the slightly acidic pH of 5.65, rather than at the strict zero-acidity pH 7.0 (see In Detail 19.1).

It is, however, too simplistic to consider atmospheric moisture to consist merely of distilled water in a pH-equilibrium with gaseous CO_2. Many other chemicals are present in precipitation in trace concentrations. For example, on windy days dusts containing calcium and magnesium are blown into the atmosphere, where precipitation containing these elements may develop pHs higher than 5.65. This is especially true of agricultural and prairie landscapes, where the ground surface is often bare of plant cover and soil particles can be easily eroded into the atmosphere.

The most abundant cations (i.e., positively charged ions) in precipitation are hydrogen ion (H^+), ammonium (NH_4^+), calcium (Ca^{2+}), magnesium (Mg^{2+}), and sodium (Na^+). The most abundant anions (negatively charged ions) are sulphate (SO_4^{2-}), chloride (Cl^-), and nitrate (NO_3^-). Other ions are also present, but in trace concentrations that have little influence on precipitation pH (see In Detail 19.2).

One of the longest-running North American records of precipitation chemistry is from a research site at Hubbard Brook, New Hampshire, in a region exposed to intense acidifying deposition. The average pH of precipitation at Hubbard Brook is about 4.2, and hydrogen ions comprise 71% of the total equivalents of cations (Table 19.1; see In Detail 19.2 for an explanation of equivalents). Sulphate and nitrate are the most important anions, occurring in a 2:1 ratio and accounting for 87% of the anion equivalents. These data suggest that most acidity in the precipitation occurs as dilute solutions of sulphuric and nitric acids. The precipitation events at Hubbard Brook that are most acidic are associated with storms that have passed over the large metropolitan regions of Boston, New York, and New Jersey. These areas have enormous emissions of SO_2 and NO_x, which are the precursor gases of much of the SO_4^{2-} and NO_3^- in acidic precipitation (as we saw in Chapter 16).

Spatial Patterns of Acidifying Deposition

Acidic precipitation is a widespread phenomenon in eastern North America (Figure 19.1), Europe, eastern Asia, and elsewhere. In eastern North America prior to the mid-1950s, precipitation with pH below 4.6 affected only local areas, mostly in southern Ontario, New York, Pennsylvania, and New England. Since then, however, this area has expanded considerably. At present, most of southeastern Canada and the eastern United States experiences acidic

IN DETAIL 19.2

CONSERVATION OF ELECTROCHEMICAL NEUTRALITY

The principle of conservation of electrochemical neutrality states that in any electrically neutral solution (i.e., one that does not carry an electrical charge), the total number of positive charges associated with cations must equal the number of negative charges of anions. For the purposes of calculating a charge balance, the concentrations of ions must be measured in units known as *equivalents*. These are calculated as the molar concentration multiplied by the number of charges on the ion. (When dealing with precipitation or surface waters, microequivalents, or µeq, are generally the units that are reported.)

This principle is relevant to the acidification of water. The concentration of H^+ can be determined as the difference in concentrations of the sum of all anion equivalents minus the sum of all cations other than H^+. Therefore, if the total equivalents of anions exceed the total equivalents of cations other than hydrogen ion, then H^+ goes into solution to balance the cation "deficit," as follows:

$$H^+ = (SO_4^{2-} + NO_3^- + Cl^-) - (Na^+ + NH_4^+ + Ca^{2+} + Mg^{2+})$$

The above equation has proven to be quite useful in studies of acidic precipitation. Prior to about 1955, the measurement of pH values was somewhat inaccurate. There were, however, reliable analyses of other important ions in surface waters and precipitation. In such cases, the above equation can be used to calculate pre-1955 pH values, providing important data for historical values of pH in waters sensitive to acidification.

TABLE 19.1 CHEMISTRY OF PRECIPITATION AT HUBBARD BROOK, NEW HAMPSHIRE

These data represent average concentrations (in microequivalents per litre) of various ions in the precipitation during a 19-year study period from 1963 to 1982, along with ± standard error (a statistical measure of variation among years). The small difference between the sums of cation and anion equivalents is due to analytical inaccuracies, which is inevitable in even the best chemical data. The average pH of the precipitation was 4.16.

CONSTITUENT	µeq/L	% OF TOTAL EQUIVALENTS
CATIONS		
H^+	69.3 ± 2.1	71.1
NH_4^+	10.6 ± 0.6	10.9
Ca^{2+}	6.5 ± 0.8	6.7
Na^+	4.8 ± 0.5	4.9
Mg^{2+}	3.0 ± 0.5	3.1
Al^{3+}	1.8 ± 0.2	1.8
K^+	1.5 ± 0.3	1.5
ANIONS		
SO_4^{2-}	54.0 ± 2.1	60.6
NO_3^-	23.5 ± 1.0	26.4
Cl^-	11.2 ± 1.2	12.6
PO_4^{3-}	0.3 ± 0.1	0.3
HCO_3^-	0.1	0.1
Sum of Cations	97.5 ± 2.5	
Sum of Anions	89.1 ± 2.6	

Source: Modified from Likens *et al.* (1984)

precipitation. It appears that the broad pattern of acidic precipitation in North America existed before the 1950s, but the phenomenon has since become more widespread and its intensity has increased. One of the most important environmental aspects of acidic precipitation is the vast size of the areas that it affects.

Precipitation chemistry varies greatly between regions (Figure 19.1). The variation reflects patterns of emission of SO_2 and NO_x, their degree of oxidation to SO_4^{2-} and NO_3^-, the prevailing direction travelled by contaminated air masses, and the amounts of acid-neutralizing dusts in the atmosphere. Atmospheric dusts are particularly important where vegetation cover is sparse, as in agricultural regions where tiny soil particles are easily carried into the atmosphere by strong winds blowing over bare fields. Dry, unpaved roads are also important sources of atmospheric dust.

Information on precipitation chemistry at four widely separated Canadian sites is summarized in Table 19.2. Dorset is located in a rural area in the Muskoka region of south-central Ontario. There, the precipitation is highly acidic, with an average pH of 4.1. The precipitation at Dorset has high concentrations of H^+, SO_4^{2-}, and NO_3^-, suggesting that its acidity is due mainly to dilute sulphuric and nitric acids. Much of the sulphate and nitrate in precipitation is derived from SO_2 and NO_x emitted by industries and automobiles to the south, and then transported

FIGURE 19.1 | CHARACTERISTICS OF PRECIPITATION IN EASTERN NORTH AMERICA

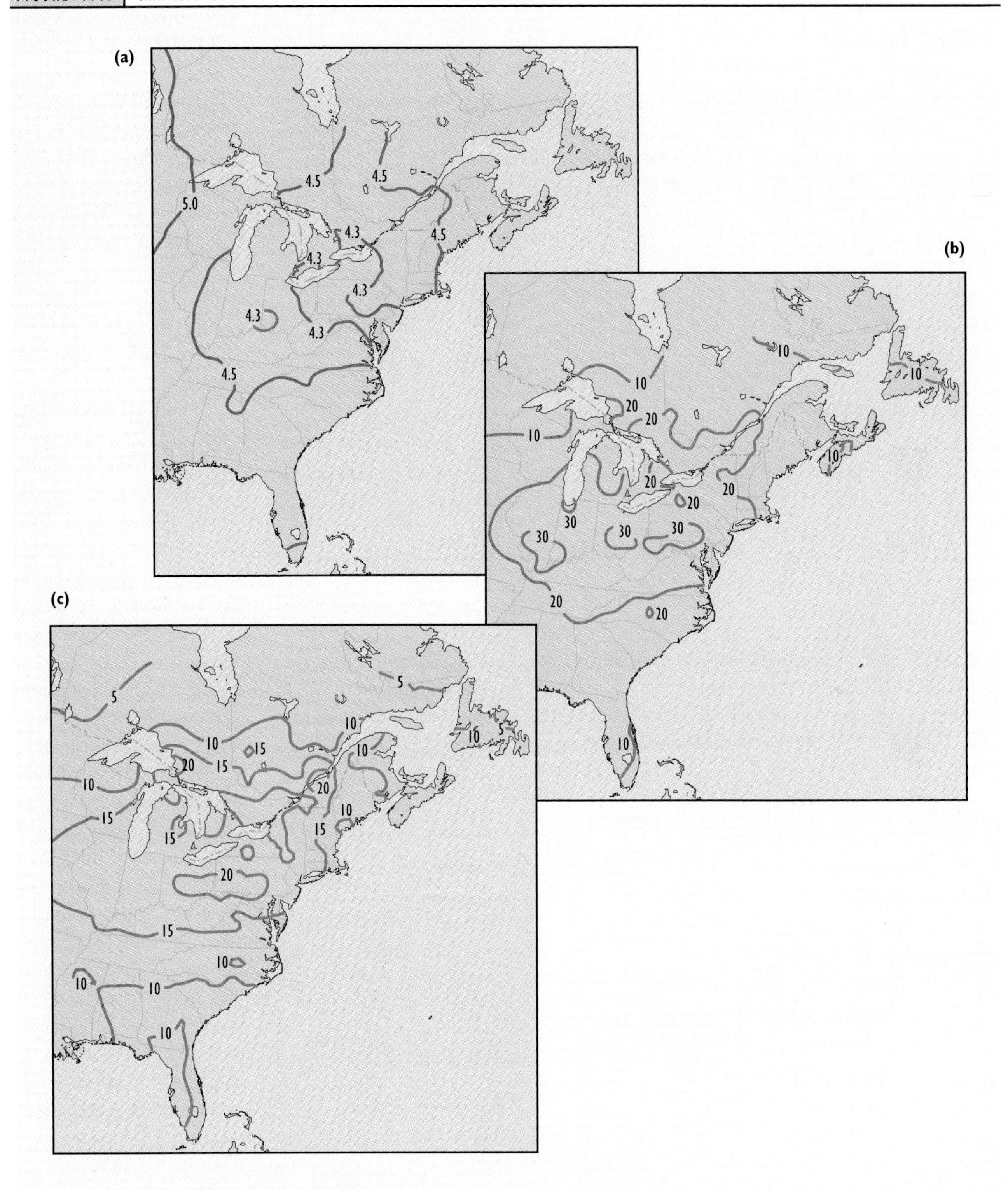

Points on the curved lines (known as isopleths) have equal annual average values of:

(a) pH in precipitation;

(b) sulphate deposition in precipitation (in kg/ha·y; corrected for sulphate of marine origin); and

(c) nitrate deposition (in kg/ha·y). Maps are for 1993.

Source: Environment Canada

TABLE 19.2 | PRECIPITATION CHEMISTRY IN VARIOUS PLACES IN CANADA

The data (in μeq/L) are average concentrations of chemicals in precipitation collected at maritime, continental, and prairie sites in Canada. The sites are described in the text. Data are for wet-only precipitation, meaning the collector was open to the atmosphere only during precipitation events.

CONSTITUENT	MARITIME: KEJIMKUJIK, N.S.	INDUSTRIAL CONTINENTAL: DORSET, ONT.	REMOTE CONTINENTAL: ELA, ONT.	PRAIRIE: LETHBRIDGE, ALTA.
CATIONS				
H^+	25.1	73.6	18.6	1.0
Ca^{2+}	4.3	10.0	12.0	112.8
Mg^{2+}	2.9	2.4	2.4	25.5
Na^+	26.1	3.9	4.3	9.6
K^+	1.1	1.0	2.0	2.3
NH_4^+	4.2	15.6	18.9	22.2
ANIONS				
SO_4^{2-}	27.5	58.3	27.1	43.5
NO_3^-	9.7	35.5	16.4	20.8
Cl^-	29.5	4.2	5.4	9.9
pH	4.6	4.1	4.7	6.0

Source: Modified from Freedman (1995)

in the atmosphere before being deposited as acidifying deposition.

The ELA (Experimental Lakes Area) site is in a remote area of northwestern Ontario, located west of Kenora (see Canadian Focus 20.1). Like Dorset, the ELA is in a landscape of Precambrian Shield, where bedrock and soil are composed of hard minerals such as granite, gneiss, and quartzite. However, the ELA site is less influenced by air masses contaminated by anthropogenic emissions; so its precipitation is less acidic (with an average pH of 4.7) than at Dorset, and has lower concentrations of nitrate and sulphate.

The Kejimkujik site in west-central Nova Scotia is also rather distant from large sources of emissions of SO_2 and NO_x. Kejimkujik often receives air masses that have moved over densely populated areas in the northeastern United States and eastern Canada. However, by the time the storm systems reach the Kejimkujik area, much of their acidic material has rained out, so the local precipitation is only moderately acidic (with an average pH of 4.6). Kejimkujik is also influenced by weather systems that have passed over the Atlantic Ocean. As a result, the precipitation at Kejimkujik has relatively high concentrations of sodium and chloride. These ions enter the precipitation from atmospheric particulates derived from sea spray. Oceanic saltwaterhas a pH of about 8 because of the presence of chemicals such as bicarbonate, so marine aerosols have an acid-neutralizing influence on precipitation in coastal regions.

Lethbridge is located in a landscape of mixed-grass prairie in southern Alberta.

TABLE 19.3 | CHEMISTRY AND DEPOSITION OF CLOUDWATER

Deposition of water and ions were studied in a conifer forest at a high-elevation site on Mount Moosilauke, New Hampshire. Cloudwater deposition occurs when fog is filtered from the atmosphere by trees. Each concentration is stated as a mean value, with standard deviation, in micro-equivalents per litre. Precipitation refers to rain and snow; % cloudwater refers to the percentage of the total deposition that was due to cloudwater.

	CONCENTRATION	DEPOSITION (kilogram/hectare·year)			
	(μeq/L; MEAN ± SD)	CLOUD	PRECIP.	TOTAL	% CLOUDWATER
H^+	288 ± 193	2.4	1.5	3.9	62
NH_4^+	108 ± 89	16.3	4.2	20.5	80
Na^+	30 ± 29	5.8	1.7	7.5	77
K^+	10 ± 4	3.3	2.1	5.4	61
SO_4^{2-}	342 ± 234	275.8	64.8	340.6	81
NO_3^-	195 ± 175	101.5	23.4	124.9	81
H_2O (cm/y)		84	180	264	32

Source: Modified from Lovett *et al.* (1982)

Precipitation at Lethbridge is non-acidic (with an average pH of 6.0) because of the acid-neutralizing influences of calcium- and magnesium-rich particulates that are relatively abundant in the atmosphere. These originated from soil dusts blown up from agricultural fields and roads. These dusts also account for the high concentrations of Ca^{2+} and Mg^{2+} in the precipitation at Lethbridge.

Rapid changes in precipitation chemistry can occur at the borders between forested landscapes and areas dominated by prairie or agricultural lands. A study in southern Ontario examined precipitation at eight places in a mostly forested area with thin soils and Precambrian Shield bedrock, and at three sites just south of the Shield in agricultural terrain with calcium-rich soil (Dillon *et al.*, 1977). The average pH of precipitation among the Shield sites was 4.1–4.2, while at sites south of the Shield it was less acidic, ranging from 4.8 to 5.8. Precipitation was less acidic in the agricultural area because of the local neutralizing influence of dusts blown up from fields and roads.

Another important characteristic of acidic precipitation is that, unlike SO_2 and metal particulate pollution, its intensity does not increase closer to large point sources of emissions, such as power plants and smelters. For instance, precipitation is no more acidic close to the superstack at Sudbury than in the larger region, yet the superstack is one of the world's largest point sources of SO_2 emissions. Moreover, when that smelter was closed by a strike in 1979, the acidity of local precipitation did not change — it averaged pH 4.49 during the seven-month strike, compared with pH 4.52 during the prior seven months when there were large SO_2 emissions (Scheider *et al.*, 1980).

Fog moisture may also be quite acidic in eastern North America and elsewhere. Fogwater collected at high elevations and at coastal locations is commonly more acidic than pH 4.0, and it can be as acidic as pH 2.5–3.0. At forested sites where fog occurs frequently, large amounts of acidity and other chemicals can be filtered out of the atmosphere by trees. This phenomenon is illustrated in Table 19.3 for a conifer forest on a foggy mountain in New Hampshire. The total inputs of atmospheric moisture to that forest are about 264 cm/y, of which rain and snow account for 68% and cloudwater deposition the other 32% (fog occurred 40% of the time). However, the concentrations of many chemicals are much higher in cloudwater than in precipitation, so their rates of deposition to forests are also higher than from precipitation. Fogwater accounted for 62% of H^+ deposition and 81% of the inputs of SO_4^{2-} and NO_3^- to the conifer forest (Table 19.3).

Terrain in the La Cloche highlands of Ontario is extremely sensitive to acidification because of thin soils and hard, quartzitic bedrock (whitish colour in the photo). Lakes in this kind of terrain have very little capability of neutralizing inputs of acidifying substances from the atmosphere, and this is why they acidify rather easily.

Source: B. Freedman

Transboundary Air Pollution

Acidifying substances and their gaseous precursors are often transported over long distances in the atmosphere, far from their sources of emission. The acidifying chemicals do not respect political boundaries, so emissions occurring in one country can degrade ecosystems and valuable resources in other countries. This transboundary context has helped to focus the attention of governments on the problem of acidifying deposition from the atmosphere.

In western Europe, for example, Scandinavians have justifiably argued that most of the acidifying deposition that has affected parts of their landscape has resulted from emissions of SO_2 and NO_x in Germany and England. This international European context was the first well-demonstrated case of so-called **LRTAP** — an acronym for the **l**ong-**r**ange **t**ransport of **a**tmospheric **p**ollutants.

Similar transboundary circumstances occur elsewhere. In eastern North America, for example, there are large populations and industrial centres in the northeastern United States. Emissions of SO_2 and NO_x from those areas waft into eastern Canada, worsening damages caused there by local emissions. It has been estimated that United States emissions are responsible for 90% of the wet deposition of acidifying nitrogen compounds in eastern Canada, along with 63% of the wet deposition of sulphur compounds, 43% of the dry deposition of nitrogen, and 24% of the dry deposition of sulphur (Shannon and Lecht, 1986). Canada also exports some of its emissions to the United States, although Canadian emissions account for less than 5% of the total deposition of sulphur and nitrogen compounds in the eastern United States.

Dry Deposition of Acidifying Substances

Dry deposition occurs during intervals between precipitation events and includes:

- uptake of the gases SO_2 and NO_x by vegetation, soil, and water
- gravitational settling of larger particles
- filtering of finer atmospheric particulates by vegetation

Forests are particularly effective at absorbing gases and particles from the atmosphere. This happens because trees have a very large and complex surface foliage and bark, which greatly enhances rates of dry deposition.

FIGURE 19.2 | ACIDIFICATION CAUSED BY SULPHUR AND NITROGEN TRANSFORMATIONS

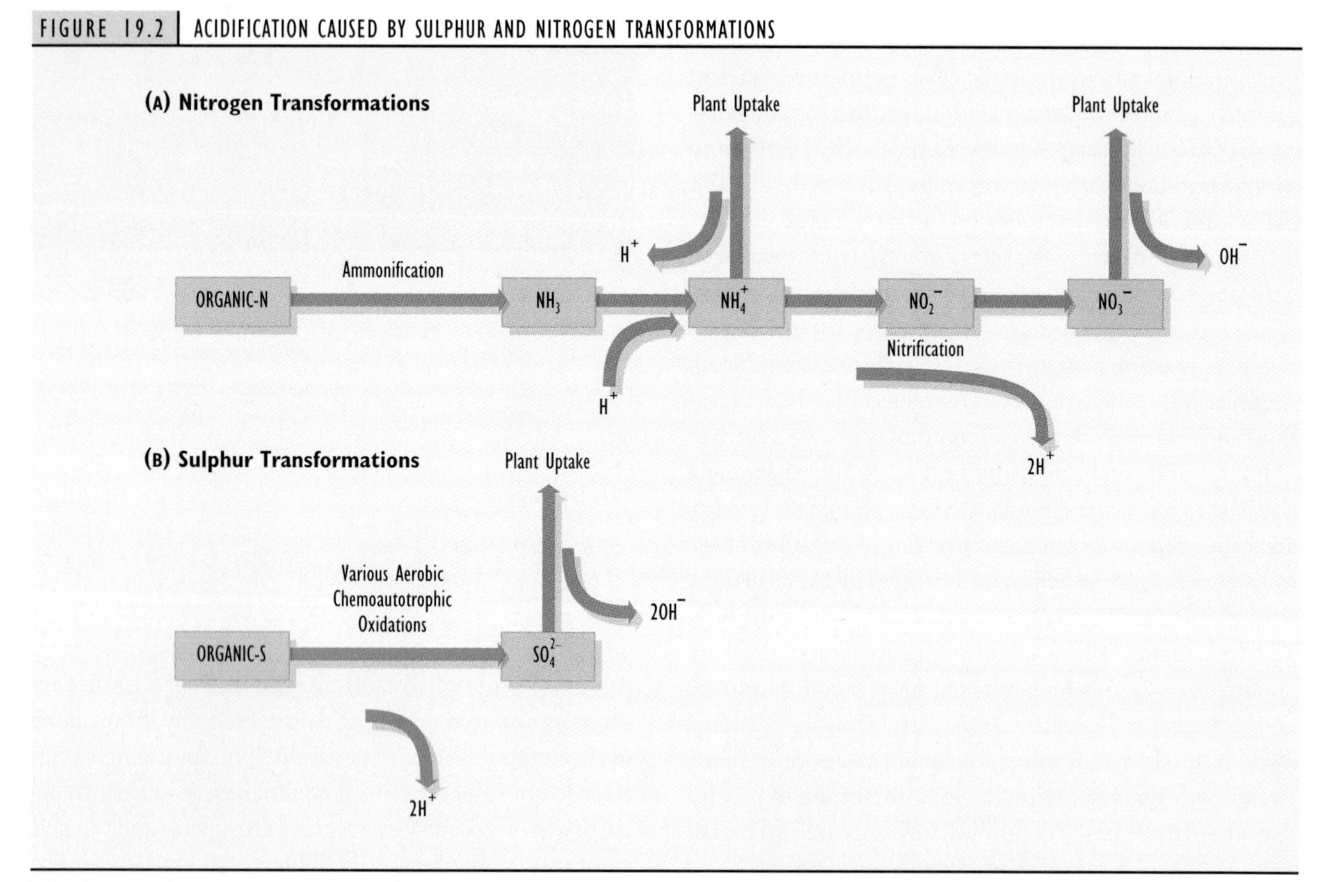

This diagram describes the acidifying effects of some important transformations of (a) nitrogen and (b) sulphur in soil or water.

Source: Modified from Reuss (1975)

Dry deposition can result in large inputs of substances from the atmosphere, including some that generate acidity when they are chemically transformed within the ecosystem. For instance, atmospheric SO_2 readily dissolves in the surface water of lakes and streams. This gas is also rather freely absorbed by plants, entering foliage through tiny, abundant pores on their surface known as stomata, and dissolving in the moist film of water that covers the internal cell surfaces. The absorbed SO_2 is oxidized to the anion sulphite (SO_3^{2-}), which is then rapidly oxidized to sulphate (SO_4^{2-}). Because the sulphate is mostly balanced electrochemically by hydrogen ions, acidity is generated by the transformation of SO_2 into SO_4^{2-} (see Figure 19.2).

NO_x gas can be similarly dry-deposited and then oxidized to nitrate (NO_3^-), which also generates an equivalent amount of H^+. The gas ammonia (NH_3) and the cation ammonium (NH_4^+) can also be dry-deposited to soil or water, where they can be oxidized by bacteria to nitrate plus equivalent quantities of H^+.

The rates of dry deposition of sulphur and nitrogen compounds are greatest when high concentrations of gaseous NO_x and SO_2 are present in the atmosphere. Such conditions typically occur in urban areas and close to large industrial sources of emissions of those gases. In these places, dry deposition accounts for larger inputs of acidifying substances than does wet deposition. In more remote, less contaminated environments, far from sources of emission, inputs with precipitation are typically larger than dry deposition.

Within 40 km of the largest smelter at Sudbury, about 55% of the atmospheric sulphur deposition occurs as dry deposition (Chan *et al.*, 1984). About 91% of the dry deposition involves SO_2, while the rest involves sulphate particulates. However, the superstack at that smelter, being extremely tall (380 m), is effective at dispersing its emissions of SO_2. Consequently, less than 1% of the emitted SO_2 is deposited close to the source (i.e., within 40 km)—almost all of the SO_2 emission is transported much farther away before it is deposited to the surface of the landscape.

Precipitation and Forest Canopies

The first surfaces in a forest that precipitation hits are those of tree foliage and bark. Most incoming rainwater drips through the canopy as so-called "throughfall," while a smaller fraction runs down tree trunks as "stemflow." In addition, some precipitation wets the canopy foliage and then evaporates back to the atmosphere.

Studies of conifer forests in Nova Scotia found that, during the growing season, about 65% of the rainfall reached the ground as throughfall, less than 1% was stemflow, and 34% evaporated back to the atmosphere from wet foliage (Freedman and Prager, 1986). Comparable data for hardwood forests were 75% throughfall, 3% stemflow, and 22% evaporation. The chemistry of throughfall and stemflow is markedly different from the chemistry of ambient precipitation. This is partly because the evaporation of water from a wetted canopy is a distillation that concentrates the remaining chemicals in the throughfall and stemflow. In addition, certain chemicals are readily leached out of foliage, increasing their concentrations in the throughfall and stemflow. In the Nova Scotia study, the concentration of K^+ was 10 times higher in throughfall and stemflow than in rainwater, while the concentrations of calcium (Ca^{2+}) and magnesium (Mg^{2+}) were 3–4 times higher. These changes in ion concentrations affect the acidity of the solutions, because H^+ is removed from solution and exchanged for Ca^{2+}, Mg^{2+}, and K^+ (these are known as *base cations*). The exchange reactions reduced H^+ quantities by 42–66% in throughfall and stemflow, compared with ambient rainfall.

In areas where the atmosphere is contaminated by SO_2, there are large increases in the sulphate concentration in throughfall and stemflow, compared with precipitation. This is caused by the washoff and leaching of sulphate that was previously dry deposited on the canopy as SO_4-containing particulates and SO_2 gas. Studies at Hubbard Brook found that sulphate deposition was about four times higher in forests than in ambient precipitation (Eaton *et al.*, 1973). Hubbard Brook is located in remote mountainous terrain, but in a region of the northeastern U.S. where the atmosphere is significantly contaminated by particulate sulphate and gaseous SO_2. At the Nova Scotian sites, which have a cleaner atmosphere, sulphate deposition was only 45% higher in forests than in precipitation (Freedman and Prager, 1986).

Soil Acidity

Soil acidity is an extremely important factor affecting the growth of vegetation. Soil acidification is a common natural process, which has been demonstrated in ecosystems by studies of succession.

One well-known study was made at Glacier Bay in southern Alaska. Glacial meltback has exposed a till substrate with a pH of about 8.0, and containing concentrations as high as 10% of carbonate minerals of calcium and magnesium (Crocker and Major, 1955). This material becomes modified over time by colonizing vegetation and climatic factors, especially by rainfall, which percolates through the soil and carries away dissolved chemicals. This results in an increased acidity of the developing soil, reaching about pH 4.8 after 70 years of succession. By that time, a conifer forest is established. The acidification is accompanied by large declines in the concentrations of Ca, Mg, and carbonates in the soil during succession.

Natural acidification is caused partly by uptake of the nutrients Ca, Mg, and K by vegetation, a process accompanied by excretion of H^+ and a decrease in the soil's **buffering capacity** (that is, the ability of the soil to resist further acidification). Another cause of acidification is the leaching of calcium and other cations out of the soil by rainwater.

After precipitation reaches the ground in a terrestrial ecosystem, it percolates downward into the soil. Various chemical changes occur as the water interacts with soil minerals, organic matter, microbes, and plant roots, including the following:

- Roots and microorganisms selectively absorb, release, and transform chemicals.
- Ions are exchanged at the surfaces of clay and organic matter.
- Insoluble minerals are made soluble by so-called weathering processes, including reactions with acids.
- Secondary minerals are formed, such as certain kinds of clays, and insoluble precipitates of iron and aluminum oxides.

These reactions contribute to important changes in the soil such as acidification, leaching of calcium and magnesium, and the solubilization of toxic metals, particularly ions of aluminum (such as Al^{3+}; see Chapter 18). All these processes occur naturally wherever inputs of water from precipitation exceed the amounts returned to the atmosphere by evapotranspiration, leaving a surplus of water to move downward through the soil. These reactions are also strongly influenced by the kind of vegetation growing on the site. For instance, pines, spruces, and oaks tend to cause soil acidification. Acidifying depositions from the atmosphere can potentially increase the rates of some of these soil processes, thereby increasing the leaching of toxic Al^{3+} and H^+ into surface waters.

Factors Affecting Soil Acidity

Soil acidity is influenced by numerous chemical transformations and ion exchanges. Some of these are carried out by organisms, while others are nonbiological reactions. The most important factors affecting soil acidity are discussed in the following sections.

Carbonic Acid

In many terrestrial ecosystems, such as grasslands and forests, the surface litter and upper soil are rich in organic matter and plant roots. Decomposition and respiration result in high concentrations of CO_2 (frequently exceeding 1%) in the atmosphere within the soil. The high CO_2 concentrations result in carbonic acid (H_2CO_3) forming in soil water, contributing to acidification. This effect is strongest in soil with a pH greater than about 6, and it is unimportant in acidic soil with a pH less than about 5.5.

The Nitrogen Cycle

Soil acidity can be affected by microbial transformations of nitrogen compounds, and by the uptake and release of these chemicals by plants (Figure 19.2). Ammonium (NH_4^+) and nitrate (NO_3^-) are especially important in this regard, because plants must take up one or both of these essential nutrients, the choice depending largely on soil acidity. In acidic soils (with pH less than about 5.5), almost all inorganic nitrogen occurs as NH_4^+. The NH_4^+ may originate from the *ammonification* of organic-N to form ammonia (NH_3) (this process is carried out by many species of soil microorganisms, as we saw in Chapter 5). Ammonia absorbs one H^+ to form NH_4^+. If the NH_4^+ ion is absorbed by a plant root, an ion of H^+ is excreted into the soil to maintain electrochemical neutrality, so there is no net change in soil acidity. However, if NH_4^+ is added directly to soil (e.g., by atmospheric deposition or fertilization), then plant uptake of NH_4^+, accompanied by the release of H^+, has an acidifying effect.

In soils with pH greater than 5.5, most inorganic-N occurs as NO_3^-, which is produced by oxidation of NH_4^+

through the process of *nitrification* (Chapter 5). Nitrification is carried out by the bacteria *Nitrosomonas* and *Nitrobacter*, both of which are intolerant of acidity. The oxidation of NH_4^+ to NO_3^- generates two H^+ ions (Figure 19.2). If the NH_4^+ originated from ammonification of organic-N (which consumes one H^+ for each NH_4^+ produced), the net effect is the release of one H^+ for each NO_3^- ion produced from organic nitrogen. However, if the NO_3^- is absorbed by a plant root, then one OH^- ion is excreted to the soil to maintain electrochemical neutrality, which is equivalent to the consumption of one H^+. In that case, the net effect on soil acidity is zero.

It is well known to farmers and agronomists that the addition of ammonium to soil can have a severely acidifying influence. This happens because the NH_4^+ becomes nitrified into NO_3^-, generating large amounts of acidity. There are two main types of ammonium inputs: the treatment of agricultural fields with fertilizers containing inorganic nitrogen (such as urea or ammonium nitrate), and the deposition of NH_3 gas and NH_4^+ ions from the atmosphere.

The Sulphur Cycle

Much of the sulphur in soil occurs in organically bound forms. Various microbial processes can transform this organic sulphur into more highly oxidized compounds. These include sulphides and elemental sulphur, but in environments well supplied with oxygen these ultimately become oxidized to sulphate. Overall, the oxidation of organic sulphur to SO_4^{2-} results in the release of one equivalent of H^+ per equivalent of SO_4^{2-} produced (this is the same as two ions of H^+ per ion of SO_4^{2-} produced; see Figure 19.2). If the SO_4^{2-} is absorbed by a plant root, an equivalent quantity of OH^- is excreted to conserve electrochemical neutrality, so there is no net effect on soil acidity. However, if atmospheric deposition causes a direct input of SO_4^{2-} to the soil, followed by uptake by plant roots, the net effect is a reduction of soil acidity. In addition, if the soil is deficient in oxygen (i.e., anaerobic), as commonly occurs in wet sites, then the SO_4^{2-} can be transformed by microbes into a sulphide compound. This transformation results in the consumption of an equivalent quantity of H^+ and a reduction in soil acidity.

Reactions associated with the sulphur cycle usually have a much smaller effect on soil acidity than those involving the nitrogen cycle. In certain situations, however, the sulphur cycle can be quite important. For instance, when wetlands are drained, their previously anaerobic soils become aerobic. This change allows bacteria to oxidize reduced sulphur compounds, such as sulphides, into sulphate. Some drained wetlands develop an extremely acidic condition known as **acid sulphate soil**. For ten or more years after these sites are drained, usually to develop new agricultural lands, the soil pH can be lower than 3. This severely impairs crop growth.

Sometimes, sulphide minerals such as pyrites (iron sulphides) become exposed to atmospheric oxygen. This allows specialized bacteria in the genus *Thiobacillus* to oxidize the sulphides, producing sulphate and oxidized iron ions, according to the reaction

$$4\,FeS_2 + 15\,O_2 + 14\,H_2O \longrightarrow 4\,Fe(OH)_3 + 16\,H^+ + 8\,SO_4^{2-}$$

This phenomenon is known as **acid-mine drainage**. It causes severe acidification of soil and surface waters, sometimes causing pHs less than 2 to develop, along with high concentrations of sulphate and toxic ions of aluminum and iron. Acid-mine drainage is an important problem in areas where disturbances associated with coal and metal mining and processing have resulted in the exposure of mineral sulphides to the atmosphere.

Uptake of Basic Cations by Plants

Terrestrial plants obtain many of their nutrients by absorbing ions from the soil in which they are growing. (Some nutrients, however, are absorbed mainly from the atmosphere, particularly CO_2). Calcium, magnesium, and potassium are important nutrients that are absorbed from soil as positively charged ions (Ca^{2+}, Mg^{2+}, and K^+, respectively). Absorption of these cations is accompanied by a release of H^+ ions into the soil environment, increasing acidity. In natural ecosystems, the absorbed Ca^{2+}, Mg^{2+}, and K^+ are ultimately returned to the soil with plant litter, so there is no long-term effect on soil acidity. However, if plant biomass is removed from the site, as it is in agriculture and forestry, these basic cations are removed along with the harvested biomass, resulting in a net acidification of the soil.

Leaching of Ions

In most soils, the anions nitrate and chloride readily leach downward into the groundwater. They may eventually reach surface waters such as streams and lakes. This is also true of sulphate, particularly in relatively young soils of glaciated regions, which include most of Canada (older soils of more southern regions often have a larger capacity for retaining sulphate). In areas that are subject to large inputs of acidifying substances from the atmosphere, the concentrations of NO_3^- and SO_4^{2-} in soil may be high enough to result in substantial rates of leaching of those ions. As these anions leach out of the soil they are accompanied by cations such as Ca^{2+}, Mg^{2+}, H^+, and Al^{3+}. This results in acidification, nutrient losses, and toxicity (associated with the aluminum ions) in both terrestrial and aquatic ecosystems.

Soil Buffering Systems

Compared with precipitation and surface waters, soil is relatively strongly buffered (or resistant to changes in pH). However, different buffering systems are important at different ranges of soil pH:

- Carbonate minerals of calcium and magnesium provide most of the buffering capacity within the pH range of >8 to 6.2.
- Silicates are important from pH 6.2 to 5.0.
- Cation-exchange capacity buffers from pH 6.2 to 4.2.
- Aluminum transformations buffer from pH 5.0 to 3.0.
- Iron transformations are important from pH 3.8 to 2.4.
- Humic substances buffer from pH 5 to 3.

Many soils in the so-called "circumneutral" pH range of 6–8 are relatively sensitive to acidification, especially if only small amounts of calcium and magnesium carbonates are present. When these kinds of soils are exposed to high inputs of acidifying substances, they can rapidly acidify to pH 3.5–4.5.

Atmospheric Deposition and Soil Acidity

The potential effects of atmospheric deposition on soil acidity have been studied in experiments in which simulated acidic "rainwater" is added to soil contained in plastic cylinders, known as lysimeters. These experiments have demonstrated that extremely acidic solutions can cause several changes in soil chemistry:

- an increase in acidity
- increased leaching of calcium, magnesium, and potassium, resulting in the loss of these nutrients and greater vulnerability of the soil to acidification
- increased solubilization of toxic metal ions — especially of aluminum, but also of iron, manganese, and others
- overwhelming of the ability of soil to absorb sulphate — after which this ion leaches freely, at a rate about equal to its rate of input. Because SO_4^{2-} is an anion, its leaching is accompanied by base cations and toxic Al^{3+} and H^+. These may eventually contribute to the acidification and toxicity of surface waters that receive the water that percolates through the soil.

One experiment involved treatment of sandy soils collected from jack pine (*Pinus banksiana*) stands in northern Ontario with simulated rainwater, adjusted with dilute sulphuric acid to a pH of 5.7, 4, 3, or 2 (Table 19.4). Even treatment with the extremely acidic pH of 2 had little effect on soil acidity, with the pH of percolating solutions being higher than 6.5 in all treatments. However, the leaching

TABLE 19.4 EXPERIMENTAL LEACHING OF SOILS BY ACIDIC SOLUTIONS

The data are concentrations, in milliequivalents per litre, of chemicals in water that drained from the bottom of lysimeters containing sandy soils collected from two jack pine (*Pinus banksiana*) stands in northern Ontario. The experiment ran for three years, with simulated rainfall being added at a rate of 100 cm/y as weekly 1–2 hour events. The data are averages of three replicates per treatment.

	pH OF LEACHING SOLUTION	CONCENTRATION IN PERCOLATE (meq/L)			
		Ca	Mg	ALL BASES	SO_4
Soil A	5.7	0.74	0.05	0.99	0.23
	4	0.67	0.05	0.99	0.22
	3	0.75	0.07	1.10	0.15
	2	0.87	0.20	1.35	0.53
Soil B	5.7	0.42	0.17	0.73	0.15
	4	0.45	0.18	0.77	0.15
	3	0.40	0.17	0.70	0.28
	2	3.72	1.47	5.50	5.43

Source: Morrison (1983)

rates of Ca, Mg, total bases (Ca + Mg + K + Na), and sulphate were much higher in the extremely acidic, pH 2 treatment. Overall, this experiment found that the soil was resistant to the effects of acid loading. Eventually, however, the resistance can be overcome by treatment with highly acidic solutions, or perhaps by long-term exposures to more moderate acidities. It must be borne in mind that experiments such as these are short-term investigations, whereas soil acidification in nature is a slow, long-term process.

Researchers monitoring soil chemistry at particular places in the field can determine whether acidification has occurred, although such studies do not necessarily identify the causes of the change. For instance, the conversion of agricultural land into conifer forest usually results in acidification of the soil. In southern Ontario, for example, the afforestation of abandoned farmland with pine or spruce caused the soil to acidify from an initial pH of 5.7 to pH 4.7 after 46 years of forest development (Brand *et al.*, 1986). It is less understood whether already forested sites will become more acidic because of atmospheric inputs of acidifying substances. A study in southern Ontario resampled forest soils after a 16-year interval, in a region where the average pH of precipitation is about 4.1. This study found no further acidification of soil over this period (Linzon and Temple, 1980).

Studies conducted elsewhere in Canada, and in the United States and Europe, have also come to rather ambiguous conclusions about the effects of atmospheric depositions on soil acidification. Except where the atmosphere is severely polluted by SO_2, e.g., near a smelter, there is no convincing evidence that atmospheric depositions have acidified soil on a wide scale. It appears that soil acidification is a potential, long-term risk associated with this type of pollution.

Acidifying Deposition and Terrestrial Vegetation

Numerous studies have demonstrated that plants can be injured by treatment with artificial solutions of "acid rain." In almost all of these studies, however, the pHs that caused injuries were considerably more acidic than is normally found in ambient precipitation.

During the 1980s, many stands of sugar maple (*Acer saccharum*) in Ontario and Quebec suffered severe thinning of their canopy. In some places, numerous trees died. Many scientists believe that these damages were caused by acid rain, or by other types of air pollution, such as ozone. Alternative hypotheses to explain the damage include climate change, severe winter weather, nutrient imbalances, and a past history of insect infestation. The controversy over the causes of the damage has not been fully resolved, but fortunately much of the damage disappeared by the mid-1990s. This photo was taken in a severely declining stand near Vineland in southern Ontario.
Source: R. Vinebrooke

For example, experiments in Norway exposed stands of young conifer trees to simulated acid rains for three years (Tveite, 1980). The "control" treatment used solutions of pH 5.6–6.1, while the "acidified" treatments used pH 4 or 3. On average, saplings of lodgepole pine (*Pinus contorta*) grew 15–20% less with the control treatment than the plants receiving the more acidic treatments! Growth of Scotch pine (*P. sylvestris*) and birch (*Betula pendula*) was also stimulated by the acidic treatments, while spruce (*Picea abies*) was unaffected. However, the moss-dominated ground vegetation of these stands was severely damaged by the most acidic treatment (pH 3).

Because laboratory experiments can be closely controlled, they are useful for determining the effects of different pHs of acidic rainwater on plants. In general, such experiments do not reduce growth until the solution pHs become more acidic than about 3.0 (for comparison, the average acidity of precipitation is about pH 4 in regions where the problem is considered "severe"). Moreover, the productivity of some particularly tolerant species can be stimulated by rainwater even more acidic than pH 3. For example, white pine seedlings (*Pinus strobus*) grew more when exposed to mists ranging from pH 2.3 to 4.0 than at pH 5.6 (Wood and Bormann, 1976). In another laboratory experiment, seedlings of 11 tree species were treated with solutions of various pHs. Acute injuries to foliage were caused only after a week of treatment at pH 2.6, an unnaturally acidic treatment (Percy, 1986).

In general, it appears that trees and other vascular plants have little risk of suffering acute injuries from exposure to ambient acidic precipitation. However, stresses associated with acidic precipitation could possibly decrease plant growth, even in the absence of acute injuries. These "hidden injuries" (see Chapter 16) might be caused by subtle disruptions of plant metabolism, or indirectly by the effects of changes in soil chemistry. Because acidic precipitation affects extensive regions, even small decreases in plant productivity could result in important economic and ecological damages. Hidden injuries, if they do occur, are most relevant to forests and other kinds of natural vegetation. Agricultural lands become acidified mostly through management practices such as cropping and the use of nitrogen fertilizers. Moreover, agricultural soils are routinely treated with liming agents to reduce their acidity (liming is discussed at the end of this chapter).

A number of studies in eastern North America and Europe have examined the possible effects of acidic precipitation on forest productivity. Although some species of trees in some regions have shown recent decreases in productivity, it has not been demonstrated that these changes were caused by acidic precipitation or other types of atmospheric pollution. Forest productivity decreases naturally as a stand matures, mainly because canopy closure intensifies competition. Forest productivity is also influenced by factors such as climate change and management practices. So far, research has not clearly separated any influences of acidic precipitation on forest productivity from effects related to succession, climate change, insect defoliation, or other factors.

Chemistry of Surface Waters

Surface waters include streams, rivers, lakes, and ponds (Chapter 14). The chemistry of surface waters is influenced by many factors, including the types of soils and vegetation in the watershed, climatic influences, and rates of deposition of chemicals from the atmosphere.

In regions where the winters are cold and a snowpack accumulates, the springtime meltwater that flows into streams, rivers, and lake surfaces tends to be relatively acidic. This so-called **acid shock** phenomenon is caused partly because snowmelt cannot percolate into the frozen, saturated soil. Therefore, the acidity of melted snow cannot be neutralized by interaction with minerals in the soil. In addition, the initial meltwaters of the snowpack are considerably more acidic than the later meltwaters. The relatively acidic conditions associated with meltwaters in the springtime are responsible for much of the toxicity of acidic surface waters.

The water chemistry of two remote lakes in a region of Nova Scotia subject to acidic precipitation is described in Table 19.5. These lakes receive a sparse input of nutrients, and consequently they are highly unproductive (or *oligotrophic*) and their waters are dilute solutions in comparison with most fresh waters (that is, their waters have low concentrations of dissolved ions). However, these lakewaters have higher concentrations of dissolved substances than does the precipitation that falls on them — their sum of cations plus anions averages 440 μeq/L, compared with 135 μeq/L in precipitation. The lakewaters contain relatively high concentrations (compared with precipitation) of calcium, magnesium, sodium, potassium, iron, aluminum, sulphate, chloride, organic carbon, and organic anions (i.e., negative charges associated with dissolved organic matter). The higher concentrations of these sub-

TABLE 19.5 | CHEMISTRY OF TWO NOVA SCOTIAN LAKES

This table summarizes data for the average concentration of chemicals in precipitation, and in the water of two remote lakes in Nova Scotia. Beaverskin Lake has clear water, while Pebbleloggitch has brown water because of the presence of a large bog in its watershed. Both are headwater lakes, which means they do not receive drainage from lakes higher in altitude. Data are in μeq/L.

CONSTITUENT	PRECIPITATION	BEAVERSKIN LAKE	PEBBLELOGGITCH LAKE
CATIONS			
Ca	4.3	20.0	18.0
Mg	2.9	32.0	30.0
Na	26.1	126.0	126.0
K	1.1	8.0	6.0
Fe	<0.1	1.0	4.0
Al	<0.1	2.2	23.4
NH_4	4.2	1.0	2.4
H	29.9	5.0	33.0
ANIONS			
SO_4	27.5	48.7	57.9
Cl	29.5	124.0	111.0
NO_3	9.7	1.0	0.9
Organic anions	<0.1	32.0	66.0
Alkalinity	0.0	0.0	0.0
Total cations	68.5	195.2	242.8
Total anions	66.7	205.7	235.8
Anions/cations	0.97	1.05	0.97
Dissolved organic C (milligram/litre)	0.0	3.9	10.0
Total carbon (mg/L)	0.0	4.5	13.8
Colour (Hazen units)	0	6	87
pH	4.6	5.3	4.5

Source: Kerekes and Freedman (1988) and Freedman and Clair (1987)

stances occur mainly because they leach from terrestrial soils and eventually migrate into the lakewater. In contrast, concentrations of ammonium and nitrate in the lakewaters are considerably lower than in precipitation, suggesting that atmospheric inputs of these nutrients are "consumed" by biological uptake within the terrestrial parts of the watershed.

Beaverskin Lake is a typical, slightly acidic, oligotrophic lake with very transparent water. Beaverskin Lake (pH 5.3) is less acidic than the local precipitation (pH 4.6) it receives. The chemistry of Pebbleloggitch Lake is greatly influenced by a bog that covers about one-third of its watershed. Large quantities of dissolved organic compounds (known as fulvic acids) leach from the bog into Pebbleloggitch Lake. The fulvic acids give the lakewater a dark-brown colour and an acidity (pH 4.5) similar to that of the precipitation. Although these two lakes are located only one km apart, they differ markedly in acidity because of the bog's influence on the concentration of organic acids in tea-coloured Pebbleloggitch Lake. In general, bog-influenced lakes and streams are naturally acidic, and usually have an acidity of pH 4–5.

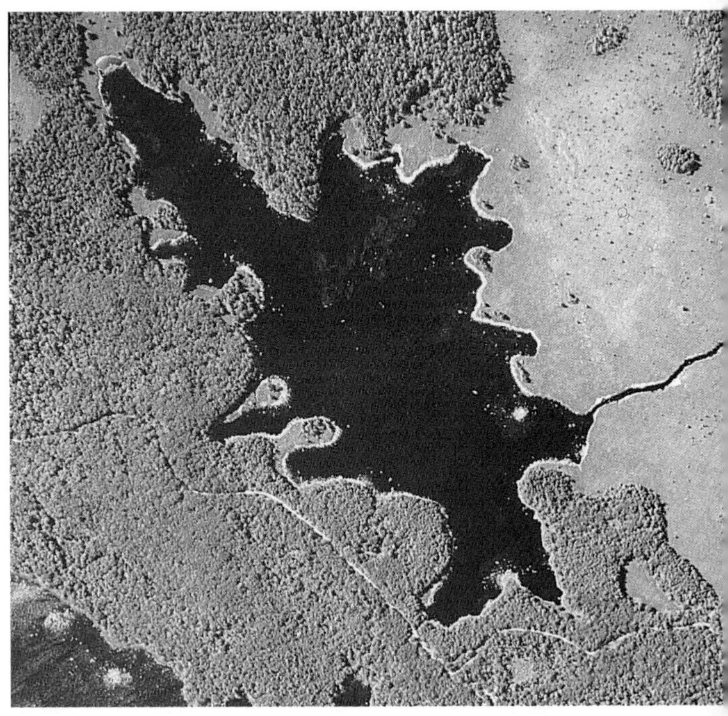

About a third of the watershed of Pebbleloggitch Lake, Nova Scotia, is an acidic, organic-rich bog (the fine, light texture in the airphoto). The rest is covered by mature conifer-dominated forest. The boggy area leaches dissolved organic acids to the lake, giving it a dark-brown colour and making it naturally acidic, with a pH of about 4.5.

Source: J. Kerekes

Acidification of Surface Waters

A widespread acidification of weakly buffered surface waters in eastern Canada has been attributed to the deposition of acidifying substances from the atmosphere. The eastern United States and Scandinavia have also been affected in this way.

A national survey of surface waters was conducted by the Environmental Protection Agency in the United States (Baker *et al.*, 1991). From a sample of 28 300 lakes around the United States, 1180 were identified as acidic. Most of these are in the eastern states. Atmospheric deposition was thought to have acidified 75% of the acidic lakes, while 3% were affected by acid-mine drainage and 22% by natural acidity draining from bogs. Of 64 300 streams sampled, 4670 were acidic, of which 47% were acidified by atmospheric deposition, 26% by acid-mine drainage, and 27% by bogs. Florida has the highest frequency of acidic lakes, mostly because of organic acids draining from natural wetlands. The influence of atmospheric deposition is most important in the northeastern United States, particularly in the Adirondack Mountains in New York State, where 10% of the lakes have a pH ≤ 5.0, and 20% a pH ≤ 5.5.

Although such a comprehensive survey of the status of individual lakes and streams has not been carried out in Canada, acidified surface waters are known to be common, particularly in the eastern part of the country. The sensitivity of Canadian surface waters to becoming acidified has been extensively surveyed (Environment Canada, 1988). Sensitivity can be described by the concentration of alkalinity in water, which is related to the amounts of calcium and magnesium in the soil and rocks of the watershed (the more calcium and magnesium present, the less sensitive the water is to acidification; see also the following section). Surface waters on 46% of Canada's land area (equivalent to about 4.0 million km^2) are considered highly sensitive to acidifying deposition, while another 21% (1.8 million km^2) are moderately sensitive. Overall, surface waters are most sensitive in eastern Canada (see Table 19.6).

TABLE 19.6 | SENSITIVITY TO ACIDIFICATION ACROSS CANADA

Data are the percentages of the total terrain with a high sensitivity to aquatic acidification.

PROVINCE/TERRITORY	% OF TERRAIN
Newfoundland and Labrador	56
Prince Edward Island	46
Nova Scotia	54
New Brunswick	31
Quebec	82
Ontario	34
Manitoba	30
Saskatchewan	37
Alberta	6
British Columbia	32
Northwest Territories	48
Yukon	43

Source: Modified from Environment Canada (1988)

Only a few studies have documented the chemical and biological changes that occur as surface waters become acidified naturally. The process has, however, been examined in important studies in which sulphuric acid was added to lakes, causing acidification (Schindler, 1990). These **whole-lake experiments** were conducted in the Experimental Lakes Area (ELA) of northwestern Ontario. The most intensively studied lake, called Lake 223, is a 27-ha, oligotrophic waterbody. Lake 223 was studied for two years before it was experimentally acidified, and for a number of years afterward. The pH of Lake 223 was initially 6.5, and its alkalinity was 80 μeq/L. Beginning in 1976, sulphuric acid was added to progressively acidify Lake 223. This reduced the pH of Lake 223 to 6.5 in 1976, 6.1 in 1977, 5.9 in 1978, 5.6 in 1979, 5.6 in 1980, and 5.0–5.1 between 1981 and 1983. The acidity of Lake 223 was then allowed to decrease, to pH 5.5–5.6 during 1984–1986 and to pH 5.8 during 1987–1988.

As expected, the concentrations of sulphate and hydrogen ions increased, because these were being added to the lake. Sulphate averaged 35 μmol/L in 1975, compared with 115 μmol/L in 1979. Increased concentrations of manganese (a 980% increase by 1980), zinc (550% increase), aluminum (155%), and sodium (26%) occurred when these chemicals dissolved out of sediments because of the acidic conditions. Acidification also caused the water of Lake 223 to become more transparent, a change that allowed increased light penetration and deeper heating of lakewaters during the summer. Many biological changes also occurred; these are described later in this chapter.

Vulnerability to Acidification

Surface waters that are vulnerable to acidification have a low alkalinity, or **acid-neutralizing capacity**. As H^+ is added to lakewater, it is absorbed by acid-neutralizing reactions until the buffering threshold is exceeded. This is followed by a rapid decrease in pH, until a different buffering system comes into play (Figure 19.3). Bicarbonate alkalinity (HCO_3^-) is the critical buffering system within the circumneutral pH range of 6–8. When this is depleted, a waterbody acidifies rapidly. Bicarbonate alkalinity reacts with H^+ to form $H_2O + CO_2$. This reaction neutralizes added H^+, so the pH does not change until the supply of alkalinity is exhausted.

The concentration of bicarbonate in natural waters is influenced by geochemical factors. Particularly important is the presence of mineral carbonates such as limestone ($CaCO_3$) or dolomite ($CaCO_3$ and $MgCO_3$ together, sometimes shown as $Ca,MgCO_3$) in the soil or bedrock of the watershed or in the aquatic sediment. As these minerals dissolve, they provide large amounts of bicarbonate alkalinity to surface waters, giving the water a large acid-neutralizing capacity. In general, carbonate-rich watersheds can generate enough alkalinity to neutralize any acidifying inputs from the atmosphere. Surface waters in these types of watersheds are not sensitive to acidification, even if they are located where the atmosphere and precipitation are rather polluted. (There might be exceptions in cases of extremely high rates of dry deposition of SO_2.)

The situation is different, however, in watersheds in which the bedrock, soil, and sediment are composed or derived from hard, poorly soluble rocks such as granite, gneiss, and quartzite. These rocks contain very small amounts of carbonate minerals. Watersheds of this type have little capacity for generating alkalinity, and are readily acidified by wet and dry depositions from the atmosphere. As was previously described, vulnerable watersheds are especially common in eastern Canada, where thin soils of carbonate-poor glacial till overlie hard bedrock of granite, gneiss, and quartzite, especially on the Precambrian Shield.

Headwater lakes and streams are at particular risk of acidification. Headwater systems receive no drainage from waterbodies at higher elevation, and their watersheds are usually quite small. Consequently, there is relatively little opportunity for rainwater to interact with soil and bedrock, so little of the acidity in precipitation can be neutralized before the water reaches the lakes and streams.

At high elevation in many mountainous regions, where crustal granite is exposed by erosion, the terrain is often vulnerable to acidification. This is the case in the Rocky Mountains of western North America and the Appalachians of the eastern United States. Again, the vulnerability occurs because granite contributes little alkalinity to surface waters.

FIGURE 19.3 | TITRATION CURVES FOR FRESHWATER

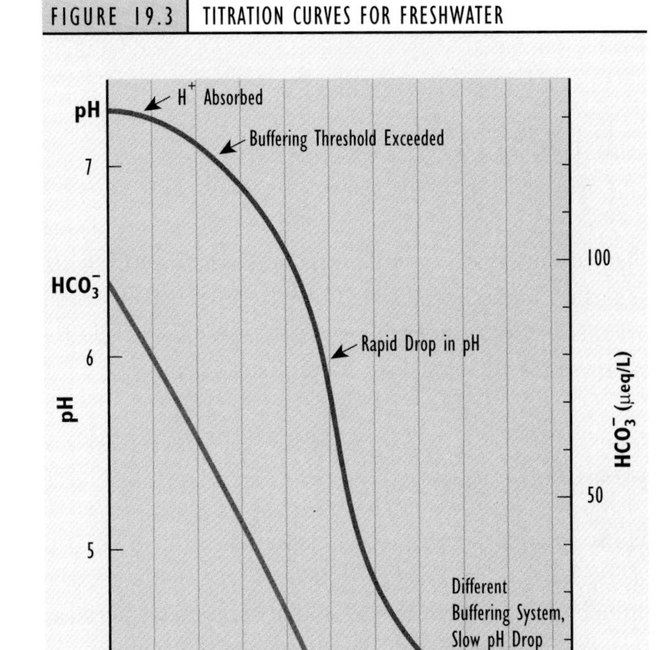

This diagram shows titration curves for a typical freshwater lake. The initial concentration of alkalinity (HCO_3^-) was 100 μeq/L. These curves are similar to what is observed when dilute acid is slowly added to a clear lakewater with a low concentration of alkalinity. Once the alkalinity is exhausted, the water acidifies rapidly.

Source: Modified from Henriksen (1980)

Within vulnerable regions, however, even small pockets of calcareous (i.e., calcium-rich) soil in the watershed can yield enough alkalinity to allow waterbodies to resist acidification by atmospheric depositions. For example, 15 lakes were surveyed in an area of Precambrian Shield in southern Ontario where precipitation is quite acidic (Dillon *et al.*, 1977). Fourteen of the lakes had little alkalinity (range of 95–175 μeq/L), were slightly acidic (pH 5.8–6.7), and were considered to be vulnerable to further acidification. One lake, however, had some calcium-rich till in its watershed. That lake had a high alkalinity (1200 μeq/L) and a high pH (7.1), and is unlikely to acidify through atmospheric influences.

Effects of Acidification on Freshwater Organisms

Many changes can occur in the biota of aquatic ecosystems as surface waters become acidified. In general, freshwater organisms are considerably more sensitive to the acidification of their habitat than are terrestrial plants.

Freshwater Algae

Many species of microscopic, single-celled algae (or *phytoplankton*) live in lakes. The water chemistry greatly influences the particular species that are present. Because species of diatoms (family Bacilliarophyceae) and golden-brown algae (Chrysophyceae) are particularly sensitive to differences in water chemistry, they are useful indicators of environmental conditions. For example, a study of 72 lakes in the Sudbury area found that certain diatom species indicated particular aspects of water chemistry (Dixit *et al.*, 1991), as follows:

This lake near a smelter at Sudbury, Ontario, was acidified to a pH less than 4, mostly by the dry deposition of SO_2 gas from the atmosphere. Since 1972, SO_2 pollution in the vicinity has been greatly abated, allowing the lake to become less acidic over time. Today the lake again provides habitat for some species that are intolerant of severe acidity.

Source: B. Freedman

1. Indicators of acidic water: *Eunotia pectinatus, Fragilaria acidobiontica, Pinnularia subcapitata, Tabellaria quadriseptata*
2. Indicators of acidic water with high concentrations of metals (Cu, Ni): *Eunotia exigua, E. tenella, Frustulina rhomboides saxonica, Pinnularia hilseana*
3. Indicators of non-acidic water: *Achnanthes lewisiana, Cyclotella meneghiniana, Fragilaria construens, F. crotonensis.*

The silica-rich cell walls of diatoms differ in shape for each species. Since these cell walls persist in lake sediment long after death of the cell, cell-wall fossils of these algae can be extracted from dated layers of lake sediment cores. As well, the water-chemistry requirements of many diatom species are known. Consequently, their presence and abundance in dated layers of sediment can be used to infer historical communities and water-chemistry conditions. This technique has been used to demonstrate that some presently acidic lakes were not acidic as recently as several decades ago.

During the acidification of Lake 223 in the Experimental Lakes Area, the phytoplankton community changed markedly. Initially, it was dominated by species of golden-brown algae; this changed to domination by green algae (family Chlorophyceae; Findlay and Kasian, 1986). Although species composition changed substantially, there was little change in the overall number of species or their diversity. A small increase in algal biomass occurred after acidification. This was apparently caused by the increased clarity of the water, which allowed more algal productivity. The acidification of Lake 223 reached pH 5.0–5.1 during 1981–1983, but was allowed to go back to pH 5.5–5.6 during 1984–1986. When this happened, algal species typical of the pre-acidification community quickly reappeared.

The phytoplankton community of lakes and ponds is much more sensitive to changes in the fertility of the water than to changes in its acidity, being especially sensitive to changes in the supply of phosphorus (see Chapter 20). The primary productivity of almost all freshwaters increases greatly if they are fertilized with phosphorus. This ecological change, known as *eutrophication*, also occurs in acidic waterbodies. This is illustrated by studies on two adjacent lakes in Nova Scotia, Little Springfield Lake and Drain Lake. After construction activities in their watersheds exposed pyrite-containing minerals to the atmosphere, both lakes became very acidic. The acidity was generated through a process similar to acid-mine drainage (Kerekes *et al.*, 1984). Little Springfield Lake had an average pH of 3.7 and supported little algal productivity; it became oligotrophic. However, Drain Lake (pH 4.0) received inputs of phosphorus-rich sewage, and became eutrophic and highly productive in spite of its extreme acidity.

Periphytes are microscopic algae that live on the surfaces of sediment, rocks, woody debris, and foliage of aquatic plants. The periphyton community can include hundreds of species, even in acidic lakes. Periphytes are especially abundant in lakes with clear water, where their late-summer biomass may develop cloudy or felt-like mats. During the experimental acidification of Lake 223 (in the ELA), a benthic mat of the filamentous green alga *Mougeotia* developed in shallow water after the pH decreased below 5.6. Mats of filamentous algae have been found in many other acidified lakes. The reasons for the growth of algal mats are not known, but may be related to a decrease in herbivorous aquatic invertebrates, resulting in reduced grazing.

Aquatic Plants

Aquatic plants (or macrophytes) can be quite abundant in shallow lakes and ponds. Acidification of some lakes has resulted in an increased abundance of aquatic mosses, especially species of *Sphagnum*. In some cases this has been accompanied by declines of other plants, such as reed (*Phragmites communis* spp.), water lobelia (*Lobelia dortmanna*), and quillwort (*Isoetes* spp.). Moreover, the invasion of lakes by *Sphagnum* mosses may accelerate the process of acidification, because these plants are extremely efficient at absorbing Ca^{2+}, Mg^{2+}, and other cations from water, which they exchange for H^+. In addition, mats of *Sphagnum* moss interfere with chemical reactions at the sediment/water interface, hindering the neutralization of acidity that might otherwise occur there.

Communities of aquatic plants differ greatly between clear-water acidic lakes and those with brown, organically stained water. For example, Pebbleloggitch Lake in Nova Scotia (see Table 19.5) has dark-brown, acidic (pH 4.5) water. The colour prevents the penetration of light into deeper water. Consequently, macrophytes can grow only within a shallow fringe of water around the edge of the lake, and only 15% of the bottom is vegetated (Stewart and Freedman, 1989). The most abundant aquatic plants, including the yellow-flowered spatter-dock (*Nuphar variegatum*), have floating leaves. Beaverskin Lake, nearby, has extremely clear water and is somewhat less acidic (pH 5.3). Because of the clarity of its water, virtually all of the bottom of Beaverskin Lake receives enough light to support aquatic plants, even to a depth of 6.5 m. Many of the macrophytes, including mats of *Sphagnum* moss, are species that maintain all of their foliage underwater.

Even in acidic lakes, the productivity of macrophytes is stimulated by nutrient additions. Drain Lake (mentioned above), an extremely acidic (pH 4.0) but eutrophic lake, has a lush productivity of aquatic plants, including species of aquatic knotweeds (*Potamogeton* spp.), which do not normally occur in acidic lakes. As with phytoplankton, the fertility of lakewaters affects the productivity of aquatic plants much more than acidity does.

Zooplankton

Zooplankton are tiny animals, mostly crustaceans, that live in the water column of lakes and ponds. Most zooplankters filter-feed on phytoplankton cells, but some are predators of other species of zooplankton. Some species of zooplankton are tolerant of acidity, and may occur in waters as acidic as pH 4 or less. The effects of acidification on zooplankton are complex, because several factors are involved:

- the toxicity of H^+ and associated metals, such as Al^{3+}
- changes in the quantities of algae available as food
- changes in predation, especially if zooplankton-eating fish are eliminated from the acidified lake

A survey of 47 lakes in Ontario found that species of zooplankton can be good indicators of water chemistry (Sprules, 1975). This is similar to the use of phytoplankton species as indicators. Some zooplankton species are indicators of acidic lakes, occurring at pH <5: *Polyphemus pediculus*, *Daphnia catawba*, and *D. pulicaria*. Others live only in lakes with pH >5: *Tropocyclops prasinus mexicanus*, *Epischura lacustris*, *Diaptomus oregonensis*, *Leptodora kindtii*, *Daphnia galeata mendotae*, *D. retrocurva*, *D. ambigua*, and *D. longiremis*. Some species are apparently indifferent to acidity, occurring over a wide range of pHs. *Diaptomus minutus*, a particularly tiny species, was the most frequently observed species in the survey, occurring over the pH range 3.8–7.0. Lakes with a pH exceeding 5 had 9–16 species of zooplankton, with 3 or 4 being dominant. More acidic lakes (pH below5) had 1–7 species, with only 1 or 2 dominants.

The experimental acidification of Lake 223 resulted in an increased abundance of zooplankton, an effect attributed to the previously described increase in phytoplankton biomass (Malley *et al.*, 1982). Throughout the acidification of Lake 223, *Diaptomus minutus* and *Cyclops bicuspidatus* remained the most abundant zooplankters. Some other species, however, were intolerant of acidification. This included the opossum shrimp (*Mysis relicta*), a predator that disappeared after the pH decreased below 5.6.

Benthic Invertebrates

Benthic invertebrates live in the sediment of lakes, streams, and other waterbodies. The number of species of these small animals tends to be lower in acidic waters. However, they can still be rather abundant, especially in acidified waterbodies from which predatory fish have disappeared. The most common benthic invertebrates in acidic lakes are certain species of insects and crustaceans (although other species of these same groups are intolerant of acidity). Mollusks do not occur in strongly acidic conditions, likely because it is difficult for them to grow their shells of calcium carbonate. A study of mollusks in more than 1000 lakes in Norway found that no species of clams could tolerate a pH below 6.0, and no snails could tolerate a pH below 5.2 (Okland and Okland, 1986).

Because sediments are relatively strongly buffered, they are much less vulnerable to acidification than the overlying water. For instance, acidity of the sediment did not change much during the experimental acidification of Lake 223 (Kelly *et al.*, 1984). In 1981, when the pH of water just above the sediment was about 5.3, at 0.5 cm into the sediment the pH was 6.0. At a depth of 2.0 cm, the pH was 6.7, virtually unchanged from the pre-acidification condition. Because the habitat of benthic invertebrates is resistant to acidification, some species are not catastrophically affected by acidification of the overlying lake. During the acidification of Lake 223, the populations of chironomid midges increased and peaked at a water pH of about 5.6 (Mills, 1984). However, the initially abundant populations of mayfly larvae disappeared at about pH 5. In addition, the crayfish *Orconectes virilis* became extirpated in Lake 223 because of reproductive failure after the pH decreased below 5.6.

Fish

Fish populations are the best-known victims of aquatic acidification. Losses of populations of trout, salmon, and other economically important fish have occurred in acidified surface waters in Canada, the United States, and Eurasia.

Studies in south-central Ontario have documented losses of fish populations from acidified lakes in the Killarney region (Beamish and Harvey, 1972; Harvey and Lee, 1982). That area is subject to severely acidic precipitation (pH 4.0–4.5), as well as to dry deposition of acidifying SO_2, because of its relative proximity to the smelters at Sudbury. A survey during the early 1970s found that 33 of 150 lakes in the Killarney area had a pH of below 4.5.

Local extirpations of several species of fish were actually observed by ecologists working on Lumsden and George Lakes in the Killarney area. There was also evidence for the losses of other fish populations, thanks to eyewitness accounts of historical sport fisheries in presently fishless lakes. The Killarney area has had 17 known extirpations of lake trout (*Salvelinus namaycush*), an important sportfish that cannot breed successfully at a pH below 5.5. There are also extirpations of smallmouth bass (*Micropterus dolomieiu*) from 12 lakes in the region, of largemouth bass (*M. salmoides*) and walleye (*Stizostedion vitreum*) from four lakes, and yellow perch (*Perca flavescens*) and rock bass (*Ambloplites rupestris*) from two others.

The pH of Lumsden Lake was 6.8 in 1961, but only 4.4 in 1971. This change caused reproductive failures and extirpations of lake trout, lake herring (*Coregonus artedii*), and white sucker (*Catostomus commersoni*). When George Lake reached pH 4.8–5.3, lake trout, walleye (*Stizostedion vitreum*), burbot (*Lota lota*), and smallmouth bass disappeared. As acidification progressed further, there were losses of northern pike (*Esox lucius*), rock bass, pumpkinseed sunfish (*Lepomis gibbosus*), brown bullhead (*Ictalurus nebulosus*), and white sucker. All of these extirpations resulted from persistent failures of the fish to reproduce in the acidified lakes.

Losses of sportfish populations have also occurred in Nova Scotia, where Atlantic salmon (*Salmo salar*) has been lost from seven highly acidic (pH below 4.7) rivers, but not from rivers with higher pHs (Figure 19.4). Lacroix and Townsend (1987) penned juvenile salmon in four acidic streams in Nova Scotia. No fish survived in streams with pH below 4.7, but all survived if the pH exceeded 4.8.

FIGURE 19.4 | CATCH OF ATLANTIC SALMON IN NOVA SCOTIAN RIVERS

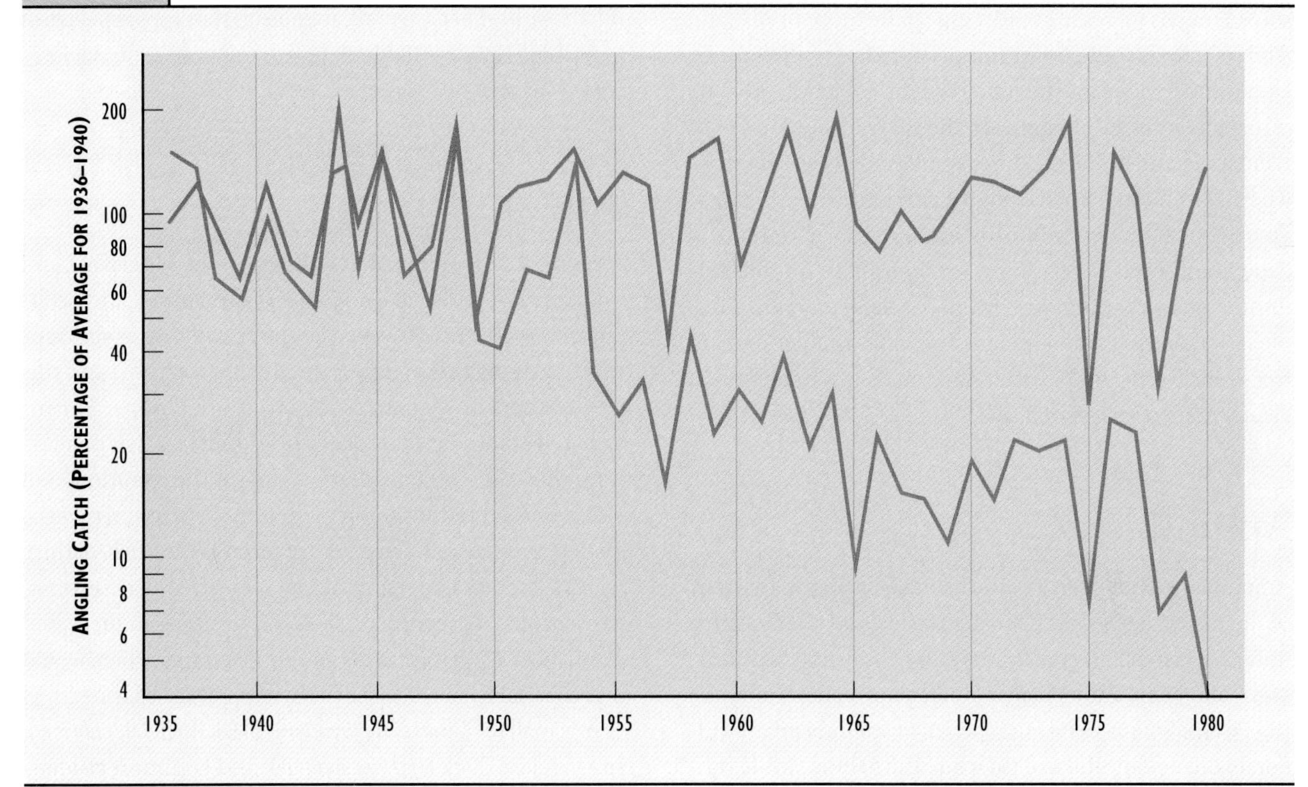

These sport-fishing data are standardized to facilitate comparisons among rivers. The top line shows data for rivers with pH greater than 5.0. The bottom line is for rivers with pH 5.0 or less.

Source: Modified from Watt *et al.* (1983)

Elsewhere, early surveys (from the 1930s) of lakes in the Adirondack Mountains of New York State found brook trout (*Salvelinus fontinalis*) in 82% of the lakes. However, in the 1970s this species was absent in 43% of 215 lakes in the same region (Schofield, 1982).

An extensive survey of 700 Norwegian lakes in the 1970s found that brown trout (*Salmo trutta*) was absent from 40% of the waterbodies and sparse in another 40% (Wright and Snekvik, 1978). Almost all of those lakes had supported trout populations before the 1950s. Losses of fish populations have been most extensive in southern Norway, where acidifying deposition is most intense.

In eastern North America, the fish species that are most tolerant of acidification are yellow perch, rock bass, central mudminnow (*Umbra limni*), largemouth bass, bluegill (*Lepomis macrochirus*), black bullhead (*Ictalurus melas*), brown bullhead, golden shiner (*Notemigonus crysoleucas*), and American eel (*Anguilla rostrata*). All of these species are known to occur in some waterbodies with pH more acidic than 4.6. All other species of fish are relatively sensitive to acidification of their habitat.

Prior to its experimental acidification, Lake 223 supported five species of fish: lake trout, white sucker, fathead minnow (*Pimephales promelas*), pearl dace (*Semotilus margarita*), and slimy sculpin (*Cottus cognatus*) (Mills *et al.*, 1987). The fathead minnow was most sensitive to acidification, declining rapidly after the pH reached 5.6. The first reproductive failure of lake trout occurred at pH 5.4, and of white sucker at pH 5.1. This led to population declines of these species as older fish died.

The physiological effects of acidity on fish have been studied extensively. In general, younger life-history stages (fry and juveniles) are more sensitive to acidity than are adult animals. This is why most losses of fish populations have been attributed to reproductive failure, rather than to mortality of adults. There are, however, many observations of adult fish being killed by exposure to acid-shock events during snowmelt in the springtime.

As surface waters acidify, the concentrations of dissolved metals increase greatly, especially those of Al^{3+} and $AlOH^{2+}$, toxic ions of aluminum. In many acidic waters, aluminum toxicity is sufficient to kill fish, regardless of any direct effects of H^+. In general, the survival and growth of fish larvae and older life-history stages become reduced when concentrations of dissolved aluminum exceed 0.1 ppm, an exposure that commonly occurs in acidic waters. In brown-coloured waters, however, almost all soluble aluminum and other metals occur as organo-metallic complexes. Metals in this state are less bioavailable than are free-ionic forms, and are therefore much less toxic than in clear waters with a similar pH.

Amphibians

Amphibians depend on aquatic habitats during at least part of their life cycle. Most Canadian species lay their eggs in water, where their larvae live until metamorphosis occurs, after which the adults utilize nearby terrestrial habitats. Research suggests that some species of amphibians are vulnerable to acidification of their aquatic habitat, while others appear to be indifferent to this environmental change.

A study of amphibian breeding sites in Nova Scotia, covering a range of pHs from 3.9 to 9.0, found that the distributions of some species were not obviously influenced by acidity (Dale *et al.*, 1985). The bullfrog (*Rana catesbeiana*) occurred from pH 4.0 to 9.0; spring peeper (*Hyla crucifer*) and yellow-spotted salamander (*Ambystoma maculatum*) from pH 3.9 to 7.8; green frog (*R. clamitans*) from pH 3.9 to 7.3; and wood frog (*R. sylvatica*) from pH 4.3 to 7.8. Yellow-spotted salamander and green, bull, wood, and pickerel (*R. palustris*) frogs all had eggs, developing larvae, and adults present in some habitats at pH 4, suggesting that reproduction was occurring at that extreme acidity. Studies in other regions, however, have shown that some species of amphibians are intolerant of acidity.

Laboratory experiments with 14 species of amphibians have shown that exposing developing eggs to pHs from 3.7 to 3.9 causes more than 85% embryonic mortality, while prolonged exposure to pH 4.0 causes mortality greater than 50% (Freda *et al.*, 1991). It must be borne in mind, however, that aquatic pHs as acidic as pH 4 are uncommon in nature. Moreover, such acidic conditions are usually associated with acid-mine drainage or natural bogs, rather than with acidification caused by atmospheric deposition (which has an acidification threshold of about pH 4.5 or higher). Overall, it appears that most species of amphibians are less liable than fish to population declines caused by acidification.

Aquatic Birds

Directly toxic effects of acidification on waterfowl and other aquatic birds have not been documented, and probably do not occur. However, acidification causes substantial changes to the aquatic habitats of waterfowl, and this has indirect consequences for bird populations. For instance, if acidification reduces or eliminates fish populations, then fish-eating waterfowl such as the common loon (*Gavia immer*) and common merganser (*Mergus merganser*) are detrimentally affected, as are piscivorous (fish-eating) raptors such as osprey (*Pandion haliaetus*). In contrast, extirpations of predatory fish could result in an increased abundance of aquatic insects and zooplankton, which would improve the food resource for other species of waterfowl such as mallard (*Anas platyrhynchos*), black duck (*A. rubripes*), ring-necked duck (*Aythya collaris*), and common goldeneye (*Bucephala clangula*).

The breeding success of common loons was studied on 84 lakes in central Ontario (Alvo *et al.*, 1988). Only 9% of breeding attempts were successful on low-alkalinity lakes (below 40 μeq/L), which are either acidic or vulnerable to acidification because the water has little acid-neutralizing capacity. In contrast, 57% of breeding attempts were successful on lakes with alkalinity of 40–200 μeq/L, and 59% on lakes exceeding 200 μeq/L. These observations likely reflect the sizes of the fish populations in these lakes.

A study of 79 small lakes and ponds in New Brunswick found that the biomass of aquatic invertebrates in the littoral (or near-shore) zone was greater in acidic waterbodies with pH 4.5–4.9 than in waterbodies with a pH greater than 5.5 (Parker *et al.*, 1992). Five species of ducks that feed on invertebrates had an average of 3.5 broods/ha on waterbodies with pH below 5.5, compared with only 0.65 broods/ha at higher pHs. The higher biomass of invertebrates in the more acidic waterbodies was likely due to decreased predation because of the reduced fish community.

Drain Lake in Nova Scotia was previously described as a highly acidic (pH 4.0) but eutrophic lake. Drain Lake is fishless and has large populations of aquatic invertebrates and plants. This habitat allows black ducks and ring-necked ducks to be more productive than is typical for lakes in the region (Kerekes *et al.*, 1984).

Reclamation of Acidified Waterbodies

Even before acidification became a high-profile issue, wildlife managers knew that habitat for sportfish in brown-water lakes could be improved by treating the acidic water with powdered limestone (calcium carbonate, $CaCO_3$) or lime (calcium hydroxide, $Ca(OH)_2$). These treatments, known as **liming**, serve to reduce acidity, clarify water, and improve aquatic productivity, particularly of fish. Not surprisingly, much research has been done on the usefulness of liming treatments in reclaiming surface waters that have been degraded by atmospheric depositions of acidifying substances.

Effects of liming on lake-water pH are illustrated in Figure 19.5 for three limed lakes and one reference (i.e., non-limed but acidic) lake in south-central Ontario (Dillon *et al.*, 1979; Yan *et al.*, 1979). Initially, the treated lakes had a water pH of 4–5, but this was increased by liming to pH 7–8. Middle and Hannah Lakes had fairly stable pHs after treatment, but Lohi quickly drifted back toward an acidic condition. This reflects the sizes of the lakes' watersheds — Lohi drains a relatively large area and flushes quickly because it receives large inputs of water. This means that the neutralization treatments are rather short-lived in Lohi Lake.

Note that fertilization of the lakes with phosphate, which stimulates the productivity of phytoplankton and macrophytes, also has an acid-neutralizing effect, although this is much smaller than the effect associated with liming.

FIGURE 19.5 | EFFECTS OF LAKE LIMING

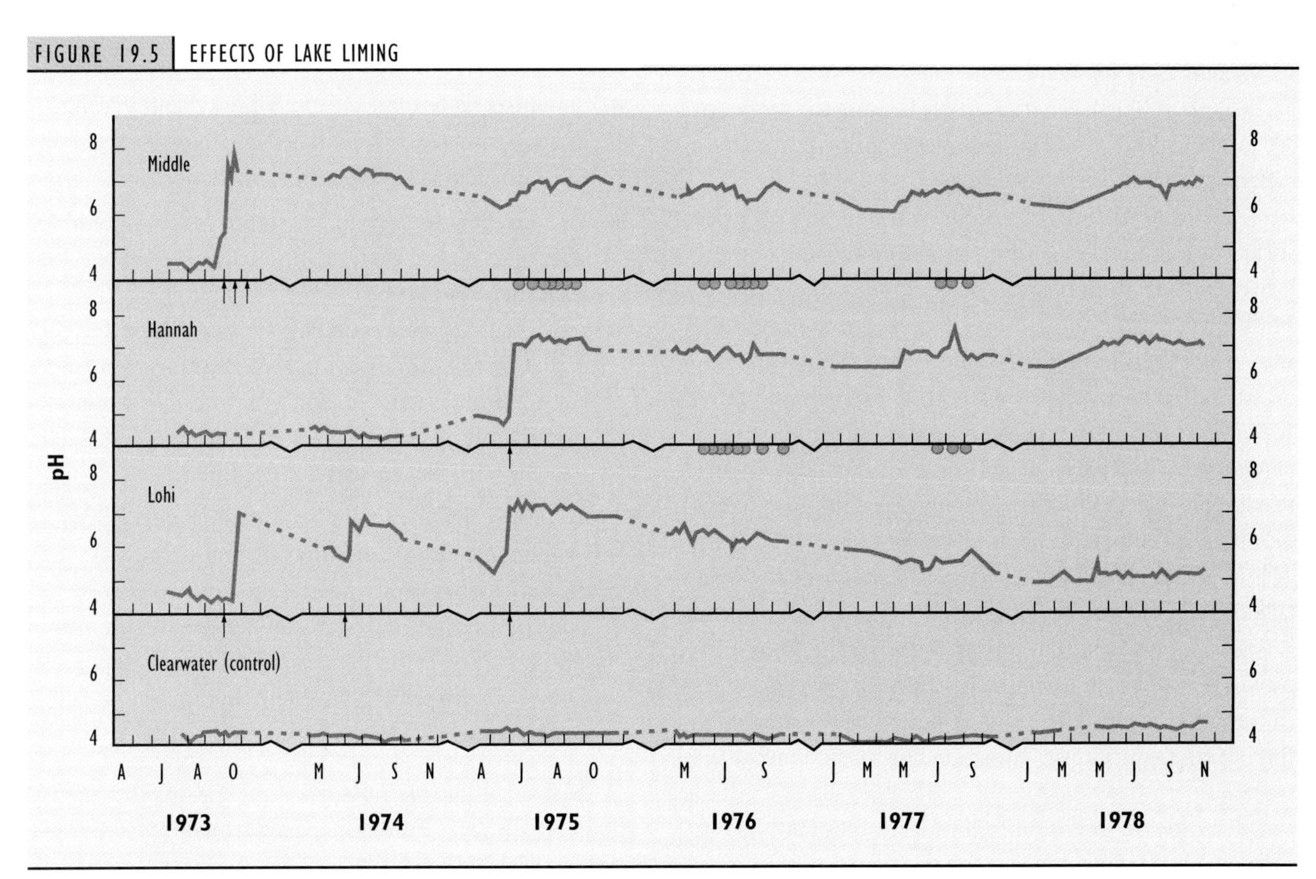

Middle, Hannah, Lohi, and Clearwater Lakes are located in the Sudbury region of Ontario. These lakes were acidified by a combination of dry and wet atmospheric depositions. The times of treatment with neutralizing agents ($CaCO_3$ or $Ca(OH)_2$) are indicated by arrows, and the times of treatment with additions of phosphate by solid dots.

Source: Modified from Dillon *et al.* (1979)

Initially, the limed lakes in the Ontario study exhibited a drastic decline in the biomass and productivity of phytoplankton and zooplankton. However, the biomass of phytoplankton rather quickly returned to the pre-liming condition, with lingering changes in species composition. Zooplankton recovered more slowly, and even after three years had not returned to their pre-liming abundance. In addition, fish kept in experimental cages in the limed lakes suffered high rates of mortality. This result was likely related to metal toxicity, since the experimental lakes were affected by particulate fallout from the Sudbury smelters. The aqueous concentrations of Cu, Ni, Zn, and Al all decreased after the lakes were limed because these metals have a lower solubility in less-acidic water. However, the metal concentrations still remained rather high, and continued to exert toxic stress on fish and other biota.

In some regions of Scandinavia, liming is used routinely to treat large numbers of acidified lakes and rivers. This is being done to mitigate the biological damages caused by acidification. Thousands of lakes have been limed in Sweden and Norway. Liming has been conducted less extensively in North America, partly because the programs are rather expensive and environmental priorities here are different from those in Scandinavia.

Research on liming has demonstrated that acidified surface waters can be neutralized. It must be understood, however, that liming treats the symptoms of damage in acidified ecosystems, but not the causes of the acidification. Moreover, liming itself represents an environmental stress that transforms a waterbody from one polluted condition to another that is still polluted, but less toxic. Liming causes severe damages to acid-adapted ecosystems, resulting in changes in species composition until new communities develop. In general, the most important benefit of liming is that less-acidic waters can support fish populations, whereas acidic waters cannot. It is important to recognize, however, that liming is not a long-term solution to the acidification of fresh waters. In part, this is because acidic waterbodies must be periodically retreated as the liming materials are consumed or flushed out of the system.

To some degree, acidified waterbodies may also be managed by fertilization, a treatment that greatly stimulates productivity, as was previously noted for Drain Lake. Acidic lakes that are fertilized can sustain a large biomass of phytoplankton, macrophytes, and invertebrates. These conditions can allow some species of waterfowl to breed in large numbers, even if fish cannot be sustained because of the acidic conditions. However, it is not always appropriate to create large numbers of highly productive lakes. For example, where recreational swimming is an important activity, large quantities of algae and macrophytes are considered a nuisance. Nevertheless, fertilization is an effective and relatively inexpensive management tool that can enhance ecological conditions in acidified waterbodies.

Reducing Emissions of Acidifying Substances

Ultimately, the extensive damages caused by acidifying depositions from the atmosphere can be resolved only by reducing the emissions of acid-forming gases. Although this fact is intuitively clear, the issue of emissions reductions remains highly controversial, for reasons such as the following:

- Scientists do not know exactly how much to reduce the emissions of SO_2 and/or NO_x in order to avoid the damages caused by acidifying deposition.
- Various emissions-reduction strategies are possible, and these vary in their economic consequences. For example, would it be more effective to target large point sources of emissions, such as smelters and power plants, while paying less attention to smaller, yet numerous, sources such as automobiles and oil-burning home furnaces? Or should large and small sources both be aggressively curtailed?
- To be effective, emissions reductions must be coordinated among neighbouring countries. For example, what will happen if the government of one country (perhaps one with large emissions of SO_2 and NO_x) does not regard acidifying deposition as a high-priority problem, but neighbouring countries do?

Not surprisingly, industries and regions that are responsible for large emissions of acid-forming gases have tended to lobby forcefully against any imposition of substantial legislated reductions of their emissions. In general, they argue that scientific justifications for these reductions are not yet convincing, while the economic costs of controls are known to be large and disruptive.

In addition, how low should the rates of atmospheric deposition of sulphur and nitrogen compounds be, in order to avoid further acidification of sensitive surface waters or to allow their recovery? The critical rates of deposition of acidifying compounds are influenced, in part, by the vulnerability of the receiving ecosystems — areas with shallow, nutrient-poor soils can sustain much lower inputs of acidifying substances than can areas with soils rich in calcium.

For the purposes of negotiations about transboundary movements of acidic precursors between Canada and the United States, the Canadian government has advocated a "desirable" rate of sulphate deposition of 20 kg/ha·y (this is sometimes referred to as a **critical load**). This particular intensity of sulphate deposition cannot yet, however, be fully justified, because the scientific understanding is still inadequate. In fact, some ecologists have suggested that many sensitive waterbodies would acidify and suffer ecological damages even at sulphate loadings of 10 kg/ha·y (Schindler, 1988). In Scandinavia, the critical sulphate loading is generally estimated to be about 15 kg/ha·y, except for particularly vulnerable landscapes, where it may be as low 6 to 12 kg/ha·y (Brodin and Kuylenstierna, 1992).

Critical rates of deposition of nitrogen compounds may be only 3–14 kg/ha·y on shallow, nutrient-poor soils, but considerably more on productive sites with calcium-rich soils.

Although there are many uncertainties about the specific causes and magnitudes of the damages caused by the atmospheric deposition of acidifying substances, it is obvious that what goes up (i.e., emissions of acid-precursor gases) must eventually come down (i.e., as acidifying depositions). This common-sense notion is supported by a great deal of scientific evidence. This knowledge, combined with public awareness and concern about acidification in many countries, has spurred politicians to begin to take effective action. This is resulting in reductions in emissions of SO_2 and NO_x, particularly in relatively wealthy countries in North America and western Europe.

In 1992, the governments of Canada and the United States signed a binational air-quality treaty aimed at reducing acidifying depositions in both countries. This agreement calls for enormous expenditures by industries and governments to substantially reduce the emissions of air pollutants, especially SO_2, during the 1990s. These cutbacks are on top of reductions of emissions that both countries had already achieved during the 1980s. In the United States, for example, emissions of SO_2 decreased from 23.1 million t in 1980 to 21.0 million t in 1990, while emissions of NO_x decreased from 20.7 million t to 19.4 million t (Anonymous, 1993).

The built environment can also be damaged by acidifying depositions from the atmosphere. For example, structures made of limestone, marble, or sandstone become eroded and chemically destabilized by the dry deposition of SO_2 and NO_x gases and by acidic precipitation. Acidifying air pollutants are seriously damaging famous artifacts of cultural heritage, such as this ancient temple known as the Parthenon in Athens, Greece.

Source: Michael Townsend/Tony Stone Images

The 1992 air-pollution treaty between Canada and the U.S. is a constructive accomplishment. It remains to be seen, however, whether emissions will be reduced enough to reverse the ecological damages caused by acidification. As well, the policies favouring reduced emissions may not be able to survive continuous challenges from politicians and economists who do not believe that such policies are necessary.

So far, actions to reduce the emissions of SO_2 and NO_x have been vigorous only in the relatively wealthy regions of North America and western Europe. In other countries, the political focus is mostly on industrial and economic growth. Air pollution and other environmental damages "subsidize" that economic growth and are often paid little heed. As soon as possible, much more political and scientific attention must be devoted to the problems of acidifying deposition and other kinds of pollution in eastern Europe, the former Soviet Union, China, India, southeast Asia, Mexico, and other rapidly developing countries. In these countries, emissions of SO_2, NO_x, and other important air-borne pollutants are galloping out of control.

Key Terms

acidification
acid rain
wet deposition
dry deposition
acidifying deposition (or the deposition of acidifying substances from the atmosphere)
LRTAP (long-range transport of atmospheric pollutants)
buffering capacity
acid sulphate soils
acid-mine drainage
acid shock
whole-lake experiments
acid-neutralizing capacity
liming
critical load

Questions for Discussion

1. Compare the chemistries of rainwater and lakewater in a region vulnerable to acidification, and discuss reasons for the differences.
2. Explain how wet and dry depositions of acidifying substances both contribute to acidification. Why do the rates and relative importance of these processes differ between urban and rural areas?
3. Are surface waters in your area acidic or likely to become so? Explain your answer in terms of the factors that influence vulnerability to acidification.
4. What environmental influences can cause soil to be come acidic, and how can this problem be managed?
5. Two options for dealing with acidifying deposition are emissions reductions to prevent the problem, and liming of water and soil to treat the symptoms. Some people have called emissions reductions the "billion dollar solution" and liming the "million dollar solution." This is because of the potentially greater capital costs of some technologies for efficient reductions of SO_2 and NO_x emissions. Which of these options (if not both) do you think is most appropriate for dealing with acidification as an environmental problem?

References

Alvo, R., D.J.J. Hussell, and M. Berrill. 1988. The breeding success of common loons (*Gavia immer*) in relation to alkalinity and other lake characteristics in Ontario. *Can. J. Zool.*, **66**: 746–752.

Anonymous. 1991. *National Acid Precipitation Assessment Program: Integrated Assessment Report*. Washington: U.S. Government Printing Office, Superintendent of Documents.

Anonymous. 1993. *1992 Report to Congress*. Washington: National Acid Deposition Assessment Program.

Baker, J.P. (principal author). 1990. Biological effects of changes in surface water acid-base chemistry. In: *Acidic Deposition: State of Science and Technology. Vol. II. Aquatic Processes and Effects*. Washington:U.S. Government Printing Office, Superintendent of Documents. pp. 13-1 to 13-381.

Baker, L.A. (principal author). 1990. Current status of surface water acid-base chemistry. In: *Acidic Deposition: State of Science and Technology. Vol. II. Aquatic Processes and Effects* Washington: U.S. Government Printing Office, Superintendent of Documents. pp. 9-1 to 9-367.

Baker, L.A., A.T. Herlihy, P.R., Kaufmann, and J.M. Eilers. 1991. Acidic lakes and streams in the United States: the role of acidic deposition. *Science*, **252**: 1151–1154.

Beamish, R.J. and H.H. Harvey. 1972. Acidification of the La Cloche Mountain Lakes, Ontario, and resulting fish mortalities. *J. Fish. Res. Board Can.*, **29**: 1131–1143.

Brand, D.G., P. Kehoe, and M. Connors. 1986. Coniferous afforestation leads to soil acidification in central Ontario. *Can. J. For. Res.*, **16**: 1289–1391.

Brodin, Y.W. and J.C.I. Kuylenstierna. 1992. Acidification and critical loads in Nordic countries: a background. *Ambio*, **21**: 332–338.

Chan, W.S., R.J. Vet, C.U. Ro, A.J.S. Tang, and M.A. Lusis. 1984. Long-term precipitation quality and wet deposition fields in the Sudbury basin. *Atmos. Environ.*, **18**: 1175–1188.

Crocker, R.L. and J. Major. 1955. Soil development in relation to vegetation and surface age at Glacier Bay, Alaska. *J. Ecol.*, **43**: 427–448.

Dale, J.M., B. Freedman, and J. Kerekes. 1985. Acidity and associated water chemistry of amphibian habitats in Nova Scotia. *Can. J. Zool.*, **63**: 97–105.

Dillon, P.J., D.S. Jefferies, W. Snyder, R. Reid, N.D. Yan, D. Evans, J. Moss, and W.A. Scheider. 1977. *Acid Rain in South-Central Ontario: Present Conditions and Future Consequences*. Rexdale, ON: Ontario Ministry of the Environment.

Dillon, P.J., N.D. Yan, W.A. Schieder, and N. Conroy. 1979. Acidic lakes in Ontario, Canada: characterization, extent, and responses to base and nutrient additions. *Arch. Hydrobiol.*, Suppl. **13**: 317–336.

Dixit, S.S., A.S. Dixit, and J.P. Smol. 1991. Multivariate environmental inferences based on diatom assemblages from Sudbury (Canada) lakes. *Freshwater Biol.*, **26**: 251–266.

Environment Canada. 1988. *Acid Rain: A National Sensitivity Assessment*. Ottawa: Environment Canada, Inland Waters Directorate.

Findlay, D.L. and S.E.M. Kasian. 1986. Phytoplankton community responses to acidification of Lake 223, Experimental Lakes Area, northwestern Ontario. *Water, Air, Soil Pollut.*, **30**: 719–726.

Freda, J., W.J. Sadinski, and W.A. Dunson. 1991. Long term monitoring of amphibians populations with respect to the effects of acidic deposition. *Water, Air, Soil Pollut.*, **55**: 445–462.

Freedman, B. 1995. *Environmental Ecology, 2nd ed.* San Diego, CA: Academic.

Freedman, B. and T.A. Clair. 1987. Ion mass balances and seasonal fluxes from four acidic brownwater streams in Nova Scotia. *Can. J. Fish. Aquat. Sci.*, **44**: 538–548.

Harvey, H.H. and C. Lee. 1982. Historical fisheries changes related to surface water pH changes in Canada. *Acid Rain/Fish., Proc. Int. Symp.*, pp. 45–55.

Henriksen, A. 1980. Acidification of freshwaters — a large scale titration. In: *Ecological Impacts of Acid Precipitation*, (D. Drablos and A. Tollan, eds.), Oslo: SNSF Project. pp. 68–74.

Kelly, C.A., J.W.M. Rudd, A. Furutani, and D.W. Schindler. 1984. Effects of lake acidification on rates of organic matter decomposition in sediments. *Limnol. Oceanogr.*, **29**: 687–694.

Kerekes, J. and B. Freedman. 1988. Physical, chemical, and biological characteristics of three watersheds in Kejimkujik National Park, Nova Scotia. *Arch. Environ. Contam. Toxicol.*, **18**: 183–200.

Kerekes, J., B. Freedman, G. Howell, and P. Clifford. 1984. Comparison of the characteristics of an acidic eutrophic and an acidic oligotrophic lake near Halifax, Nova Scotia. *Water Poll. Res. J. Canada*, **19**: 1–10.

Lacroix, G.L. and D.R. Townsend. 1987. Responses of juvenile Atlantic salmon (*Salmo salar*) to episodic increases in acidity of Nova Scotia rivers. *Can. J. Fish. Aquat. Sci.*, **44**: 1475–1484.

Likens, G.E., F.H. Bormann, R.S. Pierce, J.S. Eaton, and R.E. Munn. 1984. Long-term trends in precipitation chemistry at Hubbard Brook, New Hampshire. *Atmos. Environ.*, **18**: 2641–2647.

Lovett, G.M., W.A. Reiners, and R.K. Olson. 1982. Cloud droplet deposition in subalpine balsam fir forests: hydrological and chemical budgets. *Science*, **218**: 1303–1304.

Malley, D.F., D.L. Findlay, and P.S.S. Chang. 1982. Ecological effects of acid precipitation on zooplankton. In: *Acid Precipitation Effects on Ecological Systems*, (F.M. D'Itri, ed.), Ann Arbor, MI: Ann Arbor Science. pp. 297–327.

Mills, K.H. 1984. Fish population responses to experimental acidification of a small Ontario lake. In: *Early Biotic Responses to Advancing Lake Acidification*, (G.R. Hendry, ed.), Toronto: Butterworth. pp. 117–131.

Mills, K.H., S.M. Chalanchuk, L.C. Mohr, and I.J. Davies. 1987. Responses of fish populations in Lake 223 to 8 years of experimental acidification. *Can. J. Fish. Aquat. Sci.*, **44**, Suppl. 1, 114–125.

Morrison, I.K. 1983. Composition of percolate from reconstructed profiles of jack pine forest soils as influenced by acid input. In: *Effects of Accumulations of Air Pollutants in Forest Ecosystems*. (B. Ulrich and J. Pankrath, eds.). Berlin: Reidel. pp. 195–206.

Parker, G.R., M.J. Petrie, and D.T. Sears. 1992. Waterfowl distribution relative to wetland acidity. *J. Wildl. Manage.*, **56**: 268–274.

Percy, K.E. 1986. The effects of simulated acid rain on germinative capacity, growth, and morphology of forest tree seedlings. *New Phytol.*, **104**: 473–484.

Reuss, J.R. 1975. *Chemical/biological Relationships Relevant to Ecological Effects of Acid Rainfall.* Corvallis, OR: U.S. Environmental Protection Agency. Pub. EPA-660/3-75-032.

Reuss, J.R. and D.W. Johnson. 1985. Effect of soil processes on the acidification of water by acid precipitation. *J. Environ. Qual.*, **14**: 26–31.

Schindler, D.W. 1988. Effects of acid rain on freshwater ecosystems. *Science*, **239**: 149–157.

Schindler, D.W. 1990. Experimental perturbations of whole lakes as tests of hypotheses concerning ecosystem structure and function. *Oikos*, **57**: 25–41.

Schindler, D.W., K.H. Mills, D.F. Malley, D.L. Findlay, J.A. Shearer, I.J. Davies, M.A. Turner, G.A. Linsey, and D.A. Cruikshank. 1985. Long-term ecosystem stress: the effects of years of experimental acidification on a small lake. *Science*, **228**: 1395–1401.

Schofield, C.L. 1982. Historical fisheries changes in the United States related to decreases in surface water pH. In: *Proc. Int. Symp. Acid. Rain/Fish*. Bethesda, MD: American Fisheries Society. pp. 57–67.

Shannon, J.D. and B.M. Lesht. 1986. Estimation of source-receptor matrices for deposition of NO_x–N. *Water, Air, Soil Pollut.*, **30**: 815–824.

Sprules, W.G. 1975. Midsummer crustaceun zeoplankton communities in acid-stressed lakes. J. Fish. Res. Board Can., **32**: 389–395.

Stewart, C.C. and B. Freedman. 1989. Comparison of the macrophyte communities of a clearwater and a brown-water oligotrophic lake in Kejimkujik National Park, Nova Scotia. *Water, Air, Soil Pollut.*, **46**: 335–341.

Tveite, B. 1980. Effects of acid precipitation on soil and forest. 9. Tree growth in field experiments. In: *Ecological Impact of Acid Precipitation*, (D. Drablos and A. Tollan, eds.) Oslo: SNSF Project. pp. 206–207.

Watt, W.D., C.D. Scott, and W.J. White. 1983. Evidence of acidification of some Nova Scotia rivers and its impact on Atlantic salmon. *Can. J. Fish. Aquat. Sci.*, **40**: 462–473.

Wood, T. and F.H. Bormann. 1976. Short-term effects of a simulated acid rain upon the growth and nutrient relations of *Pinus strobus*. Upper Darby, PA: Northeastern Forest Experiment Station. U.S.D.A. For. Serv., Gen. Tech. Rep. NE-23. pp. 815–824.

Wright, R.F. and E. Snekvik. 1978. Acid precipitation: chemistry and fish populations in 700 lakes in southernmost Norway. *Verh.-Int. Ver. Theor. Angew. Limnol.*, **20**: 765–775.

Yan, N.D., R.E. Girard, and C.J. Lafrance. 1979. *Survival of Rainbow Trout,* Salmo gairdneri, *in Submerged Enclosures in Lakes Treated with Neutralizing Agents near Sudbury, Ontario.* Rexdale, ON: Ontario Ministry of the Environment.

20

EUTROPHICATION AND RELATED PROBLEMS OF WATER QUALITY

CHAPTER OUTLINE

CHAPTER OBJECTIVES

After completing this chapter, you will be able to:

1. Compare and contrast the causes of eutrophication of fresh and marine waters.
2. Explain the evidence which indicates that phosphorus is the limiting nutrient for eutrophication of freshwaters.
3. Outline how the trophic structure of an aquatic ecosystem can affect its response to eutrophication.
4. Describe the objectives and technologies used in primary, secondary, and tertiary sewage treatment.
5. Explain the role of eutrophication in the damage caused to the ecosystem of Lake Erie.

INTRODUCTION

Eutrophic waters contain large concentrations of nutrients, and as a result are highly productive. In contrast, **oligotrophic** waters are much less productive, because of a restricted availability of nutrients. **Mesotrophic** waters are intermediate between these two conditions.

Some waterbodies occur in inherently fertile watersheds, and are *naturally eutrophic*. So-called **cultural eutrophication**, however, is caused by anthropogenic nutrient additions, usually through the dumping of sewage wastes or the runoff of fertilizers from agricultural land. Both fresh and marine waters can become eutrophic through increases in their nutrient supply, although the problem is more usually found in inland freshwaters.

The most conspicuous symptom of eutrophication is a large increase in primary productivity, particularly of phytoplankton, which can develop dense populations known as an **algal bloom.** Shallow waterbodies may also experience vigorous growths of aquatic plants. Since the increased productivity of algae and macrophytes can allow the higher trophic levels to be more productive, aquatic invertebrates, fish, and waterfowl may be abundant in some eutrophic waterbodies.

Extremely eutrophic (or *hypertrophic*) waters may, however, become severely degraded. Hypertrophic waterbodies develop noxious blooms of cyanobacteria (i.e., blue-green bacteria) and algae during the summer, which may cause an off-flavour in water used for drinking, and may synthesize and release toxic organic compounds into the water. In addition, the decomposition of the algal biomass requires a large amount of oxygen, severely depleting its concentration in water. The resulting anoxic conditions are extremely stressful to many aquatic animals such as fish. Such conditions also facilitate the production and release of hydrogen sulphide (H_2S) and other noxious gases.

Cultural eutrophication, which can degrade both water quality and ecological conditions, is an important environmental problem in many areas. Severe eutrophication can impair the use of a waterbody as a source of drinking water, to support a fishery, or for recreation. Eutrophication can also degrade the ecological values of natural waters. In this chapter we will examine the causes and consequences of this environmental problem.

Lake-of-the-Woods is a large lake in northwestern Ontario and southeastern Manitoba. The watershed of the lake is largely wild and forested, but Lake-of-the-Woods supports numerous cottages and a great deal of tourism. There are also nutrient inputs with rivers draining agricultural areas in the United States to the south, and pulp mills from the east. Shallow waters and bays of Lake-of-the-Woods are subject to eutrophication caused by nutrient enrichment. This aerial photograph shows a bloom of algae, indicated by a whitish colour of the water.
Source: P. Dillon

CAUSES OF EUTROPHICATION

Most lakes in Canada are geologically "young" because they occur on landscapes released from glacial ice only about 8 000–12 000 years ago (depending on the area). Many lakes are relatively deep, having had little time to accumulate sediment in their basins. They also tend to have low rates of nutrient input. Over many centuries, however, lakes of this sort gradually increase in productivity as they accumulate nutrients. They also become shallower due to sedimentation, which results in increased rates of nutrient cycling. The slowly increasing productivity of many lakes over time is a natural expression of eutrophication.

Surface waters may also be naturally eutrophic if they occur in watersheds with fertile soils. This is the case with many lakes and ponds in the prairie regions of Canada, particularly shallow waterbodies that recycle their nutrients relatively quickly.

Anthropogenic (or cultural) eutrophication involves more rapid increases in the productivity of surface waters.

This is most often caused by nutrient loading associated with sewage dumping, or by agricultural runoff contaminated by fertilizers. Wherever humans live in large populations or engage in intensive agriculture, there are large inputs of nutrients into lakes, rivers, and other freshwaters. In coastal areas, estuaries and shallow near-shore habitats may also be affected by anthropogenic nutrient inputs.

A theory called the **Principle of Limiting Factors** states that certain ecological processes are controlled by whichever environmental factor is present in the least supply relative to demand. According to this theory, the primary production of waterbodies is limited by whichever nutrient is present in least supply relative to its demand (assuming that light, temperature, and oxygen supply are all adequate). Research has shown that, in almost all freshwaters, primary production is limited by the availability of phosphorus, occurring as the phosphate ion (PO_4^{3-}). In contrast, marine waters are usually limited by the availability of inorganic nitrogen, particularly in the form of nitrate (NO_3^-).

Phosphorus as the Limiting Nutrient in Freshwater

During the 1960s and early 1970s, there was much controversy about eutrophication, including which nutrients were the limiting factors to primary production in lakes and other surface waters. Many scientists believed that phosphorus was most important in this regard. Others, however, suggested that dissolved nitrogen, in the forms of nitrate or ammonium, was the critical limiting factor in many freshwaters. Plants and algae have relatively large demands for nitrate and ammonium, which often occur in low concentrations in water.

It was also suggested that dissolved inorganic carbon (as bicarbonate, HCO_3^-) could be limiting primary productivity. Inorganic carbon is needed in large quantities by autotrophs. In aquatic systems, HCO_3^- is replenished mainly by the diffusion of atmospheric CO_2 into the surface water, which is a rather slow process.

Eventually, several lines of scientific evidence resolved the controversy about limiting nutrients. Studies have convincingly demonstrated that phosphorus is the key nutrient that limits primary production in freshwaters. Consequently, controlling the rate of phosphorus supply is now known to be critical in the management of eutrophication.

The essential nature of the role of phosphorus is suggested in Table 20.1, which shows its typical concentration in freshwater (an index of "supply") and its concentration in plants (an index of "demand"). Because the ratio of demand to supply for phosphorus is considerably larger than for other important nutrients, it is a likely candidate to be the primary limiting factor for the productivity of algae and plants in freshwaters. The data also suggest that the second most important nutrient is inorganic nitrogen, in the form of nitrate or ammonium.

Other information is consistent with phosphorus being the controlling nutrient for eutrophication in freshwaters. Perhaps the most convincing data come from a famous series of experiments conducted in an area of northwestern Ontario known as the Experimental Lakes Area (ELA; see Canadian Focus 20.1). Most of the research applicable to eutrophication has involved the addition of nutrients at various rates and in various combinations to selected lakes, followed by careful monitoring of the ecological responses. In terms of identifying phosphorus as the key nutrient causing eutrophication, the most important **"whole-lake" experiments** are the following (Schindler, 1978, 1990; Levine and Schindler, 1989):

- For two years, Lake 304 was fertilized with phosphorus, nitrogen, and carbon. It responded by becoming eutrophic. After the fertilization with phosphorus was stopped, Lake 304 returned to its

TABLE 20.1 | DEMAND AND SUPPLY OF ESSENTIAL NUTRIENTS IN WATER

These are indexed by the typical concentrations in freshwater plants and in freshwater.

NUTRIENT	CONCENTRATION IN PLANTS (%)	CONCENTRATION IN WATER (%)	RATIO OF CONCENTRATIONS PLANTS:WATER (APPROX.)
Carbon	6.5	0.012	5 000
Silicon	1.3	0.00065	2 000
Nitrogen	0.7	0.000023	30 000
Potassium	0.3	0.00023	1 300
Phosphorus	0.08	0.000001	80 000

Source: Modified from Vallentyne (1974)

CANADIAN FOCUS 20.1

THE EXPERIMENTAL LAKES AREA

Some of the most famous Canadian research in ecology and environmental science has been conducted in the Experimental Lakes Area (ELA). Research at the ELA, funded mainly by the federal Department of Fisheries and Oceans (DFO), began during the late 1960s. The first two decades of work are closely identified with David Schindler, but hundreds of other Canadian and international scientists from government, universities, and the private sector have been involved in research in the area. Several of the leading scientists working at the ELA, notably Schindler, have won international awards for their outstanding contributions to ecology and environmental science.

The ELA is located in a remote region of northwestern Ontario, near the town of Kenora. The area contains a large number of natural lakes and wetlands. The watersheds are largely forested, with thin soils that overlie hard, Precambrian bedrock within the extensive geological formation known as the Canadian Shield. The ELA is not a protected area, so commercial forestry is actively pursued, as are mining exploration and ecotourism. However, there is an understanding among the various interests in the ELA that some of the lakes and their watersheds are being used for scientific research and should not be disturbed.

Aerial view of Lake 226 in the Experimental Lakes Area of northwestern Ontario. This lake was divided into two separate basins with a heavy vinyl curtain. The upper basin in the photograph was fertilized with phosphorus, nitrogen, and carbon, while the lower basin received nitrogen and carbon. Only the basin receiving phosphorus became eutrophic and developed blooms of phytoplankton — a response seen as a whitish hue in the photo.
Source: D. Schindler

One of the most important experimental procedures used at the ELA has been the controlled perturbation of entire lakes or wetlands. This is done to investigate the ecological effects of stressors associated with human influences, such as eutrophication, acidification, metal pollution, and flooding of wetlands. Such projects are known as "whole-lake" experiments (and some as "whole-wetland" experiments). The experimental design includes an initial study of the lake or wetland for several years to determine the baseline conditions. The system is then perturbed in some way — for example, by causing it to become eutrophic through the addition of nutrients. The experimental lakes are carefully monitored for a wide range of ecological responses, such as changes in the abundance and productivity of species and communities, changes in nutrient cycling, and changes in chemical and physical factors.

The experimental lakes are paired with reference (or nonperturbed) lakes, which are monitored to provide information on natural changes unrelated to the experimental manipulation. Because the reference lakes are monitored for long periods of time, they also provide extremely useful information relevant to the detection of changes in the ambient environment, such as climate change.

Unfortunately, research at the Experimental Lakes Area has been severely threatened by cutbacks in funding, and the future of this world-class facility and its programs is uncertain. Throughout the history of the ELA, the Department of Fisheries and Oceans provided most of the funding to maintain the research infrastructure. Most of the experiments were also funded by DFO and conducted by its personnel, or by university scientists working with them. DFO is now, however, focusing its activities on marine issues, particularly in relation to commercial fish stocks. Although negotiations are underway with

other agencies of the federal and provincial governments, notably Environment Canada, a new champion of long-term research at the ELA has not emerged, and the facility is threatened with closure. This would be a severe blow to research on aquatic ecosystems in Canada.

original oligotrophic condition, even though nitrogen and carbon additions were continued.

- The two basins of Lake 226, an hourglass-shaped lake whose two sections are connected by a constricted channel, were isolated with a vinyl curtain. One basin was fertilized with C + N + P in a weight ratio of 10:5:1, while the other received only C + N at 10:5. Only the basin receiving P developed algal blooms. After five years, the nutrient additions were stopped, and the original oligotrophic condition returned within just one year. A similarly rapid recovery was observed in another experiment, involving Lake 303, after fertilization with P was stopped.
- Lake 302N was fertilized with P + N + C. However, these nutrients were injected directly into deep waters during the summer, at a time when lakes in the ELA develop a thermal stratification (see In Detail 20.2). Because the nutrients were injected into deep waters, primary production in the surface water was not affected and eutrophication did not occur.

Research at ELA also demonstrated that when phosphorus, but not nitrogen and carbon, was deliberately added to lakes, the supply of N and C already in the lake was still capable of supporting P-induced eutrophication. Eutrophication induced a greater rate of fixation of atmospheric dinitrogen (N_2) by blue-green bacteria and an increased rate of diffusion of atmospheric CO_2 into the

In Detail 20.2

Lake Stratification

During most of the year, lakes have a rather uniform distribution of temperature throughout their depth. This allows their bottom and surface waters to mix easily under the influence of strong winds. During the summer, however, lakes often develop a persistent *stratified* condition. This is characterized by a surface layer of relatively warm water, several metres thick in most lakes (but thicker in some large lakes), lying above deeper, cooler waters. Because the density of warm water is less than that of cool water, the two layers remain physically discrete and do not mix.

The relatively warm, upper waters of a stratified lake are known as the *epilimnion*, while the cooler, deeper waters are known as the *hypolimnion*. These layers are separated by a zone of rapid change in temperature, known as the *thermocline*. During the autumn, as the epilimnion cools, the stratification diminishes and the two layers are eventually mixed by strong winds.

During stratification, oxygen and other dissolved substances can enter the hypolimnion, but mainly by diffusion across the thermocline. Because this is a slow process, hypolimnetic oxygen can be easily depleted in stratified lakes whose deeper waters receive large inputs of organic materials. The organic inputs may be associated with sewage, agricultural runoff, or algal biomass sinking from the epilimnion. The development of anoxic (no-oxygen) conditions represents an important degradation of water quality, because fish and most other animals cannot live in such an environment. Foul gases such as hydrogen sulphide (H_2S) are also generated.

Sometimes, density gradients associated with dissolved salts may also cause lakes to stratify. In these cases, a surface layer of freshwater sits on top of more saline, deeper water, with the layers separated by a steep chemical and density gradient known as a *halocline*. Lakes that stratify because of temperature gradients during the summer are common throughout Canada, while salt-stratified lakes are rare.

lakewater. These processes were sufficient to sustain eutrophication.

Eutrophication also causes pronounced changes in the species composition, relative abundance, and biomass of phytoplankton, the most important primary producers in the ELA lakes. Changes in the phytoplankton resulted in effects on organisms at higher levels of the trophic web, such as zooplankton and fish. For example, the productivity of whitefish (*Coregonus clupeaformis*) in hourglass-shaped Lake 226 was greater in the experimentally eutrophied basin than in the oligotrophic basin. The increased fish productivity occurred in response to a greater abundance of their prey of zooplankton and aquatic insects, itself a trophic response to the increased productivity of algae.

Sources and Control of Phosphorus Loading to Surface Waters

In North America, eutrophication, considered as an environmental problem, was most severe during the 1960s and 1970s. At that time, the average discharge of phosphorus to inland waters was about 2 kg/person-year. About 84% of the phosphorus loading was associated with dumping of municipal sewage, and the rest was due to agricultural sewage and fertilizers. In addition, the total nitrogen discharge at the time was 12.5 kg/person-year, of which 36% was from municipal and 64% from agricultural sources.

Because scientists have convincingly demonstrated that phosphorus is the primary limiting nutrient for eutrophication in freshwaters, control strategies have focused on reducing the input rate of that nutrient to surface waters.

Phosphorus in Detergents

One of the first targets was domestic detergents. During the 1960s and early 1970s, detergents contained large concentrations of phosphate compounds, particularly sodium tripolyphosphate, which typically accounted for 50–65% of the weight of the product (12–16% if expressed as P). The phosphates were added as so-called "builders" in the detergent formulation. Builders reduce the activities of calcium, magnesium, and other cations in the wash water, thereby allowing the actual cleaning agents in the detergent (these are known as surfactants) to work more efficiently. During the 1960s and early 1970s, as much as 3 million kg of high-phosphate detergents were used in North America each year, and virtually all of this was eventually flushed into surface waters through the sewerage system. Detergents accounted for as much as one-half of the phosphorus content of wastewater discharges in North America during the early 1970s.

Fortunately, the use of detergents is a highly discrete activity, and good substitute "builders" are available to replace phosphates. Consequently, it was relatively easy to achieve large and rapid decreases in phosphorus loading by regulating the use of high-phosphate detergents. In 1970, detergents sold in Canada could contain as much as 16% phosphorus; this was decreased to 2.2% by 1973. Some areas of North America have gone further, virtually banning the sale and use of detergents containing any phosphorus.

Sewage Treatment

In most places, the principal objective of sewage treatment is to reduce inputs of pathogenic microorganisms and oxygen-consuming organic matter to receiving waters. However, in places where surface waters are vulnerable to eutrophication, sewage wastes may also be treated to reduce the quantities of phosphorus in the effluent.

All sizeable towns and cities in Canada have facilities to collect the sewage effluent from homes, businesses, institutions, and factories. This infrastructure consists of complex webs of underground pipes and other collection devices. Some municipalities have separate systems for the collection of domestic and industrial wastes, because the latter often contain large amounts of toxic chemicals which should be treated separately. Some municipalities may also have a separate system of pipes to handle the large volumes of storm flows, which result from surface runoff of rainwater and snow meltwater. Eventually, these large quantities of waste water must be discharged into the ambient environment, usually into a nearby lake, river, or ocean. Wherever possible, however, it is highly desirable to treat the waste water to reduce its concentrations of pollutants, rather than discharge them into the environment.

Regrettably, many municipalities in Canada continue to dump their raw, untreated sewage into a nearby aquatic environment. This practice is especially common for cities

Even extremely acidic lakes can become eutrophic if they are fertilized. Drain Lake near Halifax became extremely acidic (pH 4.0) after pyritic minerals in its watershed were exposed to the atmosphere by construction activities. However, the lake also received inputs of sewage, which caused it to become quite eutrophic, allowing it to support a lush productivity of aquatic plants, algae, zooplankton, insects, and waterfowl. The lake was too acidic, however, to support fish.

Source: B. Freedman

and towns located beside an ocean, because well-flushed marine ecosystems have a huge capacity for diluting and biodegrading organic pollutants in sewage. Metropolitan areas that treat little of their sewage before dumping it into the ocean include Halifax, Saint John, and St. John's on the Atlantic coast, Montreal and Quebec City on the St. Lawrence River, and Victoria on the Pacific coast. Consequently, places like Halifax Harbour have become severely degraded by the aesthetic, hygiene-related, and ecological damages that are associated with environments receiving large quantities of raw, untreated sewage. Although the worst of these damages are relatively local, restricted to the vicinity of the sewage outfalls, they are significant. The situations should be responsibly addressed by the construction of sewage-treatment facilities.

Compared with many oceanic environments, inland waters such as lakes and rivers have much smaller capacities for diluting and biodegrading sewage wastes. Consequently, most Canadian municipalities that are located beside inland waters treat their sewage before discharging the effluent into the nearest river or lake. **Sewage treatment** can, however, vary greatly in degree and in the technologies used, as is described below.

Primary sewage treatment is relatively simple. It usually involves the screening of raw sewage to remove larger materials, and then allowing the remainder to settle, thereby reducing the quantity of suspended organic materials. The resulting effluent is then discharged into the environment, although it may first be treated with a disinfectant (usually a chlorine compound) to kill pathogenic microorganisms, particularly bacteria. Primary treatment typically removes about 60% of the suspended solids of raw sewage, and 5–15% of the phosphorus, while reducing the **biological oxygen demand** (BOD) by 25%. (BOD is the capacity of organic materials in wastewaters to consume oxygen during decomposition.)

Secondary sewage treatment may be applied to the effluent of primary treatment, mostly to further reduce the BOD level. Secondary treatment usually involves the use of a biological technology, in which aerobic decomposition of organic wastes is accomplished by enhancing the activities of microbial communities in an engineered environment. Two such biotechnologies in common use are activated sludge, generated through vigorous aeration of sewage waters to enhance decomposition of their organic content, and trickling filters, in which sewage wastes pass slowly through a complex physical substrate supporting large populations of microorganisms. These biotechnologies, along with primary treatment, produce large quantities of a humus-like product known as sludge, which can be composted and then spread on agricultural land as an organic-rich conditioner. Sludge may also be incinerated or dumped in a landfill (see Chapter 18). Primary and secondary treatments together typically remove about 30–50% of the phosphorus from sewage, and reduce the BOD by about 80%.

Tertiary sewage treatment includes processes designed to remove most of remaining dissolved nutrients from the sewage effluent. Phosphorus removal may be achieved by adding aluminum, iron, calcium, or other chemicals that develop insoluble precipitates with phosphate, which then settle out of the water. These processes can remove 90% or more of the phosphate present in the wastewater. Tertiary processes may also be used to remove inorganic nitrogen compounds, such as ammonium and nitrate.

Artificial wetlands are sometimes constructed and used to provide advanced treatment of sewage wastes. The wetlands are engineered to develop highly productive ecosystems, in which vigorous microbial communities decompose organic wastes, while algae and macrophytes decrease nutrient concentrations in the water. Most sewage-treatment wetlands are constructed outdoors, but some are developed inside greenhouses, allowing the system to work during the winter. The efficiency of these systems depends on climate, the rate of flowthrough of the sewage wastes, and the nature of the engineered wetland. Nevertheless, artificial wetlands are capable of removing as much as 30% of the phosphorus from raw sewage, while reducing BOD by as much as 90%.

Tertiary treatment to reduce phosphorus concentrations in municipal sewage effluents requires expensive investments in technology and operating costs. Consequently, this practice is only pursued under certain conditions. In Canada, tertiary treatment is used mainly by communities around the Great Lakes, especially Lakes Erie and Ontario, and Lake St. Clair, all of which have been substantially affected by eutrophication and other types of water pollution. Moreover, because the Great Lakes are affected by effluents originating from sources in both Canada and the United States, bilateral agreements have been negotiated concerning the loading of phosphorus and other pollutants. To meet the target loadings under these agreements, it is necessary to use tertiary systems to achieve a high degree of phosphorus removal from municipal sewage effluents produced by municipalities in both countries.

Elsewhere in Canada, less attention is generally paid to the removal of phosphorus from municipal sewage effluents. Although municipalities may treat their sewage, only primary or secondary systems are generally used, mostly to reduce the quantities of pathogenic microorganisms and to lower the BOD in the effluents.

In addition, agricultural livestock produce enormous quantities of faecal materials. However, their sewage effluents are rarely treated before they are disposed into the environment. Treatment facilities for agricultural sewage are considered too expensive, and are therefore not often required by regulators. This happens even though some intensive rearing systems, such as agro-industrial feedlots and factory farms, may produce huge quantities of relatively concentrated manures, equivalent to the effluents of small cities.

Case Studies of Eutrophication in Canada

Canada has an abundance of lakes and other surface waters, many of which receive substantial loadings of nutrients through sewage dumping and the runoff of agricultural fertilizers. Consequently, a great deal of research has examined the causes and consequences of eutrophication as an environmental problem in Canada. The following sections discuss two case studies of eutrophication in Canada.

Eutrophication of an Arctic Lake

Meretta and Char Lakes are two small lakes located on Cornwallis Island in northern Canada. Because of the severe climate in the High Arctic, tundra lakes are relatively simple ecosystems. Moreover, nutrient inputs to these lakes are naturally small and nutrient cycling is slow, so tundra lakes are naturally oligotrophic. Char Lake is a typical oligotrophic polar lake, with extremely clear, nutrient-poor water, low rates of primary production, and low productivity of zooplankton and fish.

Meretta Lake, in contrast, receives sewage from a small community. Because of the nutrient inputs, it has become moderately eutrophic — a development studied by scientists during the early 1970s (Schindler *et al.*, 1974).

The sewage dumping resulted in a phosphorus loading about 13 times larger in Meretta than in Char Lake, while the nitrogen loading was 19 times higher. Consequently, during the growing season, Meretta Lake developed a phytoplankton biomass averaging 12 times greater than in Char Lake, and as much as 40 times greater during the summer algal bloom, when biomass is highest. During the winter, these lakes are covered with ice, which restricts the rate at which atmospheric oxygen can enter the water. During this time the decomposition of organic materials, mostly sewage wastes but also algal biomass, exerts a great demand for oxygen in the bottom waters of Meretta Lake, resulting in the development of anoxic conditions. The oxygen depletion causes severe stress to aquatic animals, and results in reduced reproduction of arctic char (*Salvelinus alpinus*), a type of trout.

This case study demonstrates that even polar lakes, which are relatively simple ecosystems because of their severe climatic regime, can exhibit a strong eutrophication response to fertilization with limiting nutrients.

Lake Erie: Eutrophication in Combination with Other Stressors

The Great Lakes of North America are one of the world's outstanding freshwater systems. Lake Superior sits at the top of this chain of lakes, with Lakes Michigan, Huron, Erie, and Ontario located below. The lakes and their watersheds drain to the Atlantic Ocean through the St. Lawrence River, itself one of the world's great rivers (there is also some flow to the Mississippi River, through a canal near Chicago). The aggregate surface area of the Great Lakes is about 245 000 km^2, and the system drains a watershed of 539 000 km^2 (Table 20.2).

All of the Great Lakes except Lake Michigan form part of the border between Canada and the United States. Consequently, issues concerning the resources and water quality of the Great Lakes are binational in character. Waters within Canadian jurisdiction are affected by actions in the United States and vice versa; especially important are the dumping of sewage and industrial wastes, the conversion of natural forests and wetlands into agricultural and residential land uses, commercial and sport fishing, and other potentially damaging activities. Recognizing this, the governments of Canada and the United States have entered into a number of cooperative agreements regarding the management of resources, the emission of pollutants, and the research and monitoring of the condition of their shared Great Lakes ecosystem. Much of the integrated, binational activity is coordinated by the *International Joint Commission*, an administrative body with equal representation from both countries.

One of the Great Lakes, Lake Erie, has a relatively small volume, and is located in a drainage basin that has rather fertile soils. Consequently, Lake Erie has always been the most productive of the Great Lakes. However, the natural productivity of Lake Erie has been greatly increased as a result of nutrient inputs associated with urban sewage and agricultural drainage. This has created eutrophic conditions in shallower regions of the lake. The western basin of Lake Erie is particularly vulnerable to eutrophication, because it is relatively shallow and warm, and receives large inputs of sewage and agricultural runoff.

In addition to experiencing nutrient loading, Lake Erie has also been affected by other important stressors. These include contamination by potentially toxic chemicals, commercial and recreational fisheries, conversions of most of the natural ecosystems in its watershed into agricultural and urban land uses, and introductions of non-indigenous species of plants and animals. This complex of environmental stressors has greatly degraded Lake Erie's water quality and ecosystem. The damages were particularly acute during the late 1960s and early 1970s, when pollution was relatively uncontrolled. Although some of the earlier problems have been alleviated, Lake Erie is still in a highly degraded condition. In the following sections we examine the most important of the ecological changes in Lake Erie, as a case study of the ecological effects of eutrophication occurring in combination with other stressors (Freedman, 1995).

TABLE 20.2 | SIZE AND WATERSHED CHARACTERISTICS OF THE GREAT LAKES OF NORTH AMERICA

	SUPERIOR	MICHIGAN	HURON	ERIE	ONTARIO
Lake surface area (kilometre^2)	82 100	57 800	59 600	25 700	18 960
Average depth (metres)	147	85	59	19	86
Water retention time (years)	191	99	22	2.6	6
Watershed area (km^2)	127 700	118 000	134 100	78 000	64 030
Land use in watershed (%)					
forest	95	50	66	17	56
agriculture	1	23	22	59	32
urban	<1	4	2	9	4
wetland, other	4	23	10	15	8
Population in watershed ($\times 10^3$; 1986)	663.5	12 051	2 461	12 533	7 375
Population density (persons/km^2)	5.2	102.1	18.4	160.7	115.3
Phosphorus loading (10^3 tonne P/y)					
1976–1978	4	6	5	10	17
1988–1990	2	4	3	7	18
Phosphorus concentration in open water (microgram/litre)					
1983–1985	3	6	4	23	14
1988–1989	3	5	4	10	6

Source: Gregor and Johnson (1980) and Government of Canada (1991)

Oxygen Depletion

The waters of Lake Erie develop a stratified condition during the summer, which makes it difficult for oxygen to penetrate to deepwater habitats (see In Detail 20.2). If the deeper waters are subject to large demands for oxygen to decompose organic materials, oxygen depletion can result. During most summers from the 1950s to 1970s, this condition occurred widely in Lake Erie, especially in the shallow, western end of the lake. Sewage dumping and algal biomass sinking from surface waters resulted in intense demands for oxygen, causing deoxygenation of bottom waters.

Deoxygenation is extremely harmful to aquatic animals, most of which require free access to oxygen in order to live. The episodes of anoxia in Lake Erie caused great changes in the community of invertebrates living in the sediment (benthos) of the lake. The benthic community was dominated by larvae of species of mayflies (aquatic insects in the order Ephemeroptera). The most common species were *Hexagenia rigida* and *H. limbata*, which lived in surface mud in an abundance of about 400/m^2. However, following a series of severe oxygen depletions during the 1950s, the density of these insects decreased to about 40/m^2, and by 1961 the mayflies had virtually disappeared, occurring in an abundance of less than 1/m^2.

The virtual extirpation of benthic mayflies was widely reported by the popular media, which tended to sensationalize the phenomenon by suggesting that Lake Erie was "dead." This was by no means the case, because the mayflies had been replaced by a benthic fauna that is relatively tolerant of deoxygenation. These included aquatic worms known as tubificids (*Limnodrilus* spp.), insect larvae of the midge family (order Diptera, family Chironomidae), and small species of mollusks (snails in the order Gastropoda and clams in the family Sphaeriidae). The worm-dominated benthos is, however, considered to indicate a substantial degradation of ecological conditions, compared with the mayfly-dominated community that normally occurs in well-oxygenated sediments.

Algal Blooms

Because Lake Erie has a greater supply of nutrients, it supports a much larger biomass of phytoplankton than the other Great Lakes. When its eutrophication was most severe, Lake Erie's western basin supported about twice as much algal biomass as Lake Ontario (per unit of surface area), and 11 times more than oligotrophic Lake Superior.

The communities of phytoplankton vary greatly among the sub-basins of Lake Erie and between its nearshore and offshore waters. The eastern and central basins are relatively deep and mesotrophic, while the shallower western basin is eutrophic. In all three basins, however, shallow nearshore waters are more productive than offshore waters. The algal bloom that occurs during springtime in the eutrophic waters is typically dominated by diatoms of the genus *Melosira*, while the bloom occurring in late summer is dominated by the blue-green bacteria *Anabaena*, *Microcystis*, and *Aphanizomenon*, the diatom *Fragilaria*, and the green alga *Pediastrum*.

In addition, the colonial green alga *Cladophora glomerata* occurs as filamentous mats attached to rocks in shallow habitats along shores. This alga grows in locally fertile habitats in Lake Erie and some of the other Great Lakes. It was especially abundant during the 1960s and 1970s, when storms often caused mats of its biomass to detach from rocky substrates, eventually washing ashore as a malodorous mass, or sinking to deeper waters to contribute to the development of anoxic conditions.

Studies have shown that the western basin of Lake Erie has always been relatively productive, sustaining lush growths of aquatic plants and algae and large populations of fish. However, the huge nutrient inputs associated with sewage dumping and agricultural runoff increased the intensity of eutrophication throughout Lake Erie. Fortunately, these problems have been alleviated substantially since the 1970s. This is because inputs of phosphorus to the lake have decreased, mainly because of the ban on high-phosphate detergents and the construction of tertiary sewage-treatment plants to service cities and towns.

Changes in Zooplankton

The zooplankton of Lake Erie used to be dominated by relatively large species, such as *Limnocalanus macrurus* and species of *Daphnia*. By the 1960s, however, these had been replaced — mainly by smaller, previously rare species, particularly *Diaptomus siciloides*, which is considered to be an indicator of eutrophic conditions. The greatest changes occurred in the shallow, western basin of the lake, where the midsummer zooplankton density increased from less than about 7000/m^3 prior to 1940 to as much as 110 000/m^3 in 1959. However, even at that time, zooplankton species typical of oligotrophic conditions survived in the deeper, eastern basin of the lake.

Changes in the zooplankton community of Lake Erie were caused partly by the lake's increasing primary productivity, because single-celled phytoplankton are the food-base of these small crustaceans. In addition, at about the same time that Lake Erie was becoming more eutrophic, its commercial fishery was overexploiting relatively large species of fish, which are typically piscivorous (that is, they eat fish). After the demise of the larger piscivorous species, the fish community became dominated by smaller species that feed on zooplankton (these are known as planktivorous fish). These fish selectively predate on larger species of zooplankton. Therefore, smaller species are favoured and their abundance increases.

Changes in the Fishery

Lake Erie has always supported a large fishery, which typically exceeds the combined landings of all the other Great Lakes. Remarkably, the total catch by the commercial fishery on Lake Erie has been quite stable over the years. This has occurred despite enormous changes in the species of fish present, fishing technology, the intensity of eutrophication, pollution by toxic chemicals, and other aspects of habitat. Factors of habitat include the damming of rivers and streams required by some fish species for spawning, and sedimentation of shallow-water habitats by soil eroded from deforested parts of the watershed.

Although the yield of fish from Lake Erie has not declined, the nature of the fish community has changed greatly during the past several centuries. These changes illustrate a severe degradation of the fishery resource and of the natural ecosystem. When the commercial fishery on Lake Erie began in the 19th century, the prime targets of exploitation were the largest, most valuable species, especially lake whitefish (*Coregonus clupeaformis*), lake trout (*Salvelinus namaycush*), and lake herring (*Leucichthys artedi*). This is a common pattern whenever previously unexploited fishery or forest resources are initially harvested — take the best and leave the rest (see Chapter 14).

Unfortunately, populations of the initially most desirable species in Lake Erie were rapidly depleted. This happened because the fishing pressure was excessive and could not be sustained by the resource. Also, severe habitat degradation occurred in the lake, caused mainly by siltation associated with extensive deforestation of the Lake Erie watershed. As the most desirable species disappeared, the fishing industry switched to "second-choice" species, such as blue pike (*Stizostedion vitreum glaucum*), walleye (*S. v.*

vitreum), sauger (*S. canadense*), and yellow perch (*Perca flavescens*). Because of overexploitation and habitat degradation, the species of *Stizostedion* became extirpated or rare by the early 1970s. The fishery then became dominated by relatively small, low-value species such as yellow perch, and by nonindigenous fish such as rainbow smelt (*Osmerus mordax*), freshwater drum (*Aplodinotus grunniens*), and carp (*Cyprinus carpio*).

Although the total yields of fish caught in Lake Erie have remained fairly large and consistent over time, the quality of the economic resource and the integrity of the fish community have been badly degraded by overfishing and habitat changes.

Recent Changes in Environmental Quality

For a number of reasons, ecological conditions have improved markedly in Lake Erie since the late 1970s. This has largely been achieved by the construction of sewage-treatment facilities in cities and towns along the lakeshore in Canada and the United States (as well as upstream, especially on the Detroit River and Lake St. Clair). Many of these facilities include technology to reduce phosphorus inputs. More than $7.5 billion (U.S.) has been spent on improving sewage-treatment facilities since 1972, reducing the annual loading of phosphorus to Lake Erie from about 27 000 t in 1972, to 6000 t in 1985, to 3000 t in 1990 (Dolan, 1993).

These huge reductions of phosphorus inputs have alleviated eutrophication in Lake Erie (Makarewicz and Bertram, 1991). The average biomass of phytoplankton, for instance, has decreased from 3.4 g/m^3 in 1970 to 1.2 g/m^3 during 1983–1985. The largest standing crops of phytoplankton still occur in the western basin, where they averaged 1.9 g/m^3 during 1983–1987, compared with 1.0 g/m^3 in the central basin and 0.6 g/m^3 in the deepest, eastern basin. Blooms of nuisance algae associated with eutrophication have also decreased in intensity. For example, the blue-green alga *Aphanizomenon flos-aquae* had a standing crop as high as 2.0 g/m^3 in 1970, but averaged 0.22 g/m^3 during 1983–1985. Similarly, diatoms that indicate eutrophic conditions have decreased greatly in abundance, by 85% in the case of *Stephanodiscus binderanus* in the western basin, and by 94% for *Fragilaria capucina*. At the same time, diatoms that are indicators of mesotrophic or oligotrophic conditions have become more abundant, notably *Asterionella formosa* and *Rhizosolenia eriensis*. Overall, open waters of the previously eutrophic, western basin of Lake Erie are now considered to be in a mesotrophic condition, while the eastern basin is oligotrophic.

The animal communities of Lake Erie have also undergone changes since the 1970s. Species of zooplankton that are indicators of oligotrophic conditions have become more abundant, while indicators of eutrophication are fewer, and are now mostly restricted to the western basin. Since 1972, the populations of relatively large species of fish have increased greatly, particularly walleye and introduced species of Pacific salmon (*Oncorhynchus* spp.). These are fish-eating species, and their predation has greatly decreased the abundance of smaller, zooplankton-eating fish such as smelt, alewife (*Alosa pseudoharengus*), and shiners (*Notropis* spp.). The decreases of planktivorous fishes have allowed secondary increases to occur in the abundance of larger-bodied zooplankton, such as the waterflea *Daphnia pulicaria*.

The zebra mussel (*Dreissena polymorpha*) is another cause of important ecological changes in Lake Erie. This bivalve mollusk, a native of Eurasia, was accidentally introduced to the Great Lakes by the discharge of ballast water from transoceanic ships. The zebra mussel can rapidly attain extremely dense populations (up to 50 000/m^2) on hard, underwater surfaces such as rock, wood, metal, and concrete. Zebra mussels are filter-feeders, and their huge populations have an enormous capability for removing algal cells from water. Consequently, they may be responsible for some of the recent clarification of Lake Erie and eutrophic parts of other Great Lakes. In addition, the dense populations of zebra mussels have benefitted some species of ducks that winter on the Great Lakes, where they feed on benthic mollusks and other invertebrates. However, the invasion of the Great Lakes by zebra mussels has also caused serious damage, including the displacement of native mollusks that cannot compete with the dense shoals of zebra mussels. Industries and water utilities have also suffered damage from the clogging of their water-intake pipes.

Overview of the Lake Erie Case Study

Lake Erie is an important example of the cumulative detrimental effects of a variety of anthropogenic stressors on the ecological health of a large lake. The stressors that degraded Lake Erie include eutrophication caused by nutrient loading, degradation of habitats through siltation resulting from watershed deforestation, overexploitation of a potentially renewable natural resource (the fishery), pollution by oxygen-consuming sewage and toxic chemicals, and introductions of nonindigenous species. Fortunately, Lake Erie is also beginning to demonstrate that a highly degraded ecosystem can be induced to recover somewhat, assuming that the causes of the damage can be managed effectively. In the case of Lake Erie, the partial recovery was achieved substantially through reduced organic and phosphorus inputs from sewage.

Key Terms

eutrophic waters
oligotrophic waters
mesotrophic waters
cultural eutrophication
algal bloom
Principle of Limiting Factors
whole-lake experiments
sewage treatment (primary, secondary, tertiary)
biological oxygen demand (BOD)
artificial wetlands

Questions for Discussion

1. Outline the evidence showing that phosphorus is the limiting nutrient for algal productivity in lakes.
2. How is sewage treated in the community where you live? Describe the environmental benefits of treating sewage in your community and the environmental cost of not doing so.
3. Lake Erie has been affected by a variety of environmental stressors. Which of these have caused the most damage to the ecosystems of the lake?

References

Dolan, D. N. 1993. Point source loadings of phosphorus to Lake Erie: 1986–1990. *J. Great Lakes Res.*, **19**: 212–223.

Freedman, B. 1995. *Environmental Ecology, 2nd ed.* San Diego, CA: Academic.

Government of Canada. 1991. *The State of Canada's Environment*. Ottawa: Government of Canada.

Gregor, D.J. and M.G. Johnson. 1980. Non-point source phosphorus inputs to the Great Lakes. In: *Phosphorus Management Strategies for Lakes*. (R.C. Loehr, C.S. Martin, and W. Rast, eds.), Ann Arbor, MI: Ann Arbor Science. pp. 37–59.

Harper, D. 1992. *Eutrophication of Freshwaters: Principles, Problems, and Restoration*. New York: Chapman & Hall.

Hartman, W.L. 1988. Historical changes in the major fish resources of the Great Lakes. In: *Toxic Contaminants and Ecosystem Health: A Great Lakes Perspective*. (M.S. Evans, ed.). New York: Wiley.

Hebert, P.D.N., C.C. Wilson, M.H. Murdoch, and R. Lazar. 1991. Demography and ecological impact of the invading mollusc *Dreissena polymorpha*. *Can. J. Zool.*, **69**: 405–409.

Levine, S.N. and D.W. Schindler. 1989. Phosphorus, nitrogen, and carbon dynamics of Experimental Lake 303, during recovery from eutrophication. *Can. J. Fish. & Aquat. Sci.*, **46**: 2–10.

Ludyanskiy, M.L., D. McDonald, and D. MacNeil. 1993. Impact of the zebra mussel, a bivalve indicator. *BioScience*, **43**: 533–544.

Mackie, G.L. 1991. Biology of exotic zebra mussel, *Dreissena polymorpha*, in relation to native bivalves and its potential impact in Lake St. Clair. *Hydrobiologia*, **219**: 251–268.

Makarewicz, J.C. and P. Bertram. 1991. Evidence for the restoration of the Lake Erie ecosystem. *BioScience*, **41**: 216–223.

Mills, E.L., J.H. Leach, J.T. Carlton, and C.L. Secar. 1993. Exotic species in the Great Lakes: a history of biotic crises and anthropogenic introductions. *J. Great Lakes Res.*, **19**: 1–54.

Regier, H.A. and W.L. Hartman. 1973. Lake Erie's fish community: 150 years of cultural stresses. *Science*, **180**: 1248–1255.

Schindler, D.W. 1978. Factors regulating phytoplankton production and standing crop in the world's freshwaters. *Limnol. Oceanogr.*, **23**: 478–486.

Schindler, D.W. 1990. Experimental perturbations of whole lakes as tests of hypotheses concerning ecosystem structure and function. *Oikos*, **57**: 25–41.

Schindler, D.W., J. Kalff, H.E. Welch, G.J. Brunskill, H. Kling, and N. Kritsch. 1974. Eutrophication in the high arctic — Meretta Lake, Cornwallis Island (75°N Lat.). *J. Fish. Res. Bd. Canada*, **31**: 647–662.

Sonzogni, W.C., A. Robertson, and A.M. Beeton. 1983. Great Lakes management: ecological factors. *Environ. Manage.*, **7**: 531–542.

Vallentyne, J.R. 1974. *The Algal Bowl. Lakes and Man*. Ottawa: Department of the Environment. Special Publication 22.

Wetzel, R.G. 1975. *Limnology*. Toronto: Saunders.

21 PESTICIDES

CHAPTER OUTLINE

CHAPTER OBJECTIVES

After completing this chapter, you will be able to:

1. Explain the notion of "pest" and "weed," and give three reasons why it may be useful to decrease pest abundance under certain circumstances.
2. Differentiate pesticides by their target pests.
3. Classify pesticides by their major chemical groups, including inorganics, natural organics, organochlorines, organophosphates, carbamates, and biologicals.
4. Outline the environmental risks and benefits of pesticide use in sanitation, agriculture, forestry, and horticulture.
5. Explain why there is a global contamination of organisms with DDT and related organochlorines.
6. Describe the ecological damage caused by DDT and other organochlorine insecticides.
7. Outline the ecological damage caused by carbofuran, and discuss why it has taken so long for most uses of this insecticide to be banned in North America.
8. Describe the economic benefits and ecological risks of pesticide use in forestry.
9. Outline the concept of integrated pest management, and discuss whether it is applicable to all pest-management problems.

INTRODUCTION

Humans are constantly engaged in struggles against potentially debilitating competitors and diseases. **Pesticides** are substances used to gain an advantage in many of those ecological interactions. Pesticides are commonly used to protect crop plants, livestock, domestic animals, and humans from damages and diseases caused by microorganisms, fungi, insects, rodents, and other "*pests*," and to defend crops from competition with unwanted but abundant plants (i.e., "*weeds*").

It is important to understand that the words "pest" and "weed" are highly contextual in their use. In most situations, for example, white-tailed deer are highly valued for their natural, wild beauty, and they provide economic and subsistence benefits through hunting. However, this same species may be considered a serious pest when it feeds in gardens, agricultural fields, or forestry plantations. The same is true, to a greater or lesser degree, of other species that are considered pests or weeds.

Humans have been using pesticides for a long time (Hayes, 1991). There are records of unspecified chemicals being used by Egyptians to drive fleas from their homes about 3500 years ago. Arsenic has been used as an insecticide in China for at least 2900 years. In his epic poem, the *Odyssey*, the Greek poet Homer (writing about 2800 years ago) referred to the burning of sulphur (which generates toxic SO_2 gas) to purge homes and buildings of fleas and other unwanted vermin.

Pesticide use has, however, become much more common in modern times, and an enormously wider variety of substances is being used. About 300 insecticides, 290 herbicides, 165 fungicides, and many other pesticidal chemicals are available in more than 3000 different formulations (Hayes, 1991). Even more "commercial products" are available, because many of these involve similar formulations manufactured by different companies. In all cases, the successful use of a pesticide requires the choice of an appropriate substance and its proper application.

Almost all pesticides are chemicals. Some are natural biochemicals that are extracted from plants grown for that purpose, while other pesticides are inorganic chemicals based on toxic metals or compounds of arsenic. Most modern pesticides, however, are organic chemicals that have been synthesized by chemists. The costs of developing a new pesticide and testing it for its efficacy (or effectiveness against pests), toxicological properties, and environmental effects are quite large, equivalent to $20–30 million per chemical (in 1983; Hayes, 1991). However, if an effective pesticide against an important pest is discovered, the profits are potentially huge, and therefore industry willingly pays the high development costs.

Humans have acquired important benefits from many uses of pesticides, including:

- increased yields of crops, because of protection from diseases, competition, defoliation, and parasites
- prevention of much spoilage and destruction of stored foods
- prevention of certain diseases, conserving health and saving the lives of millions of people and domestic animals

It has been estimated that pests destroy 37% of the potential yield of plant crops in North America (Pimentel *et al.*, 1992). It has also been reckoned that for every dollar spent on pesticides in North American agriculture, there is a four-dollar gain in benefits through increased crop yield and prevention of damage to stored products. (These data include only "conventional" economic costs and benefits — the costs of environmental damage caused by pesticide use were not considered. See Chapter 12 for an explanation of ecological economics.)

This is not to say, however, that more pesticide use would achieve even better results. In fact, it has been argued that pesticide use in North America could be decreased by one-half without greatly decreasing crop yields. Cutting pesticide use in half would achieve important environmental benefits through reduced ecological damage (Pimentel *et al.*, 1991). In fact, three European countries (Sweden, Denmark, and the Netherlands) have recently passed legislation requiring at least a 50% reduction in agricultural pesticide use by the year 2000, and similar actions may eventually be required in North America (Matteson, 1995).

Because of the substantial benefits derived from the use of pesticides, their utilization has increased enormously during the past half-century. For example, pesticide usage increased 10-fold in North America between 1945 and 1989, although it has levelled off during the past decade or so (Pimentel *et al.*, 1992). Pesticide usage is now a firmly integrated component of the technological systems used in modern agriculture, forestry, horticulture, and public health management in most parts of the world.

Unfortunately, the considerable benefits from pesticides are partially offset by damage caused to ecosystems and sometimes to human health. Each year, about one million people are poisoned by pesticides, with 20 000 fatalities (Pimentel *et al.*, 1992). Although developing countries account for only about 20% of global pesticide use, they sustain about half of the poisonings. This is because relatively toxic insecticides are used in many developing countries, by a work force whose widespread illiteracy hinders the comprehension of instructions, and whose safety is compromised both by poor enforcement of regulations and by inadequate use of protective equipment and clothing. The most tragic case of pesticide-related poisoning occurred in 1984 at Bhopal, India. About 2800 people were killed and 20 000 seriously poisoned when a factory accidentally released 40 t of methyl isocyanate vapours to the atmosphere. Methyl isocyanate is a precursor chemical of carbamate insecticides (Rozencranz, 1988).

In addition, many pesticide applications cause ecological damages by killing **nontarget organisms** (that is, organisms that are not considered pests). These damages are particularly important when **broad-spectrum pesticides** are used. These pesticides are toxic to organisms other than the specific pest. Pesticides used as *broadcast sprays* are sprayed over a large area, such as an agricultural field, a lawn, or a stand of forest. If a broad-spectrum pesticide is broadcast-sprayed, many nontarget organisms are exposed. Many may be unintentionally damaged or killed. For example, in a typical agricultural field or forestry plantation, only a few plant species are abundant enough to compete significantly with crops and reduce their productivity. These are the "weeds" that are the targets of a broadcast herbicide application. However, many other plant species are also affected. Many of the nontarget plants provide habitat and food for animals, and help prevent erosion and nutrient losses. These ecological benefits are degraded by the nontarget damage. Similarly, broadcast insecticide spraying causes nontarget mortality to numerous arthropods (such as beneficial insects) other than the pest species, and birds, mammals, and other creatures can also be poisoned. The nontarget mortality may include predators and competitors of the pest species. This may cause secondary damage by releasing the pest from some of its ecological controls. The great challenges of pest control are to develop more effective, pest-specific pesticides and to invent non-pesticidal methods.

Pesticide use is expanding rapidly in extent and intensity. It now occurs in all countries, although to greatly varying degrees. Although much is known about the environmental damage caused by pesticide use, not all of the potential effects are understood. In this chapter we examine the nature of pesticides, their most important uses, and the characteristics of commonly used chemicals. We then examine cases of ecological damage caused by the routine use of these chemicals to deal with pest-management problems.

The Nature of Pesticides

Classification of Pesticides by Their Target

Pesticides are defined by their usefulness in causing mortality to or otherwise managing the abundance of species that are deemed to be "pests." Pesticides are, however, an extremely diverse group of chemicals. To better understand the usefulness and toxicity of pesticides, and the damage they cause, it is necessary to categorize them in various ways. One classification of pesticides is based on their intended targets.

- **Fungicides** are used to protect crop plants and animals from fungi that cause diseases and other damage.
- **Herbicides** are used to kill weeds, that is, unwanted plants that interfere with some human purpose. Most herbicide use in agriculture and forestry is intended to release crop plants from competition, while horticultural use is mostly for aesthetics.
- **Insecticides** are used to kill insects that are pests in agriculture, horticulture, or forestry, or that spread deadly diseases, such as the mosquito vectors of malaria, yellow fever, and encephalitis.
- *Acaricides* are used to kill mites that are pests in agriculture, and ticks that are vectors of certain diseases such as Lyme disease and typhus.
- *Molluscicides* are used to kill snails and slugs in agriculture and gardens, and aquatic snails which can be vectors of diseases such as schistosomiasis.

- *Nematicides* are used against nematodes, which can cause important damage to the roots of agricultural plants.
- *Rodenticides* are used to control populations of small mammals, such as mice, rats, gophers, and other rodents that are pests in agriculture or around the home.
- *Avicides* are used to kill birds, which are sometimes considered to be pests in agriculture.
- *Piscicides* are used to kill fish, which are sometimes considered pests in aquaculture.
- *Algicides* are used to kill unwanted growths of algae: for example, in swimming pools.
- *Bactericides, disinfectants, and antibiotics* are used to control infections and diseases caused by bacteria. (Note that antibiotics are not actually classified as "pesticides" under the Pest Control Products Act.)

Chemical Classification of Pesticides

Because almost all pesticides are chemicals, they can be categorized according to similarities in chemical structure. The most important groups of pesticides are described below (see Table 21.1 for a summary). (A few pesticide formulations are based on microbes; these are discussed below under "Biological Pesticides.")

Inorganic Pesticides

Inorganic pesticides are compounds that contain arsenic, copper, lead, or mercury. They are highly persistent in terrestrial environments, being only slowly dispersed by leaching and erosion by wind and water. Recently, inorganic pesticides have been widely replaced by synthetic organics and are now being used much less than in the past. Prominent examples of inorganic pesticides include

TABLE 21.1 | SOME IMPORTANT PESTICIDES

See Freedman (1995) for chemical names of these pesticides.

CLASS	MAJOR USE	IMPORTANT EXAMPLES
1. INORGANIC PESTICIDES		
a) Bordeaux mixture	fungicide	tetracupric + pentacupric sulphate (copper sulphates)
b) arsenicals	herbicides and insecticides	arsenic trioxide, sodium arsenite, calcium arsenate, Paris green, lead arsenate
2. ORGANIC PESTICIDES		
a) natural organics	mostly insecticides	nicotine, nicotine sulphate, pyrethroids, red squill, rotenone, strychnine
b) organo-mercurials	fungicides	phenyl mercuric acetate, methyl mercury, methoxyethyl mercuric chloride
c) phenols	fungicides	trichlorophenol, pentachlorophenol
d) chlorinated hydrocarbons		
DDT and relatives	insecticides	DDT, DDD or TDE, methoxychlor
lindane	insecticide	lindane
cyclodienes	insecticides	chlordane, heptachlor, aldrin, dieldrin
chlorophenoxy acids	herbicides	2,4-D, 2,4,5-T, MCPA, silvex
e) organic phosphorus compounds	mostly insecticides	parathion, methyl parathion, fenitrothion, malathion, phosphamidon, monocrotophos, glyphosate (a herbicide)
f) carbamates	mostly insecticides	carbaryl, aminocarb, carbofuran, aldicarb, methiocarb
g) triazines	herbicides	simazine, atrazine, hexazinone, cynazine, metribuzin
h) amides	herbicides	alachlor, metolachlor
i) thiocarbamates	herbicides	butylate
j) dinitroaniline	herbicide	trifluralin
k) acetaldehyde polymer	molluscicide	metaldehyde
l) pyrethroids (synthetic)	insecticides	cypermethrin, deltamethrin, permethrin, tetramethrin

Bordeaux mixture, a mixture of copper-based compounds, which is used as a fungicide to protect fruit and vegetable crops; and *arsenicals* such as arsenic trioxide, sodium arsenite, and calcium arsenate, which are used as herbicides and soil sterilants. Paris green, lead arsenate, and calcium arsenate are used as insecticides.

Organic Pesticides

Most organic pesticides are artificially synthesized chemicals, but some are natural toxins produced by certain species of plants and extracted and used as pesticides. Important examples of organic pesticides include the following:

Natural organic pesticides are extracted from plants. Nicotine and related alkaloids, for example, are extracted from tobacco (*Nicotiana tabacum*) and used as insecticides, usually applied as nicotine sulphate. Pyrethrum is a complex of chemicals extracted from flowers of daisy-like plants (*Chrysanthemum cinerariaefolium* and *C. coccinium*) and used as an insecticide. Rotenone is extracted from several tropical shrubs (*Derris elliptica* and *Lonchocarpus utilis*) and used as an insecticide, rodenticide, or piscicide. Red squill, extracted from the sea onion (*Scilla maritima*), is another rodenticide. Strychnine is a rodenticide extracted from the tropical plant *Strychnos nux-vomica*.

Synthetic organo-metallic pesticides are used mainly as fungicides. They include organomercurials such as methylmercury and phenylmercuric acetate.

Phenols include trichlorophenols, tetrachlorophenol, and pentachlorophenol, all of which are fungicides that are widely used to preserve wood and other organic materials.

Chlorinated hydrocarbons (or organochlorines) are a diverse group of synthetic pesticides. Residues of most organochlorines are quite persistent, having a half-life of about 10 years in soils. **Persistent** chemicals can remain in the environment for many years, because they are not easily degraded by microorganisms or physical agents such as sunlight or heat. The persistence of organochlorines, coupled with their strongly lipophilic nature (i.e., they are virtually insoluble in water, but highly soluble in fats and lipids), means that they become *bioconcentrated* and food-web magnified, with the highest concentrations occurring in top predators (bioconcentration and food-web magnification are explained in In Detail 18.1, on page 310).

Organochlorines include the following:

- *DDT and its insecticidal relatives*, such as DDD and methoxychlor, were once widely used as insecticides. Because of bans in North America and Europe in the early 1970s, use of these chemicals is now mainly confined to tropical countries. A persistent metabolite of DDT and DDD, the non-insecticidal chemical DDE, can accumulate in organisms.
- *Lindane* is the active constituent of hexachlorocyclohexane, an insecticide.
- *Cyclodienes* are highly chlorinated cyclic hydrocarbons, such as chlordane, heptachlor, aldrin, and dieldrin, all of which are used as insecticides.
- *Chlorophenoxy acids* have growth-regulating influences on plants, and are used as herbicides against broad-leafed angiosperm weeds. The most important compound is 2,4-D, but others include 2,4,5-T, MCPA, and silvex.
- *Other organochlorines*, such as polychlorinated biphenyls (PCBs), dioxins, and furans, are also produced by human activities and released into the environment. These chlorinated hydrocarbons are not pesticides, but are mentioned here because their ecotoxicological properties are similar to those of pesticidal organochlorines: they are persistent in the environment, and their lipophilic nature means that they can bioconcentrate and food-web magnify.

Organic phosphorus pesticides are used mostly as insecticides, acaricides, and nematicides. These chemicals are not very persistent in the environment, but most are extremely toxic to arthropods, and also to nontarget fish, birds, and mammals. Parathion, fenitrothion, malathion, and phosphamidon are prominent examples of organophosphate insecticides. Glyphosate, a phosphonoalkyl compound, is an important herbicide (this chemical is not very toxic to animals).

Carbamate pesticides have a moderate persistence in the environment, but are highly toxic to arthropods, and in some cases to vertebrates. Aminocarb, carbaryl, and carbofuran are important carbamate insecticides.

Triazine pesticides are used mostly as herbicides, and sometimes as soil sterilants. Prominent examples are atrazine, simazine, and hexazinone.

Synthetic pyrethroids are analogues of natural pyrethrum, and are used mostly as insecticides and acaricides in agriculture. The pyrethroids are highly toxic to invertebrates and fish, but are of variable toxicity to mammals, and of low toxicity to birds. Important examples are cypermethrin, deltamethrin, permethrin, synthetic pyrethrum and pyrethrins, and tetramethrin.

Biological Pesticides

Biological pesticides are formulations of microbes that are pathogenic to specific pests, and consequently have a relatively narrow spectrum of activity in ecosystems. The best examples are insecticides based on the bacterium *Bacillus thuringiensis* (or *B.t.*), types of which are used against moths, flies, and beetles. Insecticides based on nuclear polyhedrosis virus (NPV) and insect hormones have also been developed.

PESTICIDE USE

The global use of pesticides is about 2.5–2.9 million t/y, a total that includes insecticides, herbicides, fungicides, preservatives, and disinfectants (Briggs, 1992; Pimentel *et al.*, 1992). During the early 1990s, trade in these chemicals had a value of about $20 billion/y (US dollars; Pimentel *et al.*, 1992).

About 61% of all pesticides used in the United States in 1989 were herbicides, while insecticides accounted for 21%, fungicides for 10%, and others for 7% (Briggs, 1992). Data for Canada are not available, but would be similar to those for the United States (Canada accounts for about one-ninth of the North American market for pesticides.) Total expenditures for pesticides in the United States were $7.6 billion in 1989. The following is a top-10 list of pesticides used in the United States, with quantities used per year (Briggs, 1992):

1. alachlor, a herbicide; 45 million kg
2. atrazine, a herbicide; 45 million kg
3. 2,4-D, a herbicide; 24 million kg
4. butylate, a herbicide; 20 million kg
5. metolachlor, a herbicide; 20 million kg
6. trifluralin, a herbicide; 14 million kg
7. cynazine, a herbicide; 9.2 million kg
8. malathion, an insecticide; 6.9 million kg
9. metribuzin, a herbicide; 6.0 million kg
10. carbaryl, an insecticide; 5.6 million kg

The most important uses of pesticides are in agriculture and forestry, around the home, and in human health and sanitation programs. We examine these uses in the following sections.

Pesticide Use for Human Health

Various species of insects and ticks are **vectors** that transmit pathogens between organisms of the same species, or from alternate hosts either to people or to domestic and wild animals. Often, alternate hosts are not affected by the pathogens, but the final host can be seriously afflicted. The most important vectored human diseases are the following:

- malaria, caused by the protozoan *Plasmodium* and spread among humans by mosquitoes, most commonly species of *Anopheles*
- yellow fever and encephalitis, caused by viruses and spread by mosquitoes
- sleeping sickness, caused by the protozoan *Trypanosoma* and spread by the tsetse fly *Glossina*
- plague or black death, caused by the bacterium *Pasteurella pestis* and transmitted by the rat flea *Xenopsylla cheops*
- typhoid fever, caused by the bacterium *Rickettsia prowazeki* and transmitted by the body louse *Pediculus humanus*
- schistosomiasis or bilharziasis, caused by the blood fluke *Schistosoma*, with freshwater snails as the alternate host

To varying degrees, the incidences of these debilitating maladies can be controlled by using pesticides against the invertebrate vectors or the alternate hosts. The abundance of mosquitoes, for example, can be reduced by spraying insecticide in their aquatic breeding habitat and by applying a persistent insecticide to the interior walls of buildings, on which these insects commonly rest. Similarly, people infested with body lice may receive a surface dusting with an insecticide — this was an early use of DDT.

Plague can be controlled by using rodenticides along with sanitation programs to reduce rat populations. Over the past half-century, pesticides have decreased the abundance of vectors and alternate hosts, and have spared hundreds of millions of humans from the deadly or debilitating effects of certain diseases, particularly in tropical countries. (This has been an important factor in reducing death rates and allowing rapid population growth, particularly in less-developed countries.)

In fact, the first important use of the insecticide DDT was in Naples, Italy, during the Second World War, to prevent a potentially deadly plague of typhus that could have decimated Allied troops and the civilian population. Because of the enormous success of this use of DDT, and its contribution to the victorious war effort, the British Prime Minister at the time, Winston Churchill, referred to the insecticide as "that miraculous DDT powder."

Malaria has long been an important disease in tropical countries. During the 1950s, about 5% of the world's population was infected with malaria. The use of insecticides to reduce the abundance of mosquitoes has achieved huge reductions in the incidence of malaria in some countries. For instance, during 1933–1935, India recorded about 100 million cases of malaria per year and 750 000 deaths. However, the incidence of malaria was reduced to 150 000 cases and 1500 deaths in 1966, mostly because of spraying with DDT and draining mosquito-breeding wetlands (McEwen and Stephenson, 1979). Similarly, 2.9 million cases of malaria occurred in Sri Lanka in 1934, and 2.8 million in 1946. DDT use reduced this incidence to only 17 cases in 1963 (Hayes, 1991). However, malaria has recently been resurging in some tropical countries, partly because mosquitos have developed a genetically based tolerance of insecticides. Many people are again being exposed to the malarial parasite, although the disease itself can today still be controlled by drugs that prevent *Plasmodium* from multiplying in the blood. (There are also signs that the *Plasmodium* is becoming resistant to these drugs.)

Pesticides and Agriculture

Modern agriculture is a highly technological activity. Machines, energy, fertilizers, pesticides, and high-yield crop varieties are used in intensive management systems to grow crops (see Chapter 14). The roles of pesticides are to control the abundance of:

- weeds that compete with crop plants
- arthropods, nematodes, and rodents that feed on the crops or the produce
- microbial diseases that can kill the crop or diminish its yield

Undeniably, these uses of pesticides are important factors in modern agriculture. Even with pesticide use, the damages caused by pests and diseases around the world are equivalent to about 24% of the potential crop of wheat, 46% of rice, 35% of corn or maize, 55% of sugar cane, 37% of grapes, and 28% of vegetables (McEwen and Stephenson, 1979). In North America alone, pests destroy about 37% of the potential production of food and fibre crops (Pimentel *et al.*, 1992).

Management practices in agriculture have intensified greatly, particularly during the 20th century. This change has resulted in substantial increases in crop productivity. The gains in agricultural yield have been largely achieved by the combined influences of:

1. increased use of fossil-fuelled mechanization
2. applications of fertilizers
3. cultivation of improved crop varieties grown in monocultural systems
4. use of pesticides

The revolutionary changes in agrotechnology are sometimes referred to as the "*green revolution*." Although agricultural yields have increased greatly, it must be recognized that the gains are highly subsidized by:

1. intense use of nonrenewable resources such as fossil fuels, metals, and soil (which is commonly lost through erosion)
2. depletions of potentially renewable resources such as soil fertility and tilth, and groundwater and surface waters required for irrigation
3. ecological damage associated with conversions of natural ecosystems into agricultural systems
4. ecotoxicological damage caused by the use of pesticides (see also Chapter 14)

In the United States, for example, the yield of corn was typically about 1.4 t/ha·y in 1933, but this increased to 4.2 t/ha·y in 1963 and 5.1–7.1 t/ha·y during 1978–1984. In Mexico, wheat yields increased from 0.75 t/ha·y in 1945 to 2.6 t/ha·y in 1964. The typical yield of rice in Japan increased from a prewar level of 1.8 t/ha·y to 4.0 t/ha·y in 1963, while in the U.S. rice yields have reached 4.9–5.5 t/ha·y (Hayes, 1991). Similar gains in agricultural yields have been realized in Canada (see Figures 14.1 and 14.2 in Chapter 14).

Of course, it remains to be seen whether the intensified agricultural systems will be sustainable over the long term. Several factors threaten sustainability: intensive agriculture's reliance on nonrenewable energy and material resources, the potential vulnerability of genetically narrow crop varieties to pests and diseases, and the ecological damage that is caused by intensive agriculture.

Almost all intensively managed agricultural systems depend to some degree on pest control. High-yield crop varieties are often vulnerable to infestations by insect pests, to diseases, and to competition from weeds. Pesticides are routinely prescribed to manage those problems. Moreover, monocultural systems result in reduced populations of natural predators and parasites, an ecological change that can exacerbate existing pest problems and allow new ones to develop. Some ecologists have described intensively managed agricultural systems as **pesticide "treadmills,"** because they rely on pesticides, often in increasing quantities, to deal with unanticipated pest problems that emerge as "surprises."

The use of pesticides in Canadian agriculture has increased greatly in recent decades. Herbicides were applied to 21.6 million ha of farmland in 1991, representing a 2.5-fold increase over 1971 (Table 14.11, Chapter 14). Insecticides and fungicides were applied to 2.8 million ha in 1991, a 3-fold increase over 1971. Overall, pesticide use in North American agriculture increased by about 10-fold between 1945 and 1989 (Pimentel *et al.*, 1992). Interestingly, during that same period, crop losses (to insects only) actually increased somewhat, from about 7% in 1941–1951 to 13% during 1951–1974 (Hayes, 1991). These trends, which seem to contradict each other, may be due to a combination of factors: the development of increased tolerance by some pests to previously effective pesticides; the emergence of new pests through accidental introductions; changes in predator-prey relationships caused by pesticide use; and the introduction of new crop varieties which happen to be vulnerable to pests.

Agricultural damages caused by arthropod pests vary greatly. Sometimes there is direct competition with humans for a food resource, as when insects defoliate crops in fields or attack stored grain and other foods. Such depredations can sometimes obliterate agricultural yields, as can happen during acute infestations by locusts such as the spur-throated grasshopper (*Melanoplus* spp.) in North America and the desert locust (*Schistocerca gregaria*) of Eurasia. More commonly, however, insects only reduce crop yields somewhat. The corn borer (*Ostrinia nubilalis*), for instance, caused an average loss of corn yield of 9% in North America between 1963 and 1973 (McEwen and Stephenson, 1979).

In some cases, however, even minor consumption of crop biomass by insects can make the produce unsalable. This is the situation with damage caused to apples by codling moth (*Carpocapsa pomonella*). "Wormy" apples containing larvae of this insect are not very saleable to consumers, yet as many as 90% of the apples in unsprayed orchards can be infested (McEwen and Stephenson, 1979). Even minor discolorations of fruit, such as those caused by apple scab and russeting of oranges (which do not significantly affect crop productivity or nutritional quality), are treated as unacceptable by many consumers, greatly reducing the economic value of the crops. Therefore, seemingly unimportant crop damage can be a critical economic consideration for farmers and food industries. Frequently, pesticides are used to combat economic problems, even if the use of pesticides to deal with such damage is unnecessary from the ecological perspective. As with some potentially life-saving drugs, pesticides are over-prescribed for some uses in modern societies.

Much pesticide use in agriculture is targeted against weeds. "Weed" plants interfere with the productivity of crop plants by competing for limited resources of light, water, and nutrients. (Of course, "weediness" is partly a matter of context — in other situations, such as in a flower garden, some weed species may have positive attributes.) In North America, weeds can reduce the yield of corn by 50–80%, and of wheat and barley by 25–50% (McEwen and Stephenson, 1979). Consequently, fields are sprayed with a herbicide to reduce the damaging influences of weeds. Of course, the weeds must be vulnerable to toxicity by the herbicide, while the crop plant must be tolerant.

A number of herbicides are toxic to broad-leaved weeds (i.e., to dicotyledonous plants) but not to corn, wheat, barley, or other crops in the grass family (these are monocotyledonous plants). Consequently, herbicides are widely used in grain agriculture in North America. For instance, more than 83% of the corn acreage is treated. Percentages are similar for other grains (McEwen and Stephenson, 1979; Freedman, 1995).

Some important diseases of agricultural plants can also be managed with pesticides. Sometimes, insecticides are sprayed to control arthropod vectors of microbial diseases. More commonly, fungicides are used to control pathogenic fungi such as late blight (*Phytophthora infestans*) of potato, apple scab (*Venturia inequalis*), powdery mildew (*Sphaerotheca pannosa*) of peach, and pythium (*Pythium* spp.) seedrot and damping-off of many crop species. Fungicides are also used to help prevent the spoilage of stored crops by fungi such as *Aspergillus flavus*, which can grow in stored legumes, grains, and nuts, producing deadly aflatoxins that make those foods poisonous to humans and livestock.

Pesticides in Forestry

Pesticide use in agriculture is much greater than in forestry. Nevertheless, pesticide use in forestry is important, because extensive areas of natural and seminatural ecosystems, supporting mainly indigenous elements of biodiversity, are sprayed. (In contrast, pesticide use in agriculture involves the more intense disturbance of agroecosystems dominated by nonnative species.)

The use of pesticides in forestry raises a great deal of controversy, sometimes more than when the same chemicals are used in agriculture. The controversy partly concerns damage that may be caused to the many native species of wildlife that are exposed to forestry sprays. In addition, spraying in forestry is mostly undertaken by government agencies and large companies, while pesticide use in agriculture commonly involves individual farmers working a family farm. Most people have greater empathy for individuals than for big government or big business, and this can influence their opinions about pesticide use.

Pesticides are used in forestry mainly to control epidemics of defoliating insects and to manage weeds in reforested areas and plantations. If uncontrolled, these pests can decrease tree productivity. The largest insecticide spraying campaigns in North American forestry have been undertaken against spruce budworm (*Choristoneura fumiferana*), particularly in New Brunswick, where a cumulative area of about 49 million ha was sprayed between 1952 and 1992 (we examine this as a case study later in this chapter). Other important spray programs include those against gypsy moth (*Lymantria dispar*), an introduced pest that defoliates many tree species and causes extensive damage in the eastern United States and southeastern Canada; hemlock looper (*Lambdina fiscellaria*); bark beetles (especially species of *Ambrosia* and *Ips*); and some other insect pests.

Herbicide use in forestry started to become common during the 1950s. Herbicides are used mainly to reduce competition from weeds, in order to increase the growth of conifer seedlings (a case study is presented later).

Pesticide Use around the Home and in Horticulture

Pesticides are also commonly used in and around homes. For instance, insecticides are routinely used to kill household insects such as cockroaches and flies, and baits containing rodenticide are used to poison rats and mice. In addition, large quantities of pesticides are used in horticulture. Herbicides are especially widely applied, mostly to achieve the monocultural, grassy-lawn aesthetic that is sought by many homeowners. To this end, herbicides are used to kill broad-leaved plants such as dandelion and plantain. The "weed" in common "weed-and-feed" lawn preparations refers to the herbicidal action of 2,4-D, dicamba, and mecoprop.

Some of the most intensive pesticide use occurs in the management of golf courses, particularly on the putting greens, where the quality of the lawn must be extremely consistent. Fungicides are used in especially large amounts, mostly to prevent turf-grass diseases caused by fungal pathogens. On a per-unit-area basis, the use of pesticides on golf-course putting greens is more intensive than most applications in agriculture.

Environmental Effects of Pesticide Use

Pesticide applications are intended to manage the impacts of a pest, or a community of pests, by reducing their abundance and ecological influence to below an economically or aesthetically acceptable threshold. This objective can sometimes be achieved selectively, avoiding nontarget damage. For example, rodenticides can be used judiciously to kill rats and mice around homes, while minimizing toxic exposures to nontarget mammals such as cats, dogs, and children (although never eliminating the risk).

More typically, however, pesticide use involves less-judicious, broadcast applications, usually by spraying. Crop-dusting aircraft or tractor-drawn sprayers are often used, resulting in the exposure of many nontarget species to the spray. The nontarget organisms may live on the sprayed site, or they may live off-site, suffering exposure through aerial or aquatic **drift** of the pesticide. Nontarget exposures include both direct contact with the sprayed pesticide and indirect exposure through the food web.

The ecotoxicological risks inherent in nontarget exposures to pesticides (and other chemicals) are influenced by a complex of variables, as we previously examined in Chapter 15. Several points should be considered when interpreting exposures of nontarget organisms, including people, to pesticides and other chemicals:

- All chemicals are potentially toxic.
- Not all exposures to potentially toxic chemicals result in poisoning (because organisms are to some degree tolerant to exposures to pesticides and other chemicals).
- Not only some pesticides, but also some naturally occurring chemicals, are extremely toxic to many organisms, including humans.
- Humans are subject to both involuntary and — as with cigarette smoke, prescription and recreational drugs, and automobile exhaust — voluntary exposures to certain toxic chemicals.

Pesticides vary enormously in their toxicity. Herbicides, for example, are extremely toxic to at least some types of plants, but are not necessarily poisonous to animals, which differ in many important physiological respects from plants. In contrast, many insecticides and rodenticides are toxic to a wide range of animals, and can cause nontarget poisoning of many species, including humans.

The acute toxicity of a chemical to animals is defined by its LD_{50}, or the dose required to kill one-half of a test population that is exposed through food, water, or air (see also Chapter 15). The oral LD_{50} for rats is a commonly used indicator of acute toxicity to mammals. Rats are widely used in toxicological research, and are similar to humans in many aspects of their physiology. Table 21.2 compares the acute toxicities of a wide range of pesticides and some other chemicals, using rat oral LD_{50}s (see also Table 15.3, Chapter 15). Note that some of the most poisonous chemicals listed are natural biochemicals (e.g., saxitoxin, a neurotoxin produced by marine algae, and the cause of paralytic shellfish poisoning in humans and other mammals). Others are chemicals to which some people expose themselves in their pursuit of pleasure (e.g., nicotine, the addicting alkaloid in tobacco).

By poisoning organisms, pesticides may cause changes to habitats. This can exert indirect ecotoxicological stresses on some species. For example, herbicides kill plants and thereby change the habitat of animals, perhaps depriving herbivores of their preferred foods. Similarly, broad-spectrum insecticides kill large numbers of arthropods, reducing the availability of food for birds and other animals. These and other indirect effects of pesticide use can result in significant ecological damage, and are additive to the direct, toxic effects of these chemicals.

In the remainder of this chapter, we will examine several case studies of particular uses of pesticides. These are useful in illustrating the broader principles and patterns of the ecological damage caused by these chemicals.

Environmental Effects of DDT and Related Organochlorines

The first case study involves DDT and related organochlorine insecticides, such as DDD, dieldrin, and aldrin. These chemicals were once widely used in Canada and most other industrialized countries. Although these chemicals were banned there in the early 1970s, they continue to be important insecticides in some less-developed nations.

DDT and its relatives are persistent in the environment. Consequently, there are still substantial residues in the ecosystems of Canada and other developed countries, even though these chemicals have not been used there for several decades. In part, this results from the continued use of these chemicals in some tropical countries, as some residues from that use are transported to high-latitude countries by global cycling. In addition, organochlorines are more persistent in cooler environments then in warmer ones. As a result, these and some noninsecticidal organochlorines (such as PCBs and dioxins) are still important pollutants in Canada.

DDT was first synthesized in 1874, although its insecticidal properties were not discovered until 1939. It was first used successfully during the Second World War,

TABLE 21.2 | ACUTE TOXICITY OF VARIOUS CHEMICALS TO RATS

The oral LD_{50}, stated here in milligrams of chemical per kilogram of body weight, is the amount of chemical required to kill 50% of a trial population, exposed through their food in a controlled laboratory test.

CHEMICAL	RAT ORAL LD_{50} (mg/kg)
TCDD (dioxin isomer)	0.01
tetrodotoxin (globefish toxin)	0.01
saxitoxin (paralytic shellfish neurotoxin)	0.3
aldicarb (insecticide)	0.8
TEPP (insecticide)	6.8
carbofuran (insecticide)	10
parathion (insecticide)	13
methylparathion (insecticide)	14
phosphamidon (insecticide)	24
strychnine (rodenticide)	30
deltamethrin (insecticide)	31
aminocarb (insecticide)	39
nicotine (alkaloid in tobacco)	50
methiocarb (molluscicide)	65
lindane (insecticide)	88
diazinon (insecticide)	108
paraquat (herbicide)	150
caffeine (alkaloid in coffee and tea)	200
DDT (insecticide)	200
fenitrothion (insecticide)	250
2,4-D (herbicide)	370
2,4,5-T (herbicide)	500
carbaryl (insecticide)	500
triclopyr (herbicide)	650
mirex (insecticide)	740
DDE (DDT, DDD metabolite)	880
tetramethrin (insecticide)	1 000
hexazinone (herbicide)	1 690
acetylsalicylic acid (aspirin)	1 700
atrazine (herbicide)	1 750
malathion (insecticide)	2 000
sodium hypochlorite (household bleach)	2 000
sodium bicarbonate (baking soda)	3 500
sodium chloride (table salt)	3 750
permethrin (insecticide)	3 800
DDD (insecticide)	4 000
glyphosate (herbicide)	5 600
picloram (herbicide)	8 200
captan (fungicide)	9 000
ethanol (drinking alcohol)	13 700
fosamine (herbicide)	24 000
sucrose (table sugar)	30 000

Source: Modified from Freedman (1995)

especially in programs to control body lice, mosquitoes, and other disease vectors. DDT was quickly recognized as an extremely effective insecticide, and immediately after the war became widely used in agriculture, forestry, and spray programs against malaria. The use of DDT peaked in 1970, when 175 million kg were manufactured globally.

At about that time, developed countries began to ban most applications of DDT. Its use was causing ecological damage, including the contamination of humans and their agricultural food web. Some researchers thought that this contamination could be resulting in human diseases. Unfortunately, DDT use has continued in other countries, especially in the tropics, mostly in programs against mosquito vectors of diseases.

Even in those tropical countries, the use of DDT and other organochlorine insecticides has been diminishing. This is partly because many pests have developed a genetically based *tolerance* of these chemicals (also sometimes known as *resistance*), decreasing the effectiveness of the pesticides. The development of tolerance is an evolutionary process. Exposure to a toxic substance selects for tolerant individuals within genetically variable populations (see also Chapter 15). Although tolerant individuals are normally rare in unsprayed populations, they can rapidly become dominant in sprayed habitats. They are not killed by the insecticide, so they survive to reproduce, passing on the genes for tolerance to their offspring. One review of pesticide tolerance found at least 447 species of insects and mites with populations tolerant to at least one insecticide, along with more than 100 examples of fungicide-resistant plant pathogens and 48 herbicide-tolerant weeds (NRC, 1986). Tolerance is especially common among flies (order Diptera), with 156 tolerant species. Of the 51 tolerant species of malaria-carrying *Anopheles* mosquito, 47 are tolerant to dieldrin, 34 to DDT, 10 to organophosphates, and 4 to carbamates.

Two physical and chemical properties of organochlorines have an important influence on their capacity for causing ecological damage. Firstly, chlorinated hydrocarbons are highly **persistent** in the environment, because they are not easily degraded by microorganisms, or by physical agents such as sunlight or heat. For example, DDT has a typical half-life in soil of about 3–10 y. DDE is the primary breakdown product of DDT, and it has a similar persistence.

In addition, DDT and related organochlorines are extremely insoluble in water, and so cannot be "diluted" into this abundant solvent. On the other hand, these chemicals are highly soluble in fats or lipids, which occur mostly in organisms. Therefore, DDT and related organochlorines have a powerful affinity for organisms, and they **bioconcentrate** in living things in strong preference to the nonliving environment. Moreover, organisms are efficient at assimilating any organochlorines present in their food. As a result, predators at the top of the food web develop the highest concentrations of organochlorine residues, particularly in their fatty tissues (this is known as **food-web magnification** or **food-web concentration**). Both bioconcentration and food-web magnification tend to be progressive with age, that is, the oldest individuals in a population are the most contaminated (see In Detail 18.1, page 310).

The bioconcentration and food-web magnification properties of DDT are illustrated in Figure 21.1. Note that concentrations of DDT are minute in air, water, and nonagricultural soil, compared with the much higher residues in organisms. Note also that concentrations in plants are lower than in herbivores, and that residues are highest at top of the food web, for instance, in predatory birds and humans.

Another characteristic of DDT and other organochlorines is their ubiquity — their residues are present in organisms throughout the biosphere. This extraordinarily widespread contamination with organochlorines occurs because these chemicals enter a global cycle, becoming widely dispersed in the bodies of migrating organisms and after entering the atmosphere by evaporation and in wind-eroded dusts. Residues of DDT are found even in organisms in Antarctica, a place far remote from areas where DDT was ever used. In one study, the concentration of "total DDT" (see In Detail 21.1) in the fat of skuas (a marine bird; *Catharacta maccormicki*) was 5 μg/g. Smaller residues (<0.44 μg/g; 1 μg/g = 1 ppm) were found in birds feeding lower in the oceanic food web, such as fulmar (*Fulmarus glacialoides*) and macaroni penguin (*Eudyptes chrysolophus*) (Norheim *et al.*, 1982).

Although organochlorine residues are ubiquitous in the biosphere, much higher concentrations are found in animals that live close to areas where these chemicals have been used, such as North America. Marine mammals feed at or near the top of their food web and are long-lived, and for these reasons can have extremely large residues of organochlorines. For example, harbour porpoises (*Phocoena phocoena*) in Atlantic Canada have had DDT residues as high as 520 ppm in their fat (Edwards, 1975).

Large residues of organochlorines also occur in some top-predator birds, especially raptors (that is, predators such as falcons, hawks, eagles, and owls). In North America, prior to the ban of DDT use, residues averaged 12 ppm (with a maximum of 356 ppm) in a sample of 69 bald eagles (*Haliaeetus leucocephalus*), up to 460 ppm among 11 western grebes (*Aechmophorus occidentalis*), and up to 131 ppm among 13 herring gulls (*Larus argentatus*) (Edwards, 1975).

Intense exposures to DDT and other organochlorines caused important ecological damage, including bird poisonings. During the 1950s and 1960s, bird kills resulted when DDT was sprayed in urban areas to kill beetle vectors of Dutch elm disease. This disease is caused by a fungal pathogen (*Ceratocyctis ulmi*) that was introduced to North America from Europe. The fungus is transported between trees by elm bark beetles, which can be controlled to some degree using insecticide. Spraying for this purpose was intensive, typically involving an application of 0.7–1.4 kg DDT per tree. As a result, birds feeding on invertebrates in sprayed sites were exposed to lethal doses of DDT. One study in New Hampshire found 117 dead birds of various species in a 6-ha spray area. It was estimated that 70% of the breeding robins (*Turdus migratorius*) had been killed (Wurster *et al.*, 1965). So much bird mortality occurred in sprayed neighbourhoods that bird song was markedly reduced — hence the title of Rachael

In Detail 21.1

DDT Residues

DDT and DDD, once widely used as insecticides, are found as residues in organisms throughout the biosphere, particularly in fatty tissues. However, DDT and DDD are metabolized by animals into DDE, a non-insecticidal organochlorine. Like DDT and DDD, DDE is a persistent and bioaccumulating chemical. Most accounts of DDT residues actually report "total DDT," that is, DDT + DDD + DDE. In most cases, DDE accounts for most of the total-DDT residue.

FIGURE 21.1 TYPICAL RESIDUES OF DDT IN THE ENVIRONMENT

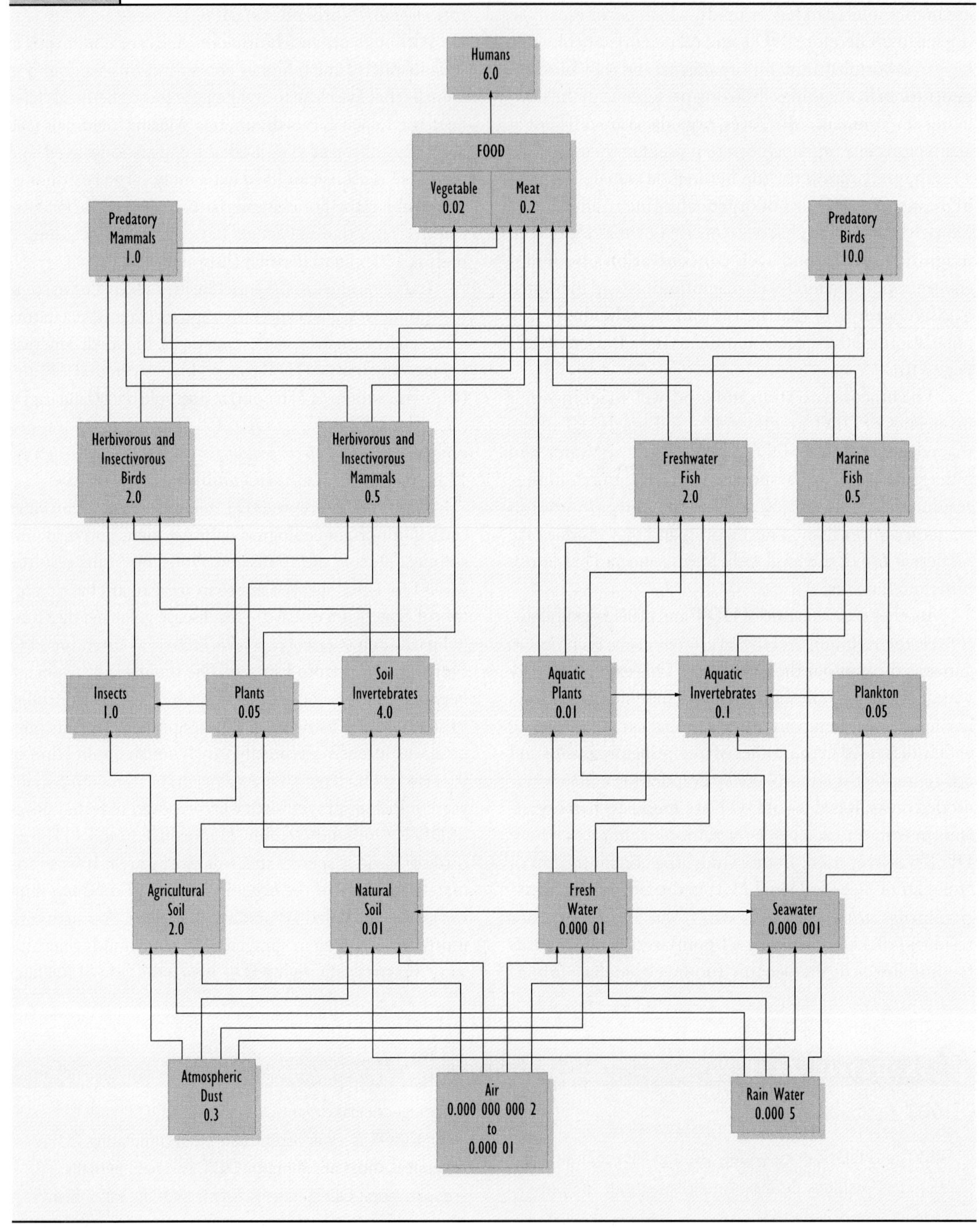

Data are in ppm.

Source: Edwards (1975)

Carson's (1962) book, *Silent Spring*, often considered a harbinger of the rapidly developing environmental movement in North America (In Detail 21.2).

In addition to the direct, acute poisoning caused by chlorinated hydrocarbons in sprayed areas, more insidious damage was also caused to birds and other wildlife over large regions. Many species experienced long-term chronic toxicity, often occurring well away from sprayed areas. It took years of population monitoring and ecotoxicological research before organochlorines were identified as the causes of this widespread damage. In fact, we can characterize the chronic poisoning of birds and other wildlife as an unanticipated ecological "surprise," occurring because scientists (and society) had no experience with the long-term effects of persistent, bioaccumulating organochlorines.

Species of raptorial birds were among the prominent victims of organochlorine insecticides. These birds are vulnerable because they feed at the top of their food web, and therefore accumulate high concentrations of

In Detail 21.2

Silent Spring

Rachael Carson, an American biologist, wrote many scientific articles and several books, the most famous of which, *Silent Spring*, was published in 1962. *Silent Spring* was aimed at a popular audience, and it is a lively and extremely controversial indictment of pesticide use as it was practised during the 1950s and early 1960s. The book pays particular attention to the use of DDT and other organochlorine insecticides. *Silent Spring* was written to warn the public at large about the known and potential dangers that these pesticides pose to wildlife, as well as to humans through contamination of their food. *Silent Spring* achieved that objective, and in fact was a literary bombshell that caused an eruption of public awareness about pesticide issues and about environmental concerns more generally.

Rachael Carson wrote about the acute and chronic effects of DDT and its organochlorine relatives on both wildlife and people. Although these insecticides are clearly useful in killing pests, Carson described how they were also causing extensive non-target mortality to non-pest arthropods, as well as to birds, mammals, and other wildlife. She also warned that humans were being widely contaminated by organochlorines, with significant residues being found, for example, in the milk of nursing mothers. She noted: "For the first time in the history of the world, every human being is now subjected to dangerous chemicals, from the moment of conception until death." Although little was known about the subject at the time, Carson warned that the chronic, low-level exposures of people to organochlorines were potentially dangerous.

A best-seller, *Silent Spring* stirred up an enormous controversy about the effects of anthropogenic chemicals in the environment. Pesticide-manufacturing companies mounted their own information and advertising programs. They attempted to discredit Carson by labelling her an irresponsible agitator and by claiming that she did not represent the views of most scientists. In fact, some of the technical details of Carson's analysis were later found to be incorrect, but this is not surprising considering the highly incomplete understanding at the time about pesticides and their environmental impacts. Nevertheless, the essential thesis of *Silent Spring* was that organochlorine insecticides can widely contaminate organisms and the environment, that they are persistent, and that they cause widespread damage. In large part these assertions have proven to be correct.

Unfortunately, Rachael Carson died an early death from cancer in 1964, just as the importance of *Silent Spring* and its message was becoming widely recognized. Today, Carson is well known as one of the most influential environmentalists in history; a pioneer who deserves much of the credit for the birth of the environmental movement during the mid-1960s. Like all environmentalists, Rachael Carson promoted an ethic of human responsibility for taking care of the biosphere and its species.

organochlorines. Breeding populations of various raptors suffered large declines. Severely affected species in North America included the peregrine falcon (*Falco peregrinus*), osprey (*Pandion haliaetus*), bald eagle (*Haliaeetus leucocephalus*), and golden eagle (*Aquila chrysaetos*).

In all cases, these species were exposed to a "cocktail" of organochlorines. This mixture included the insecticides DDT, DDD (both of which are metabolized to DDE in organisms), aldrin, dieldrin, and heptachlor, as well as PCBs, a group of noninsecticides with many industrial uses. Researchers have investigated the relative importance of these different organochlorines in causing the population declines of raptors. It appears that DDT was the more important toxin to birds in North America, while cyclodienes (particularly dieldrin) were more important in Britain (Moriarty, 1988; Cooper, 1991).

Damage caused to predatory birds was mainly associated with chronic effects on reproduction, rather than with toxicity to adults. Reproductive damage included the production of unusually thin eggshells that could break under the weight of an incubating parent, high death rates of embryos and nestlings, and abnormal adult reproductive behaviour. The number of chicks declined, resulting in rapid reductions in population sizes. The

CANADIAN FOCUS 21.3

ORGANOCHLORINES AND THE PEREGRINE FALCON

The most famous avian decline caused by organochlorines involved the peregrine falcon. Its decreasing populations were first noticed in the early 1950s in North America and western Europe (Peakall, 1990; Freedman, 1995). By 1970, the population of peregrines breeding in eastern North America (known as the *anatum* subspecies, that is, *Falco peregrinus anatum*) had stopped reproducing and was critically endangered. At the same time, populations of the *tundrius* subspecies of the arctic were declining rapidly. Only the *pealei* subspecies that breeds on the Pacific islands off British Columbia and Alaska had normal breeding success and stable populations.

Pealei falcons do not migrate. Moreover, they live in a region where pesticides are not used, and they feed mainly on nonmigratory seabirds. In contrast, *anatum* peregrines bred in a region where organochlorine pesticides were widely used, and fed on contaminated prey. The arctic *tundrius* peregrines breed in a remote wilderness where pesticides are not used, but they winter in Central and South America where their food is contaminated by organochlorines, as is some of their prey of migratory waterfowl and shorebirds in the Arctic. Studies by wildlife toxicologists demonstrated that populations of peregrine falcons with high residues of organochlorines were laying thin-shelled eggs and suffering other kinds of reproductive impairment. This damage was causing the populations of peregrines to collapse.

By 1975, the *anatum* peregrines had become extirpated in eastern North America, while the arctic *tundrius* subspecies had declined to only about 450 pairs from historical levels of 5000–8000 pairs.

Fortunately, many countries (including Canada, the United States, and most other industrialized nations) banned further uses of DDT and most other organochlorines in the early 1970s. The residues in the food of peregrines and other raptors have decreased, allowing the populations to stabilize or recover. By 1985, arctic populations of peregrines were stable or increasing, and small breeding populations had reestablished in more southern regions.

The recovery of peregrines has been enhanced by a program (funded in Canada by the Canadian Wildlife Service) in which peregrines are bred in captivity to provide young birds for release into the former range of the eastern *anatum* subspecies. Several thousand peregrines have been released in Canada and the United States, and many of these birds have survived and established breeding populations. Some of the peregrines have been released in cities, where tall buildings provide cliff-like nesting habitat. The abundant pigeons (or rock dove, *Columba livia*) and other urban birds are a suitable prey. Thanks to declining residues of organochlorine pesticides and PCBs in North America, resulting from legislated bans on the use of these chemicals, the peregrine falcon is on the way back.

case of the peregrine falcon is described in Canadian Focus 21.3.

Since the ban on most uses of DDT and other organochlorines in North America, their residues in wildlife have been declining. Some of the best data illustrating this decrease have been obtained by analyzing eggs of herring gulls (*Larus argentatus*) breeding on islands in the Great Lakes (Table 21.3). Although eggs from the various lakes differ in their organochlorine residues (depending on local sources of contamination), they all exhibited large decreases in DDE and PCB concentrations between 1971 to 1989. Well-documented decreases in residues of organochlorines have also occurred in eggs of double-crested cormorants (*Phalacrocorax auritus*) in other regions of Canada.

Modern Insecticides and Birds

The Case of Carbofuran

The most important replacements for DDT and related organochlorine insecticides have been organophosphate and carbamate chemicals. These poison insects and other arthropods by inhibiting a specific enzyme, acetylcholine esterase (AChE), which is critical in the transmission of nerve impulses. Vertebrates such as amphibians, fish, birds, and mammals are also highly sensitive to poisoning of their AChE enzyme system. In all of these animals, acute poisoning by organophosphate and carbamate insecticides can cause tremors, convulsions, and ultimately death.

Carbofuran, a carbamate insecticide, is used for many purposes in agriculture. Available as a liquid suspension, carbofuran is diluted in water and broadcast-sprayed against pests such as grasshoppers and leaf beetles. It is also available in a granular formulation, in which the insecticide coats the surface of grit particles and is sown along with the seeds to protect tender seedlings from insect damage. In Canada, the granular formulation has been commonly used in the sowing of canola (or oilseed rape) and corn.

Unfortunately, wildlife is exposed to toxic doses of carbofuran when it is used in either of these formulations. For example, if not all of the carbofuran granules are buried within the planting furrows, they can remain exposed on the ground surface (Mineau, 1993). In one method of seeding, commonly used when planting corn in Ontario,

TABLE 21.3 | CHANGES IN THE RESIDUES OF DDE AND PCBs IN BIRD EGGS IN CANADA

Eggs of herring gull and double-crested cormorant have shown decreasing residues since DDT, PCBs, and other persistent organochlorines were banned in North America in the early 1970s. Residues are measured in ppm.

(A) HERRING GULL EGGS

	BIG SISTER ISLAND, ON		GRANITE ISLAND, ON		MUGGS ISLAND, ON	
YEAR	DDE	PCBs	DDE	PCBs	DDE	PCBs
1971–1974	58	151	28	74	23	139
1975–1978	29	105	16	60	16	123
1979–1982	14	65	6	39	11	66
1983–1986	9	30	3	16	5	40
1987–1989	5	28	2	12	4	21

(B) DOUBLE-CRESTED CORMORANT EGGS

	STRAIT OF GEORGIA, BC		GREAT LAKES, ON		ST. LAWRENCE, QC	
YEAR	DDE	PCBs	DDE	PCBs	DDE	PCBs
1971–1974	4	8	16	16	6	8
1975–1978	–	–	5	6	2	8
1979–1982	1	4	2	10	2	7
1983–1986	0.5	2	–	–	2	6
1987–1989	0.5	2	2	3	1	4

Source: Modified from Bishop and Weseloh (1990) and Environment Canada (1993)

15–31% of the carbofuran granules remain exposed on the surface, equivalent to 515–1065 exposed granules per metre of furrow. Methods used to plant canola in western Canada often leave about 5% of the applied granules on the surface. These granules are avidly ingested by seed-eating birds, which require hard particles in that size range as "grit" for grinding hard seeds in their muscular gizzard. The carbofuran is extremely toxic — consumption of a mere 1–5 granules is enough to kill a small bird. Raptors and mammals may then be secondarily poisoned when they scavenge the birds' dead bodies.

In addition, if carbofuran-treated fields become flooded, as often happens during the spring and autumn, the surface water may contain large residues of the insecticide. This is particularly the case if the soil and water are acidic, because acidity greatly reduces the breakdown rate of carbofuran into less-toxic chemicals.

Of all the pesticides used recently in North American agriculture, carbofuran has probably causesd the most non-target mortality to birds and other wild animals. Even though there is no systematic program for reporting bird kills caused by pesticide use in Canada or the United States, an astonishingly large number of toxic incidents have been documented for carbofuran (Mineau, 1993). A few of these documented incidents are described below.

Granular carbofuran:

- More than 2000 Lapland longspurs (*Calcarius lapponicus*), a seed-eating finch, were killed in a freshly planted canola field in Saskatchewan in May, 1984.
- 500–1200 birds, mostly savannah sparrows (*Passerculus sandwichensis*), were killed in turnip and radish fields in British Columbia in September, 1986.

Flooded fields polluted by carbofuran:

- More than 1000 green-winged teal (*Anas carolinensis*) were killed within hours of landing in a flooded turnip field in British Columbia in the autumn of 1975.
- About 50 mallards (*Anas platyrhynchos*) and pintails (*A. acuta*) were poisoned in flooded fields in British Columbia in December, 1973.
- A total of 2450 dead widgeon (*Mareca americana*) were found one day after the spraying of an alfalfa field in California in March, 1974.

These examples represent only a small fraction of the known bird kills caused by routine uses of carbofuran in North American agriculture. There are also, of course, larger numbers of unreported incidents.

Because carbofuran usage in agriculture carries such a well-known risk of poisoning birds and other wildlife, ecologists and environmentalists have lobbied intensely to have its registration for agricultural uses withdrawn, or at least more tightly controlled. In 1993, the United States Environmental Protection Agency announced that the sand-based granular formulation of carbofuran was banned, except for some relatively minor uses and one major use (with rice crops) for which there was no suitable alternative. In 1996, Agriculture Canada, the federal agency that regulates pesticide use in Canada, announced a ban on most agricultural uses of carbofuran, which took effect in 1997. This ban prohibits most uses of carbofuran in liquid suspension, as well as the use of granular formulations containing 10% carbofuran. Unfortunately, all uses of granular formulations containing 5% carbofuran are still permitted, pending further studies and review.

The Case of Monocrotophos

Other modern insecticides are also poisoning Canadian birds. In 1996, it was discovered that agricultural use of the organophosphate monocrotophos against grasshoppers in Argentina was killing large numbers of Swainson's hawks (*Buteo swainsoni*). This raptor breeds in western Canada and the United States, and winters on the pampas of South America.

Populations of Swainson's hawks have been declining for several years. However, it was not until some birds were fitted with satellite transmitters and followed to Argentina that wildlife toxicologists discovered a probable cause of the decline — poorly regulated use of monocrotophos on the wintering grounds of the birds. Field studies in Argentina in 1996 discovered that as many as 20 000 Swainson's hawks had been killed in just one agricultural area (other regions were not surveyed), out of a total breeding population of only 400 000 birds (of which 40 000–100 000 breed in Canada).

Monocrotophos is extremely toxic to birds, although it is not very persistent in the environment. Because of the risk of ecological damage, monocrotophos has been banned in the United States, and was never registered for use in Canada. In Argentina, however, the insecticide may legally be used. Swainson's hawks are exposed to lethal doses of monocrotophos when they flutter behind spray tractors to feed on grasshoppers as they are flushed by the machinery, and also when they later feed on insecticide-contaminated prey.

Pest Problems in Forestry: Spruce Budworm and Weeds

Pesticides are used more extensively in agriculture than in forestry. However, forestry case studies better illustrate many of the ecological effects of pesticide use because forests (and clear-cuts) support mainly native species and natural or seminatural ecosystems. In contrast, agricultural ecosystems are typically dominated by nonnative species and are intensively managed, making them less amenable to the examination of ecological effects due to pesticides.

Spraying against Spruce Budworm

The largest insecticides spray programs in North American forestry have been mounted against spruce budworm (*Choristoneura fumiferana*), particularly in New Brunswick. Spruce budworm is a moth whose larvae are an important pest of fir- and spruce-dominated forests. Infestations can affect forests over tens of millions of hectares, and trees are killed after a number of years of defoliation. Mature stands dominated by balsam fir (*Abies balsamea*) are particularly vulnerable to spruce budworm damage. Spruces are also a preferred food, especially white spruce (*Picea glauca*). Red spruce (*P. rubens*) and black spruce (*P. mariana*) are less apt to suffer lethal damage.

Spruce budworm is an indigenous forest insect, which means it is always present in small populations in its fir-spruce habitat. However, it occasionally irrupts to attain an enormous abundance, and this is when it becomes a pest. Under normal conditions, only about five larvae occur on each conifer tree. This can increase to perhaps 2000 per tree at the beginning of an irruption, and more than 20 000 per tree during the peak of an outbreak. A local outbreak is typically sustained for 6 to 10 years, and then collapses. Studies in Quebec have demonstrated that historical outbreaks have occurred at an average interval of about 35 years (Blais, 1985). Outbreaks of spruce budworm are typically synchronous (occurring at the same time) over extensive areas of susceptible fir-spruce forest, although there are great variations in the abundance of budworm among the stands in different areas. The reasons for the irruptions are not well known, but likely involve several years of warm, dry weather in the springtime, which favours larval survival.

Forest Damage

It appears that the extent of the damage caused by spruce budworm may have increased during the three outbreaks of the 20th century. The outbreak that began in 1910 affected about 10 million ha, one that started in 1940 involved 25 million ha, and one beginning in 1970 affected more than 55 million ha (Figures 21.2 and 21.3). The enlarging areas of budworm infestations may be related to an increase in the extent of vulnerable fir-spruce forest. This may have been due to:

- forestry practices such as clear-cutting
- protection of forests from wildfire
- regeneration of conifer stands on abandoned farmlands
- spraying of infested stands with insecticide — a practice which may help maintain the habitat in a condition suitable for budworm

Enormous damage has been caused by budworm to economically important forests. During the most recent outbreak (1971–1984), tree mortality was equivalent to more than 38 million m^3/y (this is m^3 of saleable timber, which is useful for manufacturing into pulp or lumber). During the peak of the infestation, substantial tree mortality occurred over about 26.5 million ha in eastern Canada (Ostaff, 1985) (see Figure 21.3).

The rapid development of budworm infestations can be illustrated by the case of Cape Breton Island, Nova Scotia (Ostaff and MacLean, 1989). No defoliation caused by budworm was observed in 1973, but in 1974 moderate-to-severe defoliation occurred over 165 000 ha. This increased to 486 000 ha in 1975, and 1.22 million ha in 1976, when virtually all of the vulnerable fir-spruce forest was infested.

Irruptions of budworm can last for ten or more years, with the damage to trees increasing over time. During the first two years of severe defoliation of fir-dominated stands on Cape Breton Island, 4% of the balsam fir trees died. The cumulative mortality increased to 9% after 4 y of heavy defoliation, 37% after 6 y, 48% after 8 y, 75% after 10 y, and 95% after 12 y (Ostaff and MacLean, 1989). Across eastern Canada, tree mortality averaged 85% in mature fir-dominated stands, 42% in immature fir stands, and 36% in mature spruce stands (MacLean, 1990).

FIGURE 21.2 | FOREST AREA IN CANADA DEFOLIATED BY SPRUCE BUDWORM IN THE TWENTIETH CENTURY

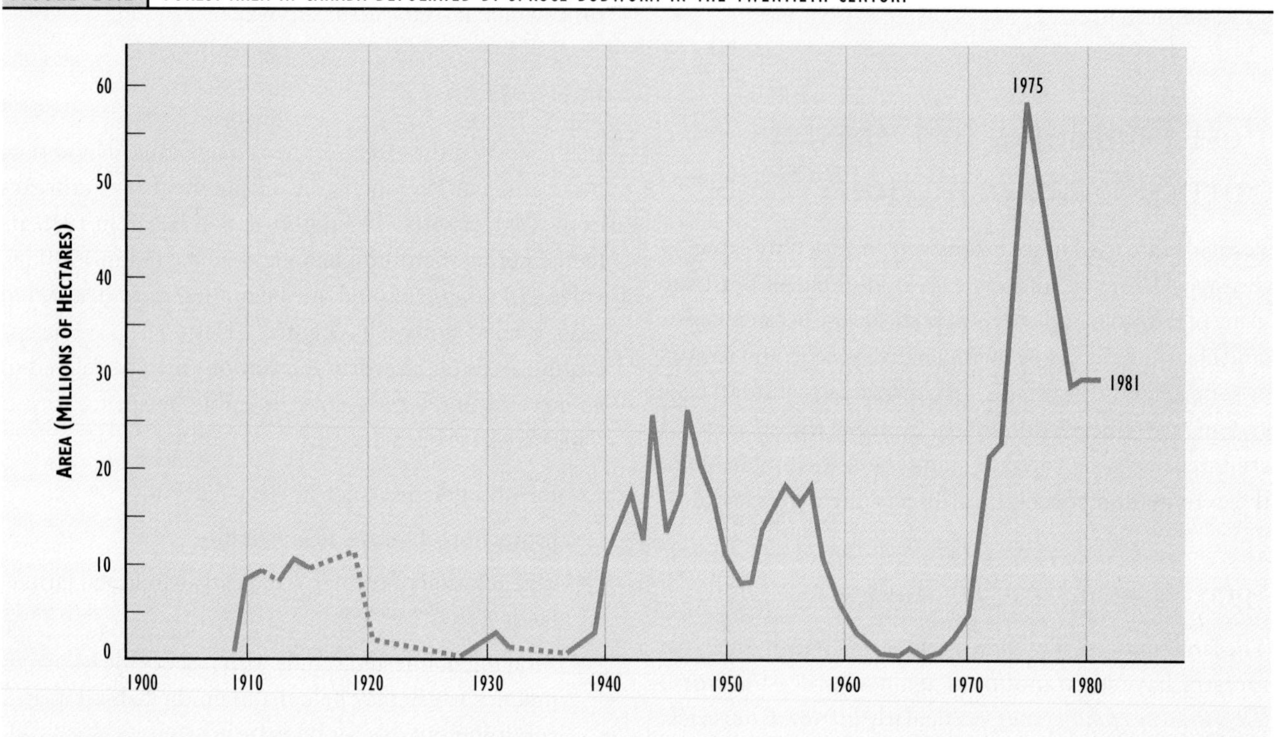

The areas correspond to stands suffering severe or moderate intensities of defoliation.

Source: Modified from Kettela (1983)

FIGURE 21.3 | AREAS OF MORTALITY CAUSED BY SPRUCE BUDWORM

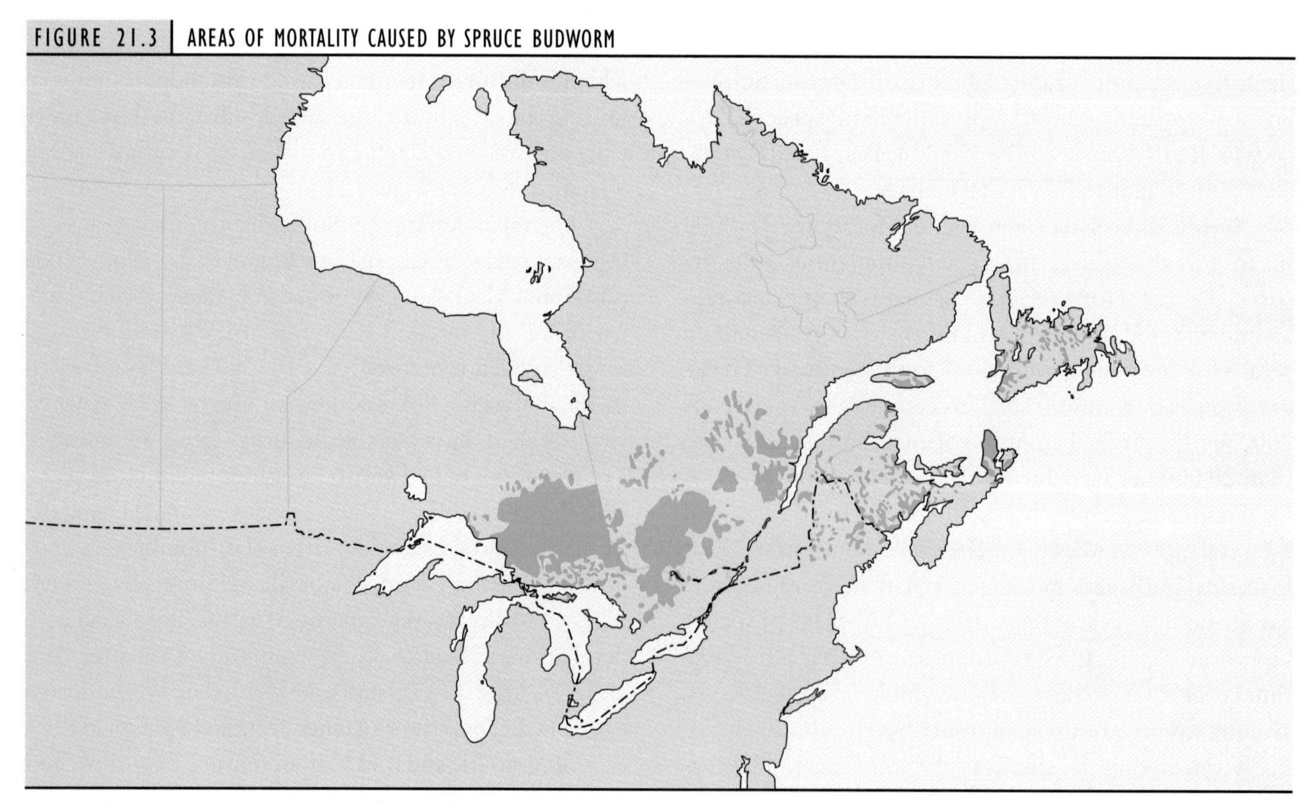

Areas of forest suffering mortality of balsam fir during the most recent irruption of spruce budworm.

Source: Modified from Kettela (1983)

An area of intensive forest damage caused by defoliation by spruce budworm on Cape Breton Island, Nova Scotia. The living trees fringing the bog are black spruce, which is relatively resistant to this insect. The extensive area of dead trees was dominated by balsam fir, a vulnerable species. This photo was taken about 8 years after the collapse of the budworm epidemic.

Source: B. Freedman

Large, mature fir and spruce trees are much more vulnerable to budworm than are smaller immature trees, which commonly survive the outbreaks. Consequently, the understorey of badly damaged stands typically contains a dense population of small fir and spruce. Known as "advance regeneration," this is important in reestablishing the next fir-spruce forest after the infestation collapses. On Cape Breton Island, for instance, severely damaged stands typically had an advance regeneration of 45 000 small balsam fir plus 3000 spruce per ha, almost all of which survived the infestation (MacLean, 1988). After the mature trees die and the canopy opens up, the small conifers grow rapidly and establish another fir-spruce forest, which becomes vulnerable decades later to developing a budworm irruption.

These observations suggest that over the long term, the budworm/forest interaction can be viewed as an ecologically stable, cyclic succession. The natural cycle of disturbance and recovery has probably been continuing for thousands of years, although it is possible that human influences have increased its scale during the present century.

Of course, spruce budworm causes great economic instability in the forest products industry, which is in competition with this insect for the fir-spruce resource. The periodic irruptions of budworm severely damage the fir-spruce forest, making it difficult for humans to plan their own orderly, long-term harvesting and management of the timber resource. One way of dealing with this resource so as to limit defoliation and prevent some tree mortality.

A ground-level view of a stand damaged by spruce budworm on Cape Breton Island. Although the mature balsam fir trees have been killed, dense regeneration of young fir is occurring in the understorey. After about 40–50 years, another mature forest will have regenerated, ready for harvesting by spruce budworm or by humans.

Source: B. Freedman

The objective of spraying is not to eradicate the budworm, but to decrease the damages it causes, and thereby maintain the forest resource and its dependent industries.

Insecticide Spraying

After the Second World War, DDT was used against spruce budworm. In 1953 alone, 804 000 ha of forest were sprayed with the insecticide in Quebec and New Brunswick. By 1968, when further use of DDT for this purpose was banned in Canada, a total of about 15 million ha had been sprayed at least once (Ennis and Caldwell, 1991). In New Brunswick alone, 5.75 million kg of DDT were sprayed onto budworm-infested stands (Armstrong, 1985).

After the use of DDT against budworm was banned in 1968, other insecticides were used. DDT was replaced by fenitrothion and phosphamidon, both organophosphate insecticides, and aminocarb, a carbamate. Of these, fenitrothion was used most widely. By 1985, phosphamidon had been sprayed over 8.1 million ha, aminocarb on 19 million ha, and fenitrothion on 64 million ha (these are the cumulative sums of annual sprayed areas — most stands were treated a number of times; Ennis and Caldwell, 1991). New Brunswick has always had the biggest spray programs for "protection" of the forest resource against budworm — up to 1985, a cumulative total of 69 million ha was treated in that province, while 37 million ha were sprayed in Quebec, 10 million ha in Maine, and 1.7 million ha in Newfoundland. The most extensive spraying in New Brunswick was in 1976, when 4.2 million ha were treated. There has been less spraying since then, declining to less than one million ha after 1985, to less than 0.5 million ha after 1990, and to zero after 1993, due to the collapse of the budworm outbreak.

During the late 1980s, a biological insecticide based on the bacterium *Bacillus thuringiensis* var. *kurstaki* (abbreviated as *B.t.*) began to displace most of the synthetic organic pesticides in budworm spray programs. *B.t.* is toxic to most lepidopterans (i.e., moths and butterflies) and to a few other insects, including blackflies and mosquitoes. Otherwise, this insecticide causes little nontarget damage. Initially, budworm control using *B.t.* was quite variable in efficacy (i.e., in effectiveness), and the costs were relatively high in comparison with insecticides such as fenitrothion. More recently, however, the technology for spraying *B.t.* has been improved. This, along with concerns about ecological damage caused by fenitrothion, has resulted in *B.t.* becoming the insecticide of choice in spray programs against budworm.

Nontarget Damage

The synthetic organic insecticides used against spruce budworm caused a great deal of nontarget mortality. Typical spray rates and toxicities for the most important insecticides used against spruce budworm are summarized in

TABLE 21.4 | ACUTE TOXICITY OF INSECTICIDES USED AGAINST SPRUCE BUDWORM

Typical application rates for forestry purposes are stated in kilograms per hectare. Acute toxicity to spruce budworm larvae and selected vertebrate species was determined under controlled laboratory conditions. The units for toxicity to budworm are in micrograms per centimetre^2 of body surface; for the trout, data are ppm in water; and for pheasant and rat, data are ppm in food. 96-h LC_{50} is the concentration in water that killed 50% of the population of trout after a 96-hour exposure.

INSECTICIDE	TYPICAL APPLICATION RATE (kg/ha)	CONTACT TOXICITY TO BUDWORM (LD_{50}; $\mu g/cm^2$)	RAINBOW TROUT 96-h LC_{50} (ppm)	PHEASANT ORAL LD_{50} (ppm)	RAT ORAL LD_{50} (ppm)
DDT	0.3–2.2	1.3	0.009	1334	87–500
Phosphamidon	0.3	0.39	7.8	4.2	15–33
Fenitrothion	0.21	0.31	2.4	56	250–600
Aminocarb	0.07	0.04	13.5	42	30
Mexacarbate	0.07	0.04	12.0	4.6	15–63

Source: Freedman (1995)

Table 21.4. Among these insecticides, DDT is the least toxic to the budworm itself, and is the most toxic to salmonid fish. Perhaps the most important reason for the 1968 ban on using DDT against spruce budworm was the mortality caused to economically important sportfish, particularly to Atlantic salmon (*Salmo salar*) and brook trout (*Salvelinus fontinalis*).

The insecticides that replaced DDT are more toxic to budworm, and can therefore be sprayed in smaller quantities while achieving the same degree of pest control. Although these insecticides are also less toxic to fish than is DDT, they can be very poisonous to other animals. Fenitrothion and aminocarb, for example, are extremely toxic to all arthropods. Spraying forests with these insecticides resulted in enormous kills of nontarget insects and spiders, including many predators of budworm. One study estimated that a typical fenitrothion spray killed as many as 7.5 million individuals of hundreds of species of arthropods per ha, although more than 90% of the killed biomass was budworm (Varty, 1975). Overall, a typical spray with fenitrothion caused a short-term decrease in arthropod biomass of 35%, and a decrease in total numbers of 50% (Agriculture Canada, 1993). Studies showed, however, that the nontarget damage to arthropods was a temporary phenomenon, and that long-term decreases in abundance were not detectable (Varty, 1975; Millikin, 1990). The postspray recovery was due mainly to recolonization from nonsprayed forest, as well as to recovery by arthropods that survived the spraying.

Bird populations are unusually abundant in infested forests because spruce budworm is an abundant and nutritious food resource for insect-eating animals. In fact, some species of birds are uncommon except in budworm-infested forests. In one study, for example, breeding populations of bay-breasted warbler increased from only 0.25 pairs/ha in noninfested forests to 30 pairs/ha during a budworm outbreak, while Tennessee warbler increased from 0 pairs/ha to 12.5 pairs/ha (Morris *et al.*, 1958).

During outbreaks of spruce budworm, almost all species of birds rely heavily on budworm larvae as food for raising nestlings. One study estimated that birds eat about 89 000 budworm larvae and pupae per ha in infested stands, compared with only 6000/ha in stands without an irruption (Crawford *et al.*, 1983). In spite of their enthusiastic efforts, bird predation has no substantial effect on the abundance of budworm during an outbreak — birds manage to consume only about 2% or less of the budworm larvae in outbreak stands. In essence, insectivorous birds are satiated by the superabundant budworm resource, and are consequently incapable of controlling the huge populations. Predation by birds may, however, be more important in reducing less-abundant budworm populations, and perhaps in helping to lengthen the intervals between serious pest outbreaks.

Because birds are so abundant in budworm-infested forests, they are exposed to insecticide sprays. Although some of the insecticides used against budworm are extremely toxic to birds, for several reasons it has proved difficult to document actual damage to avian populations. First, it is extremely difficult to find dead or dying birds in forest habitats because they occur in a small density and are quickly scavenged by other animals. Even with all the insecticide spraying in New Brunswick between 1965 and 1987, the Canadian Wildlife Service has records of only 125 dead birds (Busby *et al.*, 1989) — this is a gross underestimate of the actual mortality from spraying.

In addition, it is difficult to detect population-level changes resulting from mortality of breeding birds in forests. A census of forest birds is taken by mapping the locations from which male birds sing; this information is used to determine the boundaries of their territories. The song-censuses are conducted in the springtime over a period of 4–6 weeks, during which bird populations are quite variable. Migratory species are returning from their wintering grounds, and the different species arrive at different times. Moreover, if an individual bird happens to be killed by an insecticide spray, it is usually quickly replaced from a surplus of nonbreeding individuals that wander extensively, searching for breeding habitat not occupied by another of their species. Because of the temporal variations in bird abundance and the rapid replacement of killed individuals, it is difficult to document mortality or population changes caused by insecticide spraying.

Phosphamidon is extremely poisonous to birds (Table 21.4), and budworm spraying may have caused severe mortality to some species. One study suggested that as many as 376 000 ruby-crowned kinglets were killed in New Brunswick during the 1975 spray season, mostly by phosphamidon poisoning (Pearce and Peakall, 1977). Because they forage high in the canopy, ruby-crowned kinglets are relatively vulnerable to insecticide exposure during an aerial spray. The damage caused to birds was the most important reason why phosphamidon was banned for use in budworm spray programs after 1975.

Fenitrothion is less poisonous to birds, but it has a small margin of toxicological safety during operational sprays. While a normal application appears to cause little avian mortality, exposure to a double spray, as commonly occurs with overlapping spray swaths, can be lethal. Studies of white-throated sparrows in New Brunswick found much greater mortality and behavioural impairment

TABLE 21.5 | EFFECTS OF FOREST SPRAYING ON BIRD POPULATIONS

Effects of forest spraying with fenitrothion on populations of breeding birds in the Gaspé region of Quebec. The operational spraying procedure is to treat stands twice, about one week apart. In this study, birds were censused for 5 days before spraying, then for 7 days after the initial spray on May 21, 1976, and again for 5 days after the second spray on May 30. The sprayed stand ("Spr") was treated with fenitrothion at 0.56 kilograms per hectare, while the unsprayed reference stand ("Ref") was used to monitor population changes unrelated to spraying. Bird density is expressed as numbers/10 hectares. Only prominent species are listed.

	PRE-SPRAY		FIRST SPRAY		SECOND SPRAY	
SPECIES	**SPR**	**REF**	**SPR**	**REF**	**SPR**	**REF**
Total Bird Abundance	46.1	39.1	52.1	57.7	68.8	75.8
Number of Species	23	23	41	35	41	39
Boreal chickadee (*Parus hudsonicus*)	2.5	4.3	1.5	1.5	1.4	1.1
Winter wren (*Troglodytes troglodytes*)	1.0	3.9	2.3	4.6	1.7	3.0
American robin (*Turdus migratorius*)	3.5	3.6	4.8	3.3	5.3	1.8
Ruby-crowned kinglet (*Regulus calendula*)	8.6	4.5	5.9	3.8	4.4	6.4
Tennessee warbler (*Vermivora peregrina*)	0.0	0.0	0.3	1.5	2.3	5.8
Magnolia warbler (*Dendroica magnolia*)	0.0	0.0	0.1	1.5	0.8	3.1
Cape May warbler (*Dendroica tigrina*)	0.0	0.0	2.0	3.1	4.3	4.0
Yellow-rumped warbler (*Dendroica coronata*)	0.5	0.3	7.7	6.3	10.8	8.2
Bay-breasted warbler (*Dendroica castanea*)	0.0	0.0	0.3	0.3	2.5	4.5
Dark-eyed junco (*Junco hyemalis*)	9.6	4.1	4.2	1.8	4.5	0.5
White-throated sparrow (*Zonotrichia albicollis*)	8.9	9.6	9.7	10.5	10.4	8.4
Fox sparrow (*Paserella iliaca*)	2.1	3.8	3.3	4.2	2.5	3.1

Source: Modified from Kingsbury and McLeod (1981)

after an experimental double application of fenitrothion, compared with the normal spray rate (Busby *et al.*, 1989).

Table 21.5 shows the effects of a fenitrothion spray on the abundance of breeding birds in a forest in Quebec. When interpreting these census data, the trends in the sprayed habitat should be compared with those of the non-sprayed (or reference) forest. The comparison is necessary because the prespray census was conducted in mid-May, when many migratory birds (for example, most warbler species) had not yet returned to their breeding habitat. Consequently, the total avian abundance in the "prespray" census was much smaller than occurs later in the breeding season. If these data are considered in this relative sense, they suggest that the fenitrothion spraying had no obvious effects on the abundance or species composition of the bird community. It is likely, however, that birds were poisoned by fenitrothion during the spray, and that the damage was not reflected in the population census for the reasons discussed earlier.

It should also be recognized that budworm outbreaks cause severe damage to forests, with substantial effects on the habitat of wildlife. Table 21.6 illustrates the effects of budworm-induced habitat changes on bird populations. Defoliation by budworm caused relatively little damage to Stand A. The decreased populations of some birds in this stand (e.g., solitary vireo, many warbler species) were caused by decreased food availability, because the outbreak had collapsed and budworm is a critical food resource. In contrast, Stand B was intensely damaged by defoliation. This resulted in great habitat changes, with many dead trees (called "snags") and a lush growth of understorey plants, including a vigorous regeneration of small fir trees. In this stand, the decreased abundance of some bird species (e.g., solitary vireo, Tennessee warbler, black-throated green warbler, blackburnian warbler, and bay-breasted warbler) resulted from the changes in vegetation as well as from reduced availability of budworm larvae as food. Note in addition, that some birds (e.g., least flycatcher,

TABLE 21.6 | ABUNDANCE OF SELECTED BIRD SPECIES DURING AND AFTER A SPRUCE BUDWORM OUTBREAK

Data from a study in New Brunswick. Stand A was monitored for eight years during an infestation (1952–1959), and then for six post-outbreak years (to 1965). Stand A suffered little damage to its trees. (The average age of balsam fir trees was 120 y in both sampling periods.) Stand B was censused for five infestation years (1955–1959) and for five post-outbreak years (to 1964). Stand B suffered intense damage to its trees. (The balsam fir averaged older than 80 y when the infestation began, but less than 10 y old afterward, because so many mature trees had died.) The bird data are in numbers/40 hectares; pr = present. Only prominent species are listed.

	STAND A		STAND B	
SPECIES	DURING OUTBREAK	POST-OUTBREAK	DURING OUTBREAK	POST-OUTBREAK
Total Bird Abundance	192	95	201	84
Yellow-bellied flycatcher (*Empidonax flavifrons*)	4.6	pr	5.1	0.0
Least flycatcher (*Empidonax minimus*)	3.9	2.8	2.1	15.9
Winter wren (*Troglodytes troglodytes*)	5.9	8.2	2.2	3.3
Swainson's thrush (*Catharus ustulatus*)	15.1	11.9	14.0	8.7
Golden-crowned kinglet (*Regulus satrapa*)	8.9	3.0	2.3	0.0
Ruby-crowned kinglet (*Regulus calendula*)	pr	10.1	1.1	2.7
Solitary vireo (*Vireo solitarius*)	8.3	pr	7.7	0.0
Tennessee warbler (*Vermivora peregrina*)	9.9	0.0	25.7	0.0
Magnolia warbler (*Dendroica magnolia*)	22.0	7.9	1.3	14.9
Yellow-rumped warbler (*Dendroica coronata*)	7.1	1.6	5.1	2.1
Black-throated green warbler (*Dendroica virens*)	11.9	4.2	11.8	4.2
Blackburnian warbler (*Dendroica fusca*)	17.1	5.3	19.5	1.5
Bay-breasted warbler (*Dendroica castanea*)	55.8	16.4	78.2	2.0
Dark-eyed junco (*Junco hyemalis*)	5.9	3.5	8.4	4.1
White-throated sparrow (*Zonotrichia albicollis*)	9.3	5.0	11.4	19.5

Source: Modified from Gage and Miller (1978)

magnolia warbler, and white-throated sparrow) do well in recently disturbed stands, benefitting from the habitat associated with budworm damage.

Overall, it appears that the demonstrated ecological effects of post-DDT spray programs against spruce budworm have been relatively short-term in duration and moderate in intensity (with the exception of the effects of phosphamidon on birds). Residues of such chemicals as fenitrothion and aminocarb are not long lived, and no accumulation occurs within the food web. Although substantial mortality has been caused to many nontarget species, long-term decreases in their populations have not been documented (bearing in mind such effects are difficult to demonstrate, particularly at larger spatial scales). For example, although many individual birds have undoubtedly been poisoned by insecticide toxicity, measurable damages to their breeding populations have not been demonstrated.

Spray Policies

Spraying insecticides on forests infested by spruce budworm has various economic benefits, associated with protection of conifer forest, an important natural resource. As a result, decision-makers and regulators have considered spray programs necessary. Many ecologists, however, come to different conclusions about the benefits and costs of spraying, because they value ecological costs more highly than do resource managers and regulators. Ecologists and environmental activists are not, however, the people who make the decisions about undertaking insecticide spray programs to manage populations of budworm or other pests.

Since 1986, fenitrothion has been the only synthetic insecticide used against budworm. In 1993, a **risk assessment** of this use (Pauli et al., 1993) concluded that many of its ecotoxicological damages were significant:

> *The weight of evidence accumulated with respect to the identified and potential negative impacts caused by the forestry use of fenitrothion on non-target fauna,... and their potential ecological implications,*
> *supports the conclusion that the large-scale spraying of fenitrothion for forest pest control, as currently practised operationally, is environmentally unacceptable.*

Partly because of the strong conclusions of this risk assessment, the registration of fenitrothion for use in budworm spray programs in Canada was withdrawn in 1995. This action has left *B.t.*, a bacterial insecticide that causes little nontarget damage, as the major insecticide available for spraying against spruce budworm. (Tebufenozide, a synthetic insecticide that works by interfering with larval moulting, has also received a temporary registration for use against budworm.)

Possible Alternatives to Spraying Insecticides

In addition to the ecotoxicity caused by insecticides, spray programs against budworm have drawbacks from a forest-management perspective. Within limits, treating infested stands with insecticide maintains fir-spruce forests "alive and green," and therefore available as a resource for the economically important forest industry. However, spraying also maintains good habitat for budworm, and consequently may prolong its population outbreaks. Agencies that spray insecticides can become "locked" into this pest-management strategy, and must continue to spray if the forest resource is to be maintained in an economically viable condition. In the absence of any alternatives, spraying is perceived to be the best available short-term strategy. Clearly, however, annual spray programs over extensive areas are not desirable.

One alternative to spraying insecticides may be the use of management practices aimed at reducing the vulnerability of stands to budworm infestation. For example, relatively tolerant species such as black spruce might be planted extensively. Alternatively, the landscape could be structured so that stands of mature but vulnerable tree species are kept small. If the landscape mosaic included less-vulnerable stands that were harvested by industry at about the rate at which the trees matured, a smaller area would be vulnerable to budworm infestation. It must be recognized, however, that such actions would represent a huge intensification of forest management, and would cause enormous changes in the ecological character of the landscape. Little is known about the long-term economic viability or ecological consequences of such changes in forest-management strategies.

Further research into alternative methods of budworm control may yield novel methods that are effective. Some promise has been demonstrated by the release of large numbers of tiny wasps that are parasites of budworm, by the use of synthetic hormones that disrupt mating, and by several other innovative biotechnologies. So far, however, none of these have proven sufficiently effective for use in operational programs to reduce epidemic populations of budworm.

If regulators decide that the damage caused by spruce budworm must be controlled, it is highly desirable to utilize the least damaging, but still effective, options. At present, effective control appears to require insecticides, with the most desirable of the available insecticides being *B.t.* In fact, since the Canadian registration of fenitrothion for budworm spraying was withdrawn in 1995, *B.t.* is the only insecticide available for this purpose. If *B.t.* proves to be sufficiently effective, the use of synthetic organic insecticides in budworm spray programs, with their attendant ecotoxicological damage, will be history.

A four-year-old clear-cut of a coniferous forest in Nova Scotia. Virtually all of the vegetation is considered (by foresters) to be noncommercial "weeds," which compete with desired conifer seedlings for space, water, and nutrients. The objective of a silvicultural herbicide treatment is to reduce the abundance of weeds, allowing the conifers to grow more quickly.
Source: B. Freedman

Use of Herbicides in Forestry

The most common use for herbicides in Canadian forestry is to keep weeds from competing with young conifers. This use allows economically desired conifers to grow more rapidly so that harvests can be made more frequently (Newton, and Knight, 1981; Freedman, 1995). Minor uses of herbicides in forestry are to prepare sites for planting, to increase visibility along roads by reducing the amounts of shrubby vegetation, and to reduce the habitat available for animals, such as rabbits and hares, that impede conifer regeneration.

The forestry use of herbicides accounts for a small fraction of the total use of these chemicals, less than about 5% in Canada. However, in forestry, herbicide use affects the habitat of a diversity of indigenous species, whereas this is much less the case in agricultural and horticultural usage. In 1993, about 139 000 ha of forest land were treated with herbicides in Canada (Canadian Council of Forest Ministers, 1995).

As in agriculture and horticulture, there are alternatives to the use of herbicides in forestry. These options include crushing competing vegetation with large machines, manual cutting of weeds with brush saws, prescribed burning, and even letting sheep selectively browse weedy plants. All of these alternatives can provide a useful degree of weed control, but foresters generally regard them as more expensive and less effective than herbicide use. Therefore, these alternative methods are not widely utilized in Canada.

There are also some disadvantages of herbicide use in forestry. As with any pesticide, the successful use of herbicides for management requires the selection of an appropriate chemical and its proper application. If the right choices are not made, the weeds will not be adequately controlled, and the conifer crop may be injured. In addition, excessive suppression of "weeds" will affect the useful ecological services that they provide, such as helping to control erosion and preventing losses of soluble nutrients by leaching. Using herbicides will also reduce employment opportunities available in manual weed-management programs. For these and other reasons, including the significant fears that many people have about the potential toxicological risks of pesticides in the environment, the use of herbicides (and insecticides) in Canadian forestry has been very controversial.

Weeds in Forestry

Any disturbance of a forest is followed by a vigorous regeneration involving many species of plants which compete for access to space, nutrients, and moisture. This is true whether the disturbance is caused by a natural fire or insect attack, or by clear-cutting. During the first 10 to 15 years of succession, the plant community is typically dominated by many species other than the conifers desired by the forest industry. This can be illustrated by examining vegetation data for young clear-cuts in Nova Scotia, in which conifers contribute only 3–8% of the vegetation cover (Table 21.7). Other plants that are not economically desirable (at least not from the forestry perspective) are much more abundant in the regeneration. These "weeds" include ferns, monocotyledonous plants such as sedges and grasses, dicotyledonous herbs such as asters and goldenrods, low shrubs such as raspberries and blackberries, and taller shrubs such as birches, maples, and cherries. The domination of the site by "undesirable" plants inhibits the growth of commercially desirable conifers, and provides the economic justification for a weed-management treatment.

The effects of competing vegetation on conifer productivity are illustrated by a study of a site in New Brunswick that had been treated with herbicides 28 years previously (MacLean and Morgan, 1983). Prior to the herbicide treatment, the vigorously growing vegetation had been dominated by angiosperm shrubs, which formed a dense, 2-m-tall canopy that overtopped the shorter conifers. The herbicide spray had released the conifers from some of the stresses of competition. The study consequently found, 28 years after the herbicide treatment, that the biomass of balsam fir on the sprayed plots was about three times larger than on an adjacent unsprayed plot. From the forestry perspective, this means that the herbicide treatment allowed a conifer-dominated stand to develop more quickly, shortening the time to the next harvest.

TABLE 21.7 DOMINANT CLASSES OF VEGETATION IN DISTURBED FOREST AREAS

The vegetation was examined in four 4–6-year old clear-cuts of conifer forest in Nova Scotia. The data represent plant cover, expressed as the average percentage of the ground that is obscured by foliage. Because of overlap, values can exceed 100%.

	SITE 1	SITE 2	SITE 3	SITE 4
Mosses and liverworts	10%	8%	1%	3%
Lichens	1	1	1	1
Ferns and clubmosses	30	44	17	1
Conifers	4	8	9	5
Monocots (grasses, sedges)	12	8	6	29
Dicots – herbs	23	29	15	44
– woody shrubs	29	26	34	11
– raspberry	27	32	12	19
Total plant cover	137	153	103	113

Source: Modified from Freedman (1995)

Toxicological and Ecological Effects

The silvicultural objective of herbicide spraying is to manage the vegetation by changing its character. After a herbicide treatment in forestry, the abundance of competing vegetation is initially decreased, although it then recovers rapidly. In essence, a herbicide treatment returns the post-harvest regeneration (usually post-clear-cutting) to an earlier successional stage, while for several years releasing small conifer plants from some of the effects of competition. The overall changes in vegetation are illustrated using data from a study in Nova Scotia, in which a substantial recovery occurred within one growing season after a herbicide treatment (Table 21.8). The postspraying regeneration involved species whose seeds colonize sprayed sites, as well as many plants that were not killed by the herbicide. This study found that no species of plants were eliminated from herbicided clear-cuts, although there were

TABLE 21.8 RECOVERY OF VEGETATION AFTER HERBICIDE TREATMENT

Regenerating clear-cuts were treated with the herbicide glyphosate at four sites in Nova Scotia. The reference ("Ref") plots were not sprayed, and illustrate the normal recovery of vegetation after clear-cutting. The sprayed ("Spr") plots were sampled for one year before the herbicide treatment, and the postspray recovery was then monitored for up to 6 years. The data are percent plant cover, measured at the end of the summer.

	PRESPRAY	POSTSPRAY			
		YEAR 1	YEAR 2	YEAR 3	YEAR 6
Site 1					
Ref	79	117	142	155	174
Spr	95	64	105	126	131
Site 2					
Ref	74	102	140	156	174
Spr	97	36	77	118	145
Site 3					
Ref	97	113	139	153	159
Spr	99	43	124	134	132
Site 4					
Ref	153	186	194	201	–
Spr	149	78	111	132	–

Source: Modified from Freedman *et al.* (1993)

large differences in the relative abundance of species between the sprayed and reference plots. Different species vary in susceptibility to herbicides and in their ability to recover from disturbance.

Animals that utilize regenerating clear-cuts as habitat are also affected by a herbicide treatment. These effects can be caused by two types of influences: direct toxicity of the pesticide and changes in the character of the habitat.

Compared with many insecticides, herbicides used in forestry (such as 2,4,5-T, 2,4-D, glyphosate, hexazinone, triclopyr, and atrazine) are not very toxic to animals (see Table 21.2 for data on acute toxicity to a mammal). At exposures encountered during typical forestry uses, the direct toxicological risks from these chemicals are small and unimportant. This is particularly true of glyphosate, the most commonly used herbicide in forestry in Canada.

Glyphosate is extremely toxic to most plants, acting by blocking the metabolic synthesis of several essential amino acids. Although all plants, and some microorganisms, have this metabolic pathway, other organisms, including all animals, obtain these amino acids in the foods they eat. Consequently, glyphosate is relatively nontoxic to animals (Tables 15.2 and 21.2).

Since the toxicity of glyphosate to animals is relatively low, it is unlikely that animals inhabiting a sprayed clear-cut would be poisoned by exposure to this herbicide. The fact remains, however, that glyphosate causes large changes in their habitat, mostly by affecting the productivity and biomass of plants. Birds and other animals can be affected by a decreased availability of berries and other plant foods. In addition, the smaller foliage biomass on sprayed clear-cuts sustains smaller populations of insects and spiders, which are important foods for most birds. These indirect effects of herbicide spraying affect birds and other wildlife, even if they are not directly poisoned by the herbicide.

A study in Nova Scotia found only small changes in the abundance of birds that were breeding on clear-cuts treated with glyphosate (Table 21.9). The data indicate that the avian abundance decreased between the prespray and first postspray years. However, this change also occurred on the reference (i.e., nonsprayed) plot, suggesting it was caused by a factor, such as bad weather, unrelated to the herbicide treatment. In the second year after spraying, the abundance of birds on the sprayed plots was similar to the abundance in the first postspray year, while

A silvicultural application of the herbicide glyphosate to a clear-cut in Nova Scotia.

Source: B. Freedman

the abundance on the unsprayed plot increased to about the prespray value.

The most common species on the clear-cuts were white-throated sparrow and common yellowthroat (Table 21.9). These had a decreased abundance on both the sprayed and reference plots up to the second year after spraying, recovering by the fourth postspray year. On the reference plot, song sparrow and Lincoln's sparrow declined in abundance during the course of the study, whereas on the sprayed plots these birds were most abundant in the second and fourth years after spraying. The reference plot became colonized by some new species, including black-and-white warbler, red-eyed vireo, ruby-throated hummingbird, and palm warbler. These species did not invade the sprayed plots, because the herbicide treatment had caused the habitat to revert to a younger successional stage, which was less favourable to these birds.

Most studies of the effects of herbicides on deer and moose have examined the availability of their foods. Broad-leaved shrubs are a preferred food (known as *browse*) for most species of deer, but they are also important weeds in forestry, and are therefore a target of herbicide treatments. The quantities of browse, although initially reduced by herbicide spraying, are often increased in the longer term through the regeneration of shrubs. For example, studies in Maine found that several years after spraying, the

TABLE 21.9 | POPULATIONS OF BREEDING BIRDS ON HERBICIDE-TREATED CLEAR-CUTS

Average data are presented for four sprayed plots and one reference plot from a study in Nova Scotia. Sprayed plots were treated with glyphosate. The data are pairs of breeding birds per kilometre^2, surveyed for one year before the herbicide treatment (year 0), and then for four postspray years. Only abundant species are listed here.

SPECIES	SPRAYED PLOTS				REFERENCE PLOT			
YEAR:	0	1	2	4	0	1	2	4
Alder flycatcher	36	7	17	63	20	40	41	102
American robin	14	21	30	31	10	10	20	10
Red-eyed vireo	0	0	0	4	0	10	31	41
Magnolia warbler	5	5	5	102	0	20	20	143
Palm warbler	0	4	18	53	0	10	51	31
Mourning warbler	50	13	12	19	71	41	20	31
Common yellowthroat	151	140	90	136	122	112	122	163
Lincoln's sparrow	23	20	41	44	20	<1	<1	0
White-throated sparrow	203	118	89	155	143	71	93	163
Dark-eyed junco	42	62	61	69	31	41	61	102
Song sparrow	43	28	60	86	41	20	10	10
American goldfinch	52	24	15	13	61	41	20	20
Total Bird Abundance	623	447	444	805	539	396	528	836

Source: Modified from MacKinnon and Freedman (1993)

availability of browse was greater on treated clear-cuts, partly because the height of the shrub canopy was lower, giving white-tailed deer easier access to this food (Newton *et al.*, 1989). This is not always the case, however, and some studies have shown that herbicide spraying may decrease habitat quality for deer.

Overall, field research suggests that herbicide use in forestry has relatively small effects on birds and other wildlife that utilize clear-cuts as habitat. Other disturbances associated with forestry cause much larger effects on wildlife, particularly clear-cutting and plantation establishment (see Chapter 23).

The hazards to humans possibly inherent in herbicide use in forestry have also been intensely scrutinized. The concerns include occupational risks for people engaged in spraying or working in recently sprayed areas, as well as possible risks for the general population. These issues are highly controversial and are not fully resolved. Much of the concern has focused on phenoxy herbicides such as 2,4,5-T and 2,4-D, partly because 2,4,5-T contains a trace contamination of TCDD, an extremely toxic type of dioxin. However, the use of other herbicides, such as glyphosate, is also controversial.

It is somewhat reassuring to know, however, that many scientists believe herbicides can be used safely in forestry (and in agriculture and horticulture), provided the instructions for their use are followed carefully. Most scientists also believe that the herbicides do not cause undue risks to sprayers or people living near the treated areas. These are among the reasons why governments have registered these pesticides for uses that are economically beneficial — they allow greater productivity of both agricultural and forest crops. It must be remembered, however, that scientists have not achieved a full consensus. Therefore, the use of pesticides in forestry and for other purposes continues to be controversial.

INTEGRATED PEST MANAGEMENT

Pesticides are commonly used in agriculture, horticulture, and forestry. It is clear from this fact that most politicians, bureaucrats, and resource managers — and many scientists — have decided that the environmental "costs" associated with the use of pesticides are "acceptable," in view of economic benefits that are achieved. It is debatable, however, whether reliance on pesticide use is desirable over the long term. This is particularly true for those pesticides that are toxic to a broad spectrum of organisms. Most people would prefer less reliance to be placed on such nonspecific methods of pest control.

A much preferable approach is known as **integrated pest management** (or IPM), which employs an array of complementary tactics toward achieving pest control. Elements of an IPM system can include the following:

- use of natural predators, parasites, and other biological agents that can contribute to controlling a pest, while causing few nontarget damages
- use of crop varieties that are resistant to pests
- management of habitat to make it less suitable for pests
- careful monitoring of pest abundance, so control measures are undertaken only when necessary
- use of pesticides — but only if absolutely required as a component of an integrated strategy of pest management

A successful IPM program can greatly reduce, but is not necessarily able to eliminate, reliance on pesticides. For example, for many years the cultivation of cotton in the southern United States has relied on the intensive application of insecticides against pests such as the boll weevil. Widespread use of an IPM system to control this insect in Texas cottonfields reduced insecticide use from 8.8 million kg in 1964 to 1.05 million kg in 1976 (Bottrell and Smith, 1982). However, some use of insecticide against this pest remained necessary.

Wherever possible, IPM systems utilize control methods that are as specific as possible to the pest, avoiding or greatly reducing nontarget damage. Some of the best examples of such specific methods involve **biological control** (i.e., the use of a biological agent). The usefulness of biological control can be illustrated with the following control successes of introduced agricultural pests (Freedman, 1995):

The cottony-cushion scale (*Icyera purchasi*; a sap-sucking insect) was accidentally introduced to the United States, where it became a serious threat to citrus agriculture. Research in its native Australia discovered that populations of the pest were naturally controlled by certain insect predators and parasites. In 1888 two of its predators, a lady beetle and a parasitic fly, were introduced to California. This allowed virtually total control over this potentially disastrous pest. Unfortunately, this biological control was disrupted when DDT was used to deal with other orchard pests in the late 1940s.

St. John's wort (*Hypericum perforatum*), a common weed, is toxic to cattle. This plant became a serious pest in North American pastures after it was introduced from Europe. In 1943, two leaf beetles that feed on this plant were released in North America, and this pest is no longer an important problem.

The prickly pear cactus (*Opuntia spp.*) was imported to Australia from North America and grown as an ornamental plant. It escaped and became a serious weed in rangelands. This pest was controlled by the introduction of one of its herbivores, a species of moth whose larvae feed on the cactus.

Ragwort (*Senecio jacobea*) is a Eurasian plant that has been introduced to North America, South America, and Australia. It has become an important weed in rangelands because it crowds out native plants and is toxic to cattle. Several of the Eurasian herbivores of ragwort are now being used to control its abundance, including the cinnabar moth, ragwort flea beetle, and ragwort seed fly.

The screw-worm fly (*Callitroga hominivorax*) causes serious damage to cattle when its larvae feed on open wounds. This pest has been controlled in some areas through the release of large numbers of male flies that were reared in laboratories and sterilized by irradiation. Because female screw-worm flies mate only once, copulation with a sterile male results in unsuccessful reproduction. If this happens to enough females, the abundance of the pest decreases to an acceptable level.

Unfortunately, biological control may not be suitable for all pest problems. In fact, biological control has not succeeded in controlling most of the pests at which it has been directed. The list of failures includes forest pests such as spruce budworm. Researchers have investigated the potential for controlling budworm using pest-specific biological methods, such as bacterial, viral, and other agents of budworm-specific diseases; wasps that parasitize and kill budworm larvae; and the use of sex and developmental hormones to disrupt the mating and growth of the pest.

Some of these biological control methods have shown promise. However, they do not yet achieve a consistent kill of budworm, and are relatively expensive. For these reasons they are not considered ready for routine use against this important forest pest.

One viable alternative to the broadcast spraying of synthetic insecticides to control budworm has been found — an insecticide derived from bacterium *B.t.* (discussed earlier). Research into other biological controls continues, and some methods may prove successful. These would allow managers to develop an effective, integrated pest-management system that does not rely on broadcast spraying of insecticides, even relatively specific insecticides such as *B.t.*

At the present time, however, it appears that society will continue to rely heavily on pesticide use in intensively managed systems in agriculture and forestry. This will happen even though the intensive systems cause ecological damages, (partly because these damages are not properly accounted for as economic "costs").

It is important that vigorous research is undertaken to develop viable methods of biological control and other elements of IPM systems. This is necessary if the present reliance on the pesticide treadmill is to be replaced with less damaging methods of pest management. Such a change would deliver substantial benefits to society, because the agricultural and forestry systems that we require for sustenance could be managed on a more sustainable basis. In part, this will required that more attention is directed to the ecological damage caused by the intensive use of pesticides.

Because of this damage, it is highly desirable that effective, nonpesticidal alternatives to pest control are discovered as quickly as possible. Until this happens, pesticide use should be reduced to the lowest levels that continue to control the pests effectively. Some researchers have recently argued that pesticide use in North America is much greater than necessary, and that it could be halved without causing a significant adverse effect on crop yields (Pimentel *et al.*, 1991). In fact, several European countries have already passed legislation requiring that agricultural pesticide use be reduced by 50% by the year 2000 (Matteson, 1995). Serious consideration could also be given to such an action in Canada.

Key Terms

pesticides
nontarget organisms (or nontarget damage)
broad-spectrum pesticides
fungicide
herbicide
insecticide
vectors
pesticide "treadmill"
drift
persistence
bioconcentration
food-web accumulation (or food-web magnification)
risk assessment
integrated pest management
biological control

Questions for Discussion

1. Classify pesticides according to their intended targets, and also by their major chemical groups.
2. What are "pests"? Why do people consider it necessary to manage their abundance in agriculture, horticulture, forestry, and public health?
3. What characteristics of organochlorines have caused them to become global contaminants? Why do they pose special ecotoxicological risks for top predators?
4. Identify the benefits and environmental risks associated with pesticide use in forestry.
5. Are there effective alternatives to the continued use of pesticides? Consider the roles of integrated pest management, biological controls, and other options.

REFERENCES

Agriculture Canada. 1993. *Registration Status of Fenitrothion Insecticide*. Ottawa: Agriculture Canada, Food Production and Inspection Branch. Discussion Report 093-01.

Armstrong, J.A. 1985. Spruce budworm control program in eastern Canada. In: *Advances in Spruce Budworms Research*. Ottawa: Canadian Forestry Service. pp. 384–385.

Bishop, C. and D.V. Weseloh. 1990. *Contaminants in Herring Gull Eggs from the Great Lakes*. Ottawa: Environment Canada, State of the Environment Reporting. SOE Fact Sheet 90-2.

Blais, J.R. 1985. The ecology of the eastern spruce budworm: a review and discussion. In: *Recent Advances in Spruce Budworms Research*. Ottawa: Canadian Forestry Service. pp. 49–59.

Bottrell, D.G., and R.F. Smith. 1982. Integrated pest management. *Environ. Sci. Technol.*, **16**: 282A–288A.

Briggs, S.A. 1992. *Basic Guide to Pesticides: Their Characteristics and Hazards*. Washington: Taylor & Francis.

Busby, D.G., L.M. White, P.A. Pearce, and P. Mineau. 1989. Fenitrothion effects on forest songbirds: a critical new look. In: *Environmental Effects of Fenitrothion Use in Forestry*. Dartmouth, NS: Environment Canada, Conservation and Protection. pp. 43–108.

Canadian Council of Forest Ministers. 1995. *Compendium of Canadian Forestry Statistics, 1994*. Ottawa: Canadian Council of Forest Ministers.

Carson, R. 1962. *Silent Spring*. Boston: Houghton-Mifflin.

Cooper, K. 1991. Effects of pesticides on wildlife. In: W.C. Hayes and E.R. Laws, eds. *Handbook of Pesticide Toxicology. Vol. 1, General Principles*. San Diego, CA: Academic. pp. 463–496.

Crawford, H.S., R.W. Titterington, and D.T. Jennings. 1983. Bird predation and spruce budworm populations. *J. Forestry*, **81**: 433–435.

Edwards, C.A. 1975. *Persistent Pesticides in the Environment*. Cleveland: CRC Press.

Ennis, T. and E.T.N. Caldwell. 1991. Spruce budworm, chemical and biological control. In: L.P.S. van der Geest and H.H. Evenhuis, eds. *Tortricid Pests, Their Biology, Natural Enemies, and Control*. Amsterdam: Elsevier. pp. 621–641.

Environment Canada. 1993. *Toxic Contaminants in the Environment: Persistent Organochlorines*. Ottawa: Environment Canada. State of the Environment Reporting.

Freedman, B. 1995. *Environmental Ecology, 2nd ed.* San Diego, CA: Academic.

Freedman, B., R. Morash, and D. MacKinnon. 1993. Short-term changes in vegetation after the silvicultural spraying of glyphosate herbicide onto regenerating clear-cuts in Nova Scotia, Canada. *Can. J. For. Res.*, **23**: 2300–2311.

Gage, S.H., and C.A. Miller. 1978. *A Long-Term Bird Census in Spruce Budworm-Prone Balsam Fir Habitats in Northwestern New Brunswick*. Fredericton, NB: Maritimes Forest Research Centre. Inf. Rep. M-X-84.

Grossbard, E. and D. Atkinson (eds.). 1985. *The Herbicide Glyphosate*. London: Butterworths.

Hayes, W.J. 1991. Introduction. In: W.C. Hayes and E.R. Laws, eds. *Handbook of Pesticide Toxicology. Vol. 1, General Principles*. San Diego, CA: Academic. pp. 1–37.

Kettela, E. 1983. *A Cartographic History of Spruce Budworm Defoliation from 1967 to 1981 in Eastern North America*. Fredericton, NB: Canadian Forestry Service. Maritimes Forest Research Centre. Inf. Rep. DPC-X-14.

Kingsbury, P.D., and B.B. McLeod. 1981. *Fenitrothion and Forest Avifauna Studies on the Effects of High Dosage Applications*. Sault Ste. Marie, ON: Canadian Forestry Service. Forest Pest Management Institute. Rep. FPM-X-43.

Mackinnon, D. and B. Freedman. 1993. Effects of silvicultural use of the herbicide glyphosate on breeding birds of regenerating clear-cuts in Nova Scotia, Canada. *J. Appl. Ecol.*, **30**: 395–406.

MacLean, D.A. 1988. Effects of spruce budworm outbreaks on vegetation, structure, and succession of balsam fir forests on Cape Breton Island, Canada. In: M.J.A. Werger, P.J.M. van der Aart, H.J. During, and J.J.A. Verhoeven, eds. *Plant Form and Vegetation Structure*. The Hague, The Netherlands: SPB Academic. pp. 253–261.

MacLean, D.A. 1990. Impact of forest pests and fire on stand growth and timber yield: implications for forest management planning. *Can. J. For. Res*, **20**: 391–404.

MacLean, D.A. and M.G. Morgan. 1983. Long-term growth and yield response of young fir to manual and chemical release from shrub competition. *For. Chron.*, **59**: 177–183.

Matteson, P.C. 1995. The "50% pesticide cuts" in Europe: a glimpse of our future? *Amer. Entomol.*, **1995**: 210–220.

McEwen, F.L., and G.R. Stephenson. 1979. *The Use and Significance of Pesticides in the Environment*. New York: Wiley.

Millikin, R.L. 1990. Effects of fenitrothion on the arthropod food of tree-foraging forest songbirds. *Can. J. Zool.*, **68**: 2235–2242.

Mineau, P. 1993. *The Hazard of Carbofuran to Birds and Other Vertebrate Wildlife*. Ottawa: Environment Canada, Canadian Wildlife Service, Wildlife Toxicology Section. Tech. Rep. No. 177.

Moriarty, F. 1988. *Ecotoxicology: The Study of Pollutants in Ecosystems*. London: Academic.

Morris, R.F., W.F. Cheshire, C.A. Miller, and D.G. Mott. 1958. The numerical response of avian and mammalian predators during a gradation of the spruce budworm. *Ecology*, **39**: 487–494.

National Research Council (NRC). 1986. *Pesticide Resistance*. NRC. Washington: National Academy Press.

Newton, M., and F.B. Knight. 1981. *Handbook of Weed and Insect Control Chemicals for Forest Resource Managers*. Beaverton, OR: Timber.

Newton, M., E.C. Cole, R.A. Lautenschlager, D.E. White, and M.L. McCormack. 1989. Browse availability after conifer release in Maine's spruce-fir forests. *J. Wildl. Manage.*, **53**: 643–649.

Norheim, G., L. Somme, and G. Holt. 1982. Mercury and persistent chlorinated hydrocarbons in Antarctic birds from Bouvetoya and Dronning Maud Land. *Environ. Pollut., (Ser. A)*, **28**: 233–240.

Ostaff, D.P. 1985. Quantifying effects of spruce budworm damage in eastern Canada. In: *Recent Advances in Spruce Budworms Research*. Ottawa: Canadian Forestry Service. pp. 247–248.

Ostaff, D.P. and D.A. MacLean. 1989. Spruce budworm populations, defoliation, and changes in stand condition during an uncontrolled spruce budworm outbreak on Cape Breton Island, Nova Scotia. *Can. J. For. Res.*, **19**: 1077–1086.

Pauli, B.D., S.B. Holmes, R.J. Sebastien, and G.P. Rawn. 1993. *Fenitrothion Risk Assessment*. Ottawa: Canadian Wildlife Service. Tech. Rep. Ser. No. 165.

Peakall, D.B. 1990. Prospects for the peregrine falcon, *Falco peregrinus*, in the nineties. *Can. Field-Nat.*, **104**: 168–173.

Pearce, P.A. and D.B. Peakall. 1977. The impact of fenitrothion on bird populations in New Brunswick. NRCC **16073**: 299–305. Ottawa: National Research Council of Canada.

Pimentel, D., H. Acquay, M. Biltonen, P. Rice, M. Silva, J. Nelson, V. Lipner, S. Giordano, A. Horowitz, and M. D'Amare. 1992. Environmental and economic costs of pesticide use. *Bioscience*, **42**: 750–760.

Pimentel, D. and H. Lehman (eds.). 1992. *The Pesticide Question: Environment, Economics, and Ethics*. New York: Chapman & Hall.

Pimentel, D., L. McLaughlin, A. Zepp, B. Lakitan, T. Kraus, P. Kleinman, F. Vancini, W.J. Roach, E. Graap, W.S. Keeton, and S. Selig. 1991. Environmental and economic effects of reducing pesticide use. *Bioscience*, **41**: 402–409.

Rozencranz, A. 1988. Bhopal, transnational corporations, and hazardous technologies. *Ambio*, **17**: 336–341.

Sassman, J., R. Pienta, M. Jacobs, and J. Cioffi. 1984. *Pesticide Background Statements. Vol. 1, Herbicides*. Washington: USDA Forest Service.

Varty, I.W. 1975. Side effects of pest control projects on terrestrial arthropods other than the target species. In: M.L. Prebble, ed. *Aerial Control of Forest Insects in Canada*. Ottawa: Department of the Environment. pp. 266–275.

Wurster, D.H., C.F. Wurster, and W.N. Strickland. 1965. Bird mortality following DDT spray for Dutch elm disease. *Ecology*, **46**: 488–499.

22

OIL SPILLS

CHAPTER OUTLINE

CHAPTER OBJECTIVES

After completing this chapter, you will be able to:

1. Outline the most common causes of oil spills on land and at sea.
2. Describe how spilled oil becomes dispersed in the environment through evaporation, spreading, and other processes.
3. Explain how hydrocarbons cause toxicity to organisms.
4. Explain how petroleum kills birds and how oiled birds may be rehabilitated.
5. Describe case studies of the ecological effects of oil spills at sea and on land.
6. Discuss the potential consequences of petroleum resource development in the Arctic.

INTRODUCTION

Petroleum (or **crude oil**) is a nonrenewable natural resource (Chapter 13) that is used mainly as a source of energy. It is also used to manufacture a diverse array of petrochemicals, including synthetic materials such as plastics. Petroleum is mined in huge quantities. Pipelines and ships transport most of this volume, plus its refined products, around the globe. The risks of spillage are always present, and oil spills cause severe ecological damages.

Petroleum accounts for about 39% of the global production of commercial energy (32% in Canada). Moreover, the use of petroleum is increasing rapidly, by about 3%/y between the late 1980s and early 1990s. The fastest increases are in regions with rapidly growing economies, such as Southeast Asia (which experienced an increase of 14% in 1988). Relatively wealthy, developed countries support about 14% of the human population, but they account for 49% of the global use of petroleum — 18% in North America, 20% in Western Europe, and 11% in Japan (World Resources Institute, 1994).

The global reserves of petroleum are about 135 billion t (in 1990), of which 63% occur in the Middle East, 8% in North America (17% of this 8% is in Canada), and 6% in the former Soviet Union (Table 22.1). Almost all mining of petroleum takes place far from the places where it is consumed. The Middle East, for example, is a huge exporter of petroleum and refined products; the amount shipped abroad is about 5.4 times larger than domestic usage in that region (in 1988). In contrast, Western Europe produces only about 33% of the petroleum it consumes, while Asia and Australasia produce 35%, and the United States 48%.

Canada produces approximately as much petroleum as it consumes (Table 22.1). However, about 83% of Canadian production is in sparsely populated areas of Alberta, and 12% in similar places in Saskatchewan, while most consumption occurs in densely populated areas throughout the country. Consequently, enormous quantities of petroleum and its refined products are transported over great distances within Canada, mostly by overland pipelines, railroads, and trucks. In addition, western Canada exports large volumes of petroleum and refined products to the central and western United States and to Asia, while eastern Canada imports from the Middle East and Latin America. Therefore, even though Canada is self-sufficient in its net production and consumption of petroleum, huge quantities move within, out of, and into the country and its regions.

On the global scale, most petroleum is transported by means of oceanic tankers and overland pipelines. Local distribution systems typically involve smaller tankers, pipelines, railroads, and trucks. There is a risk of accidental spillage from any of these means of transportation. Some of these spills have been spectacular in their volumes and in the environmental damages caused. In addition, petroleum is discharged into the environment during the normal operation of tankers, especially when the crew

TABLE 22.1 PETROLEUM PRODUCTION AND USE IN SELECTED COUNTRIES IN 1990

COUNTRY	PRODUCTION OF PETROLEUM (10^6 tonne/year)	PROVEN RESERVES OF PETROLEUM (10^6 t)	MARINE SHIPPING OF PETROLEUM AND PRODUCTS (10^6 t/y)	CONSUMPTION OF PETROLEUM (10^6 t/y)
Russia	558	5 132	121	403
United States	376	3 560	350	779
Saudi Arabia	261	35 650	180	–
Iran	140	12 700	102	–
Iraq	139	13 600	–	–
China	136	3 264	36	113
Mexico	123	6 079	79	–
United Arab Emirates	96	1 300	71	–
Venezuela	93	8 604	82	–
Kuwait	90	12 785	59	–
Nigeria	84	2 400	74	–
Canada	77	720	26	75
Indonesia	67	726	73	31
Libya	54	3 151	53	–
World	2 924	134 792	3 360	3 104

Source: Colombo (1993) and World Resources Institute (1994)

cleans the storage tanks and releases oily bilge waters into the ocean.

Petroleum refineries also cause chronic pollution of their local environment, due to frequent but small spills, and to routine discharges of waste waters contaminated both with hydrocarbons and with trace quantities of metals, sulphides, phenols, and other chemicals. Small-scale dumping of used motor oil and other wastes is another important, although poorly documented, source of oil pollution.

In this chapter we examine the causes of oil spills from these various sources, and the ecological damage that can be inflicted on aquatic and terrestrial environments.

Petroleum and Its Refined Products

Petroleum is a naturally occurring mixture of liquid organic compounds, almost all of which are *hydrocarbons* (this term is explained below). Petroleum is a fossil fuel, as are coal, oil shale, oil sand, and natural gas. Fossil fuels are derived from ancient plant biomass that became buried in deep sedimentary formations. Over geologically long periods of time, this biomass was subjected to high pressure, high temperature, and anoxia. Chemical reactions resulted that eventually produced a rich mixture of gaseous, liquid, and solid hydrocarbons. Naturally occurring hydrocarbons can range in complexity from gaseous methane with a weight of only 16 g/mole, to solid substances found in coal, with molecular weights exceeding 20 000 g/mole. (In chemistry, a "mole" is a standard quantity of a substance, equivalent to the amount contained in 6.02×10^{23} atoms or molecules of the substance.)

Hydrocarbons are molecules composed entirely of hydrogen and carbon atoms. Hydrocarbons can be classified into three groups: aliphatic, alicyclic, and aromatic hydrocarbons.

Aliphatic hydrocarbons are compounds in which the carbon atoms are organized into a simple chain. Saturated aliphatics (also called paraffins or alkanes) have a single bond between adjacent carbon atoms, while unsaturated molecules have one or more double or triple bonds. These bonds are illustrated by the two-carbon aliphatic hydrocarbons ethane, $H_3C{-}CH_3$; ethylene, $H_2C{=}CH_2$; and acetylene, $HC{\equiv}CH$. Unsaturated aliphatics are relatively unstable and do not occur naturally in petroleum. They are produced during industrial refining, as well as photochemically in the environment after crude oil is spilled.

Alicyclic hydrocarbons have some or all of their carbon atoms arranged in a ring structure, which can be saturated or unsaturated.

Aromatic hydrocarbons contain one or more five- or six-carbon rings in their molecular structure. The simplest ring, C_6H_6, is known as benzene.

Crude petroleums vary greatly in their specific mixtures of hydrocarbons and other chemicals. They typically consist of about 98% liquid hydrocarbons, less than 1–2% sulphur, and less than 1% nitrogen, plus the metals vanadium and nickel in concentrations up to 0.15%. When petroleum is refined, various hydrocarbon fractions are separated by distillation at different temperatures. This is done to produce such products as natural gas, gasoline, kerosene, heating oil, jet fuel, lubricating oils, waxes, and residual fuel oil (also known as bunker fuel). In addition, a process known as catalytic cracking is used to convert some of the heavier fractions into lighter, more valuable hydrocarbons such as those in gasoline.

Oil Spills

Oil pollution can be caused by any spillage of petroleum or its refined products. The largest spills typically involve the discharge of petroleum or bunker fuel to the ocean from a disabled tanker or drilling platform, to an inland waterway from a barge or ship, or to land or freshwater from a well blowout or broken pipeline. In addition, some enormous oil spills have resulted from acts of warfare.

Terrestrial Oil Spills

Oil spills onto land are relatively common. Between 1985 and 1988, an average of 2282 oil spills per year were reported in Canada. About 42% of these occurred in the immediate vicinity of production wells, while 29% were from pipelines and 16% from tanker trucks (Government of Canada, 1991).

Most large terrestrial spills involve a ruptured pipeline, either above or below ground. Canada has about 35 000 km of pipeline for transporting petroleum and another

227 000 km of natural gas pipeline. Pipeline breaks may be caused by faulty welding, corrosion, or pump malfunctions, as well as by sabotage, earthquakes, and even armed vandals engaged in target practice.

Fortunately, the extensive Canadian network of pipelines incorporates spill sensors and other technologies that allow workers to shut down sections of pipeline rapidly. Individual accidents can then be kept relatively small and confined. Some other countries use fewer of these technologies, and consequently may suffer huge petroleum spills from overland pipelines. Such accidents occur too commonly in northern Russia, for example. Some of the pipelines there have corroded, and few effective countermeasures have been instituted for the prevention or containment of oil spills.

In general, oil spilled on land affects relatively localized areas of terrain, because most soils absorb petroleum quickly. However, much larger areas of habitat can be affected if spilled petroleum reaches a watercourse. Oil spilled onto water, particularly at sea, can affect very extensive areas, because winds and currents cause slicks to spread and disperse widely.

Marine Oil Spills

During the late 1970s and early 1980s, petroleum spills into the world's oceans amounted to about 3–6 million t/y (Table 22.2). More recent estimates suggest a substantial reduction in spillage, to about 0.6 million t in 1989. In addition to petroleum spillage, there is a substantial natural emission to the oceans of hydrocarbons not derived from petroleum. These chemicals are synthesized and released by marine phytoplankton, at a rate estimated at 26 million t/y. These huge biological releases contribute to the background concentration of hydrocarbons in marine ecosystems, equivalent to about 1 ppb (or 1 μg/L) in seawater. Because they are well dispersed and do not result in known biological damage, the biogenic emissions should be viewed as natural contamination.

In addition, there are natural emissions of petroleum to the marine environment, oozing from underwater oil seeps. These natural seeps amount to an estimated 0.2–0.6 million t/y, and can sometimes cause local ecological damage.

Massive spills associated with wrecked supertankers or well platforms at sea attract a great deal of attention, and deservedly so. It is important to recognize, however, that relatively small but frequent discharges to the environment are associated with urban runoff, oil refineries, "normal" tanker discharges, and other coastal effluents. Because these discharges are frequent, they account for a larger aggregate volume of petroleum than do the rare but spectacular massive spills (Table 22.2). The chronic coastal discharges are responsible for the local contamination and pollution by hydrocarbons that is typical of coastal cities and harbours everywhere.

Discharges of oily washings from tanker storage tanks are another important source of petroleum inputs to the oceans. After a tanker delivers a load of petroleum to a refinery, it fills its storage tanks with seawater, which acts as stabilizing ballast while the ship travels to get its next load. As the tanker approaches its destination, the ballast is often discharged into the ocean. These waste waters contain hydrocarbon residues which are equivalent to as much as 1.5% of the tanker's capacity in the case of bunker fuel, less than 1% for petroleum, and about 0.1% for light refined products such as gasoline.

This large, operational source of marine pollution has been decreasing greatly since the 1970s, due to widespread adoption of a procedure called the **load-on-top (LOT) method**. LOT separates and contains most of the oily residues before ballast waters are discharged to the marine environment (the residual oil is then combined with the next load). If used in calm seas, the LOT technique

TABLE 22.2 ESTIMATES OF PETROLEUM HYDROCARBON INPUTS TO THE WORLD'S OCEANS

Data, in units of 10^3 tonne/year, are for 1983.

SOURCE	AMOUNT (10^3 t/y)
Urban runoff and discharges	1080
Operational discharges from tankers	700
Accidents involving tankers at sea	400
Losses from non-tanker shipping	320
Atmospheric deposition	300
Natural seeps	200
Coastal refineries	100
Other coastal discharges	50
Offshore production losses	50
Total discharges	3200

Source: Koons (1984)

can recover 99% of the oily residues, although the efficiency is commonly 90% or less if the tanker has had a turbulent passage. Thanks to the widespread use of LOT by tankers, petroleum spillage with ballast waters has been reduced from about 1.1 million t in 1973 to annual levels of 0.25 million t in 1989 and subsequent years. Although LOT is widely used, some tankers continue to discharge oily wastes at sea illegally. This pollution is still an important cause of seabird mortality off the coasts of Canada and other countries.

The most disastrous marine spills of petroleum (several of which are described later in this chapter) include the following supertanker accidents:

- In 1967, the *Torrey Canyon* spilled 117 000 t of petroleum off southern England.
- In 1973, the *Metula* spilled 53 000 t in the Strait of Magellan.
- In 1978, the *Amoco Cadiz* spilled 230 000 t in the English Channel.
- In 1989, the *Exxon Valdez* spilled 36 000 t in southern Alaska.
- In 1993, the *Braer* spilled 84 000 t off the Shetland Islands of Scotland.

Canada has had its share of notable tanker spills, three of which we list here:

- The *Arrow* ran aground in Chedabucto Bay, Nova Scotia, in 1970, and spilled 11 000 t of bunker-C fuel (bunker-C is a common ship fuel). About 300 km of shoreline were polluted by the spill, and many seabirds were killed (about 2000 dead birds were collected from Chedabucto Bay, and another 5000 from Sable Island, 320 km away).
- The *Kurdistan* spilled 7500 t of bunker fuel in Cabot Strait between Newfoundland and Nova Scotia in 1979.
- The *Nestucca* spilled 875 t of bunker fuel in 1988 off Washington State, and extensively polluted shorelines on the west side of Vancouver Island, British Columbia. This spill killed many seabirds. About 3600 dead birds of 31 species were collected on the western beaches of Vancouver Island, although the total mortality is estimated at more than 10 000 birds.

Dead aquatic birds are among the most evocative and tragic victims of oils spills. This blue-winged teal (*Anas discors*) was killed by a spill of heavy fuel oil on the St. Lawrence River.
Source: B. Freedman

Some large accidental spills have occurred from offshore oil drilling platforms. The enormous blowout of the *IXTOC–I* exploration well in the Gulf of Mexico in 1979 resulted in the spillage of about 500 000 t of petroleum, representing the largest-ever accidental spill. Other platform spills include a blowout in 1969 off the coast of Santa Barbara in southern California (10 000 t), and the *Ekofisk* blowout in 1977 in the North Sea off Norway (30 000 t).

Oil Spills through Warfare

Large quantities of petroleum and refined products have been spilled during warfare. During the Second World War, for example, German submarines sank 42 tankers off eastern North America, resulting in the spillage of about 417 000 t of oil. Attacks on oil tankers were particularly common during the Iran-Iraq War of 1981–1987. In total, there were 314 attacks on oil tankers, 70% of them by Iraqi forces. That war's largest spill occurred in 1983, when Iraq damaged five tankers and three production wells at the Iranian *Nowruz* offshore facility, causing more than 260 000 t of petroleum to spill into the Gulf of Arabia.

The world's largest-ever marine spill occurred during the brief Gulf War of 1991. Iraqi forces deliberately released huge quantities of petroleum (perhaps about

0.8 million t, although estimates range as high as 2 million t) into the Gulf of Arabia from a Kuwaiti offshore loading facility. In part, this spill was made as a tactic of warfare. It was an attempt to make it difficult for Allied forces to execute an amphibious landing during the liberation of Kuwait. The spill was also, however, an act of economic and ecological terrorism.

The Iraqis also caused an extremely large spillage on land during that war, by sabotaging and igniting virtually all of the more than 700 production wells in Kuwait. An estimated 2–6 million t of petroleum per day were emitted from the burning wells. After the Gulf War was over, it took 11 months to control and cap the blowouts. By that time, an estimated 42–126 million t of petroleum had been spilled. About 5–21 million t of the petroleum accumulated as crude-oil lakes on the desert around the blowouts, while most of the rest burned in the atmosphere or evaporated.

Fate of Spilled Oil

Various natural processes affect petroleum and refined hydrocarbons after they are spilled into the environment (Figure 22.1). Depending on their chemical and physical characteristics, the various hydrocarbon fractions can selectively evaporate, spread, dissolve into water, accumulate as persistent residues, be degraded by microorganisms, or undergo some combination of these precesses.

Evaporation of fumes and vapours is important in reducing the amount of spillage remaining in the aquatic or terrestrial environment. Evaporation typically dissipates as much as 100% of gasoline spilled at sea, 30–50% of crude oil, and 10% of bunker fuel. In other words, the relatively light, volatile hydrocarbon fractions are selectively evaporated, leaving heavier residues behind. Rates of evaporation are increased by warm ambient temperatures and vigorous winds.

Spreading involves the movement of an oil slick over the surface of water or land. Spreading can occur over extremely large areas on water, but it is much more restricted on land because of the high absorptive capacity of soil. Slicks on water are moved about by currents and wind, and may eventually wash onto a shore. The degree of spreading on water is influenced by the viscosity of the spilled material, and by environmental factors such as wind speed, water turbulence, and the presence of surface ice. One experimental spillage of 1 m^3 of petroleum onto calm seawater created a slick 0.1 mm thick, with a diameter of 100 m, after 100 minutes. A petroleum slick only 0.3 μm thick or less is visible as a sheen on calm water.

Dissolution causes pollution of the water beneath an oil slick. Lighter hydrocarbon fractions are generally more soluble in water than heavier ones, while aromatics are much more soluble than alkanes (Table 22.3). After a petroleum spill at sea, the hydrocarbon concentration in water several metres beneath the slick is as much as 4–5 ppm (or g/m^3), thousands of times greater than the 1 ppb (or mg/m^3) of total hydrocarbons in ambient seawater.

FIGURE 22.1 | FATE OF SPILLED PETROLEUM ON WATER

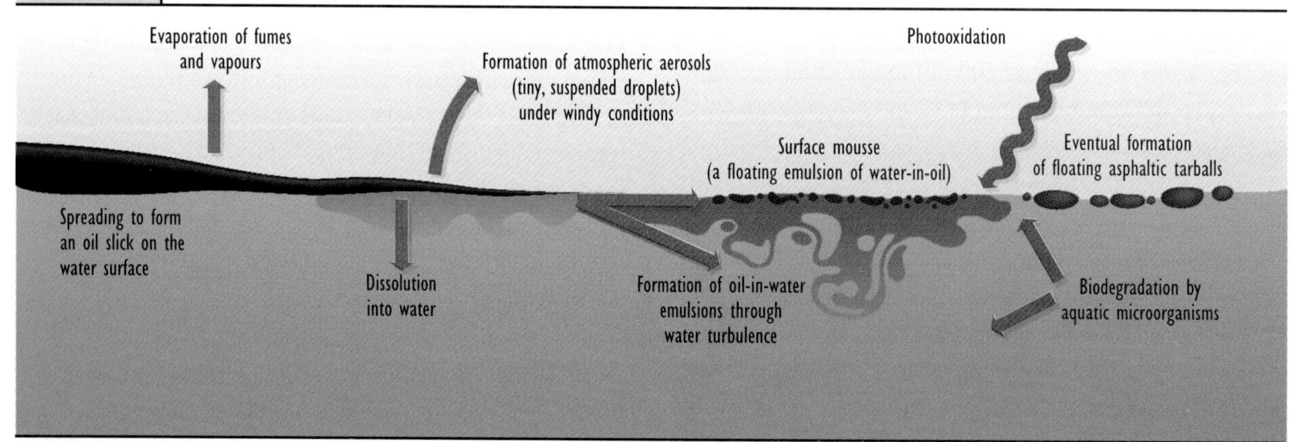

Source: Modified from Clark and MacLeod (1977)

TABLE 22.3 SOLUBILITIES OF ALKANE AND AROMATIC HYDROCARBONS

Solubilities are reported in grams per metre^3 (ppm) in fresh water. Within the aromatics, aqueous solubility decreases with increasing molecular size and with the number of aromatic rings.

HYDROCARBON	SOLUBILITY
ALKANES	
Gases (1–4 carbons)	24–62
Liquids (5–9 carbons)	0.05–39
Kerosenes (10–17 carbons)	$1–2 \times 10^{-4}$
Lubricating oils (23–37 carbons)	$10^{-7}–10^{-14}$
Residual hydrocarbons (>37 carbons)	$<10^{-14}$
AROMATICS	
Benzene (C_6H_6)	1780
Toluene (C_7H_6)	515
Naphthalene ($C_{10}H_8$, two rings)	31

Residual materials remain after the lighter fractions of the spilled petroleum have evaporated or dissolved. At sea, residual materials typically form a rather gelatinous, water-in-oil emulsion, known as "mousse" because of its vague resemblance to the whipped chocolate desert. Oil spilled offshore usually washes onto shorelines as mousse. Mousse stranded on the shore may weather to form a long-lasting tarry residue on rocks. Alternatively, mousse may eventually combine with particles of sediment on the beach to form sticky, tar-like patties that subsequently become buried, or may be washed back to sea during a storm. Mousse that does not wash ashore eventually weathers into dense, semisolid, floating asphaltic residues known as "tar balls."

Degradation refers to the slow decomposition of spilled materials by organisms and photooxidation by solar ultraviolet radiation. Many species of fungi, bacteria, and other microorganisms can utilize hydrocarbons as an energy source for metabolism. The rate of **biodegradation** varies greatly, however, depending on ambient temperature, the availability of oxygen, and the availability of key nutrients such as nitrogen and phosphorus. In general, lighter fractions of petroleum are relatively easily decomposed by biological and inorganic oxidations, while heavier fractions resist degradation and can be quite persistent in the environment.

Toxicity of Petroleum and Hydrocarbons

Acute toxicity caused by petroleum, refined products, or pure hydrocarbons is typically associated with the destruction of cellular membranes, a form of damage that results in tissue death. The toxic effects are influenced by several factors, especially the following:

- the chemical composition of the spilled material, including its component hydrocarbons
- the intensity of exposure, that is, the quantity or concentration of specific hydrocarbons or petroleum
- the frequency of exposure events, i.e., whether the pollution is a single event, chronic (continuous), or episodic (a series of repeated events)
- the timing of the exposure, e.g., whether it occurs during a critical time of the year for a species or ecosystem
- the condition of the spilled material, including the thickness of the slick, the nature of the emulsion, the degree of weathering, and the persistence of the residues
- any environmental influences on exposure and toxicity, including weather conditions, oxygen status, and the presence of other pollutants
- any toxicity associated with chemical dispersants that may be used during a cleanup
- the sensitivity of the species in the affected ecosystem to the toxic effects of hydrocarbons — a circumstance which is itself influenced by a number of biological factors

In addition, ecosystem-level damages may be influenced by physical disturbances associated with a cleanup, such as the use of dispersants and emulsifiers, hot-water washings, the removal of substrates, burning, and the tilling of oiled materials to improve aeration. Ecological effects are also influenced by any damage caused to "keystone" species. Damage to these important species results in disproportionate effects on their community.

Effects on Birds

Seabirds such as cormorants, sea ducks (e.g., eiders, mergansers, scaup, scoters), alcids (e.g., auklets, murres, puffins, and razorbills), and penguins are extremely vulnerable to oil spills. During the nonbreeding seasons, a spill can cause enormous mortality to these birds, because most species congregate in large, seasonal flocks known as "rafts." Moreover, since alcids and penguins have low reproductive rates, their abundance can take a long time to recover from mass mortality caused by an oil spill. Murres, for example, begin to breed only when they are five years old, lay only a one-egg clutch, and fledge only about one young per four breeding adults per year.

Seabirds are killed because their feathers become fouled with oil when they dive through or swim in oil-polluted water. The birds lose critical insulation and buoyancy, and die from excessive heat loss leading to hypothermia or by drowning. They also ingest highly toxic oil while attempting to clean their feathers by preening. In addition, bird eggs can be killed by even a light oiling from the feathers of a contaminated parent.

The size of a petroleum spill is not necessarily an accurate indicator of its potential for causing damage to bird populations. The ecological context is also critically important: even a small spill in sensitive habitats can wreak havoc. For example, in 1981, a relatively small discharge of oily bilge water from the tanker *Stylis* off Norway killed about 30 000 seabirds. This happened because the spill affected a critical habitat where seabirds are abundant during the winter. In another case, more than 16 000 Magellanic penguins (*Spheniscus magellanicus*) were discovered oiled on beaches in Argentina, even though no offshore slick could be found. The oil likely came from the bilge washings of a passing tanker. Similar damage occurs routinely off Newfoundland, through illegal discharges of oily bilge waters. In January, 1997, about 30 000 murres and other seabirds were killed by oiling in this way, near Cape St. Mary's.

Ecological Effects of Oil Pollution

We will use case studies of several oil spills to investigate not only ecological damage from oil pollution, but also damage from some of the cleanup methods. We will examine spills from wrecked supertankers and offshore drilling platforms, chronic emissions near petroleum refineries, and experimental oiling of terrestrial vegetation.

Oil Spills from Wrecked Tankers

The *Torrey Canyon*

The *Torrey Canyon* wreck in 1967 is one of the most famous supertanker accidents. The *Torrey Canyon* was bound for a refinery in Wales with 117 000 t of crude oil when it ran aground. The entire cargo was spilled, polluting hundreds of kilometres of coastline. Seabirds were among the most tragic victims of this spill, with at least 30 000 killed. Although almost 8000 oiled birds were captured and cleaned, the rehabilitation methods of the time were not very successful. Only a few of these birds survived long enough to be released, and post-release survival can be poor (In Detail 22.1).

In Detail 22.1

Cleaning Oiled Birds

Birds can easily become oiled if they swim or dive in water polluted by oil. Because of the great empathy that people have for these tragic victims of oil pollution, intense efforts are often made to collect oiled birds and rehabilitate them by cleaning them of oily residues and treating their poisoning (Clark, 1984; Holmes, 1984; Harvey-Clark, 1990).

The first significant effort to do this occurred after the *Torrey Canyon* spill of 1967, when about 8000 oiled birds, mostly common murres (*Uria aalge*) and razorbills (*Alca torda*), were captured and cleaned. Unfortunately, the techniques available at that time for rehabilitating oiled birds were not very effective, and only 6% of the treated animals survived for more than one month. Similarly, more than 1600 oiled birds were cleaned after the Santa Barbara spill in 1969, mostly western grebes (*Aechmophorus occidentalis*) and loons (*Gavia immer*, *G. stellata*, and *G. arctica*). Only 15% of the treated birds survived to be released.

These early rehabilitation treatments were not successful because scientists did not yet understand that more is needed than just removing oily residues from birds — their poisoning must also be addressed. Scientists determined several reasons for these problems:

- Since birds were not captured and treated soon enough after being oiled, they became hypothermic (excessively cooled). Also, the animals were ingesting petroleum residues while attempting to clean themselves.
- The means of removing oily residues from birds were inappropriately harsh. The solvents and emulsifiers used were themselves toxic, caused irreparable damage to feather structure, or did not clean the feathers sufficiently.
- Most oiled birds are hypoglycemic to some degree. This condition is characterized by a negative energy balance (that is, by low blood sugar and weight loss), and requires immediate treatment with a glucose solution.
- An important effect of hydrocarbon poisoning in birds, particularly with aromatic hydrocarbons, is disruption of the ability to regulate ion concentrations (especially sodium and potassium) in the blood plasma. This condition requires the oral administration of an electrolyte solution.
- Aromatic hydrocarbons are toxic to red blood cells, resulting in a hemolytic anemia that requires several weeks of treatment through appropriate nutrition.

Today, much better procedures are available for capturing, cleaning, and rehabilitating oiled birds. These improved techniques have been developed largely through trial and error while treating accidentally oiled birds, coupled with laboratory research involving experimentally oiled animals. Because we now know that oiled birds must be treated as soon as possible, spill-response teams try to capture oiled birds quickly. This reduces the amount of stress and toxicity the animals experience before they can be cleaned. In addition, relatively gentle cleaning solutions, known as polysorbates, are used to de-oil birds, and electrolyte solutions and glucose are routinely administered to treat dehydration and hypoglycemia.

Finally, the methods of postcleaning rehabilitation and release have been improved. Typically, birds are kept for 7–10 days after cleaning. They are released as soon as the waterproofing of their feathers has been restored, their salt-excreting metabolism has recovered, their anemia is corrected, and they have started to regain previously lost weight.

As a result of these improved methods, up to 75% of oiled birds can be released after timely cleaning and rehabilitation. However, the success rate still varies greatly, depending on bird species, the type of oil, and other factors — particularly how much time has passed between the oiling event and the capture and treatment of the animal.

Unfortunately, in spite of today's relatively successful cleaning techniques, recent studies have shown that the post-release survival of the birds is poor. It appears that fewer than 1% of treated and released seabirds survive for one year (Sharp, 1996). With such poor post-release survival rates, it is questionable whether any substantial ecological benefits are gained from the cleaning programs. Moreover, it is expensive to treat oiled birds. Large numbers of volunteers are needed, including specialists such as wildlife veterinarians. It is, of course, much better to avoid the oil spill altogether than to try to deal with the terrible damages caused to species and ecosystems.

Immediately after the wreck of the *Torrey Canyon*, an intensive cleanup of the oiled beaches began. This effort used large amounts of detergents and dispersants to create oil-in-water emulsions on polluted shorelines. The emulsions were then rinsed back to sea using pressurized water streams from hoses. Unfortunately, the chemicals used as emulsifiers were extremely toxic to marine organisms. The enthusiastic use of these materials greatly increased the toxic damage already caused by the petroleum to flora and fauna on beaches and in intertidal and subtidal habitats.

Emulsifiers were not used during the cleanup of rocky beaches. There, the marine algae, while damaged by oily residues, preserved some of their regenerative tissues, and regrew relatively quickly. Some species of intertidal invertebrates also proved rather tolerant to oiling. Many limpets (*Patella* spp.), for example, were able to survive being covered by oily residues, and were later able to graze on algae on oil-covered rocks. In all cases, damages to marine species were much more severe wherever detergent or dispersant had been used during the cleanup.

The unanticipated damages caused by toxic emulsifiers were a critical lesson from the cleanup of the *Torrey Canyon* oil spill. Soon after, less toxic dispersants were developed for use in oil-spill emergencies. Techniques improved too, so that these chemicals were used more judiciously, mainly to clean sites of high value for industrial or recreational purposes and to treat offshore locations where ecological damages would be minimized.

A post-oiling succession occurred after the *Torrey Canyon* disaster, which eventually restored ecosystems natural to the region. Oiled habitats in the rocky intertidal zone were initially colonized by the opportunistic green alga *Enteromorpha*. As invertebrate herbivores recovered in abundance, this alga was grazed and replaced by species of seaweeds, which are the typical algae of rocky intertidal habitats. Except for lingering effects on seabird populations, ecological damages caused by the *Torrey Canyon* spill turned out to be relatively short term, because recovery was vigorous.

In habitats cleaned with emulsifiers, the recovery was slower. Some areas took up to ten years to recover communities similar to those present before the spill.

The *Amoco Cadiz*

About a decade after the *Torrey Canyon* accident, in 1978, another supertanker ran aground in the same general area, but closer to France. This was the *Amoco Cadiz*, whose wreck spilled 233 000 t of petroleum and fouled about 360 km of shoreline. A 140 km portion of the fouled coast was oiled heavily. The intensive cleanup of many of the polluted beaches involved physically digging up and removing oily sand, sediment, and petroleum residues. Detergents and low-toxicity dispersants were used only to remove fouling residues in harbours, and in offshore habitats to disperse floating masses of mousse. Many of the ecological damages caused by oil pollution and the cleanup were less severe than in the case of the *Torrey Canyon*. Recovery from the *Amoco Cadiz* spill was relatively rapid, and substantially complete within several years. Some effects on invertebrates lasted for a decade, and there were lingering damages to local colonies of alcid seabirds.

The *Exxon Valdez*

The most damaging tanker accident in North American waters was the wreck of the *Exxon Valdez* in southern Alaska. About one-quarter of the petroleum produced in the United States is mined in northern Alaska. The crude oil is transported by a 1280-km overland pipeline, from the oil fields on the North Slope south to the port of Valdez in southeastern Alaska. The petroleum is then transported to markets in the western United States by a fleet of supertankers. The first part of the oceanic passage runs through Prince William Sound, using a narrow but well-charted shipping channel.

Before the *Exxon Valdez* accident on March 23, 1989, tankers had successfully navigated that passage about 16 000 times, carrying 1.25 billion t of petroleum. However, the *Exxon Valdez*, the newest ship in the Exxon tanker fleet, was incompetently steered onto a submerged reef, where it grounded. This resulted in the spillage of 36 000 t of its load of 176 000 t of petroleum into Prince William Sound, the largest-ever oceanic spill in North America. About 40% of the spill washed onto shoreline habitats of Prince William Sound, while 25% was carried out of the sound by currents, and 35% evaporated at sea. Less than 10% of the spill was recovered or burned, mostly from oiled shores.

The grounding of the *Exxon Valdez* could have been avoided by more sensible operation of the tanker and better control of its personnel. At the time that the ship went aground, its bridge was under the command of an unqualified mate. Unaccountably, the captain was in his cabin. Only some ten minutes after assuming control of the *Exxon Valdez*, the mate, who was not sufficiently familiar with the shipping channel and its aids to navigation, had run the huge supertanker securely onto an unforgiving reef.

The damages caused by the grounding of the *Exxon Valdez* were compounded by a stunning lack of preparedness by industry and government for dealing with an oil-spill emergency at the port of Valdez. Essential equipment for containment and oil recovery was not immediately available, and it took much too long to mobilize trained staff. Consequently, despite favourable weather and sea conditions during the first critical days after the grounding, few effective oil-spill countermeasures were mounted. Not until

CANADIAN FOCUS 22.2

RAISING THE *IRVING WHALE*

The *Irving Whale* is a ocean-going barge that was used to haul residual fuel oil (also known as bunker-C) from a refinery in Saint John, New Brunswick, to various industrial customers in eastern Canada. On September 7, 1970, the *Irving Whale* was caught in a fierce gale and sank in the Gulf of St. Lawrence, about 100 km north of Prince Edward Island.

For 26 years, the *Irving Whale* lay on the bottom, 67 m deep. The barge still contained about 3000 t of its shipment of 4300 t of bunker-C. Occasionally, part of its cargo leaked, floating to the surface as sticky globs of pollution. In addition to its cargo of heavy fuel oil, the *Irving Whale* also contained 7000 L of PCBs (polychlorinated biphenyls). The PCBs were used as a heat-exchange fluid within an internal system that heated the bunker-C oil to make it fluid enough for pumping. Because the wreck was sitting on the bottom of the Gulf of St. Lawrence and still contained bunker-C and PCBs within a damaged hull, the *Irving Whale* was deemed an "environmental time-bomb." As a result, the Government of Canada decided that the barge should be raised and salvaged.

The salvage operation was somewhat risky, because it was not known how much damage the hull had suffered at the time of sinking, nor how much it had corroded in 26 year. It was therefore feared that disturbing the wreck might cause an enormous spillage of oil and PCBs. The plan generated much interest and controversy, mainly because people were afraid of the pollution that might ensue. Several court actions were initiated, and the resulting injunctions caused delays in the lifting of the barge. Extended periods of bad weather also hindered the salvage operation. Eventually, however, the legal uncertainties and weather cleared sufficiently to allow the *Irving Whale* to be lifted from the depths.

Both the sunken barge and its environment were carefully examined. Next, the wreck was fitted with an apparatus that allowed it to be lifted a short distance by a heavy marine crane. Several enormous slings were then installed beneath the barge, allowing the *Irving Whale* to be brought to the surface. This was done successfully, with no significant spillage of oil or PCBs. The barge was taken to Halifax, where its remaining bunker-C cargo was off-loaded.

Remarkably, the salvaged barge was in such good shape that it may be refurbished and again used to transport fuel oil in Atlantic Canada.

The salvage of the *Irving Whale* cost the federal government at least $35 million. However, the operation removed a substantial pollution hazard from the environment, and for this reason should be regarded as an environmental success story.

---

the second day of the spill was it possible to begin to off-load unspilled petroleum from the *Exxon Valdez* to another tanker, and not until the third day were floating booms deployed to contain part of the spill. Unfortunately, a gale developed on the fourth day. It then became impossible to contain or recover spilled petroleum, which subsequently became widely dispersed. Much of the spillage eventually fouled the western shoreline of Prince William Sound and other areas downcurrent along the Alaskan coast.

The region around Prince William Sound is famous for its spectacular scenery and ecologically important populations of wildlife. Some ecological communities and species populations were severely damaged by the oil spill, although not all of the damages nor their long-term effects have been well quantified. Significant controversies have arisen both from the poor understanding of some of the ecological effects of the *Exxon Valdez* oil spill, and from the role of science and scientists in sorting out some of the legal and political aspects of the disaster (Holloway, 1996; Weins 1996). For a long time, scientists were prohibited from sharing some of their information, because of legal needs for confidentiality. Controversies arose among scientists, environmental advocates, and other interest groups about the scale and intensity of some of the reported ecological impacts.

About 1900 km of shoreline habitats in Prince William Sound and its surrounding area were oiled to some degree. A survey of 1437 km of shoreline found that 140 km were "heavily oiled," meaning that there was at least a 6-m width of oiled substrate. Another 93 km were "moderately oiled" (with a 3–6-m width of oiled beach), 323 km were "lightly oiled" (with a width of less than 3 m oiled), and the rest

The top photo shows a heavily oiled beach on Green Island, Prince William Sound, Alaska, soon after the *Exxon Valdez* disaster in 1989. The site was cleaned with warm-water washing in 1989, and then cleaned manually in 1990. In 1990 and 1991, it was fertilized to enhance microbial breakdown of the petroleum residues.

The bottom photo shows the greatly improved condition of the same beach in 1992, as a result of both natural and managed cleanups. Although there is little visible damage on the surface, there are still hydrocarbon residues deeper in the substrate.

Source: Exxon Corporation

was "very lightly oiled" (with less than a 10% cover of mousse). Overall, about 20% of the shoreline of Prince William Sound, plus an additional 14% of the beaches on the nearby Kenai Peninsula and Kodiak Island, suffered some degree of oiling.

A heroic and extremely expensive effort was undertaken to recover and clean up some of the petroleum residues of the *Exxon Valdez* spill, particularly from oiled beaches. About 11 000 people were involved in that cleanup, which cost the Exxon corporation about $2.5 billion (U.S.). The United States federal government spent an additional $154 million (U.S.). Petroleum residues were removed from many heavily oiled beaches with machines and by people wielding shovels and bags. Other places were cleaned by rinsing with sprays of hot or cold water. On some beaches, people actually wiped oiled rocks with absorbent cloths, a procedure that was semi-humorously referred to as "rock polishing."

These cleanup efforts helped greatly. They were subsequently aided by natural cleanup processes, especially winter storms and microbial degradation of residues. Consequently, the amounts of petroleum residues on rocks and beaches declined rapidly in the years following the spill. One survey of 28 oiled sites found an average of 37% surface oil cover in the first post-spill summer of 1989, but less than 2% cover in 1990. Another survey in 1991, after two post-spill winters and three summers, found that fewer than 2% of the beaches in Prince William Sound still had visible surface residues of oil, compared with 20% in the first summer after the spill. Subsurface residues of petroleum still exist in many places, and the long-term ecological implications of this less visible, lingering pollution are not fully known.

Initially, severe damages were caused to the seaweed-dominated intertidal communities of oiled beaches. These effects were made worse by certain cleanup methods, particularly those involving washing with pressurized hot water. Fortunately, much of this damage proved to be short-term. By the end of the summer of 1991, a substantial recovery of intertidal seaweeds and invertebrates had begun on many beaches, although there were lingering effects on the community structure.

Prince William Sound supports large populations of salmon and herring, with economically important fisheries for both species. In 1988, before the *Exxon Valdez* oil spill, the catch had a value of more than $90 million (U.S.). The fishery was closed in 1989 because of the oil spill. As a result, Exxon paid compensation of $302 million (U.S.) to the displaced fishers and processors (many of whom were also employed in the cleanup, earning another $105 million (U.S.) in wages and vessel charters).

In 1990, the harvest of pink salmon (*Oncorhynchus gorbuscha*) in Prince William Sound was 44 million fish, much larger than the previous record high catch of 29 million fish. These were 2-year-old fish, about one-quarter of which would have passed through Prince William Sound during their migration from rivers to the sea in 1989, the year of the spill. The rest of those juvenile salmon had been released from hatcheries in the Sound. The 1991 catch of pink salmon was also large, at 37 million fish. There was also a large harvest of Pacific herring (*Clupea harengus pallasi*) in 1990, when about 7500 t were landed. The largest catch of herring in a decade was made in 1991, when 10 800 t were taken. These data suggest that the salmon and herring fisheries were not severely degraded by the *Exxon Valdez* oil spill.

Sea otters (*Enhydra lutris*) were the hardest-hit of the marine mammals. More than 1000 individuals were killed by oiling out of a total population of 5000–10 000 in Prince William Sound. A total of 357 oiled sea otters were captured and treated, and 223 of these animals survived and were later released or placed in zoos.

Seabirds can be extremely abundant in the region, particularly in the autumn when some species "stage" to prepare for their southern migration. At that time, about 10 million seabirds inhabit the Sound area. Fortunately, the *Exxon Valdez* disaster happened in late winter, but there were still about 600 000 seabirds in the sound. About 36 000 dead birds were actually counted as beached corpses. However, many additional corpses sank or drifted out to sea, and the estimated total mortality was 375 000–435 000 seabirds. In addition, at least 153 bald eagles (*Haliaeetus leucocephalus*) were killed when they became oiled while scavenging seabird carcasses.

About 400 people, 140 boats, and 5 aircraft were hired by Exxon, the owner of the stricken supertanker, to capture and rehabilitate oiled birds. In total, they managed to treat 1600 birds of 71 species. About one-half of the captured birds died of their oil-caused injuries. The rest were treated and eventually released back to the wild. The lingering effects of hydrocarbon poisoning probably prevented most of them from surviving for long.

Oil Spills from Offshore Drill Platforms

The world's largest accidental spill was a blowout (an uncontrolled discharge from a wellhead) in 1979 from the *IXTOC-I*, a drilling platform being used for petroleum exploration off the east coast of Mexico. This blowout remained uncontrolled for more than nine months, resulting in a continuous spillage that amounted to an estimated 476 000 t of petroleum. About 50% of the spill is thought to have evaporated into the atmosphere, while 25% sank to the bottom of the Gulf of Mexico, 12% was degraded photochemically or by microorganisms, 6% was burned or recovered near the spill site, and 7% fouled about 600 km of shoreline in Mexico and Texas.

This enormous blowout caused great economic damage by fouling beaches important to tourism. In addition, the fishing industry in the Gulf of Mexico was damaged by the oiling of boats and fishing gear, by the prevention of fishing in the vicinity of slicks, and by the tainting of commercial species of fish and invertebrates with foul-tasting hydrocarbons. In addition, many birds, sea mammals, sea turtles, and other wildlife were oiled and died, although these and other ecological damages were not well documented.

Another offshore blowout occurred in 1969 off Santa Barbara in southern California. This spill involved about 10 000 t of petroleum and fouled about 230 km of coastline. Birds were the most obviously tragic victims of this blowout, with about 9000 killed, or approximately half of the population present at the time of the spill. About 60% of the dead birds were western grebes and loons which winter in the area in large numbers. Although attempts were made to capture and clean oiled birds, the efforts were not very successful.

Coastal ecosystems were also severely damaged, especially in rocky intertidal habitats, but recovery was fairly rapid. Within only one year, barnacles began recolonizing intertidal habitats, even establishing themselves on rocks that were still covered with asphaltic residues. Beaches used for recreation were cleaned by the physical removal of petroleum and oily sand, by blasting with water or steam, or by spraying with solvents to wash the oily residues back to sea. As with the *Torrey Canyon* cleanup, these cleanup methods greatly exacerbated the ecological damages. Organisms were removed along with petroleum residues, and some of the cleaning agents were toxic.

Oil Spills in the Arctic Ocean

The consequences of a petroleum spill in the Arctic Ocean are potentially catastrophic. Such a spill could result from a tanker accident in ice-choked Arctic waters, or from operations at offshore oil wells.

Climatic conditions in the Arctic are severe — a factor that greatly increases the likelihood of spills from offshore oil wells as a result of equipment failure or human error. Furthermore, the cold, icy conditions of the long, arctic winter would make it difficult to quickly drill an offshore relief well, a necessary step in controlling a blowout. Containing or cleaning up a spill into arctic seas would also be difficult.

In fact, some scientists have suggested that an offshore blowout of a drilling or production well in the Arctic Ocean could remain uncontrolled for several years. Such an event would result in an enormous, uncontainable spillage of petroleum. Because of entrapment under sea ice and the cold, nutrient-poor conditions, spilled oil would not evaporate or dissolve into seawater very effectively, and microbial biodegradation would be extremely slow. Consequently, the amounts of spilled petroleum would not decrease much over time, and most of the initial toxicity of the oil would persist. (Note that the spill from the *Exxon Valdez* occurred in boreal waters of southern Alaska, which are subject to much less severe temperature and ice conditions than occur in the Arctic Ocean.)

The ecological damage from an offshore spill in the Arctic Ocean could be catastrophic. Most of the marine wildlife of the Arctic, particularly migratory seabirds and marine mammals, is extremely vulnerable to the effects of petroleum spills. When they return to their northern breeding habitats in the early summer, marine birds and mammals commonly aggregate in dense populations in patches of ice-free water, known as leads and polynyas, within the greater expanse of sea ice. Unfortunately, these open-water habitats are places where spilled petroleum would accumulate in large quantities. Enormous mortality of migrating sea ducks, murres, seals, whales, polar bears, and other species would result as they became oiled by sticky residues. Because of the persistence of petroleum residues in the cold Arctic Ocean, this threat could persist for years, and long-term, debilitating damages to populations of these animals would likely result. Potential damages to fish, zooplankton, and other components of the marine ecosystem are little known, but would probably be less intensive than the effects on marine birds and mammals.

A number of exploration wells have been drilled on the continental shelf of the Arctic Ocean off northern Canada and Alaska (and also in boreal and temperate waters off Newfoundland and Nova Scotia). Fortunately, there have not been any large spills of petroleum from the offshore drilling activities in the Arctic Ocean of North America (although there have been several blowouts involving natural gas). However, in spite of the adoption of the most modern spill-prevention technologies, a severe spill may be inevitable in the exploration and production activity offshore in the Arctic. Such an accident would cause enormous ecological damage, from which recovery would be extremely slow.

Effects of Chronic Oil Pollution

Environments around tanker terminals and coastal petroleum refineries are chronically exposed to pollution by small but frequent oil spills, discharges of contaminated waste waters, and airborne contaminants from industrial stacks. Similarly, coastal ecosystems near cities and towns, both marine and freshwater, are chronically affected by oils and fuels that are dumped into sewers, which often discharge these wastes directly into the aquatic environment. Chronic exposures such as these are much less intense than the severe pollution associated with wrecked tankers, but ecological damages are still inflicted.

Chronic exposures to hydrocarbons and other pollutants have been blamed for unusually high frequencies of cancers and other diseases in fish and shellfish. Although the exact causes of many of these wildlife diseases have not yet been determined, many scientists believe that they are somehow caused by chronic pollution. One study, for example, found an unusually large incidence of gonadal tumours in fish (up to 100% among older male fish) sampled from a polluted river near Detroit, Michigan. However, epidemics of wildlife diseases are not always observed in chronically polluted environments.

Ecological damage at the community level has been observed near the effluent discharges from some petroleum refineries. Extensive studies in Britain, for example, have shown a deterioration of salt-marsh vegetation near oil refineries. Exposed bare mud was found where well-vegetated, grassy marshes had occurred previously. In places where industry made serious efforts to reduce the emissions of hydrocarbon pollutants, new vegetation was able to recolonize the mud and redevelop a salt marsh.

Terrestrial Oil Spills

Although oil spills cause severe damage to terrestrial vegetation, only relatively local areas are usually affected (except in the case of extremely large spills). This is because soil, particularly organic soil, has a great absorptive capacity for petroleum. In addition, much of the oil spilled on land tends to accumulate in low spots and does not spread widely. This is particularly true in much of northern Canada, where deep infiltration of petroleum into the soil may be prevented by impenetrable bedrock or permafrost. The relatively localized impacts of most terrestrial spills of petroleum are very different from the effects in aquatic environments, in which spilled oil spreads widely and can affect enormous areas.

Research has also shown that a wide range of natural, soil-dwelling microorganisms are able to utilize petroleum residues as a metabolic substrate. These oil-degrading bacteria, fungi, and other microbes are widespread in natural soils and waters. After soil becomes polluted by an oil spill, these microorganisms proliferate rapidly, responding to the presence of hydrocarbons that can be utilized as a source of metabolic energy.

Petroleum, although a carbon-rich substrate, is highly deficient in other nutrients such as nitrogen and phosphorus. Consequently, the vigour of microbial responses to oiling, and the rate of decomposition of the petroleum residues, can be increased greatly by fertilizing oiled soils. Microbial decomposition of residues can also be increased by occasionally tilling the soil to increase the availability of oxygen. In general, fertilization is a relatively inexpensive but effective way of speeding up the rate of degradation of petroleum residues, while avoiding the severe damage associated with physical cleanups. In agricultural areas in temperate regions of Canada, fertilization plus tillage can be used effectively to biodegrade petroleum residues in soil.

Of course, any spilled oil that reaches groundwater or surface waters causes severe damage there. Petroleum spills into high-energy streams and rivers become extensively dispersed, and some of the residues eventually flow into lakes or the ocean. Oil that finds its way into ponds and lakes can be quite persistent. It accumulates around the margins of those waterbodies, damaging vegetation and wildlife habitat. Studies have shown, however, that after spilled petroleum has weathered for a year or more, the toxicity of the residues has decreased to a point at which aquatic plants can grow through surface slicks without suffering much damage. The phytoplankton and zooplankton communities of ponds and lakes are also somewhat resistant to the effects of weathered petroleum; although there is much initial damage, recovery can occur after several years have passed. However, any waterfowl that attempt to feed on oiled lakes or ponds quickly become fouled with residues, and this is usually fatal.

Terrestrial Spills in the Arctic

Much exploration for petroleum has occurred in the Arctic, mostly since the late 1960s. Discoveries in Alaska were large enough to allow the construction of a pipeline to carry petroleum from the North Slope to the southern port of Valdez, for shipment to the lower American states by tankers (such as the *Exxon Valdez*). There have also been major discoveries of petroleum in the Canadian Arctic, but these are not large enough to make their commercial development economically feasible at present.

Future exploration in Canada may discover reserves of petroleum large enough to justify construction of a southward pipeline, most likely along the valley of the Mackenzie River. In fact, a pipeline along this route was seriously considered during the early 1970s, to carry both Alaskan petroleum and the smaller quantities of oil that had been discovered in and near the Mackenzie River delta in Canada.

During the *environmental impact assessments* of petroleum exploration and pipeline construction in the Arctic, researchers studied the potential effects of accidental oil spills on tundra and boreal forest ecosystems. Some of the Canadian research involved experimental spills of petroleum onto plots of vegetation. The initial damages were documented and the subsequent recovery monitored over time. These studies found that crude oil behaved as a contact herbicide to terrestrial vegetation, killing foliage and woody tissues upon direct exposure. In many species of plants, however, the perennating (regenerating) tissues were not all killed, allowing rapid regrowth after the experimental oiling.

These general observations are illustrated by a study conducted in the western Canadian Arctic (Table 22.4). The experimental oiling caused a rapid defoliation of plants. This damage is reflected by the small values of foliar cover after the oiling, in contrast to the nonoiled, reference

A one-year-old petroleum spill near Norman Wells, Northwest Territories. The dominant shoreline plant is the sedge *Carex aquatilis*. The foliage of this plant was killed by oiling, but its rhizomes in the sediment survived. The plants grew new foliage that penetrated the floating residues without suffering much damage.

Source: B. Freedman

vegetation, whose cover did not change substantially. Black spruce (*Picea mariana*) trees are the dominant plants in the two boreal forest sites. Although the spruce trees did not die immediately after oiling, they became more vulnerable to physiological stresses associated with the hard arctic winter. As a result, the oiled trees eventually died, taking as long as several years to do so.

After the initial post-oiling damages, many plant species of the forest and tundra began to recover. Black spruce was a notable exception, as no seedlings of this tree were observed during the five-year period of the study. Lichens and mosses also recovered slowly from the oiling.

Of course, the environmental consequences of northern oil development are much broader than the ecological effects of petroleum spills on land or in water. The construction of infrastructure such as roads and pipelines in remote and inhospitable terrain would have a variety of environmental consequences. In addition, the influx of large sums of money and wage employment into small northern towns would have huge socioeconomic impacts, some of them positive, others potentially disruptive. As with all other industrial activities, any potential damage to the ecological and socioeconomic environments must be identified and, as far as possible, minimized. The residual damage must then be balanced against the economic and social benefits accruing from the development of the fossil-fuel resources.

TABLE 22.4 | EFFECTS OF EXPERIMENTAL SPILLS OF CRUDE OIL ON ARCTIC VEGETATION

The studied plant communities in the western Canadian Arctic were (1) mature black spruce (*Picea mariana*) boreal forest, (2) 40-year-old spruce forest, (3) cotton-grass (*Eriophorum vaginatum*) wet-meadow tundra, and (4) dwarf-shrub tundra. The oiled vegetation was treated with petroleum at 9 litres/metre^2, while the reference vegetation was not oiled. The forest study area is near Norman Wells, and the tundra is near Tuktoyaktuk, both in the Northwest Territories.

COMMUNITY TYPE	TREATMENT	COVER (%) OF PLANT FOLIAGE			
			POST-SPILL		
		PRE-SPILL	1 y	2 y	5y
Mature boreal forest	Reference	195	215	255	240
	Oiled	350	18	10	20
40-year-old forest	Reference	355	360	260	235
	Oiled	420	20	20	95
Cotton-grass tundra	Reference	339	284	268	–
	Oiled	358	26	34	–
Dwarf-shrub tundra	Reference	339	338	292	–
	Oiled	322	55	82	–

Source: Modified from Freedman and Hutchinson (1976) and Hutchinson and Freedman (1978)

KEY TERMS

petroleum (or crude oil)
hydrocarbons
load-on-top (LOT) method
biodegradation

QUESTIONS FOR DISCUSSION

1. What are the causes of petroleum spills to the oceans?
2. Why were the ecological effects of the *Amoco Cadiz* oil spill generally fewer and shorter-lasting than those of the *Torrey Canyon*?
3. Considering the poor survival of aquatic birds after they have been "rehabilitated" and returned to the ocean, do you think that it is worthwhile to attempt to treat these victims of oil spills?
4. In view of the ecological risks, do you think that oil exploration and extraction should be allowed in the Canadian Arctic?

REFERENCES

Alexander, V. and K. Van Cleve. 1983. The Alaska pipeline: a success story. *Ann. Rev. Ecol. System.*, **14**: 443–463.

Baker, B., B. Campbell, R. Gist, L. Lowry, S. Nickerson, C. Schwartz, and L. Stratton. 1989. *Exxon Valdez* oil spill: the first eight weeks. *Alaska Fish & Game*, **21 (4)**: 2–37.

Baker, J.M. (ed.) 1976. *Marine Ecology and Oil Pollution*. New York: Wiley.

Berger, A.E. 1993. *Effects of the Nestucca oil spill on seabirds along the coast of Vancouver Island in 1989*. Vancouver: Canadian Wildlife Service. Tech. Rep. Ser. No. 179.

Bourne, W.R.P. 1976. Seabirds and pollution. In: R. Johnston, ed. *Marine Pollution*. London: Academic. pp. 403–502.

Boesch, D.F. and N.N. Robelais (eds.). 1987. *Long-term Environmental Effects of Offshore Oil and Gas Development*. London: Elsevier Science Publishers.

Cairns, J.L. and A.L. Buikema (eds.). 1984. *Restoration of Habitats Impacted by Oil Spills*. Boston: Butterworth.

Clark, R.B. 1984. Impact of oil pollution on seabirds. *Environ. Pollut., Ser. A*, **33**: 1–22.

Clark, R.C. and W.D. MacLeod. 1977. Inputs, transport mechanisms, and observed concentrations of petroleum in the marine environment. In: D.C. Malins, ed. *Effects of Petroleum on Arctic and Subarctic Marine Environments and Organisms, Vol. 1*. New York: Academic. pp. 91–224.

Colombo, J.R. (ed.). 1993. *The Canadian Global Almanac 1992.* Toronto: Global.

Cormack, D. 1983. *Response to Oil and Chemical Marine Pollution.* London: Applied Science.

Davidson, A. 1990. *In the Wake of the Exxon Valdez.* Toronto: Douglas & McIntyre.

Earle, S. 1992. Assessing the damage one year later (after the Gulf oil spill). *National Geographic,* **179 (2)**: 122–134.

Engelhardt, F.R. (ed.) 1985. *Petroleum Effects in the Arctic Environment.* New York: Elsevier Press.

Foster, M.S. and R.W. Holmes. 1977. The Santa Barbara oil spill: an ecological disaster In: J. Cairns, K.L. Dickson, and E.E. Herricks, eds. *Recovery and Restoration of Damaged Ecosystems.* Charlottesville, VA: University Press of Virginia. pp. 166–190.

Freedman, B. 1995. *Environmental Ecology, 2nd ed.* San Diego, CA: Academic.

Freedman, B. and T.C. Hutchinson. 1976. Physical and biological effects of experimental crude oil spills on low arctic tundra in the vicinity of Tuktoyaktuk, NWT, Canada. *Can. J. Bot.,* **54**: 2219–2230.

GESAMP. 1991. *Carcinogens: Their Significance as Marine Pollutants.* London: International Marine Organization. Joint Group of Experts on the Scientific Aspects of Marine Pollution (GESAMP), Report 46.

GESAMP. 1993. *Impact of Oil and Related Chemicals and Wastes in the Marine Environment.* London: International Marine Organization. Joint Group of Experts on the Scientific Aspects of Marine Pollution (GESAMP), Report 50.

Government of Canada. 1991. *The State of Canada's Environment.* Ottawa: Government of Canada.

Harvey-Clark, C. 1990. Veterinary treatment of oiled seabirds. *Bull. B.C. Veterinary Medical Assoc.,* **28 (4)**: 24–33.

Holloway, M. 1996. Sounding out science. *Sci. Amer.,* **275 (4)**: 106–112.

Holloway, M. and J. Horgan. 1991. Soiled shores. *Sci. Amer.,* **265 (4)**: 103–116.

Holmes, W.N. 1984. Petroleum pollutants in the marine environment and their possible effects on seabirds. *Rev. Environ. Toxicol.,* **1**: 251–317.

Hutchinson, T.C. and B. Freedman. 1978. Effects of experimental crude oil spills on subarctic boreal forest vegetation near Norman Wells, NWT, Canada. *Can. J. Bot.,* **56**: 2424–2433.

Jenssen, B.M. 1994. Effects of oil pollution, chemically treated oil, and cleaning on the thermal balance of birds. *Environ. Pollut.,* **86**: 207–215.

Koons, C.B. 1984. Input of petroleum to the marine environment. *Marine Tech. Soc. J.,* **18**: 97–112.

Koons, C.B. and H.O. Jahns. 1992. The fate of oil from the *Exxon Valdez* — a perspective. *Marine Tech. Soc. J.,* **26**: 61–69.

Malins, D.C. (ed.). 1977. *Effects of Petroleum on Arctic and Subarctic Marine Environments and Organisms, 2 vol.* New York: Academic.

Milne, A.R. and R.H. Herlinveaux. 1977. *Crude Oil in Cold Water.* Sidney, BC: Department of Fisheries and Department of the Environment.

National Research Council (NRC). 1985. *Oil in the Sea: Inputs, Fates, and Effects.* Washington: National Academy Press.

National Research Council (NRC). 1989. *Using Oil Spill Dispersants in the Sea.* Washington: National Academy Press.

Neff, J.M. and J.W. Anderson. 1981. *Response of Marine Mammals to Petroleum and Specific Petroleum Hydrocarbons.* London: Applied Science.

NOAA. 1982. *Ecological Study of the Amoco Cadiz Oil Spill.* U.S. Department of Commerce. Washington: National Oceanic and Atmospheric Administration (NOAA).

Paine, R.T., J.L. Ruesink, A. Sun, E.L. Soulanille, M.L. Wonham, C.D.G. Harley, D.R. Brumbaugh, and D.L. Secord. 1996. Trouble on oiled waters: lessons from the *Exxon Valdez* oil spill. *Annu. Rev. Ecol. System.,* **27**: 197–235.

Piatt, J.F., H.R. Carter, and D.N. Nettleship. 1991. Effects of oil pollution on marine bird populations. In: J. White, ed. *The Effects of Oil on Wildlife: Research, Rehabilitation, and General Concerns.* Hanover, NH: Sheridan. pp. 126–141.

Pimlott, D., D. Brown, and K. Sam. 1976. *Oil Under the Ice.* Ottawa: Canadian Arctic Resources Committee.

Sharp, B.E. 1996. Post-release survival of oiled, cleaned seabirds in North America. *Ibis,* **138**: 222–228.

Southward, A.J. and E.C. Southward. 1978. Recolonization of rocky shores in Cornwall after use of toxic dispersants to clean up the *Torrey Canyon* spill. *J. Fish. Res. Bd. Canada,* **35**: 682–706.

Sprague, J.B., J.H. Vandermeulen, and P.G. Wells. 1982. *Oil and Dispersants in Canadian Seas — Research Appraisal and Recommendations.* Environment Canada, Environmental Protection Service, Economic and Technical Review Report EPS.3-EC-82-2.

Steinhart, C.E. and J.S. Steinhart. 1972. *Blowout: A Case Study of the Santa Barbara Oil Spill.* Belmont, CA: Duxbury.

Warner, F. 1991. The environmental consequences of the Gulf War. *Environment,* **33 (5)**: 7–26.

Weins, J.A. 1996. Oil, seabirds, and science: the effects of the *Exxon Valdez* oil spill. *BioScience,* **46**: 587–597.

Wells, P.G., J.N. Butler, and J.S. Hughes. 1995. *Exxon Valdez Oil Spill: Fate and Effects in Alaskan Waters.* Philadelphia: American Society for Testing and Materials.

World Resources Institute (WRI). 1994. *World Resources 1994–95: A Guide to the Global Environment.* New York: Oxford University Press.

23

ECOLOGICAL EFFECTS OF FORESTRY

Chapter Outline

Chapter Objectives

After completing this chapter, you will be able to:

1. Explain how forest harvesting removes nutrient capital from the site.
2. Outline how forestry can damage aquatic ecosystems, and how many of those damages can be avoided.
3. Describe how clear-cutting affects biodiversity.
4. Discuss the ecological consequences of conversions of natural forests into intensively managed plantations.
5. Explain the concept of integrated forest management.

INTRODUCTION

Forestry includes both the harvesting of trees and the management of post-harvest succession to foster the regeneration of another forest. Forest science guides these activities by providing understanding of the environmental factors that affect forest productivity and regeneration. The broad goal of commercial forestry is to provide sustainable harvests of tree biomass that can be used to manufacture lumber, paper, and other industrial products.

Forestry is an extremely important economic activity in many parts of the world. This is particularly true in Canada, where an enormous industrial enterprise depends on a continuous supply of tree biomass for the production of products that have a great economic value ($49.3 billion in 1994; see Chapter 14 for forestry-related economics).

Of course, to achieve the great economic benefits of commercial forestry, trees must be harvested from extensive areas of mature forest. In Canada, **clear-cutting** is by far the most common method of forest harvesting. In clear-cutting, all of the economically useful trees are harvested at the same time. In recent years, clear-cutting accounted for 87% of the annual harvest in Canada (see Table 14.13).

It is important to understand, however, that clear-cutting is not the same as **deforestation**. Deforestation involves the permanent conversion of a forest into some other kind of ecosystem, such as an agricultural or urbanized land use. In Canada and most other industrialized countries, clear-cutting is generally followed by the regeneration of another forest. In fact, it is common practice in Canada to manage the post-harvest succession to speed up the rate of regeneration. This means that the next harvest can be made after a relatively short time, allowing more profit to be made. (This period of time is known as a harvest *rotation*.) In this sense, commercial forestry as it is usually practised does not result in a net deforestation — if appropriately managed, the forest resource is not depleted. Even though almost one million ha of forest are harvested each year in Canada, the rate of net deforestation is only about 0.1% (see Table 14.12).

Harvesting can be viewed as a type of disturbance of forest ecosystems, followed by regeneration. Other forestry-related disturbances include those associated with **silvicultural** management, such as preparing the site for planting, thinning overly dense stands, and applying herbicide or insecticide to deal with pest problems. Silviculture is practised over an extensive area in Canada. For instance, about 656 million tree seedlings were planted on 418 000 ha of clear-cuts in 1993, equivalent to 43% of the area harvested. (The other 57% regenerated naturally, i.e., without planting.) Some of the planted areas are managed quite intensively to develop **plantations**. These are highly productive tree farms, which can be viewed as agroforestry systems in which trees are intensively cultivated. Because plantations are tree-dominated ecosystems, they are considered to be forests, although they lack many of the ecological and aesthetic values of natural forests.

Even though the normal practice of forestry in Canada and other developed countries does not result in deforestation, important ecological changes still result, some of them damaging to the environment. These changes can have important consequences for the renewability of the forest resource, for other economic resources such as hunted animals and recreational opportunities, and for ecological values such as indigenous biodiversity.

In this chapter we examine some of the ecological damages associated with the harvesting and management of forests, with a focus on effects on site capability, hunted animals, and biodiversity values. Some additional effects of forestry are discussed in other chapters. (We examined effects of insecticide and herbicide spraying in Chapter 21, implications of forest disturbances for carbon storage and CO_2 emissions in Chapter 17, and effects of tropical deforestation on global biodiversity in Chapter 24.)

This chapter differs in its approach from the previous chapters in this section on Environmental Damages. Here, forestry is considered as a discrete group of industrial activities. This chapter examines some of the diverse environmental stressors associated with these industrial activities, and describes their ecological effects. A similarly integrated approach could have been used to examine the ecological effects of agriculture or urbanization, but these affect less natural and more anthropogenic ecosystems than forestry does.

FOREST HARVESTING AND SITE CAPABILITY

In Chapter 14, we defined *site capability* as the potential of land to sustain the productivity of agricultural crops. Site capability is also relevant to forestry, being an indicator of the ability of land to sustain the productivity of crops of trees. Site capability is a complex attribute, involving the levels of nutrients and organic matter in soil, the availability of moisture, and other factors influencing plant growth. These environmental factors are influenced by soil type, climate, drainage, rates of nutrient cycling and decomposition, and the nature of the plant and microbial communities present.

The ability of soil to supply plants with nutrients is a critical aspect of site capability in forestry. In large part, this ecological function depends on the **nutrient capital** of the site, or the amounts of nutrients present in the soil, in living vegetation, and in dead organic matter. When trees are harvested from an area, the large amount of nutrients in their biomass is also removed. This can deplete the nutrient capital of the site.

Forests may be harvested using a variety of methods, which vary in the amount of biomass and nutrients removed from the site. **Selection harvesting** is a relatively "soft," non-intensive system, because it involves the harvest of only some of the trees from a stand, leaving others behind, with the forest remaining substantially intact. The most intensive kinds of harvests are *clear-cuts*, in which all economically useful trees are removed from an area. The smallest clear-cuts, typically involving a hectare or so of forest, are known as *group-selection harvests*. More typically, clear-cuts entail the harvesting of trees from larger areas, on the order of 20 to 100 hectares. The largest clear-cuts can extend over hundreds, and even thousands, of hectares. Such extensive operations, however, are unusual, being generally associated with the salvaging of trees damaged by wildfire, windstorm, or insect infestation.

There are also some less intensive methods of clear-cutting. A *shelterwood harvest* is a staged clear-cut, in which some larger trees of the economically most desirable species are left standing during the initial cut. These provide a seed source and a partially shaded environment that encourages natural regeneration of their own species. Once regeneration is well underway, the large "leave" trees are harvested. *Strip-cuts* are another staged harvest, in which long, narrow clear-cuts are made at intervals, with uncut forest left in between to provide a source of seed to regenerate desirable tree species in the cut strips. Once the regeneration is established, another strip-cut is made, again leaving intact forest on one of the sides. This system of progressive strip-cutting continues until all the forest in the management block (i.e., in the specific area being managed this way) has been harvested. Typically, areas are harvested in three to four strips. To regenerate trees on the final strips, foresters rely on so-called *advance regeneration* — that is, on small individuals of tree species that existed in the stand prior to harvesting and survive the disturbance of clear-cutting.

A three-year-old shelterwood cut of hardwood forest in Nova Scotia. About 40–60% of the commercial-sized trees were removed during the harvest, leaving many of the "best" trees to grow into relatively high-quality sawlogs, and to shed seeds to promote natural regeneration of the site. This silvicultural treatment produces a complex habitat, which supports a mixture of birds typical of clear-cuts and mature forest.

Source: B. Freedman

Clear-cutting systems also vary in how intensely the biomass of individual trees is harvested. The usual *stem-only* harvest involves the removal of tree trunks, leaving the roots, stumps, and logging "slash" (i.e., cut branches and foliage) on the site. The harvested logs can then be processed into lumber, plywood, or pulp for manufacturing paper. A *whole-tree harvest* is more intensive, because it takes all of the above-ground biomass of the trees. Whole-tree harvests recover considerably more biomass than stem-only harvests. This can be an advantage if the wood will be used as a source of energy (i.e., as fuel). Rarely, a complete-tree clear-cut may be made, which involves harvesting both the above- and below-ground biomass, including stumps and large roots.

Nutrient Removals During Harvesting

Although intensive harvests such as whole-tree clear-cuts increase the yield of biomass, they also increase the removal of nutrients. Some ecologists have suggested that the nutrient removals associated with successive whole-tree harvests could degrade the capability of sites to sustain tree productivity. The problem would be especially severe if intensive harvests were conducted over short rotations. These might not allow sufficient time for the nutrient capital to recover through natural inputs, such as precipitation, nitrogen fixation, and weathering of minerals in rocks (Figure 23.1).

Site impoverishment caused by intensive cropping is a well-known problem in farming, in which severely degraded land may have to be abandoned for some or all agricultural purposes. Usually this problem can be managed, to a greater or lesser degree, by applying fertilizers to the land. Sometimes, however, the degradation of site capability, especially tilth, is too severe, and this simple mitigation is not successful. Of course, the harvest rotation in agriculture is usually annual, whereas the rotation period in forestry typically ranges from about 40 to 100 years. However, each forest harvest involves the removal of a huge quantity of biomass, and thus of nutrients.

Compare, for example, the quantities of biomass and nutrients removed by stem-only and whole-tree clear-cuts of a conifer forest in Nova Scotia (Table 23.1). In this case, the whole-tree clear-cut yielded 30% more biomass than the stem-only harvest. The increased yield is an advantage of the whole-tree method, particularly if the harvest is to be used for energy production, for which biomass quality is not important. The increased harvest of biomass is, however, mostly due to the removal of such nutrient-rich tissues as foliage and small branches. Consequently, the whole-tree harvest removed up to twice as many nutrients

FIGURE 23.1 EFFECTS OF HARVEST INTENSITY AND LENGTH OF ROTATION ON SITE NUTRIENT CAPITAL

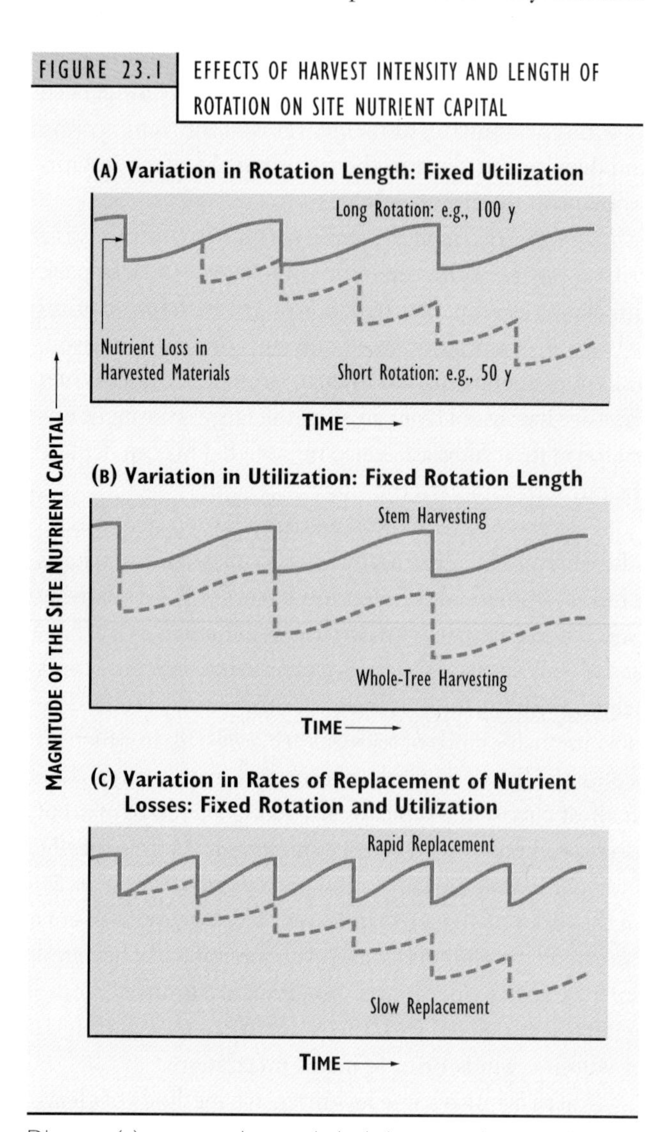

Diagram (a) suggests that a relatively long rotation can allow harvested nutrients to be replenished by inputs through rainfall, weathering of minerals, nitrogen fixation, and other means. An adequate postharvest recovery period means that harvesting is sustainable with respect to nutrient capital. Under the shorter rotation in the diagram, the harvested nutrients are not totally replenished between successive clear-cuts, resulting in a degradation of site nutrient capial.

Site capability degradation can also result from an increase in the intensity of the harvest. This is illustrated in diagram (b), in which a whole-tree clear-cut removes twice as much nutrient as a stem-only clear-cut. Diagram (c) indicates that fertile sites (solid line) are less likely to be degraded by intensive harvests over short rotations, compared with less fertile sites (dashed line).

Source: Modified from Kimmins (1996)

as did the stem-only clear-cut. In effect, a 30% increase in biomass yield by the whole-tree method was "purchased" at the ecological "expense" of 54–99% increases in the rates of removal of nutrients.

Unfortunately, there are few useful studies that allow foresters to compare the productivity of subsequent forest rotations on the same site. Such studies would take well over 100 years, requiring several generations of foresters! Therefore, it is difficult to evaluate the implications of nutrient removals in intensive forest harvesting, by either stem-only or whole-tree clear-cutting. Overall, however, it appears that an impoverishment of site nutrient capital is a less severe problem in forest harvesting than in agriculture. Consequently, nutrient removals by harvesting should be viewed as a potential, long-term problem. Because forestry is an economically important activity, and because the maintenance of site capability is critical to the sustainability of the enterprise, scientists must continue to study the effects of intensive harvesting on nutrient capital. In the short term, however, forestry causes more immediate damages to site capability and biodiversity that deserve our attention.

Leaching of Nutrients

The disturbance of forested lands can increase the rate at which dissolved nutrients are transported downward into the soil with percolating rainwater (a process known as *leaching*). If the nutrients leach deeper into the soil than tree roots can penetrate, they are effectively lost from the "working" nutrient capital of the site. Eventually, leaching nutrients can find their way into groundwater and surface waters.

The nutrients with the greatest tendency to leach are nitrate and potassium, both of which are highly water soluble. However, calcium, magnesium, and sulphate may also leach in significant amounts. Of course, following a clear-cut, any nutrient losses by leaching are in addition to the nutrients removed with tree biomass. Both these sources of nutrient loss reduce site nutrient capital.

The best-known study of nutrient leaching caused by forest disturbance was done at Hubbard Brook, New Hampshire. This large-scale experiment involved felling all the trees on a 16-ha watershed, but without removing any biomass — the cut trees were left lying on the ground. The entire watershed was then treated with herbicide for 3 y to suppress any regeneration. This severe experiment was designed to examine the effects of intense disturbance, by devegetation, on the biological control of watershed functions such as nutrient cycling and hydrology. The research was not intended to examine the effects of typical forestry practices. Still, important insights were gained into the effects of disturbance on ecological processes.

TABLE 23.1 | REMOVALS OF BIOMASS AND NUTRIENTS BY CLEAR-CUTS OF A CONIFER FOREST IN NOVA SCOTIA

This study involved weighing the biomass harvested by a stem-only clear-cut and a whole-tree clear-cut (both 0.5 hectares). Nutrient concentrations were determined for subsamples of the harvested biomass, and were used to calculate the amounts of nutrients removed. Biomass is stated in tonnes/hectare, and nutrients in kilograms/hectare. "Percentage Increase" refers to the whole-tree removals, compared with stem-only removals.

BIOMASS COMPONENT	BIOMASS (t/ha)	N (kg/ha)	P (kg/ha)	K (kg/ha)	Ca (kg/ha)	Mg (kg/ha)
STEM-ONLY CLEAR-CUT						
Tree stems	105.2	98	16	92	181	17
WHOLE-TREE CLEAR-CUT						
Tree stems	117.7	120	18	76	219	20
Branches and foliage	34.8	119	17	57	118	17
Total harvest	152.5	239	35	133	337	37
Percentage increase	30%	99%	93%	74%	54%	81%
Nutrients in the forest floor		900	62	110	290	32
Nutrients in the mineral soil (surface 30 cm)		3 860	1 220	13 300	5 460	1 740

Source: Freedman *et al.* (1981)

TABLE 23.2 NUTRIENT LOSSES IN STREAMFLOW AFTER DEFORESTATION OF A WATERSHED

The watershed is located at Hubbard Brook, New Hampshire. The data represent *net flux*, or the difference between inputs from precipitation and outputs due to streamflow. The data are reported for the first 3 y following experimental deforestation of a 16-ha watershed. The deforested area and an uncut, reference watershed of similar size are compared. Losses of material are stated in kilograms per hectare per 3-year period.

	LOSSES OF MATERIAL (kg/ha·3y)	
SUBSTANCE	DEFORESTED WATERSHED	REFERENCE WATERSHED
Nitrate (as NO_3-N)	−114.1	+2.3
Calcium	−77.7	−9.0
Silicate (as SiO_2-Si)	−30.6	−15.9
Potassium	−30.3	−1.5
Aluminum	−21.1	−3.0
Magnesium	−15.6	−2.6
Sodium	−15.4	−6.1
Sulphate (as SO_4-S)	−2.8	−4.1
Chloride	−1.7	+1.2
Bicarbonate (as HCO_3-C)	−0.1	−0.4
Ammonium (as NH_4-N)	+1.6	+2.2
Total dissolved substances	−307.8	−36.9

Source: Modified from Bormann and Likens (1979)

Overall, during a 10-year period following the cutting, the devegetated watershed lost 50 kg/ha·y of NO_3-N (that is, nitrogen in the form of nitrate), 45 kg/ha·y of Ca, and 17 kg/ha·y of K in streamflow (Bormann and Likens, 1979; Table 23.2 shows data for the first three years, which were the most dramatic). The 10-year losses were much larger from the disturbed watershed than from an undisturbed watershed monitored as a reference treatment: 4.3 kg/ha·y of NO_3-N, 13 kg/ha·y of Ca, and 2.2 kg/ha·y of K. In part, the increased losses of nutrients in streamwater were due to a 31% annual increase in the yield of water from the devegetated watershed (this was the average during the first 3 y after cutting). The increased amount of streamwater was caused by the disruption of transpiration from plant foliage. However, increases of nutrient concentrations in the streamwater were more important: during the first three years of the study, NO_3 increased by an average factor of 40, K by 11, Ca by 5.2, and Mg by 3.9. The total losses of N, Ca, and Mg in the streamwater were similar to the amounts of those nutrients in the above-ground biomass of the forest.

Because this experiment in devegetation did not involve a typical forest management practice, the measured effects were unrealistically large. However, other watershed-level studies of clear-cutting and other typical forestry practices have also found that these disturbances can increase nutrient leaching, although to a lesser degree than in the case of the devegetation studied at Hubbard Brook. For example, in the first 3 y after the clear-cutting of a 391-ha watershed in New Brunswick, there was an increased loss of nitrate in streamwater of 7 kg NO_3-N/ha·y (Krause, 1982). A study of nine clear-cut watersheds in New Hampshire found an average nitrate loss of 18 kg NO_3-N/ha·y during the first 4 y, compared with 3.5 kg/ha·y for five uncut watersheds (Martin *et al.*, 1986). In addition, calcium losses from the clear-cuts averaged 28 kg/ha·y, compared with 13 kg/ha·y for the reference watersheds, while potassium losses were 6 kg/ha·y compared with 2 kg/ha·y. However, other studies have found smaller effects of clear-cutting on nutrient losses with streamflow, especially when only a portion of the watershed was cut.

Nitrate and other highly soluble ions are leached from watersheds after clear-cutting (and after other disturbances, such as wildfire) for several reasons. First, disturbance stimulates the activity of microbes involved in the decomposition of organic matter. Clearing the forest canopy results in warmer surface soils, while decreased uptake by plants leads to an increased availability of inorganic nutrients and moisture. Second, disturbance often stimulates the microbial processes of ammonification and nitrification (see Chapter 5), leading to increased rates of production of nitrate, which is extremely soluble and readily lost from soils.

Forestry and Erosion

Forestry activities can cause severe losses of soil, called **erosion**, particularly in terrain with steep slopes. In most cases, erosion is triggered by improper practices such as constructing logging roads poorly, using streams as trails to haul logs, running log-removal trails down slopes instead of along them, and harvesting trees from steep slopes that are extremely sensitive to soil loss. Overall, road building is the most important cause of erosion in forestry lands, particularly where culverts (i.e., channelled stream crossings) are not sufficiently large or numerous, or are poorly installed or maintained.

Severe erosion causes many environmental damages, such as loss of mineral soil. In extreme cases, this may expose bedrock, making forest regeneration very difficult. Soil loss also represents a depletion of site nutrient capital. In addition, erosion causes secondary damages to aquatic habitats. These include silt deposition (or *siltation*), which degrades aquatic habitats by covering gravel substrates important to spawning fish. Also, the shallower water increases the risk of flooding.

Erosion is an important, yet largely avoidable, environmental effect of forestry in Canada. The irresponsible practices that can cause erosion have been restricted by provincial regulations, and occur far less frequently now than in the past. Practices that help to reduce erosion include the following:

- planning the routes of forest roads to avoid stream crossings as much as possible
- installing a sufficient number of adequately sized culverts
- avoiding the disturbance of stream channels by heavy equipment
- leaving buffer strips of uncut forest beside watercourses
- using log-removal practices that minimize disturbance of the forest floor (such as cable yarding — a procedure in which a tall spar anchors cables radiating into the clear-cut, allowing logs to be dragged to a central place without the use of a heavy, wheeled machine)
- allowing vegetation to regenerate quickly, to speed the reestablishment of biological moderation of erosion
- deciding to selectively harvest, or not to harvest, steep sites that are extremely sensitive to erosion

It has become a common operational practice in Canada to leave strips of uncut forest beside streams, rivers, and lakes. These buffer zones greatly reduce the erosion of streambanks, eliminate temperature increases in the water, maintain riparian (lake- and stream-bank) habitat for wildlife, and mitigate some of the aesthetic damage from forest harvesting. While it is widely accepted that buffer strips provide important benefits, there is no consensus about how wide the uncut strips should be. This is an economically important consideration, because large areas of valuable timber are withdrawn from the potential harvest when buffer strips are left. The requirements in New Brunswick, for example, are for a 15-m buffer on each side of watercourses draining watersheds larger than 600 ha, with wider buffers recommended beside well-travelled roads and beside surface waters that are commonly used for recreation. Narrower buffers may be allowed on smaller watersheds. New regulations being considered in New Brunswick would require wider riparian buffers on sites that have relatively steep terrain or are otherwise vulnerable to erosion. In some cases, selective harvesting of some of the trees may be allowed within riparian buffers, as long as this does not reduce the ecological services provided by these *special management zones*. Similar regulations have been enacted or are being considered by other provinces.

Forestry and Hydrology

Forests have a strong influence on the hydrology of watersheds. Large amounts of water are evaporated into the atmosphere by forest vegetation, especially by trees (this is known as *transpiration*, which along with evaporation of water from non-living surfaces is called *evapotranspiration*). In the absence of transpiration, an equivalent quantity of water would leave the watershed as streamflow or seepage to deep groundwater.

For example, studies of four forested watersheds in Nova Scotia found that evapotranspiration was equivalent to 15–29% of the annual inputs of water from precipitation, with runoff by streams accounting for the other 71–85% (these watersheds had no substantial drainage to deep groundwater; Freedman *et al.*, 1985).

The hydrologic budget of forested watersheds is extremely seasonal, particularly in the temperate and boreal climates typical of the forested regions of Canada. This seasonality can be illustrated by the hydrology of the Mersey River watershed in Nova Scotia (Figure 23.2). The annual input of water from precipitation is about 146 cm/y, with 82% arriving as rain and 18% as snow. About 62% of the annual input is dispersed by riverflow and 38% by evapotranspiration. The months from November to January have somewhat higher amounts of precipitation, although these seasonal variations are small in comparison with some other regions of Canada. The seasonal variations of evapotranspiration, runoff, and groundwater storage in the watershed are much more substantial. Evapotranspiration is highest during the growing season of May to September, resulting in sparse runoff. Runoff is greater during late autumn and early winter when there is little transpiration. However,

FIGURE 23.2 | SEASONAL HYDROLOGY OF THE MERSEY RIVER IN NOVA SCOTIA

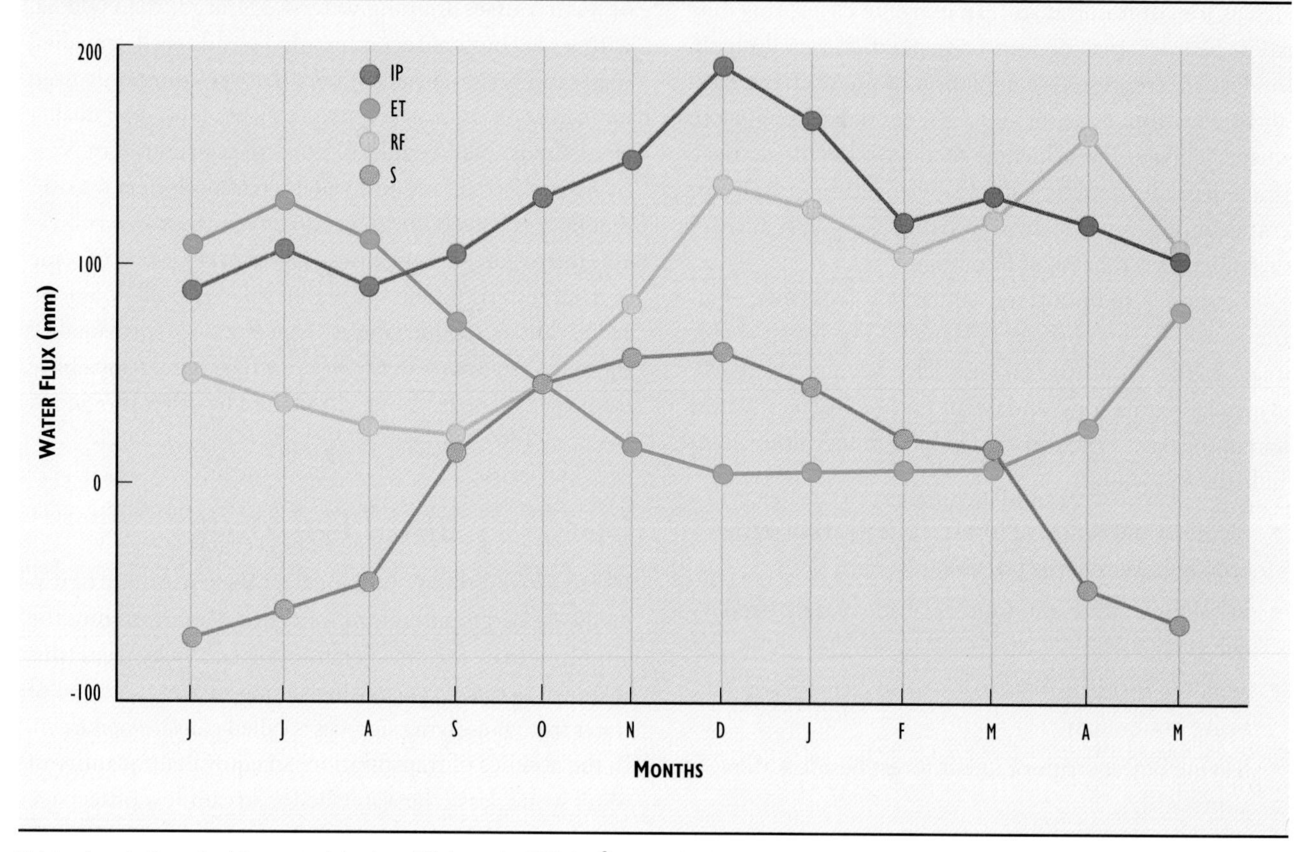

Water inputs from incident precipitation (IP) into the 723 km^2 watershed, and riverflow (RF) from the watershed, are displayed as monthly averages for the period 1968–1982. Evapotranspiration (ET) was estimated with a climatic model, and groundwater storage (S) was calculated as IP – RF – ET.

Source: Modified from Freedman *et al.* (1985)

much of the precipitation input during this period serves to recharge groundwater storage, which becomes severely depleted by the uptake of water by vegetation during the growing season. Runoff is greatest during the springtime, when the accumulated snowpack melts quickly, resulting in a burst of riverflow.

Disturbances such as wildfire and forest harvesting can substantially change the hydrology of watersheds. The seasonality and quantities of flow can change, and erosion, flooding, and other damages can occur downstream. In addition, some poorly drained sites may become wetter, because reduced transpiration can raise the height of the water table. In general, an increase in streamflow is related to the proportion of the watershed that was harvested. After an entire watershed is clear-cut, the increase in streamflow can be as much as 40% in the first year. The increase is proportionately less in the case of partial cuts.

Clear-cuts usually regenerate quickly, and the vigorous regrowth of shrubs and herbs can restore most of the original foliage area in as few as four to six years. Consequently, the biggest increases in streamflow occur in the first year after cutting, followed by rapid recovery to the pre-harvest condition. In the temperate climates prevalent in much of Canada, the largest increases in streamflow occur during late spring, summer, and early autumn, these being the seasons in which transpiration is normally most important.

Hydrology can also be affected by a change in the type of forest on the watershed. For example, if an area of hardwood forest is converted into conifer plantations, the annual streamflow may decrease. This happens because the conifers maintain foliage throughout the year, extending their transpiration season into periods when angiosperm trees lack foliage.

Weeds, Regeneration, and Reorganization

Clear-cuts and other disturbances caused by forestry usually regenerate rather quickly through the ecological process of succession. Initially, however, most naturally regenerating biomass involves plants other than the tree species that foresters consider economically desirable. As a result, the vigorous regrowth is often regarded as detrimental to silvicultural objectives. Such noncrop plants may be viewed as "weeds," and their abundance may be controlled by herbicide applications (see Chapter 21).

However, a rapid revegetation of clear-cuts and other disturbed lands confers some important ecological benefits. The regenerating plants influence the ecological "reorganization" of disturbed lands. They re-establish a measure of biological control over nutrient cycling, erosion, and hydrology, while also restoring habitat for animals. These values, degraded by disturbance of the site, are restored during the successional recovery.

For example, during the first few years after clear-cutting, fast-growing vegetation restores a high rate of nutrient uptake from the soil. In performing this uptake, the regenerating vegetation acts as a biological "sponge," tying up some of the soluble nutrients that might otherwise be leached from the site. Eventually, the early successional plants die, and their nutrients become recycled by decomposition. The absorbed nutrients are then made available to trees. In addition, a rapid regeneration of plants on sites disturbed by forestry helps to re-establish biological control over soil losses from erosion. The revegetation of clear-cuts also restores habitat for birds, mammals, and other wildlife. Clearly, the reorganization phase of succession is enhanced by the rapid regeneration of many plant species on recent clear-cuts, including those considered to be economic weeds by foresters.

Forestry and Biodiversity

Clear-cutting and other forestry practices inflict intense disturbances on forests. They cause dramatic changes in the habitat available to support species of plants, animals, and microbes, and their various communities. Some species benefit from habitat changes that occur because of forestry, but others suffer damage.

Biodiversity was previously defined (in Chapter 7) as "the richness of biological variation." Biodiversity is often considered within three levels of organization:

1. genetic variation within populations and species
2. numbers of species (or *species richness*)
3. the variety and dynamics of ecological communities on larger scales, such as landscapes

In the following sections, we shall examine the effects of forestry on aspects of Canadian biodiversity. The effects of clear-cutting on plants, mammals, birds, and fish will be discussed, largely because these groups have been relatively well studied in the context of forestry. The effects on other elements of indigenous biodiversity, such as insects, spiders, fungi, and microorganisms, are also important, but they have not yet been examined in detail.

Vegetation

Any severe forest disturbance results in changes in the dominant species and sizes of plants living on the site. Because they have such a great influence on local environmental conditions, trees are the dominant organisms in forests. When the dominance of tree-sized plants is reduced by clear-cutting, many species of plants take advantage of the relatively uncompetitive conditions that occur after any severe disturbance. These plants dominate the initial stages of post-harvest succession. They are reduced in abundance, however, or even eliminated from the community, once several decades of regeneration have gone by, and tree-size plants have been able to re-establish their dominance.

Many plants of early post-cutting succession can only be successful in open environments of the sort created by disturbances — they are intolerant of the shade and other conditions beneath a closed forest canopy. These relatively short-lived *ruderal* plants typically disperse readily. This propagation strategy is necessary for them because of the ephemeral nature of their habitats (see Chapter 9). Examples of ruderal plants that proliferate in clear-cuts and other recently disturbed forests include asters, goldenrods, grasses, and sedges. A specific example is the fireweed (*Epilobium angustifolium*), a purple-flowered species that is often abundant after wildfires (hence its name), and also after clear-cutting. Some woody plants are also ruderals, being most abundant during the recovery after forest disturbance. Examples are the red raspberry (*Rubus strigosus*), pin cherry (*Prunus pensylvanica*), and elderberry (*Sambucus racemosa*). Because of their requirement for open, recently disrupted habitats, ruderal plants benefit greatly from clear-cutting and other disturbances associated with forestry.

Unlike ruderal plants, some species are tolerant of the environmental stresses beneath a closed forest canopy. Examples of such plants are white trillium (*Trillium grandiflorum*), shield fern (*Dryopteris marginalis*), feather-mosses (such as *Pleurozium schreberi* and *Hylocomium splendens*), and often certain lichens, such as lungwort (*Lobaria pulmonaria*). These species are not very tolerant of open conditions, and decrease in abundance after clear-cutting. Once suitable conditions redevelop during the successional recovery, these plants may recolonize the site.

In general, once a clear-cut has had two to four years to regenerate, the plant community is actually more rich in species than the mature forest that was harvested (this is particularly true of vascular plants). The increase in species diversity occurs because recently disturbed habitats are relatively rich in resources such as light, nutrients, and water. Under these open conditions, many species of low-growing plants can be supported on the site, including a diversity of ruderal plants. In comparison, naturally stressful habitats, such as the understorey beneath mature, closed-canopied forests, generally support fewer species of plants.

Data from a study in a hardwood forest in Nova Scotia can be used to illustrate the species-rich nature of the vegetation after clear-cuts (Crowell and Freedman, 1994). That study examined a number of stands of mature forest, as well as clear-cuts of various ages. The ground vegetation (occurring within 2 m above the surface) averaged 11 plant species/m^2 on two 1-y-old clear-cuts, increasing to 14/m^2 on 6-y-old clear-cuts. Mature forest and clear-cuts older than 30 y had fewer species — typically 3–6/m^2 in stands with a closed canopy dominated by species of maple trees, but 9–11/m^2 in birch-dominated stands, which have a more open canopy. This comparison suggests that many species of plants, especially ruderals, can freely utilize open habitats associated with clear-cutting.

Deer, Moose, Elk, and Caribou

White-tailed deer (*Odocoileus virginianus*) and mule deer (*O. hemionus*) are the most common wild ungulates in Canada. These deer feed on woody stems (also known as browse) and low-growing herbaceous plants, and need brushy habitats for at least part of their yearly range.

During the 20th century, the abundance of white-tailed and mule deer has increased over most of their range. In fact, these deer are now more abundant in parts of their range than they were before the European colonization of Canada, when extensive landscapes were covered with mature and old-growth forests. In Nova Scotia, for example, white-tailed deer were uncommon when Europeans first began to settle the land. In fact, the species was soon extirpated by overhunting. However, white-tailed deer were re-established in the nineteenth century, both by deliberate introductions and by natural immigration from New Brunswick. Today, this species is probably more abundant in Nova Scotia than at any time since deglaciation.

The modern abundance of white-tailed deer in Canada is largely due to a widespread availability of early-successional, shrubby habitats. While-tailed deer have proliferated mainly because human activities have increased the areas of these habitats, and in some regions have decreased the populations of natural predators. The shrubby habitats were created by the abandonment of agricultural lands, by the harvesting of forests, and by wildfires. All three factors result in habitats dominated for several decades by shrub-sized plants, with a rich understorey of forbs (herbaceous, dicotyledonous plants) and graminoids (monocotyledonous species, especially grasses, sedges, and rushes).

These shrubby habitats tend to be distributed over the landscape as a mosaic of stands in various stages of secondary succession, within a matrix of mature forest. This spatial arrangement enhances the suitability of the landscape for white-tailed deer if the following conditions are also present:

- extensive production of nutritious and palatable browse in younger stands
- abundance of ecotonal (i.e., edge) habitat
- adequate availability of good "yarding" habitat of mature coniferous forest, which provides critical shelter in regions where winters are severe

If these habitat qualities occur within a mosaic of stands of various ages, the landscape is more favourable to deer than are either extensive clear-cuts or unbroken expanses of mature forest.

The central parts of large clear-cuts are not well used by deer, as they like to be close to protective forest cover. A study in eastern Canada found that white-tailed deer fed about seven times more intensively in the centres of clear-cuts less than 80 ha in area than in the centres of larger clear-cuts up to 410 ha in size (Drolet, 1978). In fact, clear-cuts for white-tailed deer are optimally rather small in size, although this varies regionally. For example, clear-cuts smaller than 4 ha in New Brunswick, and smaller than 2 ha in southern Ontario, have been recommended as optimal habitat for white-tailed deer

To some degree, the amount of useful habitat on larger clear-cuts can be increased if they have an irregular shape. Erratic shapes have a higher ratio of edge to surface area than do circular disks, square, or rectangulars shapes. Consequently, irregularly shaped clear-cuts provide more edge habitat, while also making the central part of the clear-cut more accessible to deer.

White-tailed and mule deer eat a variety of woody plants, forbs, and graminoids, and these foods are often much more abundant on cutover and burned sites than in mature forest. After clear-cutting, the biomass of browse and herbaceous plants typically increases to a maximum after about 8-15 y, followed by a decline as the tree canopy matures and shades the understorey vegetation. This successional pattern is illustrated in Table 23.3 for stands of various ages following clear-cutting in Nova Scotia. The quantity of browse peaked at about 8–13 y and then declined. The pattern for herbaceous plants was similar, but with the biomass peaking at 2–6 y after clear-cutting.

Browse is not only more abundant, but also of better nutritional quality in clear-cuts and burns than in mature forests. Recently sprouted, rapidly growing twigs tend to have higher concentrations of protein, nitrogen, and phosphorus, and are more succulent and more easily digested than the older browse found in mature forests.

Although foods for deer are relatively abundant and of good quality on clear-cuts, other characteristics of managed forestry land can restrict the use of these habitats. The central parts of large clear-cuts may be too far from protective forest cover to be well utilized. Clear-cuts may also have physical obstructions to deer movements, such as tangles of logging slash. They may also have excessively deep snow during winter, because snowfall is not intercepted by an overhead canopy of conifer trees. This can be important, because deer movement is severely restricted by snow depths greater than about 50–70 cm.

In areas with severe winters, it is critical that forest harvesting be planned to ensure the availability of suitable "yarding" habitats of mature conifer forest. In these places, the winter microclimate is less severe, browse is available, and snow depths are not excessive. In some regions, winter-yarding habitat may be more important to deer than the amount and quality of summer habitat. Since particular yards are often used for years by deer from an extensive area, these critical habitats should be protected from any cutting that might detract from their use.

Finally, ongoing disturbances in forest management areas can affect habitat use by deer. These include frequent traffic along logging roads, noise from harvesting operations, and the excessive hunting pressure which can result from easy access along forestry roads.

Moose (*Alces alces*) and elk (or wapiti, *Cervus elaphus*) may also benefit from some kinds of forest-harvesting practices. Moose feed primarily on browse, although they also eat aquatic and terrestrial herbs during the summer. Elk graze primarily on graminoids and forbs during the growing season, but eat browse during the winter when herbs are less available. Because the availability of browse and herbs usually increases after forest harvesting, the habitat

TABLE 23.3 | SHRUB AND HERB BIOMASS IN STANDS OF DIFFERENT AGES

The data are from a region of maple-birch forest in Nova Scotia. Stands aged 20 y and younger had been disturbed by clear-cutting. Biomass is reported in tonnes of dry weight per hectare.

STAND AGE (y)	NUMBER OF STANDS SAMPLED	SHRUB BIOMASS (t/ha)		HERB BIOMASS (t/ha)	
		AVERAGE	RANGE	AVERAGE	RANGE
1	2	0.45	0.4–0.5	1.1	0.93–1.2
2	2	1.9	1.8–2.1	1.6	1.4–1.7
3–6	4	8.2	6.6–12.2	1.7	1.5–2.1
8–13	5	17.6	16.9–18.7	0.57	0.27–1.0
20	1	10.9	–	0.14	–
30–40	3	4.8	0.8–7.9	0.25	0.10–0.47
50–75	5	2.4	1.1–3.9	0.23	0.70–0.45

Source: Modified from Crowell and Freedman (1994)

of moose and elk can be improved by some kinds of forest management. In general, however, these species are less favoured by forestry than are white-tailed and mule deer.

The woodland caribou (*Rangifer tarandus*; this species is known as reindeer in Eurasia) is another abundant species of deer in Canada, particularly in the north. Woodland caribou require an extensive habitat of mature coniferous forest, particularly during the winter, when so-called "reindeer mosses" (species of lichens in the genus *Cladonia*) make up much of their diet. These lichens grow on the forest floor, and are most abundant in relatively open conifer stands that are 40–100 y old. However, if the tree density is high enough to allow the canopy to close as the forest matures, these lichens decline and become replaced by feather mosses, which are not palatable to caribou. Disturbance of forests by wildfire and logging can regenerate the supply of reindeer lichens in areas with extensive closed-canopied stands. In general, however, caribou are not favoured by extensive logging of their habitat.

All the species of wild ungulates in Canada are important in hunting, an activity that generates significant economic value while also providing subsistence for many rural people (see Chapter 14). Increasingly, foresters and wildlife biologists are working together to develop integrated management plans that accommodate the need to harvest both timber and ungulates from landscapes. These plans can allow the maintenance of relatively large populations of white-tailed deer, mule deer, moose, and elk, even while clear-cutting and other forestry practices take place. We will examine *integrated forest management* in more detail at the end of this chapter.

Smaller Mammals

Hares and rabbits are abundant in most regions of Canada and are economically important as small game, as pests, and in recreational wildlife viewing. These animals feed by browsing and grazing, and can benefit from the increase in availability of low-growing shrubs and herbs as a result of forest harvesting, abandonment of agricultural lands, and wildfire. In fact, hares and rabbits can be abundant enough to significantly impede tree regeneration by girdling (chewing the bark around a sapling, killing it) and clipping (chewing the foliage and growing points of a young tree).

Other small mammals include mice, voles, shrews, and moles, which are important as components of terrestrial food webs. These small mammals sometimes impede forest regeneration by consuming tree seeds and by girdling young trees. In other cases, they are considered beneficial, because they can eat large quantities of potentially damaging insects.

Most studies report that forest harvesting has relatively minor effects on small mammals. For example, no substantial differences were found in the overall abundance, species richness, or diversity of small mammals when mature forest, 3–5-y-old clear-cuts, strip-cuts, and shelterwood cuts were compared in Nova Scotia (Table 23.4).

TABLE 23.4 | COMPARISON OF SMALL-MAMMAL COMMUNITIES AMONG HABITAT TYPES

Data are from different habitats in a maple-birch forest in Nova Scotia. Abundance is indexed by the numbers of animals caught per 100 days of trapping effort. Species richness is the number of species observed.

SPECIES	UNCUT FOREST	CLEAR-CUTS	STRIP-CUTS	SHELTERWOOD
Short-tailed shrew (*Blarina brevicauda*)	3.9	3.4	2.5	3.4
Masked shrew (*Sorex cinereus*)	2.2	4.9	2.2	2.3
Red-backed vole (*Clethrionomys gapperi*)	7.0	6.0	4.9	4.9
Deer mouse (*Peromyscus maniculatus*)	1.0	0.7	0.4	3.0
Meadow vole (*Microtus pennsylvanicus*)	0.2	2.8	2.6	0.6
Eastern chipmunk (*Tamias striatus*)	0.2	0.2	0.7	0.4
Woodland jumping mouse (*Napaeozapus insignis*)	2.9	0.3	2.9	4.7
Total abundance	17.4	18.3	16.2	19.3
Species richness	7	7	7	7

Source: Modified from Swan *et al.* (1984)

Pine marten (*Martes americana*) and fisher (*M. pennanti*) are small carnivores with extensive natural ranges in North America. Unfortunately, these furbearers have suffered large population declines in many areas, mainly because they have been trapped too intensively. In addition, over much of their range, marten and fisher appear to depend on the complex habitat structure of older-growth, coniferous forests. Consequently, they are considered to be at risk from forest harvesting and management.

Birds

Many species of birds require mature forest as habitat for breeding or wintering, or during migration. However, many other species need the types of habitat that occur during the early stages of forest succession, including those created through such forestry activities as clear-cutting.

Ruffed grouse (*Bonasa umbellus*) are commonly hunted in Canada. Also hunted, but to a lesser degree, are spruce grouse (*Canachites canadensis*) and blue grouse (*Dendragapus obscurus*). Populations of these so-called "upland game birds" are generally favoured by landscape mosaics that include both mature forest and younger, brushy stands.

Ruffed grouse occur in a variety of habitats, but prefer areas dominated by hardwood forest with some conifers mixed in, especially stands dominated by poplars and birches. These birds feed mainly on the foliage, young twigs, catkins, and buds of woody plants, also eating fleshy fruits when available. In Nova Scotia, ruffed grouse utilize clear-cuts of maple-birch forest that are five or more years old. Clear-cuts of aspen forest in Minnesota become suitable for ruffed grouse after 4–12 y of regeneration, and are then used as breeding habitat for 10–15 y, while older aspen stands are important as wintering habitat (Gullion, 1988). Wildlife biologists recommend that, to provide optimal habitat for ruffed grouse in regions of aspen forest, forestry be conducted in such a way as to create a mosaic of different-aged stands, each 10 ha or less in area, with adjacent stands differing in age by 10–15 y.

Wildlife managers often refer to the great diversity of bird species that are not hunted for sport as "nongame" species. These birds can, however, be economically important. For example, they may be useful predators of insects that damage trees or other crops, and they are the objects of bird-watching, an increasingly popular outdoor sport.

Forestry affects individual species of nongame birds, as well as their communities, in terms of density and species diversity. These effects are indirect, being caused by changes in the physical structure and plant-species composition of the available habitats. Important aspects of bird habitat structure include the quantities of living and dead plant biomass, their distribution in both vertical and horizontal planes, and the nature of *ecotones* that occur where different habitats meet.

An important aspect of horizontal structure is the distribution of patches of distinct habitats, either within a stand or on a landscape. The shapes of the patches affect their ratio of edge to area (and thus the amount of ecotonal habitat). Patch size is also important, because small, isolated patches cannot sustain species of birds that maintain large territories. The species composition of the vegetation affects the food types and other habitat elements occurring in a particular stand. The presence of cavity trees, dead trees (or snags), and logs on the forest floor is also critical to many species of birds and other animals (this is discussed in the next section). Finally, the abundance of many bird species often increases in stands in which there is an outbreak of insects, such as spruce budworm (see Chapter 21).

Many bird watchers have a general knowledge of the relationships between bird species and habitats. They use this understanding to predict the species they might see under certain conditions. Ecologists know enough about the specific requirements of some birds to be able to manage their habitat. The best forestry-related example of this practice is the use of prescribed fire to create even-aged stands of jack pine (*Pinus banksiana*) in Michigan. This ensures an appropriate habitat for the Kirtland's warbler (*Dendroica kirtlandii*), an endangered species.

Of course, each bird species has particular habitat needs. If the physical and botanical character of a habitat are changed by a major disturbance such as wildfire or clear-cutting, many species of birds can no longer breed in the affected stand. The same disturbance, however, will create habitat opportunities for early-successional species of birds. These changes are illustrated in Table 23.5, which compares the bird species breeding in mature stands and in clear-cuts of hardwood forest in Nova Scotia. The mature maple-birch forest supported an avian community with an average population of 663 pairs/km^2, dominated by ovenbird, least flycatcher, red-eyed vireo, black-throated green warbler, and hermit thrush. The 3–5-y-old clear-cuts supported a slightly less abundant bird population of

TABLE 23.5 | BREEDING BIRDS IN MATURE FOREST AND ADJACENT CLEAR-CUTS

The mature stands and 3–5-y-old clear-cuts were surveyed in Nova Scotia. The three stands of mature forest had a closed canopy dominated by maple and birch, while the three clear-cuts had a vigorous regeneration of shrubs and herbaceous plants. Less common bird species are not included here. Data are in pairs/kilometre^2.

SPECIES	FOREST			CLEAR-CUTS		
	A	B	C	A	B	C
Common snipe (*Capella gallinago*)	0	0	0	10	15	0
Ruby-throated hummingbird (*Archilochus colubris*)	0	0	0	25	30	15
Least flycatcher (*Empidonax minimus*)	290	120	0	0	0	0
Hermit thrush (*Catharus guttatus*)	60	40	30	0	0	0
Veery (*Catharus fuscescens*)	50	10	0	25	0	0
Solitary vireo (*Vireo solitarius*)	60	30	0	0	0	0
Red-eyed vireo (*Vireo olivaceous*)	80	50	30	0	0	0
Black-and-white warbler (*Mniotilta varia*)	15	50	40	0	0	0
Northern parula warbler (*Parula americana*)	15	30	40	0	0	0
Black-throated green warbler (*Dendroica virens*)	50	30	30	0	0	0
Chestnut-sided warbler (*Dendroica pensylvanica*)	0	0	0	100	40	190
Ovenbird (*Seiurus aurocapillus*)	150	120	200	0	0	0
Mourning warbler (*Oporornis philadelphia*)	0	0	0	0	0	90
Common yellowthroat (*Geothlypis trichas*)	0	0	0	25	300	130
American redstart (*Setophaga ruticilla*)	15	80	100	0	0	0
Rose-breasted grosbeak (*Pheucticus ludovicianus*)	15	10	0	0	0	0
Dark-eyed junco (*Junco hyemalis*)	15	20	15	50	70	30
White-throated sparrow (*Zonotrichia albicollis*)	0	20	0	90	190	100
Song sparrow (*Melospiza melodia*)	0	0	0	90	70	0
Total density (pairs/km^2)	815	660	515	435	745	585
Number of species	12	16	9	10	8	7

Source: Modified from Freedman *et al.* (1981)

588 pairs/km^2, dominated by chestnut-sided warbler, common yellowthroat, white-throated sparrow, and dark-eyed junco. The forests and the clear-cuts thus supported similar densities of breeding birds, but with almost entirely different species. This occurred because the forests and clear-cuts provide very different habitats in terms of physical structure and the species composition and productivity of vegetation. Although clear-cutting deprived mature-forest birds of habitat, it created opportunities for early-successional bird species.

Welsh and Fillman (1980) examined the effects on birds of clear-cutting spruce forests in northern Ontario. In that study, the largest populations of breeding birds (1020–1970 pairs/km^2) occurred in moderate-aged (11–24 y) clear-cuts. This was a considerably higher population level than in uncut spruce forest (561 pairs/km^2). The smallest densities of birds occurred in a 3-y-old clear-cut with about 200 pairs/km^2. By 5 y after the cutting, this had increased to 690 pairs/km^2. In general, the clear-cuts and mature spruce forest supported different species of birds, although there was some overlap.

Cavity Trees, Snags, and Coarse Woody Debris

Living trees with *cavities*, standing dead trees (i.e., *snags*), and logs lying on the forest floor (i.e., *coarse woody debris*) are critical elements of habitat for many species of animals. This is particularly true of many birds, which use these habitat features for nesting, as substrates for foraging, and as perches for hunting, resting, or singing. In fact, about one-third or more of the birds that breed in temperate and boreal forests depend on these features, particularly on cavities.

For example, all 12 species of woodpeckers that breed in Canada excavate cavities in snags or in living, heart-rotted trees. These cavities are used as nesting sites, for roosting at night, or for both purposes. In addition, most species of woodpeckers forage for their food of invertebrates by drilling and excavating into the bark and wood of dead and living trees. Many other species of birds nest in the abandoned cavities made by woodpeckers, or use natural cavities that have formed in rotten parts of trees. Some other birds nest in or beneath coarse woody debris, or build platform nests on snags or living trees that have suffered damage to their tops. Table 23.6 lists bird species that require these critical habitat elements for breeding in Canada.

Because so many species of birds depend on cavities, maintaining this habitat feature has become an important consideration in North American forest management. The issue is especially prominent in the older-growth forests of the west coast, where as many as six species of woodpeckers can co-occur in relatively large populations, along with many other cavity-dependent species. If we wish these species to continue to live in regions where forestry is being practised, part of the management area must be maintained as mature or old-growth forest. If plantations are established, these should be designed to provide habitat for birds that require snags and dead logs. For example, dead and dying trees can be left standing during the harvest. Cavity- and snag-dependent species are also better accommodated by less intensive harvesting techniques, such as selection cuts.

A two-year-old clear-cut of hardwood forest in Nova Scotia. Birds typically breeding in this habitat include dark-eyed junco (*Junco hyemalis*), white-throated sparrow (*Zonotrichia albicollis*), song sparrow (*Melospiza melodia*), and common snipe (*Capella gallinago*).

Source: B. Freedman

An eight-year-old clear-cut of hardwood forest in Nova Scotia. Birds breeding in this habitat include chestnut-sided warbler (*Dendroica pensylvanica*), common yellowthroat (*Geothlypis trichas*), and alder flycatcher (*Empidonax alnorum*).

Source: B. Freedman

Freshwater Biota

Forestry practices can degrade freshwater habitats in four major ways:

- by siltation (i.e., the settling of soil eroded from the land and streambanks)
- by increases in water temperature caused by the removal of shading vegetation from stream edges
- by blocking stream channels with logging debris
- by changes in hydrology

Damages may also be caused by chemical and fuel spills and as a result of pesticide spraying (see Chapter 21). All these assaults on the habitat can affect populations of freshwater organisms, including fish, amphibians, and aquatic invertebrates. These problems can be especially

TABLE 23.6 | CANADIAN BIRDS THAT REGULARLY NEST IN CAVITIES, SNAGS, OR COARSE WOODY DEBRIS

PRI = primary cavity excavator; SEC = secondary user of abandoned cavity of primary excavator; NAT = uses cavities created by natural decay; BARK = nests beneath loose bark; CWD = nests in or beneath coarse woody debris on the forest floor; PLAT = constructs a platform nest on a snag.

SPECIES	PRI	SEC	NAT	BARK	CWD	PLAT
Great blue heron, *Ardea herodias*						✓
Wood duck, *Aix sponsa*		✓	✓			
Common goldeneye, *Bucephala clangula*		✓	✓			
Barrow's goldeneye, *Bucephala islandica*		✓	✓			
Bufflehead *Bucephala albeota*		✓	✓			
Hooded merganser, *Lophodytes cucullatus*		✓	✓			
Red-breasted merganser, *Mergus serrator*					✓	
Turkey vulture, *Cathartes aura*					✓	
Bald eagle, *Haliaeetus leucocephalus*						✓
Osprey, *Pandion haliaetus*						✓
American kestrel, *Falco sparverius*		✓				
Eastern screech-owl, *Otus asio*		✓	✓			
Western screech-owl, *Otus kennicottii*		✓	✓			
Northern pygmy owl, *Glaucidium gnoma*		✓	✓			
Spotted owl, *Strix occidentalis*		✓	✓			
Barred owl, *Strix varia*		✓	✓			
Saw-whet owl, *Aegolius acadicus*		✓	✓			
Boreal owl, *Aegolius funereus*		✓	✓			
Marbled murrelet, *Brachyramphus marmoratus*						✓
Chimney swift, *Chaetura pelagica*			✓			
Vaux's swift, *Chaetura vauxi*			✓			
Lewis' woodpecker, *Melanerpes lewis*	✓					
Red-headed woodpecker, *Melanerpes erythrocephalus*	✓					
Yellow-shafted flicker, *Colaptes auratus*	✓					
Pileated woodpecker, *Dryocopus pileatus*	✓					
Yellow-bellied sapsucker, *Sphyrapicus varius*	✓					
Red-breasted sapsucker, *Sphyrapicus ruber*	✓					
Hairy woodpecker, *Picoides villosus*	✓					
Downy woodpecker, *Picoides pubescens*	✓					
Black-backed three-toed woodpecker, *Picoides arcticus*	✓					
Great crested flycatcher, *Myiarchus crinitus*		✓	✓			
Yellow-bellied flycatcher, *Empidonax flaviventris*					✓	
Tree swallow, *Tachycineta bicolor*		✓	✓			
Violet-green swallow, *Tachycineta bicolor*		✓	✓			
Purple martin, *Progne subis*		✓	✓			
Black-capped chickadee, *Parus atricapillus*	✓	✓	✓			
Mountain chickadee, *Parus gambeli*	✓	✓	✓			
Boreal chickadee, *Parus hudsonicus*	✓	✓	✓			
White-breasted nuthatch, *Sitta carolinensis*	✓	✓	✓			
Red-breasted nuthatch, *Sitta canadensis*		✓	✓			
Brown creeper, *Certhia familiaris*	✓		✓	✓		
House wren, *Troglodytes aedon*		✓	✓			
Winter wren, *Troglodytes troglodytes*					✓	
Eastern bluebird, *Sialia sialis*		✓	✓			
Mountain bluebird, *Sialia carrucoides*		✓	✓			

Source: Based on information in Godfrey (1986)

severe in hilly or mountainous terrain, because of the many small streams and rivers that occur there, and because the soil on steep slopes is highly vulnerable to erosion.

In most cases, it is possible to avoid or mitigate many of the damages to aquatic ecosystems. As we previously noted, rates of erosion can be greatly reduced if roads and culverts are constructed carefully, logs are hauled correctly, and riparian buffers of uncut forest are left beside water courses. Leaving buffer strips also avoids an excessive accumulation of logging debris in streams, as does avoiding the felling of trees into aquatic habitats. In addition, riparian buffers are extremely effective at preventing increases in water temperature, because they provide shade to streams even if nearby habitats have been clear-cut.

Snags, or dead standing trees, are a critical habitat element for many species of animals. This photo shows four young kestrels (*Falco sparverius*) that recently fledged from their nest site in a natural cavity in a pine snag left in a clear-cut in Nova Scotia.
Source: I.A. McLaren

Old-Growth Forest and Its Dependent Species

An old-growth forest is a late-successional (or climax) ecosystem. Old-growth forests are characterized by the presence of old trees, an uneven-aged population structure (that is, all age classes are represented, from young to old), and a complex physical structure. The structure includes multiple layers within the canopy, large trees, many large snags, and dead logs lying on the forest floor.

In some ecological contexts, the term "old-growth" has also been used to refer to senescent populations of shorter-lived species of trees, such as older stands of poplar, birch, or cherry. This is not, however, the meaning of "old-growth forests" being considered here. Old-growth forests are natural ecosystems with unique and special values that are not replicated elsewhere. For this reason, they are considered to have great intrinsic value, and to be important components of natural heritage. They support some species of plants and animals that do not occur in other Canadian habitats. (This is, admittedly, a relatively minor attribute of temperate and boreal old-growth forests — tropical old-growth forests sustain enormously larger numbers of dependent species; see Chapters 7 and 24.) In addition, old-growth forests deliver important ecological services, such as furnishing clean water and air, and they have economic value for outdoor recreation and ecotourism.

Old-growth forests were once much more extensive in Canada (and elsewhere in the world) than they are today. In eastern Canada, for example, extensive tracts of old-growth forests have been cleared for agriculture, or were converted into younger, second-growth forests by timber harvesting. As a result, there is very little of this ecosystem type left in the eastern provinces, where only one to a few percent of the total forest cover is now in an old-growth condition.

Old-growth forests are more abundant in parts of western Canada, particularly on the Pacific coast of British Columbia. There, the wet climatic regime favours the development of this kind of natural ecosystem (since wildfires are uncommon). Even in British Columbia, however, extensive tracts of old-growth forests have been logged or

A wintertime airphoto of a riparian buffer of uncut forest beside a stream in New Brunswick. The shading vegetation prevents the increases in water temperature that would be caused by clear-cutting to the stream edge. It also helps to prevent erosion, and maintains a corridor of intact, mature vegetation for continued use by wildlife.

Source: M. Sullivan

converted to urbanized land uses, especially in the vicinities of Vancouver and Victoria. And because old-growth timber is such a valuable natural resource, much of the remaining old-growth forest is threatened by forestry. It is likely that virtually all of the remaining old-growth forest will be logged during the next several decades, and converted into second-growth forests that will be harvested before they attain an old-growth condition. The only exceptions will be those tracts of old-growth forest that are protected in ecological reserves and parks.

Because old-growth forests are now rare in regions of eastern Canada where they were once widespread, the preservation of the remnants is a high priority for environmentalists and conservationists. This is also the case in western Canada, especially in coastal British Columbia, where there are enormous controversies over the continued logging of the remaining tracts of old-growth rainforests (see Canadian Focus 23.1).

Some of the key characteristics of old-growth forests, including their elements of biodiversity, can be accommodated by so-called *new forestry* harvesting systems that are relatively "soft" in the intensity of disturbance that they

CANADIAN FOCUS 23.1

CONTROVERSY OVER OLD-GROWTH FORESTS OF CLAYOQUOT SOUND

Clayoquot Sound is an embayment of the Pacific Ocean, reaching inland in central Vancouver Island to encompass a watershed of about 263 000 ha. Because of the mountainous terrain, a great variety of habitats is found in this region, including extensive forests on the interior mountains and the flatter coastal plain. The climatic regime is mild temperate, and there is an abundance of rainfall.

The high-rainfall regime has an important influence on the kinds of forests that develop in the region. The wet climate means that wildfires are rare, a factor that encourages the natural development of an old-growth rainforest. Before commercial forest harvesting began in the region, most of the forests of Clayoquot Sound were in an old-growth condition. Many of the trees are very large and old — some individuals are more than three meters in diameter, and more than one thousand years old. In addition, old-growth trees have fine-grained wood because of their slow growth rates, making them extremely valuable for manufacturing into lumber.

Because of the extensive logging of old-growth forests in Clayoquot Sound (and elsewhere), this ecosystem type is much less extensive today than it used to be. This is particularly true of the coastal plain and lower elevations in the mountains, where the most accessible old-growth forests were found. Today, about 80% of the remaining old-growth forests occur at higher elevations and on steeper slopes.

Almost all of the stands of old-growth timber that are logged on Vancouver Island and elsewhere in coastal British Columbia return to forest. In large part, the second-growth forests are dominated by the same species of trees that dominate the old-growth forests. The second-growth forests will, however, be harvested as soon as their trees are large enough to be used for manufacturing lumber or pulp, and this occurs at a much younger age than is

required to redevelop the old-growth condition. Enormous controversy has arisen over the rapid and extensive conversions of old-growth into second-growth forests. Many people believe that old-growth forests have great intrinsic value because they are a distinct natural ecosystem. These forests are also highly valued for cultural, aesthetic, and ecological reasons.

Environmental activists have targeted the remaining old-growth forests of the Clayoquot Sound area for protection, and have focused a great deal of activism to that end. In part, this has occurred because the area is rather accessible and is traversed by many people as they travel to Pacific Rim National Park farther up the west coast of Vancouver Island. In fact, there are more extensive old-growth forests than in Clayoquot Sound elsewhere on Vancouver Island and on the British Columbia mainland. However, because of the remote nature of those other forests, protests there would have been less effective in attracting media and public attention than actions in the Clayoquot Sound region.

The most intense protests began in March, 1993, soon after the government of British Columbia purchased stock in a logging company that had a licence to harvest timber in the Clayoquot Sound region. Several weeks later, the government issued permits for logging over 74% of the area. This sparked an explosion of public demonstrations, including a large gathering of concerned individuals at Clayoquot Sound. Many of these people travelled across Canada by train, as part of a media event that symbolically began in Newfoundland and ended in coastal British Columbia. In addition to the protests and publicity stunts, there were illegal blockades of logging roads, and other kinds of civil disobedience, leading to the arrests of more than 850 people.

In October, Mike Harcourt, then premier of British Columbia, announced the establishment of a Scientific Panel for Sustainable Ecosystem Management in Clayoquot Sound. The panel was made up of 23 members, including experts on ecology, biodiversity, forestry, ecotourism, and other interests relevant to old-growth forests and the regional controversy. The mandate of this highly regarded group was to "make forest practice in Clayoquot not only the best in the province, but the best in the world."

In June, 1995, after a series of public meetings and other deliberations, the panel released a report containing 120 recommendations. These were accepted by the government of British Columbia and passed into law in July. The panel recommended that sustainable ecosystem management should be the overriding objective for the Clayoquot Sound region, and that all activities, including those associated with forestry, should be conducted with that objective in mind.

This landscape in the Clayoquot Sound region of Vancouver Island used to be covered mainly by old-growth rainforests. Because the valley floor contained the largest trees and was relatively accessible, it was harvested first, in this case about 15 years before the photo was taken. The clear-cuts are regenerating well through a combination of planted seedlings and natural seeding-in, and another mixed-species forest will again develop. This forest will, however, be harvested long before it attains an old-growth condition. The stands at mid-slope were clear-cut about 3 years before the photo was taken, while the upper slopes have not yet been harvested. Note the erosion associated with the logging road at mid-slope. Erosion is a common result of logging roads built in mountainous terrain.

Source: B. Freedman

Many of the recommendations advocated silvicultural systems that would retain the most important of the ecological characteristics of old-growth rainforests, such as an uneven age structure, snags and cavity trees, coarse woody debris, and healthy aquatic ecosystems. The panel felt that, in large part, those objectives could be met by restricting the proportion of any large watershed that could be converted into younger age-classes at any time. This could be accomplished by limiting the cut rate within any watershed to no more than 1% of the area per year. That practice would ensure that sufficient areas of natural old-growth forest, or of managed forest having most of the habitat values of old-growth forest, would be present at any time to satisfy the needs of dependent species.

The panel also recommended that significant areas within the Clayoquot Sound region should be protected from any forestry, particularly those areas and sites that have great value because of their natural ecological features, aesthetics, cultural significance, or utility for recreation and ecotourism.

The recommendations of the panel, and their acceptance by government, satified most (but not all) people about the ecological sustainability of the forest-management plan for the Clayoquot Sound region. The panel's work should not, however, be regarded as a perfect model of sustainability, or as a permanent solution to the controversy over forestry in the old-growth forests of Clayoquot Sound or elsewhere in Canada. There is, for example, no guarantee that a future government will not choose to change the rules, and it is even possible that a future generation of Canadians would support such an action.

The best hope of preventing such future damage is to ensure that the social contract of Canadian forestry always includes an obligation to conduct industrial activities in a manner that does not degrade the timber resource, while maintaining other economic values such as populations of hunted species and ecotourism, as well as sustaining old-growth forests and other elements of our natural, ecological heritage.

cause. The best example of a new forestry system is selection cutting with snag and cavity-tree retention. Because only some of the economically valuable timber is removed during a selection harvest, the physical and ecological integrity of the forest is left substantially intact, conserving many of the old-growth values.

There are, however, limits to what can be achieved through new-forestry practices. If a goal of society is to preserve old-growth forests as a special type of natural ecosystem, this is best done by establishing large, landscape-scale protected areas. The size of the protected areas is a critical factor, since they must be large enough to sustain the long-term ecological dynamics that permit old-growth forests to develop, particularly the natural disturbance regime. This landscape perspective is important, because particular stands of old-growth forest cannot be preserved forever — they will inevitably become degraded by natural disturbance and/or environmental changes. Consequently, old-growth forests can only be sustained if large protected areas are designated to preserve the necessary ecological dynamics.

Old-growth forests are an extremely valuable natural resource, perhaps more so than any other types of forest in Canada, because they contain large individuals of economically desirable tree species. This kind of timber can be used to manufacture relatively valuable products, such as fine-grained, large-dimension lumber and plywood. Stands of old-growth forests are rarely, however, managed by foresters as a renewable, natural resource. Rather, they are almost invariably "mined" by harvesting, followed by silvicultural management to convert the site into a younger, second-growth forest. The second-growth forest will only be allowed to develop into a middle-aged forest before it is again harvested.

There is a strong economic rationale for this kind of management strategy. In the old-growth condition, forests do not sustain a positive net production of biomass. This happens because, at the stand level, the productivity by living trees is approximately balanced by the deaths of other individuals through disease, accident, or age. If the primary objective of forest management is to optimize the productivity of tree biomass in the stand, it is better to harvest middle-aged stands soon after their net productivity starts to decrease, that is, before they attain the old-growth condition. Of course, this kind of economic thinking does not take account of the special ecological and aesthetic values of old-growth forests, which are degraded by logging.

Because old-growth forests have particular structural characteristics, some species of wildlife can occur only in these kinds of habitats. This is especially true of tropical old-growth forests. In Canada, animals that depend substantially on old-growth forests include the marbled murrelet (*Brachyramphus marmoratus*), the northern spotted owl (*Strix occidentalis caurina*), and the American marten (*Martes americana*). In addition, some species of plants are more abundant in old-growth than in younger, mature forests. Examples include the Pacific yew (*Taxus brevifolia*) and the lungwort lichen (*Lobaria pulmonaria*). However, the species that depend on old-growth forests in Canada have not yet been well investigated. Ongoing biosystematic studies will discover additional examples, especially of insects, lichens, mosses, and other less conspicuous elements of forest biodiversity. All of these indigenous biodiversity values are endangered by the continued logging of old-growth forests in Canada.

A 28-year-old plantation of white spruce (*Picea glauca*) in New Brunswick. The plantation is a conifer forest, but it is simple in physical and biological structure. Although this habitat supports many native species of plants and animals, some species are eliminated by the scarcity of critical habitat elements, such as cavity trees and coarse woody debris.

Source: B. Freedman

Effects of Plantation Establishment

Clear-cutting of natural forests in Canada is often followed by the planting of a new crop of trees, usually of a conifer species, which may then be intensively managed to increase the productivity of the stand. This practice results in the development of a plantation or tree farm, which is an anthropogenic forest of a relatively simple character compared with the natural, mature or old-growth, mixed-species forest that originally occupied the site. This *ecological conversion* has important implications for natural biodiversity, because many indigenous species of the original, natural forest are unable to utilize the habitats available in a plantation.

The most important habitat changes are related to the differences in the tree species and physical structure of the plantation and the natural forest that is being replaced. A typical plantation is dominated by a population of trees of the same species and of similar size and age (in population ecology this is known as a *cohort*; in agriculture it is referred to as a *monoculture*). This is a greatly simplified ecosystem compared with many natural forests, which may contain trees of various species, sizes, and ages. Such changes are greatest when hardwood-dominated or mixed hardwood-conifer forests are replaced by conifer plantations. The changes are fewer if natural conifer-dominated forests are replaced with conifer plantations.

Of course, any changes in vegetation and habitat have secondary effects on the species of animals that can be sustained. Studies of intensively managed conifer plantations

TABLE 23.7 | BREEDING BIRDS IN NATURAL FORESTS AND PLANTATIONS

Natural, mixed-species forest and spruce and pine plantations were surveyed in southern New Brunswick. The plantations were surveyed at 3, 6, 7, and 15 years of age, and the natural forests were 60 years old. Only abundant species are listed here. Data are given as pairs/10 hectares; stand age is in years.

	PLANTATIONS				FORESTS	
SPECIES	3 y	6 y	7 y	15 y	60 y	60 y
Yellow-bellied flycatcher (*Empidonax flaviventris*)	0.0	0.0	0.0	6.9	4.1	1.3
Alder flycatcher (*Empidonax alnorum*)	0.0	5.3	4.3	13.4	0.0	0.0
Hermit thrush (*Hylocichla guttata*)	0.0	0.0	0.0	2.0	2.2	2.6
Magnolia warbler (*Dendroica magnolia*)	0.0	1.9	0.0	13.4	9.3	2.6
Yellow-rumped warbler (*Dendroica coronata*)	0.0	0.0	0.0	5.5	2.2	1.3
Black-throated green warbler (*Dendroica virens*)	0.0	0.0	0.0	0.0	2.2	5.5
Blackburnian warbler (*Dendroica fusca*)	0.0	0.0	0.0	0.0	4.5	6.0
Common yellowthroat (*Geothlypis trichas*)	0.9	15.5	9.2	15.3	0.0	0.0
Song sparrow (*Melospiza melodia*)	4.8	4.4	7.9	0.0	0.0	0.0
Lincoln's sparrow (*Melospiza lincolnii*)	4.4	1.2	17.2	6.0	0.0	0.0
White-throated sparrow (*Zonotrichia albicollis*)	0.0	12.6	5.3	8.4	1.5	0.4
Northern junco (*Junco hyemalis*)	3.5	1.0	0.8	2.5	2.2	0.4
Total bird density	15.7	53.9	47.2	102	57.8	50.6
Number of species	16	20	22	38	42	32

Source: Modified from Freedman (1995)

in southern New Brunswick have found that they can sustain an abundant population of breeding birds (Table 23.7). In fact, a 15-y-old spruce plantation supported a larger bird population than did nearby stands of natural forest, while species richness was similar. In this study, many bird species of the natural, conifer-dominated forest began to invade the developing plantations once the trees were at least 10 years old.

Plantations are particularly deficient in cavity trees, snags, and coarse woody debris. Consequently, they support few of the many species that require these critical habitat features. For example, in southern New Brunswick, conifer plantations are established by clear-cutting natural forests, then preparing the site for planting using large machines that crush the logging debris and topple any unharvested trees and snags. This kind of harvesting and intensive management results in the presence of almost no cavity trees or snags, and little coarse debris (Table 23.8).

If conifer plantations provide adequate winter cover and browse, snowshoe hares (*Lepus americanus*) can maintain large populations. In fact, these animals sometimes cause economic damage by feeding on the bark and shoots of small trees. As soon as the trees mature and start to produce sizable cone crops, red squirrels (*Tamiasciurus hudsonicus*) also find conifer plantations to be acceptable habitat.

Sometimes, plantations are established on previously agricultural or industrial lands (this is known as *afforestation*). Depending on the sort of habitat that results, these plantations are likely to enhance the populations of indigenous species by providing opportunities for forest species. This can be an important benefit in some regions of Canada where agriculture is the dominant land use, for example, in southern Ontario. However, even greater biodiversity benefits would be attained if an attempt were made to restore a more natural forest, rather than a monocultural plantation.

Landscape Considerations

Biodiversity at the landscape level is related to the distribution and richness of ecological communities, including their dynamics over time (see Chapter 7). If a landscape is covered uniformly with only one or a few types of com-

TABLE 23.8 | SNAGS AND COARSE WOODY DEBRIS IN NATURAL FORESTS AND PLANTATIONS

Natural mature forest and conifer plantations were surveyed in southern New Brunswick. Data are for snags (standing dead trees) and woody debris with diameter greater than 5 cm. Basal area is the cross-sectional area of snags or trees, and is a forestry measure related to biomass. Volume is also related to biomass. Snag density is stated as numbers/hectare, basal area as metre^2/hectare, debris density in units of 10^3/hectare, and volume as metres^3/hectare.

	SNAGS		COARSE WOODY DEBRIS	
	DENSITY (no./ha)	BASAL AREA (m^2/ha)	DENSITY (10^3/ha)	VOLUME (m^3/ha)
STANDS OF MATURE, NATURAL FOREST				
Hardwood-dominated	138	3.5	0.28	18.7
Mixedwood	188	3.6	0.30	19.9
Mixedwood	200	5.4	0.36	13.0
Mixedwood	270	4.1	0.27	32.7
Conifer-dominated	698	11.4	0.48	41.6
Conifer-dominated	1115	19.5	0.86	45.4
Conifer-dominated	467	12.7	1.03	56.5
STANDS OF PLANTATION FOREST				
21-y-old spruce	13	0.03	0.13	0.6
15-y-old spruce	0	0.00	0.17	8.9
13-y-old spruce	13	0.20	0.56	14.1
8-y-old spruce	50	0.20	2.73	28.0
7-y-old spruce	0	0.00	2.20	23.9
6-y-old larch	0	0.00	2.04	32.1
5-y-old spruce	0	0.00	2.18	23.7
4-y-old spruce	0	0.00	3.25	52.2

Source: Freedman *et al.* (1996)

munity, it has little biodiversity at this level. In contrast, an area with a complex and dynamic mosaic of communities has much greater landscape-level biodiversity. *Landscape ecology* involves the study of the patterns and dynamics of communities on landscapes (and seascapes).

Landscape-level biodiversity is greatly influenced by catastrophic disturbances that result in some stands or "patches" of late-successional communities being replaced with younger ones. Natural causes of these *stand-replacing disturbances* include wildfire, windstorms, volcanic eruptions, and insect irruptions. Human activities associated with agriculture and forestry can have similar effects.

Sometimes, forest harvesting is designed to mimic the natural patch-disturbance regime. For example, most pine forests are naturally disturbed by periodic wildfires, which may kill most of the mature trees. Soon afterward, a new cohort of tree seedlings establishes, and these grow into another mature forest. To some degree, foresters can emulate this natural cycle of disturbance when they prepare plans to harvest and manage pine forests.

Usually, however, forestry imposes an anthropogenic patch dynamic onto the forest landscape. For instance, this may occur if forestry creates an unnatural mosaic of clear-cuts and plantations of various ages, interspersed in checkerboard fashion within a matrix of any remaining natural forest and nonforest communities. In some cases, a landscape mosaic of this kind may even be recommended by game managers, because (as we have already seen) it can favour certain hunted species such as deer and ruffed grouse.

However, the patch dynamics created by the harvesting and management of forests have important implications for many other elements of biodiversity. For example, if the remaining patches of unharvested natural forest are too small or isolated from each other, they will not be able to sustain all of their native species and communities over the long term. These losses would have negative implications for natural values and for the ecological sustainability of the entire forestry system (see Chapter 12).

Forestry creates fragmented landscapes, containing successionally dynamic patches of silvicultural-forest and natural-forest habitats. Many species of native wildlife find the silvicultural habitats to be adequate for their purposes. Such habitats and their dynamics are, however, incapable of supporting other indigenous elements of biodiversity, whose survival may be placed at risk. If these values are to be protected, they must be accommodated within patches of natural forest which are protected as ecological reserves.

To achieve a balance between the economic needs of forestry and the need to conserve indigenous biodiversity, the size, shape, and spatial arrangement of the patches of managed landscapes, and particularly the ecological reserves, must be considered. For example, if the ecological reserves are too few, small, isolated, or young to accommodate all of the biodiversity objectives, then it will be necessary to design a landscape that is more ecologically appropriate. Some design options that have been recommended to meet the biodiversity objectives of ecological reserves are described in Chapter 24.

Some of the National Parks of Canada are among the largest ecological reserves in the world. Yet many of these are too small to sustain viable populations of some species, or to sustain the ecological dynamics required to allow old-growth forests to persist on the landscape. The species that are most at risk need extensive areas of suitable habitat to sustain their populations. Such species include grizzly bear (*Ursus arctos*), wolf (*Canis lupus*), spotted owl, and marbled murrelet. Even the largest National Parks in Canada may not be big enough to sustain these species over the centuries.

In such cases, the ecological reserves and their surrounding areas must be managed as "greater reserves" — that is, as an integrated ecosystem. If forestry continues to be an important economic activity in the area surrounding protected reserves, it must be conducted with a view to sustaining those species and natural communities that might be at risk on the greater landscape. In many cases, this will require changes in the forestry management systems. Such changes might include maintaining a network of protected areas connected by corridors, and incorporating critical habitat elements, such as cavity trees, snags, and coarse woody debris, into managed stands.

Integrated Forest Management

Increasingly, in many parts of Canada, foresters are working with other interested parties to develop **integrated forest management** plans that accommodate the need to harvest timber from forested landscapes while also sustaining other values. Usually, these plans focus on finding ways to conduct forestry while continuing to support populations of hunted species such as deer and elk, and of fished species such as trout and salmon. In some cases, significant efforts are also made to accommodate other uses and values, such as nonconsumptive recreation and ecotourism (e.g., birding and hiking). Canadian Focus 23.2 describes an example of various interest groups working to develop an integrated management plan for a region in New Brunswick known as the "Greater Fundy Ecosystem."

Canadian Focus 23.2

The Greater Fundy Ecosystem

The Greater Fundy Ecosystem (GFE) consists of Fundy National Park plus its surrounding area in southern New Brunswick. Although the National Park is a relatively large protected area (comprising 206 km^2), it is too small to fully sustain some important ecological values, such as viable populations of species that have large territories (for example, black bear and pileated woodpecker), or certain types of natural ecosystems, particularly old-growth forests.

For these reasons, some ecologists believe that the ecological reserve function of Fundy National Park is threatened by forestry and other economic activities in its surrounding landscape, including the development of intensively managed forest plantations adjacent to the park boundary. In effect, the natural ecosystems of Fundy National Park are becoming

insularized within a matrix of plantations and other intensively managed habitats. (In the sense meant here, insularization is a process in which a reserve becomes surrounded by different, hostile habitat, much like an island in an ocean.) Forestry is not the only important ecological stressor within the GFE. Local stressors include tourism, agriculture, and residential development, while regional stressors include climate change and acidifying depositions from the atmosphere. However, these stressors are considered less important within the GFE than the extensive conversions of natural forests into plantations.

Most of the natural forest in the GFE is dominated by mixed-species stands of spruce and balsam fir, with lesser amounts of birch, maple, and other tree species. Most of the conifer-dominated stands have been damaged by recent infestations of spruce budworm, which caused much mortality of mature fir and spruce. From the forestry perspective, the damaged stands represent a degraded natural resource. The residual economic value of the timber is extracted from budworm-damaged stands by clear-cutting, and the site is then converted into a single-species, intensively managed plantation of conifers. These trees will be harvested for pulpwood on an approximately 40-y rotation, or for sawlogs on a 60-y rotation.

A team of scientists and resource managers from government, universities, and industry have come together as the Greater Fundy Ecosystem Research Group to address ecological changes and issues within the GFE. The GFE Research Group is also a partner in the Fundy Model Forest, one of ten model forests in a national network funded by the Canadian Forest Service. The broad goal of the model-forest program is to demonstrate sustainable forestry in Canada. In most cases, this is pursued through work on sustainable timber harvests, on management options that increase forest productivity, and on the integrated management of timber, hunted wildlife, and water resources. These, plus indigenous biodiversity, natural ecosystems, and protected areas, are all important priorities in the Fundy Model Forest.

The larger, conceptual mission of the GFE Research Group is to design an "ecologically sustainable landscape" in the Greater Fundy Ecosystem. This must sustain two clusters of values:

- long-term harvests of economically important commodities — especially timber, but also hunted animals and agricultural crops — and opportunities for ecotourism
- acceptable levels of other ecological values, such as sustainable populations of indigenous species, and viable areas of natural ecosystems

This agenda acknowledges that the National Park cannot be managed in isolation from environmental influences originating in its surrounding area. This is an example of ecosystem-based management, because an attempt is being made to manage the entire GFE as a single, integrated ecosystem.

Most of the work of the GFE Research Group is directed at acquiring a better understanding of the ecological consequences of forest conversion and reserve insularization, particularly for biodiversity at the levels of species, community, and landscape. The program of ecological research began in 1992, but has been able to build on a considerable body of earlier work. The research results have been incorporated into an initial set of recommended changes to forestry management systems, aimed toward the use of ecologically more appropriate systems of harvesting and management in the Greater Fundy Ecosystem and the Fundy Model Forest. These recommendations have been subjected to broad, consensus-building consultations, and have been offered for integration into forestry management plans in the region.

The recommendations incorporate two approaches:

- a landscape-level approach directed toward the management of community types, age-class distributions, and connectedness of patches
- a finer-grained strategy that addresses the needs of particular threatened species and communities

The recommendations focus on:

- the use of forest-management practices that are compatible with the natural disturbance regime and the geological factors, soils, and other "enduring features" important in the landscape

- the maintenance of water quality and habitats of aquatic species — a goal to be achieved mainly through the protection of riparian buffers
- the conservation of cavity trees, snags, and coarse woody debris as critical habitat features
- the completion of a network of protected areas to represent all community types in the model forest.

The recommendations of the GFE Research Group were endorsed as "desirable" by the partnership of the Fundy Model Forest during a workshop in late 1996. Work is now progressing toward incorporating the recommendations into forestry management plans being prepared for implementation within the Fundy Model Forest. The next phase of GFE Research Group work will be to determine whether key aspects of the forestry management plan of the model forest are actually achieving their goals of sustaining indigenous biodiversity and ecosystem functions.

By cooperating in the design of integrated management plans, the forest industry is attempting to come to grips with some of the important controversies that are arising from their woodlands operations. Society expects that the vast forests of Canada will continue to deliver a wide range of ecological goods and services. These include the substantial economic benefits from timber harvesting, as well as satisfying the needs of sport hunters, fishers, hikers, and other outdoor recreationists. Even while they are used in these ways, forest landscapes are also expected to provide such ecologically important services as clean air and water, and carbon storage.

The forest industries are making great strides in the directions society expects of them, but much more has to be done. This is particularly true of the need to set aside additional areas that are protected from commercial forestry. If they are large enough, the protected areas can allow ecological processes to continue in a manner unfettered by major human influences, so that natural ecosystems can develop and indigenous species can sustain their populations. In addition, management systems will have to change to accommodate more of the habitat needs of Canadian biodiversity on harvested sites.

If the Canadian forest industries are to legitimately claim that they are conducting their operations in an ecologically sustainable manner, they must achieve several broad objectives. First, it is critical that the rate of harvesting forest biomass does not exceed the rate of forest productivity. At the same time, it is necessary that other economic values be sustained, such as viable populations of hunted animals and outdoor recreation opportunities. Finally, it is critical that no indigenous elements of Canadian biodiversity be endangered. Although the forest industry has been making substantial progress toward environmental protection in Canada, not all of these requirements of ecologically sustainable forestry are being satisfied.

Key Words

forestry
clear-cutting
deforestation
silviculture
plantations
nutrient capital
selection harvesting
erosion
biodiversity
integrated forest management

Questions for Discussion

1. Consider a typical forest in the region where you live. Make a list of the dominant species of trees, other plants, and animals that live in that forest, and try to identify some important interactions among those species. Discuss, for example, the habitat needs of certain animals, including the foods they eat.
2. Are forest products important in your life and in the functioning of your community? Compile a list that shows how trees are used for energy, lumber, paper, and other products. Also consider nontimber uses of forests,

such as hunting, recreation, and provision of clean air and water.

3. What elements should an integrated management plan include for a typical forested watershed in your region? Try to consider the needs to ensure a constant supply of timber, deer, sportfish, clean water, and habitat for nongame species.
4. How might forest resources make a larger contribution to the Canadian economy, or to the economy of your region? Would it be possible, for example, to harvest more wood without degrading the timber resource? Could more people be employed in woodland operations if harvesting and management activities were less mechanized? Consider also the prospects for reducing exports of raw materials such as logs, through increased local processing into manufactured products.
5. Describe the characteristics of a typical old-growth forest in the region in which you live. Do you think that the special values of that old-growth forest can be accommodated by forestry? Or can old-growth forests be preserved only by creating large protected areas, where trees are not harvested? What are the economic implications of setting aside such large tracts of potentially valuable timber? What are the ecological implications?

References

Bormann, F.H., and G.E. Likens. 1979. *Pattern and Process in a Forested Ecosystem*. New York: Springer.

Buskirk, S.W. 1992. Conserving circumboreal forests for martens and fisher. *Conservation Biology*, **6**: 318–320.

Crowell, M. and B. Freedman. 1994. Vegetation development during a post-clearcutting chronosequence of hardwood forest in Nova Scotia, Canada. *Canadian Journal of Forest Research*, **24**: 260–271.

Drolet, C.A. 1978. Use of forest clear-cuts by white-tailed deer in southern New Brunswick and central Nova Scotia. *Can. Field-Nat.*, **92**: 275–282.

Freedman, B. 1995. *Environmental Ecology, 2nd ed.* San Diego, CA: Academic.

Freedman, B., R. Morash, and A.J. Hanson. 1981. Biomass and nutrient removals by conventional and whole-tree clear-cutting of a red spruce-balsam fir stand in central Nova Scotia. *Canadian J. For. Res.*, **11**: 249–257.

Freedman, B., C. Stewart, and U. Prager. 1985. *Patterns of Water Chemistry of Four Drainage Basins in Central Nova Scotia*. Moncton, NB: Environment Canada, Inland Waters Directorate, Water Quality Branch. Technical Report IWD-AR-WQB-85-93.

Freedman, B., S. Woodley, and J. Loo. 1994. Forestry practices and biodiversity, with particular reference to the Maritime Provinces of eastern Canada. *Env. Rev.*, **2**: 33–77.

Freedman, B., V. Zelazny, D. Beaudette, T. Fleming, S. Flemming, G. Forbes, G. Johnson, and S. Woodley. 1996. Biodiversity implications of changes in the quantity of dead organic matter in managed forests. *Env. Rev.*, **4**: 238–265.

Gillis, A.M. 1990. The new forestry: an ecosystem approach to land management. *BioScience*, **40**: 558–562.

Godfrey, W.E. 1986. *The Birds of Canada, 2nd ed.* Ottawa: National Museums of Canada.

Gullion, G.W. 1988. Aspen management for ruffed grouse. In: *Integrating Forest Management for Wildlife and Fish*. St. Paul, MN: North Central Forest Experiment Station. U.S.D.A. For. Serv., Gen. Tech. Rep. NC-122. pp. 9–12.

Harris, L.D. 1984. *The Fragmented Forest*. Chicago: University of Chicago Press.

Hunter, M.L. 1990. *Wildlife, Forests, and Forestry: Principles of Managing Forests for Biological Diversity*. Englewood Heights, NJ: Prentice Hall.

Kimmins, J.P. 1996. *Forest Ecology, 2nd ed.* New York: Macmillan.

Krause, H. H. 1982. Nitrate formation and movement before and after clear-cutting of a monitored watershed in central New Brunswick. *Canadian Journal of Forest Research*, 12: 922–930.

Likens, G.E. and F.H. Bormann. 1995. *Biogeochemistry of a Forested Ecosystem, 2nd ed.* New York: Springer.

Martin, C. W., R. S. Pierce, G. E. Likens, and F. H. Bormann. 1986. *Clearcutting Affects Stream Chemistry in the White Mountains of New Hampshire*. Broomall, PA: Northeastern Forest Experiement Stations. USDA For. Serv. Research Paper NE-579.

Maser, C. 1990. *The Redesigned Forest*. Toronto: Stoddart.

Maser, C., R.F. Tarant, J.M. Trappe, and J.F. Franklin. 1988. *From the Forest to the Sea: A Story of Fallen Trees*. Portland, OR: Pacific Northwest Research Station.USDA For. Serv., Gen. Tech. Rep. PNW-GTR-229.

Morgan, K. and B. Freedman. 1986. Breeding bird communities in a hardwood forest succession in Nova Scotia. *Can. Field-Nat.*, **100**: 506–519.

Swan, D., B. Freedman, and T. Dilworth. 1984. Effects of various hardwood forest management practices on small mammals in central Nova Scotia. Can. Field-Nat., **98**: 362–364.

Swanson, F.J. and J.F. Franklin. 1992. New forestry principles from ecosystem analysis of Pacific Northwest forests. *Ecological Applications*, **2**: 262–274.

Welsh, D. and D.R. Fillman. 1980. The impact of forest cutting on boreal bird populations. *American Birds*, **34**: 84–94.

24

THE BIODIVERSITY CRISIS

CHAPTER OUTLINE

- Introduction
- Extinction as a natural process
- Extinctions and endangerment caused by humans
- Tropical deforestation and extinctions
- Species declines
- Back from the brink
- Canadian species and ecosystems at risk
- The importance of protected areas
- International conservation activities

CHAPTER OBJECTIVES

After completing this chapter, you will be able to:

1. Outline how humans are causing the modern crisis of extinction and endangerment.
2. Give several examples of species that have been rendered extinct through human activities, including cases of losses from Canada.
3. Explain how a system of protected areas is critical to the preservation of biodiversity.
4. Outline the roles of governments, nongovernmental organizations, and scientists in conserving species and other elements of biodiversity.

INTRODUCTION

Earlier, we defined biodiversity as the total richness of biological variation, and discussed reasons why biodiversity is important (see Chapter 7). In this chapter, we examine the many threats to biodiversity that are associated with human influences. We emphasize permanent damage caused to biodiversity values, particularly those values associated with losses of species and of distinctive, natural ecosystems.

Extinction refers to the loss of some *taxon* over all of its range on Earth. (The word taxon refers to any taxonomically defined group of organisms, such as subspecies, species, genus, family, order, etc.) **Extirpation** refers to a more local disappearance, with the taxon still surviving elsewhere. Extinction represents an irretrievable loss of a unique portion of the biological richness of Earth, whereas it may be possible to re-establish an extirpated taxon from a surviving population elsewhere.

Extinctions have always occurred as a result of natural influences. These include random catastrophes as well as the long-term effects of environmental changes: for example, in climate, or in biological interactions such as disease or predation (see Chapter 7). In modern times, however, and probably for the past 10 000 years or so, almost all extinctions have been caused by anthropogenic influences, particularly overharvesting and habitat changes. In fact, an enormous increase in the rate of extinctions has been one of the most important consequences of humans becoming the dominant species on Earth. Species are now disappearing so quickly that we refer to the phenomenon as an extinction crisis, or sometimes as a biodiversity crisis.

All species are unique, a fact that gives them great intrinsic value — this is a central tenet of the *biocentric world view* (see Chapter 1). Therefore, from an ethical perspective, irretrievable losses of biodiversity are a shameful consequence of the way humans are using their powers to exploit and manage other species and ecosystems. This is also a foolish way for humans to administer their global empowerment, because unique species are disappearing before they have been investigated for their potential usefulness in medicine or agriculture, and before we understand their importance as components of ecosystems. Human actions that result in extinctions can only be considered to be ecologically dangerous behaviour.

In this chapter, we examine the modern extinction crisis — the reasons why it is happening, ways of repairing some of the damage already done, and how to prevent further damage.

EXTINCTION AS A NATURAL PROCESS

Life has existed on Earth for approximately 3.5 billion years. Almost all of the species that have lived during that period of time are now extinct, having disappeared "naturally" for some reason or other. (Of course, the rest of those species are *extant*, or still living today.) In many cases, the extinct species could not adapt to changes that occurred in their environments, such as shifts in climate or increases in the intensity of disease, predation, or competition with other species. Many other species, however, disappeared during brief episodes of mass extinction, which may have been caused by unpredictable catastrophes such as a meteorite hitting the Earth.

The geological record clearly shows that numerous species and broader groups of organisms (such as genera, families, and phyla) have appeared and disappeared over time (see Chapter 6). For example, many phyla of invertebrate animals evolved during an evolutionary proliferation around the beginning of the Cambrian era some 570 million years ago. Subsequently, most of those phyla and their numerous species became extinct. About 15–20 extinct phyla from that period were discovered in a renowned fossil deposit known as the Burgess Shale, in Yoho National Park in southeastern British Columbia. Each of those extinct phyla represented a novel, fantastic, evolutionary "experiment" in invertebrate form and function. We only know about these ancient extinct creatures because their fragile, soft-bodied structures became fossilized under extraordinary geological circumstances (Gould, 1989).

The fossil record is filled with many other definitive examples of ancient extinctions. However, the rates of extinctions, and of the subsequent evolutionary radiations of new species, have not been uniform over time. The geological record clearly shows that relatively low and uniform rates of extinction persisted for extremely long periods of time. However, those tranquil eras were punctuated by about nine catastrophic events of mass extinction.

The most intense episode of mass extinction occurred at the end of the Permian period, about 245 million y ago. This natural catastrophe resulted in the loss of 54% of the families of marine organisms, including 84% of the existing genera and 96% of the species (Erwin, 1990). Another mass extinction occurred at the end of the Cretaceous period, about 65 million years ago. This famous extinction event involved the last of the dinosaurs and pterosaurs (the latter were flying reptiles), along with many other taxa, totalling perhaps three-quarters of the species living at the time. Many scientists believe that this crisis of paleobiodiversity was caused by a meteorite hitting Earth. Such a catastrophe would have ejected enormous quantities of dust into the atmosphere, resulting in a cooling of the climate that most plant and animal species were unable to tolerate.

During the past several centuries, Earth's existing heritage of biodiversity has been buffeted by another mass extinction. This is an ongoing catastrophe, which will certainly intensify into the foreseeable future. This ecological calamity is not a natural phenomenon. Rather, it is being caused by the influences and activities of modern humans.

Extinctions and Endangerment Caused by Humans

Numerous human activities are causing species to become **endangered** or extinct. (Species are considered endangered if, because of a small population or loss of habitat, they are at high risk of becoming extirpated or extinct.) The most important cause is the destruction of natural ecosystems and their conversion into habitats that are unsuitable for indigenous species, a problem that is especially acute in tropical countries. Excessive harvesting of some species is also significant, as are damages caused by introduced predators, diseases, and competitors. Any of these stressors can cause populations to become small and fragmented, resulting in much greater risks of extirpation or extinction (Figure 24.1).

FIGURE 24.1 | THE EXTINCTION VORTEX

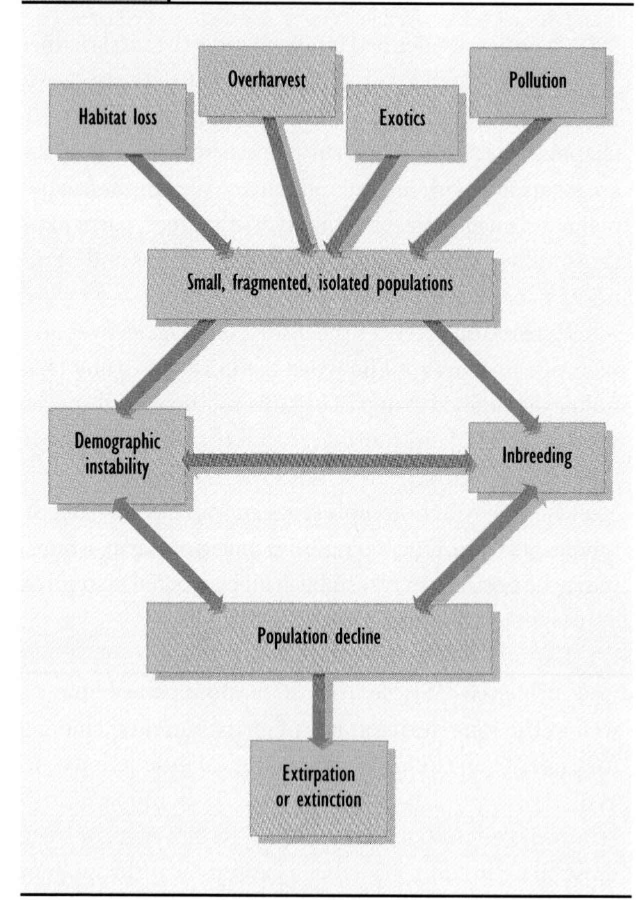

Extinction can be caused by various influences and activities of humans, such as habitat loss; excessive harvesting; and introductions of exotic diseases, predators, and competitors. As a result of these stressors, large, continuously distributed populations may fragment into small isolated units which are vulnerable to the deleterious effects of inbreeding, population instability, and random catastrophes. These can cause the endangered populations to decline further, and may ultimately result in extirpation or extinction. Conservation biologists refer to this accelerating, downward spiral of endangerment as the "extinction vortex."

Because of these anthropogenic influences, the past several centuries have witnessed huge increases in the global rate of extinction and in the number of species threatened with this catastrophe. Our knowledge of recently extinct and endangered species is relatively complete for large and conspicuous species such as vertebrate animals, particularly those of temperate and higher-latitude countries, where most biologists live. In fact, there have been more than 700 known extinctions of vertebrate animals during the past four centuries, including about 100 species of mammals and 167 species of birds. All of these extinctions were caused by human influences.

Unfortunately, we know much less about extinctions among other groups of organisms. This is particularly true of the enormous diversity of relatively small, poorly known species that live in tropical ecosystems, especially in old-growth rainforests. Undoubtedly, huge numbers of tropical species, particularly plants and invertebrates, have become extinct during the past several centuries as their natural forest habitats were converted into agricultural and other land uses. We can refer to these losses as "hidden extinctions," because so few of the extinct species had been discovered and named by taxonomists. Moreover, these hidden extinctions continue to occur rapidly, in fact at an accelerating rate, because their poorly explored tropical habitats are being destroyed so quickly.

In the following sections we will examine selected case studies of species that have been rendered extinct by humans and their activities.

Prehistoric Extinctions

Many species are useful as "resources" that humans can harvest as sources of food, medicine, timber, fuel, or for some other purpose. In many cases, the exploitation of these potentially renewable resources has been so insatiable and unsustainable that their "mining" has culminated in extinction. These once-valuable species now occur nowhere on Earth.

We previously examined extinctions caused by prehistoric hunters as they overexploited populations of large, naive animals in newly discovered lands (see In Detail 12.2, page 191). In North America, for example, it appears that hunters exterminated many species of large mammals soon after humans discovered the continent by migrating across a Beringean land-bridge around 11 or12 millennia ago, at the end of the most recent ice age. Known extinctions around that time include 77 species of mammals and several species of birds. The losses include 10 species of horse, a giant ground sloth, four species of camels, two species of bison, the mastodon and several species of mammoths, and the sabertooth tiger. Other large species of animals became extinct when South America was colonized somewhat later.

Colonizations by humans also caused prehistoric mass extinctions in other places. Australia and New Guinea were discovered about 50 000 years ago. Soon after, many species of marsupials, large flightless birds, and tortoises became extinct. New Zealand was colonized less than 1000 years ago. Within two centuries, 30 species of large birds were extinct, among them a goose, a swan, and 26 species of flightless birds known as moas. The extinctions on New Zealand proceeded as a wave that spread from northern North Island, the initial point of colonization, to South Island. Many of the moas were driven by hunters and their dogs to convenient butchering sites, where the great piles of bones were later mined by European farmers and processed into phosphate fertilizer.

In a similar fashion, the human colonization of Madagascar, about 1500 years ago, resulted in the extinction of 14 species of lemurs, 6–12 species of giant, flightless elephant birds, and various other large, edible animals. Other well-known cases of prehistoric, mass extinctions occurred on Hawaii, New Caledonia, Fiji, the West Indies, and numerous other islands.

Many of the **endemic species** that existed on small islands were particularly vulnerable to extinction. One reason for this is that island species occurred in small, isolated populations. Moreover, it is thought that many birds of remote islands had not experienced intense predation during their recent history, and consequently had evolved to be flightless, relatively large, and unafraid of predators. Such species were extremely vulnerable to hunting by humans once their islands were discovered. In addition, most island species did not co-occur with closely competing organisms. Consequently, they may have been displaced easily when more capable species were introduced. Finally, human colonization of remote islands, particularly by Europeans, resulted in great changes in natural habitats as the endemic vegetation was cleared for agricultural, urban, and tourism developments. The islands became ecologically degraded by introduced plants, animals, and diseases.

For these reasons, the biotas of remote islands have suffered particularly high rates of extinction. For example, at one time, each of the approximately 800 islands of the southern Pacific Ocean may have supported two or three endemic species of flightless rails (a family of marsh birds), plus other unique species of birds and reptiles. As these islands were discovered and colonized by prehistoric Polynesians, perhaps thousands of these endemic species became extinct through overhunting and habitat changes. For instance, a study of bird bones, recovered at an archaeological site on the island of Ua Huka, found that 14 of the 16 original bird species no longer occur there, including 10 unique species that had been rendered extinct. The extinctions worsened when Europeans secondarily colonized these and other oceanic islands, because of the great habitat losses that occurred during "development." In fact, of the 167 taxa of birds (including 95 species) around the world that are known to have become extinct since 1600, all but 9 lived on islands.

The problem of extinction-prone island biotas can be further illustrated by the case of the Hawaiian Islands, an ancient, remote archipelago of volcanic outcroppings in the Pacific Ocean. When these islands were first discovered by Polynesian seafarers, there were at least 86 species of birds, including 68 species that occurred nowhere else. Of those 68 endemics, 24 are now extinct and 29 are endangered. Similarly, the native flora of the Hawaiian archipelago at the time consisted of as many as 2000 species of flowering (or angiosperm) plants, of which 94–98% were endemic. During the past several hundred years, more than 100 of the indigenous plant species have become extinct, and more than 500 have become threatened or endangered. The extinctions and endangerment of Hawaiian species have been caused by extensive conversions of natural habitats into agricultural and urbanized landscapes, coupled with introductions of invasive species, including alien predators, competitors, virulent disease organisms, and destructive herbivores such as goats.

Species Made Extinct or Endangered by Overharvesting

Unsustainable harvesting has caused some of the most famous cases of extinction and endangerment, in some instances involving species that were initially extremely abundant. We will illustrate this phenomenon by referring to the dodo, the passenger pigeon, the great auk, and a few other notable cases. These are typical examples of the devastating effects that insatiable killing can have on vulnerable populations of wild creatures.

The dodo (*Raphus cucullatus*) was a turkey-sized, flightless bird that disappeared in 1681, making it the first well-documented extinction. The loss of this species is immortalized in everyday language by the phrase "dead as a dodo," which is used as a metaphor for an irrevocable loss. We also sometimes use the word "dodo" to describe an old-fashioned or stupid person. This etymology derives from the hapless dodo's apparent inability to adapt to the threats posed by the human colonists of Mauritius, the only place on Earth where this bird lived. Mauritius is a small island in the Indian Ocean, discovered by Portuguese sailors in 1507. It was colonized by the Dutch in 1598, who hunted the dodo for meat, gathered its eggs, cleared its habitat for agriculture, and released cats, pigs, and monkeys that preyed on dodos and destroyed their ground-level nests. These stressors caused the dodo to decline rapidly and become extinct.

The great auk (*Pinguinus impennis*), a flightless seabird, was the first well-documented anthropogenic extinction of a species whose range included North America. Early mariners knew the great auk as the original "pennegoin," although it belonged to a different family of birds (Alcidae) from the superficially similar penguins (Spheniscidae) of the Southern Hemisphere. The great auk lived throughout the north Atlantic region, breeding on a few islands off eastern Newfoundland, in the Gulf of St. Lawrence, around Iceland, and north of Scotland.

This large seabird was initially abundant in its breeding colonies. Because it was flightless, it could be killed quite easily. Consequently, the great auk had long been exploited by aboriginal people inhabiting what is now known as Newfoundland, and also by Icelanders and European fishers, as a source of fresh meat, eggs, and oil. Unfortunately, the great auk developed into a valuable economic commodity when its feathers became sought after for stuffing mattresses in the mid-1700s. This resulted in a systematic and relentless overexploitation that quickly caused the great auk to become extinct.

One of the largest breeding colonies of great auks was on Funk Island off Newfoundland. In 1785, an observer described the harvest of great auks and other seabirds on Funk Island (from Nettleship and Evans, 1985):

> *It has been customary of late years, for several crews of men to live all summer on that island, for the sole purpose of killing birds for the sake of their feathers, the destruction of which they have made is incredible. If a stop is not soon put to that practice, the whole breed will be diminished to almost nothing, particularly the penguins.*

The great auk was, in fact, extirpated from Funk Island in the early 1800s. The last two individuals of this species ever seen alive were killed in 1844 by several Icelanders who were searching for specimens to sell to a bird "collector." Because of their extreme rarity at the time, great auks and their eggs were precious to collectors — they were, unfortunately, too valuable to let live.

The passenger pigeon (*Ectopistes migratorius*) possibly numbered 3–5 billion individuals three centuries ago, when it may have been the world's most populous land bird. The passenger pigeon bred in southeastern Canada and the northeastern United States in mature forests of oak, beech, hickory, and chestnut. These trees produce large seeds known as "mast," upon which this bird fed. In the autumn, passenger pigeons migrated to the southeastern United States. They flew in enormous flocks, described as being so dense as to obscure the sun, and taking hours to pass. The birds roosted communally during winter nights, often in such large numbers that they would kill trees through the excessive deposition of guano (bird feces), while breaking stout limbs under their weight.

The naturalist John Lawson described an impressive passage of these pigeons in the Carolinas (from Feduccia, 1985):

I saw such prodigious flocks of these pigeons…in 1701–2… that they had broke down the limbs of a great many trees all over these woods, whereupon they chanced to sit and roost…These pigeons, about sun-rise…would fly by us in such vast flocks, that they would be near a quarter of an hour, before they were all passed by; and as soon as that flock was gone, another would come; and so successively one after another, for the rest of the morning.

The seeming unlimited abundance of passenger pigeons, and their habit of migrating and breeding in large and dense groups, made them an easy target for market hunters who sold the carcasses in cities and towns. During the early 1800s there was a well-organized hunt of passenger pigeons to supply urban markets with cheap, if somewhat seasonal, meat. During seasons when the hunt was on, "wagon loads of them...poured into the market...and pigeons became the order of the day at dinner, breakfast, and supper, until the very name became sickening" (A. Wilson in 1829; quoted in Feduccia, 1985).

The sizes of the harvests were staggering. For example, about one billion pigeons were taken in 1869 in breeding colonies in Michigan alone. The intensity of the commercial exploitation far exceeded sustainability, and this, along with destruction of much of the breeding habitat, caused the passenger pigeon to decline rapidly in abundance. The last known attempt at nesting was in 1894, and the last known individual died a lonely death in the Cincinnati Zoo in 1914.

The Carolina parakeet (*Conuropsis carolinensis*) used to breed widely in the southeastern United States. This parakeet was a fairly common, brightly plumaged, fruit- and seed-eating bird that foraged and roosted in groups, especially in mature hardwood forests. Carolina parakeets were not hunted as a valuable commodity. Rather, they were exterminated because they were regarded as an agricultural pest, because of damage they caused while feeding in orchards and grain fields. Unfortunately, Carolina parakeets were an easy mark for eradication because they nested and fed communally. Also, they tended to assemble around wounded colleagues, allowing an entire flock to be wiped out easily by a hunter. The last record of a flock of these parakeets was in 1904, and the last known individual died in a zoo in 1914.

Steller's sea cow (*Hydrodamalis stelleri*) was a sea mammal related to the manatees. It lived in subarctic waters around the Aleutian Islands in the Bering Sea and was hunted by aboriginal peoples of the region. Soon after this shy and inoffensive species was "discovered" by Russian explorers in 1741, it became hunted as a source of food and hides. Steller's sea cow was rendered extinct after only 26 years of exploitation.

The Caribbean monk seal (*Monachus tropicalis*) lived in warm waters of the Caribbean Sea and Gulf of Mexico. This species was encountered, and eaten, on Christopher Columbus's second voyage to the New World in 1494. Populations of the Caribbean monk seal were depleted by an 18th-century commercial hunt for its meat and blubber. The last survivors were exterminated by the subsistence hunting of local fishers.

The Eskimo curlew (*Numenius borealis*) is a large sandpiper that was quite abundant as recently as one century ago. The Eskimo curlew was exploited by market hunters during its migrations through the prairies and coasts of Canada and the United States, and also on its wintering grounds on the pampas (or grasslands) and coasts of South America. The uncontrolled hunting caused this bird to become rare by the end of the 19th century. The last observed nesting attempt was in 1866, and the last specimen "collected" (by shooting) was in Labrador in 1922. For some decades the Eskimo curlew was thought to be extinct, but recently, very small numbers of this perilously endangered bird have been seen by a few naturalists.

The right whale (*Eubalaena glacialis*) once ranged over all temperate waters of the Northern Hemisphere. Because of its slow speed, surface swimming behaviour, and rich oil content, plus the fact that it floated when dead, early whalers considered these the "right" whales to hunt. Due to commercial overhunting of right whales for their blubber, which was rendered into oil to fuel the lamps of Europe, the populations of this species collapsed over its entire range. The right whale has been extirpated from the eastern Atlantic Ocean off Europe, and is critically endangered in the western Pacific off Korea and Japan.

Only a few hundred right whales occur in the northwest Atlantic Ocean. Most of these animals spend much of the summer and autumn in the mouth of the Bay of Fundy and off southwestern Nova Scotia. They migrate south to winter along the southeastern United States and eastern Caribbean. Although not hunted for decades, the population of right whales has been slow to recover, largely because of mortality caused by collisions with ships and entanglement in fishing gear.

Species Made Extinct or Endangered by Habitat Destruction

Many species have been endangered or made extinct because their natural habitats were converted to agricultural or other land uses. We will first examine several examples of this phenomenon, and then discuss the modern destruction of tropical forests — the human activity that is causing the most extinctions today.

The American ivory-billed woodpecker (*Campephilus principalis principalis*) lived in the southeastern United States, where it bred in extensive areas of mature, bottomland hardwood forests and cypress swamps. Most of this habitat was heavily logged or converted to agriculture by the early 1900s, driving the population of ivory-billed woodpeckers into a rapid decline. There have been no sightings of this species in North America since the early 1960s. A closely related subspecies, the critically endangered Cuban ivory-billed woodpecker (*Campephilus principalis bairdii*), may still occur in very small numbers in mountain forests on Cuba.

The black-footed ferret (*Mustela nigripes*) was first "discovered" in the prairies of North America in 1851. Because of habitat losses, this predator has become extirpated in Canada and endangered in the United States. Extensive areas of its habitat of natural shortgrass and mixedgrass prairie were converted into agricultural use. Also, its principal food, the prairie dog (*Cynomys ludovicianus*), has declined in abundance. The prairie dog has been relentlessly poisoned as a perceived pest of rangelands. With little habitat or food, the black-footed ferret is unable to survive over most of its former range.

The Furbish's lousewort (*Pedicularis furbishiae*) is a herbaceous plant that grows only along a 230-km stretch of the Saint John River valley in New Brunswick and Maine. This species had been considered extinct. In 1976, it was "re-discovered" by a botanist doing field studies of the potential environmental impacts of a proposed hydroelectric reservoir on the upper St. John River in Maine. That industrial development would have obliterated the only known habitat of the lousewort. For that and other environmental and economic reasons, the dam was not constructed.

Tropical Deforestation and Extinctions

Tropical forests are Earth's most biodiverse ecosystems; their richness of species is phenomenal. Moreover, these poorly explored ecosystems are thought to contain millions of as-yet-unnamed species, particularly insects (Chapter 7). Because so many tropical-forest species have local distributions, the clearing of this ecosystem causes a disproportionate number of extinctions (that is, in comparison with extinctions due to disturbances of other kinds of natural ecosystems).

It is well known that the rates of deforestation in most tropical countries have increased alarmingly during the present century, particularly in the past several decades. This is in marked contrast to the situation in most higher-latitude countries, where forest cover has been relatively stable (see Chapter 14). In North America, for example, there was little net change (0.1%) in forest cover between 1981 and 1990 (Table 24.1). In contrast, most countries of Central and South America had substantial losses of forest cover during that period, as did most tropical countries of Africa and Asia. Globally, the rate of clearing of tropical rainforests during the 1980s was equivalent to about 1% of that biome per year — a rate which, if maintained, would imply a half-life for that biome of only about 70 y.

Most tropical deforestation is caused by conversions of forests into subsistence agriculture by poor people. These agricultural conversions are greatly increased whenever the access to the forest interior is improved. When roads are constructed for timber extraction or mineral explo-

TABLE 24.1 CHANGES IN FOREST AREA IN SELECTED COUNTRIES OF THE AMERICAS

Areas are stated in units of 10^6 kilometres^2, and deforestation rates as annual percentages ("n. d." indicates that no data are available).

COUNTRY	FOREST AREA (1990) (10^6 km^2)	DEFORESTATION (%/y; 1981–90)
NORTH AMERICA		
Canada	453.3	0.1
United States	296.0	0.1
CENTRAL AMERICA AND CARIBBEAN		
Honduras	4.6	1.9
Nicaragua	6.0	1.7
Panama	3.1	1.7
Guatemala	4.2	1.6
Mexico	48.6	1.2
Cuba	1.7	0.9
Belize	2.0	0.2
SOUTH AMERICA		
Paraguay	12.9	2.4
Ecuador	12.0	1.7
Venezuela	45.7	1.2
Bolivia	49.3	1.1
Brazil	561.1	0.6
Columbia	54.1	0.6
Peru	68.0	0.4
Guyana	18.4	0.1
Suriname	14.8	0.1
Argentina	67.0	n.d.
Chile	16.1	n.d.

Source: World Resources Institute (1994)

The greatest modern threats to biodiversity are associated with deforestation in tropical countries. This area in West Kalimantan in Indonesian Borneo was, until recently, covered in old-growth tropical rainforest. The forest was initially logged to recover its largest trees, which were used to manufacture timber and plywood for export. A secondary harvest was then made of smaller trees for local use, after which the area was converted to agricultural land use through a practice known as slash-and-burn. At the time the photo was taken, people had just moved into the area, and were engaging in subsistence agriculture. Few native species can survive in these ecologically degraded habitats.

Source: B. Freedman

ration, deforestation often follows rapidly. The complex social causes of deforestation include population growth, inequality of land ownership, and the displacement of poor people by mechanization and the commercialization of agriculture. Because of these factors, enormous numbers of poor people are seeking arable land in most of the world's less-developed countries. These people need land on which they can grow food for subsistence and some cash.

The forest conversions often involve a system of *shifting cultivation*, in which the trees are felled, the woody debris is burned, and the land is used to grow mixed crops for several years. By that time, fertility has declined and weeds have become too abundant. The land is then abandoned for a fallow period of several decades. This allows a secondary forest to regenerate, while nearby patches of forest are cleared to provide land for cultivation.

A more intensive system of subsistence agriculture, known as *slash-and-burn*, results in a more permanent conversion of the land into crop production. Slash-and-burn also involves cutting and burning the forest. After the forest is gone, however, the land is used continuously, without a fallow period during which a secondary forest could regrow.

Some tropical forests are being cleared to provide land for industrial agriculture (for example, to develop oil-palm plantations, sugarcane fields, and cattle pastures), and by commercial logging. Tropical deforestation is also caused by flooding during the development of hydroelectric reservoirs, by the harvesting of wood to manufacture charcoal, and by the harvesting of fuelwood, especially near large towns and cities. Wood is the predominant cooking fuel in many tropical countries, particularly for poorer people — for most of the world's humans, the energy crisis involves fuelwood, not fossil fuels (see Chapter 14).

Because so many species live in tropical forests, the modern rate of deforestation of this biome is having catastrophic consequences for global biodiversity. These damages will become increasingly important in the future, if the present relentless pace of tropical deforestation continues.

Fortunately, a widespread awareness and concern about this important ecological problem have developed. These have recently stimulated a great deal of research on the conservation and protection of tropical forests, and governments have started to set aside substantial areas as ecological reserves and other types of **protected areas**. By the mid-1990s, thousands of sites, comprising hundreds of millions of hectares, had received some sort of "protection" in tropical countries.

The effectiveness of this protected status varies greatly, however. It depends on factors that influence the governments' commitments to conserving forests and other indigenous ecosystems and to protecting biodiversity more generally. Political stability and priorities are particularly important considerations — these are critical to addressing the social causes of the destruction of tropical ecosystems. Social factors include poverty, population growth, inequalities of the distributions of wealth and land, and corruption. More directly, political stability and priorities determine whether enough money is available to support a system of protected areas, as well as to find effective means to control poaching of animals and lumber and to prevent other encroachments.

Poaching (i.e., illegal harvesting) of endangered wildlife is a terrible problem for species that have economic value on the international black market (see In Detail 24.1). This can be illustrated by the black rhino (*Diceros bicornis*) and the elephant (*Loxodonta africana*) in a game reserve in Zambia, Africa. In the early 1970s, the Luangwa Valley contained an estimated 100 000 elephants and 4000–12 000 black rhinos (Leader-Williams *et al.*, 1990). Unfortunately, these relatively large populations have now collapsed because of poaching, which resulted from the extremely high prices paid for rhino horns and elephant tusks in certain wealthy countries. Even though Zambian park wardens made courageous and highly motivated conservation efforts under difficult circumstances, it has so far proven impossible to control the poaching effectively. The astronomical value of horn and ivory has spawned a well-organized and effective chain of poaching, smuggling, and sale.

In spite of all of the problems, some tropical countries are developing a real commitment to the protection of threatened biodiversity. In Central America and the Caribbean, for example, Panama, Costa Rica, and the Dominican Republic have relatively progressive policies on the conservation and protection of natural ecosystems. As of 1993, 17% of the area of Panama had been given park or reserve status, as had 12% of Costa Rica, and 11% of the Dominican Republic (World Resources Institute, 1994). For perspective, we should note that the relative areas of protected land in those Latin American nations are greater than in Canada (5.5%) and the United States (4.1%), in spite of Latin America's comparative poverty. Such vigorous conservation activities are badly needed in the region; in the 1980s, the rate of deforestation was 2.6%/y in Costa Rica, 2.5%/y in the Dominican Republic, and 1.7%/y in Panama.

In other Latin American countries, conservation efforts have been disrupted by civil wars and other political instabilities and by different government and social priorities. For example, in 1993, the percentage of the national territory with protected status as parks or ecological reserves was only 0.1% in Jamaica and Honduras, 0.3% in Haiti, 0.4% in El Salvador, 0.6% in Trinidad and Tobago, and 1.0% in Mexico (World Resources Institute, 1994).

IN DETAIL 24.1

CITES and the International Trade in Endangered Species

Many endangered species are extremely valuable for one reason or another. Some are avidly sought by private collectors or by public zoos or botanical gardens, many of which may be willing to pay large sums for living specimens to add to their collections. Some animal and plant tissues are extremely valuable, a circumstance that may result in endangered species being killed for their fur, ivory, horn, fine-grained wood, or other body parts or tissues. For

example, rhinoceros horns are extremely valuable in Yemen for manufacturing into dagger handles, and in eastern Asia as an ingredient in traditional medicines. Similarly, bile from the gall bladder of bears is a precious commodity in traditional medicine in eastern Asia, as are tiger bones and the underground rhizome of ginseng, particularly if the plant was collected from a wild habitat. Elephant ivory is valued for carving and other crafts in southern Asia and elsewhere. Other costly products of endangered species include rare furs, trade in which has seriously affected large predators, especially cats such as tigers, cheetah, leopard, and jaguar.

The **C**onvention on **I**nternational **T**rade in **E**ndangered **S**pecies, often referred to by its acronym CITES, is an international treaty that commits signatory nations to controlling or preventing the trade in endangered species. CITES was established in 1973 in association with the United Nations Environment Program (UNEP). Its most important function is to monitor the international trade in endangered species and to control or prevent that trade as much as possible. For these purposes, the status of species as extinct, endangered, vulnerable, or rare is assigned by the International Union for the Conservation of Nature (IUCN). The actual trade in species is monitored by the World Wildlife Fund/IUCN "Traffic" network worldwide. The world headquarters of CITES, IUCN, and WWF are all in Switzerland.

International trade in about 400 species of animals and 150 species of plants is prohibited by CITES. In addition, the trade in about 2500 animal species and 25 000 plant species is monitored by the World Conservation Monitoring Center (WCMC) of Cambridge, UK, which publishes a series of so-called "red books" summarizing the status of, and commerce in, about 60 000 plant and 2000 animal species.

Canada is a member of CITES. Some of its responsibilities under the treaty are to monitor and report on its international trade in all species that fall under the purview of WCMC. In 1991, for example, the federal government of Canada issued 114 permits to import living CITES species or their parts (the living organisms are intended for zoological or botanical parks), and 2714 export permits (Campbell *et al.,* 1993). Legal exports of living specimens in 1991 included 6 grizzly bears, 3 polar bears, 17 otters, and certain other native species, plus many individuals of nonnative CITES species that were being traded among zoos and botanical parks. There is also a substantial legal export from Canada of body parts of some indigenous CITES species, including (in 1991) 102 narwhal tusks, 2238 wolf skins, 316 black bear skins, 349 grizzly bear skins, 266 polar bear skins, and 57 cougar skins.

Of course, these data refer only to the known, legal trade of species listed by CITES. There is also a flourishing illegal trade in Canada, particularly of animal parts such as bear gall bladders, caribou antlers, and certain kinds of furs. Further, there is an illegal trade in living animals and plants, such as some native species of orchids, and gyrfalcons and peregrine falcons, which are extremely valuable in certain countries for use in the sport of falconry. Most of the prohibited trade involves animals and plants illegally hunted or collected by poachers. In addition, there are large illegal imports of banned products into Canada. These include rare and endangered parrots, reptiles, fish, orchids, and other species that are valuable in the pet trade. There is also a burgeoning illegal trade of animal parts that are valuable in traditional medicines, particularly to service a market among North Americans interested in traditional Chinese medicine.

The illegal trade in rare and endangered species is responsible for an enormous international economy (reputedly second only to the illegal drug trade). This is why such commerce is flourishing in so many countries, including Canada. To some degree, governments can deal with the problem by more rigidly enforcing the laws governing the trade, and by imposing more severe penalties on convicted offenders. Ultimately, however, the illicit commerce is driven by a wealthy and enthusiastic marketplace. Obviously, for the sake of the endangered biodiversity, it is critical that the demand be curtailed as soon as possible. Ultimately, people's attitudes must be changed, and severe penalties must be imposed for the illegal possession of species or body parts banned by CITES.

The world's greatest expanses of tropical rainforests occur in equatorial Africa, southern and southeastern Asia, Central America, western South America, and the basin of the Amazon River. The latter region, known as Amazonia, contains the most extensive rainforests, and may support 10% of the biodiversity of Earth (Lovejoy, 1985). This rich tropical region is still extensively covered by old-growth rainforests that have been little affected by modern agriculture, lumbering, or other influences of industrial society (although virtually all of Amazonia has supported indigenous cultures for thousands of years).

The exploitation and devastation of the Amazonian forests is, however, accelerating rapidly. Great expanses of rainforest are being converted into cattle ranches. In addition, large areas of forest are being converted by poor farmers who have migrated from the heavily populated regions of Amazonian countries in search of "new" agricultural lands. Some Amazonian forests have been degraded by hydroelectric developments, by lumbering, and by the harvesting of trees to manufacture charcoal as a fuel for the production of iron.

Most of Amazonia lies in northern Brazil. The population in that region increased 10-fold between 1975 and 1986, to more than one million persons (Myers, 1988). The population continues to rise, leading to great increases in the rate of forest clearing, which in the mid-1980s was about 17 000 km^2/y. By that time, as much as 10% of Amazonian Brazil may already have been deforested (Lovejoy, 1985). For various reasons, including pressures exerted by international environmental organizations, the government of Brazil has committed itself to conserving its Amazonian biodiversity, even while vigorously "developing" the economy of the region. Up to 1993, the government of Brazil had established about 18 400 km^2 of parks and ecological reserves, another 9400 km^2 of areas with less protection (although the permitted land uses do not include deforestation), plus 115 000 km^2 of anthropological reserves, which are intended to protect the traditional homelands and cultures of indigenous peoples (World Resources Institute, 1994). Almost all of these protected areas are in Amazonia. However, as with protected areas everywhere, those established in Amazonia often suffer from poaching, illegal mining and agricultural settlements, and other kinds of prohibited activities that degrade their ecological values.

Species Declines

Recently, numerous species of certain groups of organisms have been suffering large and widespread declines in their populations, with many of them becoming endangered. These organisms include large carnivores, reptiles, amphibians, and migratory songbirds. For example, of the 42 species and geographically distinct populations of birds that are listed as endangered, threatened, or vulnerable in Canada, 38 are migratory (COSEWIC, 1996). We will examine the problem of species declines using the example of North American songbirds.

Within the past decade or so, ecologists and birdwatchers have been reporting alarming declines in the populations of many species of so-called neotropical migrants (i.e., species that spend most of the year in tropical habitats but migrate to higher-latitude regions to breed). Most of the declining species breed in mature temperate and boreal forests. Although the reasons for the songbird declines in North America are not totally understood, the most important factors are probably the following:

- extensive deforestation in the tropical wintering ranges
- disturbance of mature-forest habitat in the northern breeding ranges
- fragmentation of breeding habitats into "islands" that are too small to sustain populations over the long term, and that are easily penetrated by forest-edge predators and nest parasites (such as cowbirds — to be discussed later)
- loss of critical habitats for staging and migration
- effects of pesticides and other toxic chemicals

Bachman's warbler (*Vermivora bachmanii*) appears to have become extinct recently because of the loss of its tropical wintering habitat. Bachman's warbler used to breed in mature hardwood forests in the southeastern United States. Although suitable breeding habitat remains in this region, this warbler has not been seen since the mid-1950s and is undoubtedly extinct. Its extinction was likely caused by the clearing of its critical wintering habitats, believed to have been tropical forests on Cuba that were converted into sugarcane plantations.

Much of the evidence suggesting that populations of neotropical migrants are declining is *anecdotal* — skilled birders are not seeing as many individuals of some species as they used to, even where the habitat has not changed greatly. Unfortunately, only a few studies have closely monitored bird populations for many years in mature forest habitats. One of the best long-term data sets is for a mature forest in West Virginia, where the breeding birds, particularly the migrants, have declined substantially over a 37-y period. During 1947–1953, 25–28 species bred at that site, of which 14–16 were neotropical migrants. This decreased to only 15 species and 8 migrants breeding over 1973–1983 (Terborgh, 1989). During that same period, the total abundance of breeding birds decreased by 16%, and the abundance of neotropical migrants by 37%. In another important census of forest birds, made at Hubbard Brook, New Hampshire, 70% of the breeding species were found to have declined in population between 1969 and 1984 (Holmes *et al.*, 1986).

An extremely important data set comes from a bird observatory located at the tip of Long Point, a 32-km-long peninsula that juts into Lake Erie in southern Ontario. Because of its shape and location, Long Point attracts large numbers of birds during their migrations. During their southward autumn migration, many birds begin their crossing of Lake Erie at the tip of Long Point, having been funnelled along the landform as they move through its habitats. During the northward migration in the springtime, the tip is the first land seen during the open-water crossing of Lake Erie, and birds often landfall there in large numbers. Consequently, the Long Point observatory provides a large and dependable sample of the enormous populations of birds that breed in the extensive habitats to the north. The data from Long Point show that, between 1961 and 1988, 29 species of neotropical migrants declined in abundance, while only four species increased (Hussell *et al.*, 1992).

The causes of the declines of neotropical migrants include the reductions of their breeding habitats in North America, due to forestry activities and conversions into agriculture and urban areas. The amount of available high-quality habitat has declined greatly, while much of the remainder has been fragmented into small islands of natural habitat. This change is important because birds have less success when breeding in small fragments of forest. In part, this is because their nests are more vulnerable to predation by such species as crows, jays, magpies, skunks, and foxes.

Many migratory species have also been seriously affected by nest parasitism by the brown-headed cowbird (*Molothrus ater*), which lays its eggs in the nests of other species. The foster parents raise the voracious cowbird chick, while their own young are neglected and usually die. The brown-headed cowbird has greatly expanded its range and abundance in North America, mostly because humans have provided suitable habitats for it by disturbing the formerly extensive forests. Cowbirds feed in open areas, and they are particularly efficient at parasitizing nests near the edges of forests.

Many bird species in the northern and eastern parts of the modern range of the cowbird are extremely vulnerable to nest parasitism (Freedman, 1995). They have only recently come in contact with this parasite and have not evolved an effective defence. For example, Kirtland's warbler (*Dendroica kirtlandii*), an endangered species, can suffer a parasitism rate of 70%. Each incidence leads to reproductive failure. A study in Illinois found that two-thirds of 75 nests of various host species were parasitized by cowbirds, including 76% of 49 nests of neotropical migrants. The rate of nest parasitism of white-crowned sparrows (*Zonotrichia leucophrys*) in California increased from 5% in 1975 to 40–50% in 1990–1991, much more than the 20% rate that the sparrow population could sustain without declining.

The cowbird problem is a dilemma. This is because the only means of rescuing the endangered indigenous bird species involves killing large numbers of the brown-headed cowbird, itself an indigenous species. Although distasteful, that action is required if humans wish to deal with the severe damage that this parasite is inflicting on other species, as an indirect consequence of anthropogenic changes to its habitat.

Back From the Brink

Fortunately, dismal stories about extinctions and grievous losses do not make up all the news about biodiversity. There are also some success stories of conservation. These involve species that were taken perilously close to the brink of extinction, but have since recovered in abundance because they were given effective protection. In some cases, the recoveries have been vigorous enough that the species are no longer in danger of imminent extinction. Although these success stories are in a distinct minority (the number of endangered species is increasing much more rapidly), they are nevertheless instructive. They illustrate that positive actions can yield great benefits, both for the species in question and for the humans that can exploit them as potentially renewable resources.

The Northern Fur Seal and Some Other Seals

The northern fur seal (*Callorhinus ursinus*) lives in the northern Pacific Ocean. This seal was relentlessly exploited for its fur, and by 1920 had been reduced from a population of several million to only about 130 000. Because the northern fur seal was believed to be in danger of extinction, an international treaty was signed that strictly regulated its harvest. The seal population responded vigorously to the conservation measures, and now numbers about 900 000 individuals. The species is now sufficiently abundant to again support a commercial hunt for its fur, leather, and oil. Since this harvest is closely regulated, it should now be sustainable. Unfortunately, the northern fur seal has recently been suffering considerable mortality when individuals become entangled in drift-nets used to fish in the north Pacific. This and oil spills pose new risks to its population.

Some other species of seals were also exploited excessively, but then rebounded in abundance after the hunt was stopped, or at least sensibly regulated. Two Canadian examples are the harp seal (*Phoca groenlandica*) of the north Atlantic Ocean, which now numbers in the millions (see Chapter 14), and the grey seal (*Halichoerus gryptus*) of temperate Atlantic waters. The grey seal numbered only about 5000 individuals as recently as the mid-1960s, and was considered endangered. Since then, however, the grey seal has had remarkable population growth, and now numbers more than 100 000 animals.

The Whaling Industry

Many populations of large whales were severely depleted during the few centuries of unregulated exploitation (Chapter 14). Following protection, some whale populations have substantially recovered. The best example of such a recovery is the grey whale (*Eschrichtius robustus*) of the Pacific coast of North America, which was protected in the 1930s when its endangered population numbered only 1000–2000 animals. The grey whale now numbers about 21 000 individuals, roughly its pre-exploitation population level. Although the grey whales of the eastern Pacific are no longer endangered, the population in the eastern Atlantic was extirpated several centuries ago, and another stock in the western Pacific is critically endangered.

Other large species of whales were also depleted by commercial hunting. With few exceptions, these have been protected from exploitation since an international moratorium on whaling in 1986. Their populations are slowly recovering, although not yet to the degree achieved by the grey whale. The sperm whale (*Physeter macrocephalus*), for example, had a global pre-whaling abundance of about 2 million, but now numbers fewer than 1 million. Similarly, the fin whale (*Balaenoptera physalus*) initially numbered about 700 000, but now numbers 163 000, while the blue whale (*B. musculus*), initially numbering 250 000, now numbers only 10 000–20 000. The humpback whale (*Megaptera novaeangliae*) was about 100 000, and is now 10 000. These species will continue to recover their abundances as long as they remain protected from commercial hunting. There is intense pressure, however, for the moratorium to end for the most abundant species, particularly the minke whale (*Balaenoptera acutorostrata*).

Several other species of whales remain badly depleted, and are recovering extremely slowly, if at all. One of these is the population of right whales (*Eubalaena glacialis*) of the western Atlantic, which numbers only a few hundred individuals. Another is the right whale of the eastern Pacific Ocean, with only a hundred or so animals. Yet another is the bowhead whale (*Balaena mysticetus*) of the Arctic, with a population of about 5000. Bowhead whales are still subjected to an aboriginal hunt in northern coastal Alaska (several were also taken off Baffin Island in Canada in 1996). The most important causes of mortality of Atlantic right whales appear to be collisions with ships and entanglement in fishing nets.

The American Bison

Before the American bison or buffalo (*Bison bison*) were subjected to an intensive commercial hunt, the population was an estimated 60 million. At that time bison were the most abundant large animals in North America, ranging over most of the continent.

Some biologists believe that there are several subspecies of bison. The eastern subspecies (*B. b. pennsylvanicus*), an animal of forests and glades, ranged over much of the eastern United States. It was hunted to extinction by the mid-1800s. The plains bison (*B. b. bison*) ranged throughout the prairies of North America, and was by far the most popu-

Populations of humpback whales (*Megaptera novaeangliae*) were decimated world-wide by commercial whaling. However, this species is now protected and its numbers are increasing. Humpback whales, such as these animals off Newfoundland, spend much of the summer feeding on small fish in certain places off eastern Canada.

Source: H.Whitehead

lous subspecies. These animals migrated in enormous herds — one was described as being 80 km long and 40 km wide, another as 320 km long, and another as moving over a 160 km front! The plains bison were subjected to an intensive market hunt during the 19th century and were very nearly exterminated. Apart from the money that was made by selling meat and hides, the eradication of the plains bison may have been subtly encouraged by the United States and Canadian governments, likely for two reasons. First, the development of prairie agriculture was being disrupted by the bison herds, especially during their mass migrations. Second, since the bison were critical to the subsistence economy of the Plains Indians, extermination of these abundant animals made it easier to displace aboriginal tribes in favour of European colonists.

The most famous buffalo hunter was William F. Cody, or "Buffalo Bill," who was contracted in 1869 to provide meat for workers constructing the Union Pacific Railroad through the United States prairie. Cody reportedly killed 250 bison in a single day, and more than 4000 during an 18-month period. Once the railways were built, special excursions were organized during which a train would stop near a herd of migrating buffalo, allowing passengers to shoot animals in a leisurely fashion through the windows of the coaches. Some of the tongues (a delicacy) would be cut from the dead animals; otherwise the carcasses were left to rot. Such actions were a wanton destruction of a natural resource. More important damage was caused by the market hunts made feasible by the new railroad — the meat could be shipped quickly to urban consumers. Between 1871 and 1875, market hunters killed about 2.5 million bison per year. In addition, the Plains Indians had acquired rifles and horses by this time and were able to hunt bison much more effectively than before.

The unregulated exploitation of the plains bison was grossly unsustainable, and the species declined precipitously. By 1889 there were fewer than 1000 bison left in the United States. In addition, small herds of undetermined size survived in the Canadian prairies. Almost too late, a few closely guarded preserves were established, and some animals were captured for breeding programs. These and later actions have allowed the numbers of plains bison to increase to their present abundance of more than 50 000 animals. Of course, since almost all of the original habitat of the plains bison is gone, having been converted into agriculture, this animal will never recover its former abundance. However, the plains bison is no longer endangered.

The wood bison (*B. b. athabascae*) of the southern boreal forest of western Canada appears to be a third subspecies. It was also hunted intensely. When this subspecies became endangered, governments protected the only remaining wild population in and around Wood Buffalo National Park in northern Alberta and the southwestern Northwest Territories. Unfortunately, the genetic integrity and health of this population has been degraded by interbreeding with plains bison, which were introduced to the area by misguided wildlife managers in the late 1920s. Fortunately, a previously unknown population of wood bison was discovered in 1960 in a remote area of Wood Buffalo National Park. It appears that these have not suffered from interbreeding with the plains subspecies. Some of these "pure" wood bison were used to establish another isolated population, northwest of Great Slave Lake. Regrettably, many of the wild-ranging bison of northwestern Canada have been exposed to introduced diseases of cattle, most notably tuberculosis and brucellosis. These, along with predation by humans and wolves, have taken their toll on the wood bison. The long-term viability of their populations is cause for concern.

Agricultural interests within the federal government have proposed to exterminate virtually all of the bison in the vicinity of Wood Buffalo National Park, except for the "pure" wood bison occurring in isolated populations known to be free of bovine diseases. This slaughter is intended to prevent the spread of brucellosis and tuberculosis from bison to the cattle herds that are spreading northward in Alberta. A secondary reason for the proposed slaughter is to protect the genetic integrity of the wood bison subspecies, because it is the hybrid wood/plains animals that would be targeted for extermination, leaving the more isolated, nondiseased, pure wood bison to repopulate the cleared habitats. This proposal is highly controversial, and even if permitted would probably not be successful. An enormous effort would be required to find each and every bison in the target area, a huge wilderness of boreal forest and muskeg.

In spite of continuing problems, it appears that vigorous conservation efforts have preserved the American bison. This large herbivore will survive, although in captivity and in relatively small populations in the wild.

Some Other Recoveries from Depleted Abundance

The sea otter (*Enhydra lutris*) lives on the west coast of North America. This mammal was subjected to a devastating 18th- and 19th-century hunt for its dense and lustrous fur. In fact, for decades, the sea otter was thought to be extinct, until small residual populations were discovered in the 1930s. The sea otter has rapidly recovered its abundance over much of the west coast. This recovery has been aided by deliberate reintroductions into areas from which it had been extirpated, such as the Pacific coast of Vancouver Island. Sea otters now number more than 100 000 individuals. However, the Canadian population is still small and vulnerable enough to remain officially designated as an endangered species.

The pronghorn antelope (*Antilocapra americana*) of the western plains of North America was severely overhunted during the 19th century, and its population was reduced to about 20 000 individuals. Fortunately, strong conservation measures were implemented, and this species now numbers more than 500 000. It can again sustain a sport hunt.

The trumpeter swan (*Olor buccinator*) used to breed extensively in western North America, perhaps as far east as Ontario. Populations of this swan were devastated by hunting for its meat and skin. However, the trumpeter swan is now protected and has recovered somewhat in abundance; it now numbers more than 5000 individuals.

The wild turkey (*Meleagris gallopavo*) was widely extirpated from its natural range by hunting and habitat losses (of course, domestic varieties of this species are abundant in agriculture). Because of conservation measures and reintroductions to areas from which the species had disappeared, populations of wild turkeys have recovered substantially — for example, in some areas in southern Ontario. Many stocks of this large gamebird can again sustain a sport hunt.

The wood duck (*Aix sponsa*) was greatly overhunted for its beautiful feathers and as food. It also suffered from losses of habitat to lumbering and wetland drainage. The recovery of the wood duck has been aided greatly by the widespread provision of nest boxes in wetlands used by this cavity-nesting species. This nest-box program also benefits several other relatively uncommon cavity-nesting ducks, particularly the hooded merganser (*Lophodytes cucullatus*) and common goldeneye (*Bucephala clangula*). An unrelated program of specially designed terrestrial nest-boxes has helped to increase the abundance of eastern and western bluebirds (*Siala sialis* and *S. mexicana*), which had been declining because of habitat losses.

The American beaver (*Castor canadensis*) was one of the most sought-after species in the fur trade, a commercial activity that stimulated much of the exploration of Canada and the central and western United States. Beavers were overharvested almost everywhere, causing the species to be extirpated from most of its natural range. However, conservation measures and decreased demand for its fur have allowed the beaver to recover its populations over most of its range where the habitat is still suitable. In fact, beavers are sometimes considered to be "pests" in some of their recolonized habitats, because of the flooding they cause.

The whooping crane (*Grus americana*) is, it is hoped, an incipient success story of conservation. The whooping crane was never very abundant, even before its populations were devastated by the combined effects of hunting for its meat and feathers, losses of its breeding habitat of prairie wetlands to agriculture, deterioration of its wintering habitat along the Gulf of Mexico, and egg and specimen collecting. These stressors drove the wild population of the species down to the perilously low level of 15 individuals (in 1941). Fortunately, since then, the whooping crane has been vigorously protected from hunting, while its major breeding habitat in Wood Buffalo National Park and its wintering habitat in coastal Texas have been conserved. These measures, along with a program of captive-breeding and release, have allowed the population of whooping cranes to increase to about 250 animals (in 1993; 145 of these were in captivity). There is now cautious optimism for the survival of this species, although it is still considered endangered.

Sea otters (*Enhydra lutris*) were decimated by hunting for their fur. Their populations have since recovered over much of their range, including parts of western Vancouver Island. Although no longer hunted, sea otters are still threatened by oil spills, habitat changes caused by fishing, and illegal shooting by fishers who perceive that otters are "eating too many valuable crustaceans and shellfish."

Source: C. Harvey-Clark

Canadian Species and Ecosystems at Risk

The conservation status of species in Canada is assessed by the **C**ommittee **o**n the **S**tatus of **E**ndangered **W**ildlife **i**n **C**anada (COSEWIC). COSEWIC is a consultative body with representatives from the federal government, the provinces and territories, and interested nongovernmental organizations (NGOs), particularly the World Wildlife Fund and the Canadian Nature Federation. It makes recommendations to the federal, provincial, and territorial governments, whose responsibility it is to actually designate conservation status.

Once a species is listed as endangered in Canada, a parallel body known as "RENEW" is mandated to prepare a plan that would ensure the recovery of its population to a safer level. Because of a lack of funding and political will, however, recovery plans have been prepared for only 14 of the 65 species or populations that have been designated as endangered in Canada (as of 1995).

COSEWIC recognizes five categories of risk, each of which has a specific meaning in terms of imminent threats to the future survival of species (COSEWIC, 1996).

Extinct refers to any species of wild life that was formerly indigenous to Canada but no longer exists anywhere in the world. Canadian examples of extinct species are the great auk (*Pinguinus impennis*), passenger pigeon (*Ectopistes migratorius*), Labrador duck (*Camptorhynchus labradorium*), sea mink (*Mustela macrodon*), deepwater cisco (*Coregonus johannae*), longjaw cisco (*Coregonus alpenae*), and eelgrass

The Labrador duck (*Camptorhynchus labradorium*) was a sea-duck that used to winter on the Atlantic Coast of Canada and the northeastern United States, and probably nested in coastal Labrador. Because of unregulated hunting, the species became extinct around 1875. This is a photograph of carved models of a pair of Labrador ducks, replicated from old stuffed specimens that had been "collected" in Nova Scotia by a 19th-century naturalist.

Source: D. Josey

limpet (*Lottia alveus*). Extinct subspecies are the Queen Charlotte caribou (*Rangifer tarandus dawsoni*), blue walleye (*Stizostedion vitreum glaucum*), and Banff longnose dace (*Rhinichthys cataractae smithi*). As of 1996, ten Canadian species or subspecies were extinct.

Extirpated refers to any species or subspecies that was formerly indigenous to Canada, but now only survives in the wild elsewhere (usually in the neighbouring U.S. states). Examples include the black-footed ferret (*Mustela nigripes*), Atlantic grey whale (*Eschrichtius robustus*), Northwest Atlantic walrus (*Odobenus rosmarus*), greater prairie chicken (*Tympanuchus cupido*), pygmy short-horned lizard (*Phrynosoma douglassii douglassii*), paddlefish (*Polyodon spathula*), and the blue-eyed mary (*Collinsia verna*). As of 1996, 11 species or populations were considered extirpated in Canada.

Endangered refers to indigenous species that are threatened with imminent extinction or extirpation throughout all or a significant portion of their Canadian range. As of 1996, 65 species or populations were considered to be endangered in Canada. Examples include the Vancouver Island marmot (*Marmota vancouverensis*), bowhead whale (*Balaena mysticetus*), right whale (*Eubalaena glacialis*), whooping crane (*Grus americana*), Eskimo curlew (*Numenius borealis*), burrowing owl (*Speotyto cunicularia*), piping plover (*Charadrius melodius*), Blanchard's cricket frog (*Acris crepitans blanchardi*), blue racer snake (*Coluber constrictor*), eastern prickly pear cactus (*Opuntia humifusa*), small white ladyslipper (*Cypripedium candidum*), thread-leaved sundew (*Drosera filiformis*), and seaside centipede lichen (*Heterodema sitchensis*).

Threatened refers to any indigenous taxon that is likely to become endangered in Canada unless factors affecting its vulnerability are reversed. As of 1996, 65 species or populations were considered to be threatened in Canada. Some examples include the wood bison (*Bison bison athabascae*), sea otter (*Enhydra lutris*), Pacific humpback whale (*Megaptera novaeangliae*), marbled murrelet (*Brachyramphus marmoratus*), roseate tern (*Sterna dougallii*), white-headed woodpecker (*Picoides albolarvatus*), massasauga rattlesnake (*Sistrurus catenatus*), and American chestnut (*Castanea dentata*).

Vulnerable refers to any indigenous species that is not currently threatened but is at risk of becoming so because of small or declining numbers, occurrence at the fringe of its range or in restricted areas, habitat fragmentation, or for some other reason. As of 1996, 124 taxa or populations were considered to be vulnerable in Canada. A few examples include the grizzly bear (*Ursus arctos*), polar bear (*Ursus maritimus*), black-tailed prairie dog (*Cynomys ludovicianus*), long-billed curlew (*Numenius americanus*), ivory gull (*Pagophila eburnea*), Pacific giant salamander (*Dicamptodon tennebrosus*), spotted turtle (*Clemmys guttata*), and prairie white-fringed orchid (*Platanthera leucophaea*).

It must be recognized that the designation of species at risk is a continuing and always incomplete process. For instance, since the conservation status of only a few species of invertebrates has been investigated, endangered species in this group are enormously underrepresented in the COSEWIC list. Unfortunately, more rapid progress by COSEWIC is constrained by a shortage of funding for research into and monitoring of endangered species in Canada, and by a lack of specialists with the necessary taxonomic and ecological knowledge.

Of course, it is not sufficient merely to designate species as being at risk of extirpation or extinction. If their status in Canada is to be improved, the species and their habitats must also be protected. Remarkably, governments in Canada have not yet enacted effective legislation to protect endangered species. However, this situation is starting to change. In late 1996, the Government of Canada proposed an Endangered Species Act, which would provide protection for species occurring on federal lands. This would substantially improve the current complete lack of federally legislated protection for endangered species.

However, the proposed legislation would have little influence on the status of the many endangered species living on provincial, territorial, aboriginal land-claim, and private lands in Canada. Nor does the proposed Act address the protection of habitat of endangered species off federal lands. To some degree, this critical deficiency could be covered by legislation being prepared by provinces and territories. (In fact, such legislation already exists in Manitoba, Ontario, Quebec, and New Brunswick. However, these legislations are not very effective, particularly because they do not protect the habitat of endangered species.) However, such a piecemeal approach would likely result in uneven levels of protection for endangered species in Canada. This would be unacceptable from the conservation viewpoint.

The development of endangered species legislation in Canada is raising controversy. Governments feel the need to demonstrate that they are making rapid progress toward sustainability, and an important component of that involves the protection of endangered species and their habitats. Unfortunately, the proposals to date have been significantly lacking in that component and would not be sufficiently effective in protecting endangered biodiversity in Canada. Hopefully, the lobbying efforts of Canadian nongovernmental organizations, under the umbrella of the Canadian Endangered Species Coalition, will result in appropriate changes to the proposed federal legislation, and will also improve provincial and territorial initiatives.

Some of the natural ecosystems of Canada now occur only as small, possibly unsustainable remnants of their former extent. Because of this, they are as endangered as the species they support. The most endangered of the indigenous ecosystems are the tallgrass prairie of southwestern Ontario and southeastern Manitoba, the Carolinian forest of southern Ontario, the dry coastal Douglas-fir forest of southwestern British Columbia, the semidesert of southeastern British Columbia, and various types of old-growth forests in all parts of forested Canada, but particularly in the east (see also Chapter 8). Some of these ecosystems, particularly the tallgrass prairie and Carolinian forest, are also very rich in endangered species. It is critical that the remaining areas of these endangered ecosystems become preserved in parks, ecological reserves, and other kinds of protected areas (this is discussed further in the next section).

The Importance of Protected Areas

Protected areas are parks, ecological reserves, and other tracts of land and/or water that are set aside from intense development to conserve their natural ecological values. The intention is usually to protect threatened ecosystems, representative examples of widespread communities, or the habitat of endangered species. However, many protected areas also support certain human activities that do not severely threaten the conserved ecological values. Such activities include ecotourism, other kinds of nonconsumptive outdoor recreation, spiritual activities, education, scientific research, and in some cases even exploitative activities such as hunting, fishing, trapping, or even timber harvesting.

It is important to understand that protected areas should never be regarded as the only, or even as the most important, way to conserve endangered species and ecosystems. Native species and other natural values should be accommodated as much as possible in all areas that humans use for economic purposes, such as agriculture, forestry, fishing, or mining. The role of protected areas is to ensure that species and ecosystems that are at risk in those "working" areas still have suitable refuges where they can maintain themselves.

A national system of protected areas in Canada would involve lands controlled by federal, provincial, or territorial governments; aboriginal groups; and/or private interests. Ideally, the system of protected areas would be designed to sustain all native species and natural communities over the long term, including terrestrial, freshwater, and marine systems. To ensure that all elements of indigenous biodiversity are adequately represented within a system of protected areas, all species and community types in the country or province must be identified. This information would allow all elements of our natural-ecological heritage to be accommodated within a comprehensive *system plan* for a network of properly managed protected areas.

Of course, these are ideal criteria — no country has yet designed and implemented a comprehensive system of protected areas that contains and sustains all indigenous species and natural communities. Moreover, many existing protected areas are relatively small. Some are threatened by stressors within their boundaries or by degrading influences from the surrounding area. In addition, the environmental factors that affect the protected species and communities are often not well understood. Because of these and other problems, it is doubtful whether many of the smaller protected areas will be able to sustain their present ecological values over the long term, especially if a major disturbance occurs.

The International Union for the Conservation of Nature (IUCN) recognizes six categories of protected areas (In Detail 24.2). In 1993, there were about 8600 protected areas around the world, with a total area of 792 million ha (IUCN categories 1–5; World Resources Institute, 1994). Of this total area, about 2500 sites comprising 464 million ha were fully protected (IUCN categories 1–3), and could be considered as true ecological reserves.

Protected Areas in Canada

National parks, provincial parks, and similar kinds of protected areas are the largest and most important ecological reserves in Canada. A summary of areas protected in Canada is provided in Table 24.2. Note that the nationally protected area of about 5.5% is far smaller than recommended by conservation scientists, whose estimates range from 10 to 40% of the landmass.

As well, most parks serve additional purposes, particularly the support of economically important outdoor

In Detail 24.2

Categories of Protected Areas

The International Union for the Conservation of Nature (IUCN) recognizes six categories of protected areas. Categories 1, 2, and 3 represent strong commitments to the maintenance of natural ecosystems within the protected area, while the other categories allow some degree of resource management or extraction.

Category	Description
1.	*Scientific Reserves and Wilderness Areas* include nature, ecological, and wilderness reserves. These are managed mainly to preserve their natural condition, although some use by scientists for research and monitoring may be allowed.
2.	*National Parks and Equivalent Reserves* consist of national, state, and provincial parks plus areas under native, tribal, or other traditional ownership. These areas are managed primarily to protect ecosystems, although recreation is usually permitted.
3.	*Natural Monuments* include geological phenomena and archaeological sites, and are intended to protect natural features of aesthetic or cultural importance.
4.	*Habitat and Species Management Areas* consist of wetlands, wildlife refuges, and wildlife sanctuaries. These are intended to conserve through the protection and management of habitats. Hunting and other types of resource use may be allowed in many of these areas.
5.	*Protected Landscapes and Seascapes* include landscapes, marine areas, scenic rivers, waterways, recreational areas, trails, protected forests, and conservation areas in which interactions of humans and nature have produced areas of distinct character. These areas are managed to sustain use by both people and wild species and ecosystems.
6.	*Managed Resource Protected Areas* contain areas of primarily natural ecosystems. These are managed to ensure the protection of biodiversity, while also providing a sustainable harvest of renewable resources and ecological services.

recreation and ecotourism. To some degree, the use of parks for these purposes challenges and compromises aspects of their function as ecological reserves. For example, strictly interpreted, the ecological values of national parks are not compatible with consumptive uses of their natural resources (e.g., sport fishing), or with the development within the parks of infrastructure supporting recreation and ecotourism (e.g., campgrounds, hotels, golf courses, roads and paths, and interpretation facilities).

The ecological-reserve function of many of Canada's protected areas is also threatened by land-use and management activities in their surrounding areas. Usually, the most important external stressors are associated with forestry, agriculture, mining, tourism, or hydroelectric development. In fact, all of the national parks in more southern regions of Canada are significantly threatened in this way. We can illustrate this problem with several well-known examples.

Point Pelee National Park is a small, 15.5-km^2 park in southwestern Ontario. This park contains many species at risk and some of the few remnants of natural habitat left in the Carolinian zone of Canada (Chapter 8), most of which has been converted into agriculture or urbanized land uses. Consequently, Point Pelee National Park contains populations of many endangered species and ecosystems. However, this small park is used intensively for outdoor recreation, including birdwatching, boating, hiking, and picnicking on its beaches. To support these economically important activities, much of the park's limited area is maintained as paved roads, pathways, parking lots, campgrounds, information centres, lawns, and other infrastructure that does not directly enhance the protection of its ecological values.

In addition, the area next to the national park has been almost entirely converted into intensively managed agricultural lands, such as onion fields established on drained marshes, or into cottage and motel developments supporting tourism. These land uses have isolated the relatively natural ecosystems of Point Pelee National Park; the park is an ecological "island" surrounded by incompatible uses of the landscape. For these and other reasons, Point Pelee National Park is losing some of the natural features it is trying to protect. For example, the park has lost 10 of its original 21 species of reptiles, and 6 of 11 species of amphibians. Some of its habitats are being badly degraded by invasions of non-native plants (such as garlic mustard, *Alliaria petiolata*), which crowd out indigenous species.

TABLE 24.2 | PROTECTED AREAS IN CANADA IN 1996

PROVINCE OR TERRITORY	TOTAL AREA PROTECTED (hectares)	NUMBER OF PROTECTED AREAS	PERCENT OF JURISDICTION PROTECTED
Yukon	36 708	5	7.6
Northwest Territories	140 719	6	4.1
British Columbia	87 460	746	9.2
Alberta	61 758	53	9.3
Saskatchewan	34 101	399	5.2
Manitoba	35 903	85	5.5
Ontario	69 397	330	6.5
Quebec	64 975	185	4.2
New Brunswick	918	52	1.3
Nova Scotia	4 471	82	8.0
Prince Edward Island	128	92	2.2
Newfoundland and Labrador	7 701	66	1.7
Canada	543 609	2 101	5.5

Source: World Wildlife Fund (1996)

Fundy National Park in New Brunswick is a similar case, although its ecological values are not as severely threatened as those of Point Pelee. Fundy National Park has an area of 206 km^2, but park ecologists believe that this is not large enough to sustain viable populations of some wide-ranging species, such as black bear, pine marten, and pileated woodpecker, or certain natural ecosystems such as old-growth forests (see also Canadian Focus 23.2, pages 440–2). To some degree, these and other natural values are compromised by the development of tourism facilities within the national park, including a golf course, a salt-water swimming pool, campgrounds, interpretive facilities, and extensive lawns, roads, and trails. Also important are industrial activities in the area surrounding the park, where natural forests are being converted into forestry plantations. These do not provide habitat for many of the indigenous species of natural forests. Thus, the natural habitats of the park are becoming ecological islands within a modified landscape.

Banff National Park in southeastern Alberta was the first national park established in Canada (in 1885). The original intent was to protect some extremely scenic view-scapes and hot springs, and to develop the area to enhance the economic benefits of tourism. It was not until several decades later that the philosophical underpinning of national parks in Canada shifted strongly toward the protection of natural ecological values. In any event, the early development of Banff National Park featured the enthusiastic construction of large hotels, golf courses, skiing facilities, major highways, a transcontinental railroad, several towns, and other structures. Unfortunately, this pattern continues today with much ongoing and planned construction, coupled with the rapid development of the area east of the park for tourism, residential neighbourhoods, forestry, and other uses. These facilities severely threaten the long-term conservation of the natural values of Banff National Park. This development is engendering intense controversy, and has been the subject of a recent commission of enquiry (see Canadian Focus 24.3).

Canada's provincial and territorial governments also have a responsibility to protect natural ecological values within their domain. Some of these governments have designated numerous ecological reserves, supplemented by natural areas that are protected in provincial parks and conservation areas, which are also used for recreation, and sometimes for other purposes. British Columbia, for example, has designated about 87 460 ha as IUCN Category 2 protected areas (see In Detail 24.2).

Some municipalities also have natural-area parks that provide habitat for native species. An outstanding example is the city of Windsor, Ontario, which is protecting some of Canada's most important remnants of tallgrass prairie and numerous endangered species.

Some **e**nvironmental **n**on-**g**overnmental **o**rganizations (or ENGOs) are also active in the protection of natural areas in Canada. At the national level, the Nature Conservancy of Canada is the organization that most actively protects land to conserve Canada's indigenous biodiversity, usually by purchasing or accepting donations of properties or land-use rights. Since it began in 1962, the Nature Conservancy of Canada has secured more than 650 nature preserves, and has contributed to the protection of 500 000 ha of natural areas. At the provincial and more local levels, an increasing number of land trusts are also protecting natural areas.

Other national organizations, such as the World Wildlife Fund (Canada), the Canadian Nature Federation, the Canadian Wildlife Federation, the Sierra Club, and Greenpeace, also play important roles in protecting indigenous biodiversity. However, they mostly do this through advocacy — these ENGOs lobby governments and the private sector to pursue more effective biodiversity agendas. Also, they engage in public campaigns and conduct research toward those ends. The World Wildlife Fund (Canada), for example, has been the prime mover behind the Endangered Spaces Campaign, which has been effective in convincing governments to preserve representative areas of their natural ecosystems within protected areas. The Canadian Council of Ecological Areas is an association of conservation experts in government, ENGOs, and universities, who are working to design a strategic plan for a national system of protected areas. More information about these various ENGOs, as well as many of their provincial and territorial counterparts, is provided in Appendix II.

In spite of the diverse conservation-related activities of governments and organizations in Canada, the existing network of protected areas is highly incomplete. There are three reasons for making this statement. First, the full breadth of Canada's natural heritage is not yet represented

CANADIAN FOCUS 24.3

ECOLOGICAL INTEGRITY IN THE BOW VALLEY CORRIDOR

In 1885, Banff National Park was the first national park to be proclaimed in Canada. Banff is also the most famous of Canada's national parks, because of its spectacular scenery, easily viewed large animals, and superb infrastructure supporting tourism and outdoor recreation. All of these values combine to attract visitors from across Canada and around the world.

Banff National Park covers a rather large area (6640 km^2), and thus plays an important role in protecting the natural ecological values of its region. This role is enhanced by the fact that Banff is bordered by several other protected areas, namely Jasper, Yoho, and Kootenay National Parks and Peter Lougheed Provincial Park, which collectively comprise an area of about 26 000 km^2.

Most of the indigenous species and ecosystems of Banff National Park are well protected within its

boundaries and in surrounding lands. Some others, unfortunately, are not. These natural values are threatened by a variety of stressors, some of which exert their influence within the national park, while others make themselves felt outside its immediate boundaries.

Banff National Park hosts about five million tourists each year, generating about $6 billion in economic activity. To service its many visitors, the park contains numerous hotels, lodges, and campgrounds. To provide the tourists with interesting things to do, and to generate revenue and local employment, the park contains ski hills with associated lifts and lodges, golf courses, an extensive network of roads and trails, interpretation facilities, and two full-service settlements with more than 8000 permanent residents — the villages of Banff and Lake Louise. In addition, the transcontinental Canadian Pacific Railway passes through the national park, as does the Trans-Canada Highway. These various facilities are developed especially intensely in the so-called Banff–Bow Valley corridor, a region that encompasses the major transportation routes through the park as well as the main tourist areas.

The tourism- and transportation-related infrastructure in the Banff–Bow Valley corridor provides important support for big-business tourism and the national system of ground transportation. These facilities do not, however, enhance the natural values of Banff National Park. In fact, local populations of timber wolf, grizzly bear, and other species are thought to be at great risk in the greater Banff region, mostly because they are suffering unsustainably high death rates. The intense mortality results from collisions with vehicles on the highways, hunting outside the park, and the necessary killing of "problem" bears which become habituated to humans, particularly near campgrounds and frequently used trails. In addition, the wilderness values of extensive areas have become significantly degraded by visual and noise pollution associated with traffic, highways, railroads, ski lifts, buildings, and large numbers of people.

The various environmental challenges to the ecological integrity of Banff National Park are an increasingly serious problem. Recently, a Task Force of five independent experts, appointed by Parks Canada, studied these challenges. The Task Force was given three objectives:

1. to develop a vision for the region that would integrate ecological, social, and economic values
2. to undertake a comprehensive analysis of existing information, and to provide direction for future monitoring programs
3. to recommend changes that would allow the Banff–Bow Valley region to be used as the basis for sustainable tourism and recreation industries, while continuing to protect its heritage of ecological values

The Task Force reviewed a wealth of existing information, conducted original research, and engaged in wide-ranging public consultations. Its final report concluded that the intensive economic development in the Banff–Bow Valley region was quickly approaching an unsustainable level, and that this clearly threatened the ecological integrity of the national park (Page *et al.,* 1996). The Task Force made numerous recommendations for specific actions and policies that would help to deal with the rapidly intensifying crisis. It strongly advised that the pace and intensity of development be strictly controlled, and in some cases reversed. In essence, the Task Force concluded that Banff National Park can be an effective ecological reserve only if the use of the area by humans is kept within sustainable bounds.

The Task Force report was favourably received by the Minister responsible for Parks Canada, who affirmed that all of the major recommendations would be followed. If this is to happen, however, Parks Canada will have to take firm actions against powerful economic interests that are determined to increase the amount of recreation and transportation infrastructure in the Banff–Bow Valley corridor. Hopefully, the Minister and senior administrators in Parks Canada will have the fortitude to resist the compelling calls for additional "development," and will manage Banff National Park in a manner that makes its ecological integrity, not economic development, the bottom line.

in protected areas. Second, there are many endangered species in Canada, many of which will have to be protected in ecological reserves that do not yet exist. Third, most of the protected areas are too small to protect their ecological values over the long term.

The latter consideration is important because small areas cannot sustain viable populations of some species of wild life over the long term, even if they are protected. Small reserves also cannot sustain the ecological conditions required for certain communities to persist, particularly old-growth forests. In such cases, the protected area must be managed within the context of its surrounding landscape as a single, integrated ecosystem. Management activities in such **"greater protected areas"** should be designed to ensure the long-term viability of populations of species-at-risk, as well as natural communities-at-risk (see also Canadian Focus 23.2, pages 440–2).

FIGURE 24.2 | DESIGN OF ECOLOGICAL RESERVES

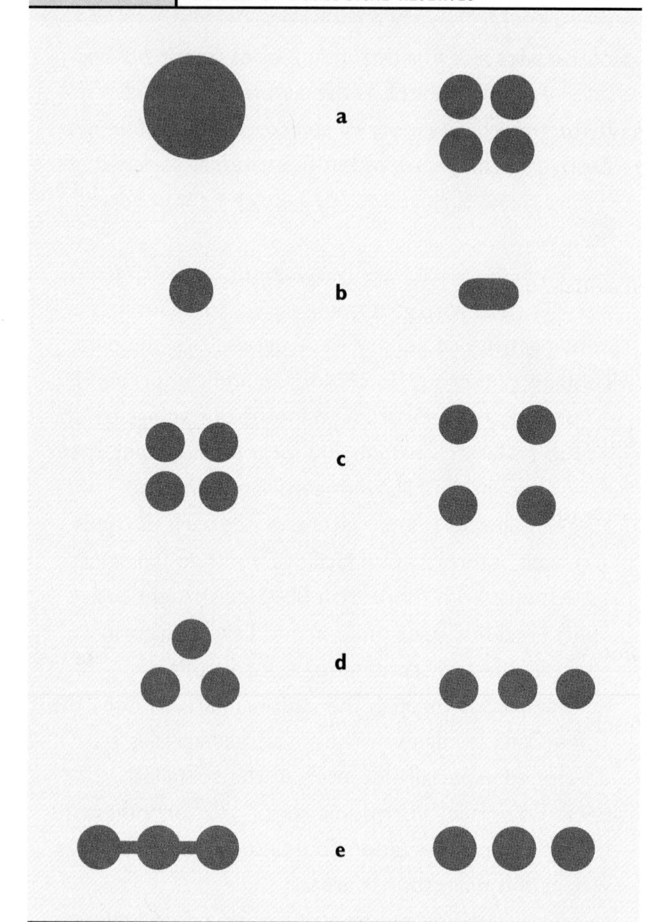

This figure summarizes the basic principles of conservation biology for the design of protected areas. In each comparison, the design on the left is better than the design on the right (total areas of each comparison are assumed to be the same). See text for discussion.

Source: Modified from Simberloff (1988)

Design and Management of Protected Areas

The design of protected areas is an important field of research in conservation biology. The essential questions involve determining the size, shape, and position of protected areas to optimize their function in protecting biodiversity, while using limited funding as efficiently as possible. The least controversial recommendations of conservation biologists are that ecological reserves should be as large and as numerous as possible. Other aspects of the design of ecological reserves are being debated actively, and the controversies will not be resolved until more research is performed. These additional aspects of reserve design include the following:

The choice between size and number of protected areas. Is it preferable to have one large reserve or a number of smaller ones with the same total area (Figure 24.2a)? Conservation biologists identify this question with the acronym SLOSS — **s**ingle **l**arge **o**r **s**everal **s**mall. According to ecological theory, populations in larger reserves are expected to have a lower risk of extinction compared with those in smaller protected areas. However, if separate populations occur in different reserves, the redundancy might protect against catastrophic losses of endangered species.

Larger reserves have more interior habitat. "Interior" habitat is uninfluenced by environmental conditions at *ecotones* (i.e., transitions between habitat types, such as a forest edge beside an adjacent field). Ecotonal habitats are often penetrated by invasive species, predators, and parasites (such as cowbirds), which can be important problems for some ecological reserves. In addition, many species require interior habitat for successful breeding. Larger reserves have proportionately more interior habitat, a factor which contributes to the protection of interior species.

Circular reserves are better. A circle has a smaller ratio of edge to area than any other two-dimensional shape. To avoid extensive edge habitat in the design of an ecological reserve, a roughly circular shape is generally preferable (Figure 24.2b).

Reserves should be close together. If a population becomes extirpated in a protected area, the chances of natural recolonization may be improved if a nearby reserve has a surviving population. Consequently, it may be better to have unconnected reserves arranged relatively close to each other, rather than far away (Figure 24.2c).

Reserves should be clumped. Similar reasoning suggests that it is better to aggregate reserves than to arrange them in a linear fashion. This would minimize the average inter-reserve distance (Figure 24.2d).

The role of corridors. A system of reserves connected by corridors of suitable habitat may provide better opportunities for gene flow and recolonization after extirpation (Figure 24.2e). (Admittedly, corridors could also make it easier for diseases, invasive species, and introduced predators to spread among reserves.)

It is important to understand that the protection of biodiversity requires more than declarations that tracts of natural areas are henceforth to be considered ecological reserves. The integrity of those reserves must also be monitored, and research on proper management is often necessary. For example, if a protected area supports populations of endangered species, their status should be monitored, as should any changes in endangered communities. If any damage is observed, the environmental causes should be determined by research, and should then, if possible, be managed to prevent or mitigate the damage. Management activities can include, among other actions, patrols to prevent poaching of animals and lumber, modification of habitats to keep them suitable for threatened species, and captive breeding and release of endangered species.

Considered together, these actions represent an integrated program of monitoring, research, and management. The application of such a system is illustrated by the case of the endangered Kirtland's warbler (*Dendroica kirtlandii*). Monitoring has shown that this species has declined rapidly in abundance, and that its global population is now only several hundred breeding pairs. Research has revealed that the breeding habitat of Kirtland's warbler consists of jack pine (*Pinus banksiana*) stands of a particular age and structure. Many such stands have now been protected in the breeding range of Kirtland's warbler. However, as these stands get older, they are no longer suitable as breeding habitat. Consequently, management is actively developing appropriate breeding habitat by prescribed burning and planting of jack pine. In addition, research has shown that the endangered warbler is heavily parasitized by the brown-headed cowbird. Consequently, management is controlling cowbird populations in the warbler's breeding habitat. Additional research and monitoring are being directed to the environmental stresses that affect Kirtland's warbler during its little-known migrations and on its wintering range. Of course, these integrated activities of monitoring, research, and management must continue as long as Kirtland's warbler remains endangered.

International Conservation Activities

The conservation of wild species of plants and animals is now regarded by almost all societies as a laudable and important objective. As a result, in most countries many people are becoming active in support of conservation. Evidence of these changes includes the fact that governments are becoming more involved in the conservation of indigenous and global biodiversity, while large numbers of nongovernmental organizations are becoming more active at local, national, and international levels (see Appendix II). In addition, more ecologists and other scientists are conducting biodiversity-related research and training in universities and other educational institutions.

All of these activities contribute to the greater agenda of biodiversity conservation, particularly by:

- identifying and protecting the habitats of rare and endangered species and ecosystems, while also conserving representative areas of natural ecosystems
- controlling illegal trade in the products of endangered species (such as elephant ivory, rhino horn, bear gall bladders, tiger bones and hides, and wild ginseng; see In Detail 24.1)
- increasing the awareness of people about biodiversity issues, and about the need to ensure adequate conservation of all of the Earth's species, ecosystems, and ecological processes
- conducting necessary research into the biology and ecology of endangered species and ecosystems
- raising or providing funds for all of the above

Not surprisingly, the intensity of these conservation activities is greatest, and increasing most quickly, in relatively wealthy, developed countries. Those countries can more easily afford to allocate significant funding and personnel to this worthwhile cause. Increasingly, however, signs of awareness of the importance of biodiversity issues, and actions to conserve those values, are also emerging in less-developed countries. This development reflects changes in the attitudes of governments in those countries and is reinforced by lobbying and funding provided by domestic and international aid agencies and nongovernmental organizations. These changes are critically important, because such a large fraction of Earth's threatened biodiversity occurs in tropical, less-developed countries.

Of course, a respect for nature has always been an integral component of most religions that developed in tropical countries, such as Buddhism, Hinduism, and Jainism. Nevertheless, this respect has not necessarily been translated into a real-world conservation ethic among the peoples of those nations. As a result, wildlife and natural habitats have suffered badly, mostly because of extensive conversions of natural ecosystems into agricultural ones, but for other reasons as well.

It is beyond the scope of this book to describe the many international agencies and organizations that are active in the protection of Earth's biodiversity. Some of the most important of these are the United Nations Environment Program (UNEP), the World Wildlife Fund, the International Union for the Conservation of Nature (or World Conservation Union; IUCN), the World Resources Institute, and Conservation International (for more information see Appendix II). All of these organizations are active in protecting species and natural lands, in education and lobbying, in research and monitoring, and in raising funds for the protection of biodiversity values.

To illustrate the rapid development of international conservation activities, we will briefly examine a global program known as the Global Biodiversity Strategy (Reid *et al.*, 1992). This is a joint program of the World Conservation Union, the World Resources Institute, and the United Nations Environment Program. The broad objectives of the Global Biodiversity Strategy are to maintain essential ecological processes and life-support systems on Earth, to preserve biodiversity, and to ensure the sustainable development of Earth's natural resources. Although these are rather general goals, they are important because they directly link the conservation of biodiversity with the sustainable development of global resources and human economies. One cannot occur without the other — a fact which must be acknowledged by any governments or agencies that support the Global Biodiversity Strategy.

Through the Global Biodiversity Strategy, all nations can initiate meaningful actions to conserve and protect their biodiversity for the benefit of present and future generations of people, as well as for reasons of intrinsic value. To achieve this end, 85 specific actions are recommended for implementation by countries commiting themselves to the Strategy. The following five actions by signatory countries are considered essential:

1. ratification and implementation of the recommendations of the international Convention on Biological Diversity, as presented in 1992 by the United Nations Environment Program at the United Nations Conference on Environment and Development, held at Rio de Janeiro, Brazil (the "Earth Summit")
2. implementation of the actions detailed in the Global Biodiversity Strategy, beginning immediately with an initial decade of intensive efforts to conserve and protect the indigenous biodiversity of signatory nations
3. creation of an international administrative mechanism to ensure broad participation in decisions concerning global biodiversity, with representation from governments, the scientific community, citizens, industry, the United Nations, and nongovernmental organizations
4. establishment of an international network, linked to the Convention on Biological Diversity, to monitor threats to biodiversity, so that individuals and organizations can be alerted and take appropriate actions
5. integration of biodiversity considerations into planning processes for national development

Other key elements of the Global Biodiversity Strategy are summarized in Table 24.3, while essential elements of the Convention on Biological Diversity are listed in Table 24.4.

It is too soon to tell whether the Global Biodiversity Strategy will be successful, because the program began only in the late 1970s (as an earlier program called the World Conservation Strategy). However, it is encourag-

TABLE 24.3 | KEY ELEMENTS OF THE GLOBAL BIODIVERSITY STRATEGY

1. Catalyze action through international cooperation and national planning
2. Establish a national policy framework for conservation of biodiversity:
   - Reform existing public policies that invite the waste or misuse of biodiversity
   - Adopt new public policies and accounting methods that promote conservation and the equitable use of biodiversity
   - Reduce demand for biological resources
3. Create an international policy environment that supports national conservation of biodiversity:
   - Integrate biodiversity conservation into international economic policy
   - Strengthen the international legal framework for conservation to complement the Convention on Biological Diversity
   - Make the development-assistance process a force for biodiversity conservation
   - Increase funding for biodiversity conservation, and develop innovative, decentralized, and accountable ways to raise funds and spend them effectively
4. Create conditions and incentives for local conservation of biodiversity:
   - Correct imbalances in the control of land and resources that cause biodiversity losses, and develop new resource-management partnerships between government and local communities
   - Expand and encourage the sustainable use of products and services from the wild for local benefits
   - Ensure that those who possess local knowledge of genetic resources benefit appropriately when it is used
5. Manage biodiversity through the human environment:
   - Create the institutional conditions for bioregional conservation and development
   - Support biodiversity conservation initiatives in the private sector
   - Incorporate biodiversity conservation into the management of biological resources
6. Strengthen protected areas:
   - Identify national and international priorities for strengthening protected areas and enhancing their role in biodiversity conservation
   - Ensure the sustainability of protected areas and their contribution to biodiversity conservation
7. Conserve species, populations, and genetic diversity:
   - Strengthen capacity to conserve species, populations, and genetic diversity in natural habitats
   - Strengthen the capacity of off-site conservation facilities to conserve biodiversity, educate the public, and contribute to sustainable development
8. Expand human capacity to conserve biodiversity:
   - Increase appreciation and awareness of biodiversity values and their importance
   - Help institutions disseminate the information needed to conserve biodiversity and mobilize its benefits
   - Promote basic and applied research on biodiversity conservation
   - Develop human capacity for biodiversity conservation

Source: Modified from Reid *et al.* (1992)

ing to know that this sort of comprehensive international effort exists and that most of Earth's nations are participating, including countries in all stages of economic development. Of course, it remains to be seen how effective the individual and collective actions will be.

If Earth's resources are to be used by humans on an ecologically sustainable basis, rare and endangered species and natural ecosystems must be protected. Clearly, an international program like the Global Biodiversity Strategy is needed to guide the process of sustainable development.

The modern predicament of extinction and endangerment of biodiversity is one of the most critical elements of the global environmental crisis. Hopefully, the increasing intensity of conservation activities worldwide will be sufficient to turn the tide, so that future generations will regard these ongoing actions as a "success story" of global conservation. Any alternative result would be catastrophically tragic.

The one process ongoing in the 1990s that will take millions of years to correct is the loss of genetic and species diversity by the destruction of natural habitats. This is the folly that our descendants are least likely to forgive us.

E.O. Wilson (cited in Reid *et al.*, 1992)

TABLE 24.4 KEY ELEMENTS OF THE CONVENTION ON BIOLOGICAL DIVERSITY

- A commitment by governments to survey their natural living resources, both domesticated and wild, and to conserve sites noted for their rich biological diversity, as well as threatened species and domesticated varieties
- Recognition that both the conservation of biodiversity in wild nature and its preservation in captivity and gene banks are key tools in any effective biodiversity conservation strategy
- A commitment by governments to ensure that any use of biodiversity is sustainable and equitable
- Recognition that conservation of biodiversity is a common concern of all humankind and that nations have the sovereign right to use their biological resources
- Recognition that access to biodiversity is contingent upon prior informed consent of the country concerned, and that those who possess traditional knowledge about genetic resources, and farmers who have contributed to and maintained diversity in crops and livestock, deserve just compensation for the use of their knowledge or their varieties
- The establishment of a financial mechanism that would provide both technical and financial assistance to developing countries in need of support for surveying, characterizing, and conserving their indigenous biodiversity
- The establishment of an administrative structure giving equal control to developed and developing countries that are Parties to the Convention in the distribution of funds under the Convention, and ensuring participation of scientists, governments, and nongovernmental organizations to advise on funding priorities
- Arrangements by which the commercial exploiters of biodiversity help finance much of its conservation in the countries that give it refuge
- Mechanisms to ensure access for developing countries to technologies for conserving and using biodiversity
- The establishment of a monitoring and early-warning system to alert governments and the public to potential threats to biodiversity

Source: Modified from Reid *et al.* (1992)

Key Words

extinction
extirpation
endangered
endemic species
protected areas
greater protected areas

Questions for Discussion

1. How are the products of biodiversity important in your life? Compile a list of ways in which you use such products, including foods, materials, and medicines, in your daily routine. Are there substitutes for any or all of these uses?
2. Do you know of any species that are rare or endangered in the region in which you live? Find several examples, and identify the habitat needs of those species. Do you think that these rare or endangered species are being adequately protected? What more could be done? How can you help?
3. Find several examples of "endangered spaces" (endangered ecosystems) in your region and decide whether they are being adequately protected. What more could be done? How can you help?
4. We learned in this chapter that the greatest modern threat to Earth's biodiversity is posed by deforestation in tropical countries. Can you think of ways in which Canadians are economically linked with deforestation in the tropics? For example, do Canadian consumers provide a demand for certain tropical-forest products? Do Canadians hold some of the foreign debt of tropical countries? How might these circumstances contribute to tropical deforestation?
5. Everyone can help to protect endangered species and spaces. List some of the ways you can contribute to solving the problems of endangered biodiversity in the region where you live, in Canada, and internationally.

References

Blockstein, D.E. and H.B. Tordoff. 1985. Gone forever: a contemporary look at the extinction of the passenger pigeon. *Amer. Birds*, **39**: 845–851.

Campbell, R.R., J.R. Robillard, and F. Pilon. 1993. *CITES Reports: 1991 Annual Report for Canada*. Ottawa: Canadian Wildlife Service.

COSEWIC. 1996. *Canadian Species at Risk*. Ottawa: Committee on the Status of Endangered Wildlife in Canada (COSEWIC).

Day, D. 1989. *The Encyclopedia of Vanished Species*. Hong Kong: McLaren.

Diamond, J.M. 1982. Man the exterminator. *Nature*, **298**: 787–789.

Ehrlich, P.R. and A. Ehrlich. 1981. *Extinction: The Causes and Consequences of the Disappearance of Species*. New York: Ballantine.

Erwin, D.A. 1990. The end-Permian mass extinction. *Annu. Rev. Ecol. Syst.*, **21**: 69–91.

Feduccia, A. 1985. *Catesby's Birds of Colonial America*. Chapel Hill, NC: University of North Carolina Press.

Finch, D.M. 1991. *Population Ecology, Habitat Requirements, and Conservation Status of Neotropical Migrating Birds*. Fort Collins, CO: Rocky Mountain Forest and Range Experiment Station. USDA For. Serv. Gen. Tech. Rep. RM-205.

Fisher, J., N. Simon, and J. Vincent. 1969. *Wildlife in Danger*. New York: Viking.

Fitter, R. 1968. *Vanishing Wild Animals of the World*. London: Kaye & Ward.

Freedman, B. 1995. *Environmental Ecology, 2nd ed.* San Diego, CA: Academic.

Gould, S.J. 1989. *Wonderful Life: The Burgess Shale and the Nature of History*. New York: Norton.

Heywood, V.H. (executive ed.). 1995. *Global Biodiversity Assessment*. Cambridge: Cambridge University Press.

Holmes, R.T., T.W. Sherry, and F.W. Sturges. 1986. Bird community dynamics in a temperate deciduous forest: long-term trends at Hubbard Brook. *Ecol. Monogr.*, **56**: 201–220.

Hussell, D.J.J., M.H. Mather, and P.H. Sinclair. 1992. Trends in numbers of tropical- and temperate-wintering land-birds in migration at Long Point, Ontario, 1961–1988. In: J.M. Hagan and D.W. Johnson, eds. *Ecology and Conservation of Neotropical Migrant Landbirds*. Washington: Smithsonian Institution Press. pp. 101–114.

Leader-Williams, N., S.D. Albou, and P.S.M. Berry. 1990. Illegal exploitation of black rhinoceros and elephant populations: patterns of decline, law enforcement, and patrol effort in Luangwa Valley, Zambia. *J. Applied Ecol.*, **27**: 1055–1087.

Lovejoy, T.E. 1985. Amazonia, people and today. In: G.T. Prance and T.E. Lovejoy, eds. *Amazonia*. New York: Pergamon. pp. 328–338.

Martin, P.S. 1984. Catastrophic extinctions and late Pleistocene blitzkrieg: two radiocarbon tests. In: M.H. Nitecki, ed. *Extinctions*. In: *Extinctions*. Chicago: University of Chicago Press. pp. 153–189.

Martin, P.S. and H.E. Wright (eds.). 1967. *Pleistocene Extinctions: The Search for a Cause*. New Haven, CT: Yale University Press.

McClung, R.M. 1969. *Lost Wild America: The Story of Our Extinct and Vanishing Wildlife*. New York: William Morrow.

Myers, N. 1979. *The Sinking Ark: A New Look at the Problem of Disappearing Species*. Oxford: Pergamon.

Myers, N. 1988. Tropical forests and their species: going… going… In: E.O. Wilson, ed. *Biodiversity*. Washington, DC: National Academy Press.

Nettleship, D.N. and P.G.H. Evans. 1985. Distribution and status of the Atlantic Alcidae. In: D.N. Nettleship and T.R. Birkhead, eds. *The Atlantic Alcidae*. New York: Academic. pp. 54–154.

Nitecki, M.H. (ed.). 1984. *Extinctions*. Chicago: University of Chicago Press.

Norton, B.G. (ed.). 1986. *The Preservation of Species*. Princeton, NJ: Princeton University Press.

Page, J.D., Bayley, S.E., Good, D.J., Green, J.E., and Ritchie, J.P.B. 1996. *Banff–Bow Valley: At the Crossroads. Summary Report*. Ottawa: Supply and Services Canada. Report of the Banff–Bow Valley Task Force.

Peters, R.L. 1991. Consequences of global warming for biological diversity. In: R.L. Wyman, ed. *Global Climate Change and Life on Earth*. New York: Routledge, Chapman, & Hall. pp. 99–118.

Primack, R.B. 1993. *Essentials of Conservation Biology*. Sunderland, MA: Sinauer.

Raup, D.M. 1986. Biological extinctions in Earth history. *Science*, **231**: 1528–1533.

Reid, W., C. Barker, and K. Miller (principal authors). 1992. *Global Biodiversity Strategy*. Washington: World Resources Institute.

Simberloff, D. 1988. The contribution of population and community biology to conservation science. *Annu. Rev. Ecol. Syst.*, **19**: 473–511.

Soule, M.E. (ed.) 1986. *Conservation Biology: The Science of Scarcity and Diversity*. Sunderland, MA: Sinauer Assoc.

Steadman, D.W. 1991. Extinction of species: Past, present, and future. In: R.C. Wyman, ed. *Global Climate Change and Life on Earth*. New York: Routledge, Chapman, & Hall. pp. 156–169.

Terborgh, J. 1989. *Where Have All the Birds Gone?* Princeton, NJ: Princeton University Press.

Vitousek, P.M. 1988. Diversity and biological invasions of oceanic islands. In: E.O. Wilson, ed. *Biodiversity*. Washington: National Academy Press. pp. 181–189.

Wilson, E.O. (ed.). 1988. *Biodiversity*. Washington: National Academy Press.

Wood, G.L. 1972. *Animal Facts and Figures*. Enfield, U.K.: Guinness Superlatives.

World Resources Institute (WRI). 1994. *World Resources 1994–95: A Guide to the Global Environment*. New York: Oxford University Press.

World Wildlife Fund. 1996. *Endangered Spaces Progress Report, 95–96*. Toronto: World Wildlife Fund.

Ziswiler, V. 1967. *Extinct and Vanishing Animals*. New York: Springer.

CBC

Canadian Case 5

Clayoquot Controversy

Old-growth forests are a special kind of natural ecosystem, characterized by large, old trees. Some contain species of plants, animals, and microbes that cannot live in younger forests. Old-growth forests are also extremely valuable as sources of timber, which is why they are being harvested rapidly in all parts of the world where they still occur.

Because of timber harvesting and the conversion of extensive areas of forest into agricultural lands, old-growth forests are now rare in eastern Canada, being limited to relatively small and isolated woodlots. Although old-growth forests have also been severely depleted on the west coast of British Columbia, the region still supports some extensive tracts.

Forests in the Clayoquot Sound region of central Vancouver Island have been harvested for decades. Intense controversy has developed, however, over the continued clear-cutting of the remaining tracts of old-growth forests in the area. In part, this has happened because large numbers of people pass through the region while travelling to Pacific Rim National Park and other recreational areas on the west coast of Vancouver Island. People do not like to see clear-cuts in places where magnificent large trees used to occur — they prefer to view big trees rather than big stumps!

However, many people earn their livelihood by working in forestry. In fact, the forest industries comprise one of Canada's largest and most important economic sectors. To some extent, all Canadians receive substantial economic benefits from the activities of the forest industries — activities chiefly based on the extensive clear-cutting of natural forests.

Resource-use conflicts associated with forestry, including those involving old-growth forests, are common in most regions of Canada. Most of the controversy pits economic interests against natural-value interests. One of the great challenges of Canadian society is to find ways to resolve these conflicts. To do this, we must find ways to use our forest resources in a sustainable manner. When this is achieved, the productivity of the forest will be maintained, and forest areas will continue to sustain their natural values, including those of indigenous species and old-growth forests as a special kind of ecosystem. However, to accomplish the latter goals, some extensive tracts will have to be protected from forestry.

Questions for Discussion

1. Make a list of products manufactured from tree biomass. Consider whether these are important in your daily life and to society more generally.
2. Do you think that we should continue to harvest old-growth forests? Or would it be better to work with only those forests that have already been incorporated into the economic system?
3. Do you believe that old-growth forests are special? Why, or why not?

Video Resource: "Clayoquot: The Sound and the Fury." News in Review (October, 1993).

CBC CANADIAN CASE 6

GLOBAL WARMING

It is well known that Earth's greenhouse effect helps to make the planet habitable, because it maintains the surface temperature within a range that organisms can tolerate (averaging about 15°C). It is also known that the greenhouse effect is due to the presence of certain trace gases in the atmosphere, especially water vapour, carbon dioxide, methane, and nitrous oxide. Because the atmospheric concentrations of some of these gases (other than water vapour) have been increasing rapidly, as a result of emissions associated with human activities, many scientists believe that Earth's greenhouse effect will be altered, and that the global climate could warm markedly.

This potential climatic influence is, however, rather controversial. Climate has always changed naturally. Some scientists believe that human influences on this process have been minor. On the other hand, there are many scientists who think that humans have already caused important changes in global climate, including a warming of surface temperatures, and an increased frequency of events of extreme weather, such as drought, hurricanes, and tornados.

The video case "Global Warming" examines these controversial issues, and particularly the evidence in support of the hypothesis that human activities are causing a marked warming of global climate. These are important issues. If it does indeed turn out that humans are responsible for this important environmental change, it will be extremely difficult to find ways of rapidly undoing the damage.

Questions for Discussion

1. Do you think that global warming is a potentially important problem? What are the benefits and damages that would likely result from this kind of environmental change?
2. Make a list of the most important anthropogenic sources of emission of greenhouse gases. Consider your own connections to these emissions. Such connections can be direct, (as when you drive an automobile) or indirect (as when large companies manufacture appliances and other machinery for use in our homes, or harvest forests to provide us with paper and lumber).
3. Do you think that society should wait until there is convincing "proof" about the possible human influence on climate change before effective control actions are undertaken? Or would it be more prudent to take a precautionary approach, and to act on the basis of reasonably anticipated effects of the well-known increases of greenhouse gases in the atmosphere?

Video Resource: Global Warming

CBC CANADIAN CASE 7

DISAPPEARING OZONE

Ozone occurs in relatively high concentrations in the upper atmospheric layer known as the stratosphere. There, it provides a critical environmental service by absorbing most of the incoming solar ultraviolet radiation, and so shielding the organisms on Earth's surface from the damaging effects of this energy.

Stratospheric ozone forms as a result of natural photochemical reactions. However, evidence is increasing that the concentrations of stratospheric ozone are being depleted by human influences. The reduction in ozone is mostly caused by reactions with substances known as chlorofluorocarbons (or CFCs; these are also known as freons). These chemicals are synthesized for use in refrigeration and the manufacturing of microelectronics. After they are released into the atmosphere, the CFCs slowly wend their way up into the stratosphere, where they break down into simpler compounds that consume ozone.

The reduced concentrations of stratospheric ozone pose important health risks to people. Exposure to ultraviolet radiation increases the probability of skin cancers, including the rapidly metastasizing cancer known as malignant melanoma. Ultraviolet radiation also causes cataracts and diminishes immune responses.

People can avoid exposure to ultraviolet radiation by staying out of the sun, wearing hats, and using sunscreens. Plants and animals, however, cannot do these things. Consequently, scientists believe that important ecological damage could be caused by the depletion of the Earth's protective veil of stratospheric ozone.

Unlike many other environmental problems, those associated with emissions of CFCs can be handled relatively effectively. An international treaty, known as the Montreal Protocol, has been negotiated. By signing this treaty, nations have committed themselves to reducing, and eventually eliminating, their emissions of CFCs. Unfortunately, CFCs are very persistent in the atmosphere. It will take decades before their influence on stratospheric ozone is significantly reduced.

In addition, the Montreal Protocol allows poorer, less-developed countries, such as China, India, and Mexico, to take a relatively long time to phase out their production and use of chlorofluorocarbons. In contrast, wealthier countries such as the United States, Canada, and nations in western Europe are expected to stop their use of these environmentally damaging chemicals much more quickly. This disparity between expectations and the regulations that govern the use of chlorofluorocarbons means that it can be extremely profitable to smuggle those chemicals from places where their use is not restricted, to those where they are closely controlled. The video "Freon" explores the highly lucrative international black market for chlorofluorocarbons, including the massive smuggling of these chemicals from Canada to the United States.

Questions for Discussion

1. The health risks associated with exposure to ultraviolet radiation are well known. In view of this, how would you explain the fact that many people continue to sunbathe and engage in other behaviours that increase their exposure to ultraviolet radiation?
2. Many environmental scientists believe that society has dealt with the problem of stratospheric ozone about as well as can be hoped. Why have similarly effective actions not been taken against other important environmental problems, such as emissions of greenhouse gases, the biodiversity crisis, and the population crisis?

Video Resource: "Freon." News in Review (September, 1997).

25

ECOLOGICALLY SUSTAINABLE DEVELOPMENT

Chapter Outline

Chapter Objectives

After completing this chapter, you will be able to:

1. Outline the process of an environmental impact assessment, and describe several recent Canadian examples.
2. Discuss how monitoring and research are crucial to understanding the causes and consequences of changes in environmental quality.
3. Explain how environmental reporting and literacy are crucial to dealing with the environmental crisis.
4. Outline the roles of governments, nongovernmental organizations, scientists, and citizens in designing and implementing an ecologically sustainable economy in Canada and internationally.

Introduction

The previous parts in this book dealt with relatively specific topics. This has allowed us to learn the subject matter of environmental science by examining certain ideas and by analyzing a body of supporting information. In this final chapter, we bring many of these topics together in an interdisciplinary fashion.

In the first sections of the chapter, we will examine topics related to environmental management and protection at the broader societal level. These topics include environmental impact assessment, environmental monitoring and research, environmental literacy, and sustainability. All of these are necessary for maintaining an acceptable level of environmental quality and healthy ecosystems — two necessary objectives for a truly sustainable socioeconomic system. We will also examine a range of actions that each of us can undertake to help prevent environmental problems.

In the concluding section of this chapter, we will briefly discuss the future prospects for industrialized economies such as that of Canada, and for spaceship Earth.

Environmental Impact Assessment

An **environmental impact assessment** is a process used to help prevent environmental problems. Environmental impact assessments do this by identifying and evaluating the potential consequences of proposed actions for environmental quality. Because they can consider ecological, physical-chemical, and other environmental effects, as well as socioeconomic consequences, environmental impact assessments are multi- and interdisciplinary activities.

Environmental impact assessments may be conducted to examine different types of activities, or planned activities, that could affect environmental quality. These include:

1. individual projects, such as proposals to construct a dam, smelter, power plant, airport, highway, or incinerator
2. integrated schemes, such as proposals to develop an industrial park, a pulp or lumber mill with its attendant wood-supply and forest-management plans, or other complex developments involving numerous projects
3. government policies that carry a risk of substantially affecting the environment

The scale of environmental assessments can vary greatly, from the examination of a relatively small proposal to construct a building near a wetland, to megaprojects associated with natural resource development.

In Canada, environmental impact assessments for proposals that involve federal funding or jurisdiction are regulated under the Canadian Environmental Assessment Act (CEAA). Most provinces and territories have similar legislated requirements, as do some local levels of government.

Any proposed project, program, or policy carries potential risks for human welfare and for other species and ecosystems. For example, a project may be proposed that will emit toxic chemicals into the environment. In such a case, it is necessary to determine if the anticipated emissions might exceed the regulated levels. The most stringent regulatory standards and guidelines relate to the maximum exposures that humans can tolerate without health risks; criteria to protect other species and ecosystems are less exacting. In addition, the project might cause disturbances, which could result in ecological damage, during its construction or operation. These disturbances should be identified and quantified, and the potential environmental damages evaluated, before permission is granted to start the project.

The process of environmental impact assessment is intended to examine these potential damages and to suggest ways of avoiding or *mitigating* them as much as possible. However, this does not necessarily mean that no damage will be caused by the proposed development. In almost all cases, some damages are inevitable.

Because most developments potentially affect a great variety of species and ecosystems, impact assessments are limited to predicting the effects on only a selection of so-called "**valued ecosystem components**" (or VECs). These components are valued because society perceives them to be important for one or more of the following reasons:

1. They are an economically important resource, such as an agricultural crop or a commercial forest, or an exploited stock of fish, mammals, or birds.
2. They are a rare or endangered species or community.
3. They are of cultural or aesthetic significance.

The initial phase of the process of environmental impact assessment is known as a "screening." The screening determines the level of assessment that a proposed activity will undergo, i.e., whether a minor review or a full assessment is appropriate. Once the level of assessment is decided, a "scoping" exercise is undertaken. This identifies any potentially important interactions between project-related activities on the one hand and human welfare or VECs on the other. In essence, the scoping compares the predicted spatial (space) and temporal (time) boundaries of the stressors associated with the proposed development, with the areas where humans and VECs are found. If potential interactions are identified, the assessors must determine whether significant damages might be caused.

Sometimes impact assessments are not well funded, or they have to be completed relatively quickly. In such cases, ecologists, toxicologists, and other environmental professionals may have to provide expert opinions about the importance of potential interactions between project-related stressors and human welfare or VECs. These professional opinions should be based on the best-available scientific information and understanding, while recognizing that such information and understanding are usually incomplete, and that differences of opinion often arise between qualified specialists. When sufficient time and funding are available, it is possible to conduct field, laboratory, and/or computer-based (or simulation) research to investigate further any risks from the potential interactions identified during the screening process. It must be understood, however, that even well-funded, properly designed, and well-executed research may yield uncertain results, particularly about damages that might occur at low intensities of exposure to project-related stressors.

Planning Options

If potentially important risks to human welfare or VECs are identified, a number of planning options must be considered. There are three broad choices.

Prevent or Avoid: One option is to avoid the predicted damages by ensuring that there are no significant exposures of people or VECs to damaging stressors related to the project. This can be done by modifying the characteristics of the development or, in cases of severe conflicts with human welfare or ecological values, by choosing to cancel the project. Because prevention and avoidance may involve substantial costs, these are sometimes considered less desirable options by the proponents of a development. Politicians and regulators may also dislike this option, since it usually involves intense controversy, and since substantial development opportunities may be cancelled. Nevertheless, there is always pressure to take as many precautions as possible before undertaking a proposed development.

Mitigation: Another option is the **mitigation** of the predicted damages — that is, to repair or offset them as much as possible. Because any direct damages to humans are considered unacceptable (and therefore to be avoided), mitigation is mainly relevant to damages inflicted on VECs, or to indirect, low-level risks caused to people. For instance, if the habitat of an endangered species is threatened, it may be possible to move the population-at-risk to a suitable habitat elsewhere, or to create or enhance a habitat at another place, so that no net damage is caused. As another example, a wetland may be destroyed by some development, with the damages offset through the creation of a comparable wetland elsewhere. Mitigations are a common way of dealing with potential conflicts between project-related stressors and VECs. However, it is important to understand that mitigations are never complete, and that there are often residual damages.

Accept the Damages: The third, and least acceptable, option (from the environmental perspective) is for decision makers to choose to allow a proposed project to cause some or all of the predicted damages to human welfare or VECs. This choice is often considered tolerable by proponents of projects and by politicians (although such a preference would rarely be explicitly stated). Their rationale is due to their perception that the socioeconomic benefits gained by proceeding with the threatening development would be greater than the costs of the environmental damage.

Environmental impact assessments often find that a proposed development carries risks of damages to environmental quality. Usually, the development is allowed to proceed in some form, with the predicted environmental damages being avoided or mitigated to the degree that is considered technologically and economically feasible. As noted previously, however, there are always residual risks that cannot be avoided or mitigated. The damages that may occur represent some of the environmental costs of development, which are real even if there is financial or some other kind of compensation (such as mitigation).

Once a project is begun, compliance monitoring is usually necessary, to ensure that regulatory criteria for pollution or health hazards are not being exceeded. It is extremely useful, although not always required, that any ecological effects also be monitored. This tests the predictions of the impact assessment and identifies any unanticipated effects or "surprises."

Ideally, monitoring programs are begun before the project actually starts. This establishes baseline conditions. The monitoring should continue for some years after the project is completed, until it is determined that important damages are not being caused by the development. If unanticipated damages are shown by the ecological effects monitoring, they may be prevented, avoided by an adaptive change in the project design, mitigated, or accepted as an ecological "cost" of development.

Some Canadian Examples of Impact Assessments

Environmental impact assessments have been conducted in all regions of Canada, examining projects varying widely in both scale and potential effects. Each of these unique cases is instructive — illustrating the environmental implications of development projects and the role played by impact assessment. In the following sections, we briefly examine selected elements of some environmental impact assessments in Canada, all from the 1990s.

Diamond Mines in the Northwest Territories: This proposal involves the development of mines to extract diamonds from five deposits discovered about 300 km northeast of Yellowknife. A variety of environmental damages may arise from this project.

First, some of the diamond-bearing rock lies beneath lakes which would have to be drained to develop the mines. These aquatic ecosystems would be destroyed.

In addition, large quantities of gravel are needed to construct roads and other infrastructure. Much of this material would be obtained from long, sinuous surface features known as eskers, which provide critical denning habitat for grizzly bear, wolf, and other high-profile species.

Further, large numbers of caribou traverse the region during their seasonal migrations. These are potentially affected by the mine and its network of roads. Substantial damage to the caribou could harm the aboriginal people in the region, who engage in a subsistence hunt for these animals. The aboriginal people might also suffer from the interference with their commercial hunt for furbearers.

Finally, the proposed mines are located in a region that is presently a huge, roadless wilderness. Conservationists, led by the World Wildlife Fund, have objected to the approval of the mine before a system of protected areas is set up for the preservation of natural ecosystems and native species, including such large carnivores as grizzly bear, wolf, and wolverine.

The diamond-mine proposal has passed its environmental impact assessment and is being allowed to proceed. It is subject to stringent requirements, however, such as the implementation of acceptable methods of disposal of mining and milling wastes, and the protection of waterbodies and rivers (other than those that must unavoidably be damaged to develop the mines and dispose of tailings). A ban has been imposed on local hunting by project personnel. As well, a monitoring program must be implemented to ensure that unanticipated damages are not caused to air or water quality, or to wildlife. The mine must also meet socioeconomic criteria, including several that deal with employment opportunities and other involvements of local people (including aboriginals) in the development. As a measure outside the scope of the formal impact assessment, the government of the Northwest Territories has committed itself to establishing protected areas in the larger region.

Destruction of Diseased Bison: Agriculture Canada proposed to slaughter almost all the bison in the southern regions of Wood Buffalo National Park and its vicinity. Some of these animals are infected with bovine tuberculosis and brucellosis, and there are concerns over the potential spread of these infectious diseases to herds of domestic cattle to the south and west of the area. The bison targeted for slaughter are hybrids between the indigenous wood bison and plains bison that were introduced to the region during the late 1920s. The proposal did not include the elimination of small populations of genetically "pure" wood bison living farther to the north, and in fact these are predicted to receive a measure of protection from the potentially harmful effects of interbreeding with hybrid animals.

This proposal was made in support of commercially important agricultural interests, but it quickly engendered intense controversy. It was opposed by virtually all conservationists, and by local aboriginal people. Although this project passed successfully through the impact assessment process, it was later suspended by the federal Minister of

The World Wildlife Fund is one of Canada's most important organizations in environmental advocacy. Its Endangered Spaces Campaign has been crucial in focusing attention on the plight of Canada's natural areas. This has resulted in greater federal and provincial efforts to designate additional ecological reserves and other protected areas. WWF is also active on the international front, for example, by supporting this project to integrate ecotourism with community development near a national park at Monteverde, Costa Rica. The "ecolodge" in the photo is community-owned and managed, and provides local families with jobs and incomes that depend on the conservation of the forests in the national park.

Source: S. Price

the Environment, largely in recognition of the intense opposition from conservation and first-nation interests.

The Hibernia Offshore Oil Development: Several decades of exploration have resulted in the discovery of large reservoirs of petroleum in undersea geological formations on the Grand Banks, east of Newfoundland. An environmental impact assessment examined a proposal from a consortium of companies to develop those valuable resources. A system of underwater wells was proposed that would feed to a central collecting system on a huge submersible platform located in 80-m-deep waters. The petroleum would be delivered periodically to onshore refineries using tanker ships.

The offshore waters of the Hibernia field are sometimes subject to intense winds, and immense icebergs pass through the area during most winters and springs. Some icebergs scour the ocean floor. These natural forces pose risks to the production and storage facilities. As well, accidents may result from equipment failure or human error. Alone or in combination, these factors could cause a massive petroleum spill. Such an accident could result in enormous damages to the fishery resource, to the abundant marine mammals and seabirds in the region, and to other ecological and economic values.

The Hibernia development includes a sophisticated system of weather and iceberg monitoring, coupled with stringent spill-prevention and control technologies. These measures have been accepted by regulators and politicians as providing an acceptable degree of environmental safety. Consequently, the Hibernia development has passed through the impact assessment process and is scheduled to be producing petroleum before the end of the 1990s.

Grande–Baleine Hydroelectric Complex: Some regions of Canada have an enormous potential for the development of hydroelectricity. One area in which this renewable source of energy is being vigorously developed is northwestern Quebec. Several large rivers flowing into James and Hudson Bays have been dammed, allowing the storage of huge reservoirs of water. Electricity is generated at times when consumer demand is greatest.

The Grande-Baleine Complex was a proposal to add to the hydroelectricity capacity of Quebec by constructing three generating stations, with a total capacity of 3212 MW, on the Grande Baleine River. The associated dams would have flooded 1667 km^2 of terrestrial habitats. Other disturbances would have included the construction of roads and transmission lines to take the power to southern markets.

This controversial development would have had important environmental impacts. The most critical of these were the ecological changes associated with the creation of such an enormous reservoir, including loss of terrestrial habitats, changes in flow regimes, effects on the aquatic ecology of rivers, and effects on water quality and ecosystems in nearshore Hudson Bay. In addition, the local populations and movements of woodland caribou and furbearing mammals would have been affected, with consequences for the livelihood of aboriginal people living in the region.

These and other potential effects were considered during a detailed environmental impact assessment, and plans were made to avoid or mitigate the damages to the degree that was possible. Ultimately, however, the proposed development did not proceed, not so much because of environmental concerns, but as a result of insufficient commitments to purchase the electricity in the northeastern United States. Without access to that foreign market, the estimated $13 billion cost of the project's construction was not considered economically feasible.

A Municipal Incinerator: This was a proposal to construct a facility to incinerate large quantities of municipal refuse from metropolitan Halifax, Nova Scotia. Some of the heat produced would have been used to generate about 16 MW of electricity (this is also known as a "waste-to-energy" facility). The incinerator would have been fitted with advanced technologies to control emissions of potentially toxic chemicals, such as metals; gases; organic particles; and organic vapours including polycyclic aromatic hydrocarbons, dioxins, and furans. Such emissions can never be totally eliminated, however, and there is controversy about the potential risks to human health inherent in even minute exposures to some chemicals, particularly dioxins and furans. As it turned out, the proposal to build the incinerator was turned down by the provincial Minister of Environment — partly because it was considered too costly in comparison with alternative methods for the disposal of municipal wastes, but partly, too, for environmental reasons.

A Peat Mine in Nova Scotia: Peat mined from bogs is used as a horticultural material, and can also be burned as a source of energy. This proposal would have developed a mine on a bog in southwestern Nova Scotia to provide peat as an industrial fuel. The impact assessment focused on the fact that the bog in question provides habitat for several rare species of plants, including a carnivorous plant called the thread-leaved sundew (*Drosera filiformis*). This species is endangered in Canada, and over most of the rest of its range in the eastern United States. Because the bog harbours the largest of the four known populations of the sundew in Canada, the provincial Minister of Environment did not allow the mine to be developed on that site. This was a controversial decision, because it cancelled a local development initiative in a region in which the economy is chronically depressed.

Environmental Monitoring and Research

For many reasons — only some of which we have examined — there are widespread and well-founded concerns about damages being caused to the quality of the environment. In response, many nations are implementing programs to monitor changes in environmental quality over time. Most of these programs are intended to document changes occurring over large regions or entire countries, and to help predict future changes. They are much larger in scale and scope than programs that monitor whether particular industrial facilities are complying with regulations and guidelines. Most large-scale monitoring is conducted by government agencies, or in some cases by nongovernmental organizations. The resulting data and knowledge are used to guide decision-making in government, to enhance the work of nongovernmental organizations, and to provide material for programs in environmental education.

In the sense meant here, **environmental monitoring** involves repeated measurements of variables related to either the inorganic environment or the structure and functioning of ecosystems. Successful programs of environmental monitoring depend on the careful choice of a limited number of representative indicators (because not everything can be monitored) and on the careful collection of reliable data. If a monitoring program detects important changes, the possible causes and consequences of those changes are usually researched.

Environmental indicators are relatively simple measurements used to represent complex aspects of environmental quality. Indicators are usually also sensitive to changes in the intensity of stressors. For example, the levels of chemical residues in species high in the food web are often used as indicators of contamination of the larger ecosystem. This is why residues of chlorinated hydrocarbons (such as DDT and PCBs) are routinely monitored in herring gulls and cormorants on the Great Lakes, and in marine mammals in nearshore waters of the Pacific, Atlantic, and Arctic Oceans (Chapter 21). Similarly, lichens are known to be very sensitive to gaseous pollutants, and are therefore sometimes monitored as indicators of air quality over large regions, including cities.

Other indicators include species that are considered to represent the general health of the ecosystem of which they are a component. For instance, the population status of grizzly bears is considered a good indicator of the quality of their extensive ecosystem, as are populations of spotted owls for western old-growth rainforests, pileated woodpecker and pine marten in some other forests, salmon and trout in certain aquatic ecosystems, and orca and other whales in their marine ecosystems.

Sometimes, *composite indicators* are monitored to track changes in environmental quality. In some respects these are environmental analogues of the composite indexes that are used to monitor complex trends in finance and economics, such as the Consumer Price Index (CPI) and the Toronto Stock Exchange (TSE) index. Because they allow complex changes to be presented in a simple manner, composite indictors are especially useful for reporting to the general public.

Composite indicators of air and water quality have been developed using data for various kinds of important pollutants, such as major gases, vapours, and particulates in the atmosphere. However, composite indicators of **environmental quality** (or of **ecosystem health** or **ecological integrity**; see In Detail 25.1) are not yet well developed. This is mostly because scientists have not yet agreed on what the component variables should be.

When a change in the indicators is measured in an environmental monitoring program, or when a change is predicted, it is necessary to understand its causes and consequences. This is generally done by using accumulated knowledge of the effects of environmental stressors on ecological structure and function, along with research that is designed to address important questions that are not yet understood. We can examine the linkages between environmental monitoring and research by considering several examples.

Suppose that environmental monitoring has detected that precipitation has become acidified in some large region (see Chapters 16 and 19). The cause(s) of the acidification might be understood by determining the concentrations of chemicals in the precipitation in addition to hydrogen ion, and by investigating local emissions of gases and particulates to the atmosphere. Researchers must also understand the consequences of the effects of increased deposition of acidifying substances on freshwater and terrestrial ecosystems, as well as the implications for buildings and other features of cities and towns. At first, the research would examine existing knowledge of the causes and ecological effects of acidification in various kinds of habitats. However, existing knowledge is always incomplete, and therefore should be supplemented with new research examining risks of acidification that are not yet understood. The accumulated information helps society to understand whether the causes of acidification can be controlled, and if so, to assess the potential environmental and economic benefits.

In another example, monitoring might indicate that the ecological character of a region is changing because natural forests are being extensively converted into plantations through industrial forestry. The ecological consequences of those changes would initially be interpreted in the light of existing knowledge of the effects of forestry, supplemented by additional research that investigates poorly

IN DETAIL 25.1

NOTIONS OF ENVIRONMENTAL QUALITY

Environmental quality, *ecosystem health*, and *ecological integrity* are important notions that help us understand the importance of changes in the environment. Like many other notions, these cannot be precisely defined. It is possible, however, to develop a common understanding of what they mean.

Because they integrate changes in many components of ecosystems and environments, all of these concepts involve complex phenomena. Environmental quality, for example, is related to the concentrations of potentially toxic chemicals and other stressors in the environment; to the frequency and intensity of disturbances; and to the effects of these on humans, other species, ecosystems, and economies. Of particular concern, of course, are stressors associated with human activities, because these have become so important in the modern world.

Ecosystem health and ecological integrity are rather similar to each other and, in many respects, to environmental quality. However, these indicators focus on changes that may be occurring in natural populations and ecosystems, rather than on effects on humans and their economy.

All of these notions involve many variables related to stressors and socioeconomic or ecological responses. As a result, they are sometimes measured with so-called composite indicators, which integrate many possible changes that are thought to be important. Composite indicators are not exact measurements of environmental quality, ecosystem health, or ecological integrity, but they do allow society to determine whether conditions are getting worse or better.

These ideas can be explained by using ecological integrity as an example. Obviously, most environmental stressors associated with human activities will enhance some species, ecosystems, and ecological processes, while at the same time damaging others. However, ecological theory suggests that systems with higher values for any or all of the following characters have a greater degree of ecological integrity:

- The ecosystem is resilient and resistant (see Glossary) to changes in the intensity of environmental stressors.
- The system is rich in biodiversity values.
- The ecosystem is complex in its structure and function.
- Large species are present.
- Top predators are present.
- The ecosystem has controlled nutrient cycling (i.e., does not "leak" nutrient capital).
- The ecosystem has a "natural" character, as opposed to being strongly affected by human influences and management.

understood issues. Specific research questions might address the effects of the ecological conversions on biodiversity, forest productivity, watershed hydrology and chemistry, and global environmental changes through effects on carbon storage (Chapter 17 and 23). This information is needed to help decision makers evaluate whether they should permit further conversions of natural forests into plantations.

Environmental monitoring and research in Canada are carried out by various agencies. Environment Canada is the most active agency at the federal level. In addition, the Department of Fisheries and Oceans deals with fisheries and oceanic environments, Natural Resources Canada provides data about nonrenewable and some renewable resources, the Canadian Forest Service provides information relevant to commercial forests, Parks Canada examines changes in national parks and their surrounding areas, and Health and Welfare Canada deals with influences of environmental quality on human health. All of the provincial and territorial governments of Canada also have comparable agencies that deal with environmental issues within their jurisdiction.

A few nongovernmental organizations also undertake a considerable amount of environmental monitoring and research. For instance, the World Wildlife Fund of Canada has several programs that fund work on endangered species and ecosystems. Canadian universities also have considerable technical expertise on environmental issues. University personnel are not usually involved in long-term monitoring programs, but many professors and graduate students undertake research into the causes and consequences of environmental changes.

Environmental monitoring programs provide society with crucial information and knowledge. Both are necessary for implementing effective programs to prevent further degradations of environmental quality and the health of ecosystems, and to repair existing damages. These actions are necessary if society is to conduct its economy in a truly sustainable manner.

Some Challenges and Successes

As we noted earlier, programs of monitoring and research should be capable of detecting recent changes in environmental quality, while also helping to predict future changes. Well-designed programs should deal with the most important stressors that are actual or potential threats to environmental quality. They should measure or predict the effects on people and on sensitive ecosystems and species, particularly those that are economically or ecologically important.

These are the simple requirements of a sensible program for monitoring and investigating environmental problems. Unfortunately, these criteria are not well met by existing programs. As a result, there are deficiencies in our understanding of some important environmental problems. Consequently, those problems are not being addressed effectively, and could become worse in the future. Some examples selected from preceding chapters include the following:

What constitutes an acceptable exposure of humans to potentially toxic chemicals? Some toxins, such as metals and many biochemicals, occur naturally. By how much can anthropogenic emissions be allowed to increase exposures beyond the natural background levels? Is any increase in exposure acceptable for non-natural toxins, such as dioxins, furans, PCBs, synthetic pesticides, and radionuclides? Or are there acceptable thresholds of exposure to these substances?

Is a widespread decline of migratory, neotropical, forest–interior songbirds occurring in North America? If so, what are the causes, and how can we manage the responsible stressors to repair the damages and prevent further losses of these native birds?

Anthropogenic emissions of CFCs (chlorofluorocarbons) may be causing a depletion of stratospheric ozone, resulting in increased ground-level exposures to ultraviolet radiation. What risks does this change have for human health, and for wild species and ecosystems? How can the damages be prevented and repaired?

What are the dimensions of the global extinction crisis? How is biodiversity important to the health of the biosphere and to human welfare? Which Canadian species and ecosystems are most at risk, and why? Should Canada expend more effort in helping to conserve tropical biodiversity, or should we focus on problems within our own boundaries? How are Canadians linked to biodiversity-depleting stressors in tropical countries?

Extensive declines and diebacks of forests have been reported in various parts of the world, including Canada. Are those damages being caused by natural environmental changes, or by stressors associated with human activities? If anthropogenic stressors are important, how can they be managed to prevent and repair the forest damages?

What are the environmental consequences and costs of conventional and nuclear warfare?

Is it possible to valuate the worth of species, communities, and ecological services (i.e., to measure their worth in dollars), so that these can be integrated into economic cost-benefit models?

How intensively can renewable resources be harvested and managed, without causing unacceptable risks to their long-term sustainability and without inflicting unacceptable damage on other species and ecosystems?

To deal properly with these and many other important issues, we must improve our understanding through better monitoring and research. This is required if we are to undertake appropriate actions to prevent and repair the environmental damages. We can illustrate the achievable benefits by examining a few "success stories" in which monitoring, research, and effective actions helped to resolve important environmental problems.

Eutrophication of freshwaters was identified as an important environmental problem during the 1960s and early 1970s. Research discovered that phosphate was the primary cause, and that the damages could largely be avoided by constructing sewage-treatment facilities and by using low-phosphorus detergents.

Contamination with persistent chlorinated hydrocarbons, such as DDT, dieldrin, and PCBs, was found to be widespread in the 1960s and 1970s. Research showed that some species, such as predatory birds, were being seriously harmed, and that there were possible effects on humans. The toxicological evidence has convinced decision makers to ban these chemicals in most countries, to the great benefit of the environment.

Acidification was recognized during the late 1970s and 1980s as an extensive phenomenon causing many ecological damages. Research showed that the problem was largely due to atmospheric depositions of sulphur and nitrogen compounds. This convinced decision makers to require reductions of industrial emissions.

Environmental Reporting and Literacy

Environmental literacy (i.e., a well-informed understanding of environmental issues) is an important social goal. Knowledge about the causes and consequences of environmental damages can influence the decisions and choices made by politicians, regulators, corporations, and individual citizens. If appropriate, these decisions and choices can influence environmental quality in a positive manner (Figure 25.1). People acquire this knowledge in various ways, the most important of which are environmental reporting and other forms of environmental education.

Environmental literacy has a pervasive influence on the attitudes that people develop. Individuals who are knowledgeable about environmental issues will make more appropriate choices of lifestyle, and will influence decision makers to ensure that proper systems of environmental management are implemented. In contrast, poorly-informed public opinion encourages less appropriate environmental choices, such as rampant consumerism and wasteful use of resources. Environmental illiteracy also fosters the development of controversial "red herrings" of environmental management, which can divert attention from more important problems.

An example of an environmental red herring is the common misunderstanding of differences between contamination and pollution. Related syndromes are known as NIMBY (i.e., "**n**ot-**i**n-**m**y-**b**ack**y**ard"), LULU ("**l**ocally **u**nacceptable **l**and **u**se"), BANANA ("**b**uild **a**bsolutely **n**othing near **an**ybody or **a**nything"), and NIMTO ("**n**ot **i**n **m**y **t**erm of **o**ffice"). NIMBY, LULU, and BANANA are common views that many people have about industrial facilities that may affect their local environment, while NIMTO is a common political response. These attitudes can result in part from a lack of information about the risks that may be associated with developments in the neighbourhood. Alternatively, NIMBY, LULU, and BANANA may result when planners and developers are insensitive to the legitimate concerns of local people.

In addition to affecting the siting of commercial and industrial facilities, NIMBY, LULU, BANANA, and NIMTO cause huge problems for planners who are attempting to build certain kinds of environmental-management facilities that society wants and needs. For example, even though all voters recognize that communities need facilities for the disposal of solid wastes and the treatment of sewage, none wish to have such works located in their own neighbourhood.

Decision makers in government and industry need objective cost-benefit analyses when dealing with environmental problems. Decision makers have the responsibility of making societal-level choices to avoid, mitigate, or accept environmental damages. Their choices are often based on their perceptions of the costs associated with environmental damages, offset by the economic benefits promised by the activity that is causing the degradation. Unfortunately, the perspective of many decision makers is that of conventional, short-term economics, rather than ecological economics (see Chapter 12). Because many social controversies have resulted from seemingly non-balanced choices, the role of decision makers is changing in Canada. In addition to, or even instead of, the actual making of choices, these people are increasingly being expected to create an appropriate climate for multilateral consultation and consensus-driven decision-making.

Environmental reporting is one process for communicating information about changes in environmental quality to various interest groups. Such reporting should involve clear and objective presentations of information, showing changes in environmental quality, and should also offer unbiased interpretations of the causes and consequences of those changes.

Environmental reporting is delivered to the broader public of Canada by various agencies, including government departments, educational institutions, nongovernmental organizations, and the mass media. A governmental instrument that has been prominent in Canada since the mid-1980s is known as state-of-the-environment reporting. The federal government has released three well-regarded, comprehensive reports on the state of the

FIGURE 25.1 | INFLUENCES ON ENVIRONMENTAL QUALITY

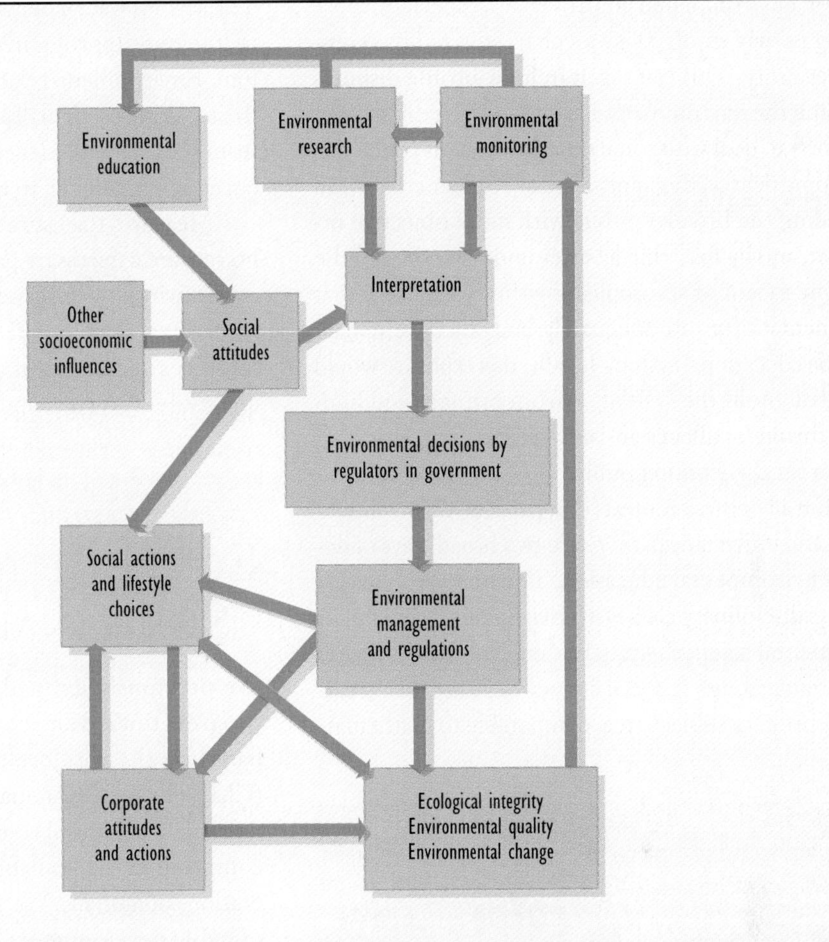

This is a conceptual model of the many influences on environmental quality, including the roles of monitoring, research, regulation, and literacy. Environmental monitoring and research provide an understanding of the causes and consequences of changes in environmental conditions. Ideally, this understanding is based on objective information from monitoring and research programs, interpreted by environmental scientists and other qualified specialists (although the interpretations may be conditioned to some degree by social and cultural influences). This knowledge is communicated to decision makers in government, who may implement regulations and undertake management activities that affect environmental quality.

Knowledge about environmental and ecological changes is also communicated to the general public, through state-of-the-environment reporting, the educational system, activities of nongovernmental organizations, and the mass media. Social attitudes regarding the environment are affected by environmental literacy, resulting in more appropriate choices of lifestyle and a public influence on the policies and actions of governments and corporations.

Source: Modified from Freedman (1995)

Canadian environment (in 1986, 1991, and 1997). Most of the provinces have also released or are preparing state-of-the-environment reports. Unfortunately, Environment Canada has now largely abandoned this function — in 1996 it closed down its division responsible for the preparation of comprehensive state-of-the-environment reports. It remains to be seen whether the provinces will maintain their commitment to the important function of environmental reporting.

Of course, most people become informed about environmental issues through the mass media, such as newspapers, magazines, radio, and television. These can be effective means of environmental education, but there are some drawbacks. Often, media presentations of issues are biased; sometimes they are inaccurate. The focus is frequently on environmental controversies, especially those involving poorly resolved issues characterized by scientific uncertainty. This can result in high-profile disputes dominating the environmental agenda, which can detract from efforts to deal with some other important problems.

To some degree, this approach can be counterbalanced by providing the broader public with more objective information, and by fostering a better understanding of the issues. One means of accomplishing this is to ensure that environmental issues are adequately, and objectively, dealt with in the educational system. Ideally, this exposure would occur throughout the system — from primary and high schools, through colleges and universities, to continuing education for the working public.

Within all of these contexts (but particularly in schools, colleges, and universities), there are two broad ways of delivering environmental education. The first involves discrete, interdisciplinary classes in environmental studies or environmental science. Arguably, environmental issues are important enough to social literacy to justify their treatment as a primary subject area, comparable to mathematics, physics, biology, languages, literature, music, and art. The second way of delivering environmental education is to integrate appropriate case material across the curriculum. Environment-related elements can be used to assist the teaching of virtually all disciplinary subjects, ranging from the physical sciences, through the other natural sciences and medicine, to the social sciences.

Environmental literacy helps people to recognize the importance of environmental problems and to support appropriate actions by government, the private sector, and nongovernmental organizations. Environmental literacy should be promoted at all levels of the educational system. This scene shows university students learning about the ecology of a bog wetland in Nova Scotia.

Source: B. Freedman

Measures to ensure that citizens are environmentally literate are a necessary part of any strategy designed to resolve environmental crises. If people understand these critical issues, they will be more willing to make personal sacrifices in support of the protection of environmental quality, biodiversity, and natural ecosystems.

Ecological Sustainability

We previously defined **sustainable development** as progress toward an economic system that is ultimately based on the wise use of renewable natural resources (Chapter 12). A sustainable economy would not deplete its capital of natural resources, and would therefore not compromise the availability of resources for future generations of humans. We also noted that **ecologically sustainable development** would allow the human enterprise to continue, but without causing unacceptable damages to other species, natural ecosystems, or to other "nonresource" values.

By these criteria, modern industrial economies are clearly nonsustainable. There are two major reasons for this alarming conclusion. The first is the obsession that politicians, economists, and other managers of national and international economies have for rapid *economic growth*, both to keep up with an expanding population and to increase the standard of living. The second reason involves the likelihood that the present size of highly industrial economies (such as those of Canada, the United States, Western Europe, and Japan) is already too large to be sustained for long. The rationale for these two statements is briefly explained in the following paragraphs (and has been further supported by more detailed discussions in earlier chapters).

Economic growth is typically achieved by forcing both nonrenewable and potentially renewable resources through an economy, thereby making the economy larger. During the 1990s, leading industrial nations such as Canada, the United States, and Japan were achieving economic growth rates of about 1–3% per year, which if maintained would double the size of their economies in only about 26–70 y. Rapidly developing economies, such as those of China, Chile, India, Indonesia, the Philippines, and Thailand, have recently been growing at 5–10% per year, sufficient to double their economies in only 7–15 y.

As we learned earlier, these economic growth rates can be achieved only as long as resources continue to be readily available. In Chapters 13 and 14, we examined numerous examples of rapidly depleting stocks of both nonrenewable and potentially renewable resources. Such examples suggest that modern economic growth rates cannot be sustained, and in fact will likely reverse themselves as crucial resources become depleted.

Moreover, many ecologists and environmentalists maintain that the present sizes of highly industrial economies (such as Canada) are already too large to be sustained. The arguments in support of this assessment are similar to those just noted — the large, "developed" economies are maintained by the forced throughput of mined resources, the supplies of which are rapidly becoming depleted.

It is common today for politicians, corporate spokespeople, and resource managers to assert publicly that they support progress toward sustainable development. However, almost all of these people are confusing genuine sustainable development, as it was defined at the beginning of this section, with "sustainable economic growth." In a resource-constrained world, unlimited economic growth can never be sustained over the long term. This is why ecologists and environmentally astute economists believe further growth to be undesirable: "Economic growth as it now goes on is more a disease of civilization than a cure for its woes" (Ehrlich, 1989).

It is important to understand that, although they are pushing society in an ill-advised direction, advocates of economic growth are not a malevolent force. Many of these people hope that growing economies will allow larger numbers of people to be productively employed, and thereby to enjoy the benefits of an industrial, material society. These are, of course, highly desirable goals. Is it prudent, however, to seek a gigantic economy that would only temporarily support a large number of people? Or would it be better to limit the scale of the human enterprise to a level that can be supported by Earth's biosphere and resources over the long term? Fundamental considerations in a sustainable human economy are the numbers of people that must be supported, the total intensity of their resource use, and the equality of standards of living among the world's peoples.

Ultimately, an ecologically sustainable economy is limited by the carrying capacity of the planet for our species and its enterprises. Important elements of a sustainable economy will include controls over the population sizes of humans and their mutualistic species (such as cattle and other domestic animals), as well as over our per-capita and total-population resource consumption. In part, the resolution of resource dilemmas will require a more equitable sharing of wealth among people living in today's poorer and richer countries. This would moderate the importance of poverty as a key factor causing environmental degradation.

Environment and Society

All levels of society have a responsibility to protect the quality of the common environment. These obligations are a central aspect of the social contract by which enlightened communities operate.

The role of government is an overarching one, because it is empowered to regulate the activities of itself, of the private sector, of nongovernmental organizations, and of individuals. Of course, many activities of government and the private sector carry risks of causing environmental degradation, and there is always a corporate obligation to avoid or mitigate damages as much as possible.

The role of nongovernmental environmental organizations is to lobby government and industry about issues, to raise the awareness of the public, and increasingly (because of fiscally driven government cutbacks) to raise funds that can be used to prevent and repair environmental damages. Finally, all individual citizens have an obligation to live their lives in an environmentally responsible manner.

In the following sections, we will briefly examine the roles and activities of key environmental organizations. Additional information about these and other environmental agencies and organizations is found in Appendix II of this book.

International Organizations

The United Nations Environment Programme (UNEP) is the principal international organization dealing with environmental matters. UNEP is responsible for coordinating global environmental efforts with other agencies of the United Nations, national governments, and nongovernmental organizations. UNEP also coordinates the development of multinational treaties and other agreements, and periodically hosts global conferences on environmental themes.

The Nature Conservancy of Canada is a nongovernmental organization whose activities focus on acquiring land or land-use rights for the protection of natural values. This project involved a donation by Shell Canada of almost 9000 hectares of terrain around Mount Broadwood in southeastern British Columbia to develop a conservation area.

Source: L.J. Nuttall

Other agencies of the United Nations also have mandates that involve environmental issues. These include the Food and Agriculture Organization of the United Nations (FAO), the United Nations Development Programme (UNDP), the United Nations Educational, Scientific, and Cultural Organization (UNESCO), the United Nations Population Fund (UNPF), the World Health Organization (WHO), and the International Labour Organization (ILO).

A wide range of nongovernmental environmental organizations are also active on the international stage:

- Organizations involved in the international conservation of biodiversity include Conservation International, the Cousteau Society, the International Union for Conservation of Nature and Natural Resources, the Nature Conservancy (U.S.), the Smithsonian Institution, and the World Wildlife Fund.
- Organizations dealing with population issues include the Population Institute, the Population Reference Bureau, and Zero Population Growth.
- Organizations with more general mandates concerning resources and other environmental issues include the Earth Island Institute, the Environmental Policy Institute, Friends of the Earth, Greenpeace International, Resources for the Future, the Sierra Club, the World Resources Institute, and the Worldwatch Institute.

Canadian Organizations

At the federal level, Environment Canada plays a central role in preserving and enhancing environmental quality. The mandate of Environment Canada includes the protection of water, air, and soil quality; renewable resources; and indigenous biodiversity. Environment Canada's institutional objective is to foster a national capacity for sustainable development, in cooperation with international, provincial, territorial, and municipal governments; other departments of the federal government; the private sector; and nongovernmental organizations.

Environment Canada is divided into three major programs (or sections):

- The mandate of the Environmental Conservation Program involves ecosystem science, wildlife habitat, biodiversity, and related issues.

- The Environmental Protection Program deals with policy and regulatory matters pertaining to chemicals in the environment, atmospheric change issues, and federal preparedness for environmental emergencies.
- The Atmospheric Environment Service is responsible for atmospheric science, weather forecasting and modelling, and watershed management issues.

Other agencies of the federal government also have important environmental mandates:

- Natural Resources Canada deals with mineral and forest resources, including aspects of the environmental impacts of mining, the use of fossil fuels, and forestry.
- Health and Welfare Canada deals with environmental issues related to human health, and also has primary jurisdiction over pesticide registrations.
- Agriculture and Agri-Food Canada deal with issues involving agricultural practices, including sustainability and pesticide registrations.
- Fisheries and Oceans Canada has a mandate to promote understanding, conservation, and beneficial use of aquatic resources.
- The Canadian Coast Guard helps to protect the marine environment through the prevention of pollution.
- Parks Canada, an agency of Heritage Canada, protects and manages national parks.
- The Department of Indian and Northern Affairs is responsible for environment and resource issues in extensive regions of northern Canada.
- The Canadian Environmental Assessment Agency conducts environmental assessments of projects involving the federal government.

All of the provincial and territorial governments of Canada have agencies similar to those listed above, for dealing with environmental responsibilities under their jurisdictions. Finally, the National Round Table on the Environment and Economy is an integrated agency of federal, provincial, and territorial governments that examines important environmental issues and provides advice on ways to achieve sustainable development in Canada.

Canada also has a wealth of nongovernmental organizations that deal with environmental issues. National

Each of us can choose to adopt a lifestyle that is less intensive in terms of its environmental impacts. Being a "green" person involves many appropriate choices, such as commuting by bicycle instead of in an automobile. You might want to make a list of other environmentally appropriate choices of lifestyle.

Source: Dick Hemingway

organizations that focus on the conservation of indigenous biodiversity include the Canadian Nature Federation, the Canadian Parks and Wilderness Society, the Canadian Wildlife Federation, the Nature Conservancy of Canada, and the World Wildlife Fund of Canada. Organizations dealing with more general mandates concerning resources and other environmental issues include the Canadian Arctic Resources Committee, Canadian Ecology Advocates, Ducks Unlimited Canada, Energy Probe Research Foundation, Friends of the Earth, Greenpeace Canada, Pollution Probe, the Royal Society of Canada, the Sierra Club (Canada), the Tree Canada Foundation, and Wildlife Habitat Canada. In addition, all of the provinces and territories have nongovernmental organizations that deal with environmental issues on a more regional basis. Many of these organizations are listed in Appendix II.

Environmental Citizenship

Although each of us individually has a relatively small effect on the environment, as a larger society our aggregate influence is enormous. If all Canadians were to pursue lifestyles that have softer effects on the environment, there would be great benefits for all of us, for future generations, and for other species. **Individual actions have relatively little effect, but many people working in concert can make a difference.**

Environmental citizenship involves actions taken by individuals and families to lessen their impact on the environment. Individual acts of environmentalism involve making lifestyle choices that include having a small family, using less energy and material resources, and causing fewer stresses to indigenous biodiversity. In addition to the many "green" actions that people can undertake, they can also give moral and financial support to organizations that deal with environmental issues at international, national, and regional levels.

Libraries and book stores stock many so-called "green" handbooks and pamphlets. These list hundreds of specific actions that individuals and families can take to lessen their effects on the environment. (Several of these books are listed in the reference section of this chapter.) The diverse possibilities include shutting off lights when leaving a room, turning thermostats down to 15° or less during the winter (while wearing warm slippers and a sweater!), avoiding wasteful travel habits (such as commuting alone in a car), avoiding the use of pesticides in lawn care, planting trees (native species being strongly preferred) to store carbon on one's property and to provide wildlife habitat, becoming a vegetarian, and giving money to environmental charities such as Greenpeace, the Nature Conservancy of Canada, or the World Wildlife Fund.

Of course, few individual Canadians will make all of the green choices possible. To do so would be to voluntarily adopt an austere lifestyle, and most people are unwilling to choose this. Instead, most will undertake some positive actions, perhaps including recycling of many household wastes, riding a bicycle to school or work, not worrying about a weedy lawn, and supporting one or two environmental organizations. This would be "*selective environmentalism*," rather than a fully green lifestyle. However, if selective environmentalism is substantial enough, and is adopted by many people, there will be positive results. Each of us is responsible for demonstrating our environmental citizenship by making as many green choices as possible, and by encouraging relatives, friends, and acquaintances to do the same.

If the citizens of Canada and other countries do not make these sensible, environmentally astute choices, the results will eventually be tragic.

Prospects for Canada and for Spaceship Earth

It is crucial that people understand how human activities cause environmental degradations, both directly and indirectly. We must also design procedures to prevent or effectively mitigate those damages. However, over the long term, human societies can only prosper if they institute sensible limitations on their populations and ensure that the harvesting and management of ecological resources are truly sustainable.

The coupling of population control with sensible strategies of environmental and ecological management will be decisive in attaining a sustainable prosperity for humans, while accommodating other species and their natural communities, on the only planet in the universe known to sustain life.

Try not to see things as they are,
but rather as how they should be.

A principle of Buddhist thought

Key Words

environmental impact assessment
valued ecosystem components (or VECs)
mitigation
environmental monitoring
environmental indicators
environmental quality
ecosystem health (or ecological integrity)
environmental literacy
environmental reporting
sustainable development
ecologically sustainable development
environmental citizenship

Questions for Discussion

1. There has not yet been a full assessment of the environmental impacts of a broad government policy, such as entering into free-trade agreements with other countries. Does this mean that important environmental considerations are not being adequately addressed when Canada's trade policies are developed?
2. Imagine that a large industrial development is being proposed for the area where you live, for example, a power plant, a sewage-treatment plant, an incinerator, a pulp mill, or a mine. Make a list of the important environmental considerations that you think should figure in an environmental impact assessment of the proposed development.
3. Consider a landscape that is being managed for the harvesting of wood for a pulp mill. What economic and ecological values would have to be accommodated by an ecological sustainable system of land use in that area?
4. Make a list of actions that you and your family could take in order to become less damaging in your environmental impact. For each action, consider the environmental benefits that would result, as well as the implications for your lifestyle.
5. Considering all that you know about environmental science, do you believe that there is a crisis in the region where you live, or in Canada, or on Earth? If you do believe that there is an environmental crisis, what are the core elements of a social strategy that would alleviate the damages?

References

Beanlands, G.E. and P.N. Duinker. 1983. *An Ecological Framework for Environmental Impact Assessment in Canada.* Halifax: Dalhousie University, Institute for Resource and Environmental Studies.

Canadian Environmental Assessment Research Council/National Research Council (CEARC/NRC). 1986. *Cumulative Environmental Effects.* Ottawa: CEARC/NRC. Board on Basic Ecology.

Ehrlich, P.R. 1989. Facing the habitat crisis. *BioScience*, **39**: 480–482.

Erickson, P.A. 1994. *A Practical Guide to Environmental Impact Assessment.* San Diego, CA: Academic.

Freedman, B. 1995. *Environmental Ecology, 2nd ed.* San Diego, CA: Academic.

Freedman, B., C. Staicer, and N. Shackell. 1993. *Recommendations for a National Ecological-Monitoring Program.* Ottawa: Environment Canada, State of the Environment Reporting Organization. Occ. Pap. Ser. No. 2.

Government of Canada. 1991. *The State of Canada's Environment.* Ottawa: Government of Canada.

Government of Canada. 1997. *The State of Canada's Environment.* Ottawa: Environment Canada, State of the Environment Reporting Organization.

Being a Green Canadian

Environment Canada. 1990. *What We Can Do for Our Environment? Hundreds of Things to Do Now.* Ottawa: Environment Canada.

Environment Canada. 1992. *Working Our Way to a Green Office: A Guide to Creating an Environmentally Friendly Office.* Ottawa: Environment Canada.

National Round Table on the Environment and Economy. 1991. *The National Waste Reduction Handbook.* Ottawa: National Round Table on the Environment and Economy.

National Round Table on the Environment and Economy. 1992. *Green Guide: A User's Guide to Sustainable Development for Canadian Colleges.* Ottawa: National Round Table on the Environment and Economy.

Pollution Probe Foundation. 1989. *The Canadian Green Consumer Guide.* Toronto: McClelland & Stewart.

CBC Canadian Case 8

Waste Not, Want Not

Many discarded materials that we commonly regard as being "wastes" or "garbage" are actually valuable resources. These can be separated from other "waste" materials and then utilized to manufacture valuable products — a process that we refer to as recycling. Examples of recyclable commodities include newspaper, metal cans, glass bottles, and many kinds of plastics. In addition, many other disused items, such as consumer electronics, furniture, and clothing, can often be repaired and reused by other people.

Many kinds of recyclable and reusable materials can be systematically collected from industrial and urban refuse and used to manufacture valuable products. In the environmental sense, these are extremely positive actions. This is because they result in great savings of energy and materials, while freeing valuable space in solid-waste disposal areas, and employing many people in the process.

The environmental benefits of reducing the use of material and energy resources in the economy can be increased by ensuring that as many people, companies, and institutions as possible are participating in recycling and reusing activities. This is something that municipalities and other levels of government can encourage in various ways: for example, by charging for garbage collection by the kilogram or the bag, or by refusing to collect some kinds of disposed materials (such as grass clippings or autumn leaves, which can be recycled by composting). Moreover, recycling, reusing, and other "green" actions can be undertaken by each of us, as efficiently as possible, in order to reduce our individual environmental impacts.

Questions for Discussion

1. List some of the ways in which you and your family can lessen your environmental impact by actions such as recycling and reusing.
2. Can companies, governments, and institutions such as schools and hospitals also lessen their impacts by recycling and reusing? List some actions that they can undertake to reduce their environmental impacts.
3. As individuals, we can choose to reduce our own environmental "footprints" by reducing our use of energy and materials, and by recycling and reusing as efficiently as possible. However, if few people do these things, do our individual actions make a significant difference at the societal level? How much freedom do you think people should have to cause environmental damage? Should destructive activities be closely regulated, or even prohibited? Consider these questions carefully: regulation represents a loss of personal freedom, even if a larger social benefit may result.

Video Resource: "Harvesting the City." *The National Magazine* (March 12, 1996).

APPENDIX I

UNITS OF MEASUREMENT AND CONVERSIONS

Standard Units

Most of the world's scientists use the *International System of Units* (or SI). The SI system is based on six primary units:

QUANTITY	NAME OF UNIT	SYMBOL
length	metre	m
mass	kilogram	kg
amount of substance	mole	mol
time	second	s
electric current	ampere	A
temperature	degree centigrade	C
luminous intensity	candela	cd
concentration of ions	equivalent	eq

All other SI units are derived from the six primary ones, and are expressed in various forms of those units. For example, the SI unit for volume is m^3, derived from the metre. Some common secondary SI units are:

QUANTITY	NAME OF UNIT	SYMBOL
area	square metre	m^2
volume	cubic metre	m^3
velocity	metre per second	m/s (or m s^{-1})
density	kilogram per cubic metre	kg/m^3 (kg m^{-3})
acceleration	metre per second squared	m/s^2 (m s^{-2})
frequency	hertz	Hz (1 Hz = 1 s^{-1})
force	newton	N (1 N = 1 kg m s^{-2})
pressure and stress	pascal	Pa (1 Pa = 1 N m^{-2})
energy or work	joule	J (1 J = 1 N m)
power	watt	W (1 W = 1 J s^{-1})
electric potential	volt	V (1 V = 1 W A^{-1})

Some multiples of SI units have been given special names, including:

QUANTITY	NAME OF UNIT	SYMBOL	DEFINITION
area	hectare	ha	1 ha = 10^4 m^2
volume	litre	L	1 L = 10^{-3} m^3
	millilitre	mL	1 mL = 1 cm^3 = 10^{-6} m^3
mass	tonne	t	1 t = 10^3 kg
amount of substance	mole	mol	6.02×10^{23} units of atoms, molecules, ions, electrons, photons, or other elementary entities
pressure	bar	bar	1 bar = 10^5 N m^{-2}
	millibar	mbar	1 mbar = 10^2 N m^{-2}
energy, work	kilowatt hour	kWh	1 kW h = 3.6×10^6 J
concentration of ions	equivalent	eg	molar concentration × number of charges on ion

The SI system specifies prefixes to denote multiples of units, in powers of ten. The most common of these prefixes are:

PREFIX	SYMBOL	EXPONENT	MULTIPLE	NUMBER
peta-	P	10^{15}	1 000 000 000 000 000	
tera-	T	10^{12}	1 000 000 000 000	one trillion
giga-	G	10^{9}	1 000 000 000	one billion
mega-	M	10^{6}	1 000 000	one million
kilo-	k	10^{3}	1 000	one thousand
hecto-	h	10^{2}	100	one hundred
deka-	da	10^{1}	10	ten
deci-	d	10^{-1}	0.1	one tenth
centi-	c	10^{-2}	0.01	one hundredth
milli-	m	10^{-3}	0.001	one thousandth
micro-	μ	10^{-6}	0.000 001	one millionth
nano-	n	10^{-9}	0.000 000 001	one billionth
pico-	p	10^{-12}	0.000 000 000 001	one trillionth

Powers of Ten

Extremely large or small numbers can be written in a compact way that does not require large numbers of zeroes. The basic rules for the "powers of ten" notation are:

1. Any figure can be represented as a number between 1 and 10 followed by ten raised to a power (or exponent).
2. If the power of 10 is positive, it means "move the decimal point this many places to the right."
3. If the power of 10 is negative, it means "move the decimal point this many places to the left."

examples: $4.26 \times 10^5 = 426\,000$
$8.82 \times 10^{-4} = 0.000\,882$

Conversions

Other types of units can be converted into SI units, and vice versa. The most common conversions are:

Length

1 inch = 25.4 mm = 2.54 cm

1 foot = 305 mm = 30.5 cm = 0.305 m

1 yard = 914 mm = 91.4 cm = 0.914 m

1 mile = 1.609 km

1 fathom = 1.83 m

1 nautical mile = 1.85 km

1 m = 39.37 in = 3.28 ft

1 km = 0.621 mile

Area

1 square inch = 645 mm^2 = 6.45 cm^2

1 square foot = 0.0929 m^2

1 square yard = 0.836 m^2

1 acre = 4050 m^2 = 0.405 ha

1 square mile = 2.59 km^2

1 m^2 = 10.8 ft^2 = 2.47×10^{-4} acre

1 ha = 2.47 acres

Volume

1 cubic inch = 16.4 cm^3

1 cubic foot = 0.0283 m^3

1 cubic yard = 0.765 m^3

1 m^3 = 35.3 ft^3 = 1.31 yd^3

Capacity

1 fluid ounce (Imp.) = 28.4 cm^3

1 fluid ounce (U.S.) = 29.5 cm^3

1 teaspoon = 4.93 cm^3

1 tablespoon = 14.8 cm^3

1 cup = 237 cm^3

1 pint (Imp.) = 568 cm^3

1 pint (U.S.) = 473 cm^3

1 quart (Imp.) = 1136 cm^3

1 quart (U.S.) = 946 cm^3

1 gallon (Imp.) = 4.55 L

1 gallon (U.S.) = 3.79 L

1 bushel (Imp.) = 0.0364 m^3

1 bushel (U.S.) = 0.0352 m^3

1 L = 0.220 imperial gallon = 0.264 U.S. gal

Mass

1 grain = 64.8 mg

1 ounce = 28.35 g

1 pound = 453.6 g = 0.454 kg

1 U.S. ton (short) = 907 kg

1 imperial ton (long) = 1002 kg

1 kg = 35.27 oz = 2.2 lbs = 1.102×10^{-3} ton

1 tonne = 10^3 kg = 0.984 imperial ton

Velocity

1 in/sec = 25.4 mm/s

1 foot/sec = 0.305 m/s

1 mile/hr = 0.447 m/s

1 knot = 0.515 m/s

1 m/s = 3.6 km/h = 3.28 ft/s = 199 ft/min = 2.24 mi/h

Pressure

1 lb/sq in = 6880 N m^{-2}

1 lb/sq ft = 47.8 N m^{-2}

1 inch of mercury (STP) = 3380 N m^{-2}

1 atm = 101325 kg/m·s^2 (Pa) = 1.01 bar = 14.7 psi = 760 mm Hg

Work

1 foot-pound = 1.36 J

Power

1 foot-pound/sec = 1.36 W

1 horsepower = 746 W

Energy

1 British thermal unit (Btu) = 1.060 kJ

1 calorie = 4.187 joule

Temperature

1 Fahrenheit degree (°) = 0.555 K

degrees Fahrenheit = 9/5 (degrees C) + 32

degrees Centigrade (or Celsius) = 5/9 (degrees F – 32)

degrees Kelvin = (degrees Celsius + 273.15)

APPENDIX II

ENVIRONMENTAL ORGANIZATIONS

This appendix contains an annotated list of organizations that may be contacted for information related to environmental issues. Organizations are listed for Canada, the Canadian provinces or regions, and international.

Canadian National Organizations

Agriculture and Agri-Food Canada, Environment Bureau, 930 Carling Avenue, Sir John Carling Bldg., Ottawa, ON, K1A 0C5. Tel: (613) 943-1611; Fax: (613) 943-1612

Purpose: To promote environmental sustainability in the Canadian agri-food industry. Also involved in public education.

Atmospheric Environment Service, 4905 Dufferin Street, Downsview, ON, M3H 5T4. Tel: (416) 739-5702; Fax: (416) 739-4212

Purpose: Concerned with weather forecasting, meteorological research, policy development, and climate and air quality monitoring and research.

Canadian Animal Rights Network, P.O. Box 687, Station Q, Toronto, ON, M4T 2N5. Tel: (416) 223-4141; Fax: (416) 730-8550

Purpose: To defend the rights of animals in research, product testing, agriculture, and hunting.

Canadian Arctic Resources Committee, 1 Nicholas Street, Suite 412, Ottawa, ON, K1N 7B7. Tel: (613) 236-7379; Fax: (613) 232-4665

Purpose: To advocate the sustainable development of arctic resources and communities.

Canadian Coast Guard, Environmental Response, Canada Building, Room 941D, 344 Slater St., Ottawa, ON, K1A 0N7. Tel: (613) 990-3400; Fax: (613) 995-4700

Purpose: To protect the marine environment through pollution prevention and countermeasures.

Canadian Ecology Advocates, 20, rue de l'Église, St-Jean, QC, G0A 3W0. Tel: (418) 829-1145; Fax: (418) 829-1276

Purpose: To advance the development of an environmentally sustainable Canadian society, where all citizens share responsibility for maintaining the diversity and integrity of the Canadian and global commons.

Canadian Environmental Assessment Agency, 200, boul. Sacré-Coeur, 14e étage, Hull, QC, K1A 0H3. Tel: (819) 997-1000; Fax: (819) 994-1469

Purpose: To conduct environmental assessments of projects involving the federal government. Facilitates public access to information concerning the project.

Canadian Environment Industry Association, 204-6 Antares Dr., Nepean, ON, K2E 8A9. Tel: (613) 723-3525; Fax: (613) 723-0060.
E-Mail: CEIAEA@capitalnet.com

Purpose: To foster environmentally sustainable economic development through the provision of state-of-the-art products, technologies and services. The CEIA comprises Canadian companies, associations, and organizations dedicated to environmentally sustainable economic development.

Canadian Environmental Law Association, 517 College Street, Suite 401, Toronto, ON, M6G 4A2. Tel: (416) 960-2284; Fax: (416) 960-9392.
E-mail: cela@web.apc.org

Purpose: To use existing laws to protect the environment and to advocate reforms of environmental law and policy.

Canadian Federation of Humane Societies, 30 Concourse Gate, Suite 102, Nepean, ON, K2E 7V7. Tel: (613) 224-8072; Fax: (613) 723-0252

Purpose: To promote a humane ethic toward animals by promoting programs, policies and laws to ensure the well-being of animals.

Canadian Nature Federation, 1 Nicholas Street, Suite 520, Ottawa, ON, K1N 7B7. Tel: (613) 562-3447; Fax: (613) 562-3371

Purpose: To promote the understanding, awareness, and enjoyment of nature and to conserve the environment so that the integrity of natural ecosystems is maintained.

Canadian Parks and Wilderness Society, 160 Bloor St. E., Suite 1335, Toronto, ON, M4W 1B9. Tel: (416) 972-0868; Fax: (416) 972-0760

Purpose: To ensure the preservation and protection of parks and wilderness areas across Canada and to promote the development of clear and effective park-management policies.

Canadian Wildlife Federation, 2740 Queensview Drive, Ottawa, ON, K2B 1A2. Tel: (1-800) 563-WILD or (613) 721-2286; Fax: (613) 721-2902
E-mail: info@cwf-fcf.org

Purpose: To educate people about wildlife and wildlife conservation, to work with governments to ensure sound environmental laws, and to finance research and recovery efforts for threatened species.

Department of Energy, Mines, and Resources, 580 Booth St., Ottawa, ON, K1A 0E4. Tel: (613) 995-0947

Purpose: Science and policy in geology and mineral resources.

Ducks Unlimited Canada, 1750 Courtwood, Suite 109, Ottawa, ON, K2C 2B5. Tel: (613) 228-0206; Fax: (613) 228-0208

Purpose: To enhance the North American waterfowl population by reclaiming, preserving, and creating quality habitats throughout Canada.

Energy Probe Research Foundation, 225 Brunswick Avenue, Toronto, ON, M5S 2M6. Tel: (416) 964-9223; Fax: (416) 964-8239

Purpose: To educate Canadians on the benefits of conservation and renewable energy and to provide business, government and the public with information on energy and related issues.

Environment Canada, Hull, QC, K1A 0H3. Tel: (819) 997-2800; Fax: (819) 953-2225

Purpose: Responsible for preservation and enhancement of the quality of the natural environment, including water, air, and soil quality; renewable resources; and nondomestic flora and fauna. Environment Canada's objective is to foster a national capacity for sustainable development in cooperation with other governments, other departments of government, the private sector, and nongovernmental organizations.

Environmental Conservation Service, Place Vincent-Massey, 351, boul St-Joseph, Hull, QC, K1A 0H3. Tel: (819) 997-2161

Purpose: To administer federal/provincial agreements and treaties regarding water resources, to conduct research on the quality and health of aquatic ecosystems, and to promote water conservation and public awareness that water is a sustainable resource.

Environmental Youth Alliance, P.O.Box 34097, Station D, Vancouver, BC, V6J 4M1. Tel: (604) 737-2258

Purpose: To educate youth to inform the community about the environment and to actively seek solutions to environmental problems and initiate action toward these solutions.

Fisheries and Oceans Canada, 200 Kent Street, Ottawa, ON, K1A 0E6. Tel: (613) 993-0600; Fax: (613) 996-9055

Purpose: To promote understanding, conservation, and beneficial use of Canadian aquatic resources, and to manage and protect these aquatic resources.

Forestry Canada, Ottawa, ON, K1A 0C5. Tel: 819-997-1107

Purpose: To promote sustainable use of Canadian forest resources.

Friends of the Earth, 251 Laurier Avenue W., Suite 701, Ottawa, ON, K1P 5J6. Tel: (613) 230-3352; Fax: (613) 232-4354. E-mail: foe@web.apc.org

Purpose: To work with others to inspire the renewal of our communities and the Earth through research, education, and advocacy.

Greenpeace Canada, 185 Spadina Ave., 6th Floor, Toronto, ON, M5T 2C6. Tel: (416) 597-8408; Fax: (416) 597-8422.
E-mail: greenpeace.toronto@green2.greenpeace.org

Purpose: Greenpeace uses nonviolent, creative confrontation to expose global environmental problems and to force solutions that are essential to a green and peaceful future. Greenpeace's goal is to ensure the ability of the Earth to nurture life in all its diversity.

Health Canada, Brooke Claxton Bldg., Tunney's Pasture, 120 Parkdale Ave., Ottawa, ON, K1A 0K9. Tel: (613) 957-2991; Fax: (613) 952-7266

Purpose: To regulate environmental contaminants and monitor environmental health issues while assessing their risk to Canadians, and to foster international, national, and regional action.

Indian and Northern Affairs Canada, Les Terrasses de la Chaudière, Ottawa, ON, K1A 0H4. Tel: (613) 997-9885

Purpose: Concerned with the environment and resources of northern Canada.

International Institute for Sustainable Development, 161 Portage Ave., 7th Floor, Winnipeg, MB, R3B 0Y4. Tel: (204) 958-7707; Fax: (204) 958-7710

Purpose: To foster the development of environmentally sustainable economies in Canada and internationally.

National Round Table on the Environment and Economy, #1500, 1 Nicholas St., Ottawa, ON, K1N 7B7. Tel: (613) 992-7189; Fax: (613) 992-7385

Purpose: To provide advice to governments on ways to achieve sustainable development in Canada.

Natural Resources Canada, 580 Booth St., Ottawa, ON, K1A 0E4. Tel: (613) 995-0947

Purpose: To promote the use of Canadian mineral resources. See also Department of Energy, Mines, and Resources and the Canadian Forestry Service, which are major divisions of Natural Resources Canada.

Nature Conservancy of Canada, 110 Eglinton Avenue W., 4th Floor, Toronto, ON, M4R 2G5. Tel: (416) 932-3202; Fax: (416) 932-3208

Purpose: To preserve biological diversity through the purchase or other means of protection of ecologically significant natural areas.

Parks Canada, Jules Légér Bldg., 25 Eddy St., Hull, QC, K1A 1K5. Tel: (819) 997-9525; Fax: (819) 953-9745

Purpose: To protect, maintain, and manage the national parks, historic sites, and other aspects of natural Canadian heritage.

Pollution Probe, 12 Madison Avenue, Toronto, ON, M5R 2S1. Tel: (416) 926-1907

Purpose: Policy research and public education focusing on hazardous materials and pollution control.

Project Preservation, RR #4, Powassan, ON, P0H 1Z0. Tel: (705) 724-5541

Purpose: To educate the public to take an active and peaceful role in environmental protection and preservation on local, regional, provincial, national, and global scales.

Royal Society of Canada, 225 Metcalfe St., Suite 308, Ottawa, ON, K2P 1P9. Tel: (613) 991-5760; Fax: (613) 991-6996

Purpose: To promote learning and research in the arts, letters, and sciences in Canada.

Solar Energy Society of Canada, 72 Robertson Rd, #26029, Nepean, ON, K2H 9R6. Tel: (613) 596-1067; Fax: (613) 596-1120

Purpose: To promote energy conservation and renewable energy sources to the public.

Tree Canada Foundation, 220 Laurier Ave. W., Suite 1550, Ottawa, ON, K1P 5Z9. Tel: (613) 567-5545; Fax: (613) 567-5270. E-mail: tcf@treecanada.ca

Purpose: To foster the planting of trees in Canada, but for nonindustrial purposes.

Wildlife Habitat Canada, 7 Hinton Avenue, Suite 200, Ottawa, ON, K1Y 4P1.

Purpose: To conserve, restore, and enhance wildlife habitat throughout the Canadian landscape.

World Wildlife Fund Canada, 90 Eglinton Avenue E., Suite 504, Toronto, ON, M4P 2Z7. Tel: (1-800) 26-PANDA, (416) 489-8800; Fax: (416) 489-3611.
E-mail: comments@wwfcanada.org

Purpose: To conserve wildlife and wild places for their own sake and for the long term benefit of people, and to protect biological diversity in Canada and around the world.

Worldwide Home Environmentalists' Network, P.O. Box 91010, West Vancouver, BC, V7V 3N3. Tel: (604) 926-5079; Fax: (604) 926-5079

Purpose: To provide educational information about environmental issues, and to support those acting to reduce waste.

Canadian Provincial or Regional Organizations

(Note that many local environmental organizations also exist. They are not listed here.)

Alberta

Alberta Fish and Game Association, 6924 - 104 Street, Edmonton, AB, T6H 2L7. Tel: (403) 437-2342; Fax: (403) 438-6872

Purpose: To promote conservation and utilization of wildlife by protecting the natural habitat.

Alberta Foundation for Environmental Excellence, 12220 Stony Plain Road, Edmonton, AB, T5N 3Y4. Tel: (403) 482-9200; Fax: (403) 482-9100

Purpose: To develop and stimulate public awareness of environmental progress.

Alberta Junior Forest Warden Association, 9920 - 108 Street, Edmonton, AB, T5K 2M4. Tel: (403) 427-3551; Fax: (403) 427-0292

Purpose: To develop appreciation, respect, and responsible use of the natural environment.

Alberta Wilderness Association, P.O. Box 6398, Station D, Calgary, AB, T2P 2E1. Tel: (403) 283-2025; Fax: (403) 270-2743

Purpose: To promote completion of a system of protected wilderness and wild rivers in Alberta, to promote sound policies which value and protect wilderness, and to promote proper management of existing protected areas.

Alberta Wildlands Trust, 61 East Whitecroft, 52313 Regional Road 232, Sherwood Park, AB, T8B 1B7.

Purpose: To conserve and protect ecologically significant lands.

Alliance for Public Wildlife, 2428 Capitol Hill Crescent NW, Calgary, AB, T2M 4C2. Tel: (403) 289-5740; Fax: (403) 284-5928. E-mail: apw@web.apc.org

Purpose: To protect North American wildlife as a resource. Opposed to privatization and commercialization.

Canadian Parks and Wilderness Society — Calgary Chapter, 1019 Fourth Ave. SW, Calgary, AB, T2P 0K8. Tel: (403) 232-6686

Purpose: To ensure the preservation and protection of parks and wilderness areas in Alberta and across Canada, and to promote the development of clear effective park management policies.

Canadian Parks and Wilderness Society — Edmonton Chapter, 11759 Groat Road, Edmonton, AB, T5M 3K6. Tel: (403) 453-8658

Purpose: (as above).

Environmental Services Association of Alberta, #250, 10508 - 82 Avenue, Edmonton, AB, T6E 2A4. Tel: (403) 439-6363; Fax: (403) 439-4249.
E-mail: info@esaa.ccinet.ab.ca

Purpose: A not-for-profit business association dedicated to building a strong environment industry through leadership in technology, education, human resources, and market development.

British Columbia

All About Us Canada, RR 3, Yellow Point Road, Ladysmith, BC, V0R 2E0. Tel: (604) 722-3349; Fax: (604) 722-3349. E-mail: eger@web.apc.org

Purpose: To provide information to grassroots environmental groups through computer networking, and to provide environmental educational materials.

Canadian Parks and Wilderness Society — B.C. Chapter, P.O. Box 33918, Station D, Vancouver, BC, V6J 4L7. Tel: (604) 886-0525

Purpose: To ensure the preservation and protection of parks and wilderness areas in British Columbia and across Canada, and to promote the development of clear and effective park management policies.

Federation of B.C. Naturalists, 321 - 1367 West Broadway, Vancouver, BC, V6H 4A9. Tel: (604) 737-3057; Fax: (604) 738-7175

Purpose: To provide natural history education, to provide naturalists with a unified voice in conservation, and to coordinate activities of natural history societies in B.C.

Greenpeace — Vancouver, 1726 Commercial Drive, Vancouver, BC, V5N 4A3. Tel: (604) 253-7701; Fax: (604) 253-0114. E-mail: greenpeace.vancouver@green2.greenpeace.org

Purpose: To promote environmental awareness and broaden understanding about the environment through high-profile, nonviolent direct actions.

Nature Conservancy of Canada — B.C., 663 Radcliffe Lane, Victoria, BC, V8S 5B8. Tel: (604) 598-7503

Purpose: To protect natural areas, particularly those with rare or endangered species or habitats.

The Nature Trust of British Columbia, 100 Park Royal South, Suite 808, West Vancouver, BC, V7T 1A2. Tel: (604) 925-1128

Purpose: To encourage the private and government sectors to protect areas of ecological significance in British Columbia.

Wildlife Rescue Association of British Columbia, 5216 Glencairn Drive, Burnaby, BC, V5B 3C1. Tel: (604) 526-7275; Fax: (604) 524-2890

Purpose: To reduce and prevent suffering of injured and pollution-damaged wildlife through rehabilitation and education.

Manitoba

Canadian Parks and Wilderness Society — Manitoba Chapter, 414 Cabana Place, Winnipeg, MB, R2H 0K4. Tel: (204) 237-5947

Purpose: To ensure the preservation and protection of parks and wilderness areas across Canada, and to promote the development of clear and effective park management policies.

Manitoba Environmental Council, 139 Tuxedo Avenue, 3rd Floor, Building 30, Winnipeg, MB, R3N 0H6. Tel: (204) 945-7031; Fax: (204) 948-2274

Purpose: To provide advice on environmental matters, to promote environmental awareness, and to conduct studies, investigations, and reviews.

Manitoba Naturalists Society, 128 James Avenue, Suite 302, Winnipeg, MB, R3B 0N8. Tel: (204) 943-9029; Fax: (204) 949-9052

Purpose: To foster an awareness and work toward preservation of the natural environment.

Manitoba Wildlife Federation, 1770 Notre Dame Avenue, Winnipeg, MB, R3E 3K2. Tel: (204) 633-5967; Fax: (204) 632-5200

Purpose: To ensure the wise use of renewable resources in Manitoba.

Manitoba Wildlife Rehabilitation Organization, P.O. Box 242, 905 Croydon Avenue, Winnipeg, MB, R3M 3S7. Tel: (204) 897-1589

Purpose: To help injured wildlife and educate the public about what humans can do to prevent wildlife injuries.

New Brunswick

Conservation Council of New Brunswick, 108 St. John Street, Fredericton, NB, E3B 4A9. Tel: (506) 458-8747; Fax: (506) 458-1047. E-mail: ccnb@web.apc.org

Purpose: To speak on environmental issues to business, industry, and government, and to educate through consultation, demonstration projects, and material.

Fundy Wilderness Coalition, P.O. Box 1281, Sussex, NB, E0E 1P0. Tel: (503) 433-5566

Purpose: To develop a wilderness area paralleling the Fundy coast from St. Martins to Fundy National Park.

Nature Trust of New Brunswick, P.O, Box 603, Station A, Fredericton, NB, E3B 5A6. Tel: (506) 453-4583

Purpose: To protect environmentally significant areas in New Brunswick and to educate people about their importance.

New Brunswick Wildlife Federation, P.O. Box 20211, Fredericton, NB, E3B 7A2. Tel: (506) 457-7468

Purpose: To conserve and protect wilderness and wildlife.

Newfoundland

Newfoundland and Labrador Conservation Corps, 20 Crosbie Place, 3rd Floor, Beothuk Building, St. John's, NF, A1B 3Y8. Tel: (709) 738-0199

Purpose: To provide training and employment opportunities in environmental enhancement and conservation.

Newfoundland and Labrador Environmental Association Inc., 140 Water Street, Suite 603, St. John's, NF, A1C 6H6. Tel: (709) 227-7099; Fax: (709) 227-7048

Purpose: To promote all aspects of environmental protection, habitat protection and conservation.

Protected Areas Association of Newfoundland and Labrador, P.O. Box 1027, Station C, St. John's, NF, A1C 5M5. Tel: (709) 726-2603; Fax: (709) 726-2603

Purpose: To work toward a network of parks and reserves to protect biodiversity in Newfoundland and Labrador.

Northwest Territories

Northern Environmental Network — Northwest Territories, 4807, 49th Street, Yellowknife, NT, X1A 3T5. Tel: (403) 920-2473; Fax: (403) 873-2654.
E-mail: enorth@web.apc.org

Purpose: To foster environmental protection and sustainable development in the Northwest Territories.

Nova Scotia

The Clean Nova Scotia Foundation, 1675 Bedford Row, P.O. Box 2528 Central, Halifax, NS, B3J 3N5. Tel: (902) 420-3474; Fax: (902) 424-5334.
E-mail: cnsf@fox.nstn.ca

Purpose: To promote and develop stewardship of the environment by changing attitudes and behaviours, and to provide information and analyses to assist Nova Scotians in making decisions, strategies, and policies toward sustainability.

Ecology Action Centre, 2nd Floor, Wallace Building, 1553 Granville Street, Halifax, NS, B3J 1W7. Tel: (902) 429-2202. E-Mail: ip-eac@ccn.cs.dal.ca

Purpose: To foster environmental protection and sustainable development in Nova Scotia.

Federation of Nova Scotia Naturalists, c/o N.S. Museum of Natural History, 1747 Summer St., Halifax, NS, B3H 3A6. E-Mail: ip-fnsn@cfn.cs.dal.ca

Purpose: To support the common interests of naturalists' clubs, and to represent those clubs at the provincial level.

The Green Web, RR #3, Saltsprings, Pictou County, NS, B0K 1P0. Tel: (902) 925-2514.
E-mail: greenweb@web.acp.org

Purpose: To serve the needs of the environmental and green movements through information distribution and exchange on local, provincial, national, and international levels.

Nova Scotia Environmental Industry Association, P.O. Box 563, Dartmouth, NS, B2Y 3Y8.

Purpose: A province-wide business organization, to contribute to achieving sustainable development through skills development, promotion, industry standards, trade, and reaching consensus.

Nova Scotia Wildlife Federation, P.O. Box 698, Halifax, NS, B3J 2T3. Tel: (902) 423-6793; Fax: (902) 422-9520

Purpose: To foster sound management and wise use of the province's renewable resources.

Ontario

Conservation Council of Ontario, Suite 202, 74 Victoria Street, Toronto, ON, M5C 2A5.

Purpose: To coordinate provincial conservation associations.

Federation of Ontario Naturalists, 355 Lesmill Road, Don Mills, ON, M3B 2W8.

Purpose: To support the common interests of naturalists' clubs, and to represent those clubs at the provincial level.

Friends of the Forest, P.O. Box 2093, Station P, Thunder Bay, ON, P7B 5E7. Tel: (807) 344-2356; Fax: (807) 343-8023. E-mail: luken@web.apc.org

Purpose: To educate the public and policy makers about long-term ecological, social, and economic effects of forestry practices.

Greenpeace — Ottawa, 126 York, #310, Ottawa, ON, K1N 5T5. Tel: (613) 562-1005; Fax: (613) 562-1006.
E-mail: greenpeace.ottawa@green2.greenpeace.org

Purpose: To promote environmental awareness and broaden understanding about the environment through high-profile, nonviolent direct actions.

Nuclear Awareness Project, P.O. Box 2331, Oshawa, ON, L1H 7V4. Tel: (905) 725-1565; Fax: (905) 725-1565.
E-mail: caware@web.apc.org

Purpose: To address issues related to nuclear power, radioactive waste, and commercial uses of radioactive materials, and to conduct research and operate a public resource centre.

Ontario Parks Association, 1220 Sheppard Avenue E., Suite 401, North York, ON, M2K 2X1. Tel: (416) 495-3440; Fax: (416) 495-3440

Purpose: To protect and conserve parks, green spaces, and the environment.

Ontario Public Interest Research Group, 455 Spadina Avenue, Suite 306, Toronto, ON, M5S 2G8. Tel: (416) 598-1576

Purpose: To provide networking and training among OPIRG Chapters, and to provide students with a socially meaningful outlet for research skills and a hands-on lesson in democracy.

Ontario Society for Environmental Education, 191 Oakhill Place, Ancaster, ON, L9G 1C8. Tel: (905) 648-1667

Purpose: To promote environmental education in Ontario.

Prince Edward Island

Environmental Coalition of Prince Edward Island, c/o Voluntary Resource Centre, 81 Prince Street, Charlottetown, PE, C1A 4R3. Tel: (902) 556-4696; Fax: (902) 368-7180

Purpose: To encourage cooperation among groups sharing environmental concerns and to promote understanding and improvement of the environment.

Island Nature Trust, P.O. Box 265, Charlottetown, PE, C1A 7K4. (902) 892-7513

Purpose: To encourage the preservation of natural areas on PEI.

Prince Edward Island Wildlife Federation, P.O. Box 753, Charlottetown, PE, C1A 7L3. Tel: (902) 566-0474

Purpose: To preserve all species of wildlife for the utilization and enjoyment of all peoples, to ensure fish and game laws are observed, and to maintain educational programs.

Quebec

Environnement jeunesse, 4545, av. Pierre-de-Coubertin, C.P. 1000, succursale M, Montréal, QC, H1V 3R2. Tel: (514) 252-3016

Purpose: To promote environmental conservation and rehabilitation, and to develop an awareness among youths on the importance of having a healthy environment.

Greenpeace — Quebec, 2444 Notre Dame Ouest, Montréal, QC, H3J 1N5. Tel: (514) 933-0021; Fax: (514) 933-1017.
E-mail: greenpeace.montreal@green2.greenpeace.org

Purpose: To promote environmental awareness and broaden understanding about the environment through high-profile, nonviolent direct actions.

Groupe de récherche et d'education sur le milieu marin, C.P. 223, 108 rue de la Cale Sêche, Tadoussac, QC, G0T 2A0. Tel: (418) 235-4646

Purpose: To educate about marine environmental protection, to research sea mammals, and to protect the St. Lawrence whales and their environment.

Groupe de récherche et d'interêt public du Québec — McGill, 3620, boul. Université, #505, Université McGill, Montréal, QC, H3A 2B2. Tel: (514) 398-7432; Fax: (514) 398-8976

Purpose: To develop students' skills in the social and environmental field.

Les amis de la vallée du Sainte-Laurent, 7734, route Marie-Victorin, Lotbinière, QC, G0S 1S0.

Purpose: To conserve natural areas in the St. Lawrence lowlands.

Regroupement ecologiste Val D'Or et environs, C.P. 605, Val d'Or, QC, J9P 4P6. Tel: (819) 825-6776.
E-mail: reve@wer.apc.org

Purpose: To develop awareness and educate on environmental issues, and to initiate and facilitate public discussions on issues affecting the environment.

Regroupment pour le jardinage amateur et ecologique, C.P. 1155, Granby, QC, J2G 9G6. Tel: (514) 372-9962

Purpose: To bring together people interested in organic gardening.

Societé pour la protection de l'environnement du college de Rosemont, 6400 - 16 ave., Montréal, QC, H1X 2S9. Tel: (514) 376-1620; Fax: (514) 376-3211

Purpose: To recycle, sort, process, and sell institutional waste material, and to conduct market studies on recycled material.

Saskatchewan

Canadian Parks and Wilderness Society — Saskatchewan Chapter, P.O. Box 914, Saskatoon, SK, S7K 3M4. Fax: (306) 665-3312

Purpose: To ensure the preservation and protection of parks and wilderness areas across Canada.

Saskatchewan Environmental Society, P.O. Box 1372, Saskatoon, SK, S7K 3N9. Tel: (306) 665-1915; Fax: (306) 665-2128. E-mail: ann@web.apc.org

Purpose: To identify issues of environmental concern in Saskatchewan, to press for responsible action on those issues, and to sponsor educational programs.

Saskatchewan Forest Conservation Network, P.O. Box 112, Big River, Sask, S0J 0E0. Tel: (306) 469-4824

Purpose: To promote forest protection through public education and direct government lobby.

Saskatchewan Natural History Society, P.O. Box 4348, Regina, SK, S4P 3W6. Tel: (306) 780-9273; Fax: (306) 781-6021

Purpose: To protect and increase awareness of Saskatchewan's natural environment.

Saskatchewan Waste Reduction Council, 101 - 219 22nd Street East, Saskatoon, SK, S7K 0G4. Tel: (306) 931-3242; Fax: (306) 665-2128

Purpose: To help Saskatchewan attain the environmental, economic, and cultural benefits that come from reducing waste, and to achieve this through consultation, communication, education, and related support services.

Saskatchewan Wildlife Federation, Box 788, 444 River Street West, Moose Jaw, SK, S6H 6J6. Tel: (306) 693-9022

Purpose: To ensure the wise use of renewable resources in Saskatchewan.

Yukon

Friends of Yukon Rivers, 21 Klondike Road, Whitehorse, YT, Y1A 3L8. Tel: (403) 668-7370

Purpose: To preserve wildlife.

Northern Environmental Network — Yukon, P.O. Box 3932, Whitehorse, YT, Y1A 5L7. Tel: (403) 668-2482; Fax: (403) 668-6637. E-mail: net@web.apc.org

Purpose: To foster environmental protection and sustainable development.

Yukon Conservation Society, P.O. Box 4163, Whitehorse, YT, Y1A 3T3. Tel: (403) 668-5678.
E-mail: ycsrick@web.apc.org

Purpose: To maintain a lasting and stable base of operation upon which YCS can support and build the programs and services it wishes to provide in the Yukon.

International Organizations

Center for Science in the Public Interest, 1501 16th Street, NW, Washington, DC, USA, 20036. Tel: (202) 332-9110

Purpose: To research and educate in food, nutrition, and health.

Conservation International, 1015 18th Street NW, Suite 1000, Washington, DC, USA, 20036. Tel: (202) 429-5660

Purpose: To purchase land or trade foreign debt for land to set aside as nature preserves in developing countries, and to provide funds and technical support for a variety of projects related to animal and plant conservation.

Cousteau Society, 930 West 21st Street, Norfolk, VA, USA, 23517.

Purpose: To research and educate on the protection and preservation of the oceans.

Ducks Unlimited, #1 Waterfowl Way, Long Grove, IL, USA, 60047.

Purpose: To purchase, restore, and protect wetland habitat for breeding and migrating waterfowl.

Earth Island Institute, 300 Broadway, Suite 28, San Francisco, CA, USA, 94133.

Purpose: Clearinghouse for international information on environmental issues.

Environmental Action, Inc., 1525 New Hampshire Avenue, NW, Washington, DC, USA, 20036. Tel: (202) 745-4870

Purpose: To lobby on a wide range of environmental issues, including transportation, solid waste, water quality, solar energy, energy conservation, toxic substances, deposit legislation.

Environmental Defense Fund, 257 Park Avenue South, New York, NY, USA, 10010. Tel: (212) 505-2100

Purpose: To protect environmental quality and public health through litigation and administrative appeals.

Environmental Law Institute, 1616 P Street NW, Suite 200, Washington, DC, USA, 20036. Tel: (202) 328-5150

Purpose: To monitor environmental legislation and conduct research and lobbying for improvements.

Environmental Policy Institute, 218 D Street, SE, Washington, DC, USA, 20036. Tel: (202) 328-5150

Purpose: To lobby on all aspects of energy development, preservation, restoration, and rational use of Earth's resources.

Food and Agriculture Organization of the United Nations (FAO), Via delle Terme di Caracalla, Rome, Italy, 00100. Tel: 57971

Purpose: International programs of research and education to improve food production, nutrition, and health.

Friends of the Earth, Inc., 218 D Street SE, Washington, DC, USA, 20003.

Purpose: To conduct research on ozone protection, tropical deforestation, global warming, and oil spill legislation.

Greenpeace International, Stichting Greenpeace Council, Keizersgracht 176, 1016 DW Amsterdam, Netherlands. Tel: 31-20-5236222; Fax: 31-20-5236200. E-mail: greenpeace.international@green2.greenpeace.org

Purpose: To use nonviolent, creative confrontation to expose global environmental problems, and to force the solutions which are essential to a green and peaceful future. Greenpeace's goal is to ensure the ability of the Earth to nurture life in all its diversity.

International Alliance for Sustainable Agriculture, 1701 University Avenue SE, Minneapolis, MN, USA, 55414.

Purpose: To promote sustainable agriculture.

International Atomic Energy Agency (IAEA), Vienna International Center, Wagramerstrasse 5, PO Box 100, A-1400 Vienna, Austria. Tel: (43-1) 2060; Fax: (43-1) 20607. E-mail: IAEO@iaea1.iaea.or.at

Purpose: To aid interested countries implement peaceful nuclear technologies for such beneficial applications as electricity production, health care, agricultural development, and industry. The IAEA also monitors civil nuclear activities to ensure that safeguarded nuclear materials are not used for military purposes.

International Fund for Animal Welfare, P.O. Box 193, Yarmouth Port, MA, USA, 02675. Tel: 508-362-4944

Purpose: To protect wild and domestic animals.

International Union for Conservation of Nature and Natural Resources (IUCN), Avenue du Mont-Blanc CH-1196, Gland, Switzerland. Tel: 022-64-71-81

Purpose: To promote scientifically based action for conserving wild animals, plants, and other resources.

National Audubon Society, 950 Third Avenue, New York, NY, USA, 10022. Tel: (212) 546-9100

Purpose: To conduct research, lobbying, and education on wildlife, wilderness, public lands, endangered species, and water resource management.

Nature Conservancy, 1815 North Lynn Street, Arlington, VA, USA, 22209. Tel: (703) 841-5300

Purpose: To conduct research and education concerning the identification, protection, and management of natural areas.

New Directions for Policy, 1101 Vermont Avenue, Suite 400, NW, Washington, DC, USA, 20005. Tel: (202) 289-3907

Purpose: To lobby on international energy policy and self-reliant economic development.

The Population Institute, 110 Maryland Avenue, NE, Washington, DC, USA, 20002. Tel: (202) 544-3300

Purpose: To educate, research, and lobby on population control.

Population Reference Bureau, 777 14th Street NW, Suite 800, Washington, DC, USA, 20005. Tel: (202) 639-8040

Purpose: To gather, interpret, and publish information on social, economic, and environmental implications of world population dynamics.

Resources for the Future, 1616 P Street NW, Washington, DC, USA, 20036. Tel: (202) 328-5000

Purpose: To research and educate about nature resource conservation issues.

Sierra Club, 730 Polk Street, San Francisco, CA, USA, 94109. Tel: (415) 776-2211.
E-mail: information@sierraclub.org

Purpose: To conduct outings, educational programs, volunteer work projects, litigation, political action, and administrative appeals on a range of environmental issues.

Smithsonian Institution, 1000 Jefferson Drive SW, Washington, DC, USA, 20560.
E-mail: nzpcr01@sivm.si.edu

Purpose: To sponsor a wide variety of environmental research and educational programs.

Student Conservation Association, Inc., Box 550, Charlestown, NH, USA, 03603. Tel: (603) 826-5206

Purpose: To coordinate environmental internships and volunteer jobs with state and federal agencies and private organizations for students and adults.

United Nations Development Program (UNDP), 1 United Nations Plaza, New York, NY, USA, 10017. Tel: (212) 906-5000

Purpose: To promote human development by helping countries develop the capacity to manage their economy, fight poverty and disease, conserve the environment, and stimulate technological innovation. UNDP forges alliances with the people and governments of developing countries, with other agencies of the United Nations, and with nongovernmental organizations.

United Nations Educational, Scientific, and Cultural Organization (UNESCO), 7 place de Fontenoy, 75352 Paris, France, Tel: 1-45-68-10-00; Fax: 1-45-67-16-90

Purpose: To research in environmental science and management of natural resources, and to provide training in environmental science and related educational programs.

United Nations Environmental Programme (UNEP), United Nations, Rm. DC2-0803, New York, NY, USA, 10017. Tel: (212) 963-8138

Purpose: To coordinate global environmental efforts with United Nations agencies, governments, and non-governmental organizations.

United Nations Population Fund (UNPF), 220 East 42nd Street, New York, NY 10017, USA. Tel: (212) 297-5000; Fax: (212) 370-0201

Purpose: To provide information and education on family planning, and to form policy and plan development using population data.

World Bank, 1818 H Street, N.W., Washington, D.C., U.S.A., 20433. Tel: (202) 477-1234; Fax: (202) 477-6391

Purpose: To lend money to developing countries for development projects, including projects for the environment, population, and health, and to research and produce publications relevant to environmental issues.

World Conservation Union (formerly International Union for the Conservation of Nature, IUCN), Rue Mauverney 28, CH-1196 Gland, Switzerland. Tel: (4122) 999 00 01; Fax: (4122) 999 00 02

Purpose: To promote nature conservation by assisting societies throughout the world to ensure that natural resource utilization is ecologically sustainable.

World Environment Center, 419 Park Avenue S, New York, NY, USA, 10016.

Purpose: To liaise between government and industry leaders to resolve environmental problems caused by industry.

World Health Organization (WHO), Pan American Sanitary Bureau, Regional Office of the World Health Organization, Washington, DC, USA.

Purpose: To deal with priority health and environment issues, including air and water pollution, hazardous wastes, water supply and sanitation in human settlements, and global environmental issues.

World Resources Institute, 1709 New York Avenue NW, Washington, DC, USA, 20006.

Purpose: Policy research centre focusing on environment and development.

World Wildlife Fund, 1255 Twenty-Third Street, NW, Washington, DC, USA, 20037. Tel: (202) 293-4800

Purpose: To research and educate on endangered species.

Worldwatch Institute, 1776 Massachusetts Avenue, NW, Washington, DC, USA, 20036. Tel: (202) 452-1999

Purpose: To research and educate on major worldwide environmental problems and environmental law and policy.

Zero Population Growth, 1400 16th Street NW, Washington, DC, USA, 20036.

Purpose: Largest nonprofit membership group concerned with impact of growing human population on resources and environment.

References

Anonymous. 1994. *The Environment Encyclopedia and Directory.* London: Europa.

Asseltine, P. and A.M. Aldighieri. 1996. *Canadian Environmental Directory 1995/96.* Toronto: Canadian Almanac & Directory.

Eiserer, L.A.C. 1991. *World Environmental Directory. 6th ed.* Silver Spring, MD: Business Publishers.

Greenpeace International World Wide Web Homepage: http://www.greenpeace.org

Gugeler, J. and T. Remkes. 1994. *The Green List.* The Canadian Environmental Network.

Nebel, B.J. 1990. *Environmental Science — The Way The World Works.* 3rd edition. Engelwood Heights, NJ: Prentice Hall.

Raven, P.H., L.R. Berg, and G.B. Johnson. 1993. *Environment.* New York: Saunders.

Trzyna, T.C. and R. Childers. 1992. *World Directory of Environmental Organizations.* World Conservation Union, Gland, Switzerland.

APPENDIX III

JOURNALS, MAGAZINES, AND PERIODICALS ON THE ENVIRONMENT

Canadian

Arctic (Journal of the Arctic Institute of North America): The Arctic Institute of North America, University of Calgary, Calgary, AB, T2N 1N4.

Biosphere: Canadian Wildlife Federation, 2740 Queensview Dr., Ottawa, ON, K2B 3A2.

Borealis: Canadian Parks and Wildlife Society, PO Box 1359, Edmonton, AB, T5J 2N2.

Canadian Environmental Control Newsletter: CCH Canadian Ltd., 6 Garamond Court, Don Mills, ON, M3C 1Z5.

Canadian Environmental Law Report: Carswell Publications, Corporate Plaza, 2075 Kennedy Rd., Scarborough, ON, M1T 3V4.

Canadian Environmental Network Bulletin: P.O. Box 1289, Station B, Ottawa, ON, K1P 5R3.

Canadian Environmental Protection: Baum Publications Ltd., 1625 Ingleton Ave., Burnaby, BC, V5C 4L8.

The Canadian Field-Naturalist: Ottawa Field Naturalists Club, P.O. Box 35069, Westgate P.O., Ottawa, ON, K1Z 1A2.

Canadian Geographic: 39 McArthur Ave., Vanier, ON, K1L 8L7.

Canadian Journal of Botany: National Research Council of Canada, Ottawa, ON, K1A 0R6.

Canadian Journal of Earth Sciences: National Research Council of Canada, Ottawa, ON, K1A 0R6.

Canadian Journal of Forest Research: National Research Council of Canada, Ottawa, ON, K1A 0R6.

Canadian Journal of Zoology: National Research Council of Canada, Ottawa, ON, K1A 0R6.

Climate Change Digest: Canadian Program Office, Canadian Climate Centre, 4905 Dufferin Street, Downsview, ON, M3H 5T4.

Conserving Canada's Natural Legacy: The State of Canada's Environment 1996: Minister of Public Works and Government Services Canada, Ottawa, ON.

Earthkeeper: 11 Oriole Crescent, PO Box 1649, Guelph, ON, N1H 6R7.

Ecoalert: Conservation Council of New Brunswick, 180 St. John's St., Fredericton, NB, E3B 4A9.

Ecodecision: Environment and Policy Society, 276 St Jacques St. W, Suite 924, Montréal, QC, H2Y 1N3.

EcoLog Canadian Pollution Legislation: Southam Business Communications Inc., 1450 Don Mills Rd., Don Mills, ON, M3B 2X7.

EcoLog Week: Southam Business Communications Inc., 1450 Don Mills Rd., Don Mills, ON., M3B 2X7.

Environment: Faculty of Environmental Studies, University of Waterloo, Waterloo, ON., N2L 3G1.

Environmental Reviews: National Research Council of Canada, Ottawa, ON, K1A 0R6.

Environment Views: Alberta Department of the Environment, 9820 106th St., Edmonton, AB, T5K 2J6.

Environmental Approvals in Canada: Butterworths Canada Ltd., 75 Clegg Rd., Markham, ON, L6G 1A1.

Environmental Eye: Businesstek Publishing Inc., P.O. Box 250, Carleton Place, ON, K7C 3P4.

Environmental Science & Engineering: Davcom Communications Inc., 10 Petch Crescent, Aurora, ON, L4G 5N7.

Forêt Conservation: Revue Foret Conservation Inc., 145 rue St Jean, 4ème étage, Québec, QC, G1H 1N4.

Franc-Vert: Union Québecoise pour la Conservation de la Nature, 160-76 rue Est, 2ème étage, Charlesbourg, QC, G1H 7H6.

From Duck Country: Ducks Unlimited Canada, 1190 Waverly St., Winnipeg, MB, R3T 2E2.

Harrowsmith: Telemedia Publishing Inc., 7 Queen Victoria, Camden E, ON, K0K 1J0.

Hazardous Materials Management: 951 Denison St., Unit #4, Markham, ON, L3R 3W9.

International Wildlife: Canadian Wildlife Federation, 2740 Queensview Dr., Ottawa, ON, K2B 3A2.

Manitoba Naturalists Society Bulletin: 302-128 James Ave., Winnipeg, MB, R3B 0N8.

Milieu: Environment Canada, 3 Buade, CP 6060, Québec, QC, G1R 4V7.

Nature Canada: Canadian Nature Federation, 453 Sussex Drive, Ottawa, ON, K1N 6Z4.

New Catalyst: Catalyst Education Society, P.O. Box 189, Gabriola, BC, V0R 1X0.

New Environment, Bedford House Publishing Corporation, 33 Draper Street, 2nd Floor, Toronto, ON, M5V 2M3.

The New Internationalist, 35 Riviera Drive, Unit 17, Markham, ON, L3R 8N4.

Northern Perspectives, Canadian Arctic Resources Committee, 1 Nicholas St., Suite 412, Ottawa, ON, K1N 7B7.

Nova Scotia Conservation: PO Box 68, Truro, NS, B2N 5B8.

Ontario Conservation News: Conservation Council of Ontario, Suite 506, 489 College St., Toronto, ON, M6G 1A5.

Ontario Recycling Update: Conservation Council of Ontario, Suite 504, 489 College St., Toronto, ON, M6G 1A5.

Prairie Forum: Canadian Plains Research Centre, University of Regina, Regina, SK, S4S 0A2.

Rapport annuel du Conseil consultatif de l'environnement et de la conservation du Québec: Ministère des Communications, Direction Generale des Publications Gouvernmentales, 2ème étage, 1279 boul. Charest Ouest, Québec, QC, G1N 4K7.

Sierra Report: Sierra Club, Western Canada Chapter, 620 View Street, Suite 314, Victoria, BC, V8W 1J6.

Silviculture: Maclean Hunter Ltd, Suite 900, 1130 W Pender St., Vancouver, BC, V6E 4A4.

STOP Press: Society To Overcome Pollution Inc., 716 St-Ferdinand, Montréal, QC, H4C 2T2

International

Ambio - A Journal of the Human Environment: Royal Swedish Academy of Sciences, Box 50005, S-104 05 Stockholm, Sweden.

Annual Review of Ecology and Systematics: Annual Reviews Inc., 4139 El Camino Way, Palo Alto, CA, 94306, USA.

Biological Conservation: Elsevier Science Publishers Ltd., Crown House, Linton Rd., Barking, Essex, IG11 8JU, U.K.

BioScience, American Institute of Biological Sciences, 730 11th St NW, Washington, DC, 20001-4521, USA.

The Bulletin, British Ecological Society, 26 Blades Court, Deodar Road, Putney, London, SW15 2NU, U.K.

Bulletin of the Ecological Society of America: Arizona State University, Box 873211, Tempe, AZ, 85287-3211, USA.

Carbon Dioxide Information Analysis Center Communications: Oak Ridge National Laboratory, Building 2001, P.O. Box X, Oak Ridge, TN 37831-6050, USA.

ChemEcology: Chemical Manufacturers Association, 1300 Wilson Boulevard, Arlington, VA, 22209, USA.

Conservation Biology: Blackwell Scientific Publications, 3 Cambridge Center, Suite 208, Cambridge, MA, 02142, USA.

Ducks Unlimited: Ducks Unlimited Inc., 1 Waterfowl Way at Gilmer Rd., Long Grove, IL, 60047-9153, USA.

Earth Island Journal: Earth Island Institute, 300 Broadway, Suite 28, San Francisco, CA, 94133-3312, USA.

Ecology: Ecological Society of America, 2010 Massachusetts Avenue, NW, Suite 400, Washington, DC, 20036, USA.

Ecological Applications: Ecological Society of America, 2010 Massachusetts Avenue, NW, Suite 400, Washington, DC, 20036, USA.

Ecological Modelling: Elsevier Science Publishers B.V., POB 181, 1000 AD Amsterdam, The Netherlands.

Ecological Monographs: Ecological Society of America, 2010 Massachusetts Avenue, NW, Suite 400, Washington, DC, 20036, USA.

Ecology Abstracts: Cambridge Scientific Abstracts, 7200 Wisconsin Ave., Bethesda, MD, 20814-4823, USA.

Econews: The Green Party, 10 Station Parade, Balham High Road, London, SW12 9AZ, U.K.

Environment: Heldref Publications Inc., 4000 Albemerle St., NW, Washington, DC, 20016-1851, USA.

Environmental Protection: Stevens Publishing Corp., 225 N New Rd., Waco, TX, 76710, USA.

Forest Ecology and Management: Elsevier Science Publishers B.V., POB 181, 1000 AD, Amsterdam, The Netherlands.

Geo: Gruner und Jahr AG und Co., 2044 Hamburg, Am Baumwall 11, Germany.

International Journal of Biometeorology: International Society of Biometeorology, Secretariat, Department of Renewable Resources, McGill University, 21 111 Lakeshore Road, Ste Anne de Bellevue, QC, H9X 1C0, Canada.

The World Conservation Bulletin: International Union for the Conservation of Nature, Rue Mauverney 28, CH-1196 Gland, Switzerland.

The Journal of Ecology: British Ecological Society, 26 Blades Court, Deodar Road, Blackwell Scientific Publications, Putney, London, SW15 2NU, U.K.

Land and Water: Land and Water Inc., 900 Central Ave., Suite 21, Box 1197, Fort Dodge, IA, 50501, USA.

National Geographic: National Geographic Society, 1145 17th Street NW, Washington, DC 20036, USA.

National Wildlife: National Wildlife Federation, 1400 16th St., NW, Washington, DC, 20036-2266 USA.

Nature: Macmillan Magazine Ltd., 4 Little Essex St., London, WC2R 3LF, U.K.

Our Planet: United Nations Environment Programme, IWSS Ltd., PO Box 119, Stevenage, Hertfordshire, SG1 4TP, England.

Populi: United Nations Population Fund, 220 East 42nd St., New York, NY, 10017, USA.

Sierra: Sierra Club, 730 Polk St., San Francisco, CA, 94109-7897, USA.

Trends in Ecology and Evolution: Elsevier Trends Journals, 68 Hills Rd., Cambridge, CB2 1LA, U.K.

GLOSSARY OF KEY TERMS

accuracy: The degree to which a measurement or observation reflects the actual value. Compare with **precision**.

acid rain: The wet deposition only of acidifying substances from the atmosphere. See also **acidifying deposition**.

acid shock: An event of relatively acidic surface water that can occur in the springtime when the snowpack melts quickly but the ground is still frozen.

acid sulphate soil: Acidic soil conditions caused when certain wetlands are drained and sulphide compounds become oxidized.

acid-mine drainage: Acidic water and soil conditions that develop when sulphide minerals become exposed to the atmosphere, allowing them to be oxidized by *Thiobacillus* bacteria.

acid-neutralizing capacity: The quantitative ability of water to neutralize inputs of acid without becoming acidified. See also **buffering capacity**.

acidification: An increasing concentration of hydrogen ions (H^+) in soil or water.

acidifying deposition: Both the wet and dry deposition of acidifying substances from the atmosphere.

acute toxicity: Toxicity associated with short-term exposures to chemicals in concentrations high enough to cause biochemical or anatomical damages, even death. Compare with **chronic toxicity**.

aerobic: Refers to an environment in which oxygen (O_2) is readily available. Compare with **anaerobic**.

afforestation: Establishment of a forest where one did not recently occur, as when trees are planted on agricultural land.

age-class structure: The proportions of individuals in various age classes of a population.

agricultural site capability: See **site capability**.

agroecosystem: An ecosystem that is managed to cultivate products for use by humans.

algal bloom: An event of high phytoplankton biomass.

ammonification: Oxidation of the organically bound nitrogen of dead biomass into ammonium (NH_4^+).

anaerobic: Refers to an environment in which oxygen (O_2) is not readily available. Compare with **aerobic**.

angiosperm: Flowering plants that have their ovules enclosed within a specialized membrane and their seeds within a seedcoat. Compare with **gymnosperm**.

anthropocentric world view: This considers humans as being more worthy than other species and uniquely disconnected from nature. The importance and worth of everything is considered in terms of the implications for human welfare. Compare with **biocentric world view** and **ecocentric world view**.

anthropogenic: Occurring as a result of a human influence.

aquaculture: The intensive cultivation of aquatic resources, sometimes in semi-domestication.

aquifer: Groundwater resources in some defined area.

artificial selection: The deliberate breeding of species to enhance traits that are viewed as desirable by humans.

artificial wetland: An engineered wetland, usually constructed to treat sewage or other organic wastes.

aspect: The direction in which a slope faces.

atmosphere: The gaseous envelope surrounding the Earth, held in place by gravity.

atmospheric inversion (or **temperature inversion**): A relatively stable atmospheric condition in which cool air is trapped beneath a layer of warmer air.

atmospheric water: Water occurring in the atmosphere, in vapour, liquid, or solid forms.

autecology: The field within ecology that deals with the study of individuals and species. Compare with **synecology**.

autotroph: An organism that synthesizes its biochemical constituents using simple inorganic compounds and an external source of energy to drive the process. See also **primary producer**, **photoautotroph**, and **chemoautotroph**.

available concentration: The concentration of metals in an aqueous extract of soil, sediment, or rocks, simulating the amount available for organisms to take up from the environment. Compare with **total concentration**.

baby boom: A period of high fecundity during 1945–1965 that occurred because of social optimism after the Second World War.

binomial: Two latinized words that are used to name a species.

bioaccumulation (or **bioconcentration** or **biomagnification**): The occurrence of chemicals in much higher concentrations in organisms than in the ambient environment.

biocentric world view: This considers all species (and individuals) as having equal intrinsic value. Humans are not considered more important or worthy than any other species. Compare with **anthropocentric world view** and **ecocentric world view**.

bioconcentration: See **bioaccumulation**.

biodegradation: The breakdown of organic molecules into simpler compounds through the metabolic actions of microorganisms.

biodiversity: The richness of biological variation, including genetic variability as well as species and community richness.

biological control: Pest-control methods that depend on biological interactions, such as diseases, predators, or herbivores.

biological oxygen demand (BOD): The capacity of organic matter and other substances in water to consume oxygen during decomposition.

biomagnification: See **bioaccumulation**.

biomass energy: The chemical potential energy of plant biomass, which can be combusted to provide thermal energy.

biome: A geographically extensive ecosystem, occurring throughout the world wherever environmental conditions are suitable.

biosphere: All life on Earth, plus their ecosystems and environments.

birth control: Methods used to control fertility and childbirth.

BOD: See **biological oxygen demand**.

bog: An infertile, acidic, unproductive wetland that develops in cool but wet climates. Compare with **fen**.

boreal forest (or **taiga**): An extensive biome occurring in environments with cold winters, short but warm growing seasons, and moist soils, and usually dominated by coniferous trees.

broad-spectrum pesticide: A pesticide that is toxic to other organisms as well as the pest.

browse: Broad-leaved shrubs that are eaten by herbivores such as hares and deer.

bryophyte: Simple plants that do not have vascular tissues nor a cuticle on their foliage.

buffering capacity: The ability of a solution to resist changes in pH as acid or base is added.

calorie: A standard unit of energy, defined as the amount of energy needed to raise the temperature of one gram of pure water from 15°C to 16°C. Compare with **joule**.

carbon credits: Actions that help reduce the atmospheric concentration of CO_2, such as fossil-fuel conservation and planting trees.

carnivore (or **secondary consumer**): An animal that hunts and eats other animals.

carrying capacity: The abundance of a species that can be sustained without the habitat becoming degraded.

chaparral: A shrub-dominated ecosystem that occurs in south-temperate environments with winter rains and summer drought.

chemoautotroph: Microorganisms that harness some of the potential energy of certain inorganic chemicals (e.g., sulphides) to drive their fixation of energy through chemosynthesis. Compare with **photoautotroph**.

chemosynthesis: Autotrophic productivity that utilizes energy released during the oxidation of certain inorganic chemicals (such as sulphides) to drive biosynthesis. Compare with **photosynthesis**.

chromosomes: Subcellular units composed of DNA and containing the genetic information of eukaryotic organisms.

chronic toxicity: Toxicity associated with exposure to small or moderate concentrations of chemicals, sometimes over a long period of time. The damages may be biochemical or anatomical, and may include the development of a lethal disease, such as cancer. Compare with **acute toxicity**.

clear-cutting: The harvesting of all economically useful trees from an area at the same time.

climate: The prevailing, long-term, meteorological conditions of a place or region, including temperature, precipitation, wind speed, and other factors. Compare with **weather**.

climate change: Long-term changes in air, soil, or water temperature; precipitation regimes; wind speed; or other climate-related factors.

coal: An organic-rich solid mined from sedimentary geological formations.

coarse woody debris: Logs lying on the forest floor.

commensalism: A symbiosis in which one of the species benefits from the interaction, while the other is not affected in either a positive or negative way.

commercial extinction: Depletion of a natural resource to below the abundance at which it can be profitably harvested.

common-property resource: A resource shared by all of society, not owned by any particular person or interest.

community: In ecology, this refers to populations of various species that are co-occurring at the same time and place.

community-replacing disturbance: A disturbance that results in the catastrophic destruction of an original community, and its replacement by another one. Compare with **microdisturbance**.

compartment: A reservoir of mass in a nutrient or material cycle.

competition: A biological interaction occurring when the demand for an ecological resource exceeds its limited supply, causing organisms to interfere with each other.

competitor: A species that is dominant in a habitat in which disturbance is rare and environmental stresses are unimportant, so competition is the major influence on evolution and community organization.

conservation: Wise use of natural resources. Conservation of nonrenewable resources involves recycling and other means of efficient use. Conservation of renewable resources includes these means, in addition to ensuring that harvesting does not exceed the rate of regeneration of the stock.

contamination: The presence of potentially damaging chemicals in the environment, but at concentrations less than those required to cause toxicity or other ecological damages. Compare with **pollution**.

control (or **control treatment**): An experimental treatment that was not manipulated, and is intended for comparison with manipulated treatments.

conventional economics: Economics as it is commonly practised, which includes not accounting for costs associated with ecological damages and resource depletion. Compare with **ecological economics**.

conversion: See **ecological conversion**.

core: Earth's massive interior, made up of hot molten metals.

Coriolis effect: An influence of Earth's west-to-east rotation, which makes winds in the Northern Hemisphere deflect to the right and those in the Southern Hemisphere to the left.

creationists: People who reject the theory of evolution in favour of a literal interpretation of Genesis, the first book of the Old Testament of the Bible. See also **scientific creationists**.

critical load: A threshold for pollutant inputs, below which it is thought ecological damages will not be caused.

crude oil: See **petroleum**.

crust: The outermost layer of Earth's sphere, overlying the lithosphere and composed mostly of crystalline rocks.

cultural eutrophication: Eutrophication caused by anthropogenic nutrient inputs, usually through sewage dumping or fertilizer runoff. See also **eutrophication**.

cultural evolution: Adaptive evolutionary change in human society, characterized by increasing sophistication in the methods, tools, and social organizations used to exploit the environment and other species. Compare with **evolution**.

decay: The decomposition or oxidation of dead biomass, mostly through the actions of microorganisms.

decomposer: See **detritivore**.

deductive logic: Logic in which initial assumptions are made and conclusions are then drawn from those assumptions. Compare with **inductive logic**.

deep drainage: Soil water that has drained to below the lower limits of plant roots.

deforestation: A permanent conversion of forest into some other kind of ecosystem, such as agriculture or urbanized land use.

demographic transition: A change in human population parameters from a condition of high birth and death rates to one of low birth and death rates.

denitrification: The microbial reduction of nitrate (NO_3^-) into gaseous N_2O or N_2.

desert: A temperate or tropical biome characterized by prolonged drought, usually receiving less than 25 cm of precipitation per year.

detritivore: A heterotroph that feeds on dead organic matter.

developed countries: Relatively wealthy countries, whose average citizens have a healthy and materialistic lifestyle, more so than in less-developed countries.

development (or **economic development**): An economic term that implies improving efficiency in the use of materials and energy in an economy, and progress toward a sustainable economic system. Compare with **economic growth**.

disturbance: An episode of destruction of some part of a community or ecosystem.

DNA: The biochemical **d**eoxyribo**n**ucleic **a**cid, the main constituent of the chromosomes of eukaryotic organisms.

dose-response relationship: The quantitative relationship between different doses of a chemical and a biological or ecological response.

doubling time: The time it takes for something to increase by a factor of two (as in population growth).

drift: Movement of applied pesticide off the intended site of deposition through atmospheric or aquatic transport.

dry deposition: Atmospheric inputs of chemicals occurring in intervals between rainfall or snowfall. Compare with **wet deposition**.

earthquake: A trembling or movement of the earth, caused by a sudden release of geological stresses at some place within the crust.

ecocentric world view: This incorporates the biocentric world view but also stresses the importance of interdependent ecological functions, such as productivity and nutrient cycling. In addition, the connections among species within ecosystems are considered to be invaluable. Compare with **anthropocentric world view** and **biocentric world view**.

ecological conversion: A long-term change in the character of the ecosystem at some place, as when a natural forest is converted into an agricultural land use.

ecological economics: A type of economics that involves a full accounting of costs associated with ecological damages and resource depletion. Compare with **conventional economics**.

ecological integrity (or **ecosystem health**): A notion related to environmental quality, but focusing on changes in natural populations and ecosystems, rather than effects on humans and their economy. See also **environmental quality**.

ecological pyramid: A model of the trophic structure of an ecosystem, organized with plant productivity on the bottom, that of herbivores above, and carnivores above the herbivores.

ecological services: Ecological functions that are useful to humans and to ecosystem stability and integrity, such as nutrient cycling, productivity, and control of erosion.

ecological stress: See **stressors**.

ecological sustainability: See **ecologically sustainable development**.

ecologically sustainable development: This considers the human need for resources within an ecological context, and includes the need to sustain all species and all components of Earth's life-support system. Compare with **sustainable development**.

ecologically sustainable economic system: An economic system that operates without a net consumption of natural resources, and without endangering biodiversity or other ecological values. Ultimately, ecologically sustainable economic systems are supported by the wise use of renewable resources.

ecologically sustainable economy: An economy in which ecological goods and services are utilized in ways that do not compromise their future availability and do not endanger the survival of species or natural ecosystems.

ecology: The study of the relationships between organisms and their environment.

economic development: See **development**.

economic growth: A term that refers to an economy that is increasing in size over time, usually due to increases in both population and per capita resource use. Compare with **development**.

ecosystem: A general term used to describe one or more communities that are interacting with their environment as a defined unit. Ecosystems range from small units occurring in microhabitats, to larger units such as landscapes and seascapes, and even the biosphere.

ecosystem approach: A holistic interpretation of the natural world that considers the web-like interconnections among the many components of ecosystems.

ecosystem health: See **ecological integrity**.

ecotoxicology: Study of the directly poisonous effects of chemicals in ecosystems, plus indirect effects such as changes in habitat or food abundance caused by toxic exposures. Compare with **toxicology** and **environmental toxicology**.

ecotype: A population specifically adapted to coping with locally stressful conditions, such as soil with high metal concentrations.

ecozone: The largest biophysical zones in the national ecological classification of Canada.

electromagnetic energy: Energy associated with photons, comprising an electromagnetic spectrum divided into components, including ultraviolet, visible, and infrared.

endangered: In Canada, this specifically refers to indigenous species threatened with imminent extinction or extirpation over all or a significant portion of their Canadian range.

endemic: An ecological term used to describe species with a local geographic distribution.

energy: The capacity of a body or system to accomplish work, and existing as electromagnetic, kinetic, and potential energies.

energy budget: An analysis of the rates of input and output of energy to a system, plus transformations of energy among its states, including changes in stored quantities.

entropy: A physical attribute related to the degree of randomness of the distributions of matter and energy.

environmental citizenship: Actions taken by individuals and families to lessen their impacts on the environment.

environmental ethics: These deal with the responsibilities of the present human generation to ensure continued access to adequate resources and livelihoods for future generations of people and other species.

environmental impact assessment: A process used to identify and evaluate the potential consequences of proposed actions or policies for environmental quality. See also **socioeconomic impact assessment**.

environmental indicators: Relatively simple measurements that are sensitive to changes in the intensity of stressors, and are considered to represent complex aspects of environmental quality.

environmental literacy: The degree to which people have a well-informed understanding of environmental issues.

environmental monitoring: Repeated measurements of indicators related to the inorganic environment or to ecosystem structure and function.

environmental quality: A notion related to the amounts of toxic chemicals and other stressors in the environment, to the frequency and intensity of disturbances, and to their effects on humans, other species, ecosystems, and economies.

environmental reporting: Communication of information about changes in environmental quality to interest groups and the general public.

environmental risk assessment: A quantitative evaluation of the risks associated with an environmental hazard.

environmental risks: Hazards or probabilities of suffering damage or misfortune because of exposure to some environmental circumstance.

environmental science: An interdisciplinary branch of science that investigates questions related to the human population, resources, and damages caused by pollution and disturbance.

environmental stressor: See **stressor**.

environmental toxicology: The study of environmental factors influencing exposures of organisms to potentially toxic levels of chemicals. Compare with **toxicology** and **ecotoxicology**.

environmental values: Perceptions of the worth of environmental components, divided into four broad classes: utilitarian, ecological, aesthetic, and moral.

environmentalist: Anyone with a significant involvement with environmental issues, usually in an advocacy sense.

erosion: The physical removal of rocks and soil through the combined actions of flowing water, wind, ice, and gravity.

estuary: A coastal, semi-enclosed ecosystem that is open to the sea and has habitats transitional between marine and freshwater conditions.

ethics: The perception of right and wrong. The proper behaviour of people toward each other and toward other species and nature.

eukaryote: Organisms in which the cells have an organized, membrane-bound nucleus containing the genetic material. Compare with **prokaryote**.

eutrophic: Pertains to waters that are highly productive because they contain a rich supply of nutrients. Compare with **oligotrophic** and **mesotrophic**.

eutrophication: Increased primary productivity of an aquatic ecosystem, resulting from nutrient inputs.

evaporation: The change of state of water from a liquid or solid to a gas.

evapotranspiration: Evaporation of water from a landscape. See also **transpiration**.

evolution: Genetically based changes in populations of organisms, occurring over successive generations.

experiment: A controlled test or investigation designed to provide evidence for, or preferably against, a hypothesis about the natural or physical world.

exposure: In ecotoxicology, this refers to the interaction of organisms with an environmental stressor at a particular place and time.

exposure assessment: An investigation of the means by which organisms may encounter a potentially toxic level of a chemical or other environmental stressor.

extinct (or **extinction**): A condition in which a species or other taxon no longer occurs anywhere on Earth.

extirpated (or **extirpation**): A condition in which a species or other taxon no longer occurs in some place or region, but still survives elsewhere.

fact: An event or thing known to have happened, to exist, or to be true. See also **hypothesis**.

fen: A wetland that develops in cool and wet climates, but is less acidic and more productive than a bog because it has a better nutrient supply. Compare with **bog**.

First Law of Thermodynamics: A physical principle stating that energy can undergo transformations among its various states, but it is never created or destroyed; thus, the energy content of the universe remains constant. See also **Second Law of Thermodynamics**.

First Nations: The aboriginal people(s) originally living in some place. This term is often used in reference to the original inhabitants of the Americas, prior to the colonization of those regions by Europeans, and their modern descendants.

fission reactions: Nuclear reactions involving the splitting of heavier, radioactive atoms into lighter ones, with the release of large quantities of energy.

fitness: The proportional contribution of an individual to the progeny of its population.

flow-through system: A system with an input and an output of energy or mass, plus temporary storage of any difference.

flux: A movement of mass or energy between compartments of a material or energy cycle.

food chain: A hierarchical model of feeding relationships among species in an ecosystem.

food web: A complex model of feeding relationships, describing the connections among all food chains within an ecosystem.

food-web magnification (or **food-web accumulation** or **food-web concentration**): The tendency for top predators in a food web to have the highest residues of certain chemicals, especially organochlorines.

forest floor: Litter and other organic debris lying on top of the mineral soil of a forest.

forestry: The harvesting of trees and management of post-harvest succession to foster the regeneration of another forest.

fossil fuel: Organic-rich geological materials, such as coal, petroleum, and natural gas.

frontier world view: This asserts that humans have a right to exploit nature by consuming natural resources in boundless quantities. See also **sustainability world view** and **spaceship world view**.

fungicide: A pesticide used to protect crop plants and animals from fungi that cause diseases or other damages.

fusion reactions: Nuclear reactions involving the combining of light nuclei, such as those of hydrogen, to make heavier ones, with the release of large quantities of energy. Fusion reactions occur under conditions of intense temperature and pressure, such as within stars and in hydrogen bombs.

Gaia hypothesis: A notion that envisions Earth's species and ecosystems as a "superorganism" that attempts to optimize environmental conditions toward enhancing its own health and survival.

gene: A region of a chromosome, containing a length of DNA that behaves as a particulate unit in inheritance and determines the development of a specific trait.

genotype: The genetic complement of an individual organism. See also **phenotype**.

geothermal energy: Heat in Earth's crust, which can sometimes be used to provide energy for heating or generation of electricity.

glaciation: An extensive geological change associated with an extended period of global climatic cooling and characterized by advancing ice sheets.

glacier: A persistent sheet of ice, occurring in the Arctic and Antarctic and at high altitude on mountains.

greater protected area: A protected area plus its immediately surrounding area, co-managed to sustain populations of indigenous species and natural communities.

green revolution: Intensive agricultural systems involving the cultivation of improved crop varieties in monoculture, and increased use of mechanization, fertilizers, and pesticides.

greenhouse effect: The physical process by which infrared-absorbing gases (such as CO_2) in Earth's atmosphere help to keep the planet warm.

greenhouse gas: See **radiatively active gas**.

gross primary production (GPP): The fixation of energy by primary producers within an ecosystem. See also **respiration**, **net primary production**, and **autotroph**.

groundwater: Water stored underground in soil and rocks.

growth: Refers to an economy or economic sector that is increasing in size over time. Compare with **development**.

gymnosperm: Vascular plants such as conifers, which have naked ovules not enclosed within a specialized membrane, and seeds without a seedcoat. Compare with **angiosperm**.

herbicide: A pesticide used to kill weeds. See also **weed**.

herbivore (or **primary consumer**): An animal that feeds on plants.

heterotroph: An organism that utilizes living or dead biomass as food.

hidden injury: A reduction in plant productivity caused by exposure to pollutants, but not accompanied by symptoms of acute tissue damages.

humidity: The actual concentration of water in the atmosphere, usually measured in mg/m^3. Compare with **relative humidity**.

hydrocarbons: Molecules composed of hydrogen and carbon only.

hydroelectric energy: The kinetic energy of flowing water, sometimes used to generate electricity.

hydrologic cycle (or **water cycle**): The movement between, and storage of water in, various compartments of the hydrosphere. See also **hydrosphere**.

hydrosphere: The parts of the planet that contain water, including the oceans, atmosphere, on land, in surface waterbodies, underground, and in organisms.

hyperaccumulator: A species that bioaccumulates metals or other chemicals to extremely high concentrations in their tissues. See also **bioaccumulation**.

hypothesis: A proposed explanation for the occurrence or causes of natural phenomena. Scientists formulate hypotheses as statements, and test them through experiments and other forms of research. See also **fact**.

igneous rocks: Rocks such as basalt and granite, formed by cooling of molten magma.

indicator: See **environmental indicator**

indigenous culture: A human culture existing in a place or region prior to its invasion, or other significant influence, by a foreign culture.

individual organism: A genetically and physically discrete living entity.

inductive logic: Logic in which conclusions are objectively developed from the accumulating evidence of experience and the results of experiments. See also **deductive logic**.

insecticide: A pesticide used to kill insects that are considered pests. See also **pesticide** and **pest**.

integrated forest management: Forest management plans that accommodate the need to harvest timber from landscapes, while also sustaining other values, such as hunted wildlife, outdoor recreation, and biodiversity.

integrated pest management (IPM): The use of a variety of complementary tactics toward pest control.

interdisciplinary: Encompassing a wide diversity of kinds of knowledge.

intrinsic (or natural) population change: Population change due only to the balance of birth and death rates.

intrinsic value: Value that exists regardless of any direct or indirect value in terms of the needs or welfare of humans.

inversion: see **atmospheric inversion**.

invertebrate: Any animal that lacks an internal skeleton, and in particular a backbone.

joule: A standard unit of energy, defined as the energy needed to accelerate 1 kg of mass at 1 m/s^2 for a distance of 1 metre. Compare with **calorie**.

K-selected: Refers to organisms that produce relatively small numbers of large offspring. A great deal of parental investment is made in each progeny, which helps to ensure their establishment and survival. Compare with **r-selected**.

keystone species: A dominant species in a community, usually a predator, with an influence on structure and function that is highly disproportionate to its biomass.

kinetic energy: Energy associated with motion, including mechanical and thermal types.

landscape: The spatial integration of ecological communities over a large terrestrial area.

laws of thermodynamics: Physical principles that govern all transformations of energy. See also **First Law of Thermodynamics** and **Second Law of Thermodynamics**.

leaching: The movement of dissolved substances through the soil with percolating rainwater.

lentic ecosystem: A freshwater ecosystem characterized by nonflowing water, such as a pond or lake. Compare with **lotic ecosystem**.

less-developed countries: Countries that are relatively poor and have not progressed far in terms of industrial and socioeconomic development.

life form: A grouping of organisms on the basis of their common morphological and physiological characteristics, regardless of their evolutionary relatedness.

life index (or **production life**): The known reserves of a resource divided by its current rate of production.

liming: Treatment of a waterbody or soil to reduce acidity, usually by adding calcium carbonate or calcium hydroxide.

limiting factor: An environmental factor that is the primary restriction on the productivity of autotrophs in an ecosystem. See also **Principle of Limiting Factors**.

lithification: A geological process in which materials are aggregated, densified, and cemented into new sedimentary rocks.

lithosphere: An approximately 80-km thick region of rigid, relatively light rocks that surround Earth's plastic mantle.

load-on-top (LOT) method: A process used in ocean-going petroleum tankers to separate and contain most oily residues before ballast waters are discharged to the marine environment.

lotic ecosystem: A freshwater ecosystem characterized by flowing water, such as a stream or river. Compare with **lentic ecosystem**.

LRTAP: The **l**ong-**r**ange **t**ransport of **a**tmospheric **p**ollutants.

macroclimate: Climatic conditions affecting an extensive area. Compare with **microclimate**.

macroevolution: The evolution of species or higher taxonomic groups, such as genera, families, or classes. Compare with **microevolution**.

management system: A variety of management practices used in a coordinated manner.

manipulative experiment: An experiment involving controlled alterations of factors hypothesized to influence phenomena, conducted to investigate whether predicted responses will occur, thereby uncovering causal relationships. See also **experiment** and **natural experiment**.

mantle: A less-dense region that encloses Earth's core, and composed of minerals in a hot, plastic state known as magma.

marsh: A productive wetland, typically dominated by species of monocotyledonous angiosperm plants that grow as tall as several metres above the water surface.

maximum sustainable yield (MSY): The largest amount of harvesting that can occur without degrading the productivity of the stock.

megacity: The world's largest urbanized areas.

mesotrophic: Pertains to aquatic ecosystems of moderate productivity, intermediate to eutrophic and oligotrophic waters. Compare with **eutrophic** and **oligotrophic**.

metal: Any relatively heavy element that in its pure state shares electrons among atoms, and has useful properties such as malleability, high conductivity of electricity and heat, and tensile strength.

metamorphic rocks: Rocks formed from igneous or sedimentary rocks that have changed in structure under the influences of geological heat and pressure.

microclimate: Climatic conditions on a local scale. Compare with **macroclimate**.

microdisturbance: Local disruptions that affect small areas within an otherwise intact community. Compare with **community-replacing disturbance**.

microevolution: Relatively subtle evolutionary changes occurring within a population or species, sometimes within only a few generations, and at most leading to the evolution of races, varieties, or subspecies. Compare with **macroevolution**.

mitigation: An action that repairs or offsets environmental damages to some degree.

monoculture: An agroecosystem managed to support only a single crop species.

montane forest: A conifer-dominated forest occurring below the alpine zone on mountains.

MSY: See **maximum sustainable yield**.

mutualism (or **mutualistic symbiosis**): A symbiosis in which both partners benefit.

natural experiment: An experiment conducted by observing variations of phenomena in nature, and then developing explanations for these through analysis of potential causal mechanisms. See also **experiment** and **manipulative experiment**.

natural gas: A gaseous, hydrocarbon-rich mixture mined from certain geological formations.

natural population change: A change in population that is due only to the difference in birth and death rates, and not to immigration or emigration.

natural selection: A mechanism of evolution, favouring individuals that, for genetically based reasons, are better adapted to coping with environmental opportunities and constraints. These more fit individuals have an improved probability of leaving descendants, ultimately leading to genetically based changes in populations, or evolution.

net primary production (NPP): Primary production that remains as biomass after primary producers have accounted for their respiratory needs. See also **respiration** and **gross primary production**.

niche: The role of a species within its community.

nitrification: The bacterial oxidation of ammonium (NH_4^+) to nitrate (NO_3^-).

nitrogen fixation: The oxidation of nitrogen gas (N_2) to ammonia (NH_3) or nitric oxide (NO).

nongovernmental organizations: Organizations that operate at arm's length from government, either to pursue advocacy or to undertake environmental, public health, or social actions.

nonrenewable resource (or **nonrenewable natural resource**): A resource present on Earth in finite quantities, so as it is used, its future stocks are diminished. Examples are metals and fossil fuels. Compare with **renewable resources**.

nontarget damage: Damage caused by a pesticide to nontarget organisms. See also **broad-spectrum pesticide** and **nontarget organism**.

nontarget organism: Organisms that are not pests, but which may be affected by a pesticide treatment. See also **broad-spectrum pesticide** and **nontarget damage**.

nuclear fuel: Unstable isotopes of uranium (^{235}U) and plutonium (^{239}Pu) that decay through fission, releasing large amounts of energy that can be used to generate electricity.

null hypothesis: A hypothesis that seeks to disprove a hypothesis.

nutrient: Any chemical required for the proper metabolism of organisms.

nutrient budget: A quantitative estimate of the rates of nutrient input and output for an ecosystem, as well as the quantities present and transferred within the system.

nutrient capital: The amount of nutrients present in a site in soil, living vegetation, and dead organic matter.

nutrient cycling: Transfers and chemical transformations of nutrients in ecosystems, including recycling through decomposition.

oceans: The largest hydrological compartment, accounting for about 97% of all water on Earth.

old-growth forest: A late-successional forest characterized by the presence of old trees, an uneven-aged population structure, and a complex physical structure.

oligotrophic: Pertains to aquatic ecosystems that are highly unproductive because of a sparse supply of nutrients. Compare with **eutrophic** and **mesotrophic**.

omnivore: An animal that feeds on both plant and animal materials.

orographic precipitation: Precipitation associated with hilly or mountainous terrain that forces moisture-laden air to rise in altitude and become cooler, causing water vapour to condense into droplets that precipitate as rain or snow.

outer space: Regions beyond the atmosphere of Earth.

overharvesting (or **overexploitation**): Unsustainable harvesting of a potentially renewable resource, leading to a decline of its stocks.

oxidizing smog: An event of air pollution rich in ozone, peroxy acetyl nitrate, and other oxidant gases.

parameter: One or more constants that determine the form of a mathematical equation. In the linear equation $Y = aX + b$, a and b are parameters, and Y and X are variables. See also **variable**.

parasitism: A biological relationship involving one species obtaining nourishment from a host, usually without causing its death.

persistence: The time it takes chemicals, especially pesticides, to be degraded in the environment by microorganisms or physical agents such as sunlight and heat.

pest: Any organism judged to be significantly interfering with some human purpose.

pesticide: A substance used to poison pests. See also **pest**, **fungicide**, **herbicide**, and **insecticide**.

pesticide tread-mill: The inherent reliance of modern agriculture and public-health programs on pesticides, often in increasing quantities, to deal with pest problems.

petroleum (or **crude oil**): A fluid, hydrocarbon-rich mixture mined from certain geological formations.

phenotype: The expressed characteristics of an individual organism, due to genetic and environmental influences on the expression of its specific genetic information. See also **genotype**.

phenotypic plasticity: The variable expression of genetic information of an individual, depending on environmental influences during development.

photoautotroph: Plants and algae that use sunlight to drive their fixation of energy through photosynthesis. See also **chemoautotroph** and **photosynthesis**.

photochemical air pollutants: Ozone, peroxy acetyl nitrate, and other strongly oxidizing gases that form in the atmosphere through complex reactions involving sunlight, hydrocarbons, oxides of nitrogen, and other chemicals.

photosynthesis: Autotrophic productivity that utilizes visible electromagnetic energy (such as sunlight) to drive biosynthesis.

phytoplankton: Microscopic, photosynthetic bacteria and algae that live suspended in the water of lakes and oceans.

plantation: In forestry, these are tree-farms managed for high productivity of wood fibre.

point source: A location where large quantities of pollutants are emitted into the environment, such as a smokestack or sewer outfall.

pollution: The exposure of organisms to chemicals or energy in quantities that exceed their tolerance, causing toxicity or other ecological damages. Compare with **contamination**.

population: In ecology, this refers to individuals of the same species that occur together in time and space.

population policy: Policy designed to influence changes in the size or structure of a human population.

potential energy: The stored ability to perform work, capable of being transformed into electromagnetic or kinetic energies. Potential energy is associated with gravity, chemicals, compressed gases, electrical potential, magnetism, and the nuclear structure of matter.

potentially renewable natural resource: An alternate phrase for renewable natural resource, highlighting the fact that these can be overexploited, and thereby treated as if they were nonrenewable resources. See also **renewable resource**.

ppb (part per billion): A unit of concentration, equivalent to 1 microgram per kilogram (μg/kg), or in aqueous solution, 1 μg per litre (μg/L).

ppm (part per million): A unit of concentration, equivalent to 1 milligram per kilogram (mg/kg), or in aqueous solution, 1 mg per litre (mg/L).

prairie: Grassland ecosystems occurring in temperate regions.

precipitation: Deposition of water from the atmosphere as liquid rain, or as solid snow or hail.

precision: The degree of repeatability of a measurement or observation. Compare with **accuracy**.

prevailing wind: Wind that blows in a dominant direction.

primary consumer: A herbivore, or a heterotrophic organism that feeds on plants or algae.

primary pollutants: Chemicals that are emitted into the environment. Compare with **secondary pollutants**.

primary producer: An autotrophic organism. Autotrophs are the biological foundation of ecological productivity. See also **primary production**.

primary production: Productivity by autotrophic organisms, such as plants or algae. Often measured as biomass accumulated over a unit of time, or sometimes by the amount of carbon fixed.

primary sewage treatment: The initial stage of sewage treatment, usually involving the filtering of larger particles from the sewage wastes, settling of suspended solids, and sometimes chlorination to kill pathogens.

Principle of Limiting Factors: A theory stating that ecological productivity (and some other functions) is controlled by whichever environmental factor is present in least supply relative to the demand.

production: An ecological term related to the total yield of biomass from some area or volume of habitat.

production life: See **life index**.

productivity: An ecological term for production standardized per unit area and time.

prokaryote: Microorganisms without an organized nucleus containing their genetic material. Compare with **eukaryote**.

protected area (or **reserve**): Parks, ecological reserves, and other tracts set aside from intense development to conserve their natural ecological values. See also **greater protected area**.

r-selected: Refers to organisms that produce relatively large numbers of small offspring. Little parental investment is made in each offspring, but having large numbers of progeny helps ensure that some will establish and survive. Compare with **K-selected**.

radiatively active gases (or **greenhouse gases** or **RAGs**): Atmospheric gases that efficiently absorb infrared radiation and then dissipate some of the thermal energy gain by reradiation.

RAGs: See **radiatively active gases**

recycling: Processing disused industrial and household goods to recover reusable materials.

relative humidity: The atmospheric concentration of water, expressed as a percentage of the saturation value for that temperature.

renewable resource (or **renewable natural resource**): These can regenerate after harvesting, and potentially can be exploited forever. Examples are fresh water, trees, agricultural plants and livestock, and hunted animals. Compare with **nonrenewable resources**.

replacement fertility rate: The fertility rate that results in the numbers of progeny replacing their parents, with no change in size of the equilibrium population.

replication: The biochemical process occurring prior to cellular division, by which information encoded in

DNA is copied to produce additional DNA with the same information.

reserve: (1) Known quantities of resources that can be economically recovered from the environment. (2) An alternative word for a protected area. See **protected area**.

residue: Lingering concentrations of pesticides and certain other chemicals in organisms and the environment.

resilience: The ability of a system to recover from disturbance.

resistance: The ability of a population or community to avoid displacement from some stage of ecological development as a result of disturbance or an intensification of environmental stress. Changes occur after thresholds of resistance to environmental stressors are exceeded.

respiration: Physiological processes needed to maintain organisms alive and healthy.

response: In ecotoxicology, this refers to biological or ecological changes caused by exposure to an environmental stressor.

risk: See **environmental risks**.

risk assessment: See **environmental risk assessment**.

RNA: The biochemical **r**ibo**n**ucleic **a**cid, which is important in translation of the genetic information of DNA into the synthesis of proteins. RNA also stores the genetic information of some viruses.

ruderal: Short-lived but highly fecund plants characteristic of frequently disturbed environments with abundant resources.

science: The systematic and quantitative study of the character and behaviour of the physical world.

scientific creationists: Creationists who attempt to explain some of the discrepancies between their beliefs (which are based on a literal interpretation of Genesis) and scientific understanding of the origin and evolution of life. See also **creationists**.

scientific method: This begins with the identification of a question involving the structure or function of the natural world, usually using inductive logic. The question is interpreted in terms of a theory, and hypotheses are formulated and tested by experiments and observations of nature.

seascape: A spatial integration of ecological communities over a large marine area.

Second Law of Thermodynamics: A physical principle stating that transformations of energy can occur spontaneously only under conditions in which there is an increase in the entropy (or randomness) of the universe. See also **First Law of Thermodynamics** and **entropy**.

secondary consumer: A carnivore that feeds on primary consumers (or herbivores).

secondary pollutants: Pollutants that are not emitted, but form in the environment by chemical reactions involving emitted chemicals. Compare with **primary pollutants**.

secondary sewage treatment: Treatment applied to the effluent of primary sewage treatment, usually involving the use of a biological technology to aerobically decompose organic wastes in an engineered environment. The resulting sludge can be used as a soil conditioner, incinerated, or dumped into a landfill. See also **primary sewage treatment**.

sedimentary rocks: Rocks formed from precipitated minerals such as calcite, or from lithified particles eroded from other rocks such as sandstone, shale, and conglomerates.

sedimentation: A process by which mass eroded from elsewhere settles to the bottom of rivers, lakes, or an ocean.

selection harvesting: Harvesting of only some trees from a stand, leaving others behind and the forest substantially intact.

sewage treatment: The use of physical filters, chemical treatment, and/or biological treatment to reduce pathogens, organic matter, and nutrients in waste waters containing sewage.

significant figures: The number of digits used when reporting data from analyses or calculations.

silvicultural management: The application of practices that increase tree productivity in a managed forest, such as planting seedlings, thinning trees, or applying herbicides to reduce the abundance of weeds.

silviculture: The branch of forestry concerned with the care and tending of trees.

site capability (or **site quality**): The potential of land to sustain the productivity of agricultural crops.

slope: The angle of inclination of land, measured in degrees (0° implies a horizontal surface, while 90° is vertical).

smog: An event of ground-level air pollution.

snag: A standing dead tree.

socioeconomic impact assessment: A process used to identify and evaluate the potential consequences of proposed actions or policies for sociological, economic, and related values. See also **environmental impact assessment**.

soil: A complex mixture of fragmented rock, organic matter, moisture, gases, and living organisms that covers almost all of Earth's terrestrial landscapes.

solar energy: Electromagnetic energy radiated by the sun.

solar system: The sun, its nine orbiting planets, miscellaneous comets, meteors, and other local materials.

spaceship Earth: An image of Earth as viewed from space, which illustrates the fact that, except for sunlight, resources needed by humans are present only on that planet.

spaceship world view: This focuses on sustaining only those resources needed by humans and their economy, and it assumes that humans can exert a great degree of control over natural processes and can pilot "spaceship Earth." See also **frontier world view** and **sustainability world view**.

species: An aggregation of individuals and populations that can potentially interbreed and produce fertile offspring, and is reproductively isolated from other such groups.

species richness: The number of species in some area or place.

stratosphere: The upper atmosphere, extending above the troposphere from 8–17 km to as high as about 50 km. See also **troposphere**.

stress-tolerator: Long-lived plants adapted to habitats that are marginal in terms of climate, moisture, or nutrient supply, but are infrequently disturbed and therefore stable, such as tundra and desert.

stressor: Environmental factors that constrain the development and productivity of organisms or ecosystems.

succession: A process of community-level recovery following disturbance.

surface flow: Water that moves over the surface of the ground.

surface water: Water that occurs in glaciers, lakes, ponds, rivers, streams, and other surface bodies of water.

sustainability world view: This acknowledges that humans must have access to vital resources, but it asserts that the exploitation of resources should be governed by appropriate ecological, aesthetic, and moral values, and should not deplete the necessary resources. See also **frontier world view** and **spaceship world view**.

sustainable development: Progress toward an economic system based on the use of natural resources in a manner that does not deplete their stocks nor compromise their availability for use by future generations of humans. Compare with **ecologically sustainable development**.

sustainable economic system (or **sustainable economy**): An economic system that can be maintained over time without any net consumption of natural resources.

swamp: A forested wetland, flooded seasonally or permanently.

symbiosis: An intimate relationship between different species. See also **mutualism**.

synecology: The study of relationships among species within communities. Compare with **autecology**.

taiga: See **boreal forest**.

tectonic forces: Forces associated with crustal movements and related geological processes that cause structural deformations of rocks and minerals.

temperate deciduous forest: A forest occurring in relatively moist, temperate climates with short and moderately cold winters and warm summers, and usually composed of a mixture of angiosperm tree species.

temperate grassland: Grass-dominated ecosystems occurring in temperate regions with an annual precipitation of 25–60 cm per year; sufficient to prevent desert from developing but insufficient to support forest.

temperate rainforest: A forest developing in a temperate climate in which winters are mild and precipitation is abundant year-round. Because wildfire is rare, old-growth forests may be common.

temperature inversion: See **atmospheric inversion**.

tertiary sewage treatment: Treatment applied to the effluent of secondary sewage treatment, usually involving a system to remove phosphorus and/or nitrogen from waste waters. See also **primary sewage treatment** and **secondary sewage treatment**.

theory: A general term that refers to a set of scientific laws, rules, and explanations supported by a large body of experimental and observational evidence, all leading to robust, internally consistent conclusions.

threatened: In Canada, this refers to any indigenous taxa likely to become endangered (in Canada) if factors affecting their vulnerability are not reversed.

tidal energy: Energy that develops in oceanic surface waters because of the gravitational attraction between Earth and the Moon, and can potentially be used to generate electricity.

tolerance: In ecotoxicology, this refers to a genetically based ability of organisms or species to not suffer toxicity when exposed to chemicals or other stressors.

total concentration: The concentrations of metals in soil, sediment, rocks, or water, as determined after dissolving samples in a strongly acidic solution. Compare with **available concentration**.

toxicology: The science of the study of poisons, including their chemical nature and their effects on the physiology of organisms. Compare with **environmental toxicology** and **ecotoxicology**.

transcription: A biochemical process by which the information of double-stranded DNA is encoded on complementary single strands of RNA, used to synthesize specific proteins.

translation: A biochemical process occurring on organelles known as ribosomes, in which information encoded in messenger RNA is used to synthesize particular proteins.

transpiration: The evaporation of water from plants. Compare with **evapotranspiration**.

trophic structure: The organization of productivity in an ecosystem, including the roles of autotrophs, herbivores, carnivores, and detritivores.

troposphere: The lower atmosphere, extending to 8–17 km.

tundra: A treeless biome occurring in environments with long, cold winters and short, cool growing seasons.

urbanization: The development of cities and towns on formerly agricultural or natural lands.

utilitarian value: The usefulness of a thing or function to humans.

valued ecosystem components (VECs): In environmental impact assessment, these are components of ecosystems perceived to be important to society as economically important resources, as rare or endangered species or communities, or for their cultural or aesthetic significance.

variable: A changeable factor believed to influence a natural phenomenon of interest or that can be manipulated during an experiment.

vascular plant: Relatively complex plants with specialized, tube-like vascular tissues in their stems for conducting water and nutrients.

VECs: See **valued ecosystem components**.

vector: Species of insects and ticks that transmit pathogens from alternate hosts to people or animals.

vertebrate: Animals with an internal skeleton, and in particular a backbone.

virgin field: In epidemiology, this is a population that is hypersensitive to one or more infectious diseases to which it has not been previously exposed.

volcano: An opening in Earth's crust from which magmic materials, such as lava, rock fragments, and gases, are ejected into the atmosphere or oceanic waters.

water cycle: See **hydrologic cycle**.

watershed: An area of land from which surface water and some groundwater flow into a stream, river, or lake.

wave energy: The kinetic energy of oceanic waves, which can be harnessed using specially designed buoys to generate electricity.

weather: The short-term, day-to-day or instantaneous meteorological conditions at a place or region. Compare with **climate**.

weathering: Physical and chemical processes by which rocks and minerals are broken down by such environmental agents as rain, wind, temperature changes, and biological influences.

weed: An unwanted plant that interferes with some human purpose.

wet deposition: Atmospheric inputs of chemicals with rain and snow. Compare with **dry deposition**.

wetland: An ecosystem that develops in wet places and is intermediate between aquatic and terrestrial ecosystems. See also **bog**, **fen**, **marsh**, and **swamp**.

whole-lake experiment: The experimental manipulation of one or more environmental factors in an entire lake.

wind: An air mass moving in Earth's atmosphere.

wind energy: The kinetic energy of moving air masses, which can be tapped and utilized in various ways, including the generation of electricity.

work: In physics, work is defined as the result of a force being applied over a distance.

working hypothesis: A hypothesis being tested in a scientific experiment or another kind of research. See also **hypothesis** and **null hypothesis**.

zero population growth (ZPG): When the birth rate plus immigration equal the death rate plus emigration.

zooplankton: Tiny animals that occur in the water column of lakes and oceans.

SUBJECT INDEX

GEOGRAPHIC INDEX

CHEMICAL INDEX

SPECIES INDEX